Studies in Classification, Data Analysis, and Knowledge Organization

Springer Japan KK

C. Hayashi · N. Ohsumi
K. Yajima · Y. Tanaka
H.-H. Bock · Y. Baba (Eds.)

Data Science, Classification, and Related Methods

Proceedings of the Fifth Conference of the International Federation of Classification Societies (IFCS-96), Kobe, Japan, March 27–30, 1996

With 240 Figures

Springer

Prof. Emeritus Chikio Hayashi
The Institute of Statistical Mathematics
4-6-7 Minami-Azabu, Minato-ku, Tokyo 106, Japan

Prof. Keiji Yajima
School of Management, Science University of Tokyo
500 Shimokiyoku, Kuki, Saitama 346, Japan

Prof. Hans-Hermann Bock
Institut für Statistik
Rheinisch-Westfälische Technische Hochschule (RWTH)
D-52056 Aachen, Germany

Prof. Noboru Ohsumi
The Institute of Statistical Mathematics
4-6-7 Minami-Azabu, Minato-ku, Tokyo 106, Japan

Prof. Yutaka Tanaka
Faculty of Environmental Science & Technology, Okayama University
2-1-1 Tsushima-naka, Okayama 700, Japan

Prof. Yasumasa Baba
The Institute of Statistical Mathematics
4-6-7 Minami-Azabu, Minato-ku, Tokyo 106, Japan

ISBN 978-4-431-70208-5   ISBN 978-4-431-65950-1 (eBook)
DOI 10.1007/978-4-431-65950-1

© Springer Japan 1998
Originally published by Springer-Verlag Tokyo Berlin Heidelberg New York in 1998

SPIN 10634047   Printed on acid-free paper

PREFACE

This volume, *Data Science, Classification, and Related Methods,* contains a selection of papers presented at the Fifth Conference of the International Federation of Classification Societies (IFCS-96), which was held in Kobe, Japan, from March 27 to 30, 1996.

The volume covers a wide range of topics and perspectives in the growing field of data science, including theoretical and methodological advances in domains relating to data gathering, classification and clustering, exploratory and multivariate data analysis, and knowledge discovery and seeking.

It gives a broad view of the state of the art and is intended for those in the scientific community who either develop new data analysis methods or gather data and use search tools for analyzing and interpreting large and complex data sets. Presenting a wide field of applications, this book is of interest not only to data analysts, mathematicians, and statisticians but also to scientists from many areas and disciplines concerned with complex data: medicine, biology, space science, geoscience, environmental science, information science, image and pattern analysis, economics, statistics, social sciences, psychology, cognitive science, behavioral science, marketing and survey research, data mining, and knowledge organization.

Data Science, Classification, and Related Methods contains 85 invited and contributed refereed papers presented during IFCS-96. Four preceding conferences were held in Aachen (Germany), Charlottesville (U.S.A.), Edinburgh (U.K.), and Paris (France). This fifth IFCS-96 conference was convened at the International Conference Center, Kobe, under the sponsorship of the Ministry of Education, Science, Sports and Culture of Japan and the Behaviormetric Society of Japan, and resulted from the close cooperation between the following ten Members of the IFCS:

British Classification Society (BCS)
Classification Society of North America (CSNA)
Gesellschaft für Klassifikation (GfKl)
Japanese Classification Society (JCS)
Jugoslovenska Sekcija za Klasifikacije (JSK)
Société Francophone de Classification (SFC)
Società Italiana di Statistica (SIS)
Vereniging voor Ordinatie en Classificatie (VOC)
Section on Classification and Data Analysis of
       Polish Statistical Society (SKAD)
Associação Portuguesa de Classificação e Análise de Dados (CLAD).

IFCS-96 was organized by the IFCS-96 Organizing Committee under the auspices of the Japanese Classification Society. In addition, the Korean Classification Society (KCS) joined the conference as a new member of the IFCS.

Moreover, the organizers appreciated very much the co-sponsorship and support of twenty-one related academic societies in Japan:

> Biogeographic Society of Japan
> Japan Association for Medical Informatics
> Japan Society for Fuzzy Theory and Systems
> Japan Society of Forest Planning
> Japan Society of Medical Electronics and Biological Engineering
> Japan Society of Plant Taxonomists
> Japan Statistical Society
> Japanese Society of Applied Statistics
> Japanese Society of Computational Statistics
> Japanese Wildlife Research Society
> The Biometric Society of Japan
> The Botanical Society of Japan
> The Ecological Society of Japan
> The Entomological Society of Japan
> The Ichthyological Society of Japan
> The Japanese Forestry Society
> The Japanese Society for Quality Control
> The Mammalogical Society of Japan
> The Operations Research Society of Japan
> The Society of Instrument and Control Engineers
> Zoological Society of Japan

A software exhibition provided the opportunity for industrial companies, research laboratories, and researchers to show their programs and data analysis application software. The various prototypes from research laboratories reflected the growing activity in this field.

The editors of this volume gratefully acknowledge the cooperation of many colleagues who selected and reviewed papers or chaired sessions during the conference. We also thank all the members of the International Scientific Committee for their help and support. We very much appreciate the active collaboration of all participants and authors, who came from more than nineteen nations and who rendered possible the scientific success of IFCS-96.

The organizers of the conference are indebted to many industrial companies and institutions that financially supported the conference:

> Central Research Services, Inc.
> Commemorative Association for the Japan World Exposition (1970)
> Dentsu, Inc.
> Government Housing Loan Corporation
> Hitachi, Ltd.
> Japan Tobacco, Inc.
> Japan Travel Bureau (Foundation)

Kansai Electric Power Co., Inc.
Labourer's Health and Welfare Association
Marketing Service Co., Ltd.
Nikkei Research, Inc.
Portpia 81 Foundation
Research and Development, Inc.
Shin Joho Center, Inc.
Tokyo Electric Power Co., Inc.
Video Research, Ltd.
Yoron Kagaku Kyokai (Public Opinion Research Center)

Finally, the editors thank the staff of Springer-Verlag Tokyo for their support
and dedication and for the opportunity for publishing this volume in the series
Studies in Classification, Data Analysis, and Knowledge Organization.

Tokyo, June 1997   Chikio Hayashi
                                                      Noboru Ohsumi

CONFERENCE COMMITTEE

Conference President

Chikio Hayashi
The Institute of Statistical Mathematics
Professor Emeritus

Scientific Program Committee

Keiji Yajima *(Chairperson)*
Yutaka Tanaka *(Co-chairperson)*

International Scientific Program Committee

S. Aivazïan, *Russian Federation*
Tomàs Aluja-Banet, *Spain*
Jean-Pierre Barthélemy, *France*
Vladimir Batagelj, *Slovenia*
Hans H. Bock, *Germany*
Srdjan Bogosavljević, *Yugoslavia*
Edwin Diday, *France*
Yadolah Dodge, *Switzerland*
Brian Everitt, *United Kingdom*
Anuška Ferligoj, *Republika Slovenia*
Wolfgang Gaul, *Germany*

Willem J. Heiser, *The Netherlands*
Dekun Hu, *People's Republic of China*
Natale Carlo Lauro, *Italy*
Jae Chang Lee, *Korea*
Fred R. McMorris, *U.S.A.*
Jacqueline J. Meulman, *The Netherlands*
Konstantin Momirović, *Yugoslavia*
Clive M. Moncrieff, *U. K.*
Fernando Costa Nicolau, *Portugal*
Alfredo Rizzi, *Italy*
Michael Windham, *U.S.A.*

Local Scientific Program Committee

Shuhei Aida
Hirotugu Akaike
Chooichiro Asano
Yasumasa Baba
Sadao Fujimura
Masashi Goto
Atsuhiro Hayashi
Chikio Hayashi
Manabu Ichino
Tadashi Imaizumi
Shuichi Iwatsubo
Sadanori Konishi
Koji Kurihara
Kazufumi Manabe
Yoshiro Matsuda

Sachiko Matsui
Nobuhiro Minaka
Hideo Miyahara
Shunichi Miyai
Masahiro Mizuta
Masakatsu Murakami
Yasuo Ohashi
Noboru Ohsumi
Sumimasa Ohtsuka
Keiichi Onoyama
Seiroku Sakai
Yoshiharu Sato
Shingo Shirahata
Masae Shiyomi
Meiko Sugiyama

Setsuo Suoh
Yoshio Takane
Kiyoshi Tanaka
Yutaka Tanaka
Tomoyuki Tarumi
Mikinori Tsuiki
Shoichi Ueda
Junzo Watada
Nagasumi Yago
Kenji Yajima
Haruo Yanai
Mitsu Yoshimura
Tadashi Yoshizawa

TABLE OF CONTENTS

Part I: General Aspects of Data Science

Part II: Methodologies in Classification

Evaluation and Assessment Procedures

Topics in Clustering and Classification

Part III: Classification and Discrimination

Statistical Approaches for Classification Problems

Discrimination and Pattern Analysis

Part IV: Related Approaches for Classification

Fuzzy and Probabilistic Modeling Methods

Spatial Clustering and Neural Networks

Symbolic and Conceptual Data Analysis

Part V: Correspondence Analysis, Quantification Methods, and Multidimensional Scaling

Correspondence Analysis and Its Application

Part VI: Multivariate and Multidimensional Data Analysis

Non-Linear Modeling and Visual Treatment

Part VII: Case Studies of Data Science

Social Science and Behavioral Science

Management Science and Marketing Science

Environmental, Ecological, Biological, and Medical Sciences

Part I

General Aspects of Data Science

Probabilistic Aspects in Classification

Hans H. Bock

Institute of Statistics, Technical University of Aachen,
Wüllnerstr. 3, D-52056 Aachen, Germany

Summary: This paper surveys various ways in which probabilistic approaches can be useful in partitional ('non-hierarchical') cluster analysis. Four basic distribution models for 'clustering structures' are described in order to derive suitable clustering strategies. They are exemplified for various special distribution cases, including dissimilarity data and random similarity relations. A special section describes statistical tests for checking the relevance of a calculated classification (e.g., the *max-F test, convex cluster tests*) and comparing it to standard clustering situations (*comparative assessment of classifications, CAC*).

1. Introduction

Consider a finite set $\mathcal{O} = \{1, ..., n\}$ of objects whose properties are characterized by some observed or recorded data (a table, a data matrix, verbal descriptions) such that 'similarities' or 'dissimilarities' which may exist among these objects can be determined by these data. Cluster analysis provides formal algorithms for subdividing the set $\mathcal{O}$ into a suitable number of homogeneous subsets, called *clusters* (classes, groups etc.) such that all objects of the same cluster show approximately the same class-specific properties while objects belonging to different classes behave differently in terms of the underlying data (separation of clusters).

A range of clustering algorithms is based on a probabilistic point of view: Data are considered as realizations of random variables, thus influenced by random errors and natural fluctuations (variations), and even the finite set of objects $\mathcal{O}$ may be considered as a random sample from an infinite universe (super-population Π). Then any class or classification of objects must necessarily be defined in terms of probability distributions for the data. This paper presents a survey on this probability-based part of cluster analysis. While we can point only briefly to various topics, a more detailed presentation with numerous references may be found in Bock (1974, 1977, 1985, 1987, 1989a, 1994, 1996a,b,c,d), Jain and Dubes (1988) and Milligan (1996); also see various other papers of this volume, e.g., by Gordon, Hardy, Lapointe and Rasson.

Let us first specify the notation: The set of objects $\mathcal{O} = \{1, ..., n\}$ shall be partitioned into a suitable number m of disjoint classes $C_1, ..., C_m \subset \mathcal{O}$ resulting in an m-partition $\mathcal{C} = (C_1, ..., C_m)$ of $\mathcal{O}$. In fact, we focus on partitional clusterings here, thus neclecting hierarchical or overlapping classifications. We will consider three types of data:

1. *A data matrix* $X = (x_{kj})_{n \times p} = (x_1, ..., x_n)'$ where for each object $k \in \mathcal{O}$ p (quantitative or qualitative) variables have been sampled and compiled into the observation vector $x_k = (x_{k1}, ..., x_{kp})'$ (' denotes the transposition of a matrix).

2. *A dissimilarity matrix* $D = (d_{kl})_{n \times n}$ where d_{kl} quantifies the dissimilarity existing between two objects k and l (e.g., $d_{kl} = ||x_k - x_l||^2$); typically, we have $0 = d_{kk} \leq d_{kl} = d_{lk} < \infty$ for all $k, l \in \mathcal{O}$.

3. *A similarity relation* $S = (s_{kl})_{n \times n}$ where $s_{kl} = 1$ or 0 if the objects k and l are considered to be 'similar' or 'dissimilar', respectively.

3

In our probabilistic context, the observed x_k, d_{kl} and s_{kl} will be realizations of suitable random variables X_k, D_{kl} and S_{kl}, respectively, whose probability distributions describe the type and extent of the underlying clustering (or non-clustering) structure. In contrast to deterministic (e.g., algorithmic or exploratory) approaches to cluster analysis, this probabilistic framework can be helpful for the following purposes:

(1) *Modeling clustering structures* allowing for various shapes of clusters.

(2) *Providing clustering criteria* under more or less specified clustering assumptions (to be distinguished from clustering *algorithms* that optimize these criteria)

(3) *Describing and quantifying the performance or optimality of clustering methods.* (e.g., in a decision-theoretic framework; see Remark 2.2).

(4) *Investigating the asymptotic behaviour of clustering methods* if, e.g., n approaches ∞ under a mixture model or under a 'randomness' hypothesis.

(5) *Testing for the 'homogeneity' or 'randomness' of the data*, either versus a general hypothesis of 'non-randomness' or versus special clustering alternatives, thus *checking for the existence of a hidden clustering of the objects.*

(6) *Testing for the relevance of a calculated classification C^** that has been obtained by a special clustering algorithm.

(7) *Determining the 'true' or an appropriate number m of classes* (see Remark 2.1).

(8) *Assessing the relevance of a special cluster C_i^* of objects* that has been obtained from a clustering algorithm (Gordon 1994).

In the following we will survey some of these topics in detail and consider several special clustering models.

2. Some probabilistic clustering models

A major and basic problem of classification theory arises by the difficulty of defining the concept of a 'class' that may be approached by philosophical, mathematical, probabilistic or heuristical tools. *Probabilistic approaches* provide a flexible way of describing the relationship between objects and classes, not just by building each class from the set of objects that share the *same* decriptor values or feature combination (as it is common, e.g., in concept theory and monothetic classification), but by allowing some variation between the objects of the same class whose properties may deviate, to some limited and random degree, from the typical class profile. In technical terms this is realized by characterizing each class (classification, clustering structure) by suitable probability distributions for the sampled data.

There are four basic probabilistic models that are often used in this context. They will be formally described for the important case when the data is provided by n random independent p-dimensional feature vectors $X_1, ..., X_n$ (for dissimilarity and similarity data see the sections 2.1.4 and 2.1.5).

2.1 The fixed-partition clustering model H_m

A *fixed-partition model H_m* assumes that there exist a fixed, but unknown partition $C = (C_1, ..., C_m)$ of the set O into m non-empty classes $C_1, ..., C_m \subset O$ and a system $\theta = (\vartheta_1, ..., \vartheta_m)$ of (unknown) parameter values $\vartheta_1, ..., \vartheta_m \in R^s$ describing the properties of these classes such that

$$X_k \quad \sim \quad f(\cdot; \vartheta_i) \quad \text{for all } k \in C_i \text{ and } i = 1, ..., m. \quad (2.1)$$

where $f(x; \vartheta)$ originates from a given parametric family of distribution densities. In this context, 'clustering' consists in estimating the unknown parameters, i.e., the number of classes $m \geq 1$, the m-partition $\mathcal{C} = (C_1, ..., C_m)$ and the parameter system $\theta = (\vartheta_1, ..., \vartheta_m)$. Note that for n data vectors we have (at least) $ms + n$ unknown parameters, *viz.* $\vartheta_1, ..., \vartheta_m$ and the class indicators $I_1, ..., I_n$ where $I_k = i$ iff $k \in C_i$.

Classical statistics provides various estimation strategies from which we consider the maximum likelihood method here (see also Remark 2.2). Assuming m to be known, maximizing the joint likelihood of the observed data vectors $x_1, ..., x_n$ is equivalent to:

$$g(\mathcal{C}, \theta) \quad := \quad \sum_{i=1}^{m} \sum_{k \in C_i} (-\log f(x_k; \vartheta_i)) \quad \rightarrow \quad \min_{\mathcal{C}, \theta} \qquad (2.2)$$

(*maximum-likelihood classification approach*). This is a combined combinatorial and analytic optimization problem whose solution(s) can be approximated by the well-known iterative *k-means algorithm* which partially minimizes g with respect to θ and $\mathcal{C}$ in turn. In fact, partial minimization with respect to the parameter θ provides the maximum likelihood (m.l.) estimates $\theta(\mathcal{C}) := (\hat{\vartheta}_1, ..., \hat{\vartheta}_m)$ for a given $\mathcal{C}$, thus reducing (2.2) to the combinatorial problem

$$g_m(\mathcal{C}) \quad := \quad \sum_{i=1}^{m} \sum_{k \in C_i} (-\log f(x_k; \hat{\vartheta}_i)) \quad \rightarrow \quad \min_{\mathcal{C}}. \qquad (2.3)$$

On the other hand, minimizing g with respect to $\mathcal{C}$ for a given parameter vector θ yields the *maximum-probability-assignment partition* $\mathcal{C}(\theta) := \hat{\mathcal{C}} := (\hat{C}_1, ..., \hat{C}_m)$ with classes

$$\hat{C}_i \quad := \quad \{k \in \mathcal{O} | \ f(x_k; \vartheta_i) = \max_{\nu=1,...,m} \{f(x_k; \vartheta_\nu)\} \ \} \qquad i = 1, ..., m \qquad (2.4)$$

(with suitable rules for avoiding ties and empty classes). $\mathcal{C}(\theta)$ can often be interpreted as a *minimum-distance partition* and will be termed in this way here. Using (2.4), the optimization problem (2.2) reduces to

$$\gamma(\theta) \quad := \quad \sum_{k=1}^{n} \min_{\nu=1,...,m} \{-\log f(x_k; \vartheta_\nu)\} \quad \rightarrow \quad \min_{\theta}. \qquad (2.5)$$

Thus, fixed-partition models provide, simultaneously, three equivalent clustering criteria (2.2), (2.3) and (2.5) and a comfortable (and fast converging) optimization strategy (*k-means algorithm*). We must remind, however, that in this formulation any resulting (optimal) class C_i^* is not necessarily a most 'homogeneous' one, described by good internal properties or even well separated from other classes, but only an element of an m-partition $\mathcal{C}^*$ that optimizes the overall-criterion (2.3), i.e., some average homogeneity of classes. This fact explains, why small classes are typically neclected in this approach (except from those far away from the rest of the data, e.g., 'outlier classes'). Empirical practice as well as theoretical investigations (Bock 1968, 1974) show that criteria of this type have the tendency to produce equally-sized classes.

We illustrate the flexibility of the fixed-partition approach by five special distribution models (see also Bock 1974, 1987, 1996a,b,c,d).

Remark 2.1: The determination of the unknown (or: a suitable) number m of classes is a conceptually and technically difficult problem that is intensively discussed, e.g., in Bock (1968, 1974, 1996a), Gordon (1997b), Milligan (1981, 1996) and Milligan and

Cooper (1985), but will not be investigated here.

2.1.1 The normal distribution case and the variance criterion

Considering the case of quantitative data vectors $X_1, ..., X_n \in R^p$, let each class C_i be described by a p-dimensional spherical normal distribution $\mathcal{N}(\mu_i, \sigma^2 E_p)$ with class-specific expectations $\mu_1, ..., \mu_m \in R^p$, $\sigma^2 > 0$ (known or unknown), and E_p the $p \times p$ unit matrix. Then the criteria (2.2) and (2.3) reduce to the equivalent well-known *SSQ* or *variance criteria*:

$$g(\mathcal{C}, \theta) := \frac{1}{n} \sum_{i=1}^{m} \sum_{k \in C_i} ||x_k - \vartheta_i||^2 \rightarrow \min_{\mathcal{C}, \mu} =: g_{mn}^* \tag{2.6}$$

$$g_m(\mathcal{C}) := \frac{1}{n} \sum_{i=1}^{m} \sum_{k \in C_i} ||x_k - \bar{x}_{C_i}||^2 \rightarrow \min_{\mathcal{C}} = g_{mn}^*. \tag{2.7}$$

Similar normal distribution models assume an $\mathcal{N}_p(\mu_i, \Sigma)$ or $\mathcal{N}_p(\mu_i, \Sigma_i)$ distribution in C_i allowing for (possibly class-specific) dependencies among the p components (Bock 1974). More general approaches characterize each class C_i by a hyperplane H_i in R^p (instead by a single point μ_i only): *principal component clustering* (Bock 1969, 1974; Diday 1973: *analyse factorielle typologique*) and *regression clustering* (Bock 1969), or constrain the class centers $\mu_1, ..., \mu_m$ to belong to the same (unknown) hyperplane $H \subset R^p$ of a small dimension s, say, such that the clustering structure is essentially low-dimensional (*projection pursuit clustering*; Bock 1987).

Remark 2.2: Clustering has been investigated in a decision theoretic framework as well, thereby looking for a clustering criterion that minimizes the expected loss incured by missing a hidden clustering structure (Binder 1968; Bock 1968, 1972, 1974; Hayashi 1974, 1993; Bernardo 1994). For example, Bock derived various Bayesian methods for normal distribution clustering models under suitable prior assumptions on the underlying parameters and showed, e.g., that (2.7) is asymptotically optimum in some cases. Similarly, Hayashi investigated the minimaxity of clustering criteria.

2.1.2 A semi-parametric convex cluster model

There are instances (e.g., in pattern recognition and image exploration) where clusters can be characterized by uniform distributions $U(D)$ concentrated on some convex set $D \subset R^p$. A corresponding fixed-partition model for $X_1, ..., X_n$ involves an m-partition $\mathcal{C}$ and m unknown non-overlapping convex domains ('parameters') $D_1, ..., D_m \subset R^p$ such that for all $k \in C_i$, $X_k \sim U(D_i)$. The resulting m.l. clustering criterion (2.3) is given by

$$g_m(\mathcal{C}) := \sum_{i=1}^{m} |C_i| \cdot \log vol_p(H(C_i)) \rightarrow \min_{\mathcal{C}} \tag{2.8}$$

where the m.l. estimate of D_i is just the convex hull $\hat{D}_i = H(C_i) := conv\{x_k | k \in C_i\}$ of the data points belonging to the class C_i, and minimization is over all m-partitions with non-overlapping $H(C_1), ..., H(C_m)$ with positive volumes (Bock 1997; see also section 3.2.3). Note that such a model is applicable only if the presumptive clusters are clearly separated by some empty space.

2.1.3 Qualitative data, contingency tables and entropy criteria

In the case of qualitative data where the j-th component X_{kj} of X_k takes its values in a finite set $\mathcal{X}_j$ of alternatives (e.g., $\mathcal{X}_j = \{0, 1\}$ in the binary case) the observed vectors $x_1, ..., x_n$ belong to the Cartesian product $\mathcal{X} := \prod_{j=1}^{p} \mathcal{X}_j$ which corresponds to the cells of the p-dimensional contingency table $\mathcal{N} = (n_\xi)_{\xi \in \mathcal{X}} = (n_{\xi_1, ..., \xi_p})$

that contains as its entries the number of objects $k \in \mathcal{O}$ with the same data vector $x_k = \xi = (\xi_1, ..., \xi_p)' \in \mathcal{X}$. Thus any clustering $\mathcal{C}$ of $\mathcal{O}$ corresponds to a decomposition of $\mathcal{N} = \mathcal{N}_1 + \cdots + \mathcal{N}_m$ into m class-specific sub-tables $\mathcal{N}_1, ..., \mathcal{N}_m$.

Loglinear models provide a convenient tool for describing the distribution of multivariate qualitative data. A loglinear model for a vector X involves various parameters (stacked into a vector ϑ) which are distinguished into *main effects* (roughly describing the size of marginal frequencies) and *interaction parameters* (of various orders) that describe the association or dependencies that might exist among the p components of X. Then the distribution density (probability function) of X takes the form $f(\xi; \vartheta) = P(X = \xi) = c(\vartheta) \cdot exp\{z(\xi)'\vartheta\}$ where $z(\xi)$ is a binary dummy vector that picks from ϑ the interaction parameters which correspond to the cell ξ of the contingency table ($c(\vartheta)$ is a norming factor). – Assuming a fixed-partition model with class-specific interaction vectors $\vartheta_1, ..., \vartheta_m$ for the m classes, the m.l. method yields the *entropy clustering criterion*

$$g_m(\mathcal{C}) \quad := \quad \sum_{i=1}^{m} |C_i| \cdot H(X, \hat{\vartheta}_i(\mathcal{C})) \quad \longrightarrow \quad \min_{\mathcal{C}} \qquad (2.9)$$

where $H(X, \vartheta_i) := - \sum_{\xi \in \mathcal{X}} f(\xi; \vartheta_i) \cdot \log f(\xi; \vartheta_i) \geq 0$ is Shannon's entropy for the distribution density $f(\cdot; \vartheta_i)$ and $\hat{\vartheta}_i = \hat{\vartheta}_i(\mathcal{C})$ the m.l. estimate for ϑ_i in the class C_i. – This and similar entropy criteria (such as *logistic regression clustering*) are considered in Bock (1986, 1993, 1994, 1996a,d) and reformulated in Celeux and Govaert (1991).

2.1.4 Clustering models for dissimilarity data

The fixed-partition approach can also be used in the case of dissimilarity-based clustering where the data is a matrix $D = (D_{kl})_{n \times n}$ of random dissimilarities D_{kl}. Recall that a basic model for describing a 'no-clustering' or 'randomness' situation assumes that all $\binom{n}{2}$ variables D_{kl} with $k < l$, whilst being independent, have the same distribution density $f(d)$, like a suitable generic variable $D^* \geq 0$, say (e.g., an exponential distribution $Exp(1)$). In this context, a clustering structure involving a partition $\mathcal{C} = (C_1, ..., C_m)$ will intuitively result if we shrink the dissimilarities between objects in the *same* class C_i by a factor $\vartheta_{ii} > 0$, and stretch the dissimilarities between objects belonging to *different* classes $C_i, C_j, i \neq j$, by another factor $\vartheta_{ij} > 0$ (typically $\vartheta_{ii} < \vartheta_{ij}$ for all i, j). The resulting *dissimilarity clustering model* reads as follows:

$$D_{kl} \quad \sim \quad \vartheta_{ij} \cdot D^* \qquad \text{for all } k \in C_i, l \in C_j. \qquad (2.10)$$

The corresponding m.l. clustering criterion is given by

$$g(\mathcal{C}, \theta) \quad := \quad \sum_{1 \leq i \leq j \leq m} \left[\sum_{k \in C_i, l \in C_j} [-\log f(d_{kl}/\vartheta_{ij})] + n_{ij} \cdot \log \vartheta_{ij} \right] \quad \longrightarrow \quad \min_{\mathcal{C}, \theta} \quad (2.11)$$

where for $i = j$ the inner sum is over $k < l$ only, and $n_{ij} = |C_i| \cdot |C_j|$ and $n_{ii} = \binom{|C_i|}{2}$ is the number of terms in the inner sum for $i \neq j$ and $i = j$, respectively. For exponentially distributed dissimilarities with $f(d) = e^{-d}$ for $d > 0$, this reduces to:

$$g(\mathcal{C}, \theta) \quad := \quad \sum_{1 \leq i \leq j \leq m} n_{ij}[\overline{D}_{C_i, C_j}/\vartheta_{ij} + \log \vartheta_{ij}] \quad \longrightarrow \quad \min_{\mathcal{C}, \theta} \quad (2.12)$$

where $\overline{D}_{C_i, C_j} = n_{ij}^{-1} \sum_{k \in C_i, l \in C_j} d_{kl}$ is the *average dissimilarity* between two classes C_i and C_j, and $\overline{D}_{C_i, C_i} = n_{ii}^{-1} \sum_{k, l \in C_i, k < l} d_{kl}$ measures the heterogeneity of C_i. – Note that the unconstrained m.l. estimate for ϑ_{ij} is given by $\hat{\vartheta}_{ij} = \overline{D}_{C_i,C_j}$ such that (2.12) reduces to the *log-distance clustering criterion*

$$g_m(\mathcal{C}) \quad := \quad g(\mathcal{C},\hat{\theta}) - \binom{n}{2} = \sum_{1 \leq i \leq j \leq m} n_{ij} \log \overline{D}_{C_i,C_j} \to \min_{\mathcal{C}}. \qquad (2.13)$$

2.1.5 Clustering models for random similarity relations

In this last example, we consider binary similarity data where two objects k and l are considered either as being 'similar' ($s_{kl} = 1$, e.g. for $k = l$) or 'dissimilar' ($s_{kl} = 0$). (Note that these data can be interpreted as a similarity or association graph with n vertices and $M := \sum_{k,l,k \neq l} s_{kl}$ links or edges). Intuitively, a corresponding clustering model should be such that links are more likely inside the clusters than between different clusters. This can be modeled by introducing linking probabilities p_{ij} between the classes C_i and C_j of $\mathcal{C}$ (typically $p_{ii} > p_{ij}$ for $i \neq j$). The resulting fixed-partition model for $\binom{n}{2}$ random and independent similarity indicators S_{kl}, $k < l$ (with $S_{kk} = 1$ and $S_{kl} = S_{lk}$) reads as follows:

$$P(S_{kl} = 1) \quad = \quad p_{ij} \qquad \text{for all } k \in C_i, l \in C_j \qquad (2.14)$$

and leads to the m.l. clustering criterion:

$$g(\mathcal{C},p) \quad := \quad \sum_{1 \leq i \leq j \leq m} (N_{ij} \log p_{ij} + (n_{ij} - N_{ij}) \log(1 - p_{ij})] \to \min_{\mathcal{C},p}. \qquad (2.15)$$

where, for $i < j$, N_{ij} (N_{ii}) is the number of observed links $s_{kl} = 1$ between C_i and C_j (inside C_i). If side constraints are neglected, $\hat{p}_{ij} := N_{ij}/n_{ij}$ is the m.l. estimate for p_{ij} and can be substituted into (2.15). – The model (2.14) has been described by Bock (1989b, 1996a,c), related models are known under the heading *'block models'* (see, e.g., Snijders and Nowicki 1996).

2.2 Random-partition clustering models and the mixture model H_m^{mix}

Clustered populations are often described by *mixture models*: The random vectors $X_1, ..., X_n$ are assumed to be independent, all with the same mixture density

$$f(x) \quad = \quad f(x;\theta,\pi) \quad := \quad \sum_{i=1}^{m} \pi_i f(x;\vartheta_i) \qquad (2.16)$$

which involves m class-specific parameters $\vartheta_1, ..., \vartheta_m$ and m unknown probabilities $\pi_1, ..., \pi_m$ (with $\sum_{i=1}^{m} \pi_i = 1$). Whilst this model incorporates *no explicit clustering of objects*, it is well known that it is obtained by a two-stage process where, in a first step, the objects of $\mathcal{O}$ are sampled independently from a superpopulation Π that is decomposed into m subpopulations $\Pi_1, ..., \Pi_m$ with relative sizes π_i and described by the parameters ϑ_i. This first step results in a non-observable *random* partition $\mathcal{C} = (C_1, ..., C_m)$ of $\mathcal{O}$ that is characterized by the (random) class indicators $I_1, ..., I_n$ according to $C_i = \{k \in \mathcal{O} | I_k = i\}$, $i = 1, ..., m$. Conditionally on $I_k = i$ (i.e., $k \in C_i$), the vector X_k is distributed with the density $f(\cdot; \vartheta_i)$.

Classical mixture analysis concentrates on the estimation of the unknown parameters $\pi_1, ..., \pi_m, \vartheta_1, ..., \vartheta_m$ for a suitable number m of components, typically by maximizing the log likelihood

$$L(\theta,\pi) := \sum_{k=1}^{n} \log(\sum_{i=1}^{m} \pi_i \cdot f(x_k; \vartheta_i)) \to \max_{\theta,pi}. \qquad (2.17)$$

(McLachlan and Basford 1988, Titterington et al. 1985). Whilst this criterion involves no classification of objects, such a classification $\hat{C} = (\hat{C}_1, ..., \hat{C}_m)$ can be constructed from the estimated parameters $\hat{\vartheta}_i, \hat{\pi}_i$ by using, in an *additional* stage, a plug-in Bayesian rule that yields the classes

$$\hat{C}_i := \{k \in \mathcal{O} | \hat{\pi}_i f(x_k; \hat{\vartheta}_i) = \max_{\nu=1,...,m} \{\hat{\pi}_\nu f(x_k; \hat{\vartheta}_\nu)\} \} \qquad i = 1, ..., m. \qquad (2.18)$$

$\hat{C} = (\hat{C}_1, ..., \hat{C}_m)$ is an 'estimate' for the random partition C.

A more appropriate approach that incorporates a partition of objects from the outset, is based on the *random-partition model* which comprizes the n independent pairs $(I_1, X_1), ..., (I_n, X_n)$ with the joint 'density' $\pi_i f(x; \vartheta_i)$ (for $x \in R^p$, $i = 1, ...m$) and maximizes the joint likelihood $l(\theta, \pi, I_1, ..., I_n; x_1, ..., x_n) = \prod_{k=1}^n \pi_{I_k} f(x_k; \vartheta_{I_k}) = \prod_{i=1}^m \prod_{k \in C_i} \pi_i f(x_k; \vartheta_i)$ of these data with respect to θ, $\pi = (\pi_1, ..., \pi_m)$ and the 'missing' values $I_1, ..., I_n$ (or C, equivalently). Using the fact that partial maximization with respect to π yields the estimates $\hat{\pi}_i = |C_i|/n$ we obtain the clustering criterion

$$g(C, \theta) := \sum_{i=1}^m \sum_{k \in C_i} (-\log f(x_k; \vartheta_i)) - n \cdot \sum_{i=1}^m (|C_i|/n) \cdot \log(|C_i|/n) \to \min_{C, \theta} \qquad (2.19)$$

(Fahrmeir, Kaufmann and Pape 1980, Symons 1981, Anderson 1985). Obviously, this adds an entropy term to the previous fixed-partition criterion (2.2). It can be shown that the classes of an optimum m-partition C^* are generated by the Bayesian rule (2.18) (after replacing $\hat{\vartheta}_i$ by the optimum values ϑ_i^*). Computationally, the likelihood $l(\theta, \pi, I_1, ..., I_n; x_1, ..., x_n)$ can be successively increased by using a *modified k-means algorithm* where, in the t-th iteration step, a new partition $C^{(t)}$ is obtained by applying the Bayesian rule (2.18) with the previous parameter estimates $\vartheta_i^{(t-1)}$ and $\pi_i^{(t-1)} = |C_i^{(t-1)}|/n$ (obtained from the previous partition $C^{(t-1)}$), in analogy to the maximum-probability-assignment partition (2.4) (Fahrmeir et al. 1980, Celeux and Diebolt 1985, Bock 1996a, Fahrmeir 1996).

2.3 Modal clusters and density-contour clusters

Another group of clustering models, designed primarily for data points $x_1, ..., x_n$ in R^p, looks for those regions of R^p where these data points are locally concentrated, or, alternatively, for the regions in which the density of points exceeds some given threshold: These regions (or the corresponding clouds of points) can be used and interpreted as 'classes' or 'clusters', especially in the context of pattern recognition and image processing.

More specifically, let $f(x)$ be the common (smooth) distribution density of $X_1, ..., X_n$ and define, for a threshold $c > 0$, by $B(c) := \{x \in R^p | f(x) \geq c\}$ the *level-c region* of f. Then the connected components $B_1(c), B_2(c), ...$ of $B(c)$ are termed *high-density clusters* (Bock 1974, 1996a) or *density-contour clusters of f* at the level c. For increasing values of c, these clusters split, but also disappear and show insofar a pseudo-hierarchical structure. The unknown density f can be approximated, e.g., by a kernel estimate $\hat{f}_n(x)$ obtained from the data $x_1, ..., x_n$ and corresponding estimates $\hat{B}_1(c), \hat{B}_2(c), ... \subset R^p$ are found from $\hat{f}_n$. From these estimated regions, a (non-exhaustive) clustering of objects or data points is obtained by defining the clusters $C_i(c) := \{k \in \mathcal{O} | \hat{f}_n(x_k) \geq c\} = \hat{B}_i(c) \cap \{x_1, ..., x_n\}$, $i = 1, 2, ...$. Note that a cluster $C_i(c)$ can show a very general (even ramified) shape in R^p, and will be particularly useful if, for a fixed sufficiently large c, it is separated by broad 'density valleys' from the rest of the data, and, for a varying c, if it is constant over a wide range of values of c.

Except for the two-dimensional case, the geometrical description of high-density clusters is difficult. Therefore many 'discretized' or modified versions of this clustering strategy have been proposed (often using a weaker or discretized version of connectivity in R^p; see Bock 1996a). From a theoretical point of view, Hartigan (1981) showed that single linkage clustering fails in detecting high-density clusters for all dimensions $p \geq 2$.

A related clustering approach focusses on *local density peaks* in R^p, i.e., on the points $\xi_1, \xi_2, \dots \in R^p$ where the underlying (smooth) density f (or its estimate $\hat{f}_n$) has its local maxima (modes): Clusters are formed by successively relocating each data point x_k into a region with a larger value of $\hat{f}_n$ (by hill-climbing algorithms, steepest ascent etc.) and then collecting all data points into the same cluster C_i, termed *mode cluster*, which finally reach the same mode of $\hat{f}_n$. Even if this approach can be criticized for its instability of the cluster concept (small local variations of f or $\hat{f}_n$ can generate an abritarily large number of modes or clusters) it is often used in image analysis and pattern recognition and there exist many algorithmic variations of this approach (cf. Bock 1996a).

2.4 Spatial clustering and clumping models

Motivated by biological and physical applications, spatial statistics provides various other models for describing a clustering tendency of points in the space R^p. A first non-parametric approach considers the data $x_1, ..., x_n$ as a realization of a *Poisson process* (restricted to a finite window $G \subset R^p$), either a *homogeneous* one with a constant intensity λ (= the average number of data points per unit square) for describing a 'homogeneous' or 'non-clustered' sample, or with a location-dependent intensity $\lambda(x)$ in the case of a *clustering structure*: Here the modes and contour regions of $\lambda(x)$ can characterize clusters similarly as in section 2.3 (when using a distribution density f), and will be determined by suitable non-parametric estimates of $\lambda(x)$ (Ripley 1981, Cressie 1991).

Another model is motivated by the spread of plants in a plane or the growing of cristals around kernels: The *Neyman-Scott process* builds clusters in three separate steps: (1) by placing random 'seed points' $\xi_1, \xi_2, \dots$ into R^p according to a homogeneous Poisson process, (2) by choosing, for each ξ_i, a random integer N_i with a Poisson distribution $\mathcal{P}(\lambda)$, and (3) by surrounding each 'parent' point ξ_i by N_i 'daughter' points $X_{i1}, ..., X_{iN_i}$ that are independently distributed according to $h((x - \xi)/\sigma)/\sigma$ (conditionally on the result of (1) and (2)) where $h(x)$ is a spherically symmetric density (typically, $h \sim \mathcal{N}(0,1)$ or $h \sim U(K(0,1))$, the uniform distribution in the unit ball $K(0,1)$). The data are then identified with the set of all daugther points X_{ik} inside a suitable window G. – There exist statistical methods for estimating the unknown parameters λ, σ etc. from these data, but the problem of reconstructing the 'clusters' (families) from the data is largely unsolved. Insofar this model is representative for a range of models (including *Cox processes, Poisson cluster processes* etc.) that focus more on the clustering *tendency* of the data than on the underlying clustering of objects.

3. Hypothesis testing in cluster analysis

A major problem in cluster analysis consists in the *interpretation* of the constructed clusters of objects and the *assessment of their relevance* for the underlying practical problem. A range of strategies can be proposed in order to solve this problem, including:

(a) *Descriptive and exploratory methods* for determining the properties of the clusters, either in terms of the observed data or by using secondary (background) information that has not yet been used in the classification process (e.g., Bock 1981);

(b) *A substance-related analysis* of the classes that looks for intuitive, 'natural' explanations or interpretations of the differences existing among the obtained classes;

(c) *A quantitative evaluation* of the benefits that can be gained by using the constructed classification in practice (e.g., in marketing, administration, official statistics, libraries etc.);

(d) *A qualitative or quantitative validation of the clusters* by comparing them to classifications obtained from other clustering methods, from alternative data (for the same objects) or from traditional systematics (see Lapointe 1997);

(e) *Inferential statistics* which is based on probabilistic models and proceeds essentially by classical hypothesis testing.

It is this latter issue (e) that will be discussed in this section. In fact, there is a long list of clustering-related questions that can be investigated by hypothesis testing. A full account is given in Bock (1996a). Here we will address only two of the major problems: Testing for homogeneity and checking the adequacy of a calculated classification (model).

3.1 Testing for homogeneity

Each clustering algorithm provides a classification of objects even if the underlying data show no cluster structure and are homogeneous in some sense. In this case, the resulting classification will typically be an artifact of the algorithm and might lead to wrong conclusions, e.g. when searching for 'natural' classes of objects. In order to avoid this error, it will be useful to check, before applying a clustering algorithm, some hypothesis of 'homogeneity' or 'randomness' and perform clustering only if this hypothesis is rejected. Depending on the type of data, the following *models for 'homogeneity'* or *'randomness'* have been considered in this context (also see Bock (1985, 1989a, 1996a) and Gordon (1996, 1997)):

H_G : $X_1, ..., X_n$ are uniformly distributed in a finite domain $G \in R^p$ (to be estimated from the data).

H^{uni}: $X_1, ..., X_n$ have the same (often unknown) *unimodal* distribution density $f_0(x)$ in R^p, with the special case:

H_1^N: $X_1, ..., X_n$ all have the same p-dimensional normal distribution $\mathcal{N}_p(\mu, \sigma^2 E_p)$.

H^D: All $\binom{n}{2}$ dissimilarities $D_{kl}, k < l$, are i.i.d., each with an arbitrary (or a specified) continuous distribution density $f(d)$; this implies the two following models:

H^{perm}: All $\binom{n}{2}$! rankings of the dissimilarities $D_{kl}, k < l$, are equally probable.

$H^{n,M}$: For each fixed number M of 'similar' pairs of objects $\{k, l\}$ (i.e. with D_{kl} smaller that a given threshold $d > 0$), these M links are purely randomly assigned to the set of all $\binom{n}{2}$ pairs of objects.

For testing one of these hypotheses versus a *general alternative of non-homogeneity* we may consider, e.g., the empirical distribution of the Euclidean distances $D_{kl} := \|X_k - X_l\|$, the nearest neighbour distances $D_k = \min_{l \neq k}\{D_{kl}\}$, maximin distances or *gap statistics* such as $T := \max_k\{D_k\}$ or T^*, the radius of the maximum ball that can be placed in the window G without containing a data point X_k, and various other test statistics. A survey of the resulting homogeneity tests is given, e.g., by Dubes and Jain (1979), Dubes and Zeng (1987) and Bock (1985, 1989a, 1996a).

A better power performance is to be expected from tests which are tailored to some *clustering alternative* of the type that has been defined in section 2. For example, there exists a range of *tests for bimodality* and *multimodality* which are suited to the concept of mode clusters or mixtures (Silverman 1981, Hartigan 1985, Sawitzki 1996) or related to single linkage clusters (i.e., the connected components of suitable similarity graphs). In particular, the graph-theoretical and combinatorial methods proposed by Ling (1973), Godehardt (1990), Godehardt and Horsch (1996), and Van Cutsem and Ycart (1996a,b, 1997) are designed to test the hypotheses H^{perm} and $H^{n,N}$, just by comparing the single linkage hierarchy calculated from the data to the one to be expected for *random* data under these hypotheses.

3.2 Testing versus parametric clustering models

The fixed-partition clustering model H_m, (2.1), and the corresponding mixture model H_m^{mix}, (2.16), are specified by a parametric density family (typically with a unimodal density $f(x; \vartheta)$), and they reduce to the same homogeneity model H_1 for $m = 1$ or $\vartheta_1 = \cdots = \vartheta_m$. Thus testing H_1 against the alternatives H_m or H_m^{mix} can, in principle, be performed with the *likelihood ratio test (LRT)* statistics

$$T_m := 2\log \frac{L_{H_m}}{L_{H_1}} = 2\log \frac{L_{H_m}(C_n^*)}{L_{H_1}} \quad \text{and} \quad T_m^{mix} := 2\log \frac{L_{H_m^{mix}}}{L_{H_1}} \quad (3.1)$$

where L_{H_m}, $L_{H_m^{mix}}$ and L_{H_1} denote the likelihood of the data maximized under the models H_m, H_m^{mix} and H_1, respectively, for a fixed number $m > 1$ of clusters, and C_n^* is the optimum m-partition resulting from (2.1) or (2.3). Unfortunately, the classical asymptotic LRT theory (yielding χ^2 distributions for T_m) fails for these clustering models, due either to the fact that H_1 is on the 'boundary' of H_m^{mix} under the parametrization (2.16) or to the discrete character of the parameter C in the fixed-partition model (2.1) (see also Hartigan 1985, Bock 1996a). However, there exist some special investigations relating to these two test criteria.

3.2.1 Testing versus the mixture model

The case of a one-dimensional normal mixture $f(x) \sim \sum_{i=1}^m \mathcal{N}(\mu_i, \sigma^2)$ has been investigated by Everitt (1981), Thode et al. (1988) and Böhning (1994) who present simulated percentiles of T_m^{mix} under $\mathcal{N}(0,1)$ for various sample sizes n and $m = 2$. The power of this LRT is investigated by Mendell et al. (1991, 1993) where it results, e.g., that $n \geq 50$ is needed to have 50% power to detect a difference $|\mu_1 - \mu_2| \geq 3\sigma$ with $0.1 \leq \pi_1 \leq 0.9$ (also see Milligan (1981, 1996), Bock (1996a)). The paper of Böhning (1994) extends these results to the case of one-dimensional exponential families and shows that inside these families, the asymptotic distribution of T_m^{mix} remains (approximately) stable. For more general cases we recommend to determine suitable percentiles of T_m^{mix} by simulations instead of recurring, e.g., to heuristic formulas.

More theoretical investigations are presented by Titterington et al. (1985), Titterington (1990), Goffinet et al. (1992), and Böhning et al. (1994). Those authors show, for two-component mixtures (with partially fixed parameters), that under H_1 the asymptotic distribution of T_2^{mix} (for $n \to \infty$) is a mixture of the unit mass at 0 and a χ_1^2 distribution. Ghosh and Sen (1985) show that the asymptotic distribution of T_2^{mix} is closely related to a suitable Gaussian process, and Bardai and Garel (1994) present the corresponding tabulations. – An alternative method for testing H_1 versus H_m^{mix} has been proposed by Bock (1977, 1985, 1996a, chap. 6.6) and uses, as test statistics, the average similarity among the sample points $x_1, ..., x_n$ which should be larger under H_1 than under the mixture alternative.

3.2.2 Testing for the fixed-classification model; the max-F test

In contrast to the mixture model, the fixed-classification model (2.1) is defined in terms of an unknown m-partition $C = (C_1, ..., C_m)$ of the n objects, for a fixed number m of classes and a given family of densities $f(x; \vartheta)$. Therefore the LRT using T_m, (3.1), can be interpreted either as

- a test for homogenity H_1 versus the clustering structure H_m;
- a test for the significance or suitability of the calculated (optimum) classification C_n^* of the n objects;
- a test for the existence of $m > 1$ 'natural' classes in the data set (versus the hypothesis of one class only).

Thus, depending on the interpretation, the analysis of the LR test has many facets.

Under the assumption that $X_1, ..., X_n$ are i.i.d, all with the same density $f(x)$ (describing either homogeneity or a mixture of distributions) the almost sure convergence and the asymptotic normality of the parameter estimates $\hat{\vartheta}_i$ has been intensively studied, e.g., in Bryant and Williamson (1978), Pollard (1982), Pärna (1986), and Bryant (1991). It appears that the asymptotic behaviour of these estimates is closely related to the solution of the 'continuous' clustering problem

$$G(\mathcal{B}, \theta) \quad := \quad \sum_{i=1}^{m} \int_{B_i} [-\log f(x; \vartheta_i)] \cdot f(x) \, dx \quad \rightarrow \quad \min_{\mathcal{B}, \theta} \qquad (3.2)$$

where minimization is over all m-partitions $\mathcal{B} = (B_1, ..., B_m)$ of R^p and all parameter vectors $\theta = (\vartheta_1, ..., \vartheta_m)$. Instead of going into details here (see Bock 1996a) we will focus on the special case of the normal distribution model described in section 2.1.1 where $X_k \sim N(\mu_i, \sigma^2 E_p)$ for $k \in C_i$ under H_m, and $X_k \sim N(\mu, \sigma^2 E_p)$ for all k in case of H_1. Here the LR test reduces to the intuitive max-F test defined by

$$k_{mn}^* \quad := \quad \max_{\mathcal{C}} \frac{\sum_{i=1}^{m} |C_i| \cdot \|\bar{x}_{C_i} - \bar{x}\|^2}{\sum_{i=1}^{m} \sum_{k \in C_i} \|x_k - \bar{x}_{C_i}\|^2} \qquad (3.3)$$

$$=: \quad \max_{\mathcal{C}} \frac{SSB(\mathcal{C})}{SSW(\mathcal{C})} = \frac{SSB(\mathcal{C}_n^*)}{SSW(\mathcal{C}_n^*)} \begin{cases} > c & \text{decide for } H_m \\ \leq c & \text{decide for } H_1 \end{cases} \qquad (3.4)$$

where C_n^* minimizes the variance criterion (2.7). In this case the continuous optimization problem (3.2) reduces to

$$G(\mathcal{B}, \mu) \quad := \quad \sum_{i=1}^{m} \int_{B_i} \|x - \mu_i\|^2 \cdot f(x) \, dx \quad \rightarrow \quad \min_{\mathcal{B}, \mu} =: G_m^* \qquad (3.5)$$

as an analogue to (2.6), and its solution $\mathcal{B}^*$ is necessarily a *stationary partition*, i.e. a minimum-distance partition of R^p generated by its own class centroids, the conditional expectations $\mu_i^* := E_f[X | X \in B_i^*]$, $i = 1, ..., m$.

For the one-dimensional case, the optimum partition $\mathcal{B}^*$ of R^1 is given by Cox (1957)

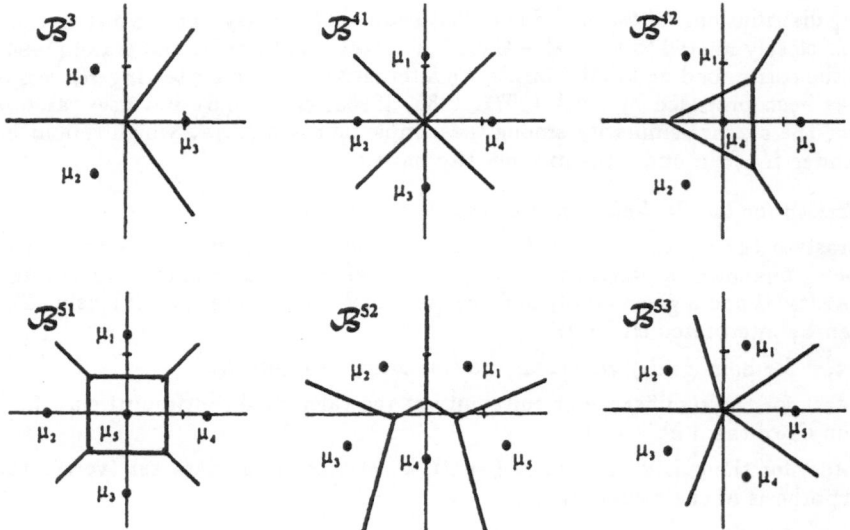

m	$\mathcal{B}$	$\mu_i = E_f[X \mid X \in B_i]$ $i = 1, ..., m$	$p_i = P(X \in B_i)$ $i = 1, ..., m$	$G_m := G(\mathcal{B}, \mu)$ $\kappa_m := (2 - G_m)/G_m$
2	$\mathcal{B}^2$	$\mu_i = \begin{pmatrix} \pm 0.79789 \\ 0 \end{pmatrix}$	$p_i \equiv 1/2$	$G_2 = 1.36338^{**}$ $\kappa_2 = 0.46694$
3	$\mathcal{B}^3$	$\mu_i = 1.03648 \cdot \begin{pmatrix} \cos(2\pi i/3) \\ \sin(2\pi i/3) \end{pmatrix}$	$p_i \equiv 1/3$	$G_3 = 0.92570^{**}$ $\kappa_3 = 1.16052$
4	$\mathcal{B}^{4,1}$	$\mu_i = 1.12838 \cdot \begin{pmatrix} \cos(\pi i/2) \\ \sin(\pi i/2) \end{pmatrix}$	$p_i \equiv 1/4$	$G_4 = 0.72676^{**}$ $\kappa_4 = 1.75194$
	$\mathcal{B}^{4,2}$	$\mu_i = 1.27910 \cdot \begin{pmatrix} \cos(2\pi i/3) \\ \sin(2\pi i/3) \end{pmatrix}$ $i = 1, 2, 3$ $\mu_4 = (0, 0)'$	$p_i = 0.24034$ $i = 1, 2, 3$ $p_4 = 0.27898$	$G_4 = 0.82034$ $\kappa_4 = 1.43801$
5	$\mathcal{B}^{5,1}$	$\mu_i = 1.36334 \cdot \begin{pmatrix} \cos(\pi i/2) \\ \sin(\pi i/2) \end{pmatrix}$ $i = 1, ..., 4,$ $\mu_5 = (0, 0)'$	$p_i = 0.18636$ $i = 1, ..., 4$ $p_5 = 0.25457$	$G_5 = 0.61448^{**}$ $\kappa_5 = 2.25477$
	$\mathcal{B}^{5,2}$	$\mu_{1,2} = \begin{pmatrix} \pm 0.70505 \\ 0.87119 \end{pmatrix}$ $\mu_{5,3} = \begin{pmatrix} \pm 1.34917 \\ -0.52106 \end{pmatrix}$ $\mu_4 = \begin{pmatrix} 0 \\ -0.89064 \end{pmatrix}$	$p_{1,2} = 0.22217$ $p_{5,3} = 0.14580$ $p_4 = 0.26405$	$G_5 = 0.62246$ $\kappa_5 = 2.21305$
	$\mathcal{B}^{5,3}$	$\mu_i = 1.17246 \cdot \begin{pmatrix} \cos(2\pi i/5) \\ \sin(2\pi i/5) \end{pmatrix}$	$p_i \equiv 1/5$	$G_5 = 0.62533$ $\kappa_5 = 2.19830$

Tab. 1: Stationary partitions $\mathcal{B} = (B_1, ..., B_m)$ of R^2 with $m = 2, ..., 5$ classes for the continuous variance criterion (3.5) if $f \sim N_2(0, E_2)$, with the class centers μ_i, class percentages p_i and criterion values $G_m := G(\mathcal{B}, \mu)$. The SSB $\kappa_m = (2 - G_m)/G_m$ is the continuous analogue to k^*_{mn}, (3.3). ** marks the optimum or best known m-partitions.

and Bock (1974, p. 179) for the cases $f \sim \mathcal{N}(0,1)$ and $f \sim U([-\sqrt{3}, \sqrt{3}])$ with variance 1. For two- and three-dimensional normals $f \sim \mathcal{N}_p(\mu, E_p)$ a range of stationary partitions $\mathcal{B}$ has been calculated by Baubkus (1985) ($m = 2, ..., 6$) and Flury (1993), the ellipsoidal case has been considered by Baubkus (1985), Flury (1993, $m = 4$), Kipper and Pärna (1992), Tarpey et al. (1995) and Jank (1996, $2 \leq m \leq 4$). For the two-dimensional normal $\mathcal{N}_2(0, E_2)$ some stationary partitions as well as their numerical characteristics are reproduced in Tab. 1; for example, the three quite distinct 5-partitions $\mathcal{B}^{5,1}$, $\mathcal{B}^{5,2}$, $\mathcal{B}^{5,3}$ differ in their G_5-values by no more than 0.013 (for other cases see Bock 1996a). It is conjectured that for $m = 2$ to 5 this list includes the optimum partitions of R^2 (see the asterisks ** in Tab. 1), but a formal proof of optimality exists only for $m = 2$ and 3 classes (Baubkus 1985).

In order to apply the max-F test, the critical threshold (percentile) c must be calculated from the null distribution of k_{mn}^* under some $f \sim H_1$. While this distribution is intractable for a finite n, the asymptotic normality of g_{mn}^* and k_{mn}^* has been proved under some regularity conditions on f, with asymptotic expectations G_m^* and $\kappa_m^* := (E_f[\|X - E_f[X]\|^2] - G_m^*)/G_m^*$, the continuous analogues of the minimum variance and max_F criteria (2.7) and (3.3), respectively (see Bryant and Williamson 1978, Hartigan 1978 (for $p = 1$), Bock 1985 (for $p \geq 1$). Since the regularity conditions include the uniqueness of the optimum partition $\mathcal{B}^*$ of (3.5), these results cannot be applied for the rotation-invariant density $f \sim \mathcal{N}_p(0, E_p)$ if $p > 1$, but suitable simulations have been conducted (see Hartigan 1978, Bock 1996a, Jank 1996). For example, Jank (1996) and Jank and Bock (1996) found that for $n \geq 100$ (in particular: $n = 1000$) the null distribution of the standardized values $(g_{mn}^* - a)/b$ and $(k_{mn}^* - a)/b$ is satisfactorily approximated by a $\mathcal{N}(0,1)$ distribution (in the range $[-2, 2]$, say) if a and b chosen to be the empirical mean and standard deviation of the optimum values g_{mn}^* and k_{mn}^*, respectively (results of the k-means algorithm) from $N = 1600$ simulations of $\{x_1, ..., x_n\}$ under $\mathcal{N}_2(0, E_2)$ and $m = 2, ..., 5$.

3.2.3 The LRT for the convex cluster case; convex cluster tests

A less investigated case is provided by clustering models where each class is characterized by a uniform distribution on a convex domain of R^p. To be specific, we consider the following three *convex clustering models* which all involve a system of m unknown non-overlapping convex sets $D_1, ..., D_m \subset R^p$ (to be estimated from the data):

H_m *Fixed-classification model:*
> $X_k \sim U(D_i)$ for all $k \in C_i$, with an unknown m-partition $\mathcal{C} = (C_1, ..., C_m)$ of $\mathcal{O}$.

H_m^{mix} *Mixture model:*
> $X_k \sim \sum_{i=1}^m \pi_i \cdot U(D_i)$ for $k = 1, ..., n$.

H_m^{uni} *Pseudo-mixture model:*
> $X_k \sim U(D_1 + \cdots + D_m)$ for $k = 1, ..., n$ (constrained to the union $D_1 + \cdots + D_m$).

These models have been introduced Bock (1997) as a generalization of some work by Rasson et al. (1988, 1994) related to H_m^{uni} (also see Rasson 1997).

Here we consider the problem of testing the hypothesis of homogeneity H_G (i.e., $X_k \sim U(G)$ for some unknown convex $G \subset R^p$) versus one of these clustering alternatives. We find the following LR tests where $G_n := con\{x_1, ..., x_n\}$ denotes the convex hull of all n data points, $D_i := H(C_i)$ the convex hull of all data points belonging to a class $C_i \subset \mathcal{O}$, and c is a critical threshold to be obtained from the H_G distribution of the test statistics:

- H_G versus H_m:

$$T_m := -\sum_{i=1}^{m} \frac{|C_i^*|}{n} \cdot \log \frac{vol_p(H(C_i^*))}{vol_p(G_n)} \quad \begin{cases} > c & \text{decide for } H_m \\ \leq c & \text{accept uniformity } H_G. \end{cases} \quad (3.6)$$

where the partition $C^* = (C_1^*, ..., C_m^*)$ minimizes the clustering criterion (2.8).

- H_G versus H_m^{mix}:

$$T_m^{mix} := \sum_{i=1}^{m} \frac{|C_i^*|}{n} \log \left[\frac{|C_i^*|}{n} \Big/ \frac{vol_p(H(C_i^*))}{vol_p(G_n)} \right] \quad \begin{cases} > c & \text{decide for } H_m^{mix} \\ \leq c & \text{accept uniformity } H_G. \end{cases} \quad (3.7)$$

where $C^* = (C_1^*, ..., C_m^*)$ is the partition that minimizes the clustering criterion

$$g_m^{mix}(C) := \sum_{i=1}^{m} |C_i| \cdot \log \frac{vol_p(H(C_i))}{|C_i|} \rightarrow \min_{C}. \quad (3.7)$$

over all partitions $C = (C_1, ..., C_m)$ with disjoint convex hulls $H(C_1), ..., H(C_m)$, all with a positive volume.

- H_G versus H_m^{uni}:
  Denote by $C^* = (C_1^*, ..., C_m^*)$ the m-partition which minimizes the *volume clustering criterion*

$$\lambda(C) := vol_p(H(C_1)) + \cdots + vol_p(H(C_m)) \rightarrow \min_{C}, \quad (3.8)$$

i.e. the sum of the volumes of the m class-specific convex hulls $\hat{D}_i := H(C_i) \subset G_n$ (supposed to be non-overlapping). Then $V_n := G_n - (H(C_1^*) + \cdots + H(C_m^*))$ is the maximum 'empty space' that is left in G_n outside of the cluster domains $\hat{D}_i$, and the LRT reduces to the following *empty space test*:

$$T_m^{uni} := \frac{vol_p(V_n)}{vol_p(G_n)} \quad \begin{cases} > c & \text{decide for clustering } H_m^{uni} \\ \leq c & \text{accept uniformity } H_G. \end{cases} \quad (3.9)$$

This test is a multivariate generalization of various one-dimensional gap tests (see Bock (1989a, 1996a, 1997), Rasson and Kubushishii (1994), Hardy (1997)).

Since all these test criteria are invariant against any regular affine transformation of the data, their distribution under H_G can be determined (e.g., by simulations) for some standardized form of G such as the unit square or the unit ball in R^p, without restriction of generality.

3.3 Power considerations and the comparative assessment of classifications

Practical experience and theoretical investigations have shown that the power of clustering tests is far from being satisfactory. For example, in normal distribution mixture cases a considerable separation of the classes is needed in order to detect a hidden clustering structure with a satisfactorily large probability (see Everitt 1981, Mendell et al. 1993, Bock 1996a). As a matter of fact, this difficulty seems to be an intrinsic feature of the classification problem, and not only a technical deficiency of our statistical tools. Insofar the result of any clustering test must be interpreted with care, more in an indicative than in a convincing sense.

Quite generally, it may be doubted if the classical hypothesis testing paradigm with only two alternative decisions: *acceptance* and *rejection* of a hypothesis H_0, will be appropriate for the clustering framework where it would be much more realistic and useful to distinguish various *grades of classifiability* which interpolate between the extreme cases of *'homogeneity or randomness'* and an *'obvious clustering structure'*. Instead of defining corresponding quantitative measures of classifiabilty here, we propose a more qualitative approach, the *comparative assessment of classifications (CAC)*: It proceeds by defining, prior to any clustering algorithm, some 'benchmark clustering situations' H_m^ϵ, i.e., data configurations or distribution models which show a more or less marked classification structure, indexed by ϵ (not necessarily a real number, but possibly a measure of class separation; see below). These configurations can be selected in cooperation of practitionners and statisticians. Then, after having calculated a special (e.g., optimum) classification C_n^* for the given data $\{x_1, ..., x_n\}$, we compare this classification with those to be expected under H_m^ϵ, for various degrees ϵ. Thus we place the observed data into a network of various different clustering situations H_m^ϵ and in order to get an idea of their underlying structure.

In the case of the fixed-classification normal model H_m, (2.1), with class-specific distributions $N_p(\mu_i, \sigma^2 E_p)$, this idea can be realized as follows:

1. Determine suitably parametrized benchmark clusterings H_m^ϵ, e.g.:

    - A partition $\mathcal{C}$ of $\mathcal{O}$ with m classes C_i of equal sizes $n_i = |C_i| = n/m$, whose class centroids μ_i are sufficiently different, e.g., $||\mu_i - \mu_j||/\sigma \geq \epsilon$ for all $i \neq j$, or $\frac{1}{m}\sum_{i=1}^m ||\mu_i - \overline{\mu}||^2/\sigma^2 \geq \epsilon$.
    - A normal mixture $\sum_{i=1}^m \pi_i \cdot \mathcal{N}_p(\mu_i, \sigma^2 E_p)$ with $\epsilon := \sum_{i=1}^m \pi_i \cdot ||\mu_i - \overline{\mu}||^2$.

2. Consider the LRT statistics T_m or, equivalently, the $max - F$ statistics k_{mn}^* for the hypothesis $H_0 : \mu_1 = \cdots = \mu_m$.

3. Determine or estimate some characteristics Q^ϵ of the (untractable) probability distribution of T_m or k_{mn}^* under the benchmark situations H_m^ϵ selected in 1., e.g., by simulating a large number of data sets $\{\tilde{X}_1, \cdots, \tilde{X}_n\}$ under H_m^ϵ and calculating the empirical mean, median or some other empirical percentile of the resulting values of T_m or k_{mn}^*.

4. Compare the values T_m or k_{mn}^* calculated from the original data $\{x_1, ..., x_n\}$ (eventually after a suitable standardization) to the characteristics Q^ϵ of the benchmark situations.

5. The *clustering tendency* or *classifiability* of the data $\{x_1, ..., x_n\}$ and the relevance of the calculated classification C_n^* is then described, illustrated and quantisized (by ϵ) by confronting them to those benchmark situations H_m^ϵ which show a weaker clustering behaviour (in the sense that, e.g., $Q^\epsilon \leq T_m$ or $\leq k_{mn}^*$) and, on the other hand, to those which describe a stronger clustering structure (i.e., where the converse inequality holds).

It is obvious that this strategy CAC is related to a formal test of a hypothesis $\epsilon \leq \epsilon_0$ versus $\epsilon > \epsilon_0$), but is more flexible due to the arbitrary selection of suitable benchmark situations. Its generalization to other clustering models is obvious.

4. Final remarks

In this paper we have described various probabilistic and inferential tools for cluster analysis. These methods provide a firm basis for deriving suitable clustering strategies and allow for a quantitative evaluation of classification results and clustering methods, including error probabilities and risk functions. In particular, various test statistics can safeguard against a too rash acceptance of a clustering structure and help to validate calculated classifications.

On the other hand, the application of these methods is not at all easy and self-evident: Problems arise, e.g., when selecting a suitable clustering model and an appropriate family of densities $f(x, \vartheta)$, or when different types of cluster shapes will simultaneously occur in the same data set. We have seen that the probability distribution of many test statistics is hard to obtain in many cases. Moreover, our analysis is always based on *only one sample from the n objects* such that we cannot evaluate the stability or variation of the resulting classification (as it would be the case if repeated samples were available for the same objects).

When comparing the risks and benefits of probability-based versus deterministic clustering approaches (which proceed, e.g., by intuitive clustering criteria or heuristic algorithms) we see that these same deficiencies exist, in some other and disguised form, for the latter methods as well. It is recommended here to combine both approaches in an exploratory way and thereby profit from both points of view. The CAC strategy presented above is an example for such an analysis.

References:

Anderson, J.J. (1985): Normal mixtures and the number of clusters problem. *Computational Statistics Quarterly* 2, 3-14.

P. Arabie, L. Hubert and G. De Soete (eds.) (1996): *Clustering and Classification.* World Science Publishers, River Edge/NJ.

Baubkus, W. (1985): *Minimizing the variance criterion in cluster analysis: Optimal configurations in the multidimensional normal case.* Diploma thesis, Institute of Statistics, Technical University of Aachen, Germany.

Berdai, A., and B. Garel (1994): Performances d'un test d'homogénéité contre une hypothèse de mélange gaussien. *Revue de Statistique Appliquée* 42 (1), 63-79.

Bernardo, J.M. (1994): Optimizing prediction with hierarchical models: Bayesian clustering. In: P.R. Freeman, A.F.M. Smith (Eds.): *Aspects of uncertainty.* Wiley, New York, 1994, 67-76.

Binder, D.A. (1978): Bayesian cluster analysis. *Biometrika* 65, 31-38.

Bock, H.H. (1968): *Statistische Modelle für die einfache und doppelte Klassifikation von normalverteilten Beobachtungen.* Dissertation, Univ. Freiburg i. Brsg., Germany.

Bock, H.H. (1969): The equivalence of two extremal problems and its application to the iterative classification of multivariate data. Report of the Conference 'Medizinische Statistik', Forschungsinstitut Oberwolfach, February 1969, 10pp.

Bock, H.H. (1972): Statistische Modelle und Bayes'sche Verfahren zur Bestimmung einer unbekannten Klassifikation normalverteilter zufälliger Vektoren. *Metrika* 18 (1972) 120-132.

Bock, H.H. (1974): *Automatische Klassifikation (Clusteranalyse).* Vandenhoeck & Ruprecht, Göttingen, 480 pp.

Bock, H.H. (1977): On tests concerning the existence of a classification. In: *Proc. First Symposium on Data Analysis and Informatics, Versailles, 1977, Vol. II.* Institut de Recherche d'Informatique et d'Automatique (IRIA), Le Chesnay, 1977, 449-464.

Bock, H.H. (1984): Statistical testing and evaluation methods in cluster analysis. In: J.K. Ghosh & J. Roy (Eds.): *Golden Jubilee Conference in Statistics: Applications and new directions.* Calcutta, December 1981. Indian Statistical Institute, Calcutta, 1984, 116-146.

Bock, H.H. (1985): On some significance tests in cluster analysis. *J. of Classification* 2, 77-108.

Bock, H.H. (1986): Loglinear models and entropy clustering methods for qualitative data. In: W. Gaul, M. Schader (Eds.), *Classification as a tool of research.* North Holland, Amsterdam, 1986, 19-26.

Bock, H.H. (1987): On the interface between cluster analysis, principal component analysis, and multidimensional scaling. In: H. Bozdogan and A.K. Gupta (eds.): *Multivariate statistical modeling and data analysis*. Reidel, Dordrecht, 1987, 17-34.

Bock, H.H. (Ed.) (1988): *Classification and related methods of data analysis*. Proc. First IFCS Conference, Aachen, 1987. North Holland, Amsterdam.

Bock, H.H. (1989a): Probabilistic aspects in cluster analysis. In: O. Opitz (Ed.): *Conceptual and numerical analysis of data*. Springer-Verlag, Heidelberg, 1989, 12-44.

Bock, H.H. (1989b): *A probabilistic clustering model for graphs and similarity relations*. Paper presented at the Fall Meeting 1989 of the Working Group 'Numerical Classification and Data Analysis' of the Gesellschaft für Klassifikation, Essen, November 1989.

Bock, H.H. (1994): Information and entropy in cluster analysis. In: H. Bozdogan et al. (Eds.): *Multivariate statistical modeling*, Vol. II. Proc. 1st US/Japan Conference on the Frontiers of Statistical Modeling: An Informational Approach. Univ. of Tennessee, Knoxville, 1992. Kluwer, Dordrecht, 1994, 115-147.

Bock, H.H. (1996a): Probability models and hypotheses testing in partitioning cluster analysis. In: P. Arabie et al. (Eds.), 1996, 377-453.

Bock, H.H. (1996b): Probabilistic models in cluster analysis. *Computational Statistics and Data Analysis* 22 (in press).

Bock, H.H. (1996c): Probabilistic models in partitional cluster analysis. In: A. Ferligoj and A. Kramberger (Eds.): *Developments in data analysis*. Metodološki zvezki, 12, Faculty of Social Sciences Press (Fakulteta za druzbene vede, FDV), Ljubljana, 1996, 3-25.

Bock, H.H. (1996d): *Probabilistic models and statistical methods in partitional classification problems*. Written version of a Tutorial Session organized by the Japanese Classification Society and the Japan Market Association, Tokyo, April 2-3, 1996, 50-68.

Bock, H.H. (1997): Probability models for convex clusters. In: R. Klar and O. Opitz (Eds.): *Classification and knowledge organization*. Springer-Verlag, Heidelberg, 1997 (to appear).

Bock, H.H., and W. Polasek (Eds.) (1996): *Data analysis and information systems: Statistical and conceptual approaches*. Springer-Verlag, Heidelberg, 1996.

Böhning, D., Dietz, E., Schaub, R., Schlattmann, P., & Lindsay, B.G. (1994): The distribution of the likelihood ratio for mixtures of densities from the one-parameter exponential family. *Annals of the Institute of Mathematical Statistics* 46, 373-388.

Bryant, P. (1988): On characterizing optimization-based clustering methods. *J. of Classification* 5, 81-84.

Bryant, P.G. (1991): Large-sample results for optimization-based clustering methods. *J. of Classification* 8, 31-44.

Bryant, P.G., and J.A. Williamson (1978): Asymptotic behaviour of classification maximum likelihood estimates. *Biometrika* 65, 273-281.

Céleux, G., & Diebolt, J. (1985): The SEM algorithm: A probabilistic teacher algorithm derived from the EM algorithm for the mixture problem. *Computational Statistics Quarterly* 2, 73-82. Cox, D.R. (1957): A note on grouping. *J. Amer. Statist. Assoc.* 52, 543-547.

Cressie, N. (1991): *Statistics for spatial data*. Wiley, New York.

Diday, E. (1973): *Introduction à l'analyse factorielle typologique*. Rapport de Recherche no. 27, IRIA, Le Chesnay, France, 13 pp.

Diday, E., Y. Lechevallier, M. Schader, P. Bertrand, and B. Burtschy (Eds.) (1994): *New approaches in classification and data analysis*. Studies in Classification, Data Analysis, and Knowledge Organization, vol. 6. Springer-Verlag, Heidelberg, 186-193.

Dubes, R., and Jain, A.K. (1979): Validity studies in clustering methodologies. *Pattern Recognition* 11, 235-254.

Dubes, R.C., and Zeng, G. (1987): A test for spatial homogeneity in cluster analysis. *J. of Classification* 4, 33-56.

Everitt, B. S. (1981): A Monte Carlo investigation of the likelihood ratio test for the number of components in a mixture of normal distributions. *Multivariate Behavioural Research* 16, 171-180.

Fahrmeir, L., Hamerle, A. and G. Tutz (Eds.) (1996): *Multivariate statistische Verfahren*. Walter de Gruyter, Berlin - New York.

Fahrmeir, L., Kaufmann, H.L., and H. Pape (1980): Eine konstruktive Eigenschaft optimaler Partitionen bei stochastischen Klassifikationsproblemen. *Methods of Operations Research* 37, 337-347.

Flury, B.D. (1993): Estimation of principal points. *Applied Statistics* 42, 139-151.

W. Gaul & D. Pfeifer (Eds.) (1996): *From data to knowledge. Theoretical and practical aspects of classification, data analysis and knowledge organization.* Springer-Verlag, Heidelberg.

Ghosh, J. K., & Sen, P. K. (1985): On the asymptotic performance of the log likelihood ratio statistic for the mixture model and related results. In: L.M. LeCam, R.A. Ohlsen (Eds.): Proc. Berkeley Conference in honor of Jerzy Neyman and Jack Kiefer. Vol.II, Wadsworth, Monterey, 1985, 789-806.

Godehardt, E. (1990): *Graphs as structural models. The application of graphs and multigraphs in cluster analysis.* Friedrich Vieweg & Sohn, Braunschweig, 240pp.

Godehardt, E., and Horsch, A. (1996): Graph-theoretic models for testing the homogeneity of data. In: W. Gaul & D. Pfeifer (Eds.), 1996, 167-176.

Goffinet, B., Loisel, P., and B. Laurent (1992): Testing in normal mixture models when the proportions are known. *Biometrika* 79, 842-846.

Gordon, A.D. (1994): Identifying genuine clusters in a classification. *Computational Statistics and Data Analysis* 18, 561-581.

Gordon, A.D. (1996): Null models in cluster validation. In: W. Gaul and D. Pfeifer (Eds.), 1996, 32-44.

Gordon, A.D. (1997a): Cluster validation. This volume.

Gordon, A.D. (1997b): How many clusters? An investigation of five procedures for detecting nested cluster structure. This volume.

Hardy, A. (1997): A split and merge algorithm for cluster analysis. This volume.

Hartigan, J.A. (1978): Asymptotic distributions for clustering criteria. *Ann. Statist.* 6, 117-131.

Hartigan, J.A. (1985): Statistical theory in clustering. *J. of Classification* 2, 63-76.

Hayashi, Ch. (19??): ......

Jain, A.K., and Dubes, R.C. (1988): *Algorithms for clustering data.* Prentice Hall, Englewood Cliffs, NJ.

Jank, W. (1996): *A study on the varaince criterion in cluster analysis: Optimum and stationary partitions of R^p and the distribution of related clustering criteria.* (In German). Diploma thesis, Institute of Statistics, Technical University of Aachen, Aachen, 204pp.

Jank, W., and Bock, H.H. (1996): *Optimal partitions of R^2 and the distribution of the variance and max-F criterion.* Paper presented at the 20th Annual Conference of the Gesellschaft für Klassifikation, Freiburg, Germany, March 1996.

Lapointe, F.-J. (1997): To validate and how to validate? That is the real question. This volume.

Ling, R.F. (1973): A probability theory of cluster analysis. *J. Amer. Statist. Assoc.* 68, 159-164.

McLachlan, G.J., and K.E. Basford (1988): *Mixture models. Inference and applications to clustering.* Marcel Dekker, New York - Basel.

Mendell, N.P., Thode, H.C., & Finch, S.J. (1991): The likelihood ratio test for the two-component normal mixture problem: power and sample-size analysis. *Biometrics* 47, 1143-1148. Correction: 48 (1992) 661.

Mendell, N.P., Finch, S.J., and Thode, H.C. (1993): Where is the likelihood ratio test powerful for detecting two-component normal mixtures? *Biometrics* 49, 907-915.

Milligan, G. W. (1981): A review of Monte Carlo tests of cluster analysis. *Multivariate Behavioural Research* 16, 379-401.

Milligan, G.W. (1996): Clustering validation: Results and implications for applied analyses. In: P. Arabie et al. (Eds.), 1996, 341-375.

Milligan, G. W., and M.C. Cooper (1985): An examination of procedures for determining the number of clusters in a data set. *Psychometrika* 50, 159-179.

Pärna, K. (1986): Strong consistency of k-means clustering criterion in separable metric spaces. Tartu Riikliku Ülikooli, TOIMEISED 733, 86-96.

Kipper, S., and Pärna, K. (1992): Optimal k-centres for a two-dimensional normal distribution. Acta et Commentationes Universitatis Tartuensis, Tartu Ülikooli TOIMEISED 942, 21-27.

Pollard, D. (1982): A central limit theorem for k-means clustering. *Ann. Probab.* 10, 919-926.

Rasson, J.-P. (1997): Convexity methods in classification. This volume.

Rasson, J.-P., Hardy, A., and Weverbergh, D. (1988): Point process, classification and data analysis.

In: H.H. Bock (Ed.), 1988, 245-256.

Rasson, J.-P., and Kubushishi, T. (1994): The gap test: an optimal method for determining the number of natural classes in cluster analysis. In: E. Diday et al. (eds.), 1994, 186-193.

Ripley, B.D. (1981): *Spatial statistics*. Wiley, New York.

Sawitzki, G. (1996): The excess-mass approach and the analysis of multi-modality. In: W. Gaul and D. Pfeifer (Eds.), 1996, 203-211.

Silverman, B.W. (1981): Using kernel density estimates to investigate multimodality. *J. Royal Statist. Soc.* B 43, 97-99.

Snijders, T.A.B. and K. Nowicki (1996): Estimation and prediction for stochastic blockmodels for graphs with latent block structure. *J. of Classification 13* (in press).

Symons, M.J. (1981): Clustering criteria and multivariate normal mixtures. *Biometrics 37*, 35-43.

Tharpey, Th., Li, L., Flury, B.D. (1995): Principal points and self-consistent points of elliptical distributions. *Annals of Statistics 23*, 103-112.

Thode, H.C., Finch, S.J., & Mendell, N.R. (1988): Simulated percentage points for the null distribution of the likelihood ratio test for a mixture of two normals. *Biometrics 44*, 1195-1201.

Titterington, D.M. (1990): Some recent research in the analysis of mixture distributions. *Statistics 21*, 619-641.

Titterington, D.M., A.F.M. Smith and U.E. Makov (1985): *Statistical analysis of finite mixture distributions*. Wiley, New York.

Van Cutsem, B., and Ycart, B. (1996a): Probability distributions on indexed dendrograms and related problems of classifiability. In H.H. Bock and W. Polasek (Eds.), 1996, 73-87.

Van Cutsem, B., and Ycart, B. (1996b): Combinatorial structures and structures for classification. *Computational Statistics and Data Analysis* (in press).

Van Cutsem, B., and Ycart, B. (1997): This volume.

Cluster Validation

A. D. Gordon

Mathematical Institute, University of St Andrews,
North Haugh, St Andrews KY16 9SS, Scotland

Summary: Clustering algorithms can provide misleading summaries of data, and attention has been devoted to investigating ways of guarding against reaching incorrect conclusions, by validating the results of a cluster analysis. The paper provides an overview of recent work in this area of cluster validation. Material covered includes: the distinction between external, internal, and relative clustering indices; types of null model, including 'data-influenced' null models; tests of the complete absence of any class structure in a data set; and ways of assessing the validity of individual clusters, partitions of data into disjoint clusters, and hierarchical classifications. A discussion indicates areas in which further research seems desirable.

1. Introduction

The topic of classification addresses the problem of summarizing the relationships within a large set of objects by representing them as a smaller number of classes (or clusters) of objects with the property that objects in the same class are similar to one another and different to objects in other classes. On occasion, it can be relevant to obtain a nested set of such partitions of the objects, providing a hierarchical classification which summarizes the class structure present at several different levels in the data.

Many different clustering algorithms have been proposed for obtaining such classifications; see, for example, Bock (1974), Hartigan (1975), Gordon (1981) and Jain and Dubes (1988). However, clustering algorithms *impose* a classification on a data set even if there is no real class structure present in it. Further, classifications of the same data set obtained using different clustering criteria can differ markedly from one another. In effect, each clustering criterion implicitly specifies a model for the data; for example, Scott and Symons (1971) demonstrated links between Ward's (1963) incremental sum-of-squares clustering criterion and a spherical normal components model (see also Binder (1978), Marriott (1982) and Bock (1989, 1996)). If this model is inappropriate for the data set under investigation, a misleading summary will be provided.

Realization of this fact has led to the consideration by investigators of how one might specify appropriate clustering strategies for data sets. Some adaptive clustering procedures have been proposed (e.g., Rohlf, 1970; Diday and Govaert, 1977; Lefkovitch, 1978, 1980; Art et al., 1982), the rationale being that the clustering procedure adapts itself in response to the structure found in the data. In effect, such procedures just involve a more general model for the data, and it seems unrealistic to expect to be able to construct a model that is sufficiently general to be appropriate for the analysis of any data set that may be encountered.

Jardine and Sibson (1971) presented a list of properties that one might require of a clustering method, and proved that the single link method uniquely satisfies these conditions. This provides a valuable characterization of the single link method, but many investigators have regarded some of the conditions as undesirable. A less prescriptive approach was provided by Fisher and Van Ness (1971) and Van Ness (1973). These authors presented a list of properties which one might expect clustering procedures or the classes obtained from them to possess, and stated whether or not these properties were possessed by each of several standard clustering criteria. From background information about the data, an investigator might be able to specify some relevant conditions; the work of Fisher and Van Ness then indicates which clustering criteria could be relevant for the analysis of that particular data set. More than one clustering criterion might be indicated as relevant, and it can be informative to analyse a data set using several different clustering criteria, and synthesize the results in a consensus classification (e.g., Diday and Simon, 1976; Gordon, 1981, Chapter 6, 1996a), the rationale being that the results are less likely to be an artifact of a clustering criterion and more likely to give an accurate representation of the structure in the data.

Investigators have commonly combined a classification of a set of objects with a low-dimensional configuration of points, in which each object is represented by a different point, points which are close together representing objects that are similar to one another. Such configurations can then be examined by eye, to establish whether or not the points fall into distinct, well-separated classes; closed curves can also be drawn around each class provided by the clustering criterion (e.g., Rohlf, 1970; Shepard, 1974) to assist in the assessment.

Further indications of the properties of classes provided by a clustering procedure are given by various plots summarizing their homogeneity and isolation from one another (e.g., Gnanadesikan et al., 1977; Rousseeuw, 1987).

It is rarely the case that investigators know with certainty which clustering procedure should be used to analyse a set of objects, and increasing attention has been paid in recent years to providing ways of testing the validity of cluster structure that are more formal than those described in the previous paragraphs. This paper presents an overview of this topic of cluster validation; earlier reviews were given by Perruchet (1983), Bock (1985), Jain and Dubes (1988, Chapter 4) and Gordon (1995).

In statistics, it is common for an exploratory analysis to be carried out on a data set, for example to assist in model formulation. Such exploratory analyses are generally followed by confirmatory analyses, which are carried out on *new* data sets. This has rarely occurred in classification, in which interest has usually resided in the set of objects which is being classified, and not in some larger population of objects, from which the classified set is regarded as comprising a representative sample. In much of the work which is described in this paper, therefore, cluster generation and cluster validation are carried out on the same data set. This has major implications for cluster validation: for example, the classes provided by a clustering algorithm are likely to be more homogeneous than those contained in a random partition of a set of objects, and care must be taken to specify appropriate reference distributions for test statistics used to assess the validity of the obtained classes.

In this context, it is relevant to distinguish three different types of validation test, involving use of *external, internal,* or *relative* cluster indices. External tests compare a classification, or part of a classification, with external information that was not used in the construction of the classification. Internal tests compare a (part of a) classification with the original data. Relative tests compare several different classifications of the same set of objects; such tests are relevant in both external and internal investigations (F.-J. Lapointe, *pers. comm.*).

Cluster validation tests can be categorized into four classes, depending on the type of cluster structure under investigation (Jain and Dubes, 1988, Chapter 4): these are tests of (1) the complete absence of class structure, (2) the validity of an individual cluster, (3) the validity of a partition into disjoint classes. and (4) the validity of a hierarchical classification.

The remainder of the paper is organized as follows: the next section describes null models for the absence of cluster structure; the following four sections provide an overview of tests, categorized as in the previous paragraph; and the final section presents a discussion, indicating areas in which it would be useful to see further research.

2. Null models

The information about objects on which a classification study is to be undertaken is usually provided in one of two formats. A *pattern matrix* is an $n \times p$ matrix $X \equiv (x_{ik})$, where x_{ik} denotes the value of the kth variable describing the ith object. If all the variables are continuous, the objects can be represented by a set of n points in p-dimensional Euclidean space. A dissimilarity matrix is an $n \times n$ matrix $D \equiv (d_{ij})$, where d_{ij} denotes the dissimilarity between the ith and j^{th} objects. Dissimilarities are symmetric $(d_{ij} = d_{ji})$ and self-dissimilarities (d_{ii}) are zero, hence it suffices to present only the $n(n-1)/2$ entries in the lower triangle of D. The relationships within a set of objects are sometimes described in terms of a symmetric *similarity matrix*, but such data can also be analysed by straightforward modifications of the theory described in this paper.

Four main classes of null model, based on either a pattern matrix or a dissimilarity matrix, have been proposed, as described in the following four subsections.

2.1. Poisson model

This model assumes that the objects can be represented by points which are uniformly distributed in some region A of p-dimensional space.

Other terms that have been used to describe this model include the Uniformity hypothesis (Bock, 1985) and the Random Position hypothesis (Jain and Dubes, 1988, Chapter 4). The region A has usually been specified to be the unit p-dimensional hypercube or hypersphere (e.g., Zeng and Dubes, 1985a; Dubes and Zeng, 1987; Hartigan and Mohanty, 1992). However, the results of tests can depend markedly on the region A that is specified (Gordon, 1996b), and it can be difficult to provide rigorous justification for choice of region. An alternative approach is to choose A to be the convex hull of the points in the data set (Smith and Jain, 1984; Bock, 1985), invoking for justification a result due to Ripley and Rasson (1977) that if points are randomly generated within a convex region of the plane, an estimate of the boundary of the region is given by a uniform dilation of the convex hull about its centroid. The rationale of such *data-influenced* null models is that it enables the construction of tests that investigate departures from lack of cluster structure and that are not influenced by unimportant differences between model and data, such as the region within which the points are located.

However, algorithms for finding the convex hull of a set of points in p-dimensional space (e.g., Chand and Kapur, 1970; Edelsbrunner, 1987, Chapter 8) and algorithms that can be used to generate data within such regions (e.g.. Dobkin and Lipton, 1976; Rubin. 1984; Chazelle, 1985) make heavy demands on computing resources. An algorithm proposed by Smith and Jain (1984) that generates points uniformly within a region approximating the convex hull can perform poorly in 'empty' regions close to the convex hull. Use of the convex hull boundary for the Poisson model is thus limited at present to data sets having small values of the dimensionality, p, though one might hope that the development of more efficient algorithms or increase in computing power might ease this restriction in the future.

2.2. Unimodal model

This model assumes that the joint distribution of the variables describing the objects is unimodal.

The assumption behind this model is that there is only one cluster in the data set. However, instead of insisting that the points are uniformly distributed within this cluster, the Unimodal model allows for the possibility that objects are closer to one another nearer the centre of the cluster than they are on the boundary.

Many different unimodal distributions exist. The most common choice for continuous variables has been the multivariate normal distribution with identity variance-covariance matrix (e.g., Rohlf and Fisher, 1968; Gower and Banfield, 1975; Hartigan and Mohanty, 1992). Data-influenced unimodal models have also been investigated, for example by allowing the data to specify the variance-covariance matrix of the multivariate normal distribution (Gordon, 1996b).

2.3. Random dissimilarity matrix model

This model assumes that the elements of the lower triangle of the dissimilarity matrix are ranked in random order, all $(n(n-1)/2)!$ rankings being equally likely (Ling, 1973a).

The model has also been referred to as the Random Graph hypothesis (Jain and Dubes, 1988, chapter 4): if each object is represented by a vertex in a graph, and an edge is present between the ith and j^{th} vertices if d_{ij} is less than some specified threshold value, then the edges are inserted in random order. In variants of the hypothesis, a specified number of (randomly-selected) edges is present in the graph, or each edge is present with a specified probability, independently of all other edges.

In that only the order of the $n(n-1)/2$ dissimilarities is taken into account, the random dissimilarity matrix model might seem to be restricted to use in conjunction with monotone admissible (Fisher and Van Ness, 1971) clustering procedures, and it appears to have been used only with the single link and complete link clustering criteria.

A major criticism of the random dissimilarity matrix model is that it ignores second- and higher-order relationships: thus, if d_{ij} were small, one might expect d_{ik} and d_{jk} to have similar ranks. Ling (1973a) argued that clusters that were not significant under this model were unlikely to be deemed significant under any other null model. However, the model seems inappropriate for objects described by a pattern matrix (Gordon, 1996b), although it might be relevant for testing for the presence of class structure when the dissimilarity matrix is provided directly, rather than being derived from a pattern matrix. Frank and Strauss (1986) proposed a more general random graph model that allows for dependence between edges that meet the same vertex.

2.4. Random permutation model

This model considers independently permuting the entries in each of the p columns of the pattern matrix. There are $(n!)^{p-1}$ essentially different matrices, each of which is regarded as equally likely under the model.

Tests based on this null model and variants of it (e.g., Harper, 1978; Strauss, 1982; Vassiliou et al., 1989) compare the class structure revealed in the analysis of the original pattern matrix with that provided by the classification of randomly-generated pseudo pattern matrices. This approach has a different rationale to use of the previously-described null models, and it is not obvious that clusters in the original data set that are more homogeneous than those found in many pseudo data sets need have a high degree of absolute homogeneity. The approach also ignores the correlation between variables, and generates pseudo-objects within a hyperrectangular region in the space of variables.

2.5. Alternative hypotheses

Alternative models, specifying the presence of some clusters in the data, have been presented. For objects described by pattern matrices, a multimodal distribution for the variables describing the objects is clearly appropriate. One relevant implementation is that the data arise from a mixture of distributions that differ only in their location parameters (Bock, 1985, 1996; Hartigan, 1985; Hartigan and Mohanty, 1992). Other models have arisen in the analysis of spatial point patterns, such as the Neyman-Scott cluster process (Diggle, 1983, Chapter 4).

In a relevant 'clustering' alternative to the random dissimilarity matrix model, the probability that an edge is present depends on whether the vertices that it links represent objects that belong to the same cluster or different clusters in a partition (Frank, 1978; Frank and Harary, 1982).

An alternative hypothesis for hierarchical classifications can be obtained by perturbing data that have perfect ultrametric structure; such data could be presented in a matrix of dissimilarities (Cunningham and Ogilvie, 1972) or their ranks (Baker, 1974), or in a pattern matrix (Gordon, 1996c).

The usefulness of tests for the absence of cluster structure can be assessed by establishing their power when data are generated under such alternative hypotheses. Such investigations have rarely been carried out.

3. Tests of the absence of class structure

It might be expected that investigators would test a set of objects to establish if they could be regarded as lacking class structure, before undertaking a classification of them. Such preliminary investigations have rarely been carried out, possibly because investigators were confident that classes existed, or because they intended to validate the results using some of the methodology described later in the paper. It should also be noted that some tests of the absence of class structure make use of the results of a classification; some of these tests can thus also be used in the validation of particular types of class structure in a data set.

Tests of the Poisson model have been based on: the number of interpoint distances less than a specified threshold (Strauss, 1975; Kelly and Ripley, 1976; Saunders and Funk, 1977); the largest nearest neighbour distance within the set of objects (Bock, 1985); and generalizations of tests for randomness of two-dimensional spatial point patterns (Ripley, 1981; Diggle, 1983) that compare distances from a randomly-selected position to the nearest point, with the distance between that (or a randomly-selected) point and its nearest neighbour (Cross and Jain, 1982; Panayirci and Dubes, 1983; Zeng and Dubes, 1985a), such tests requiring attention to be paid to edge effects.

Generalizations of a test proposed for use in two dimensions in Hopkins (1954) were found to have superior performance in comparative studies of such tests carried out by Zeng and Dubes (1985b) and Dubes and Zeng (1987).

Tests based on the distribution of the lengths of edges in the minimum spanning tree have been proposed under the Poisson model (Hoffman and Jain, 1983) and a multivariate normal model (Rohlf, 1975). Smith and Jain (1984) considered adding randomly-positioned points to the data set and using a test due to Friedman and Rafsky (1979) based on the number of edges in the minimum spanning tree of the combined data set that link original and added points. Brailovsky (1991) generated modified data sets, in which each point was either retained with a specified probability or replaced by a randomly-positioned point, and compared the strength of clustering of the original data set with that of the modified data sets.

Bock (1985) described theoretical results pertaining to the average pairwise similarity of objects and the total within-class sum of squared distances when the objects were optimally partitioned into a specified number of classes, deriving asymptotic distributions of relevant test statistics; see also Hartigan (1977, 1978). However, more work is required to establish how relevant such results are for assessing finite-sized data sets. It has been more common for tests to involve simulation studies.

Some tests are based on searching for 'gaps' or multimodality. Hartigan and Mohanty (1992) studied the properties under both Poisson and Unimodal models of a test based on a single link dendrogram: for each single link class C, the number of objects $n(C)$ in its smallest offspring class is noted, and the test statistic is the maximum value of $n(C)$ over all classes C, large values suggesting the presence of bimodality. Müller and Sawitzki (1991) described a test based on comparing the amounts of probability mass exceeding various threshold values when there are c modes in the distribution, for different values of c, but this test is at present computationally feasible only for small values of the dimensionality, p. Hartigan (1988) considered evaluating the minimum amount of probability mass required to render the distribution unimodal, but settled for estimating departures from unimodality along the edges of an optimally-rooted minimum spanning tree. Rozál and Hartigan (1994) described a test based on minimum spanning trees, constrained so that edge lengths are non-increasing on all paths to the root node(s) corresponding to the class centre(s). Sneath (1986) presented a test based on the empirical distribution function of the number of internal nodes in a dendrogram that are located at less than a specified height.

The following test statistics have been proposed for use in conjunction with the random dissimilarity matrix model to test for the absence of class structure: the number of edges required before the graph consists of a single component (Rapoport and Fillenbaum, 1972; Schultz and Hubert, 1973; Ling, 1975; Ling and Killough, 1976); the number of vertices not belonging to the largest component when a specified number of edges is present (Ogilvie, 1969); the number of components when a specified number of edges is present (Ling, 1973b; Ling and Killough, 1976); the sizes of clusters in a partition into two clusters (Van Cutsem and Ycart, 1996). Godehardt (1990) described a generalization of the random dissimilarity matrix model to *multigraphs*, in which each variable describing the objects defines a different dissimilarity matrix.

4. Assessing individual clusters

An early approach to the assessment of individual clusters was to define properties of an *ideal* cluster, and identify which clusters satisfied these conditions. For example, all within-cluster pairwise dissimilarities can be required to be less than all dissimilarities between an object in the cluster and an object outside it (van Rijsbergen, 1970); other definitions of what might constitute an ideal cluster were provided by McQuitty (1963, 1967), Jardine (1969), Ling (1972) and Hubert (1974a). However, these are fairly restrictive requirements, and few data sets possess large ideal clusters.

Various measures of the internal cohesion and external isolation of clusters have been defined (e.g., Estabrook, 1966; Bailey and Dubes, 1982). Ling (1973a) defined the 'lifetime' of a cluster to be the difference between the rank of the dissimilarity at which it is formed and the rank at which it is incorporated into a larger cluster, evaluating lifetime distributions of single link clusters under the random dissimilarity matrix model. Matula (1977) proved that the distribution of the size of the maximal complete subgraph of a random graph, in which each edge is independently present with the same probability, is highly peaked, allowing an assessment of complete link clusters. When k edges are present in a random graph, Bailey and Dubes (1982) obtained inequalities for the probabilities of obtaining indices of cohesion and isolation as extreme as the observed ones for single link and complete link clusters, plotting these bounds for different values of k. Lerman (1970, Chapter 2, 1980, 1983) defined U statistics comparing within-cluster and between-cluster dissimilarities, and assessed both partitions and individual clusters under the hypothesis of random partitions having the same cluster sizes as observed in the results. Gordon (1994, 1996b) obtained by simulation critical values of U statistics under all four types of null model described in Section 2, by reanalysing random data sets using the same clustering procedure used in the classification of the original set of objects. He noted unsatisfactory properties of the random dissimilarity matrix and random permutation models, and the dependence of the results for the random pattern matrix models on the precise specification of the null model.

Many tests of the validity of an individual cluster have involved examining the distinctness of its offspring classes (e.g., Engelman and Hartigan, 1969; Gnanadesikan et al., 1977; Sneath, 1977, 1979, 1980; Barnett et al., 1979; Lee, 1979). These tests have usually been restricted to univariate data, sometimes obtained by projection onto the line joining the centroids of the two classes. Analytical results are difficult to obtain (Hartigan, 1977), and recourse has usually been made to simulation studies.

Some tests of whether or not a cluster should be sub-divided have been used as 'local stopping rules' for deciding on the number of clusters in a data set; such tests are described in the next section.

The tests for assessing individual clusters described in this section are internal validation tests, as defined in Section 1: cluster generation and cluster validation are carried out on the same data set. They encounter the 'multiple comparison' problem that tests of different clusters in the same data set are not independent of one another, which has implications for the significance levels of the tests. By contrast, Gabriel and Sokal (1969) described a simultaneous test procedure with assured overall significance level, in which (possibly overlapping) largest homogeneous clusters are identified; their approach is also applicable to determining coarsest acceptable partitions.

5. Assessing partitions

A partition of a set of objects can be compared with an externally-specified partition of the objects by evaluating a relevant index comparing the partitions. The significance of the value of this index can be determined by comparing it with its distribution under random permutations of the labels of the objects that leave unchanged the numbers of objects in each class of the partitions. A comparative study of five indices conducted by Milligan and Cooper (1986) concluded that Hubert and Arabie's (1985) modification of Rand's (1971) index was best suited to comparing a specified partition with cluster output comprising several different numbers of clusters.

On occasion, the external information might not provide a partition of the objects to be classified, but rather describe the classes to which they are believed to belong. At the most formal level, this description could specify a statistical model comprising a mixture of a known number of classes, each with known parametric form, but with unknown mixing proportions; at the least formal level, the distribution function for each class could be provided by the empirical distribution of a class provided by a clustering algorithm (Gordon, 1996d).

However, in many classification studies, external information is not available, and it has been more common for investigators to seek answers to the following question about a given set of objects: 'How many clusters are there in the data (and what is their composition)?' There are several ways in which this question may be posed:

1. Does a partition into c (say) clusters that has been provided by a clustering algorithm comprise cohesive and isolated clusters?

2. What value(s) of c is (are) most strongly indicated as (an) informative representation(s) of the data?

These questions are addressed using, respectively, internal and relative validation tests. The multiple comparison problem is again relevant: the value of c in (1) will have been chosen by an investigator after studying the results of a classification, and most tests for determining the appropriate value(s) of c in (2) have unknown significance levels.

The first question stated above can be addressed by defining a measure of the adequacy of a partition into c classes (e.g., the total within-class sum of squared distances about the c centroids), and obtaining its distribution under a null model of the absence of class structure. Some asymptotic theoretical results have been obtained (e.g., Hartigan, 1977; Pollard, 1982; Bock, 1985), but further work is required to establish their appropriateness for finite data sets. Baker and Hubert (1976) assessed partitions into complete link clusters using the number of extraneous edges present under the random graph model. Monte Carlo tests have been carried out by evaluating the measure of partition adequacy when many randomly-generated data sets are partitioned into c classes using the same clustering procedure as employed on the original data (Arnold, 1979; Milligan and Mahajan, 1980; Milligan and Sokol, 1980).

Partitions have also been assessed by investigating their stability when slightly modified versions of the data are reanalysed (e.g., Rand, 1971; Gnanadesikan et al., 1977), and by their replicability across subsamples (e.g., McIntyre and Blashfield, 1980; Breckenridge, 1989).

A problem with internal validation tests of a partition of the set of objects into c classes is the inter-relatedness of class structure at different values of c: thus, if there were really c_0 classes in the data, the hypothesis that there were c classes in the data would probably be acceptable for a range of values of c close to c_0. The second question stated above is concerned with the problem of identifying appropriate values of c.

Procedures that seek to find the single most appropriate value of c are generally referred to as 'stopping rules'; this is because, in the absence of information about relevant values of c, investigators often obtain a complete hierarchical classification using an agglomerative algorithm, and wish to have guidance on when to stop amalgamating and regard the current partition as the optimal one. Most research on cluster validation has involved the construction of stopping rules, and this overview describes only a small selection of them.

Stopping rules can be categorized as either *global* or *local*. Global stopping rules make use of all the information contained in a partition into c clusters for each value of c. The value of c for which a specified criterion is satisfied is then identified. Many of the criteria are based on seeking the optimal (maximum or minimum) value of measures comparing within-cluster and between-cluster variability, and do not have a natural definition when $c = 1$: it is clearly a disadvantage of stopping rules if they do not allow one to reach the conclusion that the data comprise only a single cluster.

Selected global stopping rules are described in this paragraph. Caliński and Harabasz (1974) identified the value of c that maximized a scaled ratio of between-cluster to within-cluster sum of squared distances. The C-index identifies c which minimizes a standardized version of the sum of all within-cluster dissimilarities (D): if the partition has a total of r within-cluster dissimilarities, D_{min} (resp., D_{max}) is defined as the sum of the r smallest (resp., largest) dissimilarities, and $C \equiv (D - D_{min})/(D_{max} - D_{min})$. Goodman and Kruskal's (1954) γ compares all within-cluster dissimilarities with all between-cluster dissimilarities, defining a comparison to be concordant (resp., discordant) if a within-cluster dissimilarity is less (resp., greater) than a between-cluster dissimilarity; the optimal value of c is defined to be the one which maximises $(S_+ - S_-)/(S_+ + S_-)$, where S_+ (resp., $S-$) is the number of concordant (resp., discordant) comparisons. A test based on the point biserial correlation identifies the value of c which maximizes the correlation between the original dissimilarities and an $n \times n$ matrix of 1's and 0's indicating whether or not the objects belonged to the same cluster. The cubic clustering criterion (Sarle, 1983) is based on R^2, the proportion of the variance accounted for by the clusters, and identifies the value of c which maximizes a function of R^2 which includes terms derived from simulation studies under the Poisson model. Other global stopping rules were proposed by Jackson (1969), Gower (1973), Davies and Bouldin (1979), Hill (1980), Ratkowsky (1984), Krzanowski and Lai (1988), Xu et al. (1993), and many others; reviews and comparative studies were presented by Milligan (1981), Milligan and Cooper (1985), and Dubes (1987).

Local stopping rules are based on tests of whether or not a pair of clusters should be amalgamated, or a single cluster should be subdivided. Unlike global stopping rules, they are thus restricted to the analysis of hierarchically-nested sets of partitions, and use only a part of the data. They often also require the specification of a threshold, the value of which can markedly influence the properties of the stopping rule.

Selected local stopping rules are described in this paragraph. Duda and Hart (1973, Section 6.12) compared W_1, the within-cluster sum of squared distances, with W_2, the total within-cluster sum of squared distances when the cluster is optimally partitioned into two, rejecting the hypothesis of a single cluster if W_2/W_1 is sufficiently small; amalgamations cease when the hypothesis is first rejected. A test with a similar rationale proposed by Beale (1969) is based on $(W_1 - W_2)/W_2$. Legendre et al. (1985) categorized the dissimilarities in a cluster as either 'high' or 'low', and assessed whether or not it should be subdivided into two sub-clusters by carrying out a randomization test based on the proportion of high dissimilarities between the sub-clusters compared to within the cluster. Other local stopping rules have been proposed by Gnanadesikan et al. (1977) and Howe (1979), and some of the tests described in Section 4 for assessing individual clusters can also be used as local stopping rules.

For many data sets, a complete hierarchical classification has been obtained and a stopping rule has then been used to provide guidance on where to section the hierarchy to provide a partition. Such work often disregards which clustering criterion has been used to obtain the hierarchy, whereas one can expect different stopping rules – just as different clustering criteria – to be more effective in analysing different types of data. Stronger links between the processes of generating and validating partitions are provided by work that assesses the stability of a partition by reanalysis of a perturbed version of the data set (e.g., Begovich and Kane, 1982) or of bootstrap samples (Jain and Moreau, 1987); or by separately reanalysing sub-samples of the data (Overall and Magee, 1992); or by noting the value of c for which 'fuzzy partitions' are 'closest' to 'hard' partitions (e.g., Roubens, 1978; Windham, 1981, 1982; Rivera et al., 1990).

Proposals for determining the number of clusters present in a set of objects continue to be published in the research literature, often with only a cursory examination of their properties and little attempt to establish how they perform in comparison with previously-published procedures. A detailed comparative study of thirty procedures by Milligan and Cooper (1985) showed that many tests performed very poorly in detecting reasonably clear-cut clusters. The five procedures that performed best in Milligan and Cooper's (1985) simulation study were the first three global stopping rules and the first two local stopping rules described earlier in this section. It is possible that these results have been influenced by the fact that the clusters generated in Milligan and Cooper's (1985) study were sampled from mildly-truncated multivariate normal distributions.

The work described above seeks to identify the single most appropriate value for the number of clusters present in the data. However, it may be relevant to describe a set of objects in terms of partitions into two or more (possibly nested) widely-separated numbers of clusters, depicting the class structure present at several different scales in the data. Gordon (1996c) presented a study investigating the ability of modifications of Milligan and Cooper's (1985) five superior stopping rules to detect nested cluster structure.

6. Assessing hierarchical classifications

Some investigators have been interested in comparing two or more independently-derived hierarchical classifications of the same set of objects; see, for example, the references cited in Lapointe and Legendre (1995). Much of this research is considered to be outside the terms of reference of this paper, which concentrates on assessing the validity of (aspects of) a single classification; for a discussion of a general approach to synthesizing the results of several different classifications, see Lapointe (1996). However, some of this methodology is relevant for conducting external validation tests comparing a hierarchical classification of data provided by a clustering algorithm with an externally-specified hierarchical classification that does not make use of these data: thus, the methodology is not relevant for comparing the results of applying two or more different clustering algorithms to the same data set, but would be appropriate if the externally-specified classification were based on the clustering of data comprising different variables describing the same set of objects.

It is relevant to distinguish between several different ways in which the information contained in a hierarchical classification can be summarized (Gordon, 1996a). A *non-ranked tree* or *n-tree* (Bobisud and Bobisud, 1972; McMorris et al., 1983) specifies only the nested set of clusters present in the hierarchy. A *dendrogram* also specifies the height in the tree of the internal node subtending each cluster. In a *ranked tree* (Boorman and Olivier, 1973; Murtagh, 1984), only the ordering of these heights is specified. It is stressed that all trees referred to in this context are rooted.

Many different indices have been proposed for comparing two hierarchical classifications (Rohlf, 1982). The significance of the value taken by such an index can be assessed by comparing it with its distribution under a suitable null hypothesis. Hypotheses that have been considered are the *random label* hypothesis, in which labels are independently permuted, and several *random tree* hypotheses, in which distributions of various types of tree are considered (e.g., binary or multifurcating, ranked or non-ranked or dendrograms). Because the numbers of different trees in the distributions increase rapidly with n, it is common for these investigations to involve simulation studies, in which trees are randomly selected from an appropriate distribution. Furnas (1984) reviewed algorithms for the uniform generation of various types of random tree, and Lapointe and Legendre (1991) described the generation of random dendrograms. Using this theory, tests that could be used for the external validation of a complete hierarchical classification have been proposed under the random label hypothesis (Hubert and Baker, 1977) and various random tree hypotheses (e.g., Simberloff, 1987). Tests can also be carried out on hypotheses that a specified part of a hierarchical classification possesses a certain class structure (De Soete et al., 1987; Hubert, 1987, Chapter 5; Lapointe and Legendre, 1990).

It is worth stressing, however, that the distribution of trees resulting from the application of a clustering algorithm to data generated under a different null model can differ markedly from being uniform over the set of trees (Frank and Svensson, 1981).

In the cognate topic of decision trees (e.g., Breiman et al., 1984), trees have been 'pruned' so as to remove branches below uninformative internal nodes, sometimes using new data sets (e.g., Quinlan, 1987), and such methodology has been used to simplify hierarchical classifications (Fisher, 1996).

Apart from such external validation tests, little work has been carried out on assessing the validity of a hierarchical classification provided by a clustering algorithm. Information has been obtained about the distributions, under Poisson, Unimodal, and random dissimilarity matrix models of the distortion imposed on data when they are represented in a hierarchical classification (e.g., Rohlf and Fisher, 1968; Hubert, 1974b; Gower and Banfield, 1975), but if such null hypotheses are rejected, it does not follow that the complete hierarchy is validated. Lerman (1970, Chapter 4, 1981, Chapter 3) defined a measure of the extent to which dissimilarities failed to satisfy the ultrametric property defining a hierarchical classification, and investigated properties of this measure when sampling from binary pattern matrices with specified row sums. Smith and Dubes (1980) divided the set of objects into two and compared the classification of a subset of the data with the relevant part of the original classification, assessing the resemblance by reference to the random dissimilarity matrix model. Other approaches have aimed at assessing the stability of a hierarchical classification by measuring the *influence* of each object, i.e. the extent to which the results are altered if an object is deleted or differentially weighted (e.g., Gnanadesikan et al., 1977; Jambu and Lebeaux, 1983, Chapter 6; Gordon and De Cata, 1988; Jolliffe et al., 1988); separate trees, each based on $(n-1)$ objects, can be combined into a consensus tree or trees (Lanyon, 1985; Lapointe et al., 1994).

A validation test which would appear to be of considerable interest is a relative test of which of two or more hierarchical classifications provides the best summary of a given set of objects. One might consider addressing this problem by defining a suitable measure of the distortion imposed in representing the data in a hierarchical classification, and identifying the classification that has minimum distortion. However, different measures of distortion tend to favour classifications obtained using different clustering procedures (e.g., Sokal and Rohlf, 1962; Sneath, 1969; Faust and Romney, 1985): in effect, this approach simply reformulates the problem of specifying appropriate clustering procedures in terms of defining appropriate measures of distortion.

Investigators are often interested only in parts of a hierarchical classification, and assess its constituent clusters and partitions using methodology described in the previous two sections. It can then be of interest to represent the data in a *parsimonious tree* which retains only those parts of the classification that are deemed to be significant (Lerman, 1980; Lerman and Ghazzali, 1991; Gordon, 1994).

7. Discussion

The topic of cluster analysis was initially perceived as being concerned primarily with the *exploratory* analysis of sets of objects, with little attention being paid to assessing the validity of the results. Some results have been assessed solely in terms of their interpretability or usefulness, but there are clearly dangers in such an approach: the human mind is quite capable of providing *post hoc* justification for results of dubious validity. More recently, there has been an increased awareness of the importance of cluster validation. However, few studies have included a validation phase; of those that have, most have involved stopping rules, as described in Section 5.

Much research remains to be done in the field of cluster validation. Ideally, one would like to be able to specify: appropriate null hypotheses of the absence of cluster structure; alternative hypotheses describing departures from such null hypotheses which it is important to detect; test statistics with known properties, which are effective in identifying the type of class structure that is present in the data.

The precise null model that is specified can markedly influence the results of a test (Gordon, 1996b), and it would be useful to have further investigations, particularly of data-influenced null models.

Many different test procedures have been proposed, particularly for addressing the 'how many clusters?' question, and the time would seem to have come when attention should be devoted to carrying out further comparisons of these in order to determine their properties. One problem is that a 'cluster' is a vaguely-defined concept, that many different types of cluster could be present in data, and that one can expect different test procedures to be effective at detecting different types of structure. It is thus unrealistic to expect to be able to identify a single 'best' test procedure for each type of investigation. Nevertheless, Milligan and Cooper's (1985) comparative study indicated that some tests performed very poorly in detecting reasonably clear-cut structure. It thus seems useful to advocate further studies, with the aims of eliminating from further consideration tests that have poor performance, and identifying a small number of 'superior' tests; such tests could then profitably be incorporated into standard statistical and classification computer packages. It seems inevitable that such comparative studies will be largely based on assessing the performance of test procedures in the analysis of data sets that have known class structure.

Some cluster validation tests make heavy demands on computing resources, and the kind of investigation which is feasible depends on the size of the data set and the nature of the data. Nevertheless, one might hope that the future will see the development of more efficient procedures and algorithms, and an increase in the power and availability of computing facilities, thus facilitating a greater use of cluster validation methodology.

34

References:

Arnold, S. J. (1979): A test for clusters. *Journal of Marketing Research*, 16, 545–551.

Art, D., Gnanadesikan, R. and Kettenring, J. R. (1982): Data-based metrics for cluster analysis. *Utilitas Mathematica*, 21A, 75–99.

Bailey, T. A., Jr. and Dubes, R. (1982): Cluster validity profiles. *Pattern Recognition*, 15, 61–83.

Baker, F. B. (1974): Stability of two hierarchical grouping techniques case I: Sensitivity to data errors. *Journal of the American Statistical Association*, 69, 440–445.

Baker, F. B. and Hubert, L. J. (1976): A graph-theoretic approach to goodness-of-fit in complete link hierarchical clustering. *Journal of the American Statistical Association*, 71, 870–878.

Barnett, V., Kay, R. and Sneath, P. H. A. (1979): A familiar statistic in an unfamiliar guise -- A problem in clustering. *The Statistician*, 28, 185–191.

Beale, E. M. L. (1969): Euclidean cluster analysis. *Bulletin of the International Statistical Institute*, 43(2), 92–94.

Begovich, C. L. and Kane, V. E. (1982): Estimating the number of groups and group membership using simulation çluster analysis. *Pattern Recognition*, 15, 335–342.

Binder, D. A. (1978): Bayesian cluster analysis. *Biometrika*, 65, 31–38.

Bobisud, H. M. and Bobisud, L. E. (1972): A metric for classification. *Taxon*, 21, 607–613.

Bock, H. H. (1974): *Automatische Klassifikation: Theoretische und Praktische Methoden zur Gruppierung und Strukturierung von Daten (Cluster-Analyse)*. Vandenhoeck & Ruprecht, Göttingen.

Bock, H. H. (1985): On some significance tests in cluster analysis. *Journal of Classification*, 2, 77–108.

Bock, H. H. (1989): Probabilistic aspects in cluster analysis. *In Conceptual and Numerical Analysis of Data*, Opitz, O. (ed.), 12–44, Springer-Verlag, Berlin.

Bock, H. H. (1996): Probability models and hypothesis testing in partitioning cluster analysis. In *Clustering and Classification*, Arabie, P. et al. (eds.), 377–453, World Scientific Publishing, River Edge, NJ.

Boorman, S. A. and Olivier, D. C. (1973): Metrics on spaces of finite trees. *Journal of Mathematical Psychology*, 10, 26–59.

Brailovsky, V. L. (1991): A probabilistic approach to clustering. *Pattern Recognition Letters*, 12, 193–198.

Breckenridge, J. N. (1989): Replicating cluster analysis: Method, consistency and validity. *Multivariate Behavioral Research*, 24, 147–161.

Breiman, L., Friedman, J. H., Olshen, R. A. and Stone, C. J. (1984): *Classification and Regression Trees*. Wadsworth, Belmont, CA.

Caliński, T. and Harabasz, J. (1974): A dendrite method for cluster analysis. *Communications in Statistics*, 3, 1–27.

Chand, D. R. and Kapur, S. S. (1970): An algorithm for convex polytopes. *Journal of the Association for Computing Machinery*, 17, 78–86.

Chazelle, B. (1985): Fast searching in a real algebraic manifold with applications to geometric complexity. *Lecture Notes in Computer Science*, 185, 145–156.

Cross, G. C. and Jain, A. K. (1982): Measurement of clustering tendency. *In Proceedings of IFAC Symposium on Theory and Application of Digital Control (Volume 2)*, 24–29, New Delhi.

Cunningham, K. M. and Ogilvie, J. C. (1972): Evaluation of hierarchical grouping techniques: A preliminary study. *Computer Journal*, 15, 209–213.

Davies, D. L. and Bouldin, D. W. (1979): A cluster separation measure. *IEEE Transactions on Pattern Analysis and Machine Intelligence*, PAMI-1, 224–227.

De Soete, G., Carroll, J. D. and DeSarbo, W. S. (1987): Least squares algorithms for constructing constrained ultrametric and additive tree representations of symmetric proximity data. *Journal of Classification*, 4, 155–173.

Diday, E. and Govaert, G. (1977): Classification automatique avec distances adaptatives. *R. A. I. R. O. Informatique/Computer Sciences*, 11, 329–349.

Diday, E. and Simon, J. C. (1976): Clustering analysis. *In Communication and Cybernetics 10 Digital Pattern Recognition*, Fu, K. S. (ed.), 47–94, Springer-Verlag, Berlin.

Diggle, P. J. (1983): *Statistical Analysis of Spatial Point Patterns*. Academic Press, London.

Dobkin, D. and Lipton, R. J. (1976): Multidimensional searching problems. *SIAM Journal on Computing*, **5**, 181–186.

Dubes, R. C. (1987): How many clusters are best? – An experiment. *Pattern Recognition*, **20**, 645–663.

Dubes, R. C. and Zeng, G. (1987): A test for spatial homogeneity in cluster analysis. *Journal of Classification*, **4**, 33–56.

Duda, R. O. and Hart, P. E. (1973): *Pattern Classification and Scene Analysis*. Wiley, New York.

Edelsbrunner, H. (1987): *Algorithms in Combinatorial Geometry*. Springer-Verlag, Berlin.

Engelman, L. and Hartigan, J. A. (1969): Percentage points of a test for clusters. *Journal of the American Statistical Association*, **64**, 1647–1648.

Estabrook, G. F. (1966): A mathematical model in graph theory for biological classification. *Journal of Theoretical Biology*, **12**, 297–310.

Faust, K. and Romney, A. K. (1985): The effect of skewed distributions on matrix permutation tests. *British Journal of Mathematical and Statistical Psychology*, **38**, 152–160.

Fisher, D. (1996): Iterative optimization and simplification of hierarchical clusterings. *Journal of Artificial Intelligence Research*, **4**, 147–180.

Fisher, L. and Van Ness, J. W. (1971): Admissible clustering procedures. *Biometrika*, **58**, 91–104.

Frank, O. (1978): Inferences concerning cluster structure. *In COMPSTAT 1978*, Corsten, L. C. A. and Hermans, J. (eds.), 259–265, Physica-Verlag, Wien.

Frank, O. and Harary, F. (1982): Cluster inference by using transitivity indices in empirical graphs. *Journal of the American Statistical Association*, **77**, 835–840.

Frank, O. and Strauss, D. (1986): Markov graphs. *Journal of the American Statistical Association*, **81**, 832–842.

Frank, O. and Svensson, K. (1981): On probability distributions of single-linkage dendrograms. *Journal of Statistical Computation and Simulation*, **12**, 121–131.

Friedman, J. H. and Rafsky, L. C. (1979): Multivariate generalizations of the Wald-Wolfowitz and Smirnov two-sample tests. *Annals of Statistics*, **7**, 697–717.

Furnas, G. W. (1984): The generation of random, binary unordered trees. *Journal of Classification*, **1**, 187–233.

Gabriel, K. R. and Sokal, R. R. (1969): A new statistical approach to geographical variation analysis. *Systematic Zoology*, **18**, 259–278.

Gnanadesikan, R., Kettenring, J. R. and Landwehr, J. M. (1977): Interpreting and assessing the results of cluster analyses. *Bulletin of the International Statistical Institute*, **47(2)**, 451–463.

Godehardt, E. (1990): *Graphs as Structural Models: The Application of Graphs and Multigraphs in Cluster Analysis (2nd edn.)*. Friedr. Vieweg & Sohn, Braunschweig.

Goodman, L. A. and Kruskal, W. H. (1954): Measures of association for cross classifications. *Journal of the American Statistical Association*, **49**, 732–764.

Gordon, A. D. (1981): *Classification: Methods for the Exploratory Analysis of Multivariate Data*. Chapman and Hall, London.

Gordon, A. D. (1994): Identifying genuine clusters in a classification. *Computational Statistics & Data Analysis*, **18**, 561–581.

Gordon, A. D. (1995): Tests for assessing clusters. *Statistics in Transition*, **2**, 207–217.

Gordon, A. D. (1996a): Hierarchical classification. *In Clustering and Classification*, Arabie, P. et al. (eds.), 65–121, World Scientific Publishing, River Edge, NJ.

Gordon, A. D. (1996b): Null models in cluster validation. *In From Data to Knowledge: Theoretical and Practical Aspects of Classification, Data Analysis, and Knowledge Organization*, Gaul, W. and Pfeifer, D. (eds.), 32–44, Springer-Verlag, Berlin.

Gordon, A. D. (1996c): How many clusters? An investigation of five procedures for detecting nested cluster structure. *Paper presented at IFCS-96 Conference, Kobe, 27-30 March 1996*.

Gordon, A. D. (1996d): External validation in cluster analysis. *Submitted for publication*.

Gordon, A. D. and De Cata, A. (1988): Stability and influence in sum of squares clustering. *Metron*, **46**, 347–360.

Gower, J. C. (1973): Classification problems. *Bulletin of the International Statistical Institute*, **45(1)**, 471–477.

Gower, J. C. and Banfield, C. F. (1975): Goodness-of-fit criteria for hierarchical classification and their empirical distributions. *In Proceedings of the 8th International Biometric Conference*, Corsten, L. C. A. and Postelnicu, T. (eds.), 347–361, Constanţa, Romania.

Harper, C. W., Jr. (1978): Groupings by locality in community ecology and paleoecology: Tests of significance. *Lethaia*, **11**, 251–257.

Hartigan, J. A. (1975): *Clustering Algorithms*. Wiley, New York.

Hartigan, J. A. (1977): Distribution problems in clustering. *In Classification and Clustering*, Van Ryzin, J. (ed.), 45–71, Academic Press, New York.

Hartigan, J. A. (1978): Asymptotic distributions for clustering criteria. *Annals of Statistics*, **6**, 117–131.

Hartigan, J. A. (1985): Statistical theory in clustering. *Journal of Classification*, **2**, 63–76.

Hartigan, J. A. (1988): The span test for unimodality. *In Classification and Related Methods of Data Analysis*, Bock, H. H. (ed.), 229–236, North-Holland, Amsterdam.

Hartigan, J. A. and Mohanty, S. (1992): The runt test for multimodality. *Journal of Classification*, **9**, 63–70.

Hill, R. S. (1980): A stopping rule for partitioning dendrograms. *Botanical Gazette*, **141**, 321–324.

Hoffman, R. and Jain, A. K. (1983): A test of randomness based on the minimal spanning tree. *Pattern Recognition Letters*, **1**, 175–180.

Hopkins, B. (1954): A new method for determining the type of distribution of plant individuals (with an appendix by J. G. Skellam). *Annals of Botany, NS*, **18**, 213–227.

Howe, S. E. (1979): *Estimating Regions and Clustering Spatial Data: Analysis and Implementation of Methods Using the Voronoi Diagram*. Unpublished Ph.D. thesis, Brown University, Providence, RI.

Hubert, L. J. (1974a): Some applications of graph theory to clustering. *Psychometrika*, **39**, 283–309.

Hubert, L. (1974b): Approximate evaluation techniques for the single-link and complete-link hierarchical clustering procedures. *Journal of the American Statistical Association*, **69**, 698–704.

Hubert, L. J. (1987): *Assignment Methods in Combinatorial Data Analysis*. Marcel Dekker, New York.

Hubert, L. and Arabie, P. (1985): Comparing partitions. *Journal of Classification*, **2**, 193–218.

Hubert, L. J. and Baker, F. B. (1977): The comparison and fitting of given classification schemes. *Journal of Mathematical Psychology*, **16**, 233–253.

Jackson, D. M. (1969): Comparison of classifications. *In Numerical Taxonomy*, Cole, A. J. (ed.), 91–113, Academic Press, London.

Jain, A. K. and Dubes, R. C. (1988): *Algorithms for Clustering Data*. Prentice-Hall, Englewood Cliffs, NJ.

Jain, A. K. and Moreau, J. V. (1987): Bootstrap techniques in cluster analysis. *Pattern Recognition*, **20**, 547–568.

Jambu, M. and Lebeaux, M. O. (1983): *Cluster Analysis and Data Analysis*. North-Holland, Amsterdam.

Jardine, N. (1969): Towards a general theory of clustering (abstract). *Biometrics*, **25**, 609–610.

Jardine, N. and Sibson, R. (1971): *Mathematical Taxonomy*. Wiley, London.

Jolliffe, I. T., Jones, B. and Morgan, B. J. T. (1988): Stability and influence in cluster analysis. *In Data Analysis and Informatics V*, Diday, E. (ed.), 507–514, North-Holland, Amsterdam.

Kelly, F. P. and Ripley, B. D. (1976): A note on Strauss's model for clustering. *Biometrika*, **63**, 357–360.

Krzanowski, W. J. and Lai, Y. T. (1983): A criterion for determining the number of groups in a data set using sum-of-squares clustering. *Biometrics*, **44**, 23–34.

Lanyon, S. M. (1985): Detecting internal inconsistencies in distance data. *Systematic Zoology*, **34**, 397–403.

Lapointe, F.-J. (1996): To validate and how to validate? That is the real question. *Paper presented at IFCS-96 Conference, Kobe, 27–30 March 1996.*

Lapointe, F.-J., Kirsch, J. A. W. and Bleiweiss, R. (1994): Jackknifing of weighted trees: Validation of phylogenies reconstructed from distance matrices. *Molecular Phylogenetics and Evolution*, **3**, 256–267.

Lapointe, F.-J. and Legendre, P. (1990): A statistical framework to test the consensus of two nested classifications. *Systematic Zoology*, **39**, 1–13.

Lapointe, F.-J. and Legendre, P. (1991): The generation of random ultrametric matrices representing dendrograms. *Journal of Classification*, **8**, 177–200.

Lapointe, F.-J. and Legendre, P. (1995). Comparison tests for dendrograms: A comparative evaluation. *Journal of Classification*, **12**, 265–282.

Lee, K. L. (1979): Multivariate tests for clusters. *Journal of the American Statistical Association*, **74**, 708–714.

Lefkovitch, L. P. (1978): Cluster generation and grouping using mathematical programming. *Mathematical Biosciences*, **41**, 91–110.

Lefkovitch, L. P. (1980): Conditional clustering. *Biometrics*, **36**, 43–58.

Legendre, P., Dallot, S. and Legendre, L. (1985): Succession of species within a community: Chronological clustering, with applications to marine and freshwater zooplankton. *The American Naturalist*, **125**, 257–288.

Lerman, I. C. (1970: *Les Bases de la Classification Automatique*. Gauthier-Villars, Paris.

Lerman, I. C. (1980): Combinatorial analysis in the statistical treatment of behavioral data. *Quality and Quantity*, **14**, 431–469.

Lerman, I. C. (1981): *Classification et Analyse Ordinale des Données*. Dunod, Paris.

Lerman, I. C. (1983): Sur la signification des classes issues d'une classification automatique de données. *In Numerical Taxonomy*, Felsenstein, J. (ed.), 179–198, Springer-Verlag, Berlin.

Lerman, I. C. and Ghazzali, N. (1991): What do we retain from a classification tree? An experiment in image coding. *In Symbolic-Numeric Data Analysis and Learning*, Diday, E. and Lechevallier, Y. (eds.), 27–42, Nova Science, New York.

Ling, R. F. (1972): On the theory and construction of k-clusters. *Computer Journal*, **15**, 326–332.

Ling, R. F. (1973a): A probability theory for cluster analysis. *Journal of the American Statistical Association*, **68**, 159–164.

Ling, R. F. (1973b): The expected number of components in random linear graphs. *Annals of Probability*, **1**, 876–881.

Ling, R. F. (1975): An exact probability distribution on the connectivity of random graphs. *Journal of Mathematical Psychology*, **12**, 90–98.

Ling, R. F. and Killough, G. G. (1976): Probability tables for cluster analysis based on a theory of random graphs. *Journal of the American Statistical Association*, **71**, 293–300.

McIntyre, R. M. and Blashfield, R. K. (1980): A nearest-centroid technique for evaluating the minimum-variance clustering procedure. *Multivariate Behavioral Research*, **15**, 225–238.

McMorris, F. R., Meronk, D. B. and Neumann, D. A. (1983): A view of some consensus methods for trees. *In Numerical Taxonomy*, Felsenstein, J. (ed.), 122–126, Springer-Verlag, Berlin.

McQuitty, L. L. (1963): Rank order typal analysis. *Educational and Psychological Measurement*, **23**, 55–61.

McQuitty, L. L. (1967): A mutual development of some typological theories and pattern analytical methods. *Educational and Psychological Measurement*, **27**, 21–46.

Marriott, F. H. C. (1982): Optimization methods of cluster analysis. *Biometrika*, **69**, 417–422.

Matula, D. W. (1977): Graph theoretic techniques for cluster analysis algorithms. *In Classification and Clustering*, Van Ryzin, J. (ed.), 95–129, Academic Press, New York.

Milligan, G. W. (1981): A Monte Carlo study of thirty internal criterion measures for cluster analysis. *Psychometrika*, **46**, 187–199.

Milligan, G. W. and Cooper, M. C. (1985): An examination of procedures for determining the number of clusters in a data set. *Psychometrika*, **50**, 159–179.

Milligan, G. W. and Cooper, M. C. (1986): A study of the comparability of external criteria for hierarchical cluster analysis. *Multivariate Behavioral Research*, **21**, 441–458.

Milligan, G. W. and Mahajan, V. (1980): A note on procedures for testing the quality of a clustering of a set of objects. *Decision Sciences*, **11**, 669–677.

Milligan, G. W. and Sokol, L. M. (1980): A two-stage clustering algorithm with robust recovery characteristics. *Educational and Psychological Measurement*, **40**, 755–759.

Müller, D. W. and Sawitzki, G. (1991): Excess mass estimates and tests for multimodality. *Journal of the American Statistical Association*, **86**, 738–746.

Murtagh, F. (1984): Counting dendrograms: A survey. *Discrete Applied Mathematics*, **7**, 191–199.

Ogilvie, J. C. (1969): The distribution of number and size of connected components in random graphs of medium size. *Information Processing*, **68**, 1527–1530.

Overall, J. E. and Magee, K. N. (1992): Replication as a rule for determining the number of clusters in hierarchial cluster analysis. *Applied Psychological Measurement*, **16**, 119–128.

Panayirci, E. and Dubes, R. C. (1983): A test for multidimensional clustering tendency. *Pattern Recognition*, **16**, 433–444.

Perruchet, C. (1983): Une analyse bibliographique des épreuves de classifiabilité en analyse des données. *Statistiques et Analyse de Données*, **8**, 18–41.

Pollard, D. (1982): A central limit theorem for k-means clustering. *Annals of Probability*, **10**, 919–926.

Quinlan, J. R. (1987): Simplifying decision trees. *International Journal of Man-Machine Studies*, **27**, 221–234.

Rand, W. M. (1971): Objective criteria for the evaluation of clustering methods. *Journal of the American Statistical Association*, **66**, 846–850.

Rapoport, A. and Fillenbaum, S. (1972): An experimental study of semantic structures. *In Multidimensional Scaling. Theory and Applications in the Behavioral Sciences: Volume II. Applications*, Romney, A. K. et al. (eds.), 93–131, Seminar Press, New York.

Ratkowsky, D. A. (1984): A stopping rule and clustering method of wide applicability. *Botanical Gazette*, **145**, 518–523.

Ripley, B. D. (1981): *Spatial Statistics*. Wiley, New York.

Ripley, B. D. and Rasson, J.-P. (1977): Finding the edge of a Poisson forest. *Journal of Applied Probability*, **14**, 483–491.

Rivera, F. F., Zapata, E. L. and Carazo, J. M. (1990): Cluster validity based on the hard tendency of the fuzzy classification. *Pattern Recognition Letters*, **11**, 7–12.

Rohlf, F. J. (1970): Adaptive hierarchical clustering schemes. *Systematic Zoology*, **19**, 58–82.

Rohlf, F. J. (1975): Generalization of the gap test for the detection of multivariate outliers. *Biometrics*, **31**, 93–101.

Rohlf, F. J. (1982): Consensus indices for comparing classifications. *Mathematical Biosciences*, **59**, 131–144.

Rohlf, F. J. and Fisher, D. R. (1968): Tests for hierarchical structure in random data sets. *Systematic Zoology*, **17**, 407–412.

Roubens, M. (1978): Pattern classification problems and fuzzy sets. *Fuzzy Sets and Systems*, **1**, 239–253.

Rousseeuw, P. J. (1987): Silhouettes: A graphical aid to the interpretation and validation of cluster analysis. *Journal of Computational and Applied Mathematics*, **20**, 53–65.

Rozál, G. P. M. and Hartigan, J. A. (1994): The MAP test for multimodality. *Journal of Classification*, **11**, 5–36.

Rubin, P. A. (1984): Generating random points in a polytope. *Communications in Statistics: Simulation and Computation*, **B 13**, 375–396.

Sarle, W. S. (1983): *Cubic Clustering Criterion*. Technical Report A-108, SAS Institute, Cary, NC.

Saunders, R. and Funk, G. M. (1977): Poisson limits for a clustering model of Strauss. *Journal of Applied Probability*, **14**, 776–784.

Schultz, J. V. and Hubert, L. J. (1973): Data analysis and the connectivity of random graphs. *Journal of Mathematical Psychology*, **10**, 421–428.

Scott, A. J. and Symons, M. J. (1971): Clustering methods based on likelihood ratio criteria. *Biometrics*, **27**, 387–397.

Shepard, R. N. (1974): Representation of structure in similarity data: Problems and prospects. *Psychometrika*, **39**, 373–421.

Simberloff, D. (1987): Calculating probabilities that cladograms match: A method of biogeographical inference. *Systematic Zoology*, **36**, 175–195.

Smith, S. P. and Dubes, R. (1980): Stability of a hierarchical clustering. *Pattern Recognition*, **12**, 177–187.

Smith, S. P and Jain, A. K. (1984): Testing for uniformity in multidimensional data. *IEEE Transactions on Pattern Analysis and Machine Intelligence*, **PAMI-6**, 73–81.

Sneath, P. H. A. (1969): Evaluation of clustering methods (with Discussion). *In Numerical Taxonomy*, Cole, A. J. (ed.), 257–271, Academic Press, London.

Sneath, P. H. A. (1977): A method for testing the distinctness of clusters: A test of the disjunction of two clusters in Euclidean space as measured by their overlap. *Mathematical Geology*, **9**, 123–143.

Sneath, P. H. A. (1979): The sampling distribution of the W statistic of disjunction for the arbitrary division of a random rectangular distribution. *Mathematical Geology*, **11**, 423–429.

Sneath, P. H. A. (1980). Some empirical tests for significance of clusters. *In Data Analysis and Informatics*, Diday, E. et al. (eds.), 491–508, North-Holland, Amsterdam.

Sneath, P. H. A. (1986): Significance tests for multivariate normality of clusters from branching patterns in dendrograms. *Mathematical Geology*, **18**, 3–32.

Sokal, R. R. and Rohlf, F. J. (1962): The comparison of dendrograms by objective methods. *Taxon*, **11**, 33–40.

Strauss, D. J. (1975): A model for clustering. *Biometrika*, **62**, 467–475.

Strauss, R. E. (1982): Statistical significance of species clusters in association analysis. *Ecology*, **63**, 634–639.

Van Cutsem, B. and Ycart, B. (1996): *Indexed Dendrograms on Random Dissimilarities*. Rapport MAI 23, CNRS, Université Joseph Fourier Grenoble I.

Van Ness, J. W. (1973): Admissible clustering procedures. *Biometrika*, **60**, 422–424.

van Rijsbergen, C. J. (1970): A clustering algorithm. *Computer Journal*, **13**, 113–115.

Vassiliou, A., Ignatiades, L. and Karydis, M. (1989): Clustering of transect phytoplankton collections with a quick randomization algorithm. *Journal of Experimental Marine Biology and Ecology*, **130**, 135–145.

Ward, J. H., Jr. (1963): Hierarchical grouping to optimize an objective function. *Journal of the American Statistical Association*, **58**, 236–244.

Windham, M. P. (1981): Cluster validity for fuzzy clustering algorithms. *Fuzzy Sets and Systems*, **5**, 177–185.

Windham, M. P. (1982): Cluster validity for the fuzzy c-means clustering algorithm. *IEEE Transactions on Pattern Analysis and Machine Intelligence*, **PAMI-4**, 357–363.

Xu, S., Karnath, M. V. and Capson, D. W. (1993): Selection of partitions from a hierarchy. *Pattern Recognition Letters*, **14**, 7–15.

Zeng, G. and Dubes, R. C. (1985a): A test for spatial randomness based on k-NN distances. *Pattern Recognition Letters*, **3**, 85–91.

Zeng, G. and Dubes, R. C. (1985b): A comparison of tests for randomness. *Pattern Recognition*, **18**, 191–198.

What is Data Science ?*
Fundamental Concepts and a Heuristic Example

Chikio Hayashi

The Institute of Statistical Mathematics
Sakuragaoka, Birijian 304
15-8 Sakuragaoka, Shibuya-ku
Tokyo 150, Japan

Summary: Data Science is not only a synthetic concept to unify statistics, data analysis and their related methods but also comprises its results. It includes three phases, design for data, collection of data, and analysis on data. Fundamental concepts and various methods based on it are discussed with a heuristic example.

1. Introduction:

Statistics and data analysis have developed in their realms separately and contributed to the development of science, showing their unique properties. The ideas and various methods of statistics were very useful, well known and solved many problems. Mathematical statistics succeeded it and developed new frontiers with the idea of statistical inference. Thus the application of these view points brought us many useful results.
However, the development of mathematical statistics, has devoted itself only to the problems of statistical inference, an apparent rise of precision of statistical models, and to the pursuit of exactness and mathematical refinement, so mathematical statistics have been prone to be removed from reality.
On the other hand, the method of data analysis has developed in the fields disregarded by mathematical statistics and has given useful results to solve complicated problems based on mathematico-statistical methods (which are not always based on statistical inference but rather are descriptive). Some results are found in the references.
In the development of data analysis, the following tendency is often found, that is to say, data analysts have come to manipulate or handle only existing data without taking into consideration both the quality of data and the meaning of data, to cope with the methodological problem based on unrealistic artificial data with simple structure, to make efforts only for the refinement of convenient and serviceable computer software and to imitate popular ideas of mathematical statistics without considering the essential meaning.
As this differentiation proceeds with specialization, the innovation of useful methods of statistics and data analysis seem to disappear and signs of stagnation appear. The reason is that the essential aim of analysis of phenomena with data has been forgotten. For extensive and profound development of intrinsically useful methods of statistics and data analysis beyond the present state, the unification of statistics and data analysis is necessary. For this purpose, the construction of a new point of view or a new paradigm is a crucial problem. So, I will present "Data Science" as a new concept.

----

* The roundtable discussion "Perspectives in classification and the Future of IFCS" was held at the last Conference under the chairmanship of Professor H. -H. Bock. In this panel discussion, I used the phrase 'Data Science'. There was a question, "What is 'Data Science'? " I briefly answered it. This is the starting point of the present paper.

2. Fundamental Concepts of Data Science

Data Science is not only a synthetic concept to unify statistics, data analysis and their related methods, but also comprises its results. Data Science intends to analyze and understand actual phenomena with "data". In other words, the aim of data science is to reveal the features or the hidden structure of complicated natural, human and social phenomena with data from a different point of view from the established or traditional theory and method. This point of view implies multidimensional, dynamic and flexible ways of thinking.

Data Science consists of three phases : design for data, collection of data and analysis on data. It is important that the three phases are treated with the concept of unification based on the fundamental philosophy of science explained below. In these phases the methods which are fitted for the object and are valid, must be studied with a good perspective. The strategy for research in Data Science through three phases is summarized in Fig. 1.

Data Science

Design for Data

Collection of Data

Analysis on Data

The leading idea of data treatment in the research process is:

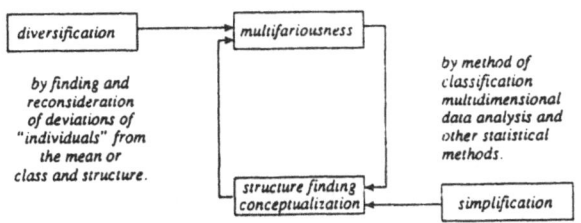

Dynamics of Both <u>Diversification</u>

and <u>Conceptualization or Simplification</u>

Fig.1 Strategy for Research

Generally speaking, phenomena are multifarious. First, these phenomena are formulated and the planning of a survey or experiment is completed, based on the ideas of Data Science (phase of design for data). Thus phenomena are expressed as multidimensional and, frequently, time-series data. The characteristics or properties of the data are necessarily made clear (phase of collection of data). The obtained data are too complicated to draw a clear conclusion. So, by methods of classification and multidimensional data analysis, and other mathematico-statistical methods, the data structure is revealed. In other words, simplification and conceptualization are carried out. However, this information generally turns out to be incomplete and unsatisfactory even though the structure finding was realized. At this stage, by finding and reconsidering the deviation of "individuals", which gives a vivid account of the roughness of conceptualization or simplification, from the mean values or class-belonging (classification) and structure, diversification of data is made. Based on this multifariousness, structure finding or conceptualization is attained, in an advanced sense, in the progressive stages. Such a circular movement of research then continues. Dynamic movement of both simplification or conceptualization and diversification begins in turn. Further, having been able to solve a problem, it is expected to discover another new problem to be solved in an advanced sense. The developmental process, in phase, design ---> collection --> analysis --->design ---> collection ---> analysis ---> design --> collection ---> analysis ---> design ---> ... and the dynamic process mentioned above, that is to say, progress and regress, are indispensable in Data Science. This shows that the methodology of Data Science develops, as it were, in the ascending-spiral-process and research proceeds as seen in spiral stairs. The main points is schematically depicted in Fig. 1.

Thus we can say that data science comprises not only the results themselves of theory and method but also all methodological results related to various processes which are necessary to work out the results mentioned above. The former is called "hard results" and the latter is called "soft results". Data Science includes simultaneously hard and soft results. It goes without saying that a useful solution emerges in coping with the complicated problem in question by the use of Data Science. It is repeatedly emphasized that the coherent idea through all items shown in Fig. 1 flows in Data Science for the purpose of analysis of phenomena with data.

3. Content of Data Science

Some concrete examples in social and medical surveys for the three phases are shown below. Before everything, it is stressed that the relevant methods are always treated with validity.

3.1 How to Design

The theory and method concerning this phase are next considered. Particularly, theoretical and systematic construction of a questionnaire is a very important problem. The problems in this phase are frequently solved using various kinds of methods of data analysis. For example,

> Sampling survey methods,
> Design of experiment,
> Evaluation of bias in quota sampling
> New systematic idea of survey planning for the solution of difficult and complicated
>   problems,
> Construction of questionnaire,
>   Theory-driven (which is an extension of hypothesis testing), Guttman's
>   Facet Theory,
>   Data-driven (exploratory approach), Hayashi's Cultural Link Analysis in
>   comparative study,
> Utilization of various types of questions, for example, dynamic use of closed and
>   open-ended questions,
> Use of various projective methods,
> Design for evaluation of data quality and data characteristics,
> Randomized response method,
> Problems of translation in international comparative study,
> and etc..

3.2 How to Collect Data

Collection of data is not only a problem of practice, but must be theoretically and concretely studied. The problems in this phase can not be solved without any information of design for data and any use of data analysis.

> Evaluation of survey bias and evaluation of experimental bias including
> question bias, interview bias, interviewer bias, observation bias, etc.
> Evaluation of non-response error,
> Evaluation of measurement error,
> Evaluation of response error, inevitably variable response data, for example,
> live data,
> Method of diminution of the relevant bias and error,
> and etc..

3.3 How to Analyze Data

The problems in this phase are, of course, closely related to the previous two phases. The main point is to obtain useful and instrumental information without any distortion or with validity. For this purpose, clear and lucid methods of analysis without unnecessary mathematical conditions only for model building and a too sophisticated style are desirable. For example,

> Various methods of scaling, quantification methods, correspondence analysis
> (analyse des données), multidimensional scaling, exploratory data
> analysis, categorical data analysis and various methods of classification and
> clustering,
> Useful data analysis suitable for the purpose,
> Useful coding of questions and their synthesis,
> Valid analysis of data including various errors,
> Evaluation of data quality and data analysis depending on data quality,
> Analysis on probabilistic response,
> Exploratory approach by data analysis,
> Method of simultaneous realization of classification and structure finding,
> Treatment of open answers in an open ended question for example, exploratory
> approach for coding or automatic processing of textual data,
> Probabilistic approach,
> Computer experiments,
> and etc..

These three phases must be synthetically treated or taken into consideration with the consistent idea in order to understand phenomena. This is the fundamental concept of Data Science. Of course, each subject will be studied separately. However, each subject must be studied in the context of Data Science. This idea will lead to the development of statistics and data analysis in a new direction. Thus the stand point of them highten and a new horizon will appear as innovative method and theory are created in three phases.

4. A Heuristic Example

As an example of the data-scientific approach, we now explain our national character survey in interchronological and international perspectives.

4.1 Fundamental Scheme of Study

What is national character ? I define it operationally as collective character on belief systems, the way of thinking and emotional attitudes, feelings or sentiments. By a survey of individuals, we can find individual response patterns on the items mentioned below.

Thus, we know that individuals have various response patterns. These are integrated in a collective through mutual and social communications in so far as individuals live in a society. This is collective character or national character (in some cases ethnic character) which is formed beyond individuals. In this situation, some principles emerge in the social environment. Receiving impacts from the exterior, social norm, customs system, paradigm, education, contemporary thought and arts, religious feelings, future course of philosophy and science, etc. are formed, as a "cultural climate" is created. Individuals are influenced by this cultural climate: the strongest influence is upon the response pattern in general social items, the second upon that in national character items and the weakest upon that in basic human feelings items. Such a perpetual circular movement continues. It is our aim to represent the collective character in terms of Data Science.

Our point of view of research is not hypothesis testing (theory-driven) but to put the emphasis on an exploratory approach (data-driven).

4.2 Time Series Data (Interchronological Approach) in Japan

First of all, we define the universe and population of the Japanese. A nation-wide sample survey is done for a sample by stratified three stage random sampling and by face to face interviewing using the same questionnaire, the contents of which cover the items shown below.

1) Fundamental Attributes, 2) Religion, 3) Family,
4) Social Life, 5) Interpersonal Relations,
6) Politics, 7) Individual Attitude toward Other Unclassified Social Issues

The outline of our survey is shown in Fig. 2.

Nation Wide Sample Survey by
Stratified Three Stage Random Sampling

Sample Spot 200 - 300
Sample Size 2000 - 4000
by face-to-face interviewing

Survey	Symbol	Year
1st Survey	I	1953
2nd Survey	II	1958
3rd Survey	III	1963
4th Survey	IV	1968
5th Survey	V	1973
6th Survey	VI	1978
7th Survey	VII	1983
8th Survey	VIII	1988
9th Survey	IX	1993
		every 5 years

[Research Committee on the Study of Japanese National Character of the Institute of Statistical Mathematics, Tokyo]

Fig. 2 Survey Design

The analysis from such time series data makes clear both enduring and changing aspects. The next step is a comparative study of national character.

4.3 Comparative Study (International Approach)

In a comparative study, the following points are indispensable,
1. How to secure comparability in a scientific sense
   Design
       Sample
       Selection of questions and construction of questionnaire
       Translation*
2. Clarification of particularity and universality (community) or speciality and generality
3. By common logic and scientific methods for easy (international understanding)

----

*   back translation, retranslation, confirmation by free question and answers, etc.

Here, in a comparative study, we present a new idea for questionnaire construction and selection of nations to be compared. This is Cultural Link Analysis (CLA in abbreviation) --Hayashi et al. 1986, 1992-- which belongs to a similar genre to Guttman's Facet Theory, (Guttman (1994)), and reveals a relational structure of collective characters of peoples in different cultural spheres (nations or ethnic groups).
   a.   A spatial link inherent in the selection of the subject culture or society.
       The connections seen in such selection may be considered along the dimensions of social environment, culture and ethnic or national characteristics.
   b.   An item structure link inherent in the commonness and differences in item response patterns within and across different cultures.
   c.   A temporal link inherent in longitudinal analysis.
An example of a. is shown in Fig.3.

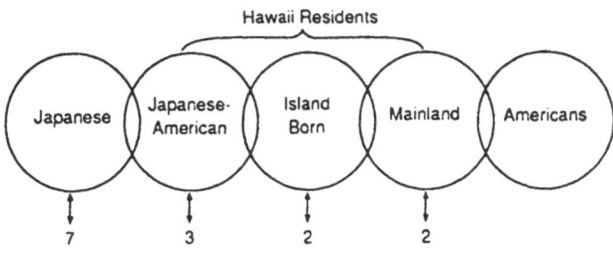

No. of Times Surveyed

(b) Multidimensional Linkage

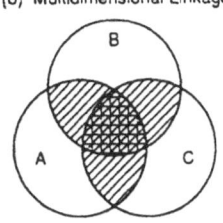

Fig.3 Cultural link survey design: selection of groups

As an example of **b** . concerning questionnaire construction, the idea is explained in Fig.4.

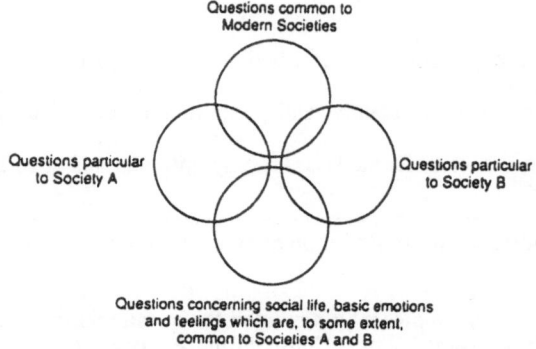

Fig.4   **Cultural Link Survey Design: Selection of Questions**

As for **c** . time series surveys in various nations or ethnic groups and their comparison are informative.

Our international comparative surveys, which consist of Americans in North America, English in UK., French in France, Germans in the past West-Germany, Dutch in the Netherlands, Italians in Italy, Japanese in Japan, Japanese-Americans in Hawaii, Japanese-Brazilians in Brazil, are described in Fig. 5 and the conjecture of link scheme is depicted in Fig.6.

1971	Japanese Americans in Hawaii (434)		
1978	Honolulu residents including JA (751)	Americans in North America (1571)	
1983	Same above (807)	------------	
1987	----------	------------	English (1043) Germans (1000) French (1013)
1988	Same above (499)	Americans in North America (1563)	
1992	Japanese Brazilians in Brazil (492)		Italians (1048)
1993			Dutch (1083)

------Nation-Wide Sampling------
(   ) Sample Size

Fig.5.   **Comparative Surveys By Cultural Link Analysis**

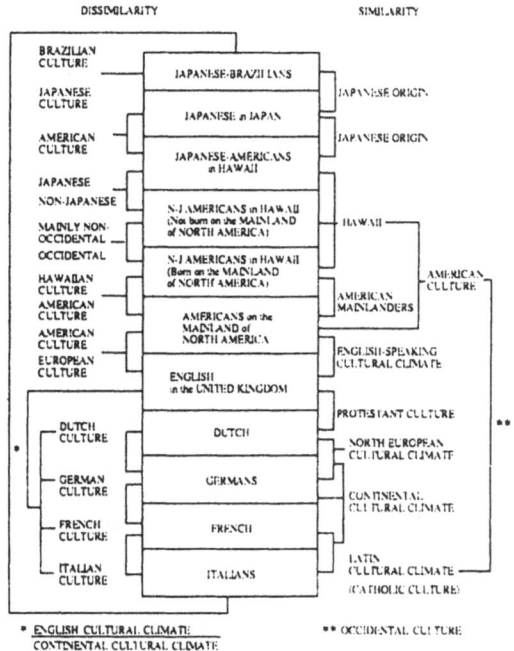

DISSIMILARITY   SIMILARITY

Fig.6 Chain in Our Study

Further Remarks : It brings us very important information to include people of Japanese origin, who settled down in foreign countries, as a linkage in order to explore and reveal the characteristics of Japanese national character.

The attitude, in which Japanese Americans are between Japanese and Americans in response distributions or data structure and, what is more, Japanese Brazilians are between Japanese and Portuguese, French and Italian in response distributions or data structure, is defined as J-attitude. The existence of J-attitude implies that J-attitude is a characteristic of Japanese national character. In other words, it may be said that J-attitude remains somewhat in Japanese Americans and Japanese Brazilians even though the tendency in them is not so strong as in the Japanese. This fact suggests that J-attitude is a Japanese characteristic. [Hayashi (1995), Hayashi C. and F. (1995)] So, it is meaningful to include people of Japanese origin in a comparative survey. However, it goes without saying that the characteristics of the Japanese are found according to the items shown in Fig.8, even though they are not J-attitude.

4.4 National Character in Statistical Terms

It is our aim to make clear and depict the following points by well-designed comparative surveys and their data analysis, i.e. "quantitative and data scientific" methods,

> "difference in some points and commonness or similarity in other points"
>   or
> "particularity in some points and universality in other points"

Since such a way of research is based on universal logic, people even in different cultural spheres can understand the results of the analysis.

Mainly considering the view of Japanese national character itself, we can summarize our study as in Fig.7. In contrast, mainly considering the comparison of national character in different cultural spheres, we can summarize our study as in Fig. 8.

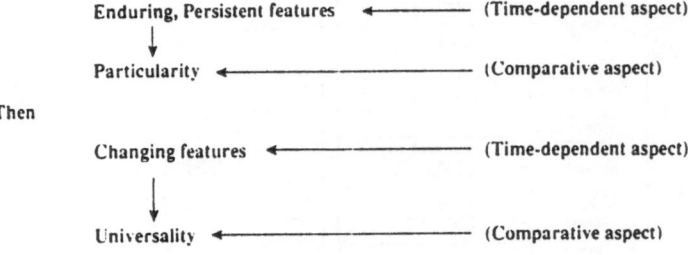

Fig.7   Japanese National Character

From these two kinds of surveys, surveys both in time and space, that is to say, continuing surveys and comparative surveys, we can define national character in statistical terms corresponding to various levels. See Fig. 8.

	Temporally Stable Consistent	Particular Characteristic compared with others
1. Majority Opinion	O	O* or O
	O	indifferent
	no datum	O or O*
2. Opinion Distributions	O	O* or O
	no datum	O or O*
3. Opinion Distributions by Various Breakdowns (for example, gender, age education, rural-urban)	O	O* or O
	no datum	O or O*
4. Changing Patterns of Opinion Distributions and Opinion Structure (including those by breakdowns, and, for example, age-cohort analysis based on time series of opinion distributions)	X	O
5-1 Opinion Structure	O or systematic change	X
5-2 Comparison of Opinion Structures		
i Existence of the Same Unidimensional Scale	O or no datum	by comparison of the scale value of nation O* or O
ii Same Structure of Opinions (more than 2 dimensional structure)	O or no datum	by comparison of the position of nation O or O*
iii Different Structure of Opinions	O or no datum	by comparison of the position of nation based on the similarity or dissimilarity analysis of structure O or O*

Fig. 8 Statistical Definition of National Character
---on various levels---

Here, majority opinion is defined as not only that supported by more than 2/3 of the individuals in the total but also that supported by more than 2/3 of the individuals in each breakdown in sex, age and education. In Fig. 8, O marks mean existence of the item, O* marks mean existence of temporally stable data and "no datum" means non existence of temporally stable evidence but existence of cross-section data. For example, as for 2. Opinion Distributions, the first line means a definition on the highest level, i.e. the opinion distribution is not only temporally stable but also particular or characteristic compared with those in different nations or ethnic groups and the second line means a definition on a lower level, i.e. temporally stable data do not exist but it is particular or characteristic compared with those in different nations or ethnic groups, in which temporally stable evidence occasionally exists. X marks mean there is no logical meaning.

4.5 Cross-Societal Surveys and Classification of Nations
--Realization of Cultural Link--

One example of data analysis of comparative surveys will be shown as below, with the following groups being taken up: Japanese, Americans, English, French, Germans, Dutch, Italians, Japanese-Americans and Japanese-Brazilians.
For example, let the opinion distribution be given in each group. Here, only one key answer category is taken up in each question item. If the number of questions is R, the number of the answer category taken up is r. Here, all questions are used except for the items of personal characteristics, for example sex, age, education, etc. We calculate the similarity index d_{ij} between i-nation and j-nation, as below.

$$d_{ij} = 1/R \sum_r^R |P_{ir} - P_{jr}|$$

where P_{ir} is the percentage of i-nation on the only one key answer category of the r-th question.
d_{ij} is a fuzzy measure of difference between i and j.

Thus we have a similarity matrix between i and j. Based on this fuzzy similarity matrix, a method of multidimensional data analysis, MDA-OR (Minimum Dimension Analysis Ordered class belonging) [Hayashi 1974, 1976], which is one kind of so-called multidimensional scaling MDS, is applied for graphical representation of groups. The quite similar configuration of groups is obtained by quantification method III or Correspondence analysis using the matrix of d's directly. The result is shown in Fig.9. This is a simple graphical summarization of the similarity relations. The degree of similarity is revealed as the distance in Euclidean space. Roughly speaking, consider that the distance corresponds to the similarity and the configuration gives a reasonable summarization of linked similarities. Here, the triangular relation mentioned above has been revealed.

The arrow means the direction of the value in the third axis in Fig.9. A line means plus direction while a dotted line means minus direction in the third dimension.
If French and Italians are deleted and the same analysis is done, Fig. 10 is obtained.
JB is found as a pole instead of French and Italians.

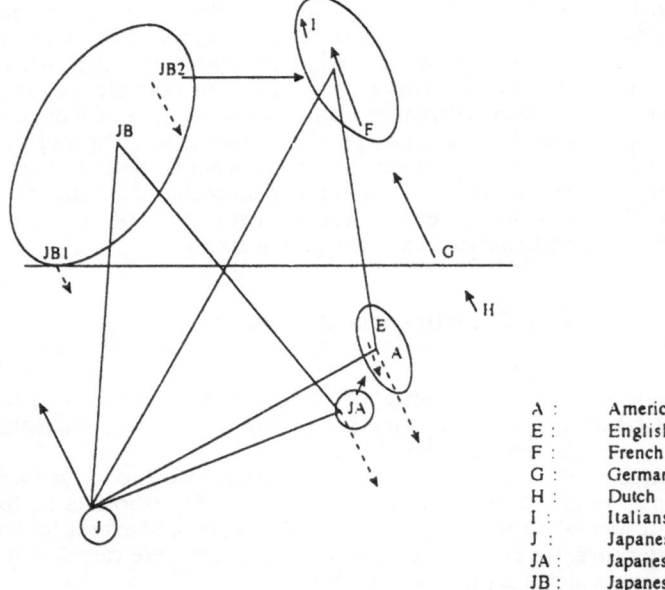

A :	Americans
E :	English
F :	French
G :	Germans
H :	Dutch
I :	Italians
J :	Japanese
JA :	Japanese-Americans in Hawaii
JB :	Japanese-Brazilians in Brazil
JB1:	First generation of JB
JB2:	Second and Third generations of JB

Fig. 9   Configurations of Nations

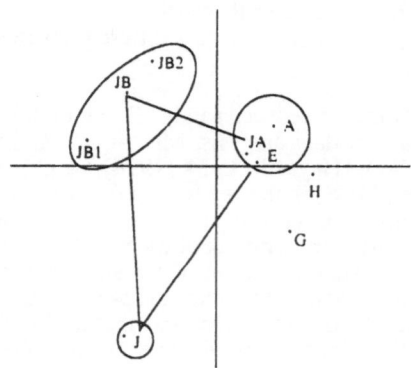

Fig.10 Configurations of Nations
(Deleting French and Italians)

The similarity between Japanese-Brazilians and French and Italians is very interesting. The positions of Japanese-Americans and Japanese-Brazilians are to be noted with J-attitude mentioned previously, the former being as a linkage between Japanese and Americans, the latter being as a linkage between Japanese and French or Italians. The link relation in Fig. 6 has been revealed in Fig. 9 by the data analysis. Thus we could clarify the entire picture of the configuration of groups.

Then, we can proceed to a detailed analysis of data without loss of sight of the whole situation. For example, the nations being different in what group of questions and common in what group of questions i.e. simultaneous classification of questions and nations or the universality and particularity of data structure across the nations.

References

The following references are relevant to the various parts of this paper.

Arabie, P., Hubert, L.J. and De Soete, G. (1996) ed.: Clustering and Classification, World Scientific.

Benzécri, J.P. (1973): *L'Analyse des Données*, Dunod.

Benzécri, J.P. (1992) : Correspondence Analysis Hand-Book, Marcel Dekker.

Bock, H.-H. and Polasek, W. (1996) ed.: Data Analysis and Information Systems, Springer.

Borg, I. & Shye, S. (1995): Facet Theory, Form and Content, Advanced Quantitative Techniques in the Social Sciences Series 5, Sage Publication.

Diday, E., J. Lemaire, J. Pouget and F. Testu (1983): *Elements d'Analyse des Données*, Dunod.

Diday, E., G. Celeux, Y. Lechevallier, G. Govaert and H. Ralambondrainy (1989): *Classification automatique et Analyse des Données: Méthodes et environment informatique*, Dunod.

Diday, E. and Y. Leschevallier (1991): Symbolic -Numeric Data Analysis and Learning, -Versailles Sept 91- Nova Science Publisher.

Gaul, W. and Pferfer, D. (1996) ed: From Data to Knowledge, Springer.

Guttman, L. (1994): *Louis Guttman on Theory and Methodology: Selected Writings*, Shlomit Levy ed, Dartmouth.

Hayashi, C. (1956): Theory and example of quantification(II). Proc. Inst. Statist. Math., 3, 69-98.

Hayashi, C. (1974): Minimum dimensional analysis MDA. *Behaviormetrika*, 1, 1-24.

Hayashi, C. (1976): Minimum dimensional analysis MDA-OR and MDA-UO, Essays in Probability and Statistics, Ikeda, S., et al. (eds.), 395-412, Shinko Tsusho Co. Ltd.

Hayashi, C. (1993): Treatise on Behaviormetrics, Asakura Shoten.

Hayashi, C. (1993): Quantification of Qualitative Data --Theory and Method --, Asakura Shoten.

Hayashi, C. (1995): *Changing and Enduring Aspects of Japanese National Character*, The Institute of Social Research, Osaka, Japan.

Hayashi, C. and Suzuki, T. (1986): *Data Analysis in Social Surveys*, Iwanami Shoten. The English version by Hayashi, C. Suzuki,T. and Sasaki, M., "*Data Analysis for Comparative Social Research: International Perspectives*" was published by Elsevier, North-Holland in 1992.

Hayashi, C. and Hayashi, F. (1995): *Comparative Study of National Character*, Proceedings of the Institute of Statistical Mathematics Vol. 43, No.1, 27-80.

Jambu, M. (1989) : *Exploration Informatique et Statistique des Données*, Dunod.

Jambu, M. (1991): Exploratory and Multivariate Data Analysis, Academic Press.

Lebart, L., Morineau, A. and Warwick, K.M. (1984): *Multivariate Descriptive Statistical Analysis*, John Wiley.

Lebart, L. and Salem, A. (1988): *Analyse Statistiques des Données*, Textuelles, Dunod.

Lebart, L. and Salem, A. (1994): Statistique Textuelle, Dunod.

Lebart, L., Morineau, A. and Piron, M. (1995): *Statistique Exploratoire Multidimensionnelle*, Dunod.

Van Cutsem, B. (1994): Classification and Dissimilarity Analysis, Springer.

Fitting Graphs and Trees with Multidimensional Scaling Methods

Willem J. Heiser

Department of Data Theory, Leiden University
P.O. Box 9555, 2300 RB Leiden
The Netherlands

Summary: The symmetric difference between sets of qualitative elements (called features) forms the basis of a distance model that can be used as a general framework for fitting a particular class of graphs, which includes additive trees, hierarchical trees and circumplex structures. It is shown how to parametrize this fitting problem in terms of a lattice of subsets, and how inclusion relations between feature sets lead to additivity of distance along paths in a graph. An algorithm based on alternating least squares and on the recent method of cluster differences scaling is described, and illustrated for the general case.

1. Introduction: Fitting distances or coordinates

Graphs and trees are increasingly considered to be attractive discrete structures for modelling general similarity or dissimilarity (or: proximity) data in the social and behavioral sciences (Arabie and Hubert, 1992; Klauer, 1994), in biology, which can built upon a conceptual tradition of long standing (Felsenstein, 1983), and in many other areas (Abdi, 1990; Barthélemy and Guénoche, 1991). Discrete structures are commonly contrasted with, and thought to be alien from the continuous spatial structures that are used in multidimensional scaling, although 'hybrid' models have been proposed (Carroll, 1976). The purpose of this paper is to take some steps towards an integrated view of these two types of models, by showing how we can deal with the problem of fitting graphs and trees with the same formalism that is the basis of least squares methods used in multidimensional scaling (MDS).

Within the framework of least squares, discrete structural representations of proximity data are usually identified by fitting a distance matrix under constraints. For example, least squares fitting of a hierarchical tree can be done by enforcing the ultrametric inequality upon a set of non-negative quantities (Hartigan, 1967; Chandon, Lemaire and Pouget, 1980), for an additive tree we can impose the four-point condition (Sattath and Tversky, 1977; Cunningham, 1978; De Soete, 1983), and for network representations one criterion that has been used is additivity of distances along every possible path (Klauer and Carroll, 1989). Most methods used in practice, while not being least squares, are nevertheless based upon classic operations on a dissimilarity matrix to transform it into a constrained distance matrix (e.g., both the single-link or minimum method and the complete-link or maximum method for hierarchical clustering can be viewed as transformations of an arbitrary dissimilarity matrix into an ultrametric distance matrix).

By contrast, an MDS model typically represents the objects in terms of points characterized by coordinates, so that distances are not parameters to be estimated, but *functions* of other parameters. These functions may be Euclidean (as is most commonly the case) or non-Euclidean (e.g., Groenen, Mathar and Heiser, 1996). The present paper will show how to set up such an indirect parametrization, which restricts the *coordinates* to be discrete (in fact, binary), for a relatively large class of graphical structures. Following Tversky (1977) and Shepard and Arabie (1979), we will use the concept of a *feature space*, in which each object of analysis is represented by some subset of features, while the features in turn are represented by subsets of objects. By restricting attention to models that can be formulated in terms of features, we are considering a particular subclass of graphical structures, to be called *feature graphs*.

The natural metric used in feature space is the city-block distance, which acquires several remarkable properties when the coordinates are restricted to be binary. Before discussing these in more detail, we need to introduce some notation.

2. Notation and reparametrization in terms of features

Let $O = \{o_1, \dots, o_i, \dots, o_n\}$ be the set of objects of analysis, and suppose that the elements of the square table $\Delta = \{\delta_{12}, \dots, \delta_{ij}, \dots, \delta_{nn}\}$ denote the values of a given dissimilarity function defined on the set of ordered pairs $O \times O$. We are looking for a graph representation of $\{O, \Delta\}$ by a *valued graph* (or network) $G = \{V, \mathcal{R}, \Lambda\}$, where the set $V = \{v_1, \dots, v_i, \dots, v_n\}$ contains the *nodes* (vertices, points), and the set $\mathcal{R}$ the *edges* (arcs, lines) of G. Thus, $\mathcal{R}$ is the collection of unordered pairs of V that defines a *relation*, that is, a subset of $V \times V$. An edge $r_{ij} = \{v_i, v_j\} \in \mathcal{R}$ is said to *join* the nodes v_i and v_j in the graph, and presence or absence of edges is indicated by the binary $n \times n$ matrix $\mathbf{A} = \{a_{ij}\}$, called the *adjacency matrix*, which has $a_{ij} = 1$ if $r_{ij} \in \mathcal{R}$, and $a_{ij} = 0$ if $r_{ij} \notin \mathcal{R}$. Finally, the graph G is *valued*: we associate with each edge present ($a_{ij} = 1$) some non-negative function value λ_{ij}, called the *edge length*, collected in the matrix $\Lambda = \{\lambda_{ij}\}$, where we define $\lambda_{ij} = 0$ when $a_{ij} = 0$. A metric on G is defined by the *path-length distance*

$$d_{ij}(\mathbf{A}, \Lambda) = d(v_i, v_j) = \sum_{(i*j*) \in \mathcal{P}(v_i, v_j)} \lambda_{i*j*}, \qquad (1)$$

in which $\mathcal{P}(v_i, v_j)$ is the set of edges on the *geodesic* (shortest path) between v_i and v_j. We write $d_{ij}(\mathbf{A}, \Lambda)$ because the distance depends not only on Λ, but also on $\mathbf{A}$ via the lists $\mathcal{P}(v_i, v_j)$. Thus, the path length is the sum of the edge lengths along the geodesic. If all λ_{ij} are equal, $d_{ij}(\mathbf{A}, \Lambda)$ is the usual graphical distance: a count of the number of edges in the shortest path from v_i to v_j.

Let us first consider the question of *embedding*, or realizability: under what conditions on Δ can the objects be mapped into a valued graph with some path-length distance? The answer is, that we may identify $\lambda_{ij} = \delta_{ij}$ for some subset $\mathcal{R}$, provided that δ_{ij} is positive-definite and satisfies the triangle inequality $\delta_{ij} \le \delta_{il} + \delta_{lj}$ (Hakimi and Yau, 1964). Note that symmetry is not required (if we allow two edges between any two nodes); but if δ_{ij} is in addition symmetric, it is a metric, and the result says that any metric can be embedded into a valued graph. In the presence of error, it is much to be preferred to optimize some loss function measuring the lack of fit of feasible model distances, rather than to rely on idealized conditions evaluated directly in terms of the data. Therefore, we study the *fitting* problem of finding G (in particular, some $\mathbf{A}$ and Λ) so that the least squares loss function

$$\sigma^2(\mathbf{A}, \Lambda) = \sum_i \sum_j (\delta_{ij} - d_{ij}(\mathbf{A}, \Lambda))^2, \qquad (2)$$

is minimal. Note that the major difficulty in (2) is finding $\mathbf{A}$, since (1) is additive in the elements of Λ, so that, once we know $\mathbf{A}$, finding Λ is just a non-negative regression problem, which can be solved by standard methods (Lawson and Hanson, 1974). How can we find out which edges to include and which to delete?

Our approach in the present paper will be to use a *reparametrization* of $d_{ij}(\mathbf{A}, \Lambda)$, which restricts attention to a certain subclass of graphs. To define the vertices of such a graph, we introduce a set of p *discrete features* $\mathcal{F} = \{F_1, \dots, F_t, \dots, F_n\}$. On the feature set $\mathcal{F}$ we define a family S of n distinct nonempty subsets $S = \{S_1, \ldots, S_i, \ldots, S_n\}$, whose union is $\mathcal{F}$. Furthermore, each feature $F_t \in \mathcal{F}$ is associated with some nonnegative *feature discriminability* parameter η_t. Every object will now be represented by some subset of features, that is, our goal will be to find a mapping $\mathcal{T}: o_i \in O \to S_i \in S$.

To rephrase the fitting problem in terms of the mapping $\mathcal{T}$, we must have a metric on (sub)sets that parallels the path-length distance. Following Goodman (1951, 1977) and Restle (1959, 1961), we may define a metric on sets, here to be called the *feature distance*

$$d(S_i, S_j) = \mu[(S_i \cup S_j) - (S_i \cap S_j)] , \tag{3}$$

where $\mu[\,\cdot\,]$ is a measure function defined over the set of features (usually, just a count), and $A - B$ is the symmetric set difference between sets A and B. Thus the feature distance measures the extent to which S_i possesses features that S_j does not have and vice versa. By elementary means it can be shown that (3) satisfies the metric axioms, and there are a number of alternative expressions of it that enable us to naturally include the feature discriminabilities η_k, which we will consider more closely in section 3. The first and foremost property of $d(S_i, S_j)$, however, is stated in the following result.

Theorem (Flament, 1963, p. 17).
Let $\mathcal{L}(S)$ be the lattice obtained from ordering the elements of S by inclusion, and consider the graph representing $\mathcal{L}(S)$ having nodes $v_i = S_i$ and an edge between v_i and v_j whenever S_i covers S_j or vice versa. Define $d_{ij}(\mathbf{A})$ as the pathlength distance (geodesic) between nodes v_i and v_j with all edge lengths λ_{ij} equal to unity. Then the feature distance $d(S_i, S_j)$ is equal to $d_{ij}(\mathbf{A})$ in the graph representation of the lattice $\mathcal{L}(S)$.

If S is an arbitrary selection of subsets, it is understood that $\mathcal{L}(S)$ includes the extension of S with all subsets that can be formed by union and intersection of its elements. In the graphical representation of this lattice of subsets there are generally several paths from S_i to S_j, but the crucial thing is that they all have equal length; hence, they are equivalent in terms of distance. Equivalence of distinct paths follows from the fact that each edge in the graphical representation of the lattice corresponds to one single element of $\mathcal{F}$, which is the feature that distinguishes the covering subset from the covered one. While the graphical distance $d_{ij}(\mathbf{A})$ is a count taken along a path of the distinguishing features in some particular order, the feature distance $d(S_i, S_j)$ is the same count of distinguishing features, taken in any order.

Another property that turns out to be crucial for the present approach is that betweenness implies additivity: that is, if S_j is inbetween S_i and S_k in the sense that either $S_i \supset S_j \supset S_k$, or $S_i \supset S_j$ and $S_k \supset S_j$, we have $d(S_i, S_k) = d(S_i, S_j) + d(S_j, S_k)$ along the path from S_i to S_k. In this case, there need not be a direct edge from S_i to S_k. This characteristic allows us to first formulate the fitting problem (2) in terms of feature distances, next sort out the additivities in the fitted distances, and finally construct the graph by excluding edges that are sums of other edges. Thus the graphs to be constructed with the present approach will always be subgraphs of the graph representation of a lattice, which forms the embedding space of the given set of objects in much the same way as Euclidean space is used to embed a finite number of points in ordinary multidimensional scaling.

3. Introducing discriminability of features: Weighted counting

In the simplest case, the Goodman-Restle feature distance (3) is a straight count of the features in the set difference between the union and the intersection of two subsets. Although there are a number of interesting re-expressions of the feature distance in terms of set operations, what we need for our MDS-like fitting problem is an expression in terms of coordinates. Let $\mathbf{E} = \{e_{it}\}$ be a binary matrix of order $n \times p$, which indicates which features of $\mathcal{F}$ are included in each of the n subsets in $\mathcal{S}$ that represent the objects. Since $\mathbf{E}$ characterizes objects as subsets of features (but also features as subsets of objects), it is (the transpose of) a *point-set incidence matrix* (see Roberts 1976, page 60). When $\mu[\cdot]$ is just a counting measure, (3) becomes

$$d(S_i, S_j) = \Sigma_t \{(1 - e_{it})e_{jt} + (1 - e_{jt})e_{it}\}$$

$$= \Sigma_t \{e_{jt} - e_{it}e_{jt} + e_{it} - e_{jt}e_{it}\}$$

$$= \Sigma_t (e_{it} - e_{jt})^2 = \Sigma_t |e_{it} - e_{jt}| , \qquad (4)$$

where the last equality follows from the binary nature of $\mathbf{E}$. Thus, the feature distance is equal to a city-block metric on a space with binary coordinates, a metric better known as the *Hamming distance*. This distance is commonly used as a dissimilarity coefficient, in a situation where the e_{it} are presence-absence data, or as a theoretical device (Boorman and Arabie, 1972), but – to the best of the author's knowledge – it has never been used as a structural model to be fitted to dissimilarity data.

Especially for fitting purposes, it is useful to take one further step and to go from a simple count to a weighted count, that is, to generalize (4) into

$$d_{ij}(\mathbf{B}) = \Sigma_t \eta_t |e_{it} - e_{jt}| = \Sigma_t |b_{it} - b_{jt}| , \qquad (5)$$

where the b_{it}s are still binary, albeit not necesarily $(0,1)$ variables, collected in the $n \times p$ matrix $\mathbf{B} = \{b_{it} = \eta_t e_{it}\}$, and where the discriminabilities η_t are nonnegative parameters to be estimated. Thus, the weighted feature distance defined in (5) allows for a differential contribution of the features to the overall length of the path from S_i to S_j. In geometrical terms, the introduction of feature discriminabilities turns the hypercube correspondng to $\mathbf{E}$ into a rectangular parallelepiped corresponding to $\mathbf{B}$.

It can be shown that the Theorem stated in the previous section still holds for the weighted feature distance, if it is adjusted to allow for unequal λ_{ij}. Since each edge in the graphical representation of the lattice $\mathcal{L}(\mathcal{S})$ corresponds to one feature in $\mathcal{F}$, we can associate exactly one feature discriminability η_t in (5) with each edge length λ_{ij} in (1). For example, if the set of edges on the shortest path between v_i and v_j would be $\mathcal{P}(v_i, v_j) = \{(v_i, v_k), (v_k, v_l), (v_l, v_j)\}$, there will be three features F_1, F_2, and F_3 on which S_i and S_j are different, with the edge lengths being related by the one-to-one mapping $\lambda_{ik} = \eta_1$, $\lambda_{kl} = \eta_2$, and $\lambda_{lj} = \eta_3$. Hence we have $d_{ij}(\mathbf{A}, \Lambda) = \lambda_{ik} + \lambda_{kl} + \lambda_{lj} = \eta_1 + \eta_2 + \eta_3 = d_{ij}(\mathbf{B})$ in this example.

4. Algorithm for fitting a feature graph

Due to the pioneering work of Hartigan (1967), Cunningham (1978), Arabie and Carroll (1980), De Soete (1983), Mirkin (1987), and others, least squares fitting of discrete models is gradually gaining ground over various *ad hoc* procedures that were once more common. After replacement of the path-length distance $d_{ij}(\mathbf{A}, \Lambda)$ by the feature distance $d_{ij}(\mathbf{B})$, the least squares loss function (2) that we are interested in becomes

$$\sigma^2(\mathbf{B}) = \Sigma_i \Sigma_j (\delta_{ij} - d_{ij}(\mathbf{B}))^2 , \qquad (6)$$

which must be minimized over all binary valued matrices $\mathbf{B}$. Because the feature distance is additive over features, it is possible to employ an alternating least squares (ALS) scheme, fitting the model one feature at a time, given some starting values $\{\underline{b}_{it}\}$. Explicitly, given the current values $\{\underline{b}_{is}\}$ for $s \neq t$, $\underline{\delta}_{ij}$ is defined as $\underline{\delta}_{ij} = \delta_{ij} - \sum_{s \neq t} |\underline{b}_{is} - \underline{b}_{js}|$, the original dissimilarity corrected for the contribution of the fixed variables. Substituting (5) into (6), and inserting $\underline{\delta}_{ij}$, we find that the ALS subproblem for feature t is to minimize, given $\underline{\delta}_{ij}$,

$$\sigma^2(\mathbf{b}_t) = \sum_i \sum_j (\underline{\delta}_{ij} - |b_{it} - b_{jt}|)^2 , \tag{7}$$

over the binary n-vector $\mathbf{b}_t$. This minimization subtask is a one-dimensional MDS problem with the coordinates restricted to form a bipartition, and therefore the cluster differences scaling (CDS) algorithm of Heiser and Groenen (1996) applies, with number of clusters equal to two. The ALS algorithm cycles over CDS subtasks until convergence.

Let us have a closer look at this particular CDS substask, by resolving $\mathbf{B}$ again in its discrete and continuous factors. Writing $|b_{it} - b_{jt}| = \eta_t \{(1 - e_{it})e_{jt} + (1 - e_{jt})e_{it}\}$, setting the partial derivative of (7) with respect to η_t equal to zero and simplifying, shows that, for any given bipartition $\{e_{it} \mid i = 1, \ldots, n\}$ the optimal value of the discriminability parameter for feature t is equal to $\max(0, \hat{\eta}_t)$, with $\hat{\eta}_t$ denoting the unconstrained minimizer

$$\hat{\eta}_t = \frac{1}{n_t(n - n_t)} \sum_i \sum_j e_{it}(1 - e_{jt})\underline{\delta}_{ij} , \tag{8}$$

where n_t is the number of objects in one group, and $n - n_t$ the number of objects in the other. Thus, the length of edge t in the fitted feature graph will be equal to the average corrected dissimilarity between the two groups of objects that constitute that particular feature. If the features are exclusive, $e_{it}(1 - e_{jt})\underline{\delta}_{ij} = e_{it}(1 - e_{jt})\delta_{ij}$, and (8) becomes just the average between-group dissimilarity. If the features are not exclusive and $\hat{\eta}_t$ is relatively large, then the corresponding bipartition must be a good discriminator by itself, on top of the contribution of the other features, since we always have $\underline{\delta}_{ij} \leq \delta_{ij}$; this justifies the name discriminability parameter.

We still have to indicate how to find $\mathbf{E}$. Loss function (7) is quadratic in one column (size n) of the binary matrix $\mathbf{E}$, so this subtask still is a hard combinatorial problem, even though its size is reduced by a factor p with repect to loss function (6). The present implementation uses a nesting of several random starts (within features and across features), together with K-means type reallocations. Heiser and Groenen (1996) have described a strategy called *Fuzzy Steps* to alleviate the local minimum problem for CDS, but it looks like the problem here is especially difficult across features (in the ALS phase), not so much within features. A more extended discussion of the algorithmic aspects of finding $\mathbf{E}$ is in preparation.

5. Example: Henley's (1969) animal terms

To see how the feature graph procedure works, dissimilarity data originally collected by Henley (1969) will be analyzed as an example. In a psychological experiment on semantic memory structure, 21 subjects were instructed to freely list from memory any animal terms they knew. From the total set of animal terms mentioned, 12 common ones were selected. Dissimilarities were computed for each pair of terms as the average (across subjects) of the proportion of items separating them in each list. Figure 1 displays an additive tree representation of the animal terms, given by Abdi (1990), with a percentage of variance accounted for of 73.0%. Many other tree representations for this example can be found in Barthélemy and Guénoche (1991).

The feature-graph method was applied with the number of features ranging from 1 to 10. The one-feature solution, which is a simple two-cluster split, had a prevalence of 7 out of 10 random initial bipartitions, and accounted for 21.9% of the variance. It splits {lion, bear, pig, sheep, goat, cow, horse} from {cat, dog, mouse, rabbit, deer}. Note that this split cannot be obtained by cutting anyone of the edges of the optimal tree, while it does

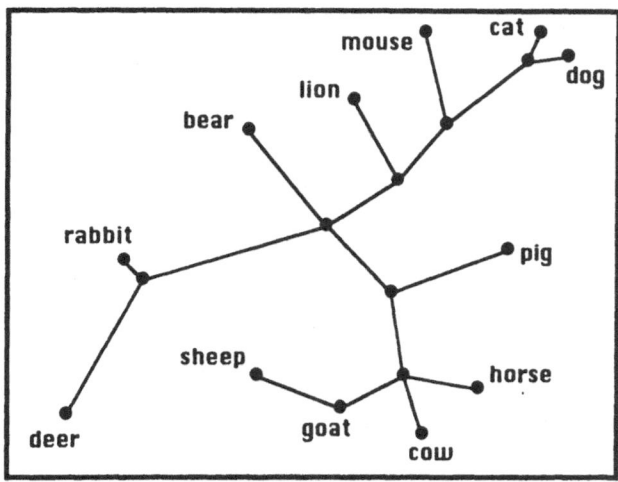

Fig. 1: Additive tree for Henley's animal terms

seem to represent the best split in terms of the dissimilarity data. The percentage of variance accounted for (VAF) for all ten solutions is given in Table 1. This table also gives for each solution the DAF (percentage of Dispersion Accounted For), defined as the sum of squared fitted distances divided by the sum of squared dissimilarities. DAF is the scale-free goodness-of-fit measure that is maximized when the badness-of-fit measure (6) is minimized.

Table 1. Goodness of fit for feature graph representations of the Henley (1969) data

# features:	1	2	3	4	5	6	7	8	9	10*
DAF	63.3	83.4	89.4	93.9	95.8	96.9	97.6	98.1	98.5	97.3
VAF	21.9	41.9	54.0	71.9	79.2	82.3	84.7	87.7	90.3	83.9

*solution with 4 unicities

We see from Table 1 that a VAF just above the percentage of the tree solution is reached with the five-feature solution (79.2%), which has a DAF of 95.8%. This solution, which does not yet discriminate all objects from each other (leaving seven objects in three small clusters), is shown in Figure 2. While the terms in the clusters {cat, dog} and {goat, cow, horse} in the feature graph are also close together in the additive tree in Figure 1, this is not the case for the cluster {bear, pig}. Another major difference is that the feature graph is not tree-like at all. As to the interpretation of the five-feature solution in Figure 2, it is clear that {deer} is an isolate (there is one feature that contrasts it with all other terms, with a discriminability of approximately 20), and that there is a "domestic" versus "wildlife" feature contrasting, from top to bottom, {sheep, cow, horse, goat, cat, dog} from {pig, bear, lion, mouse, rabbit, deer}, with a discriminability of about 14. A third important split is {cat, dog, lion, mouse} versus the rest, with discriminability 16.

A more differentiated representation arises if we look at one of the solutions with a higher number of features. How many features to take is not only a matter of amount of fit that is

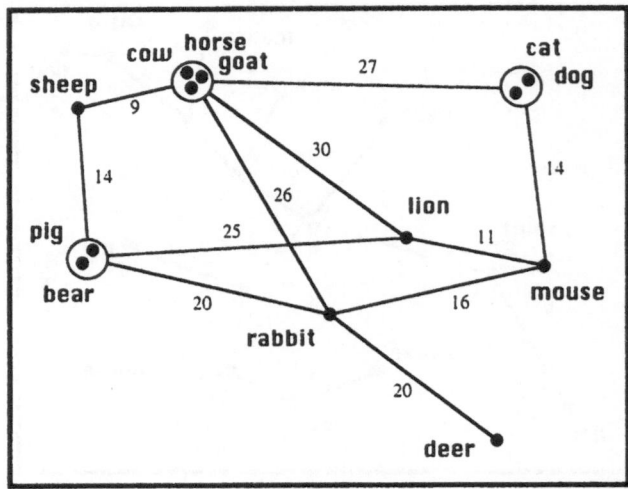

Fig. 2: Five-feature solution of Henley's data

deemed acceptable, but also depends on the issue of how many edges need to be kept, or conversely, how many additivities there are in the fitted distances. Judged by the number of edges needed, while still accounting for a reasonable amount of variance, a special ten-feature solution was selected as the best one (see the last column of Table 1; its graph with 24 edges is displayed in Figure 3). It consists of six common features (i.e., features shared by more than one object) and four unique features (not shared by any other object).

Figure 3 contains two types of nodes: the closed circles, which represent the objects of analysis, and the open circles, called *latent nodes*, which represent subsets of features that can be obtained by taking either the union or the intersection of the feature subsets characterizing two other nodes. Remember that the fitted feature graph is a subgraph of the

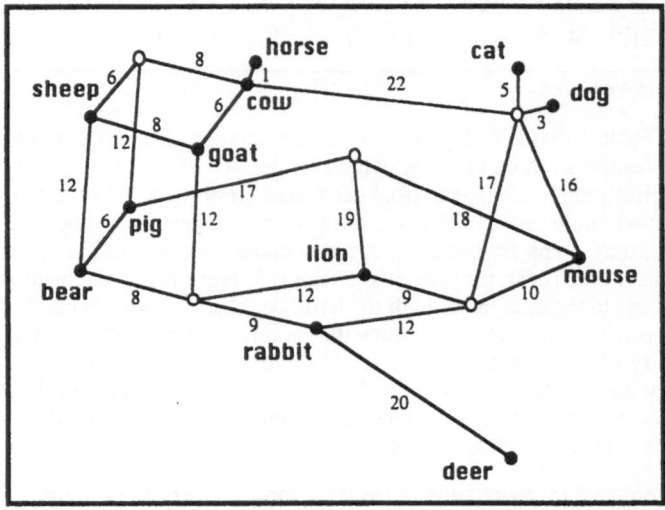

Fig. 3: Ten-feature solution of Henley's data, with 6 common features and 4 unique ones (open circles are latent nodes)

graph representation of the lattice of feature subsets, and latent nodes are other elements of this lattice that can be included afterwards, to make the graph simpler in terms of its pathways and number of edges. As a good example of the effect of the introduction of a latent node, consider four objects characterized by the subsets (BCD), (ACD), (ABD), and (ABC), and assume equal discriminability of the features. Then all distances are equal, and the objects are mapped as four points on a regular tetrahedron, with six edges. Introducing the latent node (ABCD), which is the union of each of the pairs of subsets, allows us to construct a star graph, in which there are only four edges, one between each of the manifest nodes and the latent node, and no one among the manifest nodes themselves.

The fitted edge lengths are also given in Figures 2 and 3 (rounded to integer numbers). An edge is not included in the graph if its length is the sum of two other edge lengths (a rather simple algorithm looping over all triads is sufficient to sort this out). To reconstruct the distance between two terms (and hence their dissimilarity), we just have to add the edge lengths along the shortest path between them. It will be noted that there are several instances of distinct paths with equal length. Comparing the two feature-graph solutions, it appears that there are primarily local changes: one is slightly more (less) differentiated than the other, a result that makes sense.

6. Some special cases

Modeling considerations can be formulated in terms of the lattice of subsets $L(S)$ as follows: given Δ, or some approximation of it satisfying the triangle inequality, does there exist feature distances $d(S_i, S_j)$ that arise from the feature graph of a family of subsets that has a certain property? Consequently, the fitting problem would become one of optimizing loss function (6) over a family of feature sets that have a specified structural property. In this section, we will briefly indicate some examples of structural properties that may be handled in the present framework; a more detailed treatment of them is in preparation.

Before considering the special cases, however, we discuss a technical issue that needs to be settled first. In a feature distance model, the role of presence and absence of features is symmetric, that is, we can always replace all elements e_{it} of a whole column of $\mathbf{E}$ by their complement $1 - e_{it}$ without changing the feature distance. To illustrate, the two matrices

$$\begin{bmatrix} 2 & 0 & 0 & 0 \\ 0 & 4 & 0 & 0 \\ 0 & 0 & 1 & 0 \\ 0 & 0 & 0 & 3 \end{bmatrix} \quad \text{and} \quad \begin{bmatrix} 1 & 0 & 0 & 2 & 0 & 1/2 & 0 & 3/2 \\ 0 & 1 & 2 & 0 & 0 & 1/2 & 0 & 3/2 \\ 0 & 1 & 0 & 2 & 1/2 & 0 & 0 & 3/2 \\ 0 & 1 & 0 & 2 & 0 & 1/2 & 3/2 & 0 \end{bmatrix}$$

generate the same feature distances among their rows. Thus we can freely add the complements of any column of the incidence matrix, provided that we *half* the corresponding discriminabilites. Any $n \times 2$ matrix formed by concatenating some column of $\mathbf{E}$ with its complement has the property that it has row sums equal to one, and such a matrix is called the *indicator matrix* of a feature.

Now suppose that the features are *nested*: that is, if $\mathbf{G}_t$ is the indicator matrix of feature F_t and $\mathbf{G}_s$ is the indicator matrix of feature F_s, than the matrix $\mathbf{G}_t'\mathbf{G}_s$ has at least one element equal to zero. Nestedness implies that one feature separates a subgroup from one of the two groups formed by the other feature. For instance, the bipartitions {(ABCD), (EFG)} and {(EF), (ABCDG)} are nested, since (EF) is a subset of (EFG) and (ABCD) is a subset of (ABCDG). Then, by a famous result of Buneman's (1971), the feature distance satisfies the four-point property that characterizes *additive trees* if and only if all its features are nested. Additive trees thus form an important special case of feature graphs, in which each edge corresponds to exactly one feature (or *split*).

The case of a *linear array* (Goodman, 1951, 1977; Restle, 1959, 1961), called *Guttman scale* in psychometrics, is obtained when the features are not only nested, but have an additional property. In terms of the feature incidence matrix **E**, this property implies that each column of **E** consists of either a single run of zeros followed by a single run of ones or of a single run of ones followed by a single run of zeros. When $n - 1$ distinct features have this structure, the feature graph has $n - 1$ edges, connecting the objects in a certain order, and no latent nodes. Except for the two endpoints, which have degree one, all nodes have degree two. For an exact characterization of the Guttman scale, see Holman (1995).

A *hierarchical tree* is a rooted additive tree with the extra requirement that the distance from any endpoint to the root is equal. In a feature graph, the root corresponds to the latent node that has all features, that is, $\mathcal{F}$. Then the first feature defines the first split in two groups of objects, the second feature splits one of these groups further down into subgroups, and so on. So the features are again nested. The hierarchical tree is a more parsimoneous model than the additive tree, because the requirement of equal distance to the root puts restrictions on the discriminabilities. The characterization of trees in terms of a feature model is due to Tversky (1977).

The last example of a family of subsets that satisfies a specific structural property is the *circumplex* or *radex* (Guttman, 1954). It is characterized by the *circular ones* property, which implies that each column of **E** consists of either a single run of zeros bordered by a run of ones on one or both sides, or of a single run of ones bordered by one or two run(s) of zeros. The graph of a regular circumplex is like a closed simple chain, with exactly n edges, if each feature divides the objects in equal groups (when n is even). When divisions in unequally sized groups are included in the feature set, the graph of a circumplex becomes more complicated. In the complete case, it looks like a network spanned over a (half)sphere (Heiser, 1981, chapter 4).

7. Discussion

We have seen that a metric defined on the symmetric set difference between sets of features can be used as a general framework for fitting a particular class of graphs, which includes additve trees, hierarchical trees and circumplex structures. It was shown that we can find out which edges to include in the graph by formulating the problem in terms of a lattice of subsets, using a weighted count of feature differences (the feature discriminabilities). The algorithm presented, based on alternating least squares and on cluster differences scaling, is still in its early stage of development. It always converges to a local minimum, but is as usual in this type of problem, there are an awful lot of local minima. On the positive side, it is the first systematic method to fit the Hamming distance.

A crucial ingredient of this approach to finding graph representations is the fact that inclusion relations between feature sets lead to additivity of distance along paths in a graph. In fact, Hutchinson (1989) and Klauer and Carroll (1989) used the criterion of dropping direct edges by looking at additivity of link length as their main graph construction strategy. But they applied this criterion to the dissimilarities, rather than to the fitted distances, as is proposed here. Feature graphs are similar to Corter and Tversky's (1986) extended similarity trees, but exactly how these two models are related needs further study. In any case, it seems clear that additive trees and other restricted representations do not show up spontaneously in real examples, although the method reproduces a circumplex, for example, when the data are error-free.

Choosing the number of features p is a matter that requires experience and cannot be settled yet with clear-cut rules. In most examples analyzed so far, the number of features needed to get good fit is in the neighborhood of $n/2$. Also, it appears that, as soon as p is in the range of values where the fit stabilizes, solutions with one feature more or one feature less are only different in their fine structure, as was the case in the example of the Henley (1969) data. There is a trade-off to be made with the number of links in the graph, a quantity that increases nonlinearly with p, and which we want to have as small as possible. Making a good trade-off is complicated by the fact that we can often reduce the number of links by including latent nodes, without it being clear how to do this optimally.

Unlike methods based on distance constraints, feature graph fitting can be extended without too much trouble to well-known variants of MDS, such as individual differences scaling (INDSCAL) and two-mode scaling (unfolding), possibly combined with nonlinear transformations of the data. An easy way to recognize this flexibility is to view the basic distance model (5) as a squared Euclidean distance (since the deviations $e_{it} - e_{jt}$ are zero or (minus) one, we just have to reparametrize η_t as the square of some other non-negative parameter). Then the feature graph loss function (6) is identical to Takane et al.'s (1977) SSTRESS loss function with restrictions on the configuration.

References:

Abdi, H. (1990): Additive tree representations, In: *Trees and Hierarchical Structures*, Dress, A. et al. (Eds.), 43-59, Springer Verlag, Berlin.

Arabie, P., and Carroll, J.D. (1980): MAPCLUS: A mathematical programming approach to fitting the ADCLUS model, *Psychometrika*, **45**, 211-235.

Arabie, P., and Hubert, L. (1992): Combinatorial data analysis, *Annual Review of Psychology*, **43**, 169-203.

Barthélemy, J.-P. and Guénoche, A. (1991): *Trees and Proximity Representations*, Wiley, New York.

Boorman, S.A. and Arabie, P. (1972): Structural measures and the method of sorting, In: *Multidimensional Scaling: Theory and Applications in the Behavioral Sciences*, Shepard, R.N. et al. (Eds.), 225-249, Seminar Press, New York.

Buneman, P. (1971): The recovery of trees from measures of dissimilarity, In: *Mathematics in the Archaeological and Historical Sciences*, Hodson, F.R. et al. (Eds.), 387-395, Edinburgh University Press, Edinburgh.

Carroll, J.D. (1976): Spatial, non-spatial and hybrid models for scaling, *Psychometrika*, **41**, 439-463.

Chandon, J.L., Lemaire, J., and Pouget, J. (1980): Construction de l'ultramétrique la plus proche d'une dissimilarité au sens des moindres carrés, *R.A.I.R.O. Recherche Opérationelle*, **14**, 157-170.

Corter, J.E., and Tversky, A. (1986): Extended similarity trees, *Psychometrika*, **51**, 429-451.

Cunningham, J.P. (1978): Free trees and bidirectional trees as representations of psychological distance, *Journal of Mathematical Psychology*, **17**, 165-188.

De Soete, G. (1983): A least squares algorithm for fitting additive trees to proximity data, *Psychometrika*, **48**, 621-626.

Felsenstein, J. (Ed.)(1983): *Numerical Taxonomy*, Springer Verlag, Heidelberg.

Flament, C. (1963): *Applications of Graph Theory to Group Structure*, Prentice-Hall, Englewood Cliffs, New Jersey.

Goodman, N. (1951): *The Structure of Appearence*, Bobbs-Merrill, Indianapolis, Indiana.

Goodman, N. (1977): *The Structure of Appearence* (3rd ed.), Reidel, Dordrecht, Holland.

Groenen, P.J.F., Mathar, R., and Heiser, W.J. (1995): The majorization approach to multidimensional scaling for Minkowski distances, *Journal of Classification*, **12**, 3-19.

Guttman, L. (1954): A new approach to factor analysis: The radex, In: *Mathematical thinking in the social sciences*, Lazarsfeld, P.F. (Ed.), 258-348, The Free Press, Glencoe, Illinois.

Hakimi, S.L., and Yau, S.S. (1965): Distance matrix of a graph and its realizability, *Quarterly of Applied Mathematics*, **22**, 305-317.

Hartigan, J.A. (1967): Representation of similarity matrices by trees, *Journal of the American Statistical Association*, **62**, 1140-1158.

Heiser, W.J. (1981): *Unfolding analysis of proximity data*, Unpublished doctoral dissertation, University of Leiden, The Netherlands.

Heiser, W.J., and Groenen, P.J.F. (1996): Cluster differences scaling with a within-clusters loss component and a fuzzy successive approximation strategy to avoid local minima, *Psychometrika*, **61**, in press.

Henley, N.M. (1969): A psychological study of the semantics of animal terms, *Journal of Verbal Learning and Verbal Behavior*, **8**, 176-184.

Holman, E.W. (1995): Axioms for Guttman scales with unknown polarity, *Journal of Mathematical Psychology*, **39**, 400-402.

Hutchinson, J.W. (1989): NETSCAL: A network scaling algorithm for nonsymmetric proximity data, *Psychometrika*, **54**, 25-52.

Klauer, K.C. (1994): Representing proximities by network models, In: *New Approaches in Classification and Data Analysis*, Diday, E. et al. (eds.), 493-501, Springer Verlag, Heidelberg.

Klauer, K.C., and Carroll, J.D. (1989): A mathematical programming approach to fitting general graphs, *Journal of Classification*, **6**, 247-270.

Lawson, C.L., and Hanson, R.J. (1974): *Solving least squares problems*, Prentice Hall, Englewood Cliffs, NJ.

Mirkin, B.G. (1987): Additive clustering and qualitative factor analysis methods for similarity matrices, *Journal of Classification*, **4**, 7-31.

Restle, F. (1959): A metric and an ordering on sets, *Psychometrika*, **24**, 207-220.

Restle, F. (1961): *Psychology of Judgment and Choice*, Wiley, New York.

Roberts, F.S. (1976): *Discrete Mathematical Models, with Applications to Social, Biological, and Environmental Problems*, Prentice Hall, Englewood Cliffs, New Jersey.

Sattath, S., and Tversky, A. (1977): Additive similarity trees, *Psychometrika*, **42**, 319-345.

Shepard, R.N., and Arabie, P. (1979): Additive clustering: Representation of similarities as combinations of discrete overlapping properties, *Psychological Review*, **86**, 87-123.

Takane, Y., Young, F.W., and De Leeuw, J. (1977): Nonmetric individual differences in multidimensional scaling: An alternating least quares method with optimal scaling features, *Psychometrika*, **42**, 7-67.

Tversky, A. (1977): Features of similarity, *Psychological Review*, **84**, 327-352.

Classification and data analysis in finance

Krzysztof Jajuga
Wroclaw University of Economics
ul. Komandorska 118/120
53-345 Wroclaw, Poland

Summary: The paper gives a brief review of the main areas of financial applications where classification and data analysis methods can be used. First of all, historical context is given. It is shown that the emerging of modern finance was made possible due to use of quantitative methods. The presented applications are divided into two main groups: 1) analysis of financial investments and markets, 2) corporate finance. The review is put in the framework where the relationship between dependent variable and explanatory variables is determined.

1. Historical remarks

In the paper some links between classification and data analysis on one side and the financial applications on the other side are shown. The history proved that the classification and data analysis methods are useful in financial applications. It is also clear that the usefulness of these methods will grow in future.

There are two important streams in the development of modern finance, which, on one hand, were the driving forces of this discipline and where, on the other hand, the contribution of statistics was crucial to the emerging and the development. These are:
- forecasting of financial prices;
- portfolio theory.

Forecasting of prices in financial markets (as well as in commodity markets) has been probably the most exciting issue in financial research. First of all, this is very difficult task, which has not been solved despite a lot of efforts. Secondly, people believe that by finding good forecasts of financial prices they can make a lot of money. This makes job even more exciting.

The first work on forecasting of financial prices was done by Louis Bachelier in 1900. His doctoral thesis „Theory of speculation" is considered today as seminal work, which had been unnoticed for more than fifty years. He completed thesis for degree of Doctor of Mathematical Sciences at Sorbonne.

In the dissertation he proved two statements. The first one: prices in financial markets cannot be succesfully predicted. He argued that: „contradictory opinions concerning market changes diverge so much that at the same instant buyers believe in a price increase and sellers believe in a price decrease. [...] It seems that the market, the aggregate of speculators, at a given instant can believe in neither a market rise nor a market fall, since for each quoted price, there are as many buyers as sellers" (Bachelier (1900)). The main conlusion of Bachelier is: the mathematical expectation of price changes is zero, therefore the best forecast of next price is current price. This means that the prices of financial instruments follow random walk process.

The second statement of Bachelier was: the range of the interval of prices is proportional to the square root of time. This is reflected today in many stochastic price models of ARIMA type. This was also confirmed empirically for many time series of prices.

Despite of this and some other theoretical results, financial practitioners searched for effective forecasting tools. One of such tool is so called technical analysis, the foundations of which were laid of by Charles Dow (also known as co-founder and editor of Wall Street Journal and the co-author of the most famous stock market index, Dow Jones Industrial Average). Dow claimed that financial prices (especially stock prices) are changing according to some trends. The particularly attractive (from the practical point of view) claim of the proponents of technical analysis is the following one: the directions of stock prices movements can be predicted and the forecasts can be used to develop trading strategy resulting in returns above average. The users of technical analysis try to detect regular patterns in past prices (by means of charts) which they believe will repeat in future. This is a simplified version of pattern recognition problem. Basically the idea of existence of regular patterns is used today in neural networks methodology.

The discussion between the advocates of two concepts: the first one that effective forecasts of financial prices can be made and the second one that the financial prices follow random walk process can be put in the framework of so called market efficiency. This concept was proposed by Fama (1965). According to him market is to be called efficient if current financial prices instantaneously and fully reflect all available information. For the predictability of prices so called weak form market efficiency is of particular importance. It is said that the market is a weak-form efficient if current financial prices instantaneously and fully reflect all information contained in the past history of financial prices. This means that historical prices provide no information (about future prices) that will lead to higher than average return by using trading rules based on forecasts of prices. Thus the best forecast of next stock price is current stock price. The search for the methods of financial prices forecasting (particularly stock prices forecasting) is based on the conviction that markets are not weak form efficient.

The second stream of the development of modern finance is connected to portfolio theory. Portfolio theory was laid off by Harry Markowitz. In his seminal paper (Markowitz (1952)), published in „Journal of Finance", probably in the most significant paper in the theory of finance, Markowitz introduced the concept of risk in financial investments. He was the first one to propose the use in finance the concept of a distribution of a random variable. As a measure of the return on the investment, the expected value of return was used. As a measure of investment risk, standard deviation of return was used. Then he developed the concept of risk diversification. This means that the investment risk can be reduced by forming a portfolio of stocks and this reduction depends on the correlation of the returns of between stocks belonging to portfolio.

At this time the approach proposed by Markowitz was entirely different from the traditional approach used in finance. For several years the paper of Markowitz was unnoticed. Then it caused a lot of discussions and criticisms. The weak point in Markowitz approach was the one that solving portfolio problem required very substantial amount of time by using the computers available at this time. Today portfolio theory is widely used in practice. For his contribution to economic sciences Harry Markowitz was awarded Nobel Prize in 1990.

In the beginning of seventies the substantial increase of the volatility (variability) of financial prices (particularly exchange rates and interest rates) was observed. This caused the search for the ways to cope with resulting risk. One solution was the introduction of the new financial instruments, like options and financial futures. Another solution was to look for more sophisticated mathematical methods.

In last fifteen years the enormous development in the area of computer technology occurred. This was extremely beneficial as far as the use of sophisticated mathematical and statistical methods is concerned. At present the use of these methods is not time-consuming which means that the costs of the implementation of these methods are relatively low. On the other hand, the computer software designed to solve complicated financial problems is widely available. Statistical and data analysis methods are at disposal of firms, banks and investors.

2. Statistical methods in financial applications

It is not easy to give the general framework for the presentation of applications of statistical methods in finance. The relatively simple way is to consider the financial applications through the analysis of the following function:

$$Y = f(X_1, X_2, ..., X_m) \qquad (1)$$

where:
Y - dependent variable,
$X_1, X_2, ..., X_m$ - explanatory variables.

It is not easy to systematize all quantitative methods that can be used to solve financial problems. One possible way is to classify them into two groups:
- classical multivariate data analysis methods;
- financial cybernetics methods.

This taxonomy is based on historical criterion since the second group of methods emerged in last ten years when the use of the methods requiring large amount of computer time was made possible due to the development of computer technology.

The term „financial cybernetics" was used by Thomas E. Berghage (president of company developing artificial intelligence software for finance) to describe the process of enhancing financial decision making by introducing artificial intelligence technologies. The term „cybernetics" was used for the first time by Norbert Wiener in 1948. At this time it was new science dealing with modifying or enhancing human decision systems with artificial electronic systems. In the area of finance this means to enhance financial decision making with computer systems which to some extent ressemble human systems.

One of the very first applications of financial cybernetics was the application on neural networks done by Lapedes and Farber (Lapedes and Farber (1987)). They attempted at forecasting the closing value of market index Standard and Poor 500, where as the explanatory variables the closing values from ten previous weeks were used. As an algorithm back-propagation method was used.

The other useful way to classify the methods to solve main financial problems can be done on the basis of two criteria:
1. The type of dependent variable - Y is quantitative or categorical.
2. The knowledge of the values (or the categories) of dependent variable - the values (or categories) are known or unknown.

Therefore four different classes of methods can be distinguished, leading to four research situations (cases):
1. Y is categorical variable and its categories are known.
2. Y is quantitative variable and its values are known.

3. Y is categorical variable and its categories are unknown.

4. Y is quantitative variable and its values are unknown.

There are many methods that can be used in each case. Here is a sample list:

Situation 1 - discriminant analysis, neural networks.

Situation 2 - regression analysis, neural networks.

Situation 3 - cluster analysis.

Situation 4 - principal component analysis (or any method synthetizing set of variables).

There is an opinion that financial cybernetics methods outperform classical methods (for example neural networks outperform discriminant analysis). It seems that it is still too early to give general conclusions. As a rule, artificial intelligence methods are time-consuming and difficult to interpret, therefore difficult to explain to non-experts, which are end users of these methods.

It is worth to give two very general remarks on the use of statistical methods in finance. First remark refers to the approach which can be assumed, stochastic approach or descriptive (distribution-free) approach. It is often the case that the use of stochastic approach, that is the use of methods based on distributional assumptions, is not justifiable in financial applications. Many statistical methods rely upon assumption of the normal (univariate or multivariate) distribution, but the distributions of financial variables are often heavy-tailed or asymmetric (as a rule, skewed to the right). Moreover, studied individuals often come from finite population, which also means that the stochastic approach may not be assumed. However, many statistical methods still can be used as descriptive methods, provided statistical inference is not made.

Second remark refers to the great opportunity to use classification methods in finance, because heterogeneity very often occurs in financial data. This heterogeneity may be due to the character of the studied problem - objects falling into separate classes, which we want to detect (for example: well performing companies and badly performing companies). This also may be due to the existence of different classes, known a priori (different industries, different sizes of companies). Finally this may be due to the intrinsic heterogeneity of data, for example resulting from the existence of outliers. These facts indicate the great opportunity for classification methods.

3. A review of the most important areas of financial applications

Now we give a brief review of several most important areas of financial applications, where the use of classification and data analysis methods proved to be useful. All financial applications can be divided into 2 main groups:

- analysis of financial investments and markets;
- corporate finance.

Group 1. Analysis of financial investments and markets.

A. Bond rating

Investing in bond is one of the basic financial investments. Bond gives its holder the right to receive the amount of money equal to the nominal value of the bond at maturity and to receive regular payments, so called coupons, being the interest on the loan.

One of the main types of risk occurring while investing in bonds is the so called default risk. Default risk means that the issuer of the bond does not pay back the loan and/or does not pay the interest on the loan. From the point of view of investor it is very important to evaluate default risk in order to avoid the negative consequences. This can be achieved by so called bond rating.

Bond rating consists in determination of the classes of bonds of approximately equal level of default risk. There are many institutions specializing in bond ratings (e.g. Standard and Poor's Corporation and Moody Investment Services). The usual way to determine bond rating is to ask experts to evaluate different factors influencing the default risk. As a rule past performance of the bond issuer is also taken into account while determining bond ratings. It is worth to mention that rating institutions claim that they link together financial statement data of issuers of bonds and experts' opinions.

Bond rating problem can be regarded as a determination of a function (1), where Y is the categorical variable standing for the class of bond and explanatory variables are the factors influencing the default risk. From the point of view of the statistical methods used in bond rating, we can distinguish two of the mentioned four situations, namely:

- situation 1: the categories of dependent variable are known;
- situation 3: the categories of dependent variable are unknown.

In the first case two types of data can be used:

- historical data, that is past bond ratings plus the information on the previous values of the factors influencing the default risk;
- expert opinions, obtained by assigning bond ratings to the hypothetical values of factors.

Here any of these two types of data sets can be treated as a learning set and can be used to determine a function which divides the bonds into classes. As a rule, discriminant analysis or neural network methodology can be used. The classes corresponding to particular bond ratings can be interpreted which is very important for end-users.

In the second case past data on bond ratings are not available and one uses classification methods (for example cluster analysis) to determine the classes of bonds. Here, the problem of difficulty to interpret the classes of bonds may occur.

B. Financial prices forecasting

This is by no doubt the most difficult financial problem. This task is important for different types of investors, short-term and long-term, individual and institutional. The following prices are usually predicted: commodity prices, exchange rates, interest rates and stock prices.

From the point of view of classification and data analysis this problem is regarded via a function (1). Here Y is usually quantitative variable - the price. Sometimes it can be categorical assuming one of three categories: „the price will go up", „the price will go down", „the price will stay within defined interval". To determine a function (1), historical data are used. This fits either to situation 1 or 2.

There are many approaches to financial prices forecasting. Basically they can be divided into three broad categories:

- technical analysis;
- econometric regression and time series models;
- neural networks.

Technical analysis is simple and widely used approach, which has been already mentioned. „Technicians" use different types of charts of past prices to discover regular patterns, which they believe will occur in future. Their reasoning is supported by simple indicators describing financial markets.

The large group of researchers uses econometric models which emerging from well known ARIMA approach. The development of statistical methodology and the development of computer technology allowed for implementation of these models in real world. The detailed description of these models is presented by Taylor (1986) and Mills (1993).

Neural networks as well as some other approaches (genetic algorithms, chaos theory) are relatively new approaches in financial forecasting. These models could be developped with the use of fast computers, since they require lengthy computations. The most popular are probably neural networks. Here the algorithm is used so that quite complicated nonlinear function is estimated. This function approximates the past financial prices. Then this function is used for forecasting.

As it was already mentioned, the question of market efficiency is a crucial one to the forecasting of financial prices. Those who apply the mentioned methods believe that the market is not efficient and the changes of prices do not follow random walk process.

C. Risk-return analysis

This is classical financial problem, which traces back to the origin of portfolio theory. The rationale behind this problem lies in the fact that most individual and institutional investors try to maximize their return while keeping risk as low as possible. This behaviour of investors was reflected in portfolio theory proposed by Harry Markowitz. He considered a portfolio of stocks. The portfolio problem can be regarded as a task of finding a combination of individual stocks, called portfolio, so that the expected return is as high as possible and risk is as low as possible. The main results of the classical portfolio theory are:
- expected return on a portfolio is weighted average of the expected returns on individual stocks;
- risk of a portfolio depends on the risk of individual stocks and on the correlation of returns; this means that the low, possibly negative correlations lead to low risk, while holding constant or even decreasing expected return.

Risk-return analysis can be treated as the analysis of location and spread parameters of a distribution. One possible solution is to use robust estimates of location and spread. Since building the portfolio involves multivariate distributions, estimates of multivariate location vector and multivariate scatter matrix are to be used.

Risk-return analysis can be also put in the framework of a function (1). There are two explanatory variables, return and risk, and the dependent variable is unknown and characterizes the attractiveness of the investment (for example, very attractive - high return and low risk, medium attractive - high return and high risk or low return and low risk, not attractive - low return and high risk). This problem fits to the situation 3.

D. Beta analysis

Beta coefficient is one of the most important coefficients used by financial industry. This coefficient was proposed by William Sharpe (Sharpe (1963)) in a so called single-index model. This is simple regression model which gives a linear relationship between the return on a stock (or other investment) and the return on the market (measured usually through a return on market index). The slope in this regression is beta coefficient. It is the measure of the sensitivity of a return on a stock on the changes of a return on the market. It can be also regarded as a measure of so called systematic risk or market risk. The stocks with beta higher than 1 are called aggressive stocks and the stocks with beta lower than 1 are called defensive stocks.

From the point of view of function (1), beta analysis fits to the situation 2. However, it is usually the case that people from finance industry do not pay particular attention to the justifiability of linearity of the relationship and use ordinary least squares to estimate beta coefficient. This raises the issue of the application of more advanced methods, for example nonlinear regression, robust regression or segmented regression models.

E. Factor market models

The researchers who analyze financial markets try to find a theoretical model, which explains what determine the returns on financial instruments. Among several proposed models, the widely accepted one is so called Arbitrage Pricing Theory (APT) model, proposed by Ross (Ross (1976)). This is linear model, given by the formula:

$$R = a + b_1 F_1 + b_2 F_2 + \dots + b_m F_m \qquad (2)$$

where:

R - the return on investment;

F_i - the i-th factor influencing the investment;

b_i - the sensitivity coefficient of the return with respect to i-th factor.

From the point of view of the determination of function (1), this fits to the situation 2. As a rule, historical data are used. If the factors influencing returns on the investments are known (for example: interest rates, GDP growth rate, etc.) then the regression analysis can be used. There is also a chance that one has no idea about the factors. In this case other solution can be applied, where factor analysis is used. Here the returns on different stocks are used to extract the values of unknown factors. However, it is often the case that it is very difficult (if even possible) to give useful interpretation of extracted factors.

Group 2. Corporate finance.

A. Analysis of financial condition of the firm

This is one of the very basic financial problems to be solved by investors purchasing stocks, management of the firm making decisions on the policy of the firm, banks making loans to the company.

This problem can be regarded as a determination of the function (1), where Y is the variable characterizing financial condition of the firm. This can be categorical variable, whose values correspond to the categories of the financial condition („very good", „good", „average", „bad", „very bad"). This can be also continuous variable, whose values inform on the level of the financial condition. Explanatory variables are different variables characterizing financial condition. Five groups of variables are usually used - they are called financial ratios:

- liquidity ratios - measure the firm's ability to fulfill its short-term commitments out of current or liquid assets;
- debt management ratios - measure the firm's ability to pay its debt and the interest on debt;
- activity ratios - measure the quality of management of firm's assets;
- profitability ratios - measure the firm's ability to generate its profit from available resources;
- market ratios - measure the atractiveness of firm from investor's point of view.

Therefore the analysis of financial condition of the firm can be regarded as a situation 3 or 4, depending whether the variable characterizing financial condition of the firm is categorical or quantitative one.

B. Bankruptcy prediction

Since the bankruptcy is the potential threat to all firms, the bankruptcy prediction becomes one of the most important financial tasks. It is particularly important to all investors and creditors. This problem can be regarded as a determination of a function (1), where Y is categorical variable, either binary variable („will bankupt", „will not bankrupt") or more general nominal variable (for example three categories: „will bankrupt in one year's time", „will bankrupt in the second year", „will not bankrupt in two year's time"). As the explanatory variables financial ratios, mentioned above, are usually used.

70

From the point of view of the use of statistical methods this can be regarded as a determination of a function (1) and it fits to the situation 1. Here the historical data on the bankrupt and non-bankrupt companies can be used to determine this function.

One of the first attempts to use discriminant analysis in the bankruptcy prediction was made by Altman (1968). This is classical model, called Altman model. Altman compared the financial data of 33 manufacturers who bankrupted with the data of 33 nonbankrupt firms of similar industry and asset size. From a number of avalilable financial variables he finally used 5 ratios.

References:

Altman, E.L. (1968): Financial ratios, discriminant analysis and the prediction of corporate bankruptcy, *Journal of Finance*, **23**, 589-609.

Bachelier, L. (1900): *Theory of speculation*, Gauthier-Villars, Paris.

Fama, E. (1965): The behavior of stock prices, *Journal of Business*, **37**, 34-105.

Lapedes, A. and Farber, R. (1987): Non-linear signal processing using neural networks: prediction and system modeling, Los Alamos National Laboratory Report.

Markowitz, H.M. (1952): Portfolio selection, *Journal of Finance*, **7**, 77-91.

Mills, T.C. (1993): *The econometric modelling of financial time series*, Cambridge University Press, Cambridge.

Ross, S.A. (1976): The arbitrage theory of capital asset pricing, *Journal of Economic Theory*, **13**, 341-360.

Sharpe, W.F. (1963): A simplified model for portfolio analysis, *Management Science*, **19**, 277-293.

Taylor, S. (1986): *Modelling financial time series*, Wiley, New York.

How to validate phylogenetic trees?
A stepwise procedure

François-Joseph Lapointe

Département de sciences biologiques, Université de Montréal,
C.P. 6128, Succursale centre-ville, Montréal, Québec,   H3C   3J7, Canada
e-mail: lapoint@ere.umontreal.ca

Summary: In this paper, I review some of the methods and tests currently available to validate trees, focussing on phylogenetic trees (dendrograms and cladograms). I first present some of the more commonly used techniques to compare a tree with the data it is derived from (internal validation), or compare a tree to another tree or to more than one (external validation). I also discuss some of the advantages of performing combined (total evidence) versus separate analyses (consensus) of independent data sets for validation purposes. A stepwise validation procedure defined across all levels of comparison is introduced, along with a corresponding statistical test: A phylogeny will be said to be globally validated only if it satisfies all the tests. An application to the phylogeny of kangaroos is presented to illustrate the stepwise procedure.

1. Introduction

The construction of a classification is a simpled-minded task. First, you need data, second, you need an algorithm, and then, like magic, you get a classification of your data. Indeed, the sole purpose of a classification algorithm is to do just that; i.e., return a classification (e.g., dendrogram, cladogram, pyramid, weak hierarchy, or any other type of classification). The problem with such an approach is usually that no safeguards are provided to ensure that the output is meaningful. Indeed, most algorithms (i e., clustering algorithms or phylogeny reconstruction algorithms) will return a solution, no matter what data are fed into it. This implies that a classification can even be derived from pure noise (i.e., randomly generated data). This is why validation becomes necessary.

In this paper, I will review some of the safeguards currently available to validate classifications represented in the form of trees; those trees can either be obtained by different algorithms or derived from independent data sets using the same algorithm. It is not my goal to present an exhaustive review of all validation techniques for all types of trees. I will focus my review on weighted trees such as those used in some phylogenetic studies. Furthermore, given the number of statistical papers published on the subject, I will emphasize validation methods based on permutation and/or resampling procedures. I will first show (1) how one can assess whether any phylogenetic structure was present in the data to begin with (internal validation). Then, (2) I will introduce some of the methods designed to compare trees obtained from independent data sets (external validation). I will also present (3) the rationale for combining those independent data sets before proceeding with phylogenetic reconstruction (total evidence). The combined approach will then be contrasted with (4) a consensus approach in which the trees are analyzed separately and then combined. A stepwise procedure will finally be introduced to validate phylogenetic trees using both internal and external validation methods as well as separate and combined approaches. This stepwise procedure will be used to validate the phylogeny of kangaroos.

2. What is a phylogeny ?

Biologically speaking, a phylogeny is a tree-like representation of evolutionary relationships among n different taxa (e.g., species, genera or other taxonomic units). Phylogenetic trees are usually derived from a character-state matrix representing morphological or molecular data (n species by p characters), or from a square n x n distance matrix; several algorithms and computer packages are currently available to do so (see Penny et al., 1992; Swofford et al., 1996a). In mathematical terms, a phylogeny can be defined as a connected graph without cycles. Such phylogenies are usually represented as rooted trees with labeled leaves. They can be depicted in the form of weighted trees if the branches of the phylogeny have lengths that represent the amount of evolutionary divergence between the nodes of the tree. Therefore, the sum of the lengths along the path of branches between any pair of taxa can be recorded in a path-length matrix (similarly, the number of branches on a path can be recorded in a branch-distance matrix). When the rates of change in the various branches of the phylogeny are identical, every terminal node will be equidistant from the root, and the tree can be associated with a path-length matrix that satisfies the ultrametric inequality (Hartigan, 1967):

$$d(i, j) \leq \max[d(i, k); d(k, j)], \qquad \text{for every triplet of taxa } i, j, k..$$

Such trees are usually defined as dendrograms. On the other hand, when rates vary among lineages, the path-length matrix is not ultrametric but remains additive (i.e., ultrametric trees represent a special case of additive trees with constant evolutionary rates); additive distances satisfy the four-point condition (Buneman, 1971) and apply to cladograms:

$$d(i, j) + d(k, l) \leq \max[d(i, l) + d(j, k); d(i, k) + d(j, l)], \quad \text{for any quartet of taxa } i, j, k, l.$$

The path-length matrices (ultrametric or not) are in one-to-one correspondence with a set of weighted phylogenetic trees (Jardine et al., 1967; Buneman, 1974); this is also true for branch-distance matrices (Zaretskii, 1965). Therefore, it is equivalent for validation purposes to compare phylogenetic trees or their associated path-length (or branch-distance) matrices.

3. How to test phylogenies ?

When phylogenetic trees (or their matrix representations) are to be validated, standard statistical procedures could rarely be applied (but see Li and Guoy, 1991; Li and Zharkikh, 1995). Indeed, the values in path-length (or branch-distance) matrices are not independent from one another, and thus violate the most crucial assumption of any parametric test. Furthermore, the non-independence of the observations implies that the degrees of freedom of the tests are always overestimated. If one also notes that distances seldom meet the normality assumption, and that their parameter distributions are rarely known, we have enough reason to call for permutation methods. For a given test, the probability of the null hypothesis is therefore assessed from a distribution of test statistics obtained under a permutation/randomization/resampling model. The general testing procedure follows Edgington (1995):

1. Compute a reference test statistic (i.e., REFSTAT) relevant to the question asked.
2. Permute (or resample) the data (i.e., distance, character-state, or path-length matrices).
3. Recompute the test statistic for the randomized data   (i.e., RANDSTAT).

4. Repeat steps 1 and 2 a large number of times (e.g, NPERM = 1000).

5. Compute the p-value as follows (see Dwass, 1957):

$$p = \frac{nb\left(RANDSTAT \geq REFSTAT\right) + 1}{NPERM + 1}$$

It is worth mentioning at this point that the statistical outcome of a permutation test is likely to be affected by different aspects of the procedure including, (i) the maximum possible number of random realizations of the null hypothesis, (ii) the actual number of permutations performed, (iii) the permutation model, and (iv) the test statistic selected to compute the test.

3.1 Random tree models

Depending on the type of trees compared (rooted or unrooted, labeled or unlabeled, weighted or unweighted, ultrametric or nonultrametric), different random-tree generation algorithms are distinguishable (Proskurowski, 1980; Furnas, 1984; Oden and Shao, 1984; Quiroz, 1989; Lapointe and Legendre, 1991). The sampling distribution of the trees is also important in the computations. The number of rooted trees for example, is larger than the number of unrooted ones for a given number of taxa (Phipps, 1975; Felsenstein, 1978), but smaller than the number of dendrograms (Murtagh, 1984); i.e., distinct populations of trees are considered. Still, random rooted trees can be sampled equiprobably or not; i.e., different distributions of trees with respect to the same population are considered (Lapointe and Legendre, 1995).

In the case of phylogenies, three distribution models are usually defined (Simberloff et al., 1981; Savage, 1983; Lapointe and Legendre, 1995). The first and simplest model is to generate every *topology* equiprobably. In the second model, each *tree* is equally likely; this is the "proportional-to-distinguishable-type" model of Simberloff et al. (1981). The third model implies that every branching point is equally likely when growing a tree (Harding, 1971); this is the Markovian branching model of Simberloff et al. (1981). It is interesting to note that dendrograms can be generated equiprobably under this Markovian model (Lapointe and Legendre, 1991; Page, 1991).

3.2 Random data models

When *data* matrices are used for validation purposes instead of trees, the models available differ with respect to the type of data considered. For example, character-state matrices are not randomized as distance matrices would be. In the first case, the general model is based on a permutation of the observed states within each character (Archie, 1989a; 1989c; Faith, 1991; Faith and Cranston, 1991; Källersjö et al., 1992). With such models, the phylogenetic structure of the data is destroyed by permutations (or random data generation, Klassen et al., 1991), and the probability of each state being assigned to any taxa is equally likely. This approach has been much debated (Bryant, 1992; Carpenter, 1992; Faith, 1992; Källersjö et al., 1992; Alroy, 1994; Faith and Ballard, 1994)). It nevertheless remains the model of choice in phylogenetic studies (but see Goloboff, 1991a; 1991b).

In the case of distance matrices, two types of models have been proposed by Sneath (1967). One option is to compute distance matrices from random points distributed in a multidimensional space (see Gordon, 1987); the uniform model based on a Poisson distribution and the unimodal model have been applied to generate such random distributions of points (Bock, 1985). Another option is to randomize the distances directly as if all the observations in the matrix were independent from one another; this is the random graph model of Ling (1973). In some testing procedures, the values in the matrix are held constant but the rows and columns (the labels) are permuted (Mantel, 1967). It is also possible to generate random distance matrices from permuted character-state matrices using the random-data model described above.

3.3 Test statistics

Another important aspect of validation methods is the test statistic selected to compare the actual tree to random realizations of the null hypothesis. Here again, data comparisons will differ from tree comparisons, and distances will require different statistics than character-state data. Common indices computed from a tree are its length, the consistency index (Kluge and Farris, 1969), retention index (Farris, 1989a), and homoplasy excess ratio (Archie, 1989b), among others (see also Farris, 1989b; Archie, 1990; Farris, 1991; Goloboff, 1991a; Hillis, 1991; Meier et al., 1991; Bremer, 1995). All of these statistics are used to measure how well the phylogeny fits the data; some are even used as optimality criteria for phylogenetic reconstruction (see Swofford et al., 1996a). For distance data, one usually calls for metric indices, like the matrix correlation (Rohlf, 1982), or any other measure of the fit between original and path-length distances (Rohlf, 1974; Gower, 1982; Gordon, 1987).

When trees are considered, it is always necessary to distinguish topological indices from tree metric indices; the former are designed for unweighted trees (i.e., ignoring branch lengths) whereas the latter are for weighted-tree comparisons (i.e., dendrograms or cladograms). Test statistics (i.e., consensus indices *sensu* Day and McMorris, 1985) available for topological comparisons include the partition metric (Bourque, 1978; Robinson and Foulds, 1981), the neighborhood interchange metric (Robinson, 1971; Waterman and Smith, 1978), the quartet metric (Estabrook et al., 1985; Day, 1986; Estabrook, 1992), and the triples distance metric (Critchlow et al., 1996), among many others (Bosibud and Bosibud, 1972; Margush, 1982; Hendy et al., 1984; Penny and Hendy, 1985b; Steel, 1988; Steel and Penny, 1993). When path-lengths matrices need to be compared, modified versions of the topological indices can be used (Robinson and Foulds, 1979), in addition to specific indices designed for dendrograms (Sokal and Rohlf, 1962; Day, 1983; Fowlkes and Mallows, 1983; Faith and Belbin, 1986; Lapointe and Legendre, 1990), or for any weighted trees (Williams and Clifford, 1971; Lapointe and Legendre, 1992a; Steel and Penny, 1993).

3.4 A note on resampling methods

Resampling methods (Efron, 1979; Efron and Gong, 1983; Efron and Tibshirani, 1993) are in some ways related to permutation tests. Indeed, resampling is used to assess the stability of some parts of the tree, or the tree as a whole, by comparing actual phylogenies to trees derived from resampled data. However, such methods are usually not designed as statistical tests, i.e., p-values are rarely provided and can not always be interpreted as such (Felsenstein et Kishino, 1993). Here the values in a data matrix are not permuted but sampled with or without replacement. The bootstrap (Felsenstein, 1985; Sanderson, 1989; 1995) and the jackknife (Davis, 1993; Farris et al., 1995a; 1995b) are among the most popular resampling techniques in phylogenetic studies; both methods have been used to validate phylogenies (see Hillis, 1995; Swofford et al., 1996a).

Bootstrapping proceeds by resampling the characters of a data matrix with replacement; a new phylogeny is derived from each bootstrap replicate, and the frequencies of the different clades are computed over the set of bootstrapped trees. The results are presented in the form of a consensus tree bearing the clades found in the majority of the bootstrapped trees. Character jackknifing is very similar to bootstrapping: sets of characters are sampled at random from the original matrix without replacement (i.e., a fixed number of characters are deleted in turn from the matrix) and the frequencies of the clades occuring in the jackknife replicates are used to provide support for parts of the phylogeny. Other resampling techniques (Penny and Hendy, 1985b) and generalizations of the bootstrap (Hall and Martin, 1988; Zharkikh and Li, 1995) and the jackknife (Lanyon, 1985; Lapointe et al., 1994) have been applied to phylogenetic studies.

4. Internal *versus* external validation

Phylogenetic reconstruction represents a very difficult computational problem (Graham and Foulds, 1982; Day, 1983c; 1987). When the size of the matrix becomes too large, no algorithm is guaranteed to find the optimal solution. The different methods are thus compared and assessed with respect to several criteria: efficiency, consistency, power, robustness, and falsiability (see Penny et al., 1992; Hillis, 1995). Whereas numerous simulation studies (Huelsenbeck, 1995) have shown the relative accuracy of the competing algorithms, it remains true that no single method can always converge to the correct tree. It then becomes necessary to test the trees derived by a particular method with null phylogenetic models: this is the purpose of validation techniques. Dubes and Jain (1979) distinguish between internal and external validation methods. The first type of validation is required to assess the reliability of a tree with respect to the data it is derived from and in contrast to unstructured data (Milligan, 1981). External validation, however, refers to the comparison of two or more trees derived from independent data sets. In this case, either a reference tree becomes the comparison criterion or all trees are compared to one another sequentially, or simultaneously.

4.1 Internal validation

To perform internal validation, one needs to compare a tree with the data it is derived from. When it comes to a specific phylogeny, validation proceeds by comparing the actual tree with others derived from randomly generated or permuted data (Archie, 1989a; Faith and Cranston, 1991; Källersjö et al., 1992; Alroy, 1994). Using one of the random data models described above, a distribution of the test statistic can then be computed, or tables of critical values can be generated (e.g., Klassen et al., 1991). For example, the validity of a tree as a whole can be assessed by comparing its length to a distribution of the lengths of trees derived from random data (Le Quesne, 1989; Carter et al., 1990; Faith and Cranston, 1991; Steel et al., 1992; Archie and Felsenstein, 1993). When no phylogenetic structure was present in the data to begin with, most random data sets (e.g., 95%) will lead to trees shorter than the real phylogeny. In that situation, the original tree will not be validated. Other methods based on different statistics (e.g., Alroy, 1994) test the same null hypothesis stating that a phylogeny derived from actual data is no better than what would be expected from random data. The same approach can be used to test the stability of parts of the tree (e.g., monophyletic groups) under various permutations and resampling models (Faith, 1991; Faith and Trueman, 1996; Swofford et al., 1996b).

In resampling methods (e.g., the bootstrap and the jackknife), the effect of character and/or taxonomic sampling on phylogenetic reconstruction is assessed. Mueller and Ayala

(1982) were among the first to test the validity of their trees with a resampling procedure. Since Felsenstein (1985), the bootstrap has been the most popular validation technique in phylogenetic studies (Hedges, 1992; Hillis and Bull, 1993; Rodrigo, 1993a; Dopazo, 1994, Harshman, 1994; Zharkikh and Li, 1992a; 1992b; Li and Zharkikh, 1994; Berry and Gascuel, 1996; Efron et al., 1996), with extensions for distance data (Krajewski and Dickerman, 1990; Marshall, 1991). The method remains controversial (Sanderson, 1989; 1995), and has been greatly modified by some (Zharkikh and Li, 1995). The original nonparametric bootstrap (Felsenstein, 1985) consists in resampling the characters of a data (or distance) matrix with replacement to assess the stability of a tree (the parametric bootstrap has been introduced by Huelsenbeck et al., 1995). Jackknifing can proceed in a similar fashion, except that characters are resampled without replacement (Davis, 1993). This rationale also applies to taxonomic sampling (Lecointre et al., 1993). Lanyon (1985) and Lapointe et al. (1994) have shown that deleting taxa from the analysis can be used to evaluate the stability of phylogenetic trees. The consensus of the jackknife trees is then used to evaluate the support of the tree as a whole or part of it (for an application, see Bleiweiss et al., 1994). When the trees tested with resampling models are not validated (e.g., a partially-resolved tree is obtained), the original phylogenies should be treated with caution; additional data must be gathered to improve the results.

4.2 External validation

Given two (or more) internally validated trees, the next task is to verify whether these trees tell the same story or not; that is, that they are congruent (Prager and Wilson, 1976; Mickevich, 1978; Colless, 1980; Sokal and Rohlf, 1981; Penny et al., 1982; Swofford, 1991; Bledsoe and Raikow, 1992; Patterson et al., 1993). External validation proceeds by comparing phylogenies to one another (or to a reference phylogeny) to assess whether the observed measure of congruence could be expected by chance alone. As for internal validation, permutation and resampling methods can be used for this test. In the case of the random-data model, the data from which the original trees were derived are randomized (Rodrigo et al., 1993; Farris et al., 1995b) to obtain new phylogenies which can be compared in turn to build a distribution of the test statistic. For the random-tree model, the actual pair of trees is compared to pairs of random trees (Hubert and Baker, 1977; Podani and Dickinson, 1984; Simberloff, 1987; Nemec and Brinkhurst, 1988; Page, 1988; Lapointe and Legendre, 1990; 1992a; Brown, 1994). Depending on the trees compared, the topology of the phylogeny, its branch lengths, and the taxon positions can be randomized (Lapointe and Legendre, 1995). No matter what method is used, the null hypothesis states that the phylogenies compared are not more similar than randomly generated trees would be (Lapointe and Legendre, 1990); a pair of trees is declared congruent when more similar than the majority (e.g., 95%) of the pairs of random trees. To prevent one from generating the null distribution for every test, tables of critical values for various consensus indices and different models have been produced (Day, 1983b; Shao and Rohlf, 1983; Shao and Sokal, 1986; Steel, 1988; Lapointe and Legendre, 1992b; Steel and Penny, 1993).

Even though phylogenies should always be validated, it is worth mentioning that data sets can be externally validated as well. The approach is similar to the random-data models used to compare trees; the character-state or distance matrices are randomized (or resampled) to assess their congruence. The Mantel test (1967) has been widely used to compare distance matrices. For comparing character-state data, canonical correlations can be applied with a testing procedure based on permutations (see Lapointe and Legendre,

1994). In any case, trees or data matrices that are not validated should be treated with caution. One should never rely on *ad hoc* criteria to decide which of the phylogenies is the best. It might be better to combine the data or trees to analyze them jointly.

5. Combined *versus* separate analysis

It is not unusual in phylogenetic studies (Bledsoe and Raikow, 1990; Patterson et al., 1993) to obtain different trees from different data sets (e.g., molecular *versus* molecular studies). In most cases this problem can be solved by validation methods, however. Indeed, the discrepancies between two phylogenies might disappear after internal validation (i.e., the trees might collapse to the same unresolved phylogeny). Likewise, two seemingly different trees could well turn out to be significantly more congruent than would be expected by chance alone. However, when statistically different but nonetheless valid trees are dealt with, one faces a difficult choice. How to reconcile the trees to reach a unique solution? Two options are available. The first is to combine the data sets in order to derive a phylogeny based on total evidence (Kluge, 1989). The other is to analyze the data sets separately, and then to combine the trees using a consensus method (Miyamoto and Fitch, 1995). Resolving this dilemma is currently one of the hottest debates in phylogenetics (de Queiroz et al., 1995; Huelsenbeck et al., 1996): to combine or not to combine?

5.1 Total evidence

How should one choose among different phylogenies based on independent data sets? Which one is closer to the true phylogeny? Given that different parts of the genome evolve at different rates, it is very unlikely that one would obtain identical phylogenies for slow-evolving *versus* fast-evolving genes (Russo et al., 1996), or even for morphological *versus* molecular data (Hillis, 1987). The solution, according to Kluge (1989), is to include all available data (i.e., character-state matrices) in one analysis (for a combination of distance matrices, see Lapointe and Kirsch, 1995). The rationale is that a tree based on total evidence rather than partial information will usually be more accurate as more data are added (see also Barrett et al., 1991; Eernisse and Kluge, 1993). This approach has been criticized by several authors (Huelsenbeck et al., 1996), and alternative views have been proposed (Williams, 1994; Bandelt, 1995; Miyamoto and Fitch, 1995; Nixon and Carpenter, 1996), one of which is to combine the data conditionally (Bull et al., 1993). The question then becomes one of *when* to combine data or not.

The debate on conditional data-combination (Chippindale and Wiens, 1994; Huelsenbeck et al., 1994; Wiens and Chippindale, 1994) is somewhat related to external validation. According to Bull et al. (1993), one has to make sure that the data sets are not heterogeneous before combining them. If it is shown, using randomization (Huelsenbeck and Bull, 1996) or resampling techniques (de Queiroz, 1993; Rodrigo et al., 1993; Farris et al., 1995a; 1995b), that the data are indeed more homogeneous than by chance alone, a combined analysis can be then performed; otherwise, the data matrices should be treated separately. In other words, one evaluates the congruence among the independent phylogenies to assess whether the corresponding data should be combined or not. The combined approach has been used in several applications since its publication (Miyamoto et al., 1994; Olmstead and Sweere, 1994; Omland, 1994; Mason-Gamer and Kellogg, 1996; Poe, 1996; Sullivan, 1996), and generalized to allow combination of overlapping sets of taxa (Wiens and Reeder, 1995; see also Lapointe and Kirsch, 1995). It is important to note, however, that a total-evidence tree can *never* be externally validated since all data are combined. It can only be compared with a previous total-evidence tree (i.e., one constructed before some new data set was generated). Nevertheless, a total-evidence tree must always be assessed with internal validation methods.

5.2 Consensus

Whether one decides to combine or not to combine data sets for statistical, practical, or philosophical reasons (Barrett et al., 1991; 1993; de Queiroz, 1993; Nelson, 1993), the problem remains the same: how to synthesize a profile of incongruent phylogenies? Whereas data are combined in a total-evidence approach, trees will be combined with a consensus approach (Miyamoto, 1985; Anderberg and Tehler, 1990). A consensus tree method (as opposed to consensus indices, Day and McMorris, 1985) takes as input a profile of trees and return a single solution that is in some sense representative of the entire set (Leclerc and Cucumel, 1987). Several approaches, including the strict (Sokal and Rohlf, 1981), semi-strict (Bremer, 1990), median (Barthélemy and McMorris, 1986), and majority-rule consensus (Margush and McMorris, 1981) methods have been developed to combine unweighted (see also Adams, 1972; Nelson, 1979; Stinebrickner, 1982; McMorris and Neumann, 1983; McMorris et al., 1983; Neumann, 1983; McMorris, 1985; Phillips and Warnow, 1996) or weighted trees (Stinebrickner, 1984; Lefkovitch, 1985; Lapointe and Cucumel, 1997). Other methods are designed for the construction of consensus supertrees from phylogenies bearing overlapping sets of taxa (Gordon, 1986; Baum, 1992; Ragan, 1992; Steel, 1992; Baum and Ragan, 1993; Lanyon, 1993; Rodrigo, 1993b; Purvis, 1995a; Ronquist, 1996; Lapointe and Cucumel; 1997), or the computation of common pruned trees (Finden and Gordon, 1985) and reduced consensus trees (Wilkinson, 1994; 1996).

The problem with consensus trees is that they are seldom validated. Assessing the significance of a consensus phylogeny remains problematic. As for phylogeny-reconstruction algorithms, a given consensus method will always return a solution. One then has to evaluate whether the consensus representation is pertinent or not; i.e., is it more structured than what would be expected from chance alone (Cucumel and Lapointe, 1997)? Consensus validation is somewhat related to the congruence tests used for external validation. The problem with consensus trees is that more than two phylogenies are usually considered at once; the tables of significance of consensus indices do not account for more than two trees at a time (e.g., Shao and Rohlf, 1983; Shao and Sokal, 1986; Lapointe and Legendre, 1992b). Furthermore, consensus trees are sometimes the synthesis of trees bearing nonidentical sets of taxa (Purvis, 1995b; Kirsch et al., 1997), which makes them even more difficult to test. Cucumel and Lapointe (1997) test the consensus by comparing it to the trivial classification (i.e., a bush, or a star tree); a distribution of consensus trees computed from randomly generated phylogenies (Lapointe and Legendre, 1990; 1992a) is used to assess the significance of the null hypothesis. Another approach would be to check whether the consensus falls within a confidence set (Sanderson, 1989) of the trees in the input profile.

6. Stepwise validation procedure

The debate on character congruence *versus* taxonomic congruence (*sensu* Kluge, 1989) has led to three different positions among systematists; (1) always combine data, (2) never combine data, or (3) combine conditionally. In other words, that means, (1) never use a consensus, (2) always use a consensus, or (3) sometimes use a consensus tree to represent the congruence among different studies. I am here proposing to always do both (i.e., total evidence AND consensus). The rationale is that a phylogeny is more likely to be accurate when the different approaches converge to the same solution. This is related to what Kim (1993) has shown by combining different algorithms to improve the accuracy of phylogenetic estimations. In the present case, combined and separate analyses are performed and the resulting phylogenies are assessed using a stepwise procedure (Fig. 1).

1- Initially, each and every tree produced has to be checked for internal validity (4.1).
2- Trees that satisfy the first test need to be compared to assess their congruence (4.2).
3- The congruent data sets must be combined to derive a total-evidence tree (5.1).
   That tree has to be validated.
4- The independent phylogenies must also be combined to obtain a consensus tree.
   That consensus has to be validated.
5- Finally, the trees obtained at steps 3 and 4 of the validation procedure must be compared.

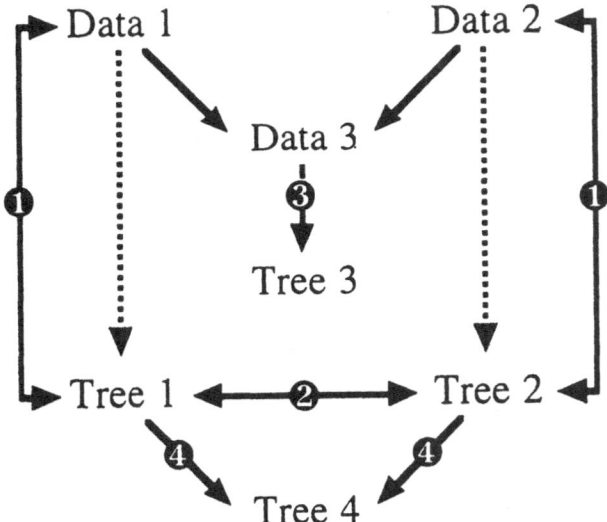

Fig. 1. Flowchart of the stepwise procedure. **Data 1** and **Data 2** represent two independent sets of data. **Tree 1** and **Tree 2** are the phylogenies derived from the corresponding data sets using any phylogenetic reconstruction algorithm. **Data 3** is obtained by combining **Data 1** and **Data 2**; **Tree 3** is the corresponding phylogeny. **Tree 4** is the consensus of **Tree 1** and **Tree 2**. Numbers refer to the different validation steps: ❶ internal validation, ❷ external validation, ❸ total evidence, and ❹ consensus.

The comparison of a total-evidence phylogeny with a consensus tree is not as simple as it seems. One could not rely on standard validation methods in this case. The problem is that those trees are not independent from one another. The testing procedure is as follows: (i) randomize (i.e., permute, resample, or simulate) the initial data sets independently, (ii) analyze these random data sets separately, and (iii) compute a consensus of the corresponding trees. At the same time, (iv) combine the random data sets and (v) derive a total-evidence phylogeny. The total-evidence and consensus trees are then (vi) compared, and the whole procedure is repeated a large number of times to build a distribution of the test statistic measuring the congruence between the two trees. If the actual trees under comparison are more similar than the majority of trees derived from random data, they are declared congruent. In that situation, the phylogenetic tree is said to be globally validated.

7. Application

To illustrate how the stepwise validation procedure works with real data, I have applied the methods to validate the kangaroo phylogeny in Kirsch et al. (1995), based on DNA-hybridization data and depicting phylogenetic relationships among 12 species. It is compared to Baverstock et al. (1989) phylogeny of 14 kangaroo species, based on immunological data. For the purpose of the demonstration, I have reduced the original data sets to only consider the nine species in common to both studies (Fig. 2).

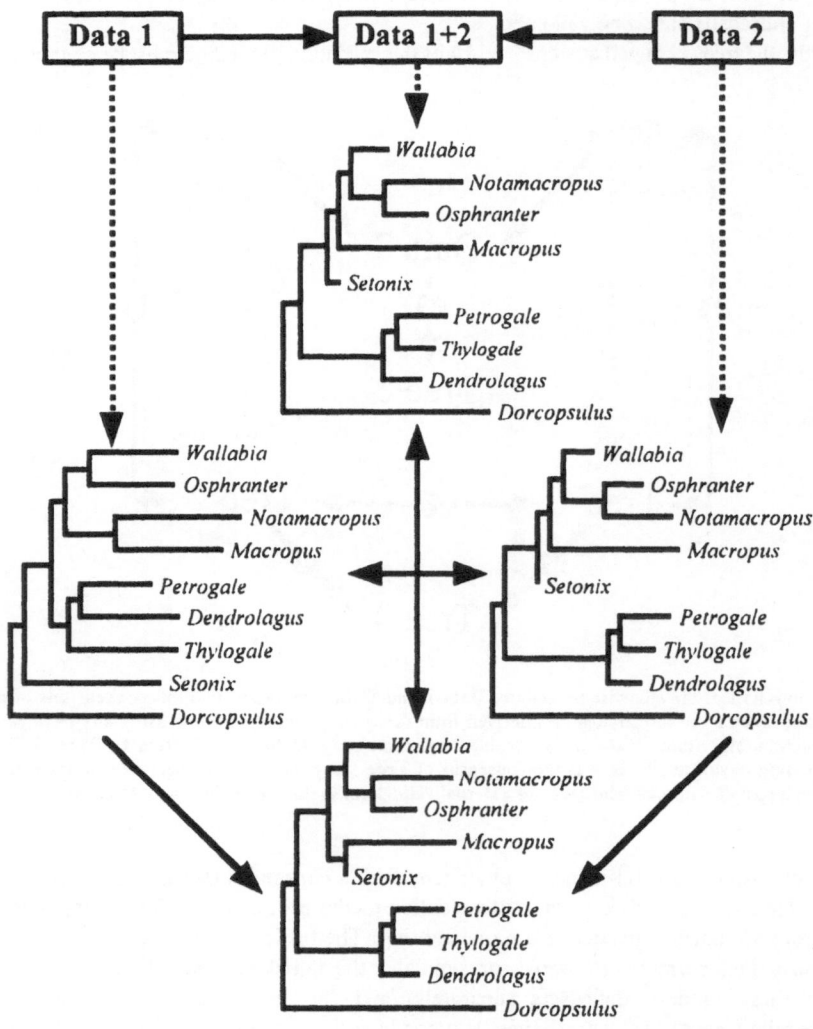

Fig. 2. Illustration of the stepwise valdation procedure. **Data 1** is Kirsch et al. (1995) DNA-hybridization data. **Data 2** is Baverstock et al. (1989) immunological data. **Data 1+2** is the average of the standardized data sets. The corresponding phylogenies were reconstructed with the FITCH algorithm (Felsenstein, 1993). The consensus was derived using the average procedure (Lapointe et al., 1994; Lapointe and Cucumel, 1997). All trees are rooted by *Dorcopsulus*.

The first step of any validation study is always to check for internal validity. In the present case, each tree had already been validated by bootstrapping and/or jackknifing in the original studies. I thus proceeded directly with external validation. Metric and topological indices were selected to compare the phylogenies depicted in the form of additive trees; the matrix correlation computed from path-length distances is 0.354, compared to 0.645 for branch distances. The latter is more extreme than would be we expected from pairs of random additive trees of the same size (Lapointe and Legendre, 1992b). That is, the two kangaroo phylogenies are topologically congruent (i.e., the data are not heterogeneous). The next step was, therefore, to combine the standardized data matrices from the different studies. I did so by a simple average of the immunological and DNA hybridization distances among the nine species (Lapointe and Kirsch, 1995). A phylogeny was derived from that total-evidence matrix (Fig. 2), and internal validation was performed with taxonomic jackknifing (Lapointe et al., 1994). Finally, the average consensus (Lapointe and Cucumel, 1997) of the original phylogenies was computed to account for branch lengths (Fig. 2). The consensus was compared to a distribution of consensus trees derived from pairs of random trees to assess its pertinence (Cucumel and Lapointe, 1997). As both the total-evidence and consensus trees were validated, the last and crucial step consisted in the comparison of those phylogenies.

The correlation between the path-length matrices is 0.996 in this case, whereas the topological correlation value is 0.927. Using the significance test described above, it was shown that this particular pair of trees is more similar than what would be expected from most consensus and total-evidence trees based on random data. The kangaroo phylogeny is thus said to be globally validated.

8. References

Adams, E. N., III. (1972): Consensus techniques and the comparison of taxonomic trees, *Systematic Zoology*, **21**, 390-397.

Alroy, J. (1994): Four permutation tests for the presence of phylogenetic structure, *Systematic Biology*, **43**, 430-437.

Anderberg, A. and Tehler, A. (1990): Consensus trees, a necessity in taxonomic practice, *Cladistics*, **6**, 399-402.

Archie, J. W. (1989a): A randomization test for phylogenetic information in systematic data, *Systematic Zoology*, **38**, 219-252.

Archie, J. W. (1989b): Homoplasy excess ratios: New indices for measuring levels of homoplasy in phylogenetic systematics and a critique of the consistency index, *Systematic Zoology*, **38**, 253-269.

Archie, J. W. (1989c): Phylogenies of plant families: A demonstration of phylogenetic randomness in DNA sequence data derived from proteins, *Evolution*, **43**, 1796-1800.

Archie, J. W. (1990): Homoplasy excess statistics and retention indices: A reply to Farris, *Systematic Zoology*, **39**, 169-174.

Archie, J. W. and Felsenstein, J. (1993): The number of evolutionary steps on random and minimum lengths trees for random evolutionary data. *Theoretical Population Biology*, **43**, 52-79.

Bandelt, H. J. (1995): Combination of data in phylogenetic analysis, *Plant Systematics and Evolution, Supplementum* **9**, 355-361.

Barrett, M. et al. (1991): Against consensus, *Systematic Zoology*, **40**, 486-493.

Barrett, M. et al. (1993): Crusade? A response to Nelson, *Systematic Biology*, **42**, 216-217.

Barthélemy, J.-P. and McMorris, F. R. (1986): The median procedure for *n*-trees, *Journal of Classification*, **3**, 329-334.

Baum, B. R. (1992): Combining trees as a way of combining data for phylogenetic inference, and the desirability of combining gene trees, *Taxon*, **41**, 3-10.

Baum, B. R. and Ragan, M. A. (1993): Reply to A. G. Rodrigo's "A comment on Baum's method for combining phylogenetic trees, *Taxon*, **42**, 637-640.

Baverstock, P. R. et al. (1989): Albumin immunologic relationships of the Macropodidae (Marsupialia), *Systematic Zoology*, **38**, 38-50.

Berry, V. and Gascuel, O. (1996): On the interpretation of bootstrap trees: Appropriate threshold of clade selection and induced gain, *Molecular Biology and Evolution*, **13**, 999-1011.

Bledsoe, A. H. and Raikow, R. J. (1990): A quantitative assessment of congruence between molecular and nonmolecular estimates of phylogeny, *Journal of Molecular Evolution*, **30**, 247-259.

Bleiweiss, R. et al. (1994): DNA-DNA hybridization-based phylogeny of "higher nonpasserines: Reevaluating a key portion of the avian family tree, *Molecular Phylogenetics and Evolution*, **3**, 248-255.

Bock, H. H. (1985): On some significance tests in cluster analysis, *Journal of Classification*, **2**, 77-108.

Bosibud, H. M. and Bosibud, L. E. (1972): A metric for classifications, *Taxon*, **21**, 607-613.

Bourque, M. (1978): Arbres de Steiner et réseaux dont varie l'emplacement de certains sommets. Ph. D. Thesis, Département d'Informatique et de Recherche Operationelle, Unversité de Montréal, Montréal.

Bremer, K. (1990): Combinable component consensus, *Cladistics*, **6**, 369-372.

Bremer, K. (1995): Branch support and tree stability, *Cladistics*, **10**, 295-304.

Brown, J. K. M. (1994): Probabilities of evolutionary trees, *Systematic Biology*, **43**, 78-91.

Bryant, H. N. (1992): The role of permutation tail probability tests in phylogenetic systematics, *Systematic Biology*, **41**, 258-263.

Bull, J. J. et al. (1993): Partitioning and combining data in phylogenetic analysis, *Systematic Biology*, **42**, 384-397.

Buneman, P. (1971): The recovery of trees from measures of dissimilarity. In: *Mathematics in Archeological and Historical Sciences*, Hodson, F. R. et al. (eds.), 387-395, Edinburgh University Press, Edinburgh.

Buneman, P. (1974): A note on the metric properties of trees, *Journal of Combinatorial Theory (B)*, **17**, 48-50.

Carpenter, J. M. (1992): Random cladistics, *Cladistics*, **8**, 147-153.

Carter, M. et al. (1990): On the distribution of lengths of evolutionary trees, *SIAM Journal of Discrete Mathematics*, **3**, 38-47.

Chippindale, P. T. and Wiens, J. J. (1994): Weighting, partitioning, and combining characters in phylogenetic analysis, *Systematic Biology*, **43**, 278-287.

Colless, D. H. (1980): Congruence between morphometric and allozyme data for *Menidia* species: A reappraisal, *Systematic Zoology*, **29**, 288-299.

Critchlow, D. E. et al. (1996): The triples distance for rooted bifurcating phylogenetic trees, *Systematic Biology*, **45**, 323-334.

Cucumel, G. and Lapointe, F.-J. (1997): Un test de la pertinence du consensus par une méthode de permutations. In: *Actes des XXIXe journées de statistique*, 299-300, Carcassonne.

Davis, J. I. (1993): Character removal as a means for assessing stability of clades, *Cladistics*, **9**, 201-210.

Day, W. H. E. (1983a): The role of complexity in comparing classifications, *Mathematical Biosciences*, **66**, 97-114.

Day, W. H. E. (1983b): Distributions of distances between pairs of classifications. In: *Numerical Taxonomy*, Felsenstein, J. (ed.), 127-131, Springer-Verlag, Berlin.

Day, W. H. E. (1983c): Computationally difficult parsimony problems in phylogenetic systematics, *Journal of Theoretical Biology*, **103**, 429-438.

Day, W. H. E. (1986): Analysis of quartet dissimilarity measures between undirected phylogenetic trees, *Systematic Zoology*, **35**, 325-333.

Day, W. H. E. (1987): Computational complexity of inferring phylogenies from dissimilarity matrices, *Bulletin of Mathematical Biology*, **49**, 461-467.

Day, W. H. E. and McMorris, F. R. (1985): A formalization of consensus index methods, *Bulletin of Mathematical Biology*, **47**, 215-229.

de Queiroz, A. (1993): For consensus (sometimes), *Systematic Biology*, **42**, 368-372.

de Queiroz, A. et al. (1995): Separate versus combined analysis of phylogenetic evidence, *Annual Review of Ecology and Systematics*, **26**, 657-681.

Dopazo, J. (1994): Estimating errors and confidence intervals for branch lengths in phylogenetic tres by a bootstrap approach. *Journal of Molecular Evolution*, **38**, 300-304.

Dubes, R. and Jain, A. K. (1979): Validity studies in clustering methodologies, *Pattern Recognition*, **11**, 235-254.

Dwass, M. (1957): Modified randomization tests for nonparametric hypotheses, *Annals of Mathematics and Statistics*, **28**, 181-187.

Edgington, E. S. (1995): *Randomization tests, 3rd Edition, Revised and Expanded.* Marcel Dekker, New York.

Eernisse, D. J. and Kluge, A. G. (1993): Taxonomic congruence versus total evidence, and the phylogeny of amniotes inferred from fossils, molecules and morphology, *Molecular Biology and Evolution*, **10**, 1170-1195.

Efron, B. (1979): Bootstrapping methods: Another look at the jackknife, *Annals of Statistics*, **7**, 1-26.

Efron, B. and Gong, G. (1983): A leisurely look at the bootstrap, the jackknife, and cross-validation, *American Statistician*, **37**, 36-48.

Efron, B. and Tibshirani, R. J. (1993): *An introduction to the bootstrap*, Chapman and Hall, New York.

Efron, B. et al. (1996): Bootstrap confidence levels for phylogenetic trees, *Proceedings of the National Academy of Sciences, USA*, **93**, 13429-13434.

Estabrook, G. F. (1992): Evaluating undirected positional congruence of individual taxa between two estimates of the phylogenetic tree for a group of taxa, *Systematic Biology*, **41**, 172-177.

Estabrook, G. F. et al. (1985): Comparison of undirected phylogenetic trees based on subtrees of four evolutionary units, *Systematic Zoology*, **34**, 193-200.

Faith, D. P. (1991): Cladistic permutation tests for monophyly and nonmonophyly, *Systematic Zoology*, **40**, 366-375.

Faith, D. P. (1992): On corroboration: A reply to Carpenter, *Cladistics*, **8**, 265-273.

Faith, D. P. and Ballard, J. W. O. (1994): Length differences topology-dependent tests: A response to Källersjö et al, *Cladistics*, **10**, 57-64.

Faith, D. P. and Belbin, L. (1986): Comparison of classifications using measures intermediate between metric dissimilarity and consensus similarity, *Journal of Classification*, **3**, 257-280.

Faith, D. P. and Cranston, P. S. (1991): Could a cladogram this short have arisen by chance alone? on permutation tests for cladistic structure, *Cladistics*, **7**, 1-28.

Faith, D. P. and Trueman, J. W. H. (1996): When the topology-dependent permutation test (T-PTP) for monophyly returns significant support for monophyly, should that be equated with (a) rejecting a null hypothesis of nonmonophyly, (b) rejecting a null hypothesis of "no structure," (c) failing to falsify a hypothesis of monophyly, or (d) none of the above? *Systematic Biology*, **45**, 580-586.

Farris, J. S. (1989a): The retention index and the rescaled consistency index, *Cladistics*, **5**, 417-419.

Farris, J. S. (1989b): The retention index and homoplasy excess, *Systematic Zoology*, **38**, 406-407.

Farris, J. S. (1991): Excess homoplasy ratios, *Cladistics*, **7**, 81-91.

Farris, J. S. et al. (1995a): Constructing a significance test for incongruence, *Systematic Biology*, **44**, 570-572.

Farris, J. S. et al. (1995b): Testing significance of incongruencies, *Cladistics*, **10**, 315-370.

Felsenstein, J. (1978): The number of evolutionary trees, *Systematic Zoology*, **27**, 27-33.

Felsenstein, J. (1985): Confidence limits on phylogenies: An approach using the bootstrap, *Evolution*, **39**, 783-791.

Felsenstein, J. (1993): *PHYLIP: Phylogeny inference package*, version 3.5c, distributed by the author, University of Washington, Seattle.

Felsenstein, J. and Kishino, H. (1993): Is there something wrong with the bootstrap on phylogenies? A reply to Hillis and Bull, *Systematic Biology*, **42**, 193-200.

Finden, C. R. and Gordon, A. D. (1985): Obtaining common pruned trees, *Journal of Classification*, **2**, 225-276.

Fowlkes, E. B. and Mallows, C. L. (1983): A method for comparing two hierarchical clusterings, *Journal of the American Statistical Association*, **78**, 553-569.

Furnas, G. W. (1984): The generation of random, binary unordered trees, *Journal of Classification*, **1**, 187-233.

Goloboff, P. (1991a): Homoplasy and the choice among cladograms, *Cladistics*, **7**, 215-232.

Goloboff, P. (1991b): Random data, homoplasy and information, *Cladistics*, **7**, 395-406.

Gordon, A. D. (1986): Consensus supertrees: the synthesis of rooted trees containing overlapping sets of labeled leaves, *Journal of Classification*, **3**, 335-348.

Gordon, A. D. (1987): A review of hierarchical classifications, *Journal of the Royal Statistical Society (A)*, **150**, 119-137.

Gower, J. C. (1983): Comparing classifications. In: *Numerical Taxonomy*, Felsenstein, J. (ed.), 137-155, Springer-Verlag, Berlin.

Graham, R. L. and Foulds, L. R. (1982): Unlikelihood that minimal phylogenies for a realistic biological study can be constructed in reasonable computational time, *Mathematical Biosciences*, **60**, 133-142.

Hall, P. and Martin, M. A. (1988): On bootstrap resampling and iterations, *Biometrika*, **75**, 661-671.

Harding, E. F. (1971): The probabilities of rooted tree-shapes generated by random bifurcations, *Advances in Applied Probability*, **4**, 44-77.

Harshman, J. (1994): The effect of irrelevant characters on bootstrap values, *Systematic Biology*, **43**, 419-424.

Hartigan, J. A. (1967): Representation of similarity matrices by trees, *Journal of the American Statistical Association*, **62**, 1140-1158.

Hedges, S. B. (1992): The number of replications needed for accurate estimation of the bootstrap *P* value in phylogenetic studies, *Molecular Biology and Evolution*, **9**, 366-369.

Hendy, M. D. et al. (1984): Comparing trees with pendant vertices labelled, *SIAM Journal in Applied Mathematics*, **44**, 1054-1065.

Hillis, D. M. (1987): Molecular versus morphological approaches to systematics, *Annual Review of Ecology and Systematics*, **18**, 23-42.

Hillis, D. M. (1991): Discriminating between phylogenetic signal and random noise in DNA sequences, In: *Phylogenetic analysis of DNA sequences*, Miyamoto, M. M. and Cracraft , J. (eds.), 278-294, Oxford University Press, New York.

Hillis, D. M. (1995): Approaches for assessing phylogenetic accuracy, *Systematic Biology*, **44**, 3-16.

Hillis, D. M. and Bull, J. J. (1993): An empirical test of bootstrapping as a method for assessing confidence in phylogenetic analysis, *Systematic Biology*, **42**, 182-192.

Hubert, L. J. and Baker, F. B. (1977): The comparison and fitting of given classification schemes, *Journal of Mathematical Psychology*, **16**, 233-253.

Huelsenbeck, J. P. (1995): Performance of phylogenetic methods in simulation, *Systematic Biology*, **44**, 17-48.

Huelsenbeck, J. P. and Bull, J. J. (1996): A likelihood ratio test for detection of conflicting phylogenetic signal, *Systematic Biology*, **45**, 92-98.

Huelsenbeck, J. P. et al. (1994): Is character weighting a panacea for the problem of data heterogeneity in phylogenetic analysis?, *Systematic Biology*, **43**, 288-291.

Huelsenbeck, J. P. et al. (1995): Parametric bootstrapping in molecular phylogenetics: Applications and performance, In: *Molecular Zoology: Strategies and Protocols*, Ferraris, J .and Palumbi, S. (eds.), Wiley, New York.

Huelsenbeck, J. P. et al. (1996): Combining data in phylogenetic analysis, *Trends in Ecology and Evolution*, **11**, 152-158.

Jardine, C. J. et al. (1967): The structure and construction of taxonomic hierarchies, *Mathematical Biosciences*, **1**, 173-179.

Källersjö, M. et al. (1992): Skewness and permutation, *Cladistics*, **8**, 275-287.

Kim, J. (1993): Improving the accuracy of phylogenetic estimation by combining different methods, *Systematic Biology*, **42**, 331-340.

Kirsch, J. A. W. et al. (1995): Resolution of portions of the kangaroo phylogeny (Marsupialia: Macropodidae) using DNA hybridization, *Biological Journal of the Linnean Society*, **55**, 309-328.

Kirsch, J. A. W. et al. (1997): DNA-hybridisation studies of marsupials and their implications for metatherian classification. Australian Journal of Zoology, in press.

Klassen, G. J. et al. (1991): Consistency indices and random data, Systematic Zoology, 40, 446-457.

Kluge, A. G. (1989): A concern for evidence and a phylogenetic hypothesis of relationships among Epicrates (Boidae, Serpentes), Systematic Biology, 38, 7-25.

Kluge, A. G. and Farris, J. S. (1969): Quantitative phyletics and the evolution of anurans. Systematic Zoology, 18, 1-32.

Krajewski, C. and Dickerman, A. W. (1990): Bootstrap analysis of phylogenetic trees derived from DNA hybridization matrices, Systematic Zoology, 39, 383-390.

Lanyon, S. (1985): Detecting internal inconsistencies in distance data, Systematic Zoology, 34, 397-403.

Lanyon, S. (1993): Phylogenetic frameworks: Towards a firmer foundation for the comparative approach, Biological Journal of the Linnean Society, 49, 45-61.

Lapointe, F.-J. and Cucumel, G. (1997): The average consensus procedure: combination of weighted trees containing identical or overlapping sets of objects, Systematic Biology, 46, 306-312.

Lapointe, F.-J. and Legendre, P. (1990): A statistical framework to test the consensus of two nested classifications, Systematic Zoology, 39, 1-13.

Lapointe, F.-J. and Legendre, P. (1991): The generation of random ultrametric matrices representing dendrograms, Journal of Classification, 8, 177-200.

Lapointe, F.-J. and Legendre, P. (1992a): A statistical framework to test the consensus among additive trees (cladograms), Systematic Biology, 41, 158-171.

Lapointe, F.-J. and Legendre, P. (1992b): Statistical significance of the matrix correlation coefficient for comparing independent phylogenetic trees, Systematic Biology, 41, 378-384.

Lapointe, F.-J. and Legendre, P. (1994): A classification of pure malt Scotch whiskies, Applied Statistics, 43, 237-257.

Lapointe, F.-J. and Kirsch, J. A. W. (1995): Estimating phylogenies from lacunose distance matrices, with special reference to DNA hybridization data, Molecular Biology and Evolution, 12, 266-284.

Lapointe, F.-J. and Legendre, P. (1995): Comparison tests for dendrograms: A comparative evaluation, Journal of Classification, 12, 265-282.

Lapointe, F.-J. et al. (1994): Jackknifing of weighted trees: Validation of phylogenies reconstructed from distances matrices, Molecular Phylogenetics and Evolution, 3, 256-267.

Leclerc, B. and Cucumel, G. (1987): Consensus en classification: Une revue bibliographique, Mathématiques et Sciences Humaines, 100, 109-128.

Lecointre, G. H. et al. (1993): Species sampling has a major impact on phylogenetic inference, Molecular Phylogenetics and Evolution, 2, 205-224.

Lefkovitch, L. P. (1985): Euclidean consensus dendrograms and other classification structures, Mathematical Biosciences, 74, 1-15.

Le Quesne, W. (1989): Frequency distributions of lengths of possible networks from a data matrix, Cladistics, 5, 395-407.

Li, W.-H. and Guoy, M. (1991): Statistical methods for testing phylogenies, In: Phylogenetic analysis of DNA sequences, Miyamoto; M. M. and Cracraft, J. (eds.), 249-277, Oxford University Press, New York.

Li, W.-H. and Zharkikh, A. (1994): What is the bootstrap technique?, Systematic Biology, 43, 424-430.

Li, W.-H. and Zharkikh, A. (1995): Statistical tests of DNA phylogenies, Systematic Biology, 44, 49-63.

Ling, R. F. (1973): A probability theory of cluster analysis, Journal of the American Statistical Association, 68, 159-164.

Mantel, N. (1967): The detection of disease clustering and a generalized regression approach, Cancer Research, 27, 209-220.

Margush, T. (1982): Distances between trees, Discrete Applied Mathematics, 4, 281-290.

Margush, T. and McMorris, F. R. (1981): Consensus n-trees, Bulletin of Mathematical Biology, 43, 239-244.

Marshall, C. R. (1991): Statistical tests and bootstrapping: Assessing the reliability of phylogenies based on distance data, Molecular Biology and Evolution, 8, 386-391.

Mason-Gamer, R. J. and Kellogg, E. K. (1996): Testing for phylogenetic conflict among molecular data sets in the tribe Triticeae (Gramineae), *Systematic Biology*, **45**, 524-545.

McMorris, F. R. (1985): Axioms for consensus functions on undirected phylogenetic trees, *Mathematical Biosciences*, **74**, 17-21.

McMorris, F. R. et al. (1983): A view of some consensus methods for trees. In: *Numerical Taxonomy*, Felsenstein, J. (ed.), 122-126, Springer-Verlag, Berlin.

McMorris, F. R. and Neumann, D. (1983): Consensus functions defined on trees, *Mathematical Social Sciences*, **4**, 131-136.

Meier, R. et al. (1991): Homoplasy slope ratio: A better measurement of observed homoplasy in cladistic analyses, *Systematic Zoology*, **40**, 74-88.

Mickevich, M. F. (1978): Taxonomic congruence, *Systematic Zoology*, **27**, 143-158.

Milligan, G. W. (1981): A Monte-Carlo study of 30 internal criterion measures for cluster-analysis, *Psychometrika*, **46**, 187-195.

Miyamoto, M. M. (1985): Consensus cladograms and general classifications, *Cladistics*, **1**, 186-189.

Miyamoto, M. M. et al. (1994): A congruence test of reliability using linked mitochondrial DNA sequences, *Systematic Biology*, **43**, 236-249.

Miyamoto, M. M. and Fitch, W. M. (1995): Testing species phylogenies and phylogenetic methods with congruence, *Systematic Biology*, **44**, 64-76.

Mueller, L. D. and Ayala, F. J. (1982): Estimation and interpretation of genetic distances in empirical studies, *Genetical Research*, **40**, 127-137.

Murtagh, F. (1984): Counting dendrograms: A survey, *Discrete Applied Mathematics*, **7**, 191-199.

Nelson, G. (1979): Cladistic analysis and synthesis: Principles and definitions, with a historical note on Adanson's *Famille des Plantes* (1763-1764), *Systematic Zoology*, **28**, 1-21.

Nelson, G. (1993): Why crusade against consensus? A reply to Barrett, Donoghue, and Sober, *Systematic Biology*, **42**, 215-216.

Nemec, A. F. L. and Brinkhurst, R. O. (1988): The Fowlkes-Mallows statistic and the comparison of two independently determined dendrograms, *Canadian Journal of Fisheries and Aquatic Sciences*, **45**, 971-975.

Neumann, D. A. (1983): Faithful consensus methods for *n*-trees, *Mathematical Biosciences*, **63**, 271-287.

Nixon, K. C. and J. M. Carpenter. (1996): On simultaneous analysis, *Cladistics*, **12**, 221-241.

Oden, N. L. and Shao, K. T. (1984): An algorithm to equiprobably generate all directed trees with k labeled terminal nodes and unlabeled interior nodes, *Bulletin of Mathematical Biology*, **46**, 379-387.

Olmstead, R. G. and Sweere, J. A. (1994): Combining data in phylogenetic systematics: An empirical approach using three molecular data sets in the Solanacae, *Systematic Biology*, **43**, 467-481.

Omland, K. E. (1994): Character congruence between a molecular and a morphological phylogeny for dabbling ducks (*Anas*), *Systematic Biology*, **43**, 369-386.

Page, R. D. M. (1988): Quantitative cladistic biogeography: Constructing and comparing area cladograms, *Systematic Zoology*, **37**, 254-270.

Page, R. D. M. (1991): Random dendrograms and null hypotheses in cladistic biogeography, *Systematic Zoology*, **40**, 54-62.

Patterson, C. et al. (1993): Congruence between molecular and morphological phylogenies, *Annual Review of Ecology and Systematics*, **24**, 153-188.

Penny, D. and Hendy, M. D. (1985a): The use of tree comparison metrics, *Systematic Zoology*, **34**, 75-82.

Penny, D. and Hendy, M. D. (1985b): Testing methods of evolutionary tree construction, *Cladistics*, **1**, 266-278.

Penny, D. et al. (1982): Testing the theory of evolution by comparing phylogenetic trees constructed from five different protein sequences, *Nature*, **297**, 197-200.

Penny, D. et al. (1992): Progress with methods for constructing evolutionary trees, *Trends in Ecology and Evolution*, **7**, 73-79.

Phillips, C. and Warnow, T. J. (1996): The asymmetric median tree - A new model for building consensus trees, *Discrete Applied Mathematics*, **71**, 311-335.

Phipps, J. B. (1975): The numbers of classifications, *Canadian Journal of Botany*, **54**, 686-688.

Podani, J. and Dickinson, T. A. (1984): Comparison of dendrograms: A multivariate approach, *Canadian Journal of Botany*, **62**, 2765-2778.

Poe, S. 1996. Data set incongruence and the phylogeny of Crocodilians, *Systematic Biology*, **45**, 393-414.

Prager, E. M. and Wilson, A. C. (1976): Congruency of phylogenies derived from different proteins, *Journal of Molecular Evolution*, **9**, 45-57.

Proskurowski, A. (1980): On the generation of binary trees, *Journal of the Association of Computing Machinery*, **27**, 1-2.

Purvis, A. (1995a): A modification to Baum and Ragan's method for combining phylogenetic trees, *Systematic Biology*, **44**, 251-255.

Purvis, A. (1995b): A composite estimate of primate phylogeny, *Philosophical Transactions of the Royal Society of London (B)*, **348**, 405-421.

Quiroz, A. J. (1989): Fast random generation of binary, t-ary and other types of trees, *Journal of Classification*, **6**, 223-231.

Ragan, M. A. (1992): Phylogenetic inference based on matrix representation of trees, *Molecular Phylogenetics and Evolution*, **1**, 53-58.

Robinson, D. F. (1971): Comparison of labeled trees with valency Three, *Journal of Combinatorial Theory*, **11**, 105-119.

Robinson, D. F. and Foulds, L. R. (1979): Comparison of weighted labelled trees. In: *Lecture Notes in Matehmatics*, Volume 748, 119-126, Springer-Verlag, Berlin.

Robinson, D. F. and Foulds, L. R. (1981): Comparison of phylogenetic trees, *Mathematical Biosciences*, **53**, 131-147.

Rodrigo, A. G. (1993a): Calibrating the bootstrap test of monophyly, *International Journal of Parasitology*, **23**, 507-514.

Rodrigo, A. G. (1993b): A comment on Baum's method for combining phylogenetic trees, *Taxon*, **42**, 631-636.

Rodrigo, A. G. et al. (1993): A randomisation test of the null hypothesis that two cladograms are sample estimates of a parametric phylogenetic tree, *New Zealand Journal of Botany*, **31**, 257-268.

Rohlf, F. J. (1974): Methods of comparing classifications, *Annual Review of Ecology and Systematics*, **5**, 101-113.

Rohlf, F. J. (1982): Consensus indices for comparing classifications, *Mathematical Biosciences*, **59**, 131-144.

Ronquist, F. (1996): Matrix representations of trees, redudancy and weighting, *Systematic Biology*, **45**, 247-253.

Russo, C. A. M. et al. (1996): Efficiencies of different genes and different tree-building methods in recovering a known vertebrate phylogeny, *Molecular Biology and Evolution*, **13**, 525-536.

Sanderson, M. J. (1989): Confidence limits on phylogenies: The bootstrap revisited, *Cladistics*, **5**, 113-129.

Sanderson, M. J. (1995): Objections of bootstrapping phylogenies: A critique, *Systematic Biology*, **44**, 299-320.

Savage, H. M. (1983): The shape of evolution: Systematic tree topology, *Biological Journal of the Linnean Society*, **20**, 225-244.

Shao, K. and Rohlf, F. J. (1983): Sampling distribution of consensus indices when all bifurcating trees are equally likely. In: *Numerical Taxonomy*, Felsenstein, J. (ed.), 132-136, Springer-Verlag, Berlin.

Shao, K. and Sokal, R. R. (1986): Significance tests of consensus indices, *Systematic Zoology*, **35**, 582-590.

Simberloff, D. (1987): Calculating probabilities that cladograms match: A method of biogeographic inference, *Systematic Zoology*, **36**, 175-195.

Simberloff, D. et al. (1981): There have been no statistical tests of cladistics biogeographical hypotheses. In: *Vicariance Biogeography: A Critique*, Nelson, G. and Rosen, D. E. (eds.), 40-63, Columbia University Press, New York.

Sneath, P. H. A. (1967): Some statistical problems in numerical taxonomy, *The Statistician*, **17**, 1-12.

Sokal R. R. and Rohlf, F. J. (1962): The comparison of dendrograms by objective methods, *Taxon*, 9, 33-40.

Sokal R. R. and Rohlf, F. J. (1981): Taxonomic congruence in the Leptopodomorpha re-examined, *Systematic Zoology*, 30, 309-325.

Steel, M. A. (1988): Distribution of the symmetric difference metric on phylogenetic trees, *SIAM Journal of Discrete Mathematics*, 1, 541-555.

Steel, M. A. (1992): The complexity of reconstructing trees from qualitative characters and subtrees, *Journal of Classification*, 9, 91-116.

Steel, M. A. and Penny, D. (1993): Distribution of tree comparison metrics—Some new results, *Systematic Biology*, 42, 126-141.

Steel., M. A. et al. (1992): Significance of the length of the shortest tree, *Journal of Classification*, 9, 63-70.

Stinebrickner, R. (1982): S-consensus trees and indices, *Bulletin of Mathematical Biology*, 46, 923-935.

Stinebrickner, R. (1984): An extension of intersection methods from trees to dendrograms, *Systematic Zoology*, 33, 381-386.

Sullivan, J. (1996): Combining data with different distributions of among-site variation, *Systematic Biology*, 45, 375-379.

Swofford, D. L. (1991): When are phylogeny estimates from molecular and morphological data incongruent?, In: *Phylogenetic analysis of DNA sequences*, Miyamoto, M. M. and Cracraft , J. (eds.), 295-333, Oxford University Press, New York.

Swofford, D. L. et al. (1996a): Phylogenetic inference, In: *Molecular Systematics, 2nd edition*, Hillis, D. M. et al. (eds.), 407-514, Sinauer, Sunderland.

Swofford, D. L. et al. (1996b): The topology-dependent permutation test for monophyly does not test for monophyly, *Systematic Biology*, 45, 575-579.

Waterman, M. S. and Smith, T. F. (1978): On the similarity of dendrograms, *Journal of Theoretical Biology*, 73, 789-800.

Wiens, J. J. and Chippindale, P. T. (1994): Combining and weighting characters and the prior agreement approach revisited, *Systematic Biology*, 43, 564-566.

Wiens, J. J. and Reeder, T. W. (1995): Combining data sets with different numbers of taxa for phylogenetic analysis, *Systematic Biology*, 44, 548-558.

Wilkinson, M. (1994): Common cladistic information and its consensus representation: Reduced Adams and reduced cladistic consensus trees and profiles, *Systematic Biology*, 43, 343-368.

Wilkinson, M. (1996): Majority-rule reduced consensus trees and their use in boostrapping, *Molecular Biology and Evolution*, 13, 437-444.

Williams, D. M. (1994): Combining trees and combining data, *Taxon*, 43, 449-453.

Williams, W. T. and Clifford, H. T. (1971): On the comparison of two classifications on the same set of elements, *Taxon*, 20, 519-522.

Zaretskii, K. (1965): Constructing a tree on the basis of a set of distances between the hanging vertices, *Uspekhi Mathematika Nauk*, 20, 90-92. (in Russian).

Zharkikh, A. and Li, W.-H. (1992a): Statistical properties of bootstrap estimation of phylogenetic variability from nucleotide sequences. I. Four taxa with a molecular clock, *Molecular Biology and Evolution*, 9, 1119-1147.

Zharkikh, A. and Li, W.-H. (1992b): Statistical properties of bootstrap estimation of phylogenetic variability from nucleotide sequences. II. Four taxa without a molecular clock, *Journal of Molecular Evolution*, 35, 356-366.

Zharkikh, A. and Li, W.-H. (1995): Estimation of confidence in phylogeny: The full-and-partial bootstrap technique, *Molecular Phylogenetics and Evolution*, 4, 44-63.

Some Trends in the Classification of Variables

F. Costa Nicolau[1],   H. Bacelar-Nicolau[2]

[1]New University of Lisbon,
Department of Mathematics,   Faculty of Sciences and Technology,
Laboratory of Statistics and Actuarial Mathematics (LEMA)

[2]University of Lisbon, Faculty of Psychology and Education,
Laboratory of Statistics and Data Analysis (LEAD),   CEA / JNICT,
Portugal

Summary: In this paper we review a class of hierarchical clustering methods based on similarity coefficients and aggregation criteria which are associated to the integral transformation by the (probabilistic) distribution function of some suitable sample statistics. Some properties of those methods we have studied are remembered and/or derived here. Applications on either simulated or real data set have shown this approach performs better than the traditional one (using empirical clustering methods) in many situations. Moreover we define some "hybrid" criteria, which we generalise in order to get some mixed or parametric hierarchical clustering methods. Inside of such parametrical families we are able to find, among different criteria those better fitting to the initial similarities,   and to search for stability and validity of those methods.

1. Introduction

The most useful hierarchical clustering methods are of agglomerative type, iteratively transforming the unit clusters partition $\{\{x\}| x \varepsilon E\}$ into the trivial one $\{E\}$, merging at each step the most similar clusters, E being the set submitted to the analysis. These methods require a previous double choice: the comparison function (dissimilarity/ similarity, represented as   cf) γ_{xy} between pairs of elements x,yεE, and the cf $\Gamma(A, B)$ between pairs of clusters of E. The generalisation of γ to Γ may be performed in several ways, as there is a certain lack of consensus on the precise definition of "cluster" and "resemble clusters". Thus the question of measuring the resemblance remains a central one.

In our approach to cluster analysis we assume the following reference hypothesis: 1- we are dealing with some similarity coefficient γ_{xy} following a unit uniform distribution; 2- the measure of resemblance between each pair of clusters A and B   is set in the sample of γ-similarities $\left[\gamma_{ab}|(a,b)\varepsilon A \times B\right]$, crossing   A and B, with size $\alpha\beta$, where α=card A, β=card B.   Moreover, we shall suppose, in the present paper that: 3-   the $\alpha\beta$ γ-similarities are i.i.d. uniform on $[0, 1]$.

This reference hypothesis is quite suitable   and natural to evaluate the global resemblance between clusters of E. Based on it some good hierarchical and non-hierarchical (Nicolau, Brito, 1989) clustering methods and fruitful ideas arise. All those methods are included in the package CLASSIF and the results obtained on either simulated or real data show very often a clear progress regarding to other traditional ones, specially in the cases of AVL, AVM and particular mixed methods (like AVB) which are generated from some parametric families of agglomerative methods.

In the sequence we briefly examine the general procedure to get the cf 's γ and Γ, in section 2. For details see Bacelar-Nicolau (1972, 1979, 1980, 1981), Costa Nicolau (1980, 1983, 1985), Bacelar-Nicolau and Costa Nicolau (1981,1985), and Lerman (1972, 1981). Section 3 concerns some properties of AVL and AVM methods, and section 4 finally presents some parametric families defining mixed aggregation criteria from the Single Linkage and the AVL methods (Bacelar-Nicolau and Costa Nicolau, 1994).

2. Similarity and distribution function: hierarchical clustering probabilistic approach

In our probabilistic approach the general procedure to get cf γ and Γ is the following:

Step 1: Defining a probabilistic similarity coefficient γ

Let $S : E \times E \rightarrow R$ be a similarity function (random variable), chosen according to the nature of date, and let s_{xy} be its particular value on the pair $[x, y]$.

Let S be transformed by the cumulative distribution function, computed under some convenient reference hypothesis (This hypothesis will be defined in each particular case. Occasionally we use an independence hypothesis, considering our lack of knowledge about the structure of data.):

$$(x, y) \xrightarrow{S} s_{xy} \xrightarrow{df} \gamma_{xy} = \text{Prob } (S \leq s_{xy})$$

$\gamma_{xy} = \text{Prob } (S \leq s_{xy})$ is a new, probabilistic, similarity coefficient: $\gamma_{xy} \in [0, 1]$ and $\gamma_{xy} = \gamma_{yx}$; on the other hand γ_{xy} being an increasing function of $s_{xy}^{\cdot}$, global invariant hierarchical criteria (Sibson (1972)) will give exactly the same clustering results either with $s_{xy}^{\cdot}$ or with γ_{xy}. The probabilistic similarity coefficient verifies the minimal properties usually required for being a comparison function.

Often γ_{xy} will be calculated approximately, assuming our data set to have a large dimension. Then $S^{\cdot} = (S - E(S))/\sigma_S$ follows an asymptotic normal unit $N (0, 1)$ distribution, and in each case we can find:

$\gamma_{xy} = \text{Prob } (S \leq s_{xy}) = \text{Prob } (S^{\cdot} \leq s_{xy}^{\cdot}) = \Phi (s_{xy}^{\cdot})$

where Φ is the cumulative distribution function (cdf) of $N (0, 1)$.

In the case of binary data, for instance, we can take $s_{xy} = \Sigma \{ I_x(i).I_y(i) \mid i \in D \} = $ number of common presences of x and y over the descriptive set D, where $I_x(i) = 1$ if i verifies x, $=0$ otherwise. Thus S will have an hypergeometric or binomial distribution (or

Poisson) as exact law, depending on the underlying reference hypothesis (concerning the marginal frequencies of the 2x2 contingency table associated to the pair (x,y)). Then γ_{xy} can be estimated by the corresponding cdf of the normal approximation of each selected discrete distribution.

Moreover, in binary case H. Bacelar Nicolau (1981, 1987, 1989) found the usual association coefficients are grouped in several distributional equivalent classes in the following sense: either they have the same exact distribution or else they share the same normal asymptotic distribution. This means that using the probabilistic coefficient, one can choose for hierarchical clustering purpose only one coefficient in each distributional equivalent class, the other ones giving (exactly or asymptotically) the same hierarchical results (dendrogram, for instance).

In the case of frequency or contingency tables we usually take the probabilistic coefficient based on the affinity coefficient (Matusita, 1951; Bacelar-Nicolau, 1988): the above reference hypothesis will then be referred either to a sampling scheme or a permutational scheme, depending on the way the data have been observed. The sampling scheme has been extended to integer data (Bacelar-Nicolau and Nicolau, 1993), while the permutation scheme usually applies to real data.

Often we refer to the probabilistic coefficient γ_{xy} as the VL similarity (V for Validity, L for Linkage), because, as Tiago de Oliveira pointed out, the probabilistic coefficient validates in a probabilistic scale the basic linkage s between each pair (x, y). On the other hand VL comes primary from "Vraisemblance du Lien", that is Likelihood Function Link. In fact much of our research on this matter find their roots in the works of Lerman (1970) and Bacelar-Nicolau (1972), where the probabilistic coefficient was for the first time used in binary case: it was called as VL, "Vraisemblance du Lien" similarity coefficient, and its first associated aggregation criterion AVL, "Algorithme de la Vraisemblance du Lien". Subsequent extensions have often kept the same label.

Step 2: Defining a probabilistic aggregation criterion

Assume that we are dealing with some similarity coefficient γ_{xy} following a unit uniform distribution and the measure of resemblance between each pair of clusters A and B is based on the set $\left[\gamma_{ab}|(a,b)\epsilon A \times B\right]$, crossing A and B, with size $\alpha\beta$ (where α=card A, β=card B). Also assume the $\alpha\beta$ γ-similarities are i.i.d. uniform on [0, 1].

This reference hypothesis is quite general and fits well to many applied situations. On the other hand it is well known the integral transformation of a continuous random variable leads to the uniform distribution on [0, 1]. The S random variable in step 1 being in fact of discrete type, the size of the descriptive set D is in general large enough in cluster analysis, to apply this property to the sample of γ-similarities.

Let's now introduce some probabilistic cf 's that will generalise γ_{xy} to the cf $\Gamma(A, B)$ between subsets of E. We have studied, in particular, the cf coefficients associated to the following statistics:

$$p_{A,B} = \max \left[\gamma_{ab} \mid (a,b) \in A \times B \right]$$

$$q_{A,B} = \min \left[\gamma_{ab} \mid (a,b) \in A \times B \right]$$

$$\overline{p}_{A,B} = \frac{1}{\alpha\beta} \Sigma \left[\gamma_{ab} \mid (a,b) \in A \times B \right]$$

$$t_{A,B} = \left[p_{A,B} + q_{A,B} \right] / 2$$

where $p_{A,B}$, $q_{A,B}$ and $\overline{p}_{A,B}$ are the basic statistics for single linkage, complete linkage and mean average linkage methods, respectively, and $t_{A,B}$ is a statistics already used (with empirical dissimilarity coefficients) by some researchers to get a compromise between the single and complete linkage methods. Here we are not interested in directly using those statistics for clustering purposes; instead we want to take their cumulative distribution functions, in order to use the probabilistic approach to hierarchical classification. As we have pointed out we assume the $\alpha\beta$ γ-similarities are i.i.d. in this work.

The first probabilistic aggregation criterion, MaxProb, derived from $p_{A,B}$, is the so-called AVL method, from Validity-Linkage Algorithm or "Algorithme de la Vraisemblance du Lien", which was first purposed by Lerman (1970), its main properties being established by Bacelar-Nicolau (1972) in the context of classification of variables. It was generalised in Lerman (1981), Bacelar-Nicolau (1981) and Nicolau and Bacelar-Nicolau (1981), for instance. The second statistics, $q_{A,B}$, was used as support to a probabilistic clustering method for frequency data introduced by Goodall, where the probabilistic similarity is defined from the χ^2 distribution (in Legendre and Legendre (1983)). In our approach the probabilistic aggregation criterion MinProb associated to $q_{A,B}$ did not perform very well. The third probabilistic criterion, AvmProb, derived from $\overline{p}_{A,B}$ is also the so-called AVM method, Mean-Validity Algorithm. Finally the probabilistic criterion generated by the "hybrid" statistics $t_{A,B}$ will not be considered now, since we are most interested in studying parametric mixed criteria. Thus, among those four statistics we shall take in the present paper only $P_{A,B}$ and $\overline{p}_{A,B}$. Under the i.i.d. assumption referred to above, we obtain for the corresponding AVL and AVM clustering criteria the following expressions:

$$\Gamma_1(A, B) = p_{A,B}^{\alpha\beta}$$

and

$$\Gamma_2(A,B) \begin{cases} \dfrac{1}{(\alpha\beta)!} \displaystyle\sum_{0 \le j < \alpha\beta \overline{p}_{A,B}} (-1)^j \binom{\alpha\beta}{j} \left[\alpha\beta\overline{p}_{A,B} - j \right]^{\alpha\beta} \dots Y \le \overline{p}_{A,B} < 5 \\[2ex] 1 - \dfrac{1}{(A,B)!} \displaystyle\sum_{0 \le j < \alpha\beta(1-\overline{p}_{A,B})} (-1)^j \binom{\alpha\beta}{j} \left[\alpha\beta(1-\overline{p}_{A,B}) - j \right]^{\alpha\beta} \dots .5 \le \overline{p}_{A,B} \le 1 \end{cases}$$

Both the AVL and AVM methods generally perform quite well on either simulated or real data. Nevertheless they differ in some specific features, which can allow us to choose each one for different particular situations. In the next section we compare the two methods in what concerns the presence/absence of inversions in the corresponding dendrograms and the type of hierarchical structure they produce.

3. Comparing MaxProb (AVL) and AvmProb (AVM) methods

3.1 - Index level or Absence/Presence of inversions

Let $\Gamma^{(k-1)}$ be the table of similarity values between clusters at the (k-1)-th step, of an agglomerative process, and $f(k) = \max \Gamma^{(k-1)}$. We call $d(k) = 1-f(k)$ a level index: it measures the lack of cohesion of the successive clusters formed at the sequential hierarchical tree levels; in this sense one usually hopes $d(k)$ to be a decreasing function of the levels k. Nevertheless we want to point out that we believe the existence of inversions in $d(k)$ can be quite compatible with the construction of good hierarchical classifications.

Concerning the level index, the following property holds:

Property - The level index of AVM method can present inversions, whereas the level index of AVL method is strictly increasing.

Proof: The absence of monotony of the AVM level index is simply a consequence of the statistical convergence of the mean statistics of an i.i.d. sample of the unit uniform distribution to the value 0.5: when the entries of Γ are updated at the end of each step of the clustering algorithm, the Γ values will increase, if the mean of VL similarities is greater than 1/2, or decrease, if that mean is lesser than 1/2. So we can naturally have $d(k+1) < d(k)$ at some levels of the AVM dendrogram.

Now in order to see that for the AVL method one has $d(k+1) > d(k)$ always, let's first suppose that $\Gamma^{(k-1)}$ has only a maximum value (binary tree), so that only a pair of clusters verifies the criterion at each step. Let (A, B) be the pair of clusters satisfying the AVL criterion (maximisation of $\Gamma^{(k-1)}$) and let P_k be the partition defined at k-th step:

$$P_k = P_{k-1} - \{A, B\} \cup \{A \cup B\}.$$

Then : $\Gamma^{(k)} = \Delta^{(k-1)} \cup \Gamma^{(K)}_{A \cup B}$ (disjoint components)

where

$$\Delta^{(k-1)} = \Gamma^{(k-1)} - \Gamma^{(k-1)}_A \cup \Gamma^{(k-1)}_B),$$

$$\Gamma^{(k-1)}_A = \{\Gamma^{(k-1)}(A, X) | X \neq A, X \in P_{k-1}\} \subset \Gamma^{(k-1)}$$

$$\Gamma^{(k-1)}_B = \{\Gamma^{(k-1)}(B, X) | X \neq B, X \in P_{k-1}\} \subset \Gamma^{(k-1)}$$

$$\Gamma^{(k)}_{A \cup B} = \{\Gamma^{(k)}(A \cup B, Y) | Y \neq A \cup B, Y \in P_k\} \subset \Gamma^{(k)}$$

Now, either: $\max \Gamma^{(k)} = \max \Delta^{(k-1)}$ and then $\max \Gamma^{(k)} < \max \Gamma^{(k-1)}$

or else: $\max \Gamma^{(k)} = \max \Gamma^{(k)}_{A \cup B}$

and then $\quad \max \Gamma^{(k)} = p_{A \cup B, C}^{(\alpha+\beta)\gamma} = p_{A, C}^{(\alpha+\beta)\gamma} < p_{A, C}^{\alpha\gamma} < \max \Gamma^{(k-1)}$,

where $\quad C \in P_k$, $\max \Gamma^{(k)} = \Gamma_{A \cup B, C}^{(k)}$, $p_{A, C} = \max(p_{A, C}, p_{B, C})$, $\gamma = \text{card C}$.

The proof is then completed in the case of binary trees.

In the general case, if h pairs of clusters, $\{A_i, B_i\}$, $i = 1, \ldots, h$, verify the AVL criterion at k-th step, then $\Gamma^{(k)}$ becomes: $\quad \Gamma^{(k)} = \Delta^{(k-1)} \cup \Delta^{(k)}$

where now,

$$\Delta^{(k-1)} = \Gamma^{(k-1)} - [(\bigcup_{i=1}^{h} \Gamma_{A_i}^{(k-1)}) \cup (\bigcup_{i=1}^{h} \Gamma_{B_i}^{(k-1)})] \subset \Gamma^{(k-1)}$$

and $\quad \Delta^{(k)} = \bigcup_{i=1}^{h} \Gamma_{A_i \cup B_i}^{(k)}$

so that the monotony property of the index level of AVL always holds.

3.2 - "Symmetry" versus "bipolarisation" effects

As successive updating of the cf Γ in AVM method tend to reinforce the strong links and, conversely, to decrease the week links one usually observes in the AVM dendrograms a sort of "bipolarisation effect". Most examples treated so far by AVM method clearly show the (same) manner how this method works: the kernel of each cluster grows by joining element after element in some kind of local chain effect (similar to the characteristic global effect of single linkage method); once a kernel is finished, another one begins to built in the same way, and the process continues until all the main clusters are formed; those clusters are then merged, without chain effect, producing some good-looking and coherent trees. This double effect, globally conserves the initial structure as expressed by the cf γ and is responsible after all by the trustworthy fitting of AVM trees to the data.

A different tendency can usually be observed in the AVL hierarchies, which seldom are associate to a chain effect. Instead they produce quite regular trees with clusters of equal size at each level of the tree. We call this the "symmetry effect" of AVL method, the responsible for that being the exponent of $p_{A,B}$ in the formula of Γ. This exponent performs in fact a sort of brake action to the chain effect at all levels of the dendrogram. One can easily understand the symmetry effect of AVL in the following example:

Suppose there are h clusters of equal size α at step k (k = 0, 1, 2, ...); let $A, B \in P_k$ the clusters which are going to be merged. Thus updating the cf between any other cluster C and the new merged cluster $A \cup B$ will give:

$$\Gamma^{(K+1)}(A \cup B, C) = p_{A \cup B, C}^{2\alpha^2}$$

while the link between C and other cluster of P_{k+1} is expressed by

$$\Gamma^{(K+1)} = \Gamma^{(k)}(C, X) = p_{C, X}^{\alpha^2}$$

So $A \cup B$ will attract C, shaping a new cluster $A \cup B \cup C$ of size 3, only if

$$p_{A \cup B, C}^{2\alpha^2} > p_{C, X}^{\alpha^2}$$

or equivalently:

$$p_{A \cup B, C} > \sqrt{p_{C, X}}, \quad \forall X \neq C, A \cup B$$

Therefore it is natural that clusters of size 2α arise before any other group of size 3α could emerge.

Concerning "symmetry" effect of AVL versus "bipolarisation" effect of AVM, recent work of Nicolau and Bacelar-Nicolau has been developed which associate them with spatial dilating and spatial contracting properties of agglomerative methods (Lance and Williams, 1967).

4. Parametric approach to adaptive clustering

Once the VL similarity is chosen as cf γ_{xy} between pairs of elements, the Single Linkage (SL) and the AVL methods being associated to the same basic aggregation function, $p_{A,B} = \max\left[\gamma_{ab}|(a,b) \in A \times B\right]$, they can be generated by a conjoined formula:

$$\Gamma(A,B) = p_{A,B}^{g(\alpha,\beta)}$$

where $\alpha = $ card A, $\beta = $ card B: $g(\alpha,\beta) = 1$ for SL and $g(\alpha,\beta) = \alpha\beta$ for AVL.

One could expect that by varying $g(\alpha,\beta)$ with $1 < g(\alpha,\beta) < \alpha\beta$, a sort of compromise will be built between SL and AVL methods: $\Gamma(A,B) = p_{A,B}^{g(\alpha,\beta)}$ will be more polluted by the chain effect when $g(\alpha,\beta)$ remains near 1, and more contaminated by the symmetry effect as long as $g(\alpha,\beta)$ is in the neighbourhood of $\alpha\beta$.

Therefore a bridge between SL and AVL methods can be established by using, for instance, the following chain of exponents:

$$1 \leq \min(\alpha,\beta) \leq \frac{2\alpha\beta}{\alpha+\beta} \leq \sqrt{\alpha\beta} \leq \frac{\alpha+\beta}{2} \leq \max(\alpha,\beta) \leq \alpha\beta$$

The correspondent agglomerative cf based on VL similarity define the following iterative clustering procedure

$$p_{A,B} \geq p_{A,B}^{\min(\alpha,\beta)} \geq p_{A,B}^{\frac{2\alpha\beta}{\alpha+\beta}} \geq p_{A,B}^{\sqrt{\alpha\beta}} \geq p_{A,B}^{\frac{\alpha+\beta}{2}} \geq p_{A,B}^{\max(\alpha,\beta)} \geq p_{A,B}^{\alpha\beta}$$

This methodology assures economic computation, invariance with respect to the initial order of the elements or the clusters to be merged, and some way of evaluating the brake action on both the chain and symmetry effects. In what concerns these aspects, the two methods associated to the geometric and the arithmetic means of the cardinal of clusters being compared perform quite well, producing fine interpretable hierarchical trees.

The above iterative clustering procedure has been generalised by defining some suitable parametric families of aggregation criteria to link the SL and AVL methods.

The first idea for finding such a family was to take a scalar transformation of $\alpha\beta$: $\alpha\beta \rightarrow \alpha\beta/\xi$, where $1 < \xi < \alpha\beta$, fixed ξ. This turned out not to be a good solution, since it is easy to prove that:

Given $\Gamma_g(A, B) = p_{A,B}^{g(\alpha,\beta)}$, then any other method $\Gamma_{g'}$, such that $g' = \delta\, g(\alpha, \beta)$, with fixed $\delta > 0$, will give exactly the same hierarchic tree built by Γ_g, differences existing only in the values of the level index.

The invariance does not hold with linear transformations of the exponent $g(\alpha,\beta)$, $\Gamma_g(A, B)$ and $\Gamma_{\delta g + \epsilon}(A, B) = p_{A,B}^{\delta g(\alpha,\beta) + \epsilon}$ generally producing different hierarchic trees. Thus a natural solution to the question of defining some appropriate parametric family of agglomerative criteria linking the SL and AVL methods appears to be the following:

$$\Gamma(A, B) = p_{A,B}^{\delta\,\alpha\beta + \epsilon}$$

where $1 \leq \delta\ \ \alpha\beta + \epsilon \leq \alpha\beta$, $\alpha = $ card A, $\beta = $ card B, the exponent function being a linear convex combination of the exponents of SL and AVL, respectively 1 and $\alpha\beta$.

Experimental work was conducted either on simulated or real data in order to study this parametric family and its role in searching for hierarchical methods which better fit the initial similarities, as well as in looking for stability and validity of those methods.

On the other hand the above parametric family has been later extended in order to include the two probabilistic methods associated to the exponents with the geometric and the arithmetic means of the cardinal of clusters being compared, in the former iterative clustering procedure. We get in this case $g(\alpha,\beta\ ;\ \epsilon,\xi) = 1 / (1 + \epsilon\, ((\ \alpha \times \beta)^{\xi} - 1))$, where ϵ and ξ both take values in the interval $[0,1]$. More recently the whole family has been included in the probabilistic similarity version of the well known Lance and Williams formula (Bacelar-Nicolau and Nicolau, 1994). The recursive Lance and Williams formula, designed for dissimilarity coefficients can in fact easily be adapted to clustering methods based on similarities and particularly on VL-similarity. One has:

$$\Gamma(A \cup B, C) = \alpha_1\, \Gamma(A, C) + \alpha_2\, \Gamma(B, C) + \beta\, \Gamma(A, B) + \gamma\, |\, \Gamma(A, C) - \Gamma(B, C)|$$

where the constants α_1, α_2, β, γ vary according to the method we want to reproduce.

The extended formula derived in order to include the probabilistic hierarchical family, needs only two more constants, which are the parameters ϵ, ξ above and can be represented as follows:

$$\Gamma(A \cup B, C) = [\ \alpha_1\, \Gamma(A, C)^{g(\alpha\eta;\epsilon,\xi)} + \alpha_2\, \Gamma(B, C)^{g(\beta\eta;\epsilon,\xi)} + \beta\, \Gamma(A, B)^{g(\alpha,\beta;\epsilon,\xi)}$$
$$+ \gamma\, |\ \Gamma(A, C)^{g(\alpha\eta;\epsilon,\xi)} - \Gamma(B, C)^{g(\beta\eta;\epsilon,\xi)}|\ ]^{1/g(\alpha,\beta;\epsilon,\xi)}$$

where $g(\alpha,\eta\ ;\ \epsilon,\xi) = 1 / (1 + \epsilon\, ((\ \alpha \times \eta\ )^{\xi} - 1))$, η being the cardinal of the cluster C.

We can easily see that making $\epsilon = 0$ we simply get the first formula above. On the other hand we find as particular cases in the family: the single linkage SL ($\alpha_1 = \alpha_2 = \gamma = 1/2$, $\beta = \epsilon = 0$), the validity-linkage AVL ($\alpha_1 = \alpha_2 = \gamma = 1/2$, $\beta = 0$, $\epsilon = \xi = 1$) and the brake-validity linkage AVB ($\alpha_1 = \alpha_2 = \gamma = 1/2$, $\beta = 0$, $\epsilon = 1$, $\xi = 1/2$) algorithms.

Using such parametric family enables us to analyse the robustness of our clustering methods: by varying ε and ξ from 0 to 1, the other coefficients remaining unchanged, we can study the stability of the AVL family of models, like before. But this procedure can of course now be generalised to the comparison among the probabilistic family and the agglomerative methods generated by the former Lance and Williams formula.

5. Conclusions

The concept of the probabilistic (VL) similarity obtained by using (the uniform transformation) the distribution function of a random variable and its extension to agglomerative criteria, allow us to establish a general consistent probabilistic approach to the hierarchical clustering methods. Validity-affinity combined with AVL, AVB and AVM methods are good examples of this approach.

Moreover the probabilistic similarity enables us to derive a parametric clustering approach to the hierarchic methodology, that is a recursive formula generating criteria which appear to be a mixture of the single linkage (chain effect) and the AVL (symmetry effect) methods. Questions so important as stability, optimisation and compatibility with the data, concerning the hierarchies produced by the parametric agglomerative methods, may in that way be enlighten.

Finally note that this concept of probabilistic similarity does not consume itself in the field of hierarchical agglomerative methods: we have also developed a non hierarchical probabilistic approach based on the distribution function of the mean of a unit uniform sample, which gives an excellent non hierarchical method of k-means type.

References:

Bacelar-Nicolau, H. (1972) - *Analyse d'un algorithme de classification automatique* - Thèse de 3ème Cycle, Univ. ParisVI (I.S.U.P.), Nov. 1972.

Bacelar-Nicolau, H. (1981) - *Contribuições ao estudo dos coeficientes de comparação em Análise Classificatória* - Doct. Thesis, FCL, Univ. de Lisboa.

Bacelar-Nicolau, H. (1987) - *On the distribution equivalence in cluster analysis*, NATO ASI Series, vol F30, Patt. Recogn. Theory and Applic., P.A.Devijver/ J.Kitler (eds), Springer-Verlag , 73-79.

Bacelar-Nicolau, H. (1988) - *Two Probabilistic Models for Classification of Variables in Frequency Tables*. Classif. and Rel. Meth. of Data Analysis, H.H.Bock (ed.). North Holland,181-186.

Bacelar-Nicolau, H. .(1989) - *Sur l'équivalence distributionnelle entre coefficients d'association*, Bulletin of the International Statistical Institute (ISI), 47th Session, Contributed Papers.Book 1. 89-90

Bacelar-Nicolau, H. ; Nicolau, F.C.(1993) - *Classifying integer scale data by the affinity coefficient: a probabilistic approach:* Proceedings of the Sixth Intern. Symp. on Applied Stochastic Models and Data Analysis (ASMDA), J.Jansen and C.H.Skiadas (ed), World Scientific, vol 1, 63-74

Bacelar-Nicolau, H. ;Nicolau ,F.C.(1994) - *Exploratory and confirmatory discrete multivariate analysis in a probabilistic approach for studying the regional distribution of AIDS in Angola.* New App. in Classif. and Data Analysis, E.Diday, Y. Lechevallier, M. Shader (ed.), Springer-Verlag, 610-618

Lance, G.N.; Williams, W.T. (1967) - *A general theory of classificatory sorting strategies. Hierarchical systems.* The Computer Journal, vol 9, no 4, 373-380

Legendre, L.; Legendre, P. (1983) - *Numerical Ecology.* Elsevier Sc. Publ. I. C. LERMAN,I.C.(1970) - *Sur l'Analyse des Données Préalable à une Classification Automatique,* Rev. Math. et Sc.Hum., vol 32, 8ème année, 5-15.

Lerman, I.C. (1973) - *Etude distributionnelle de statistiques de proximité entre structures finies de même type, application à la classification automatique.* Cahiers du BURO, Paris.

Lerman, I.C. (1981) - *Classification et Analyse Ordinale des Données.* Dunod.

Matusita, K. (1951) - *On the Theory of Statistical Decision Functions.* An. In. Stat. Math.,vol.III,1-30.

Nicolau, F. Costa (1981) - *Critérios de análise classificatória hierárquica baseados na função de distribuição* - Doct. Thesis, FCL, Univ. de Lisboa.

Nicolau, F. Costa ; Bacelar-Nicolau, H. (1981) - *Nouvelles méthods d'agrégation basées sur la fonction de répartition.* Collection Séminaires INRIA de Classification et Perception par Ordinateur 1981, INRIA, Domaine de Voluceau-Rocquencourt, France.

Nicolau, F. Costa (1983) - *Cluster Analysis and Distribution Function,* Meth. Oper. Res., vol. 45, Verlag Anton Hain, 431-433.

Nicolau, F. Costa ; Brito, M.P.(1989) - *Improvment in NHMEAN method.* Data Analysis, Learning Symbolic and Numerical Knowledge (ed. E. Diday) Nova Science Publishers

Sibson, R. (1972) - *Order invariant methods for Data Analysis* in J.R.S.S., B, vol.34, nº.3, 311-338.

CONVEXITY METHODS IN CLASSIFICATION

Jean-Paul Rasson[1]

[1]F.U.N.D.P., Département de Mathématique,
8, Rempart de la Vierge,
B–5000 Namur, Belgium
Tel.: + 32 81 724928 - Fax: + 32 81 724914
E-Mail : jpr@math.fundp.ac.be

Summary : We investigate the solutions to the clustering and the discriminant analysis Problems when the points are supposed to be distributed according to Poisson Processes on convex supports. This leads to very intuitive criteria for homogeneous Poisson Processes based on the Lebesgue measures of convex hulls. For non homogeneous Poisson Processes, the Lebesgue measures have to be replaced by intensities integrated on convex hulls.

Similar geometrical tools, based on the Lebesgue measure, are used in the context of Pattern Recognition. First, a discriminant analysis algorithm is developed for estimating a convex domain when inside and outside points are available. Generalisation to non convex domains is explored.

Introduction

This research has been initiated by the question of D.G. Kendall of how to estimate a bounded convex set observing only the realization of an homogeneous Poisson Process inside this convex set. The solution (Ripley and Rasson, 1977; Rasson, 1979) was an homothetic expansion of the convex hull of the sample from its centroid. The following question, raised by E. Diday, was naturally the problem of using these arguments in Clustering. This led us to the maximum likelihood estimation of the hypothetized support for clustering, i.e. the union of K bounded convex sets. The solution was "find the partition of the points in K subgroups of such that the sum of the Lebesgue measures of their convex hulls is minimal" (Hardy and Rasson 1982).

Then came the question of the corresponding discriminant analysis raised by A.D. Gordon. The Bayesian solution for the same hypothesis (still for an homogeneous Poisson Process) was to affect the new point to the sample for which the Lebesgue measure added by convexity to its convex hull is minimal (Baufays and Rasson, 1984a, 1984b, 1985). But this could not classify the points belonging to more than one convex hull as this is often the case of pixels in images.

Then to deal with them, we moved to non-homogeneous independent Poisson Processes with convex supports. The main change in the solution was to replace the Lebesgue measure by the integrated intensity. But the same equation gave also the solution for points lying in the intersections of the supports.

1. The Stationary Poisson Process

1.1. The model

We consider that we deal with a clustering problem where the points we observe are generated by a Poisson Process and are distributed in D where D is the union of g disjoint domains $(D_k), 1 \leq k \leq g$. Our point of view will be that if we are able to find back these domains, making some inference about them, we will, in some sense, solve the clustering problem.

1.2. The maximum likelihood solution of the clustering problem.

Let x denote the sample vector $(x_1, ..., x_n)$ with $x_i \in I\!\!R^d, i = 1, ..., n$. The indicator function of a set A at the point y is defined by : $1\!\!1_A(y) = 1$ if $y \in A$ and 0 otherwise.

So, since with our hypothesis, the points will be independently and uniformly distributed on D, the likelihood function takes the form

$$F_D(x) = \frac{1}{(m(D))^n} \cdot \prod_{i=1}^{n} 1\!\!1_D(x_i)$$

where $m(D)$, the Lebesgue measure of D, is the sum of the measures of the g subsets $D_k (1 \leq k \leq g)$. The domain D, parameter of infinite dimension, for which the likelihood is maximal, is, among all those which contain all the points, the one whose Lebesgue measure is minimal.

If we do not impose more conditions on the subsets, we can easily find g sets D_k which contain all the points and are such that the sum of their measures is zero. Thus there are many trivial solutions to the problem. Nevertheless, we can easily see that the problem of estimating a domain is not well-posed and that the weakest assumption that makes the domain D estimable is the convexity of the D_k (Baufays and Rasson, 1984).

With a partition of the set of points into g sub-domains having disjoint convex hulls, we can associate a whole class of estimators ; indeed we only have to find g disjoint convex sets, each of them containing one of the . For each partition, the likelihood has a local maximum : the convex hulls of the g subsets. The global maximum will be attained with the partition for which the sum of the Lebesgue measures of the convex hulls of the g subgroups is minimal. This is the solution we seek.

Practically, if the basic space is $I\!\!R$, we look for the g disjoint intervals containing all the points such that the sum of their lengths is minimal. In $I\!\!R^2$ (or $I\!\!R^3$), we try to find the g groups of points such that the sum of the areas (volumes) of their disjoint convex hulls is minimal.

1.3. The statistical model and the rule for the associated discriminant analysis.

The conditional distribution for the k-th population is assumed to be uniform in a convex compact domain D_k , and the a priori probability p_k that an individual belongs to population k is proportional to the Lebesgue measure of D_k. The convex domains D_k are assumed disjoint. The density of population k, $f_k(x)$, and the unconditional density $f(x)$, are respectively equal to :

$$f_k(x) = \frac{1}{m(D_k)} \cdot 1\!\!1_{D_k(x)}$$

$$f(x) = \frac{1}{m(D)} \cdot \sum_{k=1}^{g} 1\!\!1_{D_k(x)}$$

The decision rule is the Bayesian one, with the unknown parameters - the convex sets D_k - replaced by their maximum likelihood estimations. Let X_k be the labeled sample of population k, $H(X_k)$ be its convex hull, and x be the individual to be assigned to one of the g populations. If x is allocated to the k-th group, the estimates of the domains D_k are equal to :

$$\widehat{D}_j = \begin{cases} H(X_j) \text{ if } j \neq k \\ H(X_j \cup \{x\}) \text{ if } j = k \end{cases}$$

The maximum likelihood estimate of $p_k f_k(x)$ is therefore :

$$\widehat{p_k f_k}(x) = \frac{1_{\widehat{D_k}}(x)}{\sum_{j=1}^{g} m(H(X_j)) + S_k(x)}$$

with $S_k(x) = m(H(X_k \cup \{x\})) - m(H(X_k))$

As the convex sets are assumed disjoint, x can be allocated to the k-th group, if and only if $H(X_k \cup \{x\})$ and $H(X_j)$ are disjoint for all $j \neq k$. Otherwise, we define arbitrarily : $S_k(x) = +\infty$.

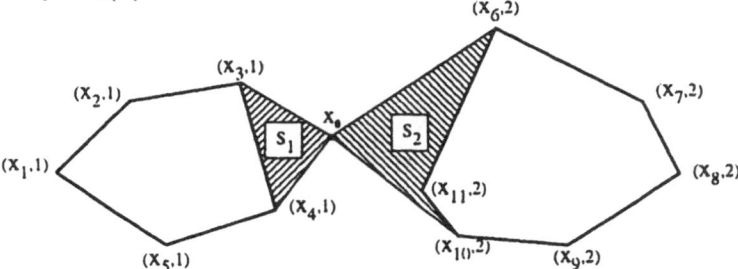

Figure 1.

The allocation rule is then :

assign x to the k-th population if and only if $S_k(x) < S_j(x), k \neq j$.

2. The Non-stationary Poisson Process

In order to avoid an heuristical part in the supervised classification procedure, we generalize now the basic hypothesis, namely the Homogeneous Poisson Process, one into the hypothesis of Non Homogeneous Poisson Process.

2.1. The Maximum Likelihood solution for the clustering problem.

We shall consider therefore that we deal with a clustering problem where the points we observe are generated by a Non Homogeneous Poisson Process with intensity $q(.)$ (satisfying $q(x) > 0$) and are observed only in D, where D is the union of g disjoint convex domains.

$$D = \bigcup_{k=1}^{g} D_k \ , \quad D_k \text{ convex disjoint.}$$

Our point of view will be again to try to find the maximum likelihood estimation of the D_k. If x denote the sample vector $(x_1, ..., x_n)$ with $x_i \in \mathbb{R}^d, i = 1, ..., n$ its likelihood will be

$$F_D(x) = \frac{1}{(m(D))^n} \cdot \prod_{i=1}^{n} 1_D(x_i).q(x_i)$$

where $m(D) = \int_D q(x)dx$ and $q(.)$ is the intensity of the process.

Thus, if the intensity is known (or, maybe, has been estimated), the maximum likelihood clustering solution, since the maximum likelihood estimation of a convex domain based on a sample of points inside this domain is still the convex hull of this sample, is then:
Find the g groups of points for which the sum of the intensities integrated on their convex hulls is minimal.

2.2. The associated discriminant analysis.

It would be easy to reconstruct the discriminant rule associated with this (unique) Non Stationary Poisson Process. We just have to replace the Lebesgue's measures of any convex set by the intensity integrated on this convex set.

Thus if $q_k(.)$ is the intensity of the Process and satisfying $q_k(x) > 0 \Leftrightarrow x \in D_k$ (e.g. $q_k(.) = q(.) \, \mathbb{1}_{D_k}(.)$ for the case of unique Process on disjoint sets), we may suppose that any point is distributed on $D = \cup_{k=1}^{g} D_k$ with respect to the density function

$$f(x) = \sum_{k=1}^{g} p_k . f_k(x)$$

where $f_k(x) = \dfrac{q_k(x)}{\int_{D_k} q_k(y)dy}$ and $p_k = \dfrac{\int_{D_k} q_k(y)dy}{\sum_{k=1}^{g} \int_{D_k} q_k(y)dy}$

As usual, the possibly convex supports D_k are estimated by the convex hulls $H(X_k)$ of the training set points. If we denote $S = \sum_{k=1}^{g} \int_{H(X_k)} q_k(y)dy$ and $S_k(x) = \int_{H(X_k \cup \{x\}) \backslash H(X_k)} q_k(y)dy$ then the Bayesian classification rule becomes :
assign the new point x to the class k such that $\widehat{p_k f_k}(x) = q_k(x)/(S + S_k(x))$ *is maximal.*

When the local intensities $q_k(x)$ do not depend on k , this rule simply consists in assigning x to the class k such that the added intensity $S_k(x)$ is minimal. Thus, in this case and when the convex hulls are disjoint, we still keep the convex admissibility property i.e. "if the point x belongs to the convex hull of only one class, x is assigned to this class".

3. Pattern Recognition applications

3.1. The inside/outside problem : presentation.

Suppose that X is a Poisson point process within a fixed finite window $F \subset \mathbb{R}^d$. In F, we have a compact convex domain D. We suppose that the Poisson process X is homogeneous on F, with density λ. We observe a fixed number $t \geq 1$ of realizations of X in F, from which n turn out to be inside the domain D and m outside of D ($t = n + m$). We want to estimate the unknown convex domain D. This problem is indeed the third problem of Grenander (1973).

The solution we propose to the problem is the use of the discriminant rule we have just described in cluster analysis. The situation here is quite similar as we have two disjoint domains D and its complementary $\bar{D} = F \setminus D$. The main difference lies in the non convexity of $\bar{D}$.

3.2. The discriminant analysis rule and the inside/outside problem

The same idea is applied to the inside/outside problem, with D and $\bar{D}$ playing the role of C_1 and C_2. The major difference here is that $\bar{D}$ is no longer a convex set.

Let us note $(x, y) = \{x_1, y_1, \ldots, x_{n+m}, y_{n+m}\}$ the realizations of the homogeneous Poisson process X and the labeling variable Y : $y_i = 1$ if $x_i \in D$ and $y_i = 2$ otherwise. Let $y_i = 1$ for $i = 1, \ldots, n$ and $y_i = 2$ for $i = n+1, \ldots, n+m$, without restricting generality.

Conditionally on n and m fixed, the likelihood function for (x, y) is

$$L_D(x, y) = \left(\frac{1}{m(D)^n} \prod_{i=1}^{n} \mathbb{1}_{[x_i \in D]} \right) \left(\frac{1}{m(\bar{D})^m} \prod_{i=n+1}^{n+m} \mathbb{1}_{[x_i \in \bar{D}]} \right)$$

$$= \left(\frac{1}{m(D)} \right)^n \left(\frac{1}{m(\bar{D})} \right)^m \mathbb{1}_{[H(x_1, \ldots, x_n) \subseteq D]} \, \mathbb{1}_{[J(x_{n+1}, \ldots, x_{n+m} | x_1, \ldots, x_n) \subseteq \bar{D}]}$$

where $J(x_{n+1}, \ldots, x_{n+m} | x_1, \ldots, x_n)$ is the "shadow" statistic defined in Hatchel, Meilijson and Nadas (1981). It is defined as:

$$J(x_{n+1}, \ldots, x_{n+m} | x_1, \ldots, x_n) = \bigcup_{i : y_i = 2} \{x_i + \lambda(x_i - b) \in \mathbb{R}^d | \lambda \geq 0, b \in H(x_1, \ldots, x_n)\}.$$

$(H(.), J(.|.))$ is a minimal sufficient statistic for the estimation of D. See Figure 2.

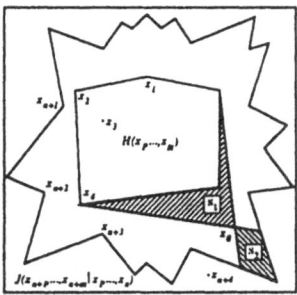

Figure 2.

It is known that $J(x_{n+1}, \ldots, x_{n+m} | x_1, \ldots, x_n)$ has similar properties as the convex hull statistic $H(x_1, \ldots, x_n)$. It is a consistent estimate of $\bar{D}$. It is robust with respect to small changes in the location of the data points. It satisfies the equivariance requirement and underestimates the Lebesgue measure of $\bar{D}$ in the same way as $m(H(x_1, \ldots, x_n))$ does for $m(D)$, ie

$$E[m(J(x_{n+1}, \ldots, x_{n+m} | x_1, \ldots, x_n))] = \left(1 - \frac{E[V_{m+1}]}{m+1} \right) m(\bar{D})$$

with $E[V_{m+1}]$ being the expected number of extreme points of $J(.|.)$ for $m+1$ observations in $\bar{D}$. See Ripley and Rasson (1977) and Rémon (1993).

The use of $H(x_1, ..., x_n)$ and $J(x_{n+1}, ..., x_{n+m}|x_1, ..., x_n)$ as estimates of the two unknown domains [here D and $\bar{D}$] in the criterion proposed by Baufays and Rasson is the key idea of our discriminant algorithm.

One gets then the following boundary of the regions allocating to D and $\bar{D}$. This is the set of points x_0 such that:

$$p_D \widehat{f_D}(x_0) = p_{\bar{D}} \widehat{f_{\bar{D}}}(x_0)$$

i.e.

$$\frac{1}{m(H(x_1, ..., x_n, x_0)) + m(J(x_{n+1}, ..., x_{n+m}|x_1, ..., x_n))}$$
$$= \frac{1}{m(H(x_1, ..., x_n)) + m(J(x_{n+1}, ..., x_{n+m}, x_0|x_1, ..., x_n))}$$

i.e.

$$S_1(x_0) = S_2(x_0)$$

where

$$S_1(x_0) \equiv m(H(x_1, ..., x_n, x_0)) - m(H(x_1, ..., x_n))$$

and

$$S_2(x_0) \equiv m(J(x_{n+1}, ..., x_{n+m}, x_0|x_1, ..., x_n)) - m(J(x_{n+1}, ..., x_{n+m}|x_1, ..., x_n)).$$

This boundary gives us a practical and easily computable estimate $\hat{D}$ for the unknown domain D. See Figure 3 for large data set results. The symmetric difference between D and $\hat{D}$ is noted by $D \triangle \hat{D} = D \cup \hat{D} \setminus D \cap \hat{D}$.

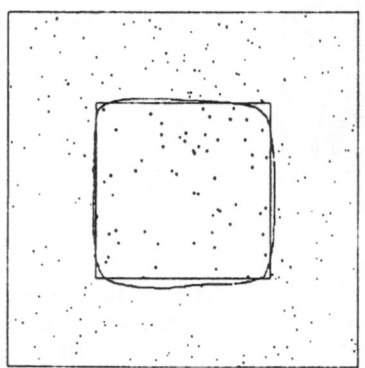 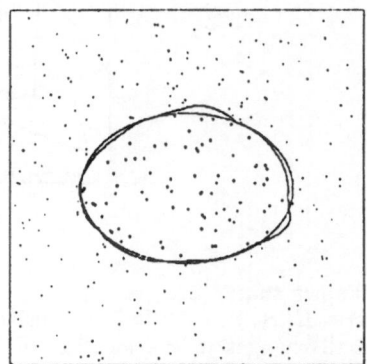

Figure 3a: Quadratic D with $m(D) = 0.25$: $t = 250$ observations with $n = 59$ from D yield $\hat{D}$ with $m(\hat{D}) = 0.26$ and $m(D \triangle \hat{D}) = 0.018$.

Figure 3b: Ellipsoidal D with $m(D) = 0.20$: $t = 300$ observations with $n = 68$ from D yield $\hat{D}$ with $m(\hat{D}) = 0.20$ and $m(D \triangle \hat{D}) = 0.011$.

3.3. Properties of $\hat{D}$

The estimator $\hat{D}$ yields a consistent estimation of D, as it is bound by two consistent estimators $H(.)$ and $J(.|.)$ of D. It has a piecewise continuous boundary. Unfortunately it happens not to be a convex set. On the other hand, this last feature turns out to be an advantage when a similar reasoning is applied to the estimation of non-convex domains.

This estimator $\hat{D}$ is robust with respect to small changes in the location of data points, as it is based only on the shapes of $H(.)$ and $J(.|.)$, which are robust in this sense. Such a property is rare in spatial statistics. For instance, the estimator $\tilde{D}$ proposed by Moore et al. (1988) can be very sensitive to small change in the location of data points.

Let us note that the time required for the computing of $\hat{D}$ does not depend on the number of points, excepted for the computation of the convex hull and shadow statistic. The amount of required cpu-time is only a function of the precision asked for the estimator $\hat{D}$.

3.4. Conclusions and future researches.

Our estimate for D, based on a well known discriminant analysis criterion, seems to be a powerful tool for pattern recognition. Moreover, it is quite straightforward to generalize it to a non-homogeneous Poisson process.

This research is currently working on recognition of non convex domains. The first results seems very encouraging as shown by the estimation of some letter A. See figure 4 where our algorithm is compared to a discriminant rule based on the distance to the nearest neighbour.

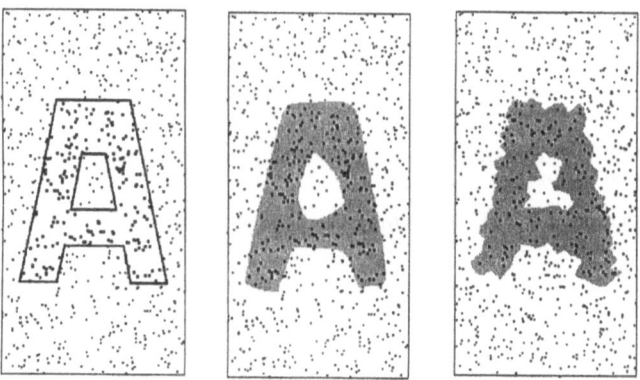

Figure 4a: A non-convex body to be estimated : the letter A.

Figure 4b: Our discriminant analysis algorithm.

Figure 4c: The estimation based on the distance to the nearest neighbour.

References :

Baufays, P., Rasson, J.-P. (1984): Une nouvelle règle de classement, utilisant l'enveloppe convexe et la mesure de Lebesgue, *Statistique et Analyse des Données*, **2**, pp. 31-47.

Baufays, P., Rasson, J.P. (1984): Propriétés théoriques et pratiques et applications d'une nouvelle règle de classement, *Statistique et Analyse des Données*, **vol.9/3**, pp.1-10.

Baufays, P., Rasson, J.P. (1985): A new geometric discriminant rule. *Computational Statistics Quaterly*, **vol. 2**, issue 1, 15-30.

Degytar, Y.U., Finkelsh Tein, M.Y. (1974): Classification Algorithms Based on Construction of Convex Hulls of Sets, *Engineering Cybermetics*,**12**, pp. 150-154.

Duda, R.O., Hart, P.E. (1973): *Pattern Recognition and Scene Analysis*, Wiley, Chichester.

Efron, B. (1965): The Convex Hull of a Random Set of Points. *Biometrika*,**52**, pp. 331-453.

Fisher, L., Van Ness, J.W. (1971): Admissible Clustering Procedures. *Biometrika*,**58**, pp. 91-104.

Fukunaga, K. (1972): *Introduction to Statistical Pattern Recognition*, Academic Press, New York.

Grenander, U. (1973): Statistical geometry: a tool for pattern analysis. *Bulletin of the American Mathematical Society*,vol. **79**, 829-856.

Hand, D.J. (1981): *Discrimination and Classification* ,Wiley, Chichester.

Hardy, A., Rasson, J.-P. (1982): Une Nouvelle Approche des Problèmes de Classification Automatique. *Statistique et Analyse des Données*, **7**, pp. 41-56.

Hartigan, J.A. (1975): *Clustering Algorithms* , Wiley, Chichester.

Mac Lachlan, G.J. (1992): *Discriminant Analysis and Statistical Pattern Recognition*, Wiley, New York.

Moore, M., Lemay, Y. and Archambault, S. (1988): Algorithms to reconstruct a convex set from sample points. Computing Science and Statistics, In: *Proceedings of the 20th Symposium on the Interface*, Eds. E.J. Wegman, D.T. Gantz and J.J. Miller, ASA, Virginia, 553-558.

Rasson, J.P. (1979): Estimation de domaines convexes du plan. *Statistique et Analyse des Données*,**1**, pp. 31-46.

Rémon, M. (1994): The estimation of a convex domain when inside and outside observations are available. *Supplemento ai Rendiconti del Circolo Matematico di Palermo*, **serie II, no 35**, 227-235.

Rémon, M. (1996): A Discriminant Analysis Algorithm for the Inside/Outside Problem. *Computational Statistics and Data Analysis*.

Ripley, B.D., Rasson, J.P. (1977): Finding the edge of a Poisson forest. *Journal of Applied Probability*,**14**, 483-491.

Toussaint, G.T. (1980): Pattern Recognition and Geometrical Complexity, In: *Proc. Fith Int. Conf. Pattern Recognition*, pp. 1324-1347, IEEE.

Part II

Methodologies in Classification

- Evaluation and Assessment Procedures
- Topics in Clustering and Classification

Part II.

Methodologies in Classification

Statistical Assessment, Clustering,
Topics in Clustering and Classification

How Many Clusters? An Investigation of Five Procedures for Detecting Nested Cluster Structure

A. D. Gordon

Mathematical Institute, University of St Andrews,
North Haugh, St Andrews KY16 9SS, Scotland

Summary: The paper addresses the problem of identifying relevant values for the number of clusters present in a data set. The problem has usually been tackled by searching for a best partition using so-called stopping rules. It is argued that it can be of interest to detect cluster structure at several different levels, and five stopping rules that performed well in a previous investigation are modified for this purpose. The rules are assessed by their performance in the analysis of simulated data sets which contain nested cluster structure.

1. Introduction

The aim of cluster analysis is to provide informative summaries of multivariate data sets, and in particular to investigate whether or not a set of n (say) objects can validly be described in terms of a smaller number of clusters of objects that have the property that objects in the same cluster are similar to one another and different from objects in other clusters. Clustering procedures provide little guidance on ways of addressing the problem of determining relevant values for the numbers of clusters, c (say), present in a data set. This has long been recognized as a challenging problem; overviews of the topic have been presented by Jain and Dubes (1988, Chapter 4), Bock (1996) and Gordon (1996).

This paper addresses the problem of determining which values of c are most strongly indicated as providing informative representations of the data. Published work to date has concentrated on identifying the single most appropriate value for c, and the test procedures and rules that have been proposed for addressing this problem are collectively usually referred to as 'stopping rules', since investigators have often obtained a complete hierarchical classification using an agglomerative algorithm, and wish to have guidance on when amalgamation should cease. Many different stopping rules have been proposed in the research literature, often with only cursory examination of their performance. The most detailed comparative study of which the author is aware was carried out by Milligan and Cooper (1985). These authors assessed the ability of thirty stopping rules to predict the correct number of clusters in a collection of different randomly-generated data sets after these had been analysed using four standard clustering criteria implemented in an agglomerative algorithm (single link, group average link, Ward's sum-of-squares criterion, and complete link). Some of the proposed stopping rules performed very poorly, and cannot be recommended for further use.

Specifying a single 'best' value for c will on occasion provide a misleading representation of the cluster structure present in data. The aim of the current study is to assess the ability of modifications of five stopping rules – those whose performance was best in Milligan and Cooper's (1985) study – to detect when *several different*, widely-separated values of c would be appropriate, that is, when structure is present in the data at several different levels. For example, it might be valid to summarize a data set in terms of two different, nested partitions: one into three clusters, and the other into twelve clusters.

The remainder of the paper is organized as follows. The stopping rules, and their modifications to allow more than one value of c to be indicated, are described in the next section. The rules are assessed by their performance in the analysis of randomly-generated data whose structure is known; the method in which the data are generated is described in the third section. The final section contains a summary of the results of analysing the data and some general comments.

2. Stopping Rules

Stopping rules can be categorized as *global* or *local*. Global rules are based on the complete data set, typically seeking the optimal value of some index that compares within-cluster and between-cluster variability. The partitions into clusters for two different values of c thus need not be hierarchically-nested, although in practice they usually will be. There is often not a natural definition of the within/between variability corresponding to the case $c = 1$, and such indices possess the unsatisfactory feature of being unable to indicate that the data comprise just a single cluster.

Local rules involve an assessment of whether or not a single cluster should be sub-divided into two sub-clusters (or a pair of clusters should be amalgamated). They are thus restricted to the assessment of hierarchically-nested sets of partitions, and are based on a subset of the data; this latter property means that the effective sample size for the test is usually much smaller than the size of the data set.

The five rules investigated in this study are defined below, in the order in which they were ranked in Milligan and Cooper's (1985) investigation.

1. CH. An index proposed by Caliński and Harabasz (1974), for assessing a partition into c clusters of a set of n objects described by numeric variables, is defined by

$$CH \equiv [B/(c-1)]/[W/(n-c)],$$

where W and B denote, respectively, the total within-cluster sum of squared distances (about the centroids), and the total between-cluster sum of squared distances.

2. DH. Duda and Hart (1973, Section 6.12) proposed a local rule for deciding if a cluster should be subdivided into two sub-clusters, based on comparing the within-cluster sum of squared distances (W_1) with the sum of within-cluster sum of squared distances when the cluster is optimally partitioned into two (W_2). The hypothesis of a single cluster is rejected if

$$W_2/W_1 < 1 - 2/(\pi p) - z[2(1 - 8/(\pi^2 p))/(mp)]^{1/2},$$

where p denotes the dimensionality of the data, m denotes the number of objects in the cluster being investigated, and z is a standard normal deviate specifying the significance level of the test. Amalgamation has generally proceeded until the hypothesis can first be rejected.

3. C. This index is based on the sum of all within-cluster pairwise dissimilarities (D). If the partition has r such dissimilarities, D_{min} (*resp.*, D_{max}) is defined as the sum of the r smallest (*resp.*, largest) pairwise dissimilarities, and

$$C \equiv (D - D_{min})/(D_{max} - D_{min}).$$

4. γ. This index, proposed by Goodman and Kruskal (1954), has been widely used for assessing cluster output (*e.g.*, Hubert, 1974). In the present instance, comparisons

are made between all within-cluster pairwise dissimilarities (d_{ij}, say) and all between-cluster pairwise dissimilarities (d_{kl}, say): a comparison is deemed concordant (*resp.*, discordant) if d_{ij} is strictly less (*resp.*, greater) than d_{kl}. The index is defined by

$$\gamma \equiv (S_+ - S_-)/(S_+ + S_-),$$

where S_+ (*resp.*, S_-) denotes the number of concordant (*resp.*, discordant) comparisons.

5. *Beale.* A test proposed by Beale (1969) has been used as a local stopping rule for assessing whether or not a single cluster should be sub-divided. The test involves comparing

$$\left(\frac{W_1 - W_2}{W_2}\right) \Big/ \left(\left(\frac{m-1}{m-2}\right) 2^{2/p} - 1\right)$$

(where W_1, W_2, m and p are defined in the DH test above) with an $F_{p,(m-2)p}$ distribution. As for the DH test, amalgamation proceeds until the hypothesis can first be rejected.

Three of these five rules are global, with the single most appropriate value for c being indicated by the maximum value of CH or γ, or the minimum value of C (with the restriction that not all values of c are investigated, as some indices can display distracting patterns for values of c close to n (Milligan and Cooper (1985)). Such global rules can readily be extended for use in the detection of nested cluster structure, by recording all *local* optima of the index. If the correct solution comprises partitions into $c_1, c_2, ..., c_k$ clusters, an ideal index would have local optima at all of, and only, these values of c. One might hope that the values taken by the index at these local optima were also the k most extreme values, but it is possible that this may not occur because of the inter-relatedness of structure at neighbouring values of c: thus, the value of the index may be more extreme at $(c_i + 1)$ clusters than at c_j clusters.

The local stopping rules proposed by Beale (1969) and Duda and Hart (1973) require the specification of a significance level α or threshold value z, and Milligan and Cooper (1985) selected values that ensured the best possible performance of these rules. Relevant threshold values depend on characteristics of the data sets under investigation, such as the values of n and p: for example, in the use of the DH stopping rule on their data, Milligan and Cooper (1985) chose z to be 3.2, whereas a similar examination for the data in the current study would specify z to be 4.0. The need to specify a threshold, whose most appropriate value varies in this way, is an unsatisfactory feature of a stopping rule.

The two local rules have been modified for use in the detection of nested cluster structure by abandoning their more formal hypothesis-testing aspects and just identifying large values of the corresponding z or F statistic. The critical values of the $F_{p,(m-2)p}$ distribution used in Beale's (1969) test depend on p and m, but for the values of p used in the current study, the variation is not large for even moderately small values of m. For neighbouring values of c, the z and F statistics will usually be evaluated on disjoint subsets of the data, and it could be argued that the relevant values of c are indicated by the k largest values of the statistics. However, it was found in this study that this approach generally provided inferior results to those indicated by the local maxima of the z and F statistics (in which neighbouring values of c cannot be indicated), and the results which are presented here are based on this latter strategy.

3. Parameters of Simulation Study

In each of the data sets that was generated, the number of objects, n, was equal to 60. Nested cluster structure was imposed at two levels: the objects were arranged in c_2 high-level classes $\{A_j \ (j = 1, ..., c_2) \}$, where $c_2 \in \{2, 3, 4\}$; each A_j was subdivided into h_j low-level classes $\{B_{js} \ (s = 1, ..., h_j) \}$, with $l_{js} \equiv card(B_{js})$ denoting the number of objects in B_{js}. Attention was restricted to cases in which the number of low-level classes $c_1 \equiv \sum_j h_j$ was equal to 12. The class-size distributions at each level were specified to be 'balanced' or 'unbalanced' as defined below. The high-level classes are said to be balanced if $h_j = 12/c_2 \ (j = 1, ..., c_2)$, and unbalanced if $(h_1, h_2) = (8, 4)$ for $c_2 = 2$; $(h_1, h_2, h_3) = (5, 4, 3)$ for $c_2 = 3$; and $(h_1, h_2, h_3, h_4) = (4, 3, 3, 2)$ for $c_2 = 4$. The low-level classes are said to be balanced if $l_{js} = 5$ for all j, s, and unbalanced if $l_{11} = 16$ and $l_{js} = 4$ for all other j, s.

The data consist of n p-dimensional normal random vectors, where $p = c_2 + 12$ and the p coordinate values are independent of each other and have the same variance σ^2. The classes are distinguished by their centres: the p-dimensional vector specifying the expected value of each object belonging to B_{js}, has only two non-zero elements, $b > 0$ in position j (characterizing A_j) and 1 in position $c_2 + \sum_{k<j} h_k + s$.

Three different pairs of values of (b, σ^2) were chosen, to allow varying degrees of spread of points within and between the classes. Let d_l denote the distance between a pair of objects belonging to the same low-level class, d_m denote the distance between a pair of objects belonging to the same high-level class but different low-level classes, and d_h denote the distance between a pair of objects belonging to different high-level classes. The aim is to have

$$prob(d_l > d_m) = \alpha = prob(d_m > d_h).$$

For each value of $c_2 \ (= 2, 3, 4)$, a large-scale simulation study established the values of b and σ^2 necessary to ensure that $\alpha = 0.00005, 0.005$, and 0.01.

For each combination of all the previous factor levels, three replications were carried out. Thus, the number of data sets generated was 3 (number of high-level classes) $\times$ 2 (low-level balance) $\times$ 2 (high-level balance) $\times$ 3 (amount of within-class variability) $\times$ 3 (number of replications) = 108. Each of these data sets was analysed using the same four clustering criteria employed by Milligan and Cooper (1985), to provide 432 analyses, each of which was assessed using the five rules described in the previous section.

4. Results and Discussion

The performances of the five rules are tabulated in Tables 1–5, in which attention has been restricted to recording the two values of c which provided the most extreme local optima of the corresponding index; values of c greater than 20 were not considered in this exercise, for the reasons given earlier. The results in the tables have been accumulated over the less variable and less interesting factors. Increasing the amount of within-cluster variability generally degraded the performance of the rules, a result also noted by Cooper and Milligan (1988). However, imposing a lack of balance in the cluster sizes did not have this effect, and on occasion led to an improvement in performance. The value of c_2 had some effect on the results, but this was greatly outweighed by differences due to the clustering criterion employed. These differences are not surprising, because each clustering criterion implicitly involves a model for the data and one should not expect them to be equally effective in detecting the clusters that have been generated. Each cell in Tables 1–5 thus contains five numbers: the four outer ones record the numbers of times that these numbers of clusters were indicated by the rule when applied to the dendrograms provided by, in clockwise order from top left: the single link (SL), group average link (AL), complete link (CL) and sum-of-squares (SSQ) criteria; the central number is the sum (Σ) of the four outer ones. When the two optimal values were equal, each of the relevant entries in a table was augmented by 0.5.

Table 1. Performance of the CH rule: the figures show the frequencies with which various values of c were indicated as the numbers of clusters, for 108 data sets analysed by four clustering criteria.

Optimal value for c	Second optimal value for c							
	c_2		12		11 or 13		other	
c_2	SL	AL	30	72	27	3	27	6
	Σ		257		34		35	
	SSQ	CL	78	77	1	3	1	1
12	23	27					0	0
	104						0	
	27	27					0	0
11 or 13	1	0					0	0
	1						0	
	0	0					0	0
other	0	0	0	0	0	0	0	0
	0		0		0		0	
	0	0	0	0	0	0	0	0

Tables 1–3 summarize the results for the three global rules; correct detections appear in one of the two cells $(12,c_2)$ or $(c_2,12)$. It should be noted that only 107 results are reported for the sum-of-squares criterion in Table 1: in one of the data sets for which $c_2 = 2$, the only local optimum occurred at $c = 2$ (i.e., no low-level clusters were indicated). Further, these accumulated results hide the fact that when $c_2 = 2$, the CH index never achieved its maximum value (but usually achieved its second-largest local maximum value) at 12 clusters; as c_2 increased, so did the frequency with which $c = 12$ was specified as the optimal solution.

Table 2. Performance of the C rule.

Optimal value for c	Second optimal value for c							
	c_2		12		11 or 13		other	
c_2			10	10	2	2	14	3
			45		9		19	
			11.5	13.5	3	2	1	1
12	32	39					2	10
	187						26	
	58.5	57.5					7	7
11 or 13	8	12					2	6
	35						19	
	7	8					5	6
other	28	12	1	2	1	3	8	9
	50		10		8		24	
	5	5	3	4	2	2	5	2

The results for the CH rule are highly encouraging: the two largest local maxima have indicated the correct values of c_1 and c_2 in more than 90 % of the cases for three of the clustering criteria; further, the numbers of times for which these are the only local maxima are 79 (group average link), 99 (sum-of-squares), and 98 (complete link), and no more than three local maxima are ever indicated for clusters provided by the latter two clustering criteria.

Table 3. Performance of the γ rule.

Optimal value for c	Second optimal value for c			
	c_2	12	11 or 13	other
c_2		9 11 46 12.5 13.5	3 2 12 4 3	15 6 24 1 2
12	33 41 185 54.5 56.5			1 7 23 9 6
11 or 13	9 10 33 7 7			1 6 16 4 5
other	28 11 51 6 6	1 5 15 4 5	1 2 6 2 1	7 7 21 4 3

The results for the C' and γ rules are summarized in Tables 2 and 3 respectively. For these indices, there is less variation in the results for different values of c_2. When one data set comprising 3 and 12 clusters was analysed using single link and group average link, both of the indices obtained their optimal value when $c = 3$, 12 and 13: these results were interpreted as a successful outcome, and the relevant numbers in the $(c_2,12)$ and $(12,c_2)$ cells were each augmented by 0.5.

Table 4. Performance of the DH rule.

Optimal value for c	Second optimal value for c			
	c_2	12	11 or 13	other
c_2		23 16 49 5 5	23 25 79 15 16	60 60 291 86 85
12	0 0 0 0 0			0 0 0 0 0
11 or 13	0 0 0 0 0			0 0 0 0 0
other	0 0 0 0 0	0 0 0 0 0	1 0 1 0 0	1 7 12 2 2

The performances of the C' and γ rules are very similar, each correctly specifying both values of c_1 and c_2 for just under two-thirds of the data sets analysed by the sum-of-squares and complete link clustering criteria; this proportion rises to nearly three-quarters when near misses (specifying $c_1 = 11$ or 13, instead of 12) are included. However, there is an increase in the mean number of local optima of the rules, and values of both c_1 and c_2 are incorrectly identified about 10 % of the time.

The results for the two local rules are summarized in Tables 4 and 5. It can be seen that the modification of Duda and Hart's (1973) rule has proved very effective in identifying the smaller number of clusters, but poor at detecting the 12 clusters that are present. The modification of Beale's (1969) rule has performed very poorly. Both of these rules have a tendency to provide a large number of local maxima. They would appear to have little to offer in this kind of investigation.

Table 5. Performance of the *Beale* rule.

Optimal value for c	Second optimal value for c			
	c_2	12	11 or 13	other
c_2		1 7 / 13 / 2 3	12 8 / 35 / 7 8	34 26 / 129 / 36 33
12	0 10 / 12 / 1 1			5 1 / 9 / 2 1
11 or 13	9 11 / 30 / 5 5			7 5 / 18 / 3 3
other	23 22 / 105 / 30 30	5 4 / 20 / 5 6	2 6 / 14 / 3 3	10 8 / 47 / 14 15

By contrast, the three global rules, and in particular the CH rule, would seem to have considerable potential for the detection of nested cluster structure. However, enthusiasm should be tempered by two observations. First, as in the Milligan and Cooper (1985) study, the cluster structure present in the simulated data sets was reasonably clear-cut. Secondly, in both the current investigation and that conducted by Milligan and Cooper (1985), the simulated clusters were generated using multivariate normal distributions (mildly truncated, in Milligan and Cooper's (1985) study). Several authors (*e.g.*, Scott and Symons, 1971) have noted reasons why the sum-of-squares criterion is particularly relevant for analysing such data, and one can speculate that a rule based on total within- and between-cluster sum-of-squares like the CH index might be better able to detect such clusters than clusters of other shapes. Nevertheless, further support has been provided for the rules that performed well in this study, and – until presented with evidence to the contrary – one can recommend their collective use to applied scientists seeking to understand the underlying cluster structure in their data.

References

Beale, E. M. L. (1969): Euclidean cluster analysis. *Bulletin of the International Statistical Institute*, **43(2)**, 92–94.

Bock, H. H. (1996): Probability models and hypotheses testing in partitioning cluster analysis. In *Clustering and Classification*, Arabie, P., Hubert, L. J. and De Soete, G. (eds.), 377–453, World Scientific, River Edge, NJ.

Caliński, T. and Harabasz, J. (1974): A dendrite method for cluster analysis. *Communications in Statistics*, **3**, 1–27.

Cooper, M. C. and Milligan, G. W. (1988): The effect of measurement error on determining the number of clusters in cluster analysis. In *Data, Expert Knowledge and Decisions*, Gaul, W. and Schader, M. (eds.), 319–328, Springer-Verlag, Berlin.

Duda, R. O. and Hart, P. E. (1973): *Pattern Classification and Scene Analysis*. Wiley, New York.

Goodman, L. A. and Kruskal, W. H. (1954): Measures of association for cross-classifications. *Journal of the American Statistical Association*, **49**, 732–764.

Gordon, A. D. (1996): Cluster validation. *Paper presented at IFCS-96 Conference, Kobe, 27-30 March, 1996*.

Hubert, L. (1974): Approximate evaluation techniques for the single-link and complete-link hierarchical clustering procedures. *Journal of the American Statistical Association*, **69**, 698–704.

Jain, A. K. and Dubes, R. C. (1988): *Algorithms for Clustering Data*. Prentice-Hall, Englewood Cliffs, NJ.

Milligan, G. W. and Cooper, M. C. (1985): An examination of procedures for determining the number of clusters in a data set. *Psychometrika*, **50**, 159–179.

Scott, A. J. and Symons, M. J. (1971): Clustering methods based on likelihood ratio criteria. *Biometrics*, **27**, 387–397.

Partitional Cluster Analysis with Genetic Algorithms:
Searching for the Number of Clusters[1]

J.A. Lozano, P. Larrañaga and M. Graña

Dept. of Computer Science and Artificial Intelligence
University of the Basque Country
P.O. Box 649, 20080 San Sebastián, Spain
e-mail: lozano@si.ehu.es
tel.: (+34 43) 218000, fax.: (+34 43) 219306

Summary: In this article we deal with the problem of searching for the number of clusters in partitional clustering in $\mathcal{R}^2$. We set up the problem as an optimization problem by giving a real function on the different partitions that is optimized when the number of clusters and the classes are the most natural. We use the Genetic Algorithm for optimizing this function. The algorithm has been applied to the well-known Ruspini data and to synthetically generated datasets, with different cluster numbers and underlying distributions. The results are encouraging.

1. Introduction

Cluster Analysis (Hartigan (1975); Everitt (1974); Jain and Dubes (1988)) is an important technique in the field of exploratory data analysis. It is a tool for grouping a set of objects into classes such that 'similar' ones are in the same class and 'different' ones in different classes. Cluster analysis explores the data known about the objects to be classified and tries to uncover the underlying structure without requiring the assumptions common to most classical statistical methods. Two main different types of clustering methods exist: hierarchical methods, which result in a nested sequence of partitions, and partitional methods, which give one single partition.

Genetic Algorithms (G.A's) (Goldberg (1989)) are probabilistic search algorithms which simulate natural evolution. They are based on the mechanics of natural selection and genetics. They combine 'survival of the fittest' among string structures with a structured yet randomized information exchange. In G.A.'s the search space of a problem is represented as a collection of individuals. The individuals are represented by character strings, which are referred to as *chromosomes*. The purpose is to find the individual from the search space with the best 'genetic material'. The quality of an individual is measured with an objective function. The part of the search space to be examined in each iteration is called the *population*. A G.A. works approximately as follows. First, the initial population is chosen at random, and the quality of each of its individuals is determined. Next, in every iteration parents are selected from the population. These parents produce children, which are added to the population. For all newly created individuals of the resulting population a probability near zero indicates that they 'mutate', i.e. they change their hereditary distinctions. The population is reduced to its initial size by removing some individuals from it according to some selection criterion. One iteration of the algorithm is referred to as a *generation*.

---

[1]This work is supported by the Diputación Foral de Gipuzkoa, under grant 95/1127 and by the Basque Government, under grant PI 94/78.

Some attempts to solve the clustering problem with G.A.'s have already been done. Krovi (1991) described the different aspects of designing a G.A. for cluster analysis and explained how a set of objects can be grouped into two clusters using binary strings. Bhuyan et al. (1991) developed a G.A. for the partitioning of n objects into k clusters, where $1 \leq k \leq n$ and k is given. They started by considering three different representations for their individuals but finally decided on the so-called *ordered* representation. Their preliminary experimental results reflected the superiority of the genetic algorithms over the known heuristic methods. Cucchiara (1993) showed the effectiveness of the G.A.'s in clustering problems in image analysis. Jones and Beltramo (1993) used integer encoding with the application of an operator used in the travelling salesman problem while Bezdek et al. (1994) used three different distances. Babu and Murty (1994) did not tackle the clustering problem with G.A.'s but with evolution strategies; another type of algorithm based on the principles of natural selection. In all of the research that is mentioned above the number of clusters into which to group the objects is supposed to be given.

Yet, some research has been carried out on the problem of the optimal number of clusters, using methods based on entropy-based statistical complexity criteria (Celeux and Soromeno (1993); Bozdogan (1994)) and statistical tests (Hardy (1994); Rasson and Kubushishi (1994); Gordon (1995)).

Our research on the use of G.A.'s for cluster analysis focuses upon the search for the optimal number of clusters. We want to develop an algorithm that automatically classifies the objects into an adequate number of clusters without this number being specified. The main problem in the development of such an algorithm is the definition of an evaluation function that makes it possible to compare the fitness of clustering consisting of a distinct number of clusters. Other difficult steps are the selection of a suitable clustering representation and the development of the operators that define the mutation and offspring production processes. An ongoing study about that using G.A.'s can be seen in Luchian et al. (1994). We have carried out experiments with five artificial data sets and with the well-known Ruspini data sets, in order to test our clustering method.

2. Setting up the problem as a combinatorial optimization problem

First of all, we set up the problem. A set $\mathcal{X} = \{x_1, x_2, \ldots, x_n\}$ of n two-dimensional real objects is given , the set $\mathcal{P}_k(\mathcal{X})$ denotes all partitions of the set $\mathcal{X}$ in nonempty k classes. We want to define a function:

$$F : \bigcup_{k=1}^{n} \mathcal{P}_k(\mathcal{X}) \longrightarrow \mathcal{R} \qquad (1)$$

such that the global optimum of the function F will be found in the number of clusters k and in the groups that are the most natural. It is important to note that the size of the search space $\bigcup_{k=1}^{n} \mathcal{P}_k(\mathcal{X})$, can be expressed by the following expression (Bhuyan et al. (1991)):

$$\sum_{k=1}^{n} \frac{1}{k!} \sum_{j=1}^{k} (-1)^{k-j} \binom{k}{j} j^n. \qquad (2)$$

In order to define this function we need to think about the characteristics that define a natural cluster. Hence, an important characteristic is that there are no big empty spaces inside a cluster (assuming that the cluster is not a ring), so we divide the space that is occupied by the objects into small squares, all of these squares having the same area. We check to see if the squares are empty or not. Each empty square is assigned a value of one, and a value of zero is assigned to a non-empty squares. With the former grid it is possible to assign a real value to every partition of the set $\mathcal{X}$ in each number of clusters. Given a partition $\{\mathcal{X}_1, \mathcal{X}_2, \ldots, \mathcal{X}_k\}$ the value given to it is the sum of the values given to each cluster (the algorithm now has the possibility of being a parallel algorithm). For a cluster we calculate its convex hull and then we sum the value of the squares whose centres are inside the convex hull. If we denote $H(\mathcal{X}_i)$ to be the convex hull of the partition $\mathcal{X}_i$, $V(x, y)$ the value assigned to a square with centre (x, y) and C the set of centres of squares, an initial aproximation to the function can be written as follows:

$$F^{\bullet}(\{\mathcal{X}_1, \mathcal{X}_2, \ldots, \mathcal{X}_k\}) = \sum_{i=1}^{k} \sum_{(x,y) \in C \cap H(\mathcal{X}_i)} V(x, y). \tag{3}$$

However this function is not capable of distinguishing between the optimal partition which gives zero to the former function, and the partition that can be constructed when splitting one of the clusters of the previous partition into two. Because of this, we need to add to the preceding function a value $\alpha \times k$, where α denotes a postive real number and k specifies the number of clusters. The final function is:

$$F(\{\mathcal{X}_1, \mathcal{X}_2, \ldots, \mathcal{X}_k\}) = F^{\bullet}(\{\mathcal{X}_1, \mathcal{X}_2, \ldots, \mathcal{X}_k\}) + \alpha \times k. \tag{4}$$

At first sight the value of α does not play an important role, and the only constraint is $\alpha < 1$. The reason is that for $\alpha \geq 1$ our function could assign a lower value to a partition that has an empty square inside rather than a partition with one cluster more and without empty squares inside which would be the natural partition. Later we will see that the value of α can be important in special cases.

Finally, there is another question that is left to answer: what is the size of the squares?. This is the key question in our approach. To calculate it we have used a simple approximation that works, as we will see later, well enough. As we do not have any information about the points, we are going to assume that the points have been generated at random, following a uniform distribution. Then if the natural structure is just one cluster and we want to discover it, we must not find an empty square in the convex hull formed for all the points. This is because it would be possible to split two or more clusters in such a way that the empty square would not be in any clusters and the optimum value of the objective function could be found in the partition in two or three clusters. Figure 1 shows that the square marked with an arrow has a value of 1 so the value of the objective function in one cluster is $1 + \alpha \times 1$ while the value of the function with two cluster is $0 + \alpha \times 2$. Hence we are going to choose the size of the square such that the probability of finding such a square will be quite small, in our case we take 0.001, i.e.

$$(1 - \frac{r^2}{S})^n \leq 0.001 \tag{5}$$

where r is the size of square, S is the area of the convex hull and n is the number of points. We have taken in each experiment the smallest r that complies with the constraint.

3. The Genetic Algorithm.

The kind of G.A. that we have used is the so-called Steady-State Genetic Algorithm

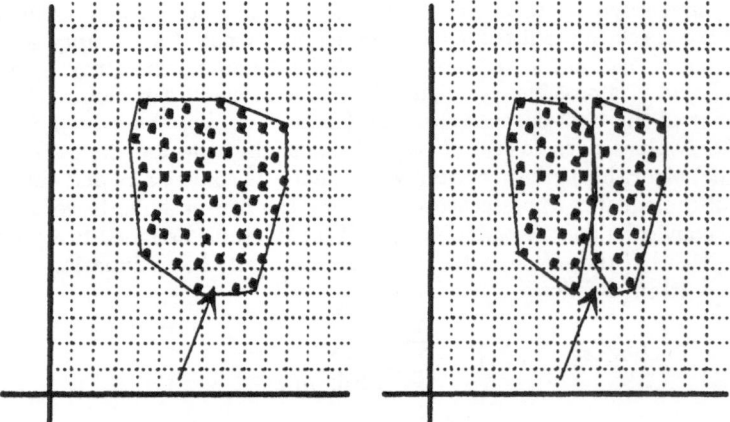

Fig. 1: Justification of the size of the square r.

(SSGA)(Whitley and Kauth (1988)). A pseudocode for this algorithm is:

begin SSGA

    Create initial population at random
    WHILE NOT stop DO

        BEGIN
        Select two parents from the population
        Let the selected parents *Produce a child*
        Mutate the child with certain probability
        Extend the population by assigning the child to it
        Reduce the extended population to the original size
        END

    Output the optimum of the population.

end SSGA

In this algorithm a population of size λ is maintained. In each step of the algorithm two individuals of the population are chosen with a certain probability proportional to their fitness function; a series of operators, usually crossover operators and mutation operators, are applied to produce a new individual. The fitness function is evaluated in the new individual and the individual is introduced into the population if its fitness function value is better than the worst (the individual with the worst fitness value in the population) individual in the population. This cycle continues until a stopping rule has critical value (convergence, number of iterations,...).

An important point about the algorithm is that the algorithm converges to the optimum , unlike the classical G.A. (Lozano et al. (1995)). Once the kind of algorithm we use is established, we need to define the parameters of the algorithm. The first point to define is the representation of the individuals. In the classic G.A. the individual used to be binary strings, however we are going to use strings of integers. An individual of the G.A. is a string of size n (number of points) which contains a permutation of the numbers $1, 2, \ldots, n$. We want an individual of the population to be a possible partition of n objects in k clusters, k being each value between 1 and n. Thus we decode each string of integers as follows: for each k the points represented by the first k integers of the permutation are taken as members and centres of a cluster and the next numbers are added in order to the cluster whose centre is nearest to the represented point. Once a point is added to a cluster the centre of this cluster changes to the centre of gravity of the points in the cluster. With this decoding we have every permutation of the n numbers representing a partition of the n points for every value of number of clusters k.

Of course we now have another problem, that is, which value to assign to a permutation of the numbers of objects. The former function assigns a value to each partition of n objects in k clusters, however every permutation represents a partition for every value of k. We solve this problem, of course, by giving to each individual the smallest value that takes the function in the different partitions for $k = 1, 2, \ldots, n$. Taking the former evaluation of each permutation into account, our algorithm can be seen as a hybrid G.A. where a local optimizer is applied in each evaluation.

The second point to note is the kind of operators that can be applied to the individuals (strings of integers) to reproduce them for getting new individuals. Hence we have studied the kind of operators that have been used in G.A.'s with permutation of number representation. Most of the work in this sort of representation has been directed to the design of genetic operator to solve the travelling salesman problem with a path representation. We have some experience in applying these operators to other research fields (Larrañaga et al. (1996)). The crossover operators that we have used are: CX (cycle crossover), ER (edge-recombination crossover), OX1 (order-crossover) and PMX (partially map crossover). As mutation operator we have used : SM (scramble mutation), SIM (insert mutation), ISM (simple inversion mutation), IVM (inverse mutation), EM (exchange mutation), DM (displace mutation).

Finally, there remain the parameters of the G.A., i.e., size of the population, probability of mutation, probability of crossover, and stopping criteria. These will be discussed in the next section.

4. The experimental results

We have carried out ten experiments with each pair of operators and each of the six datasets (see Figure 2), the Ruspini dataset and five other generated datasets. The features of the five generated data sets can be seen in Table 1.

data set	points	clusters	underlying dist
1	1000	1	uniform
2	400	1	uniform
3	800	3	uniform
4	1000	5	uniform
5	1000	5	normal

Tab. 1: The parameters of datasets.

Every experiment has been carried out with the GENITOR software (Whitley and Kauth (1988)) and in a SparcServer 630 MP. The parameters we used are a population size $\lambda = 20$, a mutation probability $p_m = 0.1$, and we used as a stopping rule the following criterion: if the mean function value of the population did not change in 20 generations we stopped the algorithm.

The previous parameters were set up when we carried out the experiments with the Ruspini and the first of the generated datasets that had a big influence on the results

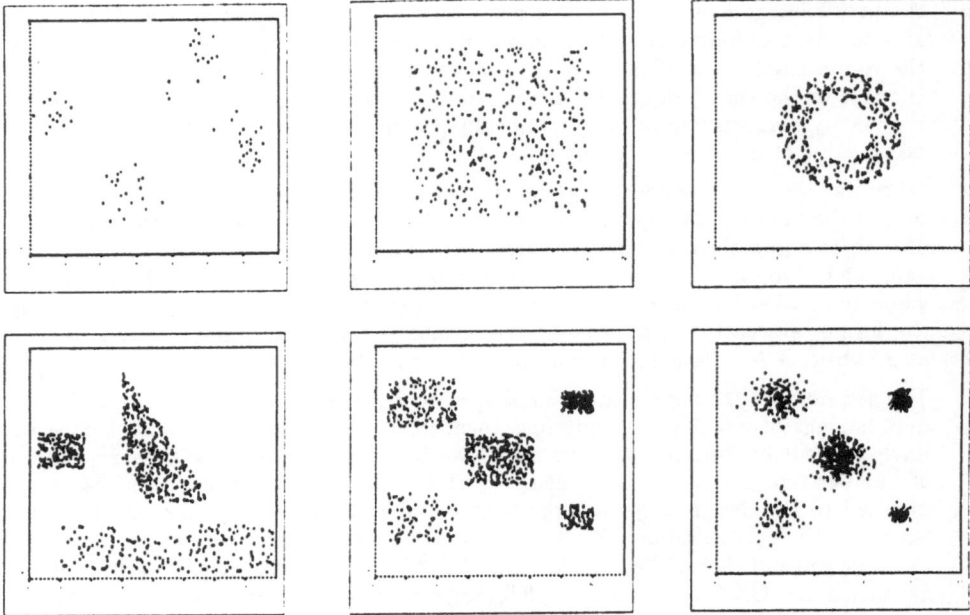

Fig. 2: The Datasets

obtained with the other datasets.

The search spaces are for the Ruspini data, dataset 1 and 2, dataset 3, and dataset 4 and 5: $1.178 \times 10^{80}, 2.755 \times 10^{393}, 2.755 \times 10^{793}, 2.755 \times 10^{993}$. It is important to keep the size of these spaces in mind to realize how quickly the G.A. finds the optimum. In the case of generated data sets we only search up to 10 clusters because in the other case the search would be impossible because of the huge space and the time to evaluate the function.

The number of clusters obtained in each evaluation, each operator and each dataset can be seen in Table 2. For instance, with the operator CX in the dataset 3, we reached 4 clusters in 4 evaluations, 5 in 56 evaluations and 6 clusters in 0 evaluations.

	Ruspini		Data 1	Data 2	Data 3			Data 4			Data 5				
Clus.	4	5	1	4	4	5	6	5	6	7	5	6	7	8	9
CX	60	0	60	60	4	56	0	59	1	0	0	56	4	0	0
ER	60	0	60	60	4	56	0	55	5	0	3	36	21	0	0
OX1	60	0	60	60	4	56	0	59	1	0	5	41	14	0	0
PMX	59	1	60	60	4	54	2	46	11	3	8	31	17	3	1
SM	40	0	40	40	2	37	1	36	3	1	4	30	5	0	1
SIM	39	1	40	40	5	34	1	36	4	0	0	27	13	0	0
ISM	40	0	40	40	0	40	0	35	4	1	1	31	8	0	0
IVM	40	0	40	40	5	35	0	36	4	0	3	27	10	0	0
EM	40	0	40	40	1	39	0	38	2	0	4	23	12	1	0
DM	40	0	40	40	3	45	0	38	1	1	4	26	8	2	0

Tab. 2: Results of the experiments.

A Kruskal-Wallis test has been applied to the results.

The following analysis has been made about the experiments:

Ruspini data. Every operator reaches the optimum without problems. It can seen that there is a statistically significant difference between the operators in relation to the number of evaluations, PMX is the fastest.

Dataset 1. Again it has been an easy problem for our approach. The operator PMX is the fastest. (84.46 function evaluations on average against the worse 154.5).

Dataset 2. It has proved a difficult problem for our approach, but there is a way of solving it. Choose a parameter r (size of the square) in such a way that only a empty square would be inside of the ring, and a parameter α bigger than 0.5. Of course, it is not a natural approach.

Dataset 3. The result is not very good, but this is a problem of the small parameters values used in the algorithm. In this problem the objective function gets the optimum in three clusters but the way in which we decode the permutation makes it difficult for the algorithm to find the optimum. However we have carried out other experiments with this dataset where the optimum was reached. If we take into account the operators, the PMX operator continues to be the fastest (96.15 function evaluations) and the slowest is ER (154.7).

Dataset 4. This is the first dataset for which we find some difference between the operator with respect to the objective function. The operator CX is the best, i.e., it finds the correct number of clusters more times than the other operators. In relation to speed PMX again is the fastest.

Dataset 5. The results are not very good because of our way of choosing r (5). Some more experiments with a small change in the parameter r allow our algorithm to reach the correct number of clusters in nearly every execution. Again the best operator in relation to the objective function is the CX and in relation to the number of function evaluation the PMX (119.7) is the best and the worst is the OX1 (227.1).

5. Conclusion and Future Work

We have given an algorithm that searches for the number of clusters in partitional cluster analysis and at the same time it gets the most natural classes. Moreover our algorithm is very flexible in the sense that it could be used to find only the classes given the number of clusters, or it could find the optimum number of clusters between given possible values. The results are encouraging but it is important to note the dependence of our algorithm on the parameter r. Some more experiments changing this parameter seem to be a good way to continue with this work. Of course other obvious steps in our future work could be to generalize our approach to objects in $\mathcal{R}^d$ spaces. This has a problem, that while the size of the search space does not change, the evaluation function is more expensive, the fundamental point, search the convex hull, is much more complicated in $\mathcal{R}^d$.

In addition we plan to apply G.A.'s to hierarchical clustering and to the most modern pyramidal clustering.

Acknowledgements

Thanks to Prof. A. Hardy for providing software and references.

References:

Babu, G.P. and Murty, M.N. (1994): Clustering with evolution strategies, *Pattern Recognition*, **27**, 2, 321-329.

Bezdek, J.C. et al. (1994): Genetic Algorithm Guided Clustering, In: *Proc. of The First IEEE Conference on Evolutionary Computation*, 34-40.

Bozdogan, H. (1994): Choosing the number of clusters, subset selection of variables, and outlier detection in the standard mixture-model cluster analysis, In: Diday E., Lechevallier Y., Schader M., Bertrand P. and Burtschy B. (eds.), *New Approaches in Classification and Data Analysis*, Springer-Verlag, 169-177.

Bhuyan, J.N. et al. (1991): Genetic Algorithms for clustering with an ordered representation, In: Belew and Booker (eds.), *Proceedings of the Fourth International Conference on Genetic Algorithms*, 408-415, Morgan Kaufmann.

Celeux, G. and Soromenho, G. (1993): An entropy criterion for assesing the number of clusters in a mixture model, Technical Report 1874 *INRIA*, France.

Cucchiara, R. (1993): Analysis and comparison of different genetic models for the clustering problem in image analysis, In: Albrecht R.F., Reeves C.R. and Steele N.C. (eds.), *Artificial Neural Networks and Genetic Algorithms*, Springer-Verlag, 423-427.

Everitt, B.S (1974): *Cluster Analysis*, John Wiley & Sons, Inc.

Goldberg, D.E. (1989): *Genetic Algorithms in Search, Optimization, Machine Learning*, Addison-Wesley.

Gordon, A.D. (1995): Test for asessing clusters, *Statistics in Transition*, **2**, 207-217.

Hardy, A. (1994): An examination of procedures for determining the number of clusters in a data set, In: Diday E., Lechevallier Y., Schader M., Bertrand P. and Burtschy B. (eds.), *New Approaches in Classification and Data Analysis*, Springer-Verlag, 178-185.

Hartigan, J.A. (1975): *Clustering Algorithms*, John Wiley & Sons, New York.

Jain, A.K. and Dubes, R.C. (1988): *Algorithms for Clustering Data*, Prentice Hall.

Jones, D.R. and Beltramo, M.A. (1993): Solving partitioning problems with Genetic Algorithms, In Albrecht R.F., Reeves C.R. and Steele N.C. (eds.). *Artificial Neural Networks and Genetic Algorithms*, Springer-Verlag, 423-427.

Krovi, R. (1991): Genetic Algorithms for clustering: A preliminary investigation, In: *Proceedings of the Twenty-Fifth International Conference on System Sciences*, 4, 540-544.

Larrañaga, P. et al. (1996): Learning Bayesian Networks Structures by Searching for the Best Ordering with Genetic Algorithms, *IEEE Transactions on Systems Man and Cybernetics*, **26**, 4. In press.

Lozano, J.A. et al. (1995): Genetic Algorithms: Bridging the Convergence Gap, submitted to *Evolutionary Computation*.

Luchian, S. et al. (1994): Evolutionary automated classification, In: *Proc. of The First IEEE Conference on Evolutionary Computation*, 585-589.

Rasson, J.P. and Kubushishi, T. (1994): The gap test: an optimal method for determining the number of natural classes in cluster analysis, In Diday E., Lechevallier Y., Schader M., Bertrand P. and Burtschy B. (eds.), *New Approaches in Classification and Data Analysis*, Springer-Verlag, 186-193.

Whitley, D. and Kauth, J. (1988): Genitor: A different Genetic Algorithm, In: *Proceedings of the Rocky Mountain Conference on Artificial Intelligence*, **2**, 189-214.

Explanatory Variables in Classifications and the Detection of the Optimum Number of Clusters

János Podani

Department of Plant Taxonomy and Ecology
Loránd Eötvös University, Ludovika tér 2
H-1083 Budapest, Hungary
Fax: +36 1 1338 764. Email: PODANI@LUDENS.ELTE.HU

Summary: An ordinal approach to the *a posteriori* evaluation of the explanatory power of variables in classifications is proposed. The contribution of each variable is assessed in a way fully compatible with the distance or dissimilarity function used in the clustering process. Then, a simple ranking-based measure is applied to express the relative agreement or disagreement of variables with a given partition. This measure treats all variables equally, no matter how influential they were when the classification was actually created. The sum of measures for all variables reflects their overall agreement and can be used to select an optimal partition from a hierarchical classification.

1. Introduction

An integral part of the interpretation of clustering results is to evaluate how the individual variables explain the classes. Finding an order of importance of variables for an existing classification is often called *a posteriori* **feature selection** (cf. Dale et al. 1986), as opposed to *a priori* feature selection, when the variables are ranked before the analysis starts (e.g., Orlóci 1973, Stephenson and Cook 1980), and to *forward* selection, in which evaluation of variables is part of the algorithm (e.g., Jancey and Wells 1987, Fowlkes et al. 1988).

Attention in *a posteriori* feature selection may be focused on two fundamental aspects of classification: cluster **cohesion** and **separation** (*sensu* Gordon 1981). The analysis can be restricted to either of these aspects (e.g., to contributions to within-cluster sum of squares only). Alternatively, the effect of variables on the distinction between clusters as well as on the internal "homogeneity" of clusters is simultaneously incorporated in the study, even if the clustering method did not actually consider both. A simple possibility which comes to the mind first is to compute for each variable the ratio of within-group and between-group sum of squares as an index of explanatory power.

It is emphasized, however, that there is no point in examining cluster cohesion and separation in terms of sum of squares when, say, the starting matrix contained chord distances or percentage dissimilarity values and the algorithm was single or complete linkage sorting. In other words, the evaluation procedure has to be **compatible** with the distance coefficient used in creating the classification. Since in many fields of science, e.g., in biological taxonomy, relatively few classifications are based on sum of squares or variance, and often the clustering models are not even Euclidean, a more generally applicable, yet flexible, criterion is required. Godehardt's (1990) multigraph approach, in which each variable is treated independently, seems to satisfy this requirement.

The third point emphasized here is that the importance of variables may be judged in two ways. The more obvious one is the measurement of the **absolute effect** of each variable upon the creation of clusters. For example, in case of Euclidean distance and centroid clustering, we can examine how far apart the cluster centroids are for each variable, and then order the variables on this basis. This ordering will emphasize variables that dominated the classification process, and may neglect others that are equally if not more interesting for the *a posteriori* interpretation of clusters. In fact, any variable supporting the given partition may prove useful in subsequent descriptions, no matter how small this support is in absolute terms. Thus, an alternative procedure free from the implicit variable weighting, i.e., measurement of the **relative importance** of the variable may prove useful. Lance and Williams (1977) are early proponents of this approach, by suggesting taking the ratio of between-cluster sum of squares to the total for each continuous variable or to compute Cramer's index (see also Anderberg 1973) for each binary or multistate variable. These criteria are thus only data-type-dependent and do not consider the manner in which the dissimilarities were calculated. The procedure described in this paper releases implicit weighting by introducing an **ordinal measure of the explanatory power** of variables in non-hierarchical classifications. This measure also satisfies the requirement of being compatible with the dissimilarities used and relies equally on both cluster separation and cohesion.

I will also provide an alternative approach to the familiar problem of detecting the optimum number of clusters. The sum of measures of explanatory power for all variables will be defined as an **overall measure of the agreement** (in a sense: consensus) among the variables regarding the partition of m objects into t clusters. Plotting the sum over a reasonable range of t values provides a graphical means to find the optimum, if any. This approach, as will be seen, is radically different from most of the methods reviewed and compared by Milligan and Cooper (1985).

2. Variable contributions

The procedure starts with evaluating the contribution of each variable to the distances or dissimilarities between objects. To ensure compatibility, the determination of this contribution must be specific to the distance or dissimilarity function used. As an example, the total contribution of variable i to all the $z=m(m-1)/2$ values in the lower semimatrix of $\mathbf{D}^2$ containing the squared Euclidean distances for m objects is computed as

$$\Phi_i = \sum_{j=1}^{m-1} \sum_{k=j+1}^{m} g_{ijk} \, ,$$

where $g_{ijk} = (x_{ij} - x_{ik})^2$ is the contribution of variable i to d^2_{jk} and is written as an element of matrix $\mathbf{G}_i$. The contributions are strictly additive, the matrix of squared distances is therefore reproduced as

$$\mathbf{D}^2 = \sum_{i=1}^{n} \mathbf{G}_i \, ,$$

with n as the number of variables. Formulae for computing contributions have been derived and are presented without proofs for 15 other distance and dissimilarity measures (Table 1). The measures themselves are not shown here, because most of them are well-known from the clustering literature. A full list is found in Podani (1994), although the

Tab. 1: Contribution of variable i to d_{jk} for several, well-known dissimilarity and distance functions. For presence/absence coefficients we assume that $x_{ij}=1$ for presence) and $x_{ij}=0$ for absence. Contributions are ranked in ascending order for most measures, except for those marked with an *, for which ranking is the reverse (see text).

Euclidean distance $\qquad\qquad\qquad (x_{ij} - x_{ik})^2$

Manhattan distance, 1-simple match. coeff. $\qquad | x_{ij} - x_{ik} |$

Penrose SIZE $\qquad\qquad\qquad\qquad x_{ij} - x_{ik}$

Chord distance * $\qquad\qquad \dfrac{x_{ij}\, x_{ik}}{\sqrt{\sum\limits_{h=1}^{n} x_{hj}^2 \sum\limits_{h=1}^{n} x_{hk}^2}}$

Canberra metric $\qquad\qquad \dfrac{|\, x_{ij} - x_{ik}\, |}{|x_{ij}| + |x_{ik}|}$

Percentage difference, 1-Sorensen $\qquad \dfrac{|\, x_{ij} - x_{ik}\, |}{\sum\limits_{h=1}^{n} x_{hj} + x_{hk}}$

1-Ruzicka, 1-Jaccard $\qquad\qquad \dfrac{|\, x_{ij} - x_{ik}\, |}{\sum\limits_{h=1}^{n} \max[x_{hj},\, x_{hk}]}$

1-Similarity ratio * $\qquad\qquad \dfrac{x_{ij}\, x_{ik}}{\sum\limits_{h} x_{hj}^2 + \sum\limits_{h} x_{hk}^2 + \sum\limits_{h} x_{hj} x_{hk}}$

1-Russell - Rao $\qquad\qquad (1 - x_{ij} x_{ik})/n$

1-Rogers - Tanimoto $\qquad\qquad \dfrac{2\,|\, x_{ij} - x_{ik}\, |}{n + \sum\limits_{h=1}^{n} 2\,|x_{hj} - x_{hk}|}$

1-Sokal - Sneath $\qquad\qquad \dfrac{2\,|\, x_{ij} - x_{ik}\, |}{\sum\limits_{h=1}^{n} \max[x_{hj}, x_{hk}] + |x_{hj} - x_{hk}|}$

1-Anderberg 1 * $\qquad \dfrac{x_{ij} x_{ik}}{\sum\limits_{h} x_{hj}} \cdot \dfrac{x_{ij} x_{ik}}{\sum\limits_{h} x_{hk}} \cdot \dfrac{(1-x_{ij})(1-x_{ik})}{n - \sum\limits_{h} x_{hj}} \cdot \dfrac{(1-x_{ij})(1-x_{ik})}{n - \sum\limits_{h} x_{hk}}$

1-Kulczynski * $\qquad \dfrac{\min[x_{ij}, x_{ik}]}{\sum\limits_{h} x_{hj}} + \dfrac{\min[x_{ij}, x_{ik}]}{\sum\limits_{h} x_{hk}}$

reader may also consult Anderberg (1973), Sneath and Sokal (1973) and Orlóci (1978). Note that some indices known generally as similarity functions are expressed as complements.

3. A new measure of explanatory power

The second part of the analysis involves determining the the **rank order** of the g_{ijk} scores for each variable i for a given partition P_t of m objects into t clusters, each with m_s objects, $s=1,...,t$. For most coefficients of distance, e.g., Manhattan and Euclidean distances, it is reasonable to state that a variable completely explains P_t if **all of its within-cluster contributions are smaller than the between-cluster contributions**, and have therefore the smallest ranks. (For functions with an asterisk in Table 1 the situation is the reverse, however. In these cases, ranking is done in descending order to keep the generality of the statement that within-cluster contributions should be ranked first in the optimal case. In the sequel, we assume the previous type to simplify discussion.) Consequently, we have the **minimum sum of ranks of within-cluster contributions** of any variable, denoted by $R_{(t)min}$. This quantity is obtained as

$$R_{(t)min} = (q^2+q)/2 \quad \text{where} \quad q= \sum_{s=1}^{t} \binom{m_s}{2}.$$

Let, further, $R_{(i,t)obs}$ be an **observed sum of ranks of within-cluster contributions** for variable i. Clearly, $R_{(t)min} \leq R_{(i,t)obs}$. Ties in the rank order can be resolved randomly which has negligible effects for large values of z. The sum of ranks for between-cluster contributions will not be used, because it conveys no extra information.

Let $R_{(t)exp}$ denote the **random expectation** for the null situation, i.e., when the variable does not make distinction as to whether a contribution is within- or between clusters (indifferent variable). In other words, contributions are arranged at random with the expected sum of within-cluster ranks given by

$$R_{(t)exp} = p \, (z^2+z)/2$$

where $p=q/z$ is the probability that a randomly chosen value in the rank order is a within-cluster contribution. The expectation will be used below as a reference basis for constructing the formula.

Then, the **explanatory power** of the variable is defined as the complement of the deviation of the actual sum of ranks from the minimum as divided by the deviation of the expectation from the minimum:

$$r(i,t) = 1.0 - \frac{R(i,t)obs - R(t)min}{R(t)exp - R(t)min}$$

$r_{(i,t)}$ values close to 1.0 indicate high explanatory power, values around zero reflect indifference, whereas negative values correspond to a situation when variable i is contradictory with P_t. (I deliberately avoid using the term discriminatory power, because it is usually coined with variance-related concepts as in discriminant analysis.) The variables may be ordered based on their r scores, to facilitate interpretation of clusters and to detect variables which happen to be indifferent or even contradictory with the given partition.

4. Detecting the optimum number of clusters

The coefficient of explanatory power can be used in turn to determine the optimum number of clusters in hierarchical classifications. The intuitive basis for this is that the more variables support a given partition the more acceptable it is. The criterion to be used is defined as the sum of coefficients of explanatory power,

$$\sigma_t = \sum_{i=1}^{n} r(i,t)$$

which will be called the **coefficent of cluster separation**. The upper bound of this coefficient is n, reached in the unanimous situation with all variables fully explaining the partition. The σ_t coefficient is computed for each level of interest in the hierarchy and the results are plotted against t. For data sets with group structure, the curve shows a peaked effect allowing to detect the number of clusters at which the **majority of variables support the same clustering** in terms of their ranked within-cluster contribution scores. For very many clusters, each with 1-2 objects only, the increase of σ_t is a necessity, but such trivial clusters attract no interest anyway (these clusters are usually excluded from such studies, cf. Milligan and Cooper 1985). Absence of clear-cut peaks is indicative of either strong disagreements between variables as to the "optimum" value of t, or complete lack of group structure, so the $r_{(i,t)}$ values must be inspected.

Computer program SYN-TAX 5.02 (Podani 1994) designed for classification purposes includes an option for computing the explanatory power of variables and the coefficient of cluster separation at several levels in a dendrogram and for plotting the graph automatically (available for PCs and Macintosh computers).

5. Example

The method will be demonstrated by an actual example coming from community ecology. A total of 80 vegetational plots (objects) represent a sample of dolomite grasslands of Sas-hill, Budapest, Hungary (for more details, see Podani 1985). The plots have been described in terms of percentage cover scores of 123 vascular plant species. For the purpose of illustration, the matrix of Euclidean distances of objects was subjected to complete linkage clustering (Fig. 1). The explanatory power of variables and the coefficient of cluster separation were computed for the top ten cut levels in the dendrogram, i.e., for $t=2$ to 11. The plot of cluster separation against the number of clusters (Fig. 2) indicates high agreement of variables for two and three clusters, with the maximum at $t=3$. When t is raised from 3 to 4, the coefficient drops by more than 50% and for more groups it remains about the same. The analysis thus suggests that the given classification is best supported by the species at the 3-cluster level. It is therefore worthwhile to examine the rank order of variables based on their explanatory power values for $t=3$ (Tab. 2). To save space, the table lists only the first ten and the last ten species from the rank order. Those at the beginning of the list are the best indicators of difference between closed (the two smaller groups) and open grasslands (the large group), whereas species with negative scores counter-support this classification because they tend to differentiate the large group even further. These species have been widely used as discriminatory species to subdivide the relatively open communities. The analysis revealed, however, that the majority of species are contradictory with this, showing that the classical syntaxonomic classification was subjectively based on a narrow subset of species.

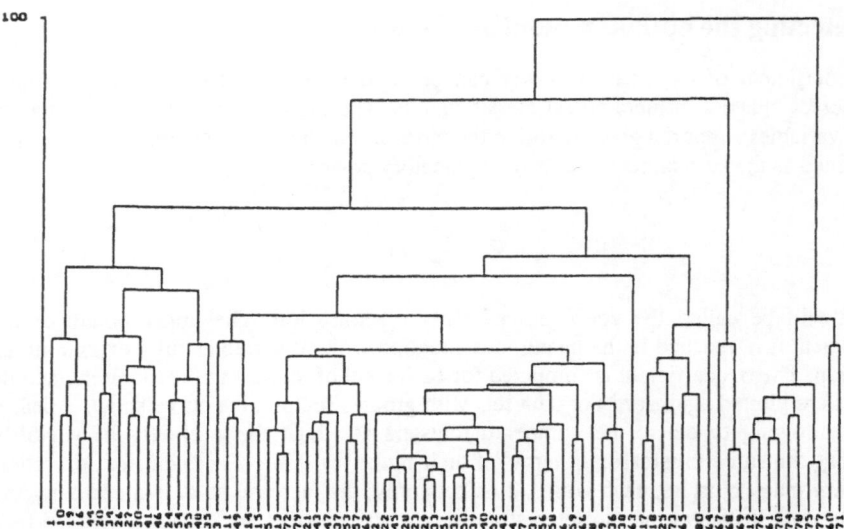

Fig. 1: Dendrogram showing complete linkage clustering of 80 vegetational plots, based on the Euclidean distances among objects using percentage cover scores of 123 species..

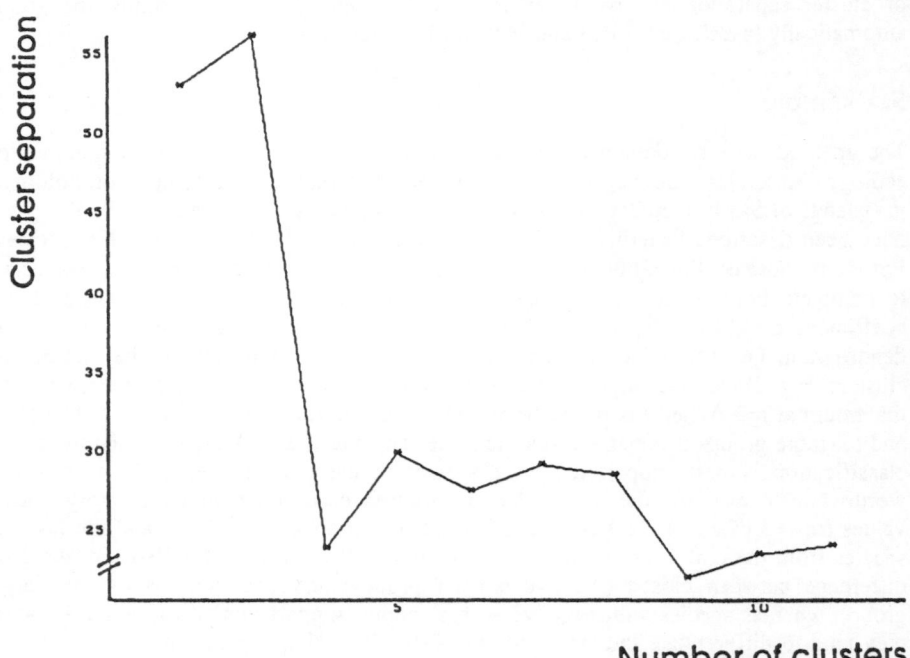

Fig. 2: Plot showing the relationship between the number of clusters and the coefficient of cluster separation for the top ten partitions obtained from the dendrogram of Fig. 1.

Tab. 2: The first ten and the last ten species in the rank order of variables for the three-cluster partition obtained from the dendrogram in Fig. 1.

Rank	Species	r_i	Rank	Species	r_i
1	Cytisus hirsutus	.865	114	Dianthus serotinus	.057
2	Festuca sulcata	.835	115	Andropogon ischaemum	.050
3	Bupleurum falcatum	.803	116	Helianthemum canum	.047
4	Pimpinella saxifraga	.771	117	Thymus praecox	.024
5	Asyneuma canescens	.763	118	Sanguisorba minor	.005
6	Veronica spicata	.758	119	Stipa eriocaulis	-.004
7	Polygonatum odoratum	.757	120	Festuca pallens	-.051
8	Campanula sibirica	.725	121	Carex liparocarpos	-.053
9	Carlina intermedia	.719	122	Chrysopogon gryllus	-.162
10	Adonis vernalis	.717	123	Seseli leucospermum	-.177

6. Discussion

The measure of explanatory power proposed in this paper is a non-metric criterion because actual differences are irrelevant: it is the rank order of contributions that matters. Thus, even if a variable had negligible effects on the distances (because of lack of commensurability, for example), it may turn out to be a good explanatory variable afterwards. Also, ranking variables based on the r values reveals an aspect rarely emphasized: the identification of variables that do not agree with the partition. Finding these variables may lead to revisions of former classifications. The possibility is also raised here that after the removal of these variables a repeated classification based on the reduced set of variables may provide a more noise-free classification. This is certainly an aspect which merits future investigations.

The coefficient of cluster separation, being based on ranked contributions, is considerably different from the currently known indices of optimum number of clusters as reviewed by Milligan and Cooper (1985). In addition to the ranking technique, the most substantial difference is that whereas the other methods are less dependent on the number of variables (so that they can be best demonstrated with a two-dimensional example), the present technique is more meaningful when there are quite a few variables. Therefore, it may perform very poorly in a low dimensional situation if compared to the other methods, and evaluation of the method proposed here along the lines of Milligan and Cooper's study would be irrelevant. The only exception seems to be the Ratkowsky and Lance (1978) criterion, which involves computation of the ratio used by Lance and Williams (1977) for each variable, and takes the average over variables. It is perhaps an explanation for the fairly poor performance of this measure in the two-dimensional case of Milligan and Cooper's study, though Ratkowsky and Lance reported high success, usually with many dimensions. It is also noted here that whereas almost all methods provide the same result after rigid rotation of the axes (rotation invariance) the Ratkowsky and Lance criterion and the one suggested in this paper are exceptions.

As with other formulae for detecting the optimum number of clusters in hierarchical classifications, the possibility to incorporate the measure directly as a clustering criterion may be examined. The coefficient of cluster separation is computationally very demanding, however. (The actual example presented in this paper took ten hours on a PC 486.) Building clusters based on a global nonmetric criterion similar to σ_f will provide a clustering procedure completely compatible with the optimality measure.

Acknowledgements:

The author expresses his sincerest thanks for receiving an OMFB Travel Grant (No. MEC 96-0176) and an OTKA Travel Grant. (No. U21456) to participate at IFCS'96, Kobe, where this contribution was presented. This study was funded by the OTKA Hungarian National Research Grant No. T19364. I am grateful to A. D. Gordon (University of St. Andrews, U.K.) for his comments on the manuscript, and to M. B. Dale (CSIRO, Australia) and Sz. Bokros (ELTE, Budapest) for discussions.

References:

Anderberg, M. R. (1973): *Cluster Analysis for Applications*. Academic, New York.

Dale, M. B., Beatrice, M., Venanzoni, R. and Ferrari, C. (1986): A comparison of some methods of selecting species in vegetation analysis. *Coenoses*, **1**, 35-52.

Fowlkes, E. B., Gnanadesikan, R. and Kettenring, J. R. (1988): Variable selection in clustering. *Journal of Classification*, **5**, 205-228.

Godehardt, E. (1990): *Graphs as Structural Models: The Application of Graphs and Multigraphs in Cluster Analysis* (2nd ed.). Vieweg & Sohn, Braunschweig.

Gordon, A. D. (1981): *Classification: Methods for the Exploratory Analysis of Multivariate Data*. Chapman and Hall, London.

Jancey, R. C. and Wells, T. C. (1987): Locality theory: the phenomenon and its significance. *Coenoses*, **2**, 31-37.

Lance, G. N. and Williams, W. T. (1977): Attribute contributions to a classification. *Australian Computer Journal*, **9**, 128-129.

Milligan, G. W. and Cooper, M. C. (1985): An examination of procedures for determining the number of clusters in a data set. *Psychometrika*, **50**, 159-179.

Orlóci, L. (1973): Ranking characters by a dispersion criterion. *Nature*, **244**, 371-373.

Orlóci, L. (1978): *Multivariate Analysis in Vegetation Research*. Junk, The Hague.

Podani, J. (1985): Syntaxonomic congruence in a small-scale vegetation survey. *Abstracta Botanica*, **9**, 99-128.

Podani, J. (1994): *Multivariate Data Analysis in Ecology and Systematics*. SPB Publishing, The Hague.

Ratkowsky, D. A. and Lance, G. N. (1978): A criterion for determining the number of groups in a classification. *Australian Computer Journal*, **10**, 115-117.

Sneath, P.H.A. and Sokal, R. R. (1973): *Numerical Taxonomy*. Freeman, San Francisco.

Stephenson, W. and Cook, S. D. (1980): Elimination of species before cluster analysis. *Australian Journal of Ecology*, **5**, 263-273.

Random dendrograms for classifiability testing

Bernard Van Cutsem, Bernard Ycart

Laboratoire de Modélisation et Calcul, Université Joseph Fourier, Grenoble
B.P. 53, F-38041 GRENOBLE Cedex 9,
Bernard.Van-Cutsem@imag.fr, Bernard.Ycart@imag.fr

Summary: We propose statistical tests to decide between an hypothesis of non classifiability of the data against the presence of a classification in some simple situations. We consider only the single link algorithm and two null hypotheses of non classifiability, according to whether the dissimilarities or the objects themselves are i.i.d. random variables. Each choice for the distribution of the input of the single link algorithm induces a different probability distribution on the output, which is a random indexed dendrogram. Certain characteristics of these random indexed dendrograms are studied, and their asymptotic distributions computed under each null hypothesis. All these random variables can be used to define statistical tests. Explicit examples of such tests are provided.

1. Introduction

Mathematical structures such as partitions, trees or dendrograms are commonly used to analyze and represent classification structures on n objects. Classical algorithms construct these structures either from dissimilarities between pairs of objects or from the values of a set of variables on the n objects.

We consider here only hierarchical classifications also called indexed dendrograms (ID), as introduced by Hartigan (1967), Johnson (1967), Jardine, Jardine and Sibson (1967). In practical situations, ID's are deduced from dissimilarities using ascending hierarchical algorithms. Descriptions of such algorithms can be found in classical books on classification such as Sneath and Sokal (1973) or Jain and Dubes (1988). We focus here on the Single Link Algorithm (SLA) which is among the most commonly used. The ID's obtained using the SLA will be called single link indexed dendrograms (SLID).

Even if the data do not present any cluster (they are homogeneous in some sense), the SLA will produce an ID. In some instances, external information on the data can be used to confirm or not the exhibited structure. If this is not possible, it is nevertheless important to be able to decide if this ID is significant or not. A method is to consider a probability distribution on the set of data which corresponds to the absence of classification structure (no clusters), and then derive the distribution of the corresponding random SLID. Then statistical tests for this null hypothesis of non classifiability of data against an hypothesis of existence of a classification structure can be constructed. Bock (1996) is a thorough review on the problem of testing partitions as structures of classification.

In Van Cutsem and Ycart (1994), we made a first attempt in this direction by describing the finite set of stratified dendrograms on n objects. This set was endowed with the equiprobability and the distributions of some characteristics were derived. We considered mainly the number of levels of such dendrograms, and the sizes of partitions. Those results can be easily extended to binary and strictly binary stratified dendrograms (see exact definitions below).

133

More realistic hypotheses bear on non classifiable data. In Van Cutsem and Ycart (1996a), we considered two other types of null hypotheses corresponding to non classifiable data.

- **Model 1.** The first hypothesis concerns dissimilarities. It supposes that the dissimilarities between objects are exchangeable random variables, and more particularly i.i.d. random variables uniformly distributed on $[0,1]$.

- **Model 2.** The second hypothesis concerns objects. It supposes that the n objects are i.i.d. points in a metric space distributed according either to a uniform distribution on a convenient domain or to a unimodal distribution. Dissimilarities between objects are then defined as their distances in the representation space.

Model 1 has been considered long ago, for instance by Ling (1973). In Van Cutsem and Ycart (1996b), a probabilistic analysis of the main ascending hierarchical algorithms was proposed. Different variables attached to indexed dendrograms were introduced, including the *sequence of levels*, the *survival time* of an object (i.e. the number of partitions in which this object is isolated) and the *ultrametric distance* between two objects. The distributions of these random variables attached to ID's produced by the SLA, (and also the average link and complete link algorithms) applied to random data corresponding to model 1 were studied, and exact as well as asymptotic results were proposed.

Our goal in the present paper is twofold. Firstly we want to extend the results obtained in Van Cutsem and Ycart (1996b) to a particular case of model 2, more precisely to i.i.d. objects uniformly distributed in the interval $[0,1]$. The main results for model 1 are recalled in theorem 3.2, where explicit asymptotic distributions for large sets of objects are given. The corresponding results for model 2 are given in theorem 4.3. Interestingly enough, the asymptotic distribution turns out to be the same (scaled Gumbel distribution) for the last index of the dendrogram. However, it is quite different for other variables. For instance the size of the smallest subset in the penultimate partition tends to 1 in probability for i.i.d. dissimilarities (model 1), whereas for i.i.d. points on $[0,1]$, it has a uniform distribution.

Our second aim is to demonstrate the applicability of our results to some problems of classifiability testing. To do this, we introduce simple classifiability hypotheses to be tested against models 1 and 2. They consist in assuming that the set of objects is partitionned into two subsets, such that the distribution of dissimilarities inside each set is stochastically smaller than that of dissimilarities between the two subsets. For both models, we derive explicit tests based on the last index of the SLID and compute the probability of detecting the given partition under the classifiability hypothesis.

The article is organized as follows. Section 2 recalls basic definitions. We summarize in section 3 the results of Van Cutsem and Ycart (1996b) concerning model 1. Section 4 contains new results for i.i.d. points uniformly distributed on $[0,1]$. Applications to classifiability testing are presented in section 5.

2. Basic definitions

A dissimilarity on a set S of n objects is as usual a function $d\colon S^2 \to \mathbb{R}^+$ such that $d(a,a) = 0$, $d(a,b) = d(b,a)$ for any pair of objects a and b in S. Moreover we suppose here that d is definite, that is: $(d(a,b) = 0) \Rightarrow (a = b)$.

An indexed dendrogram is a sequence $\{(P_\ell, \lambda_\ell)\}_{0 \leq \ell \leq \ell_{max}}$ such that

1) $\{P_\ell\}_{0 \leq \ell \leq \ell_{max}}$ is a sequence of nested partitions of S such that P_0 is the partition into singletons and $P_{\ell_{max}}$ is the partition with only one element $\{S\}$. The dendrogram is defined by $\{P_\ell\}_{0 \leq \ell \leq \ell_{max}}$.

2) $\{\lambda_\ell\}_{0 \leq \ell \leq \ell_{max}}$, the sequence of indices of the dendrogram, is strictly increasing and starts from $\lambda_0 = 0$.

A sequence $\{P_\ell\}_{0 \leq \ell}$ of partitions is nested if any subset of a partition P_ℓ is union of subsets of $P_{\ell-1}$. An ID is called strictly binary if any partition P_ℓ is obtained by joining exactly two subsets of $P_{\ell-1}$. For a strictly binary dendrogram, $\ell_{max} = n - 1$. An ID is a stratified dendrogram (SD) if, for any $\ell \in \{0, \ldots, \ell_{max}\}$, $\lambda_\ell = \ell$.

If A and B are two disjoint subsets of S, we define

$$\delta(A, B) = \min\{d(a,b) : a \in A, b \in B\}.$$

The function δ defines a dissimilarity on pairwise disjoint subsets of S. We suppose in the sequel that all dissimilarities, either between objects or subsets, are distinct. The SLA constructs an ID from a dissimilarity d by the iterative procedure which starts from $(P_0, 0)$, and for each $(P_{\ell-1}, \lambda_{\ell-1})$,

1) computes $\delta_{min} = \min\{\delta(A, B) : A \neq B \in P_{l-1}\}$.

2) defines the partition P_ℓ by grouping together the pair of elements of $P_{\ell-1}$ whose dissimilarity is δ_{min}.

3) sets $\lambda_\ell = \delta_{min}$.

The algorithm stops when $P_\ell = \{S\}$.

The associated SLID is then a strictly binary indexed dendrogram. Moreover, for each $\ell \in \{0, \ldots, n-1\}$, P_ℓ is a partition into exactly $n - \ell$ subsets.

Given a dissimilarity d on S, with no ties (except 0), we attach to d two families of undirected graphs, the vertices of which are the objects in S.

- The *discrete family* is the sequence $\{g_m\}_{0 \leq m \leq n-1}$, where for each m, g_m has m edges which correspond to the first m pairs $\{a, b\}$ of objects, ranked in increasing order of dissimilarities.

- The *continuous family* is the family $\{g(\lambda), 0 \leq \lambda\}$, where for each real $\lambda \geq 0$, $g(\lambda)$ is the graph with set of edges $\{\{a, b\} : d(a,b) \leq \lambda\}$.

Of course these two dissimilarity graph families are closely related and their use will depend on whether the interest bears more on partitions (discrete) or on the sequence of indices (continuous). The SLA can be expressed according to these families of graphs as follows.

1) Rank the $\alpha(n) = \frac{n(n-1)}{2}$ pairs (a, b) of objects according to the (strictly) increasing order of the dissimilarities,

$$d(a_1, b_1) < d(a_2, b_2) < \ldots < d(a_{\alpha(n)}, b_{\alpha(n)}).$$

2) Determine the strictly increasing sequence of integers m_ℓ by $m_0 = 0$ and, for any $\ell \in [0, n-1]$,

$$m_\ell = \min\{m \; : \; g_m \text{ has } n - \ell \text{ connected components}\} \; .$$

3) Define P_ℓ to be the partition of S in the connected components of g_{m_ℓ} and $\lambda_\ell = d(a_{m_\ell}, b_{m_\ell})$.

This point of view was introduced long ago by Ling (1973), see also Godehardt (1988) and the many references therein. Thus a (strictly binary) SLID $\{(P_\ell, \lambda_\ell)\}_{0 \le \ell \le n-1}$ can be described by a sequence of graphs indexed in three different ways according to the preference given to the *discrete level* m_ℓ, to the *continuous level* λ_ℓ, or to the *rank level* ℓ.

A SLID is a complex object and our first objective is to summarize its characteristics into certain variables which, later on, will serve as test statistics. The characteristics we consider are the following. Each of them can be expressed from the three points of view described above, according to the choice of its scale value.

1. The sequences of indices, defined either by the continuous levels, $\{\lambda_\ell\}_{0 \le \ell \le n-1}$ or by the discrete levels $\{m_\ell\}_{0 \le \ell \le n-1}$. The analysis of the gaps in the sequence of indices can suggest the presence of clusters in the ID.

2. The survival time of a singleton, which is the minimum level ℓ at which a given singleton is not an isolated point for the partition P_ℓ. This survival time can also be evaluated either by the discrete level m_ℓ or by the continuous one λ_ℓ. An object which is isolated at too high a level may be significantly different from the others.

3. The ultrametric distance between two given objects, defined as the minimum level ℓ at which both objects are in a same subset of the partition P_ℓ. This ultrametric distance can be evaluated either by the discrete level m_ℓ or by the continuous level λ_ℓ of partition P_ℓ. Two objects which are separated at too high a level may suggest the existence of two different clusters.

4. The size of subsets of partitions P_ℓ. Balanced or unbalanced subsets may provide indications on the presence or not of clusters.

These variables are respectively denoted as follows.

λ_ℓ	continuous level of the partition P_ℓ
m_ℓ	discrete level of the partition P_ℓ
$\lambda(a)$	survival time of a expressed as a continuous level
$m(a)$	survival time of a expressed as a discrete level
$\ell(a)$	survival time of a expressed as a rank of partition
$\lambda(a, b)$	ultrametric distance of a and b expressed as a continuous level
$m(a, b)$	ultrametric distance of a and b expressed as a discrete level
$\ell(a, b)$	ultrametric distance of a and b expressed as a rank of partition
$t_\ell(a)$	size of the cluster containing a in partition P_ℓ.

3. Random dissimilarities on [0,1]

Let us denote by $D(a, b)$ the random dissimilarity between objects a and b. Throughout this section, we shall suppose that the $D(a, b)$'s are independent and uniformly distributed on $[0, 1]$. The exchangeability of the $D(a, b)$'s implies that, when pairs of objects are ranked according to the increasing order of the dissimilarities, the resulting order is uniformly distributed on the set of the $\alpha(n)!$ possible orders. We denote by *Prob* the reference probability distribution.

When applying the SLA to random dissimilarities, one obtains a random SLID (RSLID) and random families of graphs denoted as usual with capital letters $\{G_m\}$, $\{G(\lambda)\}$. The variables attached to a deterministic SLID that were described in the previous section become now random variables, denoted also with capital letters,

$$\Lambda_\ell, \ M_\ell, \ \Lambda(a), \ M(a), \ L(a), \ \Lambda(a, b), \ M(a, b), \ L(a, b), \ T_\ell(a) \ .$$

In Van Cutsem and Ycart (1996b), we derived the exact and asymptotic distributions of all these variables. Exact distributions involve some combinatorics on graphs and use mainly the following numbers.

$$\begin{aligned} \gamma(n, m) \quad &= \quad \text{number of graphs with } n \text{ vertices}, m \text{ edges} \\ \gamma(n, m, k) \quad &= \quad \text{number of graphs with } n \text{ vertices}, m \text{ edges} \\ &\qquad \text{and } k \text{ connected components.} \end{aligned}$$

These numbers can be computed using recurrence relations, but the actual implementation is quickly limited by numerical explosion.

We summarize below, without proofs, the distributions of Λ_l, $\Lambda(a)$, $\Lambda(a, b)$, $M(a, b)$ and T_{n-2}. Details of the proofs and other related results can be found in Van Cutsem and Ycart (1996b).

Theorem 3.1 *With the above hypotheses and notations,*

- $\forall \ell \in \{0, 1, \ldots, n - 1\}$, $\forall \lambda \in [0, 1]$,

$$Prob(\Lambda_\ell \leq \lambda) = \sum_{m=\ell}^{\alpha(n)} \sum_{k=1}^{n-\ell} \gamma(n, m, k)\, \lambda^m\, (1 - \lambda)^{\alpha(n)-m}.$$

- $\forall \lambda \in [0, 1]$,

$$Prob(\Lambda(a) \leq \lambda) = 1 - (1 - \lambda)^{n-1} \ .$$

- $\forall \lambda \in [0, 1]$,

$$Prob(\Lambda(a, b) \leq \lambda) = \sum_{m=0}^{\alpha(n)} Prob(M(a, b) \leq m) \binom{\alpha(n)}{m} \lambda^m (1 - \lambda)^{\alpha(n)-m}.$$

where $\forall m = 1, \ldots, \alpha(n - 1) + 1$,

$$Prob(M(a, b) \leq m) = \sum_{(n_1, m_1) \in A(n, m)} \binom{n - 2}{n_1 - 2} \frac{\gamma(n_1, m_1, 1)\, \gamma(n - n_1, m - m_1)}{\gamma(n, m)}$$

with

$$A(n, m) = \{(n_1, m_1) : 2 \leq n_1 \leq n, \ n_1 - 1 \leq m_1 \leq \alpha(n_1), \ 0 \leq m - m_1 \leq \alpha(n - n_1)\} \ .$$

- $\forall n_1 = 1, \ldots, n-1,\ Prob(T_{n-2}(a) = n_1) =$

$$\sum_{(m,m_1) \in B(n,n_1)} \frac{\binom{n-1}{n_1-1} \gamma(n_1, m_1, 1)\, \gamma(n-n_1, m-m_1-1)}{\binom{\alpha(n)}{m-1}}\ \frac{n_1\,(n-n_1)}{\alpha(n)-m+1}$$

where

$$B(n, n_1) = \{(m, m_1)\ :\ n_1-1 \le m_1 \le \alpha(n_1),\ n-n_1-1 \le m-m_1-1 \le \alpha(n-n_1)\}.$$

The size of the numbers involved makes the exact computation of the probability distributions of theorem 3.1 impossible for n larger than a few tens. Fortunately, explicit asymptotic distributions are also available. These asymptotics are also derived in Van Cutsem and Ycart (1996b). The proofs are based on the theory of random graphs. The key observation is that $\{G_m\}$ and $\{G(\lambda)\}$ are random graph processes in the sense of Bollobás (1985). In particular the discrete and the continuous families of graphs are equivalent in some sense for n tending to infinity. More precisely, G_m and $G(\lambda)$ will have similar properties if $\lambda \sim m\,/\,\alpha(n)$. As an illustration, we summarize below the asymptotic distributions of some of our variables.

Theorem 3.2 *With the above hypotheses and notations,*

- $\forall \ell \ge 1,\ \forall x \in I\!R^+$,

$$\lim_{n \to \infty} Prob(\frac{n^2}{2} \Lambda_\ell < x) = 1 - e^{-x}(1 + \cdots + \frac{x^{\ell-1}}{\ell-1!}).$$

- $\forall x \in I\!R$,

$$\lim_{n \to \infty} Prob(n\Lambda_{n-1} - \log(n) \le x) = e^{-e^{-x}}.$$

- $\forall x \in I\!R^+$,

$$\lim_{n \to \infty} Prob(n\Lambda(a) \le x) = 1 - e^{-x}.$$

- $\forall x \in]1, +\infty[$,

$$\lim_{n \to \infty} Prob(n\Lambda(a, b) \le x) = \left(1 - \frac{\varphi(x)}{x}\right)^2$$

where $\varphi(x)$ *is the only solution in* $]0, 1[$ *of the equation* $\varphi e^{-\varphi} = x e^{-x}$.

-

$$\lim_{n \to \infty} Prob(\min(T_{n-2}(a), n - T_{n-2}(a)) = 1) = 1 .$$

4. Random points on [0,1]

We now consider the simplest case where the n objects are represented by n random points on the unit interval. Let $U_1, U_2, \ldots, U_n$ denote n independent random variables uniformly distributed on $[0, 1]$. As usual, we denote by $U_{(1)}, U_{(2)}, \ldots, U_{(n)}$ the ordered variables, and by $R_1, R_2, \ldots, R_n$ the sequence of ranks. The object a is then represented by the point $U_a = U_{(R_a)}$. In this model, the dissimilarity $D(a, b)$ between objects a and b is $|U_a - U_b|$.

It is easily checked that

1) the density of $U_{(.)} = (U'_{(1)}, U'_{(2)}, \ldots, U'_{(n)})$ is

$$f_{U_{(.)}}(u_1, u_2, \ldots, u_n) = n! \, \mathbb{I}_{\Omega_n}(u_1, u_2, \ldots, u_n)$$

where

$$\Omega_n = \{(u_1, u_2, \ldots, u_n) : 0 \le u_1 \le u_2 \le \ldots \le u_n \le 1\}$$

and where $\mathbb{I}_A$ denotes the indicator function of the set A.

2) the ranks $R_{(.)} = (R_1, R_2, \ldots, R_n)$ define a permutation of the first n integers which is uniformly distributed on the $n!$ permutations of these integers.

Let us define $\Delta_i = U_{(i+1)} - U_{(i)}$ to be the distance between two consecutive points in the representation. The density of $\Delta = (\Delta_1, \Delta_2, \ldots, \Delta_{n-1})$ is easily computed.

$$f_\Delta(\delta_1, \delta_2, \ldots, \delta_{n-1}) = n! \, (1 - (\delta_1 + \delta_2 + \ldots + \delta_{n-1})) \, \mathbb{I}_{\Omega'_{n-1}}(\delta_1, \delta_2, \ldots, \delta_{n-1})$$

where

$$\Omega'_{n-1} = \{(\delta_1, \delta_2, \ldots, \delta_{n-1}) : \forall i, 0 \le \delta_i \text{ and } \delta_1 + \delta_2 + \cdots + \delta_{n-1} \le 1\} \,.$$

Then, for all $i \in \{1, 2, \ldots, n-1\}$, Δ_i is distributed according to a Beta distribution $\beta(1, n)$. Moreover the variables $\Delta_1, \Delta_2, \ldots, \Delta_{n-1}$ are exchangeable.

We now associate a SLID to the dissimilarities $D(a, b)$. Since almost surely there are no ties in dissimilarities, this ID is strictly binary. We shall consider separately, the sequence of levels $\Lambda_0 = 0, \Lambda_1, \ldots, \Lambda_{n-1}$ and the stratified strictly binary dendrogram denoted by BD (see Figure 1 for an example).

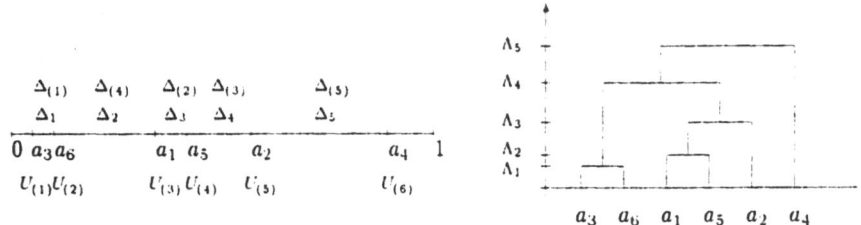

Figure 1: Six random points $[0, 1]$ and their associated SLID.

Theorem 4.1 *With the above hypotheses and notations.*

1. *The sequence of levels of the RSLID is the ordered sequence $\Delta_{(1)}, \Delta_{(2)}, \ldots \Delta_{(n-1)}$ associated to the variables $\Delta_1, \Delta_2, \ldots, \Delta_{n-1}$.*

2. *BD is uniformly distributed on the finite set $\mathcal{B}_n$ of stratified strictly binary dendrograms on n objects.*

3. *$\Delta_{(.)} = (\Delta_{(1)}, \Delta_{(2)}, \ldots, \Delta_{(n-1)})$ and BD are independent.*

The proof is elementary. Result 1 is obvious because of the properties of the SLA. To prove 2 we remark that, as the variables Δ_i are exchangeable, their ranks define a permutation which is uniformly distributed on the set of all the $(n-1)!$ permutations of $\{1, 2, \ldots, n-1\}$. Fix first a permutation of the n objects, and independently a permutation of the Δ_i's. Then the stratified dendrogram BD is determined. But 2^{n-1} such choices will lead to the same binary dendrogram, since the order of the pair of sets which are joined at each of the $n-1$ levels can be switched. Thus each binary dendrogram is obtained with probability

$$\frac{2^{n-1}}{n!\,(n-1)!}.$$

Result 3 is easy to check and uses once more the exchangeability of the variables Δ_i.

Remark. This decomposition of the RSLID into the product of two independent random structures given by the levels and the stratified binary dendrogram can be extended in a product of three independent random structures if we decompose a stratified dendrogram into the product of labels and of an unlabelled strictly binary dendrogram.

The consequences of this theorem are important. Firstly the distribution of levels can be deduced directly from that of $\Delta_{(\cdot)}$. Many results on survival times of objects, ultrametric distances between objects expressed as ranks of partitions, and consequently expressed as levels, can be directly obtained from combinatorics on the set $\mathcal{B}_n$ of strictly binary dendrograms on n objects. Here are for instance the distributions of survival times and ultrametric distances expressed as ranks of partitions.

Theorem 4.2 *Let BD denote a stratified strictly binary dendrogram uniformly distributed on the set $\mathcal{B}_n$.*
Then $\forall \ell \in \{1, \ldots, n-1\}$,

$$Prob(L(a) = \ell) = \frac{n - \ell}{\binom{n}{2}},$$

$$Prob(L(a, b) = \ell) = \frac{n + 1}{(n-1)\binom{n-\ell+2}{2}}.$$

Also, $\forall n_1 \in \{1, 2, \ldots, n-1\}$,

$$Prob(T_{n-2}(a) = n_1) = \frac{1}{n-1}.$$

The proof is easy.
The distributions of the levels Λ_k are more difficult to compute. We give only a few indications.

1) It is obvious that, using the exchangeability of the Δ_i's,

$$
\begin{aligned}
Prob(\Lambda_k \le \lambda) &= Prob(\Delta_{(k)} \le \lambda) \\
&= \sum_{j=k}^{n-1} \binom{n-1}{j} Prob(\Delta_1 \le \lambda, \ldots, \Delta_j \le \lambda, \Delta_{j+1} > \lambda, \ldots \Delta_{n-1} > \lambda).
\end{aligned}
$$

2) One can prove that, for any $k \in \{1, 2, \ldots, n-1\}$,

$$Prob(\Delta_1 > \lambda, \ldots, \Delta_k > \lambda) = ((1 - k\lambda)_+)^n ,$$

where $(x)_+ = \max\{x, 0\}$.

3) This yields

$$Prob(\Delta_1 \leq \lambda, \ldots, \Delta_h \leq \lambda, \Delta_{h+1} > \lambda, \ldots, \Delta_k > \lambda) =$$
$$((1 - (k-h)\lambda)_+)^n - \binom{h}{1}((1 - (k-h+1)\lambda)_+)^n + \cdots +$$
$$(-1)^j \binom{h}{j}((1 - (k-h+j)\lambda)_+)^n + \cdots + (-1)^{n-1} \binom{h}{j}((1 - (n-1)\lambda)_+)^n .$$

A tedious computation allows to derive $Prob(\Delta_{(k)} \leq \lambda)$. For the survival time $\Lambda(a)$ of a singleton, the exact formula is

$$Prob(\Lambda(a) \leq \lambda) = 1 - \frac{2}{n}(1 - \lambda)^n - \frac{n-2}{n}((1 - 2\lambda)_+)^n .$$

The ultrametric distance of two elements $\Lambda(a, b)$ can also be computed along the same lines. More interesting are the asymptotic distributions for continuous variables. They are obtained using the classical approximation by a Poisson process (*cf.* for instance Feller (1971)).

Theorem 4.3 *With the above hypotheses and notations,*

• $\forall x \in \mathbb{R}^+$,

$$\lim_{n \to \infty} Prob(n^2 \Lambda_1 \leq x) = 1 - e^{-x} .$$

• $\forall x \in \mathbb{R}$,

$$\lim_{n \to \infty} Prob(n\Lambda_{n-1} - \log(n) \leq x) = e^{-e^{-x}} .$$

• $\forall x \in \mathbb{R}^+$,

$$\lim_{n \to \infty} Prob(n\Lambda(a) \leq x) = 1 - e^{2x} .$$

• $\forall x \in]1, +\infty[$,

$$\lim_{n \to \infty} Prob(n\Lambda(a, b) - \log(n/3) \leq x) = e^{-e^{-x}} .$$

• $\forall x \in]0, 1/2[$,

$$\lim_{n \to \infty} Prob((1/n)\min(T_{n-2}(a), n - T_{n-2}(a)) \leq x) = 2x .$$

Thus the asymptotic distribution for the last index Λ_{n-1} is the same for i.i.d. dissimilarities (theorem 3.2) and for i.i.d. points on $[0, 1]$. The asymptotic distributions of $\Lambda(a)$ and $\Lambda(a, b)$ are different but with scalings of similar orders of magnitude. The main difference comes from the penultimate partition which is very unbalanced in the case of i.i.d. dissimilarities and corresponds to a random cut in the case of i.i.d. points on $[0, 1]$. The proof of theorem 4.3 uses the well known Poisson approximation, under the following equivalent form. Consider n i.i.d. points, uniformly distributed on $[0, n]$. Let $\Delta_1, \ldots, \Delta_k$ be a fixed number of distances between consecutive neighbors. As n tends to infinity, they are asymptotically independent, exponentially distributed with parameter 1. Recall that the infimum of n i.i.d. exponential r.v.'s is exponential with parameter n. Their supremum is asymptotically Gumbel, with location parameter $Log(n)$.

5. Examples of classifiability testing

Hartigan and Mohanty (1992) propose tests of classifiability based on the cluster sizes at different levels of a SLID. In this section, we shall consider only the last two levels. As an illustration of our tests, we define a partition $\overline{P}$ of the set $S = \{1, \ldots, n\}$ into two subsets $A = \{1, \ldots, n_1\}$ and $B = \{n_1 + 1, \ldots, n_1 + n_2\}$. Here, n_1 and $n_2 = n - n_1$ are meant as integer valued functions of n, tending to infinity with n. Our classifiability hypotheses $\mathcal{H}_1$ and $\mathcal{H}'_1$ depend on the partition $\overline{P}$ and on a positive parameter μ.

The hypothesis $\mathcal{H}_1$ is: the dissimilarities $D(a, b)$ are independent and distributed as follows

1) if $(a, b) \in A^2 \cup B^2$, $D(a, b)$ is uniformly distributed on $[0, 1]$,

2) if $(a, b) \in A \times B$, $D(a, b)$ is uniformly distributed on $[\mu, 1 + \mu]$.

The null hypothesis $\mathcal{H}_0$ is the hypothesis of non classifiability of section 3 (i.i.d. dissimilarities, model 1). We denote by $Prob_\mu$ the probability distribution corresponding to hypothesis $\mathcal{H}_1$.

Using the SLA we derive a RSLID and consider the last index Λ_{n-1}, the partition P_{n-2} and the variable $Z = \min\{D(a, b) : (a, b) \in A \times B\}$. We introduce two more variables $\Lambda^*_{n_1}$ and $\Lambda^*_{n_2}$ which are the highest levels of the two sub-ID's obtained by the SLA applied to the restrictions of D to A and B respectively.

We first consider the probability p_μ of good detection of $\overline{P}$ by the SLA:

$$p_\mu = Prob_\mu(P_{n-2} = \overline{P}) \, .$$

It is clear, using properties of the SLA, that:

$$p_\mu = Prob_\mu(\max(\Lambda^*_{n_1}, \Lambda^*_{n_2}) < Z) \, .$$

This quantity evaluates the possibility for SLA to detect the partition $\overline{P}$. It is certainly bigger than $\hat{p}_\mu$ below, which is a good approximation for large n_1 and n_2.

$$\hat{p}_\mu = Prob_\mu(\max(\Lambda^*_{n_1}, \Lambda^*_{n_2}) < \mu) \, .$$

This last probability is easy to derive since the two variables $\Lambda^*_{n_1}$ and $\Lambda^*_{n_2}$ are independent and asymptotically distributed according to a Gumbel distribution with convenient location and scale parameters (cf. theorem 3.2).

$$\hat{p}_\mu \simeq e^{-e^{-\mu n_1 + Log(n_1)}} \, e^{-e^{-\mu n_2 + Log(n_2)}} \, .$$

Table 1 gives some values of $\hat{p}_\mu$ in the case $n_1 = n_2 = 100$.

μ	0.04	0.05	0.06	0.07	0.08	0.09	0.10
$\hat{p}_\mu$	0.0257	0.2599	0.6091	0.8333	0.9351	0.9756	0.991

Table 1: Values of the probability of detection at the penultimate partition.

Let us now test $\mathcal{H}_0$ against $\mathcal{H}_1$, using the statistic Λ_{n-1}. We determine the critical region $[c_\alpha, 1]$, at significance level α, by

$$Prob_0(\Lambda_{n-1} \geq c_\alpha) = \alpha \, .$$

Since the asymptotic distribution of Λ_{n-1} is Gumbel, for n large enough, c_α is determined by

$$e^{-e^{-c_\alpha n + Log(n)}} \simeq 1 - \alpha .$$

Table 2 presents a few values for $n_1 = n_2 = 100$.

α	0.1	0.05	0.01	0.005	0.001
c_α	0.0377	0.0413	0.0495	0.0530	0.0610

Table 2: Critical values for the test on the last index.

Notice that, somewhat paradoxically, the hypothesis of non classifiability $\mathcal{H}_1$ may be rejected by the test on the last level, whereas the penultimate partition does not detect the correct structure.

The power of the test on the last level is given by

$$\pi(\mu) = Prob_\mu(\Lambda_{n-1} \geq c_\alpha) ,$$

and it is clear that $\pi(\mu) = 1$ if $\mu \geq c_\alpha$. As $\lim_{n \to +\infty} c_\alpha = 0$, we see that, for any fixed $\mu > 0$,

$$\lim_{n \to +\infty} \pi(\mu) = 1 .$$

Consider now model 2. The null hypothesis $\mathcal{H}_0'$ is: the n objects are i.i.d. points in $[0,1]$. For the hypothesis $\mathcal{H}_1'$ we choose: the objects in A are i.i.d. points uniformly distributed in $[0, \mu_1]$, and the objects in B are i.i.d. points uniformly distributed in $[\mu_2, 1]$, with

$$0 < \mu_1 < \mu_2 < 1 \quad , \quad \mu_2 - \mu_1 = \mu > 0 \quad .$$

If the statistic of test is the last level Λ_{n-1}, since its asymptotic distribution is the same for model 1 as for model 2, we obtain exactly the same critical region as before. However the probability of detection p_μ has changed. Define again $\Lambda_{n_1}^*$ and $\Lambda_{n_2}^*$ as the last levels of the sub-SLID's on the sets A and B respectively. For n_1 and n_2 large enough, the asymptotic distributions of $\Lambda_{n_1}^*$ and $\Lambda_{n_2}^*$ are still Gumbel, but with different scalings. One has

$$\hat{p}_\mu = Prob_\mu(\max(\Lambda_{n_1}^*, \Lambda_{n_2}^*) < \mu)$$
$$\simeq e^{-e^{-\frac{\mu n_1}{\mu_1} + Log(n_1)}} e^{-e^{-\frac{\mu n_2}{(1-\mu_2)} + Log(n_2)}}$$

As an example, table 3 gives some values of $\hat{p}_\mu$ in the case $n_1 = n_2 = 100$, $\mu_1 = (1 - \mu)/2$ and $\mu_2 = (1 + \mu)/2$ (the gap is centered).

μ	0.02	0.03	0.04	0.05	0.06	0.07	0.08
$\hat{p}_\mu$	0.0342	0.6625	0.9531	0.9947	0.9994	0.9999	1.

Table 3: Values of the probability of detection at the penultimate partition.

References:

Bock, H.H. (1996): Probability models and hypotheses testing in partitioning cluster analysis. In: *Clustering and classification*, Arabie, P. *et al.* (eds.), 377–453, World Scientific, Singapore.

Bollobás, B. (1985): *Random Graphs*, Academic Press, London.

Feller, W. (1971): *An introduction to probability theory and its applications*, vol. II, Wiley, London.

Godehardt, E.(1988): *Graphs as Structural Models*, Vieweg, Brauschweig/Wiesbaden.

Hartigan, J.A. (1967): Representations of similarity matrices by trees. *J. Amer. Statist. Assoc.*, **62**, 1140–1158.

Hartigan, J.A. and Mohanty, S. (1992): The RUNT test for multimodality. *J. of Classification*, **9**, 63–70.

Jain, A.K. and Dubes, R.C. (1988): *Algorithms for clustering data*. Prentice Hall, Englewood Cliffs.

Jardine, C.J., Jardine, N., and Sibson, R. (1967): The structure and the construction of taxonomic hierarchies. *Math. Biosci.*, **1**, 171–179.

Johnson, S.C. (1967): Hierarchical clustering schemes. *Psychometrika*, **32**, 241–254.

Ling, R.F. (1973): A probability theory of cluster analysis, *J. Amer. Statist. Ass.*, **68**, 159–164.

Sneath, P.H. and Sokal, R.R. (1973): *Numerical Taxonomy*, Freeman, San Francisco.

Van Cutsem, B. and Ycart, B. (1994): Renewal-type behaviour of absorption times in Markov Chains. *Adv. Appl. Probab.*, **26**, 988–1005.

Van Cutsem, B. and Ycart, B. (1996a): Probability distributions on indexed dendrograms and related problems of classifiability. In: *Data Analysis and Information Systems*, Bock, H. (ed.), 73–87. Springer-Verlag, Berlin.

Van Cutsem, B. and Ycart, B. (1996b): Indexed dendrograms on random dissimilarities. *J. of Classification*, to appear.

The L_p-product of ultrametric spaces and the corresponding product of hierarchies

Bernard Fichet

Laboratoire de Biomathématiques
Université d'Aix-Marseille II
27 Boulevard Jean Moulin
13385 Marseille, France.

Summary : The L_p-product ($1 \leq p \leq \infty$) of r indexed hierarchies is introduced in connection with the L_p-product of the corresponding r ultrametric spaces. The Cartesian product of two hierarchies appears to be a quasi-hierarchy. Endowed with an index of L_p-type ($p < \infty$). this quasi-hierarchy is in bijection with the L_p-product of two ultrametric spaces. The indexed hierarchy associated with the supremum product of r ultrametric spaces is also characterized.

1 Introduction

This paper is devoted to the L_p-product of indexed hierarchies and its relationship with the L_p-product of ultrametric spaces. In this approach, quasi-hierarchies will appear as an fundamental structure. Recall that the main axiom of a quasi-hierarchy, axiom *iii*) below, has been investigated by Batbedat (1989) and Bandelt and Dress (1989) in defining weak hierarchies. This axiom stipulates that the intersection of three clusters always is the intersection of two clusters among them. Then an indexed quasi-hierarchy is defined by adding some usual axioms. Quasi-hierarchical classification may be regarded as a unifying way for two extensions of hierarchical classification. Indeed, indexed quasi-hierarchies extend indexed (but not weakly-indexed) pseudo-hierarchies, also called "pyramids", and additive trees. For references concerning the three previous concepts, see Durand and Fichet (1988), Bertrand and Diday (1991) and Buneman (1974).

A bijection has been established between indexed quasi-hierarchies and particular dissimilarities, called quasi-ultrametrics. See Diatta and Fichet (1994) or Bandelt (1992) via a four-point characterization.

In this paper we show that the Cartesian product of two hierarchies is a quasi-hierarchy. Moreover, from two indexed hierarchies a level index of L_p-type ($p < \infty$) is produced in connection with the L_p-product of the corresponding two ultrametric spaces. Finally the supremum product of r ultrametric spaces is ultrametric and the associated indexed hierarchy is characterized as a subclass of the Cartesian product of the corresponding r hierarchies.

The reader will find an analogy with the primary approach of Benzecri and Escofier defining correspondence analysis, see Escofier (1969). Introducing the χ^2-metric on the row-set and the column-set of categories, they produce a global scattering which is nothing but the L_2-product of the two metric sets.

Let us note that the results given here have been presented by the author at the 19th Annual Conference of the Gesellschaft für Klassifikation e.V, held in Basel, March

145

1995. and the I.F.C.S. 5th Conference held in Kobe, March 1996.

2   Preliminaries and notations

Let I be a finite set. A *quasi-hierarchy* $\mathcal{H}$ on I is a collection of nonempty subsets of I obeying the following four axioms:

i) $I \in \mathcal{H}$

ii) $\forall H \in \mathcal{H}$, $\cup \{H' \in \mathcal{H} , H' \subset H\} \in \{H, \emptyset\}$

iii) $\forall H_1, H_2, H_3 \in \mathcal{H}$, $H_1 \cap H_2 \cap H_3 \in \{H_1 \cap H_2, H_2 \cap H_3, H_3 \cap H_1\}$

iv) $\forall H, H' \in \mathcal{H}$, $H \cap H' \in \mathcal{H} \cup \{\emptyset\}$

Such a definition stays valid whenever I is infinite. In the finite case as considered here, we have an equivalent definition. The class $\mathcal{H}$ is a quasi-hierarchy if and only if it obeys *i)*, *iii)*, *iv)* and

ii')   minimal elements of $\mathcal{H}$ partition I

A quasi-hierarchy $\mathcal{H}$ is said to be *total* (definite) if and only if:

v) $\forall i \in I$, $\{i\} \in \mathcal{H}$

Note that *v)* implies *ii)* (or *ii')*)

Recall that a *hierarchy* obeys *i)*, *ii)* (or *ii')*) and

vi) $\forall H, H' \in \mathcal{H}$, $H \cap H' \in \{H, H', \emptyset\}$

Thus a hierarchy is a quasi-hierarchy.

Recall that a hierarchy admits a well-known visual display, called a *dendrogram*. In a quasi-hierarchy $\mathcal{H}$, a predecessor of a cluster (element) H in $\mathcal{H}$, $H \neq I$, is any minimal element of the family $\{H' \in \mathcal{H} , H' \supset H\}$. It follows from *vi)* that in a hierarchy H has a unique predecessor. In a quasi-hierarchy, there exists a smallest cluster, say H_{ij}, containing two fixed units i and j in I. That derives from *iv)*.

A level index is defined as usually. An *indexed quasi-hierarchy* is a pair $(\mathcal{H}, f)$ where $\mathcal{H}$ is a quasi-hierarchy and f is a function mapping $\mathcal{H}$ into $\mathbb{R}_+$ and obeying:

vii) $(H \in \mathcal{H} , H \text{ minimal}) \Longrightarrow f(H) = 0$

viii) $(H, H' \in \mathcal{H} , H \subset H') \Longrightarrow f(H) < f(H')$

A *stratified quasi-hierarchy* is a pair $(\mathcal{H}, \preceq)$ where $\mathcal{H}$ is a quasi-hierarchy and $\preceq$ is a (linear) quasi-order on $\mathcal{H}$ obeying:

vii') $(H \in \mathcal{H} , H \text{ minimal}) \Longrightarrow \forall H' \in \mathcal{H} , H \preceq H'$

viii') $(H, H' \in \mathcal{H} , H \subset H') \Longrightarrow H \preceq H'$ and not $H' \preceq H$

Clearly an indexed hierarchy induces a stratified quasi-hierarchy by:

$$\forall H, H' \in \mathcal{H} , H \preceq H' \Longleftrightarrow f(H) \leq f(H')$$

A *dissimilarity* on I is a mapping $d : I^2 \mapsto \mathbb{R}_+$ obeying

$$\forall (i,j) \in I^2 , d(i,j) = d(j,i) \text{ and } \forall i \in I , d(i,i) = 0$$

Such a dissimilarity is said to be *proper* (definite) iff $d(i,j) = 0 \Rightarrow i = j$

(I, d) is called a dissimilarity space.

We note $B^d(i,r)$ the (closed) *ball* with centre i and radius $r \geq 0$, i.e. $B^d(i,r) = \{k \in I \; / \; d(i,k) \leq r\}$. We define a 2-ball as the set $B^d_{ij} = B^d(i, d(i,j)) \cap B^d(j, d(i,j))$. The family of 2-balls is denoted by $\mathcal{B}_d = \{B^d_{ij} \; . \; i, j \in I\}$.

For any $J \subseteq I$, $\mathrm{diam}_d(J)$ is the *diameter* of J, i.e.

$$\mathrm{diam}_d(J) = \max\{d(i,j) \ , \ i,j \in J\}$$

There is a well-known bijection between the set of indexed hierarchies and the set of particular dissimilarities, called ultrametrics, see Johnson (1967), Jardine et al. (1967), Benzecri (1973). An *ultrametric* obeys the ultrametric inequality: $\forall i,j,k \in I$. $d(i,j) \leq \max[d(i,k),d(k,j)]$. The set of ultrametrics on I will be denoted $D_u(I)$. In fact, there are many equivalent definitions for ultrametricity. The following five statements will be useful in this paper.

$d \in D_u(I)$

$\Leftrightarrow \ \forall(i,j) \in I^2$. $B^d(i,d(i,j)) = B^d(j,d(i,j))$

$\Leftrightarrow \ \forall(i,j) \in I^2$, $B^d_{ij} = B^d(i,d(i,j))$

$\Leftrightarrow \ \left[\forall(i,j,k,l) \in I^4 \ , \ k,l \in B^d(i,d(i,j)) \Rightarrow B^d(k,d(k,l)) \subseteq B^d(i,d(i,j))\right]$

$\Leftrightarrow \ \forall(i,j) \in I^2$. $\mathrm{diam}_d B^d(i,d(i,j)) = d(i,j)$

According to those properties, Diatta and Fichet (1994) define a *quasi-ultrametric* as a dissimilarity obeying both

$i.x$) $\ \forall(i,j,k,l) \in I^4 \ ; \ k,l \in B^d_{ij} \Rightarrow B^d_{kl} \subseteq B^d_{ij}$ (*inclusion condition*)

$.x$) $\ \forall(i,j) \in I^2$, $\mathrm{diam}_d B^d_{ij} = d(i,j)$ (*diameter condition*)

Thus an ultrametric is a quasi-ultrametric. We denote by $D_{qu}(I)$ the set of quasi-ultrametrics on I.

Now, we have the ingredients to recall a theorem extending the one-to-one correspondence between indexed hierarchies and ultrametrics, see Diatta and Fichet (1994).

Theorem 1 *Let* $d \in D_{qu}(I)$. *Then* $(\mathcal{B}_d, \mathrm{diam}_d)$ *is an indexed quasi-hierarchy. The quasi-hierarchy* $\mathcal{B}_d$ *is total iff* d *is proper. This is a hierarchy iff* $d \in D_u(I)$. *Conversely, given an indexed quasi-hierarchy* $(\mathcal{H}, f)$, *let* δ *define the mapping from* I^2 *into* $I\!R_+$ *by:* $\forall(i,j) \in I^2$, $\delta(i,j) = f(H_{ij})$. *Then* δ *is the unique quasi-ultrametric such that* $(\mathcal{B}_\delta, \mathrm{diam}_\delta) = (\mathcal{H}, f)$.

We end the section by specifying the L_p-product of dissimilarity spaces. Let (I_1, d_1), $\ldots, (I_r, d_r)$ be r dissimilarity spaces. Let $I = I_1 \times \cdots \times I_r$. For every $1 \leq p \leq \infty$, define

$$d : I^2 \mapsto I\!R_+ \ \text{by:} \ \forall i = (i_1, \ldots, i_r), \ j = (j_1, \ldots, j_r) \in I \ , \ d(i,j) = \left[\sum_l d_l^p(i_l, j_l)\right]^{\frac{1}{p}}$$

Clearly d is a dissimilarity and (I,d) is called the L_p-product of $(I_1, d_1), \ldots, (I_r, d_r)$. We note $d = d_1 \oplus_p \cdots \oplus_p d_r$. For $p = 1$, we have the *direct product* and the following simpler notation: $d = d_1 \oplus \cdots \oplus d_r$. When $p = \infty$, (I,d) is called the *supremum product*.

3   The product of two hierarchies

Given r sets I_l , $l = 1, \ldots, r$ and a collection of subsets $\mathcal{H}_l$ on every set I_l, we denote by $\mathcal{H} = \mathcal{H}_1 \times \cdots \times \mathcal{H}_r$ the set $\{H = H_1 \times \cdots \times H_r \ / \ H_l \in \mathcal{H}_l \ , \ l = 1, \ldots, r\}$. Up to a bijection, $\mathcal{H}$ is the Cartesian product of $\mathcal{H}_1, \ldots, \mathcal{H}_r$.

Proposition 1 *Let* $\mathcal{H}_1$ *and* $\mathcal{H}_2$ *be two hierarchies on* I_1 *and* I_2, *respectively. Then* $\mathcal{H} = \mathcal{H}_1 \times \mathcal{H}_2$ *is a quasi-hierarchy on* $I_1 \times I_2$, *called the product of* $\mathcal{H}_1$ *and* $\mathcal{H}_2$. *It is total if and only if both* $\mathcal{H}_1$ *and* $\mathcal{H}_2$ *are total.*

Proof. It is clear that axioms i), ii) (or ii')) and iv) of a quasi-hierarchy are fulfilled. In particular $H = H_1 \times H_2$ is minimal in $\mathcal{H}$ iff both H_1 and H_2 are minimal in $\mathcal{H}_1$ and $\mathcal{H}_2$.

Now, let $H = H_1 \times H_2$, $H' = H_1' \times H_2'$, $H'' = H_1'' \times H_2''$ be three elements of $\mathcal{H}$. Then $H \cap H' \cap H'' = (H_1 \cap H_1' \cap H_1'') \times (H_2 \cap H_2' \cap H_2'')$.

First suppose that $H_1 \cap H_1' \cap H_1'' = \emptyset$. Since $\mathcal{H}_1$ is a hierarchy, two clusters, say H_1 and H_1', are necessarily disjoint. Then:

$$H \cap H' \cap H'' = \emptyset = (H_1 \cap H_1') \times (H_2 \cap H_2') = H \cap H'$$

A similar property holds whenever $H_2 \cap H_2' \cap H_2'' = \emptyset$.

Finally, suppose that $H_1 \cap H_1' \cap H_1''$ and $H_2 \cap H_2' \cap H_2''$ are nonempty.

Without loss of generality (w.l.o.g.), let $H_1 \subseteq H_1' \subseteq H_1''$.

If $H_2 = \min(H_2, H_2', H_2'')$, then $H \cap H' \cap H'' = H = H \cap H'$, for example.

Otherwise, suppose w.l.o.g. that $H_2' = \min(H_2, H_2', H_2'')$. Then: $H \cap H' \cap H'' = H_1 \times H_2' = H \cap H'$. Axiom iii) of a quasi-hierarchy is fulfilled.

The last assertion of the proposition is obvious. ∎

Let us observe that the quasi-hierarchy $\mathcal{H}$ obtained in the previous proposition is very particular. For example, the following property is noteworthy. Every cluster $H = H_1 \times H_2$ of $\mathcal{H}$, with $H_1 \neq I_1$ and $H_2 \neq I_2$, has exactly two predecessors, specifically $H_1 \times H_2'$ and $H_1' \times H_2$, where H_1' and H_2' stand for the (unique) predecessors of H_1 in $\mathcal{H}_1$ and H_2 in $\mathcal{H}_2$.

Corollary 1 *Let $(\mathcal{H}_1, f_1)$ and $(\mathcal{H}_2, f_2)$ be two indexed hierarchies on I_1 and I_2, respectively. Let $\mathcal{H} = \mathcal{H}_1 \times \mathcal{H}_2$ and $f : \mathcal{H} \mapsto I\!\!R_+$ such that:*

$$\forall H = H_1 \times H_2 \in \mathcal{H}, \quad f(H) = [f_1^p(H_1) + f_2^p(H_2)]^{\frac{1}{p}}, \quad 1 \le p < \infty$$

Then $(\mathcal{H}, f)$ is an indexed quasi-hierarchy, called the L_p-product of $(\mathcal{H}_1, f_1)$ and $(\mathcal{H}_2, f_2)$.

We note $f = f_1 \oplus_p f_2$ or simply $f = f_1 \oplus f_2$ when $p = 1$.

For different values of p, the corresponding indexed quasi-hierarchies do not induce the same stratified quasi-hierarchy. For a counter-example define $(\mathcal{H}_1, f_1)$, $(\mathcal{H}_2, f_2)$ and clusters $H_1, H_1' \in \mathcal{H}_1$, $H_2, H_2' \in \mathcal{H}_2$ such that: $f_1(H_1) = 1 - \varepsilon$, $f_1(H_1') = 1/2$, $f_2(H_2) = \varepsilon'$, $f_2(H_2') = 1/2$, $0 < \varepsilon' < \varepsilon < 1$. Then $f_1 \oplus f_2 (H_1 \times H_2) < f_1 \oplus f_2 (H_1' \times H_2')$ whereas $f_1 \oplus_2 f_2 (H_1 \times H_2) > f_1 \oplus_2 f_2 (H_1' \times H_2')$ for ε sufficiently small.

In terms of dissimilarities, we have the following proposition.

Proposition 2 *Let (I_1, d_1) and (I_2, d_2) be two ultrametric spaces and let (I, d) be their L_p-product $(1 \le p < \infty)$. Then:*

1. $\forall i, j \in I$, $B_{ij}^d = B^{d_1}(i_1, d_1(i_1, j_1)) \times B^{d_2}(i_2, d_2(i_2, j_2))$

2. *d is quasi-ultrametric.*

Note that the family of 2-balls of (I, d) does not depend on p.

Proof. Let $k \in B^{d_1}(i_1, d_1(i_1, j_1)) \times B^{d_2}(i_2, d_2(i_2, j_2))$. Then:

$$d(i, k) = [d_1^p(i_1, k_1) + d_2^p(i_2, k_2)]^{\frac{1}{p}} \leq [d_1^p(i_1, j_1) + d_2^p(i_2, j_2)]^{\frac{1}{p}} = d(i, j)$$

Similarly, since every point of a ball in an ultrametric space is a centre of the ball, we have: $d(j, k) \leq d(i, j)$. Thus $k \in B_{ij}^d$.

Conversely, let $k \in B_{ij}^d$. Then:

$$\max[d_1^p(i_1, k_1) + d_2^p(i_2, k_2), d_1^p(j_1, k_1) + d_2^p(j_2, k_2)] \leq [d_1^p(i_1, j_1) + d_2^p(i_2, j_2)]$$

Suppose, by way of contradiction, that for example $d_1(i_1, k_1) > d_1(i_1, j_1)$. Since d_1 is ultrametric, we have $d_1(i_1, j_1) < d_1(i_1, k_1) = d_1(j_1, k_1)$

Thus, $\max[d_2(i_2, k_2), d_2(j_2, k_2)] < d_2(i_2, j_2)$. That violates the ultrametric inequality.

Consequently, Point 1. is proved.

Then, the inclusion condition and the diameter condition in terms of 2-balls for d derive from the same conditions in terms of balls for d_1 and d_2. ∎

Proposition 2 shows that the L_p-product of two ultrametric spaces is connected with an indexed quasi-hierarchy. Similarly, we deduce from Corollary 1 that the L_p-product of two indexed hierarchies is connected with a quasi-ultrametric. In fact, each mapping is the inverse of the other.

Proposition 3 *The indexed quasi-hierarchy associated with the L_p-product of two ultrametric spaces is the L_p-product of the corresponding two indexed hierarchies.*

Since the clusters of a hierarchy are the balls and the clusters of a quasi-hierarchy are the 2-balls, the proof is immediate from Proposition 2 and Corollary 1. We may also use the opposite way by observing that the smallest cluster of $\mathcal{H}_1 \times \mathcal{H}_2$ containing i and j is the Cartesian product of the smallest clusters of $\mathcal{H}_1$ and $\mathcal{H}_2$ containing the corresponding components of i and j.

The following coherent diagram summarizes the previous results.

$$\begin{array}{ccccc} [(\mathcal{H}_1, f_1) & : & (\mathcal{H}_2, f_2)] & \longrightarrow & (\mathcal{H}_1 \times \mathcal{H}_2, \ f_1 \oplus_p f_2) \\ \updownarrow & & \updownarrow & & \updownarrow \\ [d_1 \in \mathcal{D}_u(I_1) & ; & d_2 \in \mathcal{D}_u(I_2)] & \longrightarrow & d_1 \oplus_p d_2 \in \mathcal{D}_{qu}(I_1 \times I_2) \end{array}$$

Given the L_p-product $(\mathcal{H}_1 \times \mathcal{H}_2, \ f_1 \oplus_p f_2)$, it is easy to recover the hierarchical components $(\mathcal{H}_1, f_1)$ and $(\mathcal{H}_2, f_2)$. Indeed, a subset H_1 of I_1 is in $\mathcal{H}_1$ iff $H_1 \times I_2 \in \mathcal{H}_1 \times \mathcal{H}_2$. Thus we have $\mathcal{H}_1$ and $\mathcal{H}_2$ and in particular their minimal elements. Then, for every $H_1 \in \mathcal{H}_1$, $f_1(H_1) = f_1 \oplus_p f_2(H_1 \times H_2)$ where H_2 stands for any minimal element of $\mathcal{H}_2$.

A similar way stays valid for a more general problem. Is a given indexed quasi-hierarchy on $I_1 \times I_2$ the L_p-product of two indexed hierarchies $(\mathcal{H}_1, f_1)$ and $(\mathcal{H}_2, f_2)$ on I_1 and I_2 for some p? Indeed, the previous procedure gives potential candidates as clusters of $\mathcal{H}_1$ and $\mathcal{H}_2$. Then it suffices to check whether $\mathcal{H}_1$ and $\mathcal{H}_2$ are hierarchies and whether $\mathcal{H} = \mathcal{H}_1 \times \mathcal{H}_2$. Similarly we have potential level indices f_1 and f_2 and a unique potential real number p.

In terms of metric spaces, such a property remains clear still.

4   The supremum product of indexed hierarchies

This paragraph is devoted to the supremum product $(p = \infty)$ of hierarchies. As opposed to the previous case $(p < \infty)$, we need here a level index on each hierarchy.

Proposition 4 *Let $(\mathcal{H}_1, f_1), \ldots, (\mathcal{H}_r, f_r)$ be r indexed hierarchies on $I_1, \ldots, I_r$, respectively. Let $\mathcal{H}^* = \mathcal{H}_1 \times \cdots \times \mathcal{H}_r$. Define $f^* : \mathcal{H}^* \mapsto I\!R_+$ by : $\forall H = H_1 \times \cdots \times H_r \in \mathcal{H}^*$, $f^*(H) = \max f_l(H_l)$. Let $\mathcal{H}$ be the subclass of $\mathcal{H}^*$ defined by:*

$$H \in \mathcal{H} \text{ iff } (H' \in \mathcal{H}^*, \ H \subseteq H', \ f^*(H') \leq f^*(H)) \Longrightarrow H = H'$$

Let f be the restriction of f^* to $\mathcal{H}$.
Then $(\mathcal{H}, f)$ is an indexed hierarchy, called the supremum product of $(\mathcal{H}_1, f_1), \ldots , (\mathcal{H}_r, f_r)$. We note $(\mathcal{H}, f) = (\mathcal{H}_1, f_1) \otimes \cdots \otimes (\mathcal{H}_r, f_r)$.

Proof.   Clearly axiom i) of a hierarchy is fulfilled. The following equivalences are also obvious:
$H = H_1 \times \cdots \times H_r$ minimal in $\mathcal{H}^* \Leftrightarrow H_l$ minimal in $\mathcal{H}_l$ for every $l \Leftrightarrow f^*(H) = 0$.
Thus, a minimal element of $\mathcal{H}^*$ is in $\mathcal{H}$ and the set of minimal elements of $\mathcal{H}^*$ is the set of minimal elements of $\mathcal{H}$. The class $\mathcal{H}$ obeys axiom ii) (or ii')) of a hierarchy. Moreover, if H is minimal in $\mathcal{H}$, $f(H) = 0$.
Finally, let $H = H_1 \times \cdots \times H_r$, $H' = H'_1 \times \cdots \times H'_r \in \mathcal{H}$, with $H \cap H' \neq \emptyset$.
Define $H''_l = H_l \cup H'_l$ for every l, and $H'' = H''_1 \times \cdots \times H''_r$. Since $H_l \cap H'_l \neq \emptyset$ for every l, H''_l is equal to H_l or H'_l and belongs to $\mathcal{H}_l$.
Suppose w.l.o.g. that $f^*(H) \leq f^*(H')$. Then, for every l, $f_l(H''_l) \leq f^*(H')$, so that $f^*(H'') \leq f^*(H')$. From the definition of $\mathcal{H}$, we deduce $H'' = H'$.
Consequently we have: $H \subseteq H'$ and $\mathcal{H}$ is a hierarchy.
If $H, H' \in \mathcal{H}$ with $H \subseteq H'$, then clearly $f(H) \leq f(H')$ and from the definition of $\mathcal{H}$ we deduce: $f(H) = f(H') \Rightarrow H = H'$. Moreover we have seen that $f(H) = 0$ whenever H is minimal, so that the proof is complete. ■

Proposition 5 *Let $(I_1, d_1), \ldots, (I_r, d_r)$ be r ultrametric spaces and let (I, d) be their supremum product.*
Then, d is ultrametric.

Proof.   Let $i, j, k \in I_1 \times \cdots \times I_r$. There exists s such that: $d(i, j) = d_s(i_s, j_s)$.
Then: $d(i, j) = d_s(i_s, j_s) \leq \max[d_s(i_s, k_s), d_s(k_s, j_s)] \leq \max[d(i, k), d(k, j)]$. ■

As in the previous section, we may establish a coherent diagram.

Proposition 6 *The indexed hierarchy associated with the supremum product of r ultrametric spaces is the supremum product of the corresponding r indexed hierarchies.*

Proof. Denote by $(\mathcal{H}_1, f_1), \ldots, (\mathcal{H}_r, f_r)$ the indexed hierarchies defined on $I_1, \ldots, I_r$, respectively. Let $I = I_1 \times \cdots \times I_r$ and let $(\mathcal{H}, f)$ be the supremum product of $(\mathcal{H}_1, f_1), \ldots, (\mathcal{H}_r, f_r)$. For every $i = (i_1, \ldots, i_r)$, $j = (j_1, \ldots, j_r) \in I$ and for every $l = 1, \ldots, r$, let $H^l_{i_l j_l}$ be the smallest cluster of $\mathcal{H}_l$ containing i_l and j_l. Define $H'^l_{i_l j_l}$ as the greatest cluster of $\mathcal{H}_l$ containing i_l and j_l, and such that $f_l\left(H'^l_{i_l j_l}\right) \leq \max_l\left[f_l\left(H^l_{i_l j_l}\right)\right]$. Let s be an integer such that $f_s\left(H^s_{i_s j_s}\right) = \max_l\left[f_l\left(H^l_{i_l j_l}\right)\right]$. Then $H'^s_{i_s j_s} = H^s_{i_s j_s}$. We show that $H' = H'^1_{i_1 j_1} \times \cdots \times H'^r_{i_r j_r}$ is in $\mathcal{H}$ and is the smallest cluster of $\mathcal{H}$ containing i and j. Indeed the existence of H' is obvious and $H' \in \mathcal{H}$ derives from the definition of $\mathcal{H}$. Now, let $H'' = H''_1 \times \cdots \times H''_r \in \mathcal{H}$ be a cluster containing i and j. For every l, $H^l_{i_l j_l} \subseteq H''_l$. Then we have: $f(H'') \geq f_s(H''_s) \geq f_s\left(H^s_{i_s j_s}\right) = f(H')$. Thus $H' \subseteq H''$ and H' has the announced property. The result follows since $f(H') = \max_l\left[f_l\left(H^l_{i_l j_l}\right)\right]$. ∎

We may also use the opposite way. Observing that a ball $B^d(i, d(i, j))$ for the supremum product may be expressed as $B^{d_1}(i_1, d(i, j)) \times \cdots \times B^{d_r}(i_r, d(i, j))$, it suffices to show that the family of such balls coincides with the supremum product of hierarchies defined in Proposition 4.

The following diagram summarizes the previous properties.

$$
\begin{array}{ccc}
[(\mathcal{H}_1, f_1) \quad ; \cdots ; \quad (\mathcal{H}_r, f_r)] & \longrightarrow & (\mathcal{H}_1, f_1) \otimes \cdots \otimes (\mathcal{H}_r, f_r) \\
\updownarrow \qquad\qquad\qquad \updownarrow & & \updownarrow \\
[d_1 \in \mathcal{D}_u(I_1) \quad ; \cdots ; \quad d_r \in \mathcal{D}_u(I_r)] & \longrightarrow & d_1 \oplus_\infty \cdots \oplus_\infty d_r \in \mathcal{D}_u(I_1 \times \cdots \times I_r)
\end{array}
$$

Although the supremum product does not contain all elements of the Cartesian product, it is still possible to extract the hierarchical components from such a structure and to solve a more general problem as in paragraph 3. Indeed, for every $H_1 \in \mathcal{H}_1$, it always exists some $H_l \in \mathcal{H}_l$, $l = 2, \ldots, r$ such that $H = H_1 \times H_2 \times \cdots \times H_r \in \mathcal{H}$. Thus $\mathcal{H}_1, \ldots, \mathcal{H}_r$ follow. Furthermore, owning the hierarchies, there is a minimal $H_2 \times \cdots \times H_r$, $H_l \in \mathcal{H}_l$, $l = 2, \ldots, r$ such that $H = H_1 \times H_2 \times \cdots \times H_r \in \mathcal{H}$ for a fixed $H_1 \in \mathcal{H}_1$. Then $f_1(H_1) = f(H)$. One might also use the fastidious but clearer way of ultrametric spaces.

Replacing the maximum by the minimum, we may establish, with a similar proof, a property analogous to the one of Proposition 4. From $\mathcal{H}^\bullet = \mathcal{H}_1 \times \cdots \times \mathcal{H}_r$, define $f_\bullet : \mathcal{H}^\bullet \mapsto \mathbb{R}_+$ by: $\forall H = H_1 \times \cdots \times H_r \in \mathcal{H}^\bullet$, $f_\bullet(H) = \min_l f_l(H_l)$. Let $\underline{\mathcal{H}}$ be the subclass of $\mathcal{H}^\bullet$ defined by:

$$H \in \underline{\mathcal{H}} \text{ iff } (H' \in \mathcal{H}^\bullet, H' \subseteq H, f_\bullet(H') \geq f_\bullet(H)) \Longrightarrow H = H'$$

Let $\underline{f}$ be the restriction of $f_\bullet$ to $\underline{\mathcal{H}}$.

Then $\left(\underline{\mathcal{H}}, \underline{f}\right)$ is an indexed hierarchy in $I = I_1 \times \cdots \times I_r$ (not on I), i.e. $\underline{\mathcal{H}}$ obeys only axioms $ii)$ and $vi)$ of a hierarchy.

5 Example

We give here a simple and illustrative example.
Two indexed hierarchies are given via their dendrograms in Figure 1.

Figure 2 exhibits the three clusters at the level 5 for the L_1-product of the indexed hierarchies given in Figure 1: $\{i_1, i_2, i_3\} \times J$, $I \times \{j_1, j_2\}$, $I \times \{j_4, j_5\}$. We may imagine a practical procedure, with a computer, displaying in this way the clusters at a given level.

The dendrogram for the supremum product of the same indexed hierarchies is visualized in Figure 3.

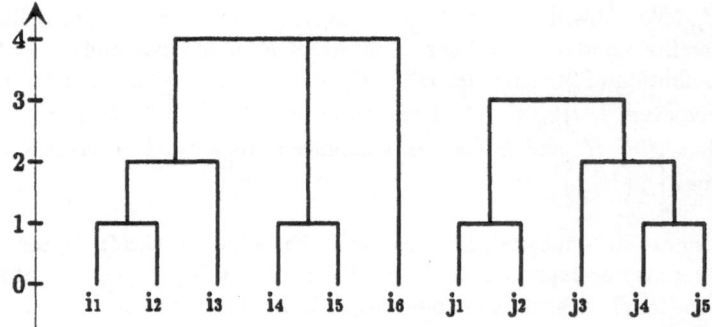

Figure 1: Two indexed hierarchies

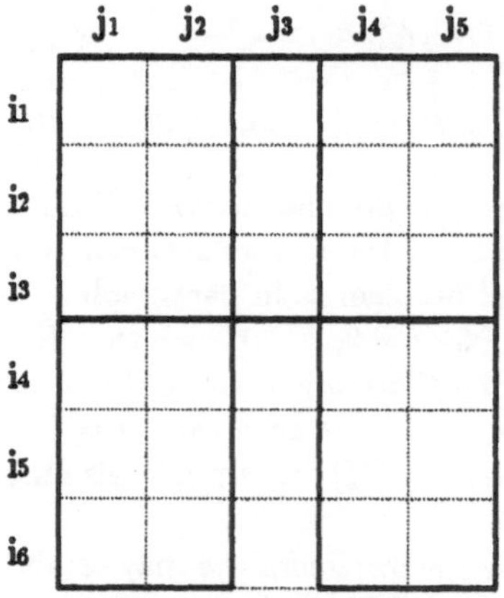

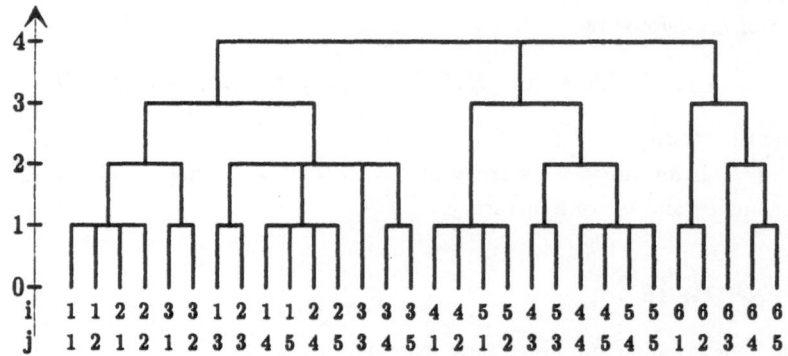

Figure 2: The three clusters at the level 5 of the L_1-product of hierarchies

Figure 3: The supremum product of the indexed hierarchies

6 References

Bandelt, H.J. (1992): Four-point characterization of the dissimilarity functions obtained from indexed closed weak hierarchies, *Math. Seminar der Universität, Hamburg.*

Bandelt, H.J. and Dress, A.W. (1989): Weak hierarchies associated with similarity measures: an additive clustering technique, *Bull. Math. Biology,* 51, 113–166.

Batbedat, A. (1989): Les dissimilarités Medas et arbas, *Statistique et Analyse des données,* 14, 3, 1–18

Benzecri J.P. (1973): *L'analyse des données. Tome 1, La Taxnomie.*, Dunod, Paris.

Bertrand, P. and Diday, E. (1991): Les pyramides classifiantes: une extension de la structure hiérarchique, *C.R. Acad. Sci. Paris, Série I,* 693–696.

Buneman, P. (1974): A note on metric properties of trees, *J. Combin. Theory, Ser. B,* 17, 48–50.

Diatta, J. and Fichet, B. (1994): From Apresjan hierarchies and Bandelt-Dress weak hierarchies to quasi-hierarchies, In:*New approaches in Classification and Data Analysis*, Diday, E. et al. (eds.), 111-118, Springer-Verlag, Berlin.

Durand, C. and Fichet, B. (1988): One-to-one correspondences in pyramidal representation: a unified approach, In: *Classification and Related Methods of Data Analysis*, Bock, H. (ed.), 80–85, North-Holland, Amsterdam.

Escofier, B. (1969): L'analyse factorielle des correspondances, *Cahiers du B.U.R.O., Université de Paris VI*, 13,25–29.

Jardine, C.J., Jardine, N. and Sibson, R. (1967): The structure and construction of taxonomic hierarchies, *Mathematical Biosciences,* 1, 465–482.

Johnson, S.C. (1967): Hierarchical clustering schemes, *Psychometrika*, 32, 241–254.

Towards Comparison of Decomposable Systems

The University of Aizu, Fukushima 965-80 Japan

Summary: The paper focuses on the comparison of decomposable systems on the base of combinatorial descriptions of systems and their parts. Our system description involves the following interconnected hierarchies: a tree-like system model; criteria and restrictions for system components (nodes of the model); design alternatives (DAs) for nodes; interconnection (Is) or compatibility between DAs of different system components; estimates of DAs and Is. A vector-like proximity for rankings is described.

1. Introduction

Usually the following basic system analytical problems have been examined to analyze complex systems: (1) to compare two system versions; (2) to classify system versions; (3) to evaluate a system version set (e.g., aggregation, construction of consensus, etc.); (4) to analyze tendencies of system changes (evolution, etc.); (5) to reveal the most significant system parameters; (6) to plan an improvement process for a system. Here we examine system representations on the base of structural or combinatorial objects. We assume that a system is decomposable one, and may have several versions. Thus the following kinds of problems are basic ones: (a) description of the system and their parts; (b) operations of the system analysis and transformation. Our description of decomposable systems consists of the following interconnected hierarchies (Levin, 1996b): (1) a system tree-like model); (2) requirements (criteria, restrictions) to system components (nodes of the model); (3) design alternatives (DAs) for nodes; (4) interconnection (Is) or compatibility between DAs of different components; (5) factors of compatibility.

The system proximity maybe examined as the following structure corresponding to the system model: proximity of hierarchical tree-like model; proximity of requirement hierarchy (criterion hierarchy; restriction hierarchy; compatibility factors hierarchy); proximity of DAs (sets of DAs, estimates on criteria, and priorities); proximity of Is (set of Is with priorities). We consider three levels of combinatorial decriptions: (1) basic combinatorial objects (points in a space; vectors; sets; partitions; rankings, strings, trees, posets, etc.); (2) elements of the system description: leaf nodes; set of DAs and/or Is; tree-like system model; criteria for DAs; etc.; (3) basic system descriptions, e.g., complete description; external requirement. So a vector-like proximity for rankings is examined.

2. Measurement of proximity

First let us consider approaches to modeling a proximity (distance, similarity, closeness, dissimilarity, etc.) for combinatorial objects. Note that these investigations have been executed in various disciplines (e.g., mathematical psychology; decision making; chemistry; linguistics; morphological schemes of systems in technological forecasting; biology; genetics; data and knowledge engineering; network engineering; architecture; combinatorics). A survey of coefficients for measures of similarity, dissimilarity, and distance from viewpoint of statistical sciences is presented in (Gower, 1995). From a system viewpoint it is reasonable to examine some functions, operations and corresponding requirements to mathematical models of the proximity

(Table 1).

Table 1. Some operations and requirements to mathematical models

Functional phase	Operations	Requirements
Development	Selection Design	Problem relevantness Completeness Univeersality Easiness Generalizability Habitulness
Representation	Mathematical description Text description Graphical presentation Animation Composite approach	Easy to visualize Understandability Operationability Relevantness to use
Study and learn	Understanding Remembering Evaluation of features Identification Processing	Ability to process and analyze Habitualness Coordination with intuition Understanability Simpleness Easy to visualize
Utilization (processing)	Processing Transformation Composing Coordination	Easy to process (by human, by computer) Composability

Formal requirements to proximity models are based on three Freshe's axioms specifying metrics, and sometimes on additional axioms (Kemeny and Snell, 1972; etc.). In some cases, the triangulation axiom is rejected, e.g., for architectural objects (Zeitoun, 1977), for rankings (Belkin and Levin, 1990). Measuring the proximity between combinatorial objects is based on the following approaches: (1) a metric in a parameter space; (2) attributes of the largest common part of objects (intersection) or an unification (the minimal covering construction); (3) minimum of changes (change path) which allows to transform an initial object into a target one.

Secondly let us consider scales of measuring. Traditionally R_1, $[0, 1]$ or an ordinal scale are applied. Huber and Arabie use measures of agreement or consensus indices, e.g., from $[-1, 1]$ (Hubert and Arabie, 1985). Recently some extensions of metric spaces have been proposed (Pouzet and Rosenberg, 1994; Barthelemy and Guenoche, 1991; etc.), for examples: (a) graphs and ordered sets (Jawhari et al., 1986); (b) conceptual lattices for complex scaling (Ganter and Wille, 1989); (c) ordered sets and semilattices for partitions (Barthelemy et al., 1986); (d) simplices for rankings (Belkin and Levin, 1990). Generally, Arabie and Hubert have examined three approaches to compare combinatorial objects (sequences, partitions, trees, graphs) through given matrices (Arabie and Hubert, 1992): (a) axiomatic approach to construct a "good" measures; (b) usage of structural representations; (c) usage of an optimization task.

Finally, in complex cases, the following approaches maybe applied: (a) multidimensional scaling (Torgenson, 1958; Kruskal, 1977; etc.); (b) usage of graphs and ordered sets as a kind of a metric space (Jawhari et al. 1986; Barthelemy et al., 1986; Barthelemy and guenoche, 1991; Ganter and Wille, 1989); (c) integrating or composing of a global proximity from distances or proximities of system components.

3. Combinatorial objects and system description

Main approaches to compare combinatorial objects are the following (Arabie and Hubert, 1992; Pouzet and Rosenberg, 1994; etc.):

1. Points in a space: traditional metrics.

2. Sets, system of representatives: metrics (Margush, 1982; etc.).

3. Partitions: metrics (Mirkin and Cherny, 1970; etc.), multidimensional scaling (Arabie and Boorman, 1973), measures of agreement or consensus indices (Hubert and Arabie, 1985), ordered sets, semilattices (Barthelemy et al, 1986).

4. Linear (ordinal) rankings: metrics (Kendall, 1962; Kemeny and Snell, 1972; Cook and Kress, 1984).

5. Strings: maximum subsequence (Wagner and Fisher, 1974; Sellers,1974), minimum supersequence (Timkovskii, 1989; etc.), metrics (Hannenhalli and Pevzner, 1995).

6. Linear rankings with values (set of numbers): metrics (Rote, 1991).

7. Group rankings (elements with ordinal priorities): metrics (Cook and Kress, 1984; Kendall, 1962), vector-like proximity (Belkin and Levin, 1990).

8. Sets of strings: measures from [0, 1] (Lemone, 1982).

9. Trees: metrics (Boorman and Oliver, 1973; Robinson and Foulds, 1981; Margush, 1982), distance as the minimum cost transformation, and the largest common substructure (Tai, 1979), proximity (Barthelemy and Guenoche, 1991; Akutsu and Halldorsson, 1994).

10. Trees with labeled leafs: metrics (Hendy et al., 1984; Day, 1985).

11. Hierarchies: metrics (Botafogo et al., 1982).

12. Graphs and posets: metrics (Bogart, 1975; etc.), structural representations of proximity (Arabie and Hubert, 1992).

An aggregation of combinatorial objects is studied in (Hubert and Arabie, 1985; Day, 1985; Barthelemy et al, 1986; Gordon, 1986; Arabie and Hubert, 1992; etc.). A relationship of basic elements of our system model and combinatorial objects above is the following: (1) points in a space: leaf nodes, DAs, Is, priorities of DAs, estimates of Is, requirements; (2) sets: leaf nodes, DAs, Is, requirements; (3) partitions: requirements; (4) ordinal rankings: DAs, Is, requirements; (5) strings: DAs, Is; (6) linear rankings: DAs, Is, requirements; (7) sets of strings: DAs, requirements; (8) trees: system model, requirements; (9) posets: estimates of DAs and Is, requirements. Clearly, it is reasonable to apply typical system descriptions of decomposable systems as follows: (1) external requirements: criteria, factors, restrictions; (2) system model: structural (tree-like) model, leaf nodes (components), DAs; (3) extended system model: structural (tree-like) model, leaf nodes (components), DAs, Is, priorities of DAs, ordinal estimates of Is.

4. Vector-like proximity for rankings

Usually the proximity measures have been used as scalar functions which satisfy to Freshe's axioms for metrics (pseudo-metrics). Kendall's proximity measure is used the most widely (Kendall, 1962). Here we describe a vector-like proximity for rankings (Levin, 1988; Belkin and Levin, 1990). In this case, a measurement scale is a simplex in which for a set of measured objects a poset maybe constructed. Let $G = (A, E)$ be a digraph, where $A = \{1, ..., i, ..., n\}$ is a set of vertices (i.e., objects, discrete information units), E is a set of edges corresponding to a preference.

By the above we may examine the following kinds of the digraphs: (1) tree (denoted as T); (2) parallel-series graph (P); (3) acyclic graph or partial ordering (R); (4) chain or linear ordering (L); (5) layered structure S (group ordering, ranking or stratification) in which the set A is divided into m subsets (layers) without intersections as follows: (a) $A(l)$, $l = 1, ..., m$, and $\forall l_1, l_2, l_1 \neq l_2, |A(l_1) \& A(l_2)| = 0$, (b) if $l_1 < l_2$ then $i_1 \succ i_2$ $\forall i_1 \in A(l_1)$ and $\forall i_2 \in A(l_2)$; (6) fuzzy layered structure S_f allowing any object to belong to the group of successive layers: the number or the interval of layers which the ith object is belonged to is defined $\forall i \in A$: $\pi(i)$ or $d(i) = [d(1, i), d(2, i)]$,

$1 \le d(1,i) \le d(2,i) \le m$ and $\pi(i) = d(1,i)$ if $d(1,i) = d(2,i)$. Thus the system of intervals $\{\delta(i)\}$ is specified. By the analogy of definitions above it is possible to specify clusters, and fuzzy clusters. Sometimes the comparison of structures representing the union of similar graphs (e.g. 'chains'- NL, layered structures - NS) has a particular interest in practice. Let $\Theta(S)$ be a set of all layered structures on A.

Definition 1. We say that

$$\delta_\pi(i,S,Q) = \pi(i,S) - \pi(i,Q),$$

$$\delta_\pi(i,j,S,Q) = \pi(i,S) - \pi(j,S) - (\pi(i,Q) - \pi(j,Q)),$$

where $\pi(i,S) = l \ \forall i \in A(l)$ in S, are the first order error $\forall i \in A$, and the second order error $\forall(i,j) \in \{A * A | i \ne j \ \forall S, Q \in \Theta(S)\}$ respectively. Thus for an estimate of a discordance between the structures $S, Q \in \Theta(S)$ with respect to i and (i,j) we obtain an integer-valued scale with the following ranges: $-(m-1) \le r \le m-1$ for $\delta_\pi(i,S,Q)$, and $-2(m-1) \le r \le 2(m-1)$ for $\delta_\pi(i,j,S,Q)$.

Definition 2. Let

$$x(S,Q) = (x[-(m-1)], ..., x[-1], x[1], ..., x[m-1]), \qquad (1)$$

$$y(S,Q) = (y[-2(m-1)], ..., y[-1], y[1], ..., y[2(m-1)]), \qquad (2)$$

be vectors of an error (proximity) $\forall S, Q \in \Theta(S)$ with respect to components i (1st order), and the pairs (i,j) (2nd order). The vector components are:

$$x[r] = |\{i \in A | \delta_\pi(i,S,Q) = r\}|/n,$$

$$y[r] = 2|\{(i,j) \in \{A * A | i \ne j\} | \delta_\pi(i,j,S,Q) = r\}|/(n(n-1)).$$

It maybe reasonable to define similar vectors of the higher order also. Moreover it is possible to examine the weighted errors of the first and the second order with taking into account dependence on corresponding number l of layer $A(l)$ for the definition of vector components.

Now denote a set of arguments for the components of vectors x and y as follows: $\Omega = \{-k, ..., k\})$, negative values as Ω^-, and positive ones as Ω^+. In addition, we use vectors x with aggregate components of the following types (similarly, for y):

$$x[k_1, k_2] = \sum_{r=k_1}^{k_2} x[r], \quad x[\le -k] = \sum_{r=-(m-1)}^{-k} x[r], \quad x[\ge k] = \sum_{r=k}^{m+1} x[r], k > 0,$$

$$x[|r|] = x[r] + x[-r].$$

Definition 3. Let $|x(S,Q)| = \sum_{r \in \Omega} x[r]$, $|y(S,Q)| = \sum_{r \in \Omega} y[r]$ be modules of the vectors.

Afterhere we will consider vector x as a basic one.

Definition 4. We will call vectors truncated ones if
(1) the part of terminal components is rejected, e.g.

$$x(S,Q) = (x[-k_1], x[-(k_1-1)], ..., x[-1], x[1], ..., x[k_2-1], x[k_2]),$$

and one or both of following conditions are satisfied: $k_1 < m-1$, $k_2 < m-1$;
(2) aggregate components are used as follows:

$$x(S,Q) = (x[\le k_1], ..., x[k_a - 1], x[k_a, k_b], x[k_b + 1], ..., x[\ge k_2]),$$

$$x(S,Q) = (x[|1|], ..., x[|r|], ..., x[|k|]).$$ (3)

Definition 5. Let us call vector x (y):
(a) the two-sided one, if $|\Omega^+| \neq 0$ and $|\Omega^-| = 0$;
(b) the one-sided one, if $|\Omega^+| = 0$ or $|\Omega^-| = 0$;
(c) the symmetrical one, if $-r \in \Omega^-$ exists $\forall r \in \Omega^+$, and vice versa;
(d) the modular one, if it is defined with respect of definition 4 (3).
So we obtain a pair of linear orders on the components of vectors x (1), and y (2):
component $1(-1) \prec ... \prec$ *component* $k(-k)$.
Clearly, if the components are aggregate ones, the orders will be analogues ones.

Definition 6. $x_1(S,Q) \succeq x_2(S,Q)$, $\Omega(x_1) = \Omega(x_2)$, $\forall S,Q \in \Theta(S)$, if any decreasing of weak components x_1 in the comparison with x_2 is compensated by corresponding increasing of it's 'strong' components ($r, p \in \Omega^+$ or $-r, -p \in \Omega^-$):

$$\sum_{r \geq u}^{r} x_1[r] - \sum_{r \geq u}^{r} x_2[r] \geq 0, \forall u \in \Omega^+ (\forall - u \in \Omega^-, -r \leq -u).$$ (4)

It is possible to force condition (4) by using a right side which is equal to a parameter $v > 0$.

Definition 7. Let $M = \{x \in X | \sum_{r \in \Omega} x[r] = 1\}$ be a marginal set (similarly for y).

Note that $\forall x$ (y) there exists a dominating subset $D(x) = \{\eta \in M | \eta \succeq x\}$.

Definition 8. Let a pair of vectors x_1, x_2 (y_1, y_2) be:
(a) comparable ones, if $x_1 \succeq x_2$ (therefore $D(x_2) \supseteq D(x_1)$ and vice versa);
(b) strongly uncomparable ones, if $|D(x_1, x_2)| = |D(x_1) \& D(x_2)| = 0$;
(c) weakly uncomparable ones, if $|D(x_1, x_2)| \neq 0$, $D(x_1, x_2)$ does not include $D(x_1)$, $D(x_2)$.

Finally let us consider properties of our vector-like proximity as follows:

1. Condition (4) defines a poset.
2. $0 \leq |x(S,Q)| \leq 1$, $0 \leq |y(S,Q)| \leq 1$ $\forall S,Q \in \Theta(S)$.
3. $x(S,Q) \succeq (0,...,0)$, $y(S,Q) \succeq (0,..,0)$ $\forall S,Q \in \Theta(S)$.
4. The following condition is true for one-sided vectors: $x(S,Q) \prec (0,0,...,0,1)$, $y(S,Q) \prec (0,0,...,0,1)$.
5. The following condition is true for any two-sided vector $x(S,Q), \forall S,Q \in \Theta(S)$: there exists such vector $e = (e[-k_1], 0,...,0, e[k_2]) \in M(k_1, k_2 > 0)$, that $x(S,Q) \succeq e$ (similarly, for y).
6. For any modular vector the following is true: $x(S,Q) = x(Q,S)$ (similarly, for y).
7. For any two-sided symmetrical vector the following is true: $x(S,Q) = x^*(Q,S)$, where $x^*[r] = x[-r]$ (similarly, for y).
8. $\forall x(S,Q)$, $\forall S,Q \in \Theta(S)$, the following is true: if $x(S,Q) = (0,..,0)$ then $S = Q$.

An assessment of proximity between fuzzy layered structures is a more complicated problem. Let us consider an example of qualitative vector-like proximity for any fuzzy structures $S_f, Q_f \in \Theta(S_f)$, where $\Theta(S_f)$ is a set of all fuzzy layered structures on A.

Definition 9. Let $z(S_f, Q_f) = (z[-(m-1)], ..., z[-1], z[1], ..., z[m-1])$, where

$$z[r] = |\{i \in A | d(2,i,S_f) - d(1,i,Q_f) = r, \ |d(i,S_f) \& d(i,Q_f)| = 0\}|/n, \ r > 0;$$

$$z[r] = |\{i \in A | d(1,i,S_f) - d(2,i,Q_f) = -r, \ |d(i,S_f) \& d(i,Q_f)| = 0\}|/n, \ r < 0;$$

be a vector of the 1st order proximity between $\forall S_f, Q_f \in \Theta(S_f)$ with respect to $\forall i \in A$.

In the same way, we may describe properties for vectors z, which are similar to those of vector x (y) (besides the 8th one).

5. Example

Let us consider an example: $A = \{1,2,3,4,5,6,7,8,9\}$; S^1 : $A_1 = \{2,4\}, A_2 = \{9\}, A_3 = \{1,3,7\}, A_4 = \{5,6,8\}$; and S^2 : $A_1 = \{7,9\}, A_2 = \{1,3\}, A_3 = \{2,5,8\}, A_4 = \{4,6\}$. Let $\|g_{ij}\|$, $(i,j \in A)$ be an adjacency matrix for graph G:

$$g_{ij} = \begin{cases} 1, & \text{if } i \succ j, \\ 0, & \text{if } i \sim j, \\ -1, & \text{if } i \prec j. \end{cases}$$

Then Kendall proximity measure (metric) for graphs G^1 and G^2 is the following (Kendall, 1962): $\rho_K(G^1, G^2) = \sum_{i<j} |g_{ij}^1 - g_{ij}^2|$, where g_{ij}^1, g_{ij}^2 are elements of adjacency matrices of graphs G^1 and G^2 respectively. Adjacency matrices, corresponding to our example, are the following:

$$|g_{ij}(S^1)| = \begin{pmatrix}
. & -1 & 0 & -1 & 1 & 1 & 0 & 1 & -1 \\
1 & . & 1 & 0 & 1 & 1 & 1 & 1 & 1 \\
0 & -1 & . & -1 & 1 & 1 & 0 & 1 & -1 \\
1 & 0 & 1 & . & 1 & 1 & 1 & 1 & 1 \\
-1 & -1 & -1 & -1 & . & 0 & -1 & 0 & -1 \\
-1 & -1 & -1 & -1 & 0 & . & -1 & 0 & -1 \\
0 & -1 & 0 & -1 & 1 & 1 & . & 1 & -1 \\
-1 & -1 & -1 & -1 & 0 & 0 & -1 & . & -1 \\
1 & -1 & 1 & -1 & 1 & 1 & 1 & 1 & .
\end{pmatrix}$$

$$|g_{ij}(S^2)| = \begin{pmatrix}
. & 1 & 0 & 1 & 1 & 1 & -1 & 1 & -1 \\
-1 & . & -1 & 1 & 0 & 1 & -1 & 0 & -1 \\
0 & 1 & . & 1 & 1 & 1 & -1 & 1 & -1 \\
-1 & -1 & -1 & . & -1 & 0 & -1 & -1 & -1 \\
-1 & 0 & -1 & 1 & . & 1 & -1 & 0 & -1 \\
-1 & -1 & -1 & 0 & -1 & . & -1 & -1 & -1 \\
1 & 1 & 1 & 1 & 1 & 1 & . & 1 & 0 \\
-1 & 0 & -1 & 1 & 0 & 1 & -1 & . & -1 \\
1 & 1 & 1 & 1 & 1 & 1 & 0 & 1 & .
\end{pmatrix}$$

So, Kendall's distance is: $\rho_K(S^1, S^2) = 31$. Proposed vector-like proximity allows to describe dissimilarity between two structures more prominent:

$$\pi(S^1) = (\pi_1(S^1), ..., \pi_i(S^1), ..., \pi_4(S^1)) = (3,1,3,1,4,4,3,4,2),$$

$$\pi(S^2) = (\pi_1(S^2), ..., \pi_i(S^2), ..., \pi_4(S^2)) = (2,3,2,4,3,4,1,3,1).$$

$$\delta_i^\pi(S^1, S^2) = (1,-2,1,-3,1,0,2,1,1),$$

$$\delta_{ij}^\pi(S^1, S^2) = \begin{pmatrix}
. & 3 & 0 & 4 & 0 & 1 & -1 & 0 & 0 \\
-3 & . & -3 & -1 & -3 & -2 & -4 & -3 & -3 \\
0 & 3 & . & -4 & 0 & 1 & 1 & 0 & 0 \\
-4 & 1 & 4 & . & -4 & -3 & -5 & -4 & -4 \\
0 & 3 & 0 & 4 & . & 1 & -1 & 0 & 0 \\
-1 & -2 & -1 & 3 & -1 & . & -1 & -1 & -1 \\
1 & 4 & -1 & 5 & 1 & 1 & . & 1 & 1 \\
0 & 3 & 0 & 4 & 0 & 1 & -1 & . & 0 \\
0 & 3 & 0 & 4 & 0 & 1 & -1 & 0 & .
\end{pmatrix}$$

$$x(S^1, S^2) = (x^{-3}, x^{-2}, x^{-1}, x^1, x^2, x^3) = (1, 1, 0, 5, 1, 0),$$
$$y(S^1, S^2) = (y^{-6}, y^{-5}, y^{-4}, y^{-3}, y^{-2}, x^{-1}, y^1, y^2), y^3, y^4, y^5, y^6) =$$
$$(0, 1, 5, 5, 1, 6, 6, 0, 1, 1, 0, 0).$$

Here we do not use in x and y the coefficients: $\frac{1}{n}$ and $\frac{2}{n(n-1)}$. Aggregate components can be applied too: $x(S^1, S^2) = (x^{\leq -1}, x^{\geq 1}) = (2, 6)$.

6. Conclusion

A combinatorial approach (e.g., description, design, transformation) to decomposable systems has been described in (Levin 1996a, Levin, 1996b). Comparison of decomposable systems can be applied in decision making, information engineering (e.g., hypertext systems), network design, etc. Clearly, that composing a global system proximity from proximities of system parts is a central problem. Finally, let us point out basic kinds of changes for our system description: (1) internal changes: microlevel (DAs), parts (subsystems, requirements), macrolevel (model); (2) external changes: requirements. Note proposed vector-like proximity can be used in aggregation of rankings.

References:

Arabie, Ph. and Boorman, S.A. (1973): Multidimensional Scaling of Measures of Distance between Partitions. *J. of Math. Psychology*, **10**, 2, 148–203.

Arabie, Ph. and Hubert, L.J. (1992): Combinatorial Data Analysis. *Annual Review of Psychology*, **43**, 169–203.

Barthelemy, J.-P., Guenoche, A. (1991): *Trees and proximity Representation*, Wiley, New York.

Barthelemy, J.-P., Leclerc, B. and Monjardet. B. (1986): On the Use of Ordered Sets in Problems of Comparison and Consensus of Classifications. *J. of Classification*, **3**, 2, 17–224.

Belkin, A.R. and Levin, M.Sh. (1990): *Decision Making: Combinatorial Models of Information Approximation*, Nauka Publishing House, Moscow (in Russian).

Bogart, K.P. (1973): Preference Structures I: Distance between Transitive Preference Relations. *J. Math. Soc.*, **3**, 49–67.

Boorman, S.A. and Oliver, D.C. (1973): Metrics on Spaces of Finite Trees. *J. of Math. Psychology*, **10**, 1, 26–59.

Botafogo, R.A., Rivlin, E. and Shneiderman, B. (1992): Structural Analysis of Hypertexts: Identifying Hierarchies and Useful Metrics. *ACM Trans. on Information Systems*, **10**, 2, 142–180.

Cook, W.D. and Kress. M. (1984): Relationships between l^1 Metrics on Linear Rankings Space. *SIAM J. on Appl. Math.*, **44**, 1, 209–220.

Day, W.H.E. (1985): Optimal Algorithms for Comparing Trees with Labeled Leafs. *J. of Classification*, **2**, 1, 7–28.

Ganter, B. and Wille, R. (1989): Conceptual Scaling. *In Applications of Combinatorics and Graph Theory to the Biological and Social Sciences*. Roberts F.S. (ed.), 140–167, Springer-Verlag, New York.

Gordon, A.D. (1986): Consensus Supertrees: The Synthesis of Rooted Trees Containing Overlapping Sets of Labeled Leafs. *J. of Classification*, **3**, 2, 335–348.

Gower, J.C. (1985): Measures of Similarity, Dissimilarity, and Distance. *In Encyclopedia of Statistical Sciences*, **5**, S. Kotz, and N.L. Johnson (eds.), 397-405, Wiley, New York.

Hannenhalli, S. and Pevzner, P. (1995): *Transforming Men into Mice*, Technical Report CSE-95-012. Dept. of Computer Science and Engineering, Pennsylvania State University.

Hendy, M.D., Little, C.H.C. and Penny, D. (1984): Comparing Tress with Pendant Vertices Labelled. *SIAM J. on Appl. Math.*, **44**, 5, 1054-1065.

Hubert, L. and Arabie, P. (1985): Comparing Partitions. *J. of Classification*,**2**, 3, 193-218.

Jawhari, E.M., Pouzet, M. and Missane, D. (1986): Retracts: Graphs and Ordered Sets From The Metrics Point of View. *In Contemporary Mathematics*, **57**, Rival, I. (ed.), AMS, Providence.

Kemeny, J.G. and Snell, J.L. (1972): *Mathematical Models in the Social Sciences*, MIT Press, Cambridge, Mass.

Kendall, M. (1962): *Rank Correlation Methods*, 3rd ed., Hafner, New York.

Kruskal, J.B. (1977): The relationship between multidimensional scaling and clustering. *In Classification and Clustering*, Van Ryzin, J. (ed.). 17-44, Academic, New York.

Lemone, K.A. (1982): Similarity Measures Between Strings Extended to Sets of Strings. *IEEE Trans. on Pattern Analysis and Machine Intelligence*, 4, 3, 345-347.

Levin, M.Sh. (1988): Vector-like Criterion for Proximity of Structures. *In Abstracts of The Third Conf. "Problems and Approaches of Decision Making in Organizational Management Systems"*, Inst. for System Analysis, Moscow, 20-21 (in Russian).

Levin, M.Sh. (1996a): Towards Combinatorial Engineering of Decomposable Systems, *In 13th European Meeting on Cybernetics and Systems*, 1, Vienna, 265-270.

Levin, M.Sh. (1996b): Hierarchical Morphological Design of Decomposable Systems. *Concurrent Engineering: Research and Applications*, 4, 2, 111-117.

Margush, T. (1982): Distance Between Trees. *Disc. Appl. Math.*, 4, 281-290.

Mirkin, B.G. and Chernyi, L.B. (1970): On Measurement of Distance Between Partitions of a Finite Set of Units. *Automation and Remote Control*, **31**, 786-792.

Pouzet, M. and Rosenberg, I.G. (1994): General Metrics and Contracting Operations. *Discrete Mathematics*, **130**, 1-3, 103-169.

Robinson, D.F. and Foulds, L.R. (1981): Comparison of Phylogenetic Trees. *Mathematical Biosciences*, **53**, 1/2, 131-147.

Rote, G. (1991): Computing the Minimum Hausdorff Distance between Two Point Sets on a Line under Translation. *Information Processing Letters*, **38**, 3, 123-127.

Sellers, P.H. (1974): An Algorithm for the Distance between two Finite Sequences. *J. of Combinatorial Theory, Ser. A*, **16**, 253-258.

Tai, K.-C. (1979): The Tree-to-Tree Correction Problem. *J. of the ACM*, **26**, 3, 422-433.

Timkovskii, V.G. (1989): Complexity of Common Subsequence and Supersequence Problems and Related Problems, *English translation from Kibernetika*, **5**, 1-13.

Torgenson, W.S. (1958): *Theory and Method of Scaling*. Wiley, New York.

Wagner, R.A. and Fisher, M.J. (1974): The String-to-String Correction Problem. *J. of the ACM*, **21**, 1, 168-173.

Zeitoun, J. (1977): *Trames Planes. Introduction a une etude architecturale des trames*. Bordas, Paris.

Performance of Eight Dissimilarity Coefficients to Cluster a Compositional Data Set

Maria Cristina Martín

Faculty of Engineering Sciences, Osaka University
Machikaneyamacho 1–3, Toyonaka-shi
Osaka 560, Japan

Summary: Concerned with the problem of clustering a compositional data set consisting of vectors of positive components subject to a unit-sum constraint, as a first step we looked for an appropriate dissimilarity coefficient or distance between two compositions. In this paper we selected eight different dissimilarity measures, and their performance was evaluated by means of graphics and cluster validity coefficients of six clustering methods applied to three compositional data sets. Almost recent criteria for measures of compositional difference are also tested for those measures emerging as the best to cluster compositions.

1. Introduction

Any compositional data set consists of N D-part compositions with x_{ri} the ith-component of the rth-composition $(r = 1, \ldots, N; i = 1, \ldots, D)$ satisfying the requirements that each component is non-negative and that the sum of all the components in each composition is 1.

It is our objective to study the problem of Cluster Analysis for compositonal data sets. As a first step, we will look for an appropriate dissimilarity coefficient or distance between two compositions. Aitchison (1992) discussed some criteria to define a measure of difference between two compositions, although no clustering problems were tried by the author. Briefly, Aitchison's proposal can be stated as follows:

The appropriate sample space for a compositional data vector $x = (x_1, \ldots, x_D)$ is the d-dimensional positive simplex (Aitchison (1986, Chapter2))

$$S^d = \{(x_1, \ldots, x_D): \ x_1 > 0, \ldots, x_D > 0 \ and \ x_1 + \cdots + x_D = 1\}. \tag{1}$$

As a way to eliminate the constant-sum constraint difficulty that each compositional data vector $x \in S^d$ must satisfy and therefore to work in a space without restrictions, Aitchison (1986, Sec.4.6) proposed the following two transformations of the data:

- the <u>logratio vector</u> $y = \log\left(\frac{x_{-D}}{x_D}\right) \in R^d$, where x_{-D} is the vector x with the last component omitted, and
- the <u>centred logratio vector</u> $z = \log\left(\frac{x}{g(x)}\right) \in R^D$, where $g(x) = (x_1 \cdots x_D)^{\frac{1}{D}}$ is the geometric mean of the components.

Aitchison's ideas are based on three postulates:

1.- A composition contains information about relative, not absolute, magnitudes of its components. In other words, by writing $r_i = \frac{x_i}{x_D}$ $(i = 1, \cdots, d)$ a composition can be completely determined by $x_i = \frac{r_i}{r_1 + \cdots + r_d + 1}$ $(i = 1, \cdots, d)$ and $x_D = \frac{1}{r_1 + \cdots + r_d + 1}$;

2.- Any discussion of the variability of a composition can be expressed in terms of ratios, such as $\frac{x_i}{x_j}$ of components;

3.- No compositional information is lost if $\log\left(\frac{x_i}{x_j}\right)$ is studied instead of ratios themselves.

Then, Aitchison(1992) (see also Aitchison(1994)) derives

$$\left[\sum_{i<j}\left(\log(\frac{x_{ri}}{x_{rj}}) - \log(\frac{x_{si}}{x_{sj}})\right)^2\right]^{\frac{1}{2}} \tag{2}$$

"as the simplest and most tractable measure of difference or distance between two compositions $x_r = (x_{r1}, \ldots, x_{rD})$ and $x_s = (x_{s1}, \ldots, x_{sD})$". The above expression, apart from a constant factor, coincides with the Euclidean distance between two centred logratio vectors, which is the first proposal of Aitchison (1986, p.193) as the inquired measure.

The fundamental point of Aitchison's criteria in order to obtain (2) is:

" Any scalar measure of difference between two compositions x_r and x_s must be expressible in terms of ratios of components in the same composition (this is in terms of $\frac{x_{ri}}{x_{rj}}$ and $\frac{x_{si}}{x_{sj}}$), and also as ratios of components in different compositions (this is as $\frac{x_{ri}}{x_{si}}$ or $\frac{x_{si}}{x_{ri}}$). In other words, the measure will be expressible as a function of the ratio $\frac{r}{R}$ (or in a symmetric way as $\frac{R}{r}$) where $r = \frac{x_{ri}}{x_{rj}}$ and $R = \frac{x_{si}}{x_{sj}}$".

In the present paper, we first review some dissimilarity coefficients or distances which are currently being used in Cluster Analysis and select those we consider capable of measuring the difference between two compositions. Then, we conduct a study of six clustering methods applied to three compositional data sets in order to explore, and compare with Aitchison's proposal, the performance of the different dissimilarities displayed. This profit is evaluated by means of graphics and coefficients of cluster validity of the many clustering techniques used. Also, Aitchison's criteria for those coefficients emerging as the best to cluster compositions are tested.

2. Dissimilarity Measures and Cluster Methods considered

As possible measures of difference between two compositions $x_r = (x_{r1}, \ldots, x_{rD})$ and $x_s = (x_{s1}, \ldots, x_{sD})$ we consider:

City Block or Manhattan Distance: $\quad \sum_{i=1}^{D} |x_{ri} - x_{si}|$

Euclidean Distance: $\quad \left[\sum_{i=1}^{D} (x_{ri} - x_{si})^2\right]^{\frac{1}{2}}$

Chebychev Distance: $\quad \max_i |x_{ri} - x_{si}|$

Jeffreys Matusita Distance: $\quad \left[\sum_{i=1}^{D} \left(\sqrt{x_{ri}} - \sqrt{x_{si}}\right)^2\right]^{\frac{1}{2}}$

Divergence Dissimilarity of Jeffreys: $\quad \sum_{i=1}^{D} \log\left(\frac{x_{ri}}{x_{si}}\right)(x_{ri} - x_{si})$

Bhattacharyya (log): $\quad -\log\left(\sum_{i=1}^{D} \sqrt{x_{ri}}\sqrt{x_{si}}\right)$

Bhattacharyya (arccos): $\quad \arccos\left(\sum_{i=1}^{D} \sqrt{x_{ri}}\sqrt{x_{si}}\right)$

Aitchison Distance: $\quad \left[\sum_{i=1}^{D} \left(\log(\frac{x_{ri}}{g(\boldsymbol{x}_r)}) - \log(\frac{x_{si}}{g(\boldsymbol{x}_s)})\right)^2\right]^{\frac{1}{2}}$

Based on the above cited measures, four Hierarchical and two Partitioning Clustering Methods were considered. Single, Complete and Average Linkage were the Agglomerative Hierarchical Clustering Methods used, and MacNoughton-Smith algorithm was the Divisive Hierarchical Clustering Method applied. Among Partitioning

Clustering Methods we have considered Partitioning around Medoid and Fuzzy Analysis. Single and Complete Linkage outputs were obtained by means of SPSS or Splus packages; whereas, programs AGNES (Average Linkage), DIANA (Divisive Analysis), PAM (Partitioning around Medoid) and FANNY (Fuzzy Analysis) of Kaufman and Rousseeuw (1990) were used in the computation of the other algorithms.

3. Criteria of Comparison

The behavior of the cited dissimilarity coefficients or distances in order to group compositions was evaluated by means of:

3.1: The usual Dendrogram when Single and Complete Linkage were applied.

3.2: Banner plots and the corresponding Agglomerative and Divisive coefficients when AGNES and DIANA were used (Kaufman and Rousseeuw (1990, p.212 and p.263)). Both coefficients range between 0 and 1. A coefficient nears 1 indicates that a strength clustering structure has been found.

3.3: Silhouette plots and the Silhouette coefficients when PAM and FANNY were considered (Kaufman and Rousseeuw (1990, p.83)). Conclusions here are based on Kaufman and Rousseeuw subjective interpretations of the silhouette coefficient (SC):

$0.75 \leq SC \leq 1.00$, a strong structure has been found,

$0.50 \leq SC < 0.75$, the structure is reasonable,

$0.25 \leq SC < 0.50$, the structure is weak and could be artificial,

$SC < 0.25$, no substantial structure has been found.

3.4: The cluster validity was also evaluated by means of the normalized version of Dunn's partition coefficient (Kaufman and Rousseeuw (1990, Chapter 4)) when FANNY was applied. Dunn's partition coefficient always varies from 0 (entirely fuzzy) to 1 (hard clustering).

4. Applications and Results

Three compositional data sets (two of them consisting of three-part compositions, and therefore, allowing a visual representation in a ternary diagram) were used in all the comparisons:

- (Sand, Silt, Clay) compositions of 39 Sediment samples in an Arctic Lake (Aitchison(1986, p.359, Data5));

- (Albite, Blandite, Cornite) compositions, abbreviated as (ABC), of 25 specimens of Hongite and 25 specimens of Kongite (Aitchison(1986, p.354-356, Data 1 and 2));

- (Albite, Blandite, Cornite, Daubite, Endite) or (ABCDE) compositions, for 25-Hongite and 25-Kongite specimens (Aitchison(1986, p.354-356, Data 1 and 2)).

In the ternary diagram of Figure 1 are represented the ABC subcompositions for Hongite and Kongite specimens. When Single Linkage was applied, the marked curvature or banana configuration displayed by the data apparently prevails over the type of dissimilarity proposed. This means that, independent of the measure used, from the dendrogram by Single Linkage we can construct a cluster with 49 elements and the remaining element in another cluster (the loop, in Figure 1,was merely drawn to indicate the isolated point). So the special criticized property of Single Linkage to give a long serpentine cluster (the so called "chaining") allowed in this example to join compositions with insignificant proportions of cornite and compositions with insignificant proportions of blandite. The same effect was observed from the dendrograms of Single Linkage for (ABCDE) compositions.

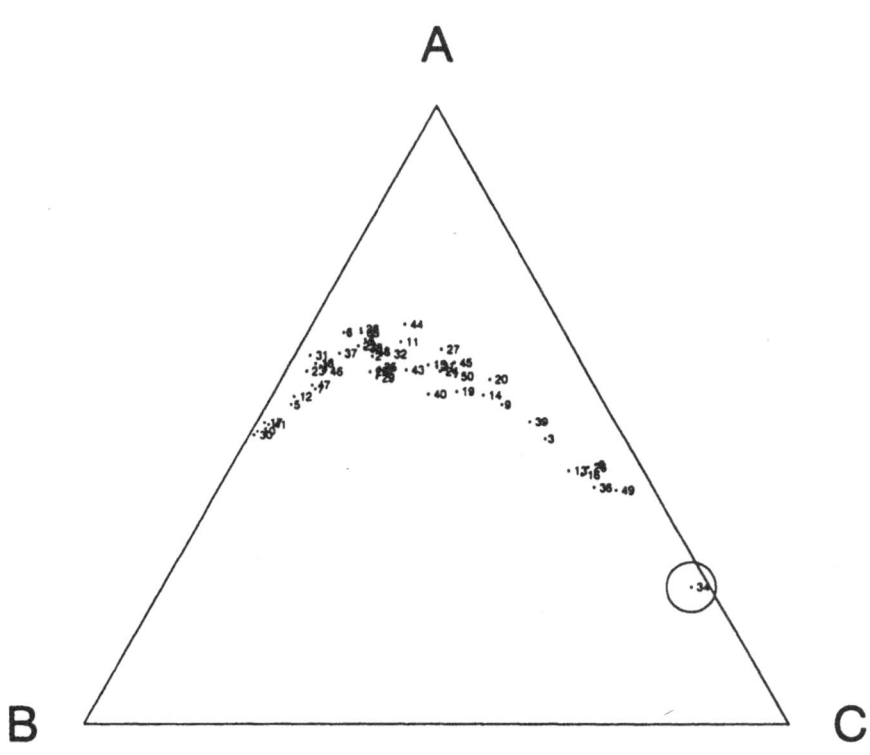

Figure 1: Ternary Diagram of (A,B,C) subcompositions of 25 specimens of Hongite and 25 specimens of Kongite

The ternary diagrams of Figure 2 summerize the observations from the dendrograms by Single Linkage of the three-part compositions of 39-samples in the Arctic Lake Sediments.

The extensive variation in the ratio of clay to sand is again represented in the shape of a banana. Even when the ternary diagrams show a clear cluster of compositions with low proportions of sand and approximately equal proportions of silt and clay, this cluster was not detected by Aitchison's approach when Single Linkage was applied. Aitchison's distance prefers to isolate compositions with a very low proportion of clay (Figure 2. (a)), whereas for the other seven coefficients the same four and six cluster solutions could be observed (Figure 2. (b)). Clearly, the last four-solutions are reasonable, but not the solutions using Aitchison's coefficient.

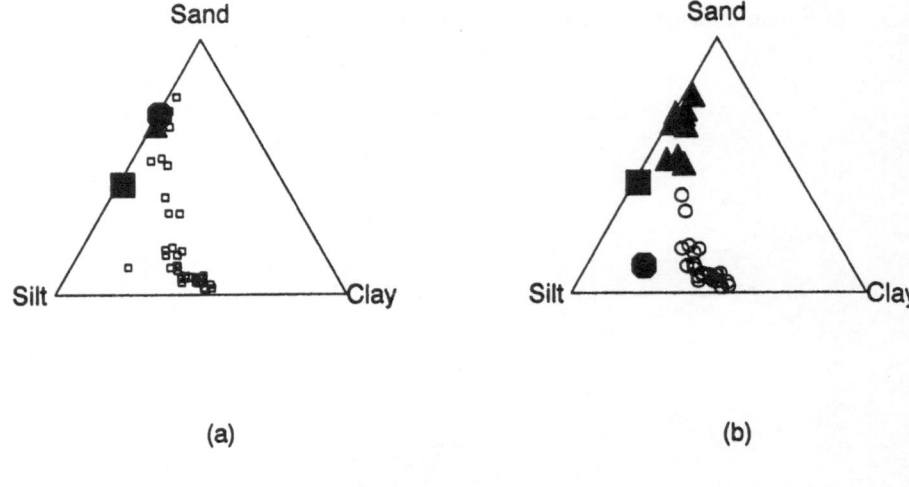

(a)

(b)

(a) 4-clusters using Aitchison's distance
(b) 4-clusters using 7-different coefficients

Figure 2: Ternary Diagrams of 39-(Sand,Silt,Clay) compositions in the Arctic Lake Sediments and the 4-cluster solutions by Single Linkage with eight different dissimilarity coefficients

No special or different behavior of dissimilarities could be observed from the dendrograms by Complete Linkage. However, the structures are more compact when Divergence of Jeffreys and Bhattacharyya(log) dissimilarities are used. However, "very compact clusters even not necessarily well separated" is the known tendency of Complete Linkage.

Tables 1 to 4 summarize the cluster coefficients of AGNES, DIANA, PAM and FANNY for the eight dissimilarities under study in each of the compositional data. Because of the amount of output, only coefficients (not plots) are shown here. Also, we reproduce silhouette coefficients by PAM (not by FANNY) and Dunn's partition coefficients by FANNY for $k = 2, \ldots, 7$ where k is the selected number of clusters (results for $k = 8, \ldots, N - 1$ are, in general, still worse).

All these coefficients (Tables 1 to 4) underline the fact that Divergence of Jeffreys and Bhattacharyya (log) dissimilarities produce the better structures of the compositional data sets considered.

Table 1: Agglomerative Coefficients of three different Compositional Data Sets using eight different Dissimilarity Coefficients

C.Data	Manhat	Euclid	Cheby.	J.Mats	Diverg	B(log)	B(acos)	Aitch.
Sedim.	0.93	0.93	0.93	0.93	0.99	0.99	0.93	0.93
HK(3)	0.94	0.94	0.94	0.92	0.99	0.99	0.93	0.94
HK(5)	0.89	0.90	0.91	0.89	0.99	0.98	0.89	0.90

Table 2: Divisive Coefficients of three different Compositional Data Sets using eight different Dissimilarity Coefficients

C.Data	Manhat	Euclid	Cheby.	J.Mats	Diverg	B(log)	B(acos)	Aitch.
Sedim.	0.95	0.94	0.95	0.94	0.99	0.99	0.94	0.96
HK(3)	0.96	0.96	0.96	0.95	1.00	1.00	0.96	0.97
HK(5)	0.93	0.94	0.95	0.94	1.00	1.00	0.94	0.95

Table 3: Silhouette Coefficients by PAM of three different Compositional Data Sets using eight different Dissimilarity Coefficients

Arctic Lake Sediments - Sand, Silt and Clay compositions								
# cl.	Manhat	Euclid	Cheby.	J.Mats	Diverg	B(log)	B(acos)	Aitch.
2	0.72	0.71	0.72	0.70	0.87	0.88	0.71	0.67
3	0.52	0.52	0.52	0.55	0.71	0.71	0.55	0.54
4	0.52	0.52	0.52	0.51	0.67	0.67	0.50	0.50
5	0.52	0.53	0.52	0.55	0.74	0.74	0.54	0.51
6	0.51	0.51	0.51	0.48	0.75	0.74	0.45	0.52
7	0.54	0.53	0.54	0.46	0.60	0.60	0.47	0.49
Hongite and Kongite - ABC subcompositions								
# cl.	Manhat	Euclid	Cheby.	J.Mats	Diverg	B(log)	B(acos)	Aitch.
2	0.64	0.63	0.64	0.55	0.74	0.75	0.57	0.54
3	0.52	0.51	0.52	0.50	0.70	0.69	0.53	0.53
4	0.54	0.54	0.54	0.51	0.73	0.73	0.55	0.51
5	0.53	0.53	0.53	0.54	0.74	0.74	0.55	0.56
6	0.53	0.53	0.53	0.47	0.74	0.74	0.52	0.52
7	0.46	0.48	0.46	0.50	0.72	0.72	0.51	0.52
Hongite and Kongite - ABCDE compositions								
# cl.	Manhat	Euclid	Cheby.	J.Mats	Diverg	B(log)	B(acos)	Aitch.
2	0.58	0.60	0.61	0.52	0.72	0.72	0.53	0.49
3	0.46	0.48	0.52	0.46	0.66	0.67	0.47	0.43
4	0.45	0.47	0.48	0.44	0.64	0.64	0.44	0.40
5	0.44	0.46	0.48	0.39	0.63	0.63	0.39	0.41
6	0.44	0.46	0.50	0.30	0.62	0.58	0.30	0.39
7	0.35	0.37	0.41	0.29	0.59	0.59	0.32	0.39

Table 4: Dunn's Partition Coefficients by FANNY of three different Compositional Data Sets using eight different Dissimilarity Coefficients

Arctic Lake Sediments - Sand, Silt and Clay compositions							
# cl.	Manhat	Euclid	Cheby.	J.Mats	Diverg	B(log)	B(acos)	Aitch.
2	0.58	0.57	0.58	0.56	0.82	0.82	0.56	0.51
3	0.42	0.42	0.42	0.42	0.72	0.72	0.42	0.41
4	0.42	0.42	0.42	0.39	0.71	0.71	0.39	0.37
5	0.36	0.36	0.36	0.32	0.64	0.64	0.32	0.34
6	0.32	0.32	0.32	0.32	0.65	0.64	0.30	0.34
7	0.34	0.33	0.34	0.33	0.67	0.66	0.32	0.35
Hongite and Kongite - ABC subcompositions							
# cl.	Manhat	Euclid	Cheby.	J.Mats	Diverg	B(log)	B(acos)	Aitch.
2	0.38	0.37	0.38	0.36	0.68	0.68	0.36	0.34
3	0.34	0.33	0.34	0.35	0.67	0.67	0.36	0.36
4	0.36	0.36	0.36	0.37	0.70	0.70	0.38	0.37
5	0.37	0.37	0.37	0.31	0.70	0.67	0.37	0.38
6	0.35	0.35	0.35	0.30	0.69	0.69	0.34	0.36
7	0.33	0.33	0.33	0.32	0.67	0.67	0.34	0.35
Hongite and Kongite - ABCDE compositions							
# cl.	Manhat	Euclid	Cheby.	J.Mats	Diverg	B(log)	B(acos)	Aitch.
2	0.26	0.29	0.31	0.28	0.63	0.64	0.28	0.24
3	0.25	0.27	0.30	0.27	0.62	0.62	0.27	0.25
4	0.24	0.26	0.30	0.25	0.61	0.60	0.25	0.21
5	0.18	0.23	0.28	0.18	0.57	0.56	0.19	0.17
6	0.18	0.20	0.23	0.16	0.55	0.55	0.16	0.17
7	0.17	0.19	0.22	0.16	0.50	0.50	0.16	0.14

5. Conclusions

The rank order performance for the dissimilarity coefficients or distances examined was replicated for the six clustering methods applied to the three compositional data sets, namely:

5.1: Divergence of Jeffreys and Bhattacharyya using logarithm produce the better (excellent in general) recovery of cluster structures;

5.2: The relative merits of the other coefficients are not straightforward. They indicate the lowest recovery values, especially when partitioning clustering methods are used.

Partitioning Clustering Methods appear more sensible than Hierarchical Clustering Methods to evaluate the performance of the different dissimilarity coefficients in order to cluster a compositional data set. For example, Agglomerative and Divisive Coefficients, independent of the measure used, always ranged between 0.90 and 1 (strong clustering structures). However, Silhouette and Dunn's partition coefficients clearly allow the separation of Divergence and Bhattacharyya(log), given reasonable or good structures, from the other measures under analysis which show weak or poor structures.

Finally, since Aitchison's proposal does not reveal advantages over the other coefficients investigated, and it also can conduce to unreasonable cluster structures, we wish to know if Divergence of Jeffreys and Bhattacharyya(log) dissimilarities satisfy the cited Aitchison's criteria for measures of compositional difference. That is, can we express Divergence and Bhattacharyya(log) dissimilarities as a function of $\frac{r}{R}$ with $r = \frac{x_{ri}}{x_{rj}}$ and $R = \frac{x_{si}}{x_{sj}}$, $(r, s = 1, \ldots, N; \ i, j = 1, \ldots, D)$?

Considering two-part compositions Divergence Dissimilarity of Jeffreys becomes:

$$\log\left(\frac{r}{R}\right)\left[\frac{(r+1) - (R+1)}{(r+1)(R+1)}\right]. \tag{3}$$

A similar computation does not allow us to express Bhattacharyya(log) coefficient as a function of the ratio $\frac{r}{R}$.

Thus, despite some complex function of $\frac{r}{R}$ but according to Aitchison's criteria, Divergence Dissimilarity of Jeffreys is an admissible measure of difference between two compositions. Additional research is essential to judge Divergence's apparently superior behavior and consequently to support it as "the" measure to cluster a compositional data set.

5. References

Aitchison, J. (1986): The Statistical Analysis of Compositional Data. *Chapman and Hall, London*

Aitchison, J. (1992): On criteria for Measures of Compositional Difference. *Mathematical Geology*, Vol. 24, No.4, p.365-379.

Aitchison, J. (1994): Principles of Compositional Data Analysis. *Multivariate Analysis and its Applications. IMS Lectures Notes-Monograph Series*, Vol.24, p.73-81.

Kaufman, L. and Rousseeuw, P.J. (1990): Finding Groups in Data. *John Wiley and Sons, Inc.*

Consensus of Hierarchical Classifications

Bruno Simeone [1]. Maurizio Vichi [2]

[1] Dipartimento di Statistica, Probabilità e Statistiche Applicate,
Università "La Sapienza" di Roma,
P.le A. Moro, 5, 00185, Roma, Italy.
e-mail: marsalis@rosd.sta.uniroma1.it

[2] Dipartimento di Metodi Quantitativi e Teoria Economica,
Università "G. D'Annunzio" di Chieti.
Viale Pindaro, 42, 65127, Pescara. Italy.
e-mail: vichi@dmqte.unich.it.

Summary: In this paper, after briefly reviewing the techniques to achieve one or more consensus dendrograms, we propose new algorithms. These perform a sequence of elementary tree operations to obtain at each step, in a greedy fashion, a dendrogram that is as close as possible to the given ones. The time and the space complexities of such procedures make them suitable for large scale applications. A numerical example is used to show the proposed technique.

Keywords: consensus classification, hierarchical classification, n-tree ultrametric matrices

1. Introduction.

Often in analyzing multivariate data, a set of hierarchical classifications can be determined as the result of: (*i*) different hierarchical clustering methods applied on the same set of n objects; (*ii*) the same clustering algorithm (aggregative or divisive) embodying different proximity measures between pairs of objects; (*iii*) a hierarchical technique applied on r *frontal slices* of a *three-way* data set, i.e., on different multivariate data sets, relative to the same n multivariate objects examined in different occasions, such as times, spaces, etc. In this paper we will consider especially the last case.

One general problem in numerical taxonomy is to compare and synthesize the given set of hierarchical classifications, determining a consensus structure (often an n-tree. less frequently a dendrogram). The intuitive reasons to search for a consensus may be different: in case (*i*) we are looking for a single and more natural classification not depending on the clustering technique considered; or, in case (*ii*), for a classification not depending on the chosen dissimilarity measure; or, in case (*iii*), we may want to synthesize several different classifications into a single one, by detecting the relevant and stable information in the individual classifications.

In consensus theory, much attention has been devoted to taxonomic models such as n-trees and three approaches have been identified (Barthélemy, Leclerc and Monjardet, 1986) to determine a consensus n-tree: (1) *constructive* ones, (Adams. 1972; Margush and McMorris, 1981; McMorris et al., 1983), where purely combinatorial methods are proposed: these generally involve heuristic algorithms with interesting properties; (2) *axiomatic* ones, (McMorris and Neumann, 1983; Neumann, 1983; Stinebrickner, 1984; Barthélemy and McMorris. 1986; Day et al., 1986; Neumann and Norton. 1986), whereby after the formulation of some general "particularly desirable" properties of consensus procedures (or functions) methods satisfying these properties are defined if possible; (3) *optimization* ones (Barthélemy and Monjardet, 1981), where after the choice of a distance index between each observed n-tree and the consensus n-tree, the consensus minimizing such distance is determined. The three approaches are not independent, since, for example, consensus methods defined via the optimization approach generally give rise to heuristic procedures satisfying some given axioms or properties. These heuristics procedures are used to define initial good solutions for the optimization algorithm.

An alternative approach searches for a common pruned tree (Rosen. 1978) by pruning the least number of leaves of the given trees in order to render them equivalent. An interesting overview on consensus methods is given in Gordon. (1996).

In general, a consensus n-tree discovers replicated classes belonging, completely or partially, to more than one classification. but the distances among the replicated classes are not evaluated. However, these distances are fundamental to help us to choose one among (at most) n-1 partitions, in the consensus n-tree. For this reason we are interested in determining a consensus dendrogram. In this field, the few available consensus procedures generally follow a constructive or axiomatic approach. In practice, a consensus dendrogram is produced on the basis of a consensus procedure for n-trees (e.g., Neumann, 1983; Stinebrickner, 1984): in this way one obtains a consensus n-tree and can use a level function for each node of the consensus n-tree. Hence, for example. we can choose an intersection method (strict consensus, cardinality intersection, *durchschnitt* rule. etc.) and a level function such as the arithmetic mean.

Among the three approaches for determining a consensus dendrogram, we prefer to consider the optimization one, which, in our opinion, is closer to the general concept of consensus, since it does not necessarily need "desirable", but in any case subjective, properties to be defined and satisfied. We will always consider also the constructive approach to write algorithms for determining good initial solutions of the optimization problems.

In this paper, after briefly reviewing the techniques to achieve a consensus dendrogram. we propose new algorithms which, starting from the given dendrograms, perform a sequence of elementary tree operations so as to obtain at each step, in a greedy fashion, a dendrogram that is as close as possible to the given ones. The time - and space - complexities of such procedures make them suitable for large scale applications. Two numerical examples are provided to show the proposed technique.

2. Notation and Basic results.

Let r hierarchical classifications of a set I of n objects be given in terms of:

(i) n-trees $T_1, T_2, ..., T_r$ where each T_h is a collection of at most $2n$-1 subsets of I such that 1); $\{i\} \in T_h$, $i=1, ... , n$ and $\emptyset \notin T_h$; 2) $J, J' \in T_h \Rightarrow$ either $J \subset J'$ or $J' \subset J$ or $J \cap J' = \emptyset$; 3) $I \in T_h$ (^{1}). Thus, the n-tree is given by the n trivial clusters (leaves) $\{i\}$, ($i \in I$) and the n-1 clusters of units (internal nodes), obtained by the n-1 steps of fusion performed by a hierarchical algorithm (hence, the last cluster is the root I).

As customary, any n-tree T is represented as a tree, generally as a binary tree, i.e., with exactly n-1 internal nodes, rooted at I, whose nodes are subsets of I belonging to T, and where there are two directed edges (J,H) and (J,K) whenever $\{H, K\}$ is a *bipartition* of J. The binary tree will still be denoted, for simplicity, by T.

If (J,H) is any edge, we say that J is a *predecessor* of H and that H is a *successor* of J. If there is a directed path from J to J', then J' is said to be a *descendant* of J, and J an *ancestor* of J'. In particular, every node is both a descendant and an ancestor of itself. A *terminal* (or *leaf*) is any node without successors. A node is *internal* if it is neither a terminal nor the root. Two nodes of T are *brothers* if they have the same predecessor. Let J be a node of T. The *sub-tree of T rooted at J* - denoted by T_J - is the rooted tree (with root J) whose nodes are the descendants of J in T and where (J, H) is an edge if it is an edge in T. We denote by T^J the rooted tree obtained from T after deletion of all the descendants of J other than J, together with all edges incident to them.

(ii) Dendrograms $\Delta_1, \Delta_2, ..., \Delta_r$, where $\Delta_h = \{\delta(I_{1h}), \delta(I_{2h}), ..., \delta(I_{2n-1 h})\}$. Here I_{jh} is the cluster obtained right after the j-th fusion, initially there are n clusters formed by the individual objects, and $\delta(I_{jh})$ is the corresponding value of fusion. One must have $\delta(I_{1h}) \le \delta(I_{2h}) \le ... \le \delta(I_{2n-1 h})$ and in particular $I_{jh} \subseteq I_{ih} \Rightarrow \delta(I_{jh}) \le \delta(I_{ih})$.

(iii) Ultrametric matrices $U_1, U_2, ..., U_r$, where $U_h = \{u_{ijh}: i.j \in I\}$ and the elements u_{ijh} satisfy the ultrametric inequality $u_{ijh} \le \max(u_{ilh}, u_{jlh}) \ \forall \ (i.l.j) \in I$, ($h = 1, ...,r$).

(^{1}) In the following we represent the n-tree by its non trivial classes: $T_h = \{I_{1h}, I_{2h}, ..., I_{n-2 h}\}$, where $1 < |I_{jh}| < n$ for all j.

3. Least Squares Consensus Dendrogram.

Lefkovitch (1985) defined a Euclidean consensus to be (usually) a dendrogram, embedding the ultrametric matrices U_1, U_2, ..., U_r into a Euclidean space with at most n-1 dimensions, via classical scaling. Then, all the r sets of points are rotated, using polar rotation, so that, each set of points is: $X_h V_h' = V_h \Lambda_h V_h'$, where X_h denote the matrix of coordinates, obtained via scaling, corresponding to the h-th ultrametric matrix, Λ_h is the diagonal matrix with non null elements the eigenvalues of $X_h X_h'$ and V_h are the corresponding eigenvectors. The polar rotation allows to represent the r sets of points in the same coordinate system, with basis the identity matrix I. Another consequence following from the polar rotation is that it is easy to obtain, by a convex combination with coefficients w_h, $C = \Sigma_h w_h V_h \Lambda_h V_h'$, a consensus set of points that minimizes $\Sigma_h \|C - V_h \Lambda_h V_h'\|^2$. Hence, a consensus classification structure, generally, but not always a dendrogram, is obtained from this convex combination of the r rotated sets of points, reversing the process to embed the ultrametric matrices into a Euclidean space. Following the optimization approach Lapointe and Cucumel (1991), Vichi (1993) determined the *Average Consensus* or *Least Squares Consensus Dendrogram*, solving the following quadratic problem:

$$\begin{cases} \min \sum_{h=1}^{r} \|U_h - U\|^2 \\ \text{subject to} \\ u_{ij} \leq \max(u_{il}, u_{jl}) \text{ for every triplet } (i,j,l) \in I \\ diag(U) = 0 \end{cases} \qquad [P1]$$

where $U = \{u_{ij}: i,j \in I\}$ is the closest least squares matrix subject to the ultrametric condition. The zero diagonal condition, together with the ultrametric constraints, implies that U is non-negative and symmetric. As $u_{ii} = 0$ for $i=1,...,n$ the variables u_{ii} are not needed in [P1].

The ultrametric constraints can be written under the equivalent disjunctive form: "either $u_{ij} \leq u_{il}$ or $u_{ij} \leq u_{jl}$". Hence [P1] is a disjunctive program with quadratic convex objective function. Disjunctive programs are known to be NP-hard even for linear objective functions. The only known exact solution methods are combinatorial, and thus their running times grow exponentially as the problem size increases. Actually, a result of Krivánek and Morávek (1986) states that [P1] is NP-hard.

For a given U, let $\Gamma(U) \equiv \{(i,j,p): 1 \leq i,j,p \leq n; i \neq j, j \neq p, i \neq p; u_{ij} \leq u_{ip}, u_{ij} \leq u_{jp}\}$.
Then [P1] is equivalent to

$$\begin{cases} \min \|U - \bar{U}\|^2 \\ \text{subject to} \\ \sum_{(i,j,p) \in \Gamma(U)} (u_{ip} - u_{jp})^2 = 0, \\ diag(U) = 0. \end{cases} \qquad [P2]$$

where $\bar{U} = (1/r)(U_1 + ... + U_r)$.

This is a consequence of the following remarks:

(i) the ultrametric constraint $u_{ij} \leq \max(u_{ip}, u_{jp})$ is equivalent to the condition that the two largest elements among u_{ij}, u_{ip}, u_{jp} be equal;

(ii) One has

$$\sum_{h=1}^{r} \|U_h - \bar{U}\|^2 + r\|\bar{U} - U\|^2 + 2tr[(\bar{U} - U)\sum_{h=1}^{r}(U_h - \bar{U})] = \text{constant} + r\|\bar{U} - U\|^2 + 0.$$

Notice that if $\bar{U}$ is ultrametric (a sufficient condition for this to happen is that the binary n-trees $T_1, T_2, ..., T_r$ associated with the ultrametric matrices are equal (Vichi 1995)) then it is obviously the optimal solution to [P2].

[P2] may be reformulated, following a classical approach, as an unconstrained quadratic minimization problem, considering the penalty function:

$$\min\| U - \bar{U} \|^2 + \lambda (\sum_{(i,j,p) \in \Gamma(t)} (u_{ip} - u_{jp})^2)$$

[P3]

Algorithms for finding good (although not necessarily optimal) solutions to the above problems have been given by Vichi (1993) ([P1], [P2] versions); Carroll and Pruzansky, (1980) and De Soete (1984), ([P3] version).

4. Algorithms.

In view of the NP-hardness of [P1], [P2], [P3], and of the exponentially growing running times of the available exact solution algorithms, it makes sense to look for fast heuristic procedures. yielding "good", although not provably optimal, solutions. Any such solution is also useful as a starting point for iterative global optimization procedures.

An initial feasible solution to [P1], [P2] and [P3] may be obtained through a hierarchical classification algorithm having in input the Euclidean matrix $\bar{U}$. The best choice is the average link method, which it is known to give, among all aggregative methods, the ultrametric matrix U at minimum distance from the given dissimilarity matrix (in our case $\bar{U}$), see for example Cunningham and Ogilvie (1972). This variant will be called Algorithm 1. The resulting dendrogram can be considered as a member of the family of consensus dendrogram methods proposed by Ghashghai et al. (1989), as a generalization of the Stinebrickner's top-down method for dendrograms (Stinebrickner, 1984a) and of Neumann's generalized intersection methods for n-trees (Neumann, 1983).

A second heuristic algorithm (Algorithm 2) for [P1], [P2] and [P3] has been proposed by Vichi (1994). Essentially, starting from U the algorithm: (i) computes a least upper bound ultrametric (via complete linkage) and the largest lower bound ultrametric (via the single linkage), (ii) finds the mean matrix of these two ultrametrics; and (iii) repeats steps (i) and (ii) on the current mean matrix until convergence is achieved. If the limit matrix (which always exists) happens to be not ultrametric, then the average link method is applied to it.

A third algorithm is actually a procedure to improve the objective function:

$$\text{o.f.} = \sum_{h=1}^{r} \| U_h - U \|^2$$

(1)

when a dendrogram approximating U is given. In our case the dendrogram is defined by one of the previous two procedures.

Before introducing the algorithm we need to define a *swap operation* and some of its properties.

Let T be an n-tree containing, among others, the following sets: (i) three pairwise disjoint sets J, H, K; (ii) $J\cup H$; (iii) $J\cup H\cup K$; then a *swap operation* transforms T into the n-tree T' or T'', where $J\cup H$ is replaced by $J\cup K$ or $H\cup K$, respectively.

The swap operation can be described by the tree representation in Fig. 1.

Fig. 1: Trees associated with a swap operation

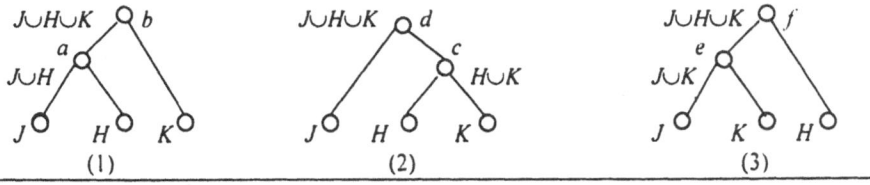

Thus, given the n-tree T with classes J, H, K, whose fusion is represented in Fig. 1.1, a swap operation is represented in Fig. 1.2, or in Fig. 1.3. These fusions give rise to the n-trees T'' and T'. respectively.

Example 1: Given the n-tree (including also the trivial classes) $T = \{1,2,3,4.5,12,34,125,I\}$, a swap on classes $\{1\}, \{2\}, \{5\}$ gives the n-trees $T' = \{1,2,3,4,5,15,34,125,I\}$, $T'' = \{1,2,3,4,5,25,34,125,I\}$.

From an algebraic viewpoint. the trees T, T' and T'' correspond to the three possible ways to obtain $J \cup H \cup K$ by successive unions starting from J, H, K, namely, $(J \cup H) \cup K$, $(J \cup K) \cup H$, $(H \cup K) \cup J$.

Theorem : Given two binary n-trees S and T on the same set I, it is always possible to obtain T from S, or vice-versa, by a finite sequence of swap operations.

Remark : If S and T are two n-trees on the same set I, if the sub-trees S_J and T_J are identical, and if T^J can be obtained from S^J by a finite sequence of swaps, then T can be obtained from S by a finite sequence of swaps as well.

We are now in position to prove the theorem.

Proof. By induction on $n = |I|$. For $n=2$ or 3 the theorem is trivial. Suppose that the theorem holds for all p-trees with $p \leq n-1$, and let S and T be any two n-trees on I ($n \geq 4$).

Case 1) There exist two identical sub-trees S_J and T_J with $|J| \geq 2$.

Then by the inductive hypothesis T^J may be obtained form S^J through a finite sequence of swaps and thus by the above remark the theorem holds also for n-trees.

Case 2) There are no two identical sub-trees S_J and T_J with $|J| \geq 2$.

Then consider any two brother terminals L and M in the tree T (two brother terminals must always exist). Both L and M are terminals also in S, but they are not brothers in S, otherwise Case 1 would occur. We are going to describe a procedure that, starting from S, produces after a finite sequence of swaps an S' where L and M are brother terminals. Then S' and T have two identical sub-trees -- namely, those rooted at $L \cup M$. Thus, we fall back into Case 1) and the theorem follows.

Fig. 2: Swaps in the procedure DOUBLE LIFTING

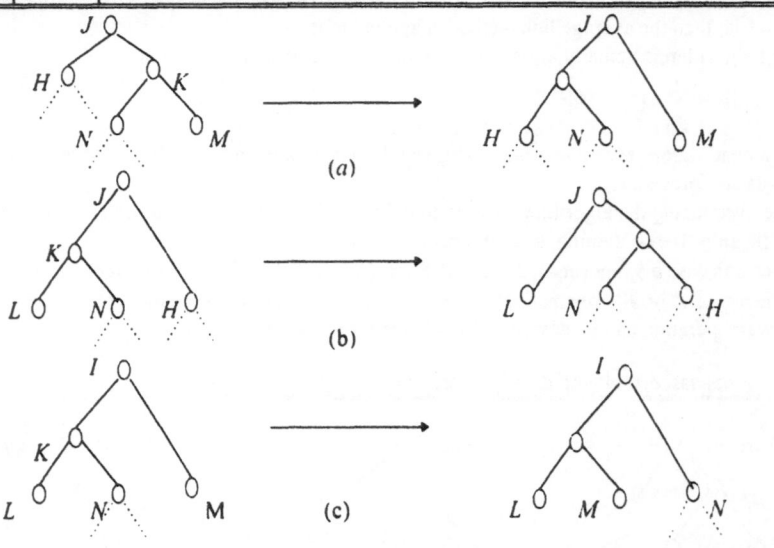

In the procedure to be described below. and with reference to the current tree S', we denote by pred(J) and brother(J) the predecessor and the brother of the node J ($J \neq I$), respectively.

PROCEDURE DOUBLE LIFTING	$\Rightarrow$

$S':=S$;
$K:=\text{pred}(M)$;
While $K \neq I$ **do**
  $N:=\text{brother}(M)$; $J:=\text{pred}(K)$; $H:=\text{brother}(K)$;
  Perform on S' the swap indicated in Fig. 2(a);
  Let S' be the tree after the swap;
  $K:=\text{pred}(M)$;
End While
$K:=\text{pred}(L)$;
While $\text{pred}(K) \neq I$ **do**
  $N:=\text{brother}(L)$; $J:=\text{pred}(K)$; $H:=\text{brother}(K)$;
  Perform on S' the swap indicated in Fig. 2(b);
  Let S' be the tree after the swap;
  $K:=\text{pred}(L)$;
End While $\Rightarrow$
	$N:=\text{brother}(L)$; $M:=\text{brother}(K)$;
	Perform on S' the swap indicated in Fig 2(c);
	Let S' be the tree after the swap;
	output S';
	end

Notice that: (*i*) After each swap in the first While-loop, the depth of M (that is, the number of edges of the path from the root I to M) decreases by one, and after the loop is exactly 1; (*ii*) After each swap in the second While-loop, the depth of L decreases by one, and after the loop is equal to 2; (*iii*) After the final swap, L and M become brother terminals; (*iv*) the overall complexity of DOUBLE LIFTING is $O(n)$.

An Example of execution of DOUBLE LIFTING is shown in Fig. 3.

Fig. 3: (*a*) Two 6-trees S and T; terminals 1 and 4 are brothers in T, but not in S. (*b*) Execution of a DOUBLE LIFTING: starting from S, after 5 swaps a tree S_5 is generated, in which 1 and 4 are brothers

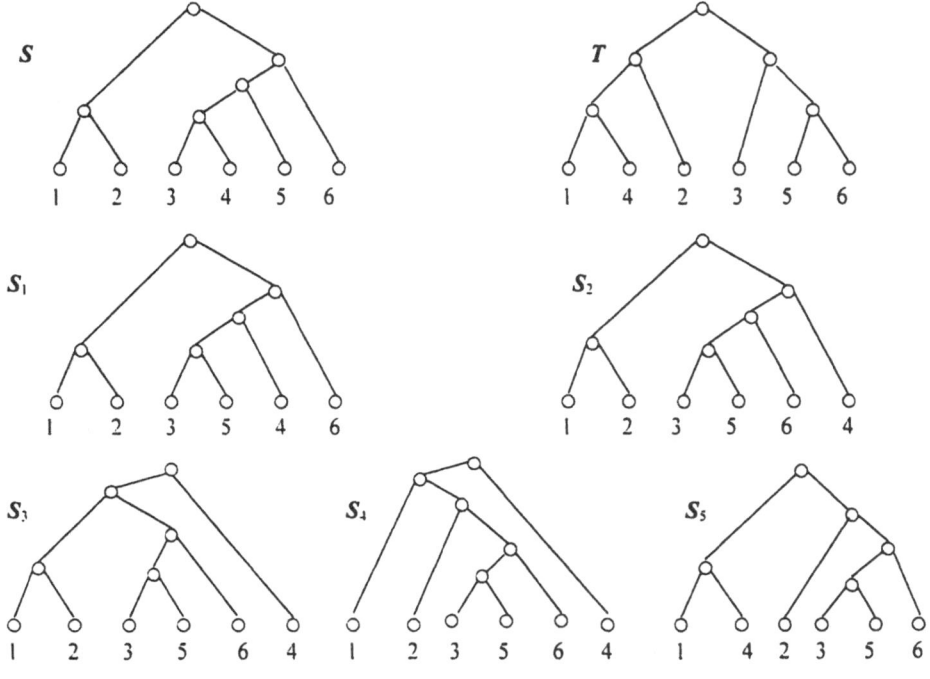

The above theorem motivates the following Algorithm 3 (although it is not enough to guarantee the optimality of the dendrogram generated by such an algorithm). For a better understanding of the algorithm, consider again the three sub-trees in Fig. 1.

The sub-trees represent three sub-dendrograms of a dendrogram Δ only if: $a \leq b$; or $c \leq d$; or $e \leq f$, where a, b, c, e, f, are the levels of fusion, as reported in Figure 1, respectively.

Furthermore, the contribution of each sub-dendrogram to the objective function (1) is given by the amount:

$$s_1 = s_a + s_b = \sum_{\substack{h=1 \\ f \in J \\ g \in H}}^{r} \sum (u_{fgh} - a)^2 + \sum_{\substack{h=1 \\ f \in J \cup H \\ g \in K}}^{r} \sum (u_{fgh} - b)^2 ; \tag{2}$$

$$s_2 = s_c + s_d = \sum_{\substack{h=1 \\ f \in H \\ g \in K}}^{r} \sum (u_{fgh} - c)^2 + \sum_{\substack{h=1 \\ f \in H \cup K \\ g \in J}}^{r} \sum (u_{fgh} - d)^2 ; \tag{3}$$

$$s_3 = s_e + s_f = \sum_{\substack{h=1 \\ f \in J \\ g \in K}}^{r} \sum (u_{fgh} - e)^2 + \sum_{\substack{h=1 \\ f \in J \cup K \\ g \in H}}^{r} \sum (u_{fgh} - f)^2 ; \tag{4}$$

respectively.

The minimum increase s_1 is given, for a sub-tree of type (1) in Fig. 1, when the level of fusion is:

$$a^* = \frac{1}{r|J| \cdot |H|} \sum_{\substack{h=1 \\ f \in J \\ g \in H \\ f < g}}^{r} \sum u_{fgh} ; \quad b^* = \frac{1}{r|J \cup H| \cdot |K|} \sum_{\substack{h=1 \\ f \in J \cup H \\ g \in K \\ f < g}}^{r} \sum u_{fgh} ; \quad \text{if } a^* < b^*. \tag{5}$$

otherwise if $a^* \geq b^*$ the minimum increase is given by

$$a^{**} = b^{**} = \frac{1}{r|J \cup H \cup K|} \sum_{\substack{h=1 \\ f,g \in J \cup H \cup K \\ f < g}}^{r} \sum u_{fgh} , \tag{6}$$

but in the last case the n-tree (1) in Fig. 1 becomes a bush, i.e., the level of fusion a is equal to the level of fusion b. In fact, the value a^* is the arithmetic mean of the dissimilarities among classes J and H, while b^* is the mean of the dissimilarities among classes $J \cup H$ and K. Thus, $a^* > b^*$ means that the best feasible solution is a tree with one level of fusion (bush) equal to the mean of dissimilarities between the elements of cluster $J \cup H \cup K$.

Similarly, the minimum increase s_2 is given, for a sub-tree of type (2), in Fig. 1, when the level of fusion is

$$c^* = \frac{1}{r|H| \cdot |K|} \sum_{\substack{h=1 \\ f \in H \\ g \in K \\ f < g}}^{r} \sum u_{fgh} ; \quad d^* = \frac{1}{r|H \cup K| \cdot |J|} \sum_{\substack{h=1 \\ f \in H \cup K \\ g \in J \\ f < g}}^{r} \sum u_{fgh} ; \quad \text{if } c^* < d^*. \tag{7}$$

otherwise if $c^* \geq d^*$ the minimum increase is given by $c^{**} = d^{**}$ as in (6).

The sub-tree of type (3), in Fig 1, has level of fusion

$$e^* = \frac{1}{r|J| \cdot |K|} \sum_{\substack{h=1 \\ f \in J \\ g \in K \\ f < g}}^{r} \sum u_{fgh} ; \quad f^* = \frac{1}{r|J \cup K| \cdot |H|} \sum_{\substack{h=1 \\ f \in J \cup K \\ g \in H \\ f < g}}^{r} \sum u_{fgh} ; \quad \text{if } e^* < f^*. \tag{8}$$

otherwise if $e^* > f^*$ the minimum increase is given by $e^{**} = f^{**}$ as in (6).

ALGORITHM 3

Given a dendrogram $\Delta = \{\delta(I_1), \delta(I_2), ..., \delta(I_{2n-1})\}$, with at most n-1 internal nodes, where $I_1, I_2, ..., I_{n-1}$ are the non singleton classes associated with the internal nodes;

Step 0: Set the iteration parameter $k:=1$;

Step1: Visit the k-th internal node according to the order given by the non-decreasing levels of fusion $\delta(I_k)$;

If cluster I_k is the root of a sub-tree with at least one internal node **then**

   If the sub-tree has more than one internal node **then**

      aggregate those clusters with the smallest level of fusion. Let J, H, K be the three leaves of this sub-tree. With a swap operation we have one of the three sub-trees in Fig. 1;

   End If;

   Step 2: Compute (a^*, b^*), (c^*, d^*), (e^*, f^*);

      If $a^* \leq b^*$ or $c^* \leq d^*$ or $e^* \leq f^*$ (i.e., at least one pair is feasible) **then**

         Compute for the feasible levels of fusion the increases s_1, s_2, s_3;
         Consider the sub-dendrogram with the smallest increase of the objective function;

      End If;

Else

Compute for the class with two elements $\{i,j\}$ the mean of u_{yh} $h=1, ..., r$.

End If;

Step 3: $k:=k+1$;

repeat step 1 to **step 3**, n-1 times.

The worst time-complexity of the algorithm is $O(rn^3)$, since processing the k-th node, which has at most k proper descendants, takes $O(k^2r)$, $k=1,..... n$-1.

A fourth algorithm is illustrated in the following table, and can be applied directly on the original dissimilarity matrices $\mathbf{D}_1$, $\mathbf{D}_2$, ..., $\mathbf{D}_r$.

ALGORITHM 4

Step 0: (initialization): let r matrices $\mathbf{D}_1$, $\mathbf{D}_2$, ..., $\mathbf{D}_r$ be given, where $\mathbf{D}_h = \{d_{yh}\ i,j \in I\}$; these may be dissimilarity or ultrametric matrices. Set the iteration parameter $k:=1$;

Step 1: For each matrix $\mathbf{D}_h$: $h=1, ..., r$, find the minimum value;

$$d_{y1} = \min\{\mathbf{D}_1\}; ...; d_{lmh} = \min\{\mathbf{D}_h\}; ...; d_{pqr} = \min\{\mathbf{D}_r\}.$$

These are the values of fusion between groups: (G_i, G_j), ..., (G_l, G_m), ..., (G_p, G_q), respectively, and represent the k-th smallest value of fusion of r dendrograms.

Step 2: Compute the means of the dissimilarities:

$$_1E\{d_{jv}, v=1,...,r\}; ...; _hE\{d_{lmv}, v=1,...,r\}; ...; _rE\{d_{pqv}, v=1,...,r\};$$

Step 3: Among the above means, choose the smallest one, let it be $_hE\{d_{lmv}, v=1,....r\}$, which is at least as large as the minimum mean detected at iteration k-1.

Step 4: The increment of the objective function in [P1] after the fusion of groups (G_l, G_m) is:

$$DEV\{d_{lmv}, v=1,...,r\} = \Sigma_v(d_{lmv} - _hE)^2$$

Thus, the fusion of groups (G_l, G_m) with cardinality n_l and n_m defines the k-th group I_k with associated value of fusion $_kE\{d_{lmv}, v=1,...,r\}$. Cluster I_k represents the k-th node of the consensus n-tree associated with the r dendrograms.

Step 5: Update the matrices $\mathbf{D}_1$, $\mathbf{D}_2$, ..., $\mathbf{D}_r$, i.e., the distances between the fused cluster I_k and a generic cluster G_v: $d_h(I_k, G_v) = (n_l / (n_l+n_m)) d_h(G_l, G_v) + (n_m / (n_l+n_m)) d_h(G_m, G_v)$

where $d_h(I_k, G_v)$ is the distance between $G_l \cup G_m$, and cluster G_v in the h-th matrix $\mathbf{D}_h$.

Step 6: $k:=k+1$;

repeat Step 1 to **Step 6**, n-1 times.

---

5. Application

In order to show the behavior of the proposed algorithms a well-known benchmark in cluster analysis, given by Michner (1970), has been used. This data set arises from a taxonomic problem on 11 types of Hoplites bees, described by 23 variables related to the form and characteristics of the bees (further details are reported in the original paper). Using the distance matrix between pairs of bees, Everitt (1993) compares the dendrograms obtained by single linkage, complete linkage and average linkage between groups (UPGMA). The ultrametric matrices U_1, U_2 and U_3 are reported in Vichi (1993). The mean matrix $\bar{U} = (1/3)(U_1 + U_2 + U_3)$, and the dendrogram of the average linkage applied on $\bar{U}$ are shown in Fig. 4.

For each of the 10 steps of UPGMA, the class and the corresponding value of fusion is reported

class	3,4	10,11	3,4,6	1,2	3,4,6,9	3,4,6,9,5	3,4,6,9,5,1,2	3,4,6,9,5,12,10,11	7,8	I
value of fusion	0.303	0.645	0.676	0.940	0.947	1.0485	1.3646	1.515	1.53	1.5782

The objective function (1) has the value: 13.9372

Algorithm 3 is applied on the dendrogram obtained by UPGMA on the mean matrix (Fig. 4)

Step 0: Given the dendrogram in Fig. 4; set $k:=1$;

- $k=1, 2$. Both nodes {3,4} and {10,11}, with level of fusions 0.303 and 0.645, have no internal node among their successors and cannot be roots of sub-trees. The values 0.303 and 0.645 are the means of the corresponding values in the three ultrametric matrices. The increase of the o.f. is 0 for both nodes;

- $k=3$. **Step 1**: The internal node {3,4,6}, with level of fusion 0.676, is the first node that can be considered as the root of the sub-tree in Fig. 5 (i). With a swap the two sub-trees in Fig. 5 (ii) and (iii) are obtained. **Step 2**: compute $a^*=0.303$, $b^*=0.676$; $c^*=0.676$, $d^*=0.489$; $e^*=0.676$, $f^*=0.489$. The only feasible solution is a^* and b^* and the increase of the o.f. is $s_h=0.0420353$.

- $k=4$. This node {1,2} with level of fusion 0.940, has no internal node among its successors and hence it cannot be the root of a sub-tree. The value 0.940 is the mean of elements between (1,2) in the three ultrametric matrices. The increase of the o.f. is 0;

Fig. 4: Mean matrix $\bar{U}$ of the ultrametric matrices associated with single linkage, complete linkage and average linkage applied on Hoplites data (Everitt, 1993). The dendrogram is obtained applying average linkage on the mean matrix.

	1	2	3	4	5	6	7	8	9	10	11
1	0.000										
2	0.940	0.000									
3	1.383	1.333	0.000								
4	1.383	1.333	0.303	0.000							
5	1.391	1.391	1.066	1.066	0.000						
6	1.383	1.333	0.676	0.676	1.066	0.000					
7	1.530	1.530	1.630	1.630	1.630	1.630	0.000				
8	1.376	1.376	1.598	1.598	1.598	1.598	1.530	0.000			
9	1.383	1.333	0.947	0.947	0.996	0.947	1.630	1.598	0.000		
10	1.570	1.570	1.493	1.493	1.493	1.493	1.630	1.598	1.493	0.000	
11	1.570	1.570	1.493	1.493	1.493	1.493	1.630	1.598	1.493	0.645	0.000

- $k=5, 6, 7, 8$. In steps 1 and 2 the sub-dendrograms (iv), (vii), (x), (xiii) are found to be optimal respectively.

- $k=9$. This node {7,8}, with level of fusion 1.53, has no internal node among its successors and it cannot be the root of a sub-tree. The value 1.53 is the mean of the elements between (7,8) in the three ultrametric matrices. The increase of the o.f. is 0.10492;

- k 10. **Step 1**: The internal node {1,2,3,4,5,6,7,8,9,10,11} is the last node that can be considered as the root of a sub-tree. The associated sub-tree with one internal node is shown in Fig. 5 (xvi). With one swap the sub-trees (xvii) and (xviii) in Fig. 5 are obtained. **Step 2**: compute $a^*=1.53033$.

Fig. 5: Steps of the algorithm 3 for which a swap operation can be executed.

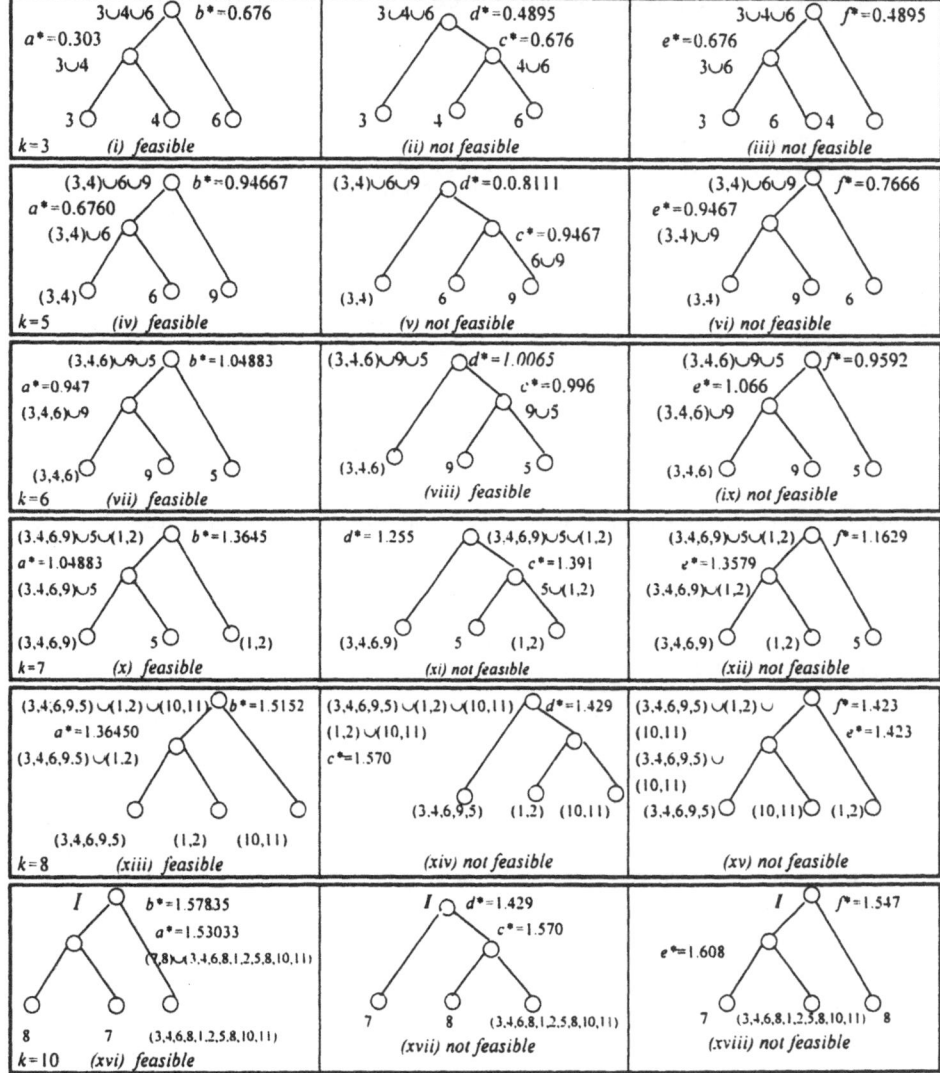

$b^*=1.57835$, $c^*=1.54885$, $d^*=1.6001$, $e^*=1.60785$, $f^*=1.54700$. The feasible solutions are a^*, b^* with $s_1=s_a+s_b = =0.10492+4.15339=4.25831$ and c^*, d^* with $s_2=s_c+s_d=2.17669+2.05085=4.22754$. Thus, the sub-dendrogram (xvii) is taken as best one. Note that in this case the sub-dendrogram changes from the original one and the increase of the o.f. is given by s_a+s_b, disregarding the increase given in the 9-th step.

For each of the 10 main iterations, the class and the corresponding value of fusion is reported below. Note that the procedure has changed some values of the levels of fusion of dendrogram in Fig. 4. In iteration 10 also the form of the dendrogram has been changed from the original.

class	3,4	10,11	3,4,6	1,2	3,4,6,9	3,4,6,9,5	3,4,6,9,5,1,2	3,4,6,9,5,1,2,10,11	7,8	1
value of fusion	0.303	0.645	0.676	0.940	0.94667	1.04883	1.3646	1.51524	1.54885	1.6001

The objective function has value: $0+0+0.0420253+0.24454+0.083308+5.9516695 + 2.820236 + 2.17669+2.05085=12.8311064$. The same solution has been obtained by Vichi (1993) solving [P3] with the truncated-Newton method.

The second data set, analyzed by Carroll *et al.* (1984), describes over-the-counter pain reliever usage in remedying three common maladies. On the three arrays of dissimilarities the average linkage algorithm has yielded the dendrograms reported in Vichi (1994).

On these data Algorithm 4 has been applied.

k=1; **step 1.** u(4;7;1)=20.29; u(3;4;2)=20.24; u(5;10;2)=20.03; **step 2, 3.** E(d(4;7);i=1,2,3))=26.76; E(d(3;4;i=1,2,3))=22.68; E(5;10;i=1,2,3))=20.96; **step 4.** DEV(d(5;10;i=1,2,3))=1.8066801. Thus, 5 with 10 at level **20.96** are aggregated, and o.f.=1.8066801.

For k=2, 3, 4, 5, 6, 7, 8, Steps 1 to 4 can be executed in a similar way.

k=9; **step 1.** u(3,4,1,7,6,9;5,10,1.8;1)=37.41; u(3,4,1,7,6,9;5,10,2,8;2)=34.65; u(3,4,1,7,6,9;5,10,2,8;3)= =33.46; **step 2,3.** E(d(3,4,1,7,6,9;5,10,2,8;i=1.2,3))=35.17; **step 4.** DEV(d(3,4,1,7,6,9 ;5,10,2,8;i=1,2, 3))= 197.09015; Thus, fuse **3, 4, 1, 7, 6, 9** with **5, 10, 2, 8** at level **35.17**; The o.f. value is 232.73522 + 197.09015=429.82538.

Thus, the dendrogram obtained can be synthetized as follows:

class	5,10	3,4	1,3,4	2,5,10	6,9	1,3,4,7	2,5,10,8	1,3,4,7,6,9	I
value of fusion	20.963	22.683	24.234	25.120	27.287	27.736	27.923	31.940	35.173

The same result has been obtained by Vichi (1994) through the solution of [P3] by the truncated-Newton method and also through the Algorithm 2 briefly outlined in this paper.

References

Adams, E. N. (1972): Consensus techniques and comparison of taxonomic trees, *Systematic Zoology*, 21, 390-397.3

Barthélemy, J. P., Leclerc, B. and Monjardet B. (1986): On the use of Ordered Sets in Problems of Comparison and Consensus of Classifications, *Journal of Classification*,3, 187-224.

Barthélemy, J. P., and Mc Morris, F.R. (1986): The median procedure for *n*-trees, *Journal of Classification*, 3, 329-334.

Barthélemy, J. P. and Monjardet B. (1981): The Median Procedure in Cluster Analysis and Social Choice Theory, *Mathematical Social Sciences*, 1, 235-267.

Carroll, J. D., Clark, L.A. and De Sarbo, W. S. (1984): The representation of three-way proximity data by single and multiple tree structure models, *Journal of Classification*, 1, 24-74.

Carroll, J.D. and Pruzansky, S. (1980): Discrete and Hybrid Scaling Models, In: *Similarity and Choice*, E. D. Lantermann and H. Feger (Eds.), Huber, Bern, 108-139.

Cunningham. K., M. and Ogilvie, J., C. (1972): Evaluation of hierarchical grouping techniques: a preliminary study. *Computer Journal*, 15, 209-213.

Day, W. E., McMorris, F.R. and Meronk, D.B. (1986): Axioms for Consensus Function Based on Lower Bounds in Posets, *Mathematical Social Sciences*, 12, 185-190.

De Soete, G. (1984): A Least Squares Algorithm for fitting an ultrametric tree to a dissimilarity matrix, *Pattern Recognition Letters*, 2, 133-137.

Everitt, B., S. (1993): *Cluster Analysis*, Edward Arnold, II edition.

Ghashghai. E., Stinebrickner, R. and Suters, W.H. (1989): A Family of Consensus Dendrograms Methods. Abstract of the paper presented at the *Second Conference of IFCS*. Charlottesville.

Gordon. A. D. (1987): A Review of Hierarchical Classification, *The Journal of the Royal Statistical Society*, A, vol. 150, 2, 119-137.

Gordon. A. D. (1996): Hierarchical Classification, In: P. Arabie *et al.* (Eds), *Clustering and Classification*, World Scientific Publishing, 65-121.

Krivánek, M. and Morávek, J. (1986): NP-Hard Problems in Hierarchical-Tree Clustering. *Acta Informatica*, 23, 311-323.

Lapointe, F. J. and Cucumel, G. (1991): The Average Consensus. Abstract of the paper presented the *Third Conference of the IFCS*, Edinburgh, Scotland.

Lefkovitch, L.P. (1985): Euclidean Consensus Dendrograms and Other Classification Structures. *Mathematical Biosciences*, 74, 1-15.

Margush, T. and Mc.Morris, F.R. (1981): Consensus *n*-tree, *Bulletin of Mathematical Biology*, 43, 239-244.

McMorris F.D., Meronk, D. B. aand Neumann, D.A. (1983): A view of Some Consensus Methods for Trees. In J. Felsestein (Ed.), *Numerical Taxonomy*, Spriger-Verlag. Berlin.

McMorris. F.D. and Neumann, D. A. (1983): Consensus Functions on Trees, *Mathematical Social Sciences*, 4, 131-136.

Neumann, D. A. (1983): Faithful consensus methods for *n*-trees, *Mathematical Biosciences*, 63, 271-287.

Neumann D. A. and Norton V. T. (1986): On Lattice Consensus Methods, *Journal of Classification*, 3, 225-255.

Powell, M.J.D. (1983): Variable Metric Methods for Constrained Optimization, *Mathematical Programming: The State of the Art*, In Bachem A. *et al*. eds, Springer Verlag, 288-311.

Rosen, D. E. (1978): Vicariant Patterns and Historical Explanation in Biogeography, *Systematic Zoology*, 27, 159-188.

Stinebrickner, R. (1984a): s-consensus trees and indices, *Bulletin of Mathematical Biology*, 46, 923-935.

Vichi M. (1993): Un algoritmo dei minimi quadrati per interpolare un insieme di classificazioni gerarchiche con una classificazione consenso, *Metron*, 51, 3-4, 139-163.

Vichi M. (1994): An algorithm for the consensus of the hierarchical classifications, *Proceedings of Italian Statistical Society*, 37, 261-268.

Vichi M. (1995): Principal Classification analysis of a three-way data set, presented at the meeting: Analisi dei dati multidimensionali, Napoli, 30-31 october.

On the Minimum Description Length (MDL) Principle for Hierarchical Classifications

Peter G. Bryant

Graduate School of Business Administration
University of Colorado at Denver
Campus Box 165
Denver, Colorado 80217-3364 U.S.A.

Summary: Hierarchical clustering procedures such as single-, average-, or complete-link procedures produce a series of groupings of the data arranged in the form of a hierarchy, or tree structure. In most cases, the choice of where to "cut" the tree is left to the user. Occasional formal guidelines have usually been based on ideas of random sampling, but that assumption is often violated in the contexts in which cluster analysis is used. This paper explores the application of Rissanen's MDL principle to derive possible guidelines for cutting the tree. These guidelines do not assume random sampling.

1. Background

1.1 Hierarchical clustering

Commonly used hierarchical clustering procedures group obervations into a nested sequence of classifications. Often that sequence is represented by a tree or dendogram.

A Simple Example: To fix ideas, let us consider a simple example consisting of the seven univariate observations

$$y = (1, 2, 5, 7, 12, 16, 20)^t, \tag{1}$$

which are to be grouped in some appropriate manner.

To use agglomerative hierarchical methods, we must specify an appropriate distance measure such as Euclidean distance, city-block distance, etc., and an aggregation criterion such as single linkage or complete linkage, which specifies how the distances between groups of observations are determined from the distances between individual observations. Such measures and aggregation criteria are discussed in standard textbooks such as Everitt (1993). The tree produced by single linkage clustering using Euclidean distance for the data in (1) is given in Fig. 1.

1.2 The problem of cutting the tree

Hierarchical methods do not require that we specify *a priori* how many groups are to be found, and this is often an advantage, but neither do they give us specific guidance on how many groups we have actually found. For those problems in which the tree is not the fundamental object of interest, but is simply a means to obtain a final grouping, the user must determine at what point it is useful to "cut" the tree. For example, in Fig. 1, if we cut the tree at (vertical) level 4.5, say, we obtain two groups, (1,2,5,7) and (12,16,20), while if we cut the tree at level 3.5, we obtain 4 groups, (1,2,5,7), (12),(16), and (20). The finer the subdivision, the more accurate (in some sense) the description of the data is, but the additional accuracy comes at the expense of a more complicated model. At what point, then, should we cut the tree?

Figure 1: Clustering for Simple Example Using Single Linkage

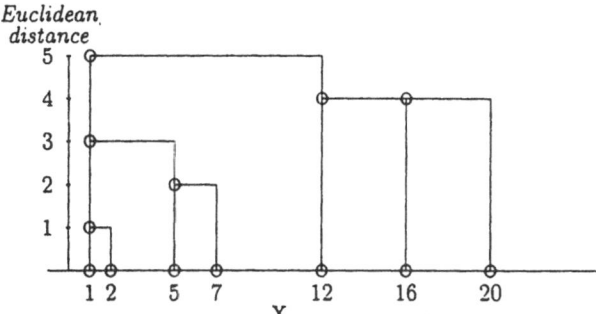

1.3 Possible approaches

At any given level of aggregation, that is, for any point at which we cut the tree, we usually obtain a corresponding figure or merit of some kind, such as the "pooled within group distance." The smaller this measure, the better the grouping describes our data. One way to determine an appropriate grouping is to plot this figure of merit versus the level of aggregation. The resulting curve often displays distinct dips at one or more points, and such points indicate clusterings which appear "significant" in some sense. How *much* of a dip is enough to be considered significant is harder to specify, though. For Euclidean distance, Duda and Hart (1973), for example, suggest referring the ratio of two successive figures of merit to some critical value, although in many cases of clustering, the sampling theory assumptions underlying classical statistical approaches to determining such critical values will be violated. In the next section, I explore an approach based on the MDL principle, an approach which doesn't depend on sampling theory.

2. The MDL Principle

2.1 General Ideas

The Minimum Description Length (MDL) principle articulated and developed by Rissanen (1987, 1989, 1996 and elsewhere) suggests a way of choosing statistical models for data when our purpose is simply to describe the given data rather than to estimate the parameters of some hypothetical population. The MDL principle is similar to a penalized maximum likelihood approach in that each proposed model has a total figure of merit combining the ability of the model to describe the data and the complexity of the model. The references cited give details of the approach. I apply it here to the case of Euclidean distance models for hierarchical clustering.

2.2 An MDL Criterion for Gaussian Models

The MDL principle requires that certain somewhat subjective choices be made by the user. Typically the result is not a single criterion but a family of criteria. For the common case of linear models $y = X\beta + e$ with Gaussian errors e, though, one appropriate form of the criterion turns out to be

$$MDL = \frac{n}{2}\ln\left(s^2\right) + \frac{n-p}{2}\ln\left(\frac{n}{2}\right) - \ln\Gamma\left(\frac{n-p}{2}\right) + \frac{p}{2}\ln\left(\frac{nR^2}{2\pi}\right). \qquad (2)$$

where $s^2 = n^{-1} \times$ (Error Sum of Squares) is the estimated error variance for y, n is the number of data points, and $p < n$ is the number of variables in the linear model. The parameter R may chosen in various ways, as long as it satisfies

$$R \geq \max_{j \leq p} \hat{\gamma}_j - \min_{j \leq p} \hat{\gamma}_j$$

where $\hat{\gamma}$ is derived from the vector of least squares coefficients, suitably standardized. The details of this derivation are given in Bryant (1996).

2.3 Application to Cutting a Hierarchical Tree

To each level of a tree representing p groups, there corresponds a Gaussian model with p independent variables. For example the design matrix X corresponding to the top level of Fig. 1 (a single group) is

$$X = (\begin{matrix} 1 & 1 & 1 & 1 & 1 & 1 & 1 \end{matrix})^t,$$

while that corresponding to the division into the groups (1,2,5,7) and (12,16,20) is

$$X = \begin{pmatrix} 1 & 1 & 1 & 1 & 0 & 0 & 0 \\ 0 & 0 & 0 & 0 & 1 & 1 & 1 \end{pmatrix}^t,$$

and so forth. To each grouping represented by the tree, there will correspond an MDL figure of merit (2), combining its ability to represent the data with the complexity of the description. The best level at which to cut the tree is the one yielding the smallest value of MDL.

The criterion given above is for univariate data. In clustering, it is more usual to have m-variate data ($m > 1$), and for such cases, we replace s^2 in (2) by the total within group squared Euclidean distance divided by $n' = nm$ and replace n and p by n' and $p' = nm$, respectively. For non-Euclidean distance measures d, the corresponding models would use a probability density measure proportional to e^{-d}, though the detailed calculations to produce an analogue of (2) will often be messy.

3. Some Numerical Examples

3.1 Simple Example (continued)

For the example in Fig. 1, the figures of merit turn out to be those given in Tab. 1.

Table 1: MDL Criteria for Simple Example ($n = 7$)

Grouping	MDL	p	s	R
(1,2,5,7,12,16,20)	16.71	1	6.676	1.348
(1,2,5,7)(12,16,20)	12.16	2	2.797	2.732
(1,2,5,7)(12,16)(20)	10.29	3	2.096	2.254
(1,2,5,7)(12)(16)(20)	10.20	4	1.803	2.621
(1,2)(5,7)(12)(16)(20)	10.05	5	.598	11.307
(1,2)(5)(7)(12)(16)(20)	10.52	6	.267	25.284
(1)(2)(5)(7)(12)(16)(20)	Undefined			

They suggest that for these data, each finer subdivision of the data is preferred at least slightly to the one which precedes it, until we reach the last: splitting the (5,7) group into two components seems to cost more than it is worth.

3.2 Johnson and Wichern's Cereal Data

We may illustrate the MDL approach further using the data from Johnson and Wichern (1988, page 587) listed in Tab. 2.

Table 2: Johnson and Wichern's Cereal Data

Id	Brand	Protein	Carbo	Fat	Calories	VitA
1	Life	6	19	1	110	0
2	Grape Nuts	3	23	0	100	25
3	Super Sugar Crisp	2	26	0	110	25
4	Special K	6	21	0	110	25
5	Rice Krispies	2	25	0	110	25
6	Raisin Bran	3	28	1	120	25
7	Product 19	2	24	0	110	100
8	Wheaties	3	23	1	110	25
9	Total	3	23	1	110	100
10	Puffed Rice	1	13	0	50	0
11	Sugar Corn Pops	1	26	0	110	25
12	Sugar Smacks	2	25	0	110	25

Five properties are given for each of twelve breakfast cereals. A clustering of these observations using standardized variables, the complete linkage criterion, and Euclidean distance is summarized in the tree in Fig. 2, and the corresponding figures of

Figure 2: Complete Linkage Clustering of Johnson and Wichern's Cereal Data

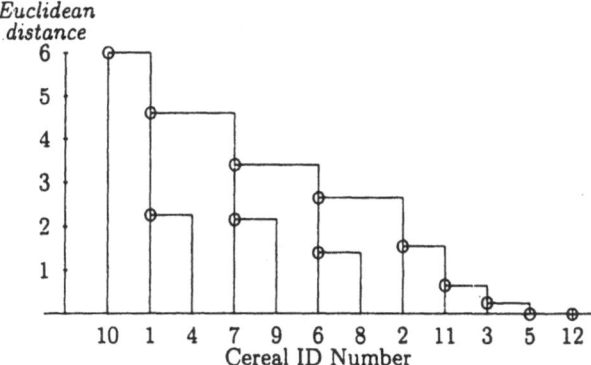

merit from the MDL criterion are listed in Tab. 3 for several groupings derived by cutting the tree.

We see, for example, that the three-group clustering is *not* preferred to that with two groups, but the proposed division into four groups *is* preferred to the division into three groups. Such "reversals" may happen, since for Euclidean distance, the optimal division into, say, $k+1$ groups is not necessarily a subdivision of the division into k groups. For these data, it seems likely that the results may be sensitive to the particular distance and aggregation criteria used, and to the choice of a hierarchical method rather than some other kind. The analyst would be well-advised to explore these issues further, rather than simply accepting the complete linkage, Euclidean distance results.

Table 3: MDL Criteria for Clustering Johnson and Wichern's Cereal Data ($n' = 60$)

Number of Groups	MDL	p'	s	R
2	19.02	10	.760	.547
3	20.92	15	.614	1.278
4	17.00	20	.479	1.642
5	20.39	25	.434	1.814

4. Remarks

The MDL approach as explored here is clearly easiest to apply in the case of mathematically tractable measures of error, such as least squares, at least in the sense that the formulae correspond naturally to the distances being used.

On the other hand, (2) could be used to assess any series of classifications whatever, hierarchical or not, without reference to the distances or other criteria used to generate them. It can thus be used as a kind of external check on the results of other methods. It seems likely that the MDL measures will be most useful when combined with other measures and results. They are intended to augment careful thought and reflection, not to replace them.

The last subdivision, in which all observations are distinct and there is no clustering, has no corresponding MDL figure of merit, as the sum of squared errors becomes 0. This will often be of little practical consequence, though is it theoretically unappealing.

Finally, note that other MDL criteria are possible, too, and they will not necessarily lead to identical conclusions. The differences among them arise from different specifications of allowable ranges for the parameters, scaling of the observations, etc. These different specifications are roughly analogous to different prior distributions in Bayesian analysis, though the exact formalisms are different. Some remarks on this are given in Bryant (1996).

References

Bryant, P. (1996): *The Minimum Description Length Principle for Gaussian Regression.* Working Paper 1996-08, University of Colorado at Denver, Graduate School of Business Administration. Denver, Colorado 80217-3364.

Duda, R. O. and Hart, P.E. (1973): *Pattern Classification and Scene Analysis.* John Wiley & Sons, New York.

Everitt, B. S. (1993): *Cluster Analysis.* Edward Arnold, London.

Johnson, R. A. and Wichern, D. W. (1988): *Applied Multivariate Statistical Analysis,* second edition, Prentice-Hall, Englewood Cliffs, N. J.

Rissanen, J. (1987): Stochastic complexity. *Journal of the Royal Statistical Society, Series B*, 49,3,223–265

Rissanen, J. (1989): *Stochastic Complexity in Statistical Inquiry.* World Scientific Publishing Co., Singapore.

Rissanen, J. (1996): Shannon-Wiener information and stochastic complexity, In: *Proceedings, N. Wiener Centenary Congress,* East Lansing, Michigan.

Consensus Methods for Pyramids and Other Hypergraphs

J. Lehel, F. R. McMorris, R. C. Powers

Department of Mathematics
University of Louisville
Louisville, KY 40292
U.S.A.

Summary: A classification can most generally be viewed as a hypergraph, which is simply a set of subsets (clusters) of the finite set S of objects being studied. In this paper, we are primarily concerned with consensus functions on tree hypergraphs such as pyramids and totally balanced hypergraphs.

1. Introduction and Definitions

Let S be a finite set of n objects. A (simple) *hypergraph* H on S is a set of non-empty subsets of S. If $A \in H$, A is called an *edge* of H. The set of all clusters of a classification of S is thus a hypergraph with the clusters as edges, and so we use the words 'cluster' and 'edge' interchangeably. Of course, the clusters of a hierarchical classification usually are structured into a tree-like relationship. In what follows, we will give an overview of several types of hypergraphs that have proved (or should prove) useful in classification studies. We then briefly investigate possibilities for consensus functions on some tree-like hypergraphs.

Throughout it is assumed that all hypergraphs H on S satisfy the following: $\{x\} \in H$ for all $x \in S$, and $S \in H$. A hypergraph H on S is a *hierarchy* (also frequently called an *n-tree*) if $A \cap B \in \{\emptyset, A, B\}$ for all clusters (edges) $A, B \in H$, and is a *weak hierarchy* if $A \cap B \cap C \in \{A \cap B, A \cap C, B \cap C\}$ for all clusters $A, B, C \in H$. Let T denote the set of all hierarchies on S and W the set of all weak hierarchies on S. A hypergraph H is a *pyramid* if $A \cap B \in H \cup \{\emptyset\}$ for all clusters $A, B \in H$, and there is a linear ordering of S so that each cluster of H is an interval in this ordering (Bertrand and Diday (1991)). Let P denote the set of all pyramids on S. It is not hard to see that $T \subseteq P \subseteq W$. There has been much interesting work relating the classes T, P and W to various dissimilarity measures on S (see Bertrand (1995), Gaul and Schader (1994), and the excellent collection of papers in Van Cutsem (1994)) but we are concerned here only with the relationships that the resulting clusters have among themselves.

A *cycle* (of length k) in a hypergraph H is an alternating sequence of vertices and edges $v_1, e_1, ..., v_k, e_k$ where $e_1, ..., e_k$ are distinct edges, $v_1, ..., v_k$ are distinct vertices, $v_i, v_{i+1} \in e_i$ for all $i = 1, ..., k-1$, and $v_k, v_1 \in e_k$. The cycle is *special* if $k \geq 3$ and $v_i \in e_j$ if and only if $j = i, j = i - 1$ or $(i, j) = (1, k)$. Thus a cycle is not special when there is an edge of the cycle that contains at least three vertices of the cycle. In Bandelt and Dress (1989) it is noted that a hypergraph H is a weak hierarchy if and only if H contains no special cycles of length 3. When a hypergraph has no special cycles at all, it is called *totally balanced*.

Before showing how totally balanced hypergraphs fit into the scheme, we need to consider hypergraphs whose clusters inherit structure from a relation on its vertices. A hypergraph H is an *interval hypergraph* if there is a path P so that every $A \in H$ is an interval of P. (See Duchet (1995) for standard terminology. Note that a *graph* is simply a hypergraph with each edge having two vertices.) Thus every pyramid is an interval hypergraph. A hypergraph H is a *tree hypergraph* if there is a tree T so that every $A \in H$ is a subtree of T. Let $\mathcal{I}$ denote the set of interval hypergraphs on S, $T\mathcal{B}$ the set of totally balanced hypergraphs on S, and $T\mathcal{H}$ the set of tree hypergraphs on S. From results in Lehel (1983) and Lehel (1985) we have that $\mathcal{I} \subseteq T\mathcal{B}$, $T\mathcal{B} = \mathcal{W} \cap T\mathcal{H}$, and $H \in T\mathcal{B}$ if and only if every subhypergraph of H is in $T\mathcal{H}$. Thus the complete list of inclusions is $T \subseteq P \subseteq \mathcal{I} \subseteq T\mathcal{B} \subseteq \mathcal{W}$ $(T\mathcal{B} \subseteq T\mathcal{H})$ with examples existing that show proper inclusions.

Because of the above characterizations of $T\mathcal{B}$ as those weak hierarchies that are also tree hierarchies, it is our opinion that totally balanced hypergraphs merit further study for possible uses in classification theory. However, in this paper our concern is with consensus methods for various hypergraphs and we now turn our attention to this topic.

2. Consensus

Let $\mathcal{H}$ denote a class of hypergraphs on S. A *consensus function* on $\mathcal{H}$ is a mapping $C : \mathcal{H}^k \to \mathcal{H}$, where k is a fixed positive integer. Elements of $\mathcal{H}^k$ are called *profiles* and are denoted by $\pi = (H_1, ..., H_k)$, $\pi' = (H'_1, ..., H'_k)$, etc. Among the general types of consensus functions are the counting rules and the intersection rules. A *counting rule* puts a cluster in $C(\pi)$ if it appears sufficiently often in the hypergraphs making up the input profile π. For example, the *majority rule* on T (Margush and McMorris (1981)) puts a cluster in the output if it appears in more than half of the input hierarchies. Counting rules from T^k into T were characterized in McMorris and Neumann (1983) and counting rules from $(T \cup \mathcal{W})^k$ into $\mathcal{W}$ were characterized in McMorris and Powers (1991). We will shortly investigate the possibilities for counting rules on P and $T\mathcal{B}$.

An *intersection rule* puts a cluster in the output $C(\pi)$ when it is the intersection of certain clusters from the input hierarchies in π. For $H \in \mathcal{H}$, let $h : H \to Z_0$ (Z_0 denotes the set of nonnegative integers) be defined by $h(A) = t$ if and only if $S = A_0 \supset A_1 \supset ... \supset A_t = A$ (proper inclusion) with each $A_i \in H$, and maximum in length. When $\mathcal{H} = T$, h is an easily visualized height function. For $\pi = (H_1, ..., H_k) \in \mathcal{H}^k$ and $j \in Z_0$ let $L_j(\pi) = \{X_1 \cap ... \cap X_k : X_i \in H_i$ and $h(X_i) = j$ for $i = 1, ..., k\}^*$. (For a set of subsets R, $R^* = \{X \in R : X \neq \emptyset\}$). Now set $C_D(\pi) = \cup_{j=0}^{\infty} L_j(\pi)$. The driving motivation for considering $C_D(\pi)$ is that by intersecting clusters at the same level across the profile π one might be able to produce a hypergraph whose clusters represent areas of partial overlap (agreement). This is precisely the type of information lost when using counting rules. A problem, of course, is that $C_D(\pi)$ might not be the same type of hypergraph as those making up the profile π. However, when $\mathcal{H} = T$ then $C_D(\pi) \in T$, and in this case C_D was studied in Neumann (1983). Intersection rules on T are further investigated in Adams (1986), Powers (1995) and Vach (1994). Problems start to arise even as we pass from T to P, and in McMorris and Powers (1996) it is noted that $C_D(\pi)$ need not be a pyramid when $\pi \in P^k$. However when each pyramid in π is based on the same linear ordering of S then $C_D(\pi)$ is an interval hypergraph, from which a pyramid can be formed. The resulting consensus function on P is characterized in McMorris and Powers (1996).

We now seek counting rules for P and $T\mathcal{B}$. Recall that a counting rule $C : \mathcal{H}^k \to \mathcal{H}$ can be described by a threshold l. It is then referred to as a $M_l - rule$ with $A \in M_l(\pi)$ if and only if $\frac{|\{i : A \in H_i\}|}{k} > l$ for $\pi \in \mathcal{H}^k$. The codomain of M_l is of concern. For example, when $\mathcal{H} = T$ then the majority rule $M_{\frac{1}{2}}(\pi) \in T$ for all $\pi \in T$ (Margush and McMorris (1981)); while if $\mathcal{H} = \mathcal{W}$, then $M_{\frac{2}{3}}(\pi) \in \mathcal{W}$ for all $\pi \in \mathcal{W}^k$ (McMorris and Powers (1991)). Clearly $M_1(\pi) \in \mathcal{H}$ for all $\pi \in \mathcal{H}^k$ where $\mathcal{H} \in \{\mathcal{T}, \mathcal{P}, \mathcal{I}, \mathcal{TB}, \mathcal{W}\}$ and is usually called the *unanimity rule*.

Surprisingly, counting rules other than the unanimity rule fail for $\mathcal{P}$ and $\mathcal{TB}$ as our example shows.

Example 1: Let $S = \{x_1, ..., x_n\}$ with $n \geq 3$. Define the hypergraphs $H_1, ..., H_n$ as follows:

$$H_1 = \{S, \{x_1\}, ..., \{x_n\}, \{x_1, x_2\}, \{x_2, x_3\}, ..., \{x_{n-1}, x_n\}\},$$
$$H_2 = \{S, \{x_1\}, ..., \{x_n\}, \{x_2, x_3\}, \{x_3, x_4\}, ..., \{x_n, x_1\}\},$$

$$H_n = \{S, \{x_1\}, ..., \{x_n\}, \{x_n, x_1\}, \{x_1, x_2\}, ..., \{x_{n-2}, x_{n-1}\}\}.$$

It is easy to see that each $H_i \in \mathcal{P}$ and thus $H_i \in \mathcal{TB}$. Letting $k = n$ and $\pi = (H_1, ..., H_k)$ we now see that $M_l(\pi)$ has a special k-cycle for all $l \in (0, 1)$ and is thus not a totally balanced hypergraph (and hence not a pyramid). Therefore the only l that works is $l = 1$.

If we try to dodge the problem pointed out in the example by requiring each pyramid in $\pi \in \mathcal{P}^k$ to be defined from the same linear order of S, then any selection of clusters from those that appear in the H_i's will give an interval hypergraph from which a pyramid is easily formed by taking intersections of intervals and adding the singletons and S. In particular, a cluster that appears in only one out of the k hypergraphs could be part of the consensus output and this is contrary to the notion of consensus.

Generalizing the $M_{\frac{2}{3}}$-rule for weak hierarchies gives the following result.

Theorem: Let $\pi = (H_1, ..., H_k)$ where each hypergraph H_i has no special cycle of length m. Set $l = \frac{m-1}{m}$ and assume that $\lceil lk \rceil < k$. Then $M_l(\pi)$ has no special cycle of length m.

Proof: Let $A_1, ..., A_m \in M_l(\pi)$. Then, for $j = 1, ..., m$, $|\{i : A_j \in H_i\}| > lk$. Since $m\lceil lk \rceil > (m-1)k$ it follows that there exists $j \in \{1, ..., k\}$ such that $A_1, ..., A_m \in H_j$. Since H_j has no special cycle of length m we have that $A_1, ..., A_m$ is not a special cycle of length m. Hence $M_l(\pi)$ has no special cycle of length m. $\square$

We point out that if $l < \frac{m-1}{m}$, then $M_l(\pi)$ might have a special cycle of length m. This leads to another interpretation as to why M_1 is the only counting rule that works on $\mathcal{TB}$. If $\pi \in (\mathcal{TB})^k$ and we are trying to eliminate special cycles of all lengths in $M_l(\pi)$, we must have $lim_{m \to \infty} l = lim_{m \to \infty}(\frac{m-1}{m}) = 1$.

We now are ready to make a proposal for an approach to consensus for hypergraphs that utilize clusters from both counting and intersection rules. This procedure is first described in general terms as follows: Let $\mathcal{H}$ be a fixed class of hypergraphs for which there is a smallest $l \in (0, 1]$ such that $M_l(\pi) \in \mathcal{H}$ for all $\pi \in \mathcal{H}^k$. For $\pi \in \mathcal{H}^k$, consider $C_D(\pi)$ and add clusters from $C_D(\pi)$ subject to preserving membership in $\mathcal{H}$ and other appropriate constraints.

To illustrate this approach consider $\mathcal{H} = \mathcal{P}$. We have seen that $l = 1$ for pyramids so for $\pi \in \mathcal{P}^k$ we first form unanimity consensus $M_1(\pi)$. Next construct $C_D(\pi)$ and sort $C_D(\pi) = \{A_1, ..., A_m\}$ according to a criterion such as size, $|A_1| \geq |A_2| \geq \cdots \geq |A_m|$.

Now consider the hypergraph $M_1(\pi) \cup \{A_1\}$. The idea is to have A_1 as a cluster in the consensus output if and only if $M_1(\pi) \cup \{A_1\}$ is a pyramid. This procedure continues until a decision is made about the last cluster A_m. Thus the final consensus pyramid gives clusters that are obtained by either counting or intersection. One should, however, consider exact algorithms and their associated complexities and we leave this for future work.

Acknowledgement

The research of F.R. McMorris was supported by the United States Office of Naval Research Grant N00014-95-1-0109.

3. References

Adams, E.N. III (1986): N-trees as nestings: Complexity, Similarity, and Consensus, *Journal of Classification*, **3**, 2, 299–317.

Bandelt, H.-J. and Dress, A. (1989): Weak hierarchies associated with similarity measures-an additive clustering technique, *Bulletin of Mathematical Biology*, **51**, 1, 133–166.

Bertrand, P. (1995): Structural Properties of Pyramidal Clustering, In: *Partitioning Data Sets*, Cox, I. et al. (eds.), DIMACS Series in Discrete Mathematics and Theoretical Computer Science, **19**, 35–53, AMS, Providence, RI.

Bertrand, P. and Diday, E. (1991): Les pyramides classifiantes: une extension de la structure hiérarchique, C. R. Acad. Sci. Paris, Série I, 693–696.

Duchet, P. (1995): Hypergraphs, In: *Handbook of Combinatorics*, Graham, R. et al. (eds.), VOL. 1, 381–432, MIT Press, Cambridge, MA.

Gaul, W. and Schader, M. (1994): Pyramidal classification based on incomplete dissimilarity data, *Journal of Classification*, **11**, 2, 171–193.

Lehel, J. (1983): Helly-hypergraphs and abstract interval structures, *ARS Combinatoria*, **16-A**, 239–253.

Lehel, J. (1985): A characterization of totally balanced hypergraphs, *Discrete Mathematics*, **57**, 59–65.

Margush, T. and McMorris, F.R. (1981): Consensus n-trees, *Bulletin of Mathematical Biology*, **43**, 239–344.

McMorris, F.R. and Neumann, D.A. (1983): Consensus functions defined on trees, *Mathematical Social Sciences*, **4**, 131–136.

McMorris, F.R. and Powers, R.C. (1991): Consensus weak hierarchies, *Bulletin of Mathematical Biology*, **53**, 679–684.

McMorris, F.R. and Powers, R.C. (1996): Intersection rules for consensus hierarchies, In: *Proceedings of the third international conference on ordinal and symbolic data analysis*, Diday, E. et al. (eds.), 301–308, Springer Verlag, Berlin.

Neumann, D.A. (1983): Faithful consensus methods for n-trees, *Mathematical Biosciences*, **63**, 271-287.

Powers, R.C. (1995): Intersection rules for consensus n-trees, *Applied Mathematics Letters*, **8**, 4, 51–55.

Vach, W. (1994): Preserving consensus hierarchies, *Journal of Classification*, **11**, 1, 59–77.

Van Cutsem, B. (1994): *Classification and dissimilarity analysis*, Lecture Notes in Statistics, New York.

On the Behavior of Splitting Criteria for Classification Trees

Roberta Siciliano. Francesco Mola

Dipartimento di Matematica e Statistica
Università di Napoli Federico II
Via Cintia - Monte S.Angelo
80126 - Naples - Italy
e–mail: r.sic@dmsna.dms.unina.it
f.mola@dmsna.dms.unina.it

Summary: In the framework of classification trees, the behavior of splitting criteria is investigated through a simulation study and applications on a real data set. Some emphasis is appointed to the strength of the dependency among variables, the choice of the splitting rule, the role played by the type of predictors, the stability of the classification rule. Alternative splitting criteria and new splitting rules are also proposed to deal with the computational effort of splitting procedures in large data sets.

1. Classification trees and splitting criteria

A classification tree procedure consists of a recursive sequential binary division of N cases belonging to J groups. A sample of N cases on which are observed a response variable Y with J classes and M predictors $(X_1, \ldots, X_M)$, not necessarily all of the same type, is assumed to be available. Such sample (called *learning sample*). is used to grow–up a binary tree. A *test sample*, with the same structure of the learning sample, or a ν–fold cross–validation can be used to find an optimal size tree and then to validate the final classification rule. As a result, a classification tree procedure provides not only a classification rule for new cases of unknown class (*decision tree*) but also an analysis of the dependence structure in large data sets (*exploratory tree*).

In a classification tree procedure a crucial moment is represented by the choice of the splitting criterion to divide a group of cases at the node t into two subgroups associated to the left node t_l and the right node t_r respectively. A splitting criterion usually satisfies the *coherence principle* in order to choose at each node the division of cases such that the two subgroups are internally most homogeneous and externally most heterogeneous (see for example Breiman *et al.*, 1984; Mingers, 1988).

The set of splits at each node is formed considering for each predictor all possible binary divisions of its categories. As an example, a numerical predictor with K distinct values produces $K - 1$ possible splits; a nominal predictor with K categories generates $2^{K-1} - 1$ possible splits. A split s, also called *dichotomous splitting variable* or *binary question*. sends a proportion p_l of cases to the left node t_l and a proportion p_r of cases to the right node t_r.

Several splitting criteria have been proposed in literature which differences rely on the type of predictors that can be considered and the adopted splitting rule; see for instance the DNP (Discrimination Non Parametrique) splitting procedure of Celeux and Lechevallier (1982), the RECPAM (RECursive Partitioning and AMalgamation) approach of Ciampi and Thiffault (1987) (for a detailed review on binary segmentation procedures see Mola, 1993). In the following, we will consider the methodology CART (Classification and Regression Trees) of Breiman *et al.* (1984) and the two-stage binary segmentation of Mola and Siciliano (1992, 1994).

1.1 CART splitting criterion

Breiman *et al.* (1984) have introduced the CART methodology providing several innovations; one of these is represented by the possibility to deal simultaneously with both numerical and categorical predictors. This was allowed by the definition of a general splitting criterion based on the concept of impurity at a given node. According to the CART splitting criterion the following *decrease in impurity* is maximized for each $s \in Q$

$$\Delta i(s|t) = i(t) - (p_l i(t_l) + p_r i(t_r)) \qquad (1)$$

where $i(\cdot)$ is the impurity function, p_l and p_r are the proportions of cases at the left node and the right node respectively. Depending on the definition of the impurity function different splitting rules can be used such as for example the Gini index of heterogeneity $G_Y(t) = 1 - \sum_j p(j|t)^2$ and the entropy index $H_Y(t) = - \sum_j p(j|t) \log p(j|t)$, where $p(j|t)$ is the proportion of cases with class j at node t.

1.2 Two-stage splitting criterion

Mola and Siciliano (1992, 1994) have introduced a two–stage splitting criterion in order to consider the "global" role played by the predictor as well as the "local" role played by any splitting variable. The first stage provides a *variable selection* of one or more significant predictors which are used to define the set of possible splits; the second stage applies a *splitting rule* to select the best split. Three strategies can be adopted with respect to the use of either (a) statistical indexes, or (b) statistical modeling, or (c) factorial methods. We briefly describe the first strategy; for the second strategy we refer to Mola, Klaschka and Siciliano (1996), for the third strategy we refer to Mola and Siciliano (1997b) in this volume.

Originally, the two-stage splitting criterion was based on the predictability index τ of Goodman and Kruskal $\tau_{Y|X}(t) = \frac{\sum_i \sum_j p^2(j|i,t)p(i|t) - \sum_j p^2(j|t)}{1 - \sum_j p^2(j|t)}$, where $p(j|i,t)$ is the proportion of cases that belong to class j given that they have category i of X at node t and $p(i|t)$ is the proportion of cases that has category i of X at node t (Mola and Siciliano, 1992). A further index can be proposed, namely the conditional entropy index of Shannon $H_{Y|X}(t) = - \sum_i \sum_j p(j|i,t) \log p(j|i,t)$.

Both the above mentioned indexes can be proved to be special cases of the following general measure for the proportional reduction in the heterogeneity of the response variable Y due to the information provided by the predictor X (globally considered):

$$\gamma_{Y|X}(t) = \frac{i_Y(t) - \sum_i p(i|t) i_{Y|i}(t)}{i_Y(t)} \qquad (2)$$

where $i_Y(t)$ is the measure of heterogeneity for the variable Y at node t and $i_{Y|i}(t)$ is the same measure for the conditional distribution of Y given the modality i of predictor X; using the Gini index yields to the predictability τ index whereas using the entropy index yields to the conditional entropy. [1]

As a result, we can describe the two stages of the splitting criterion using such general index $\gamma_{Y|\cdot}(t)$ to be defined for both the predictor and the splits generated by a given predictor.

---

[1] Notice that the impurity measure in CART is nothing else than an heterogeneity index and for this reason we have adopted the same notation for the heterogeneity index here as for the impurity measure in section 1.1.

At the first stage, we maximize for each predictor $m \in M$

$$\max_{m \in M} \gamma_{Y|X_m}(t) = \gamma_{Y|X^*}(t) \tag{3}$$

and, at the second stage, we maximize for each split $s \in S$ of the predictor X^*

$$\max_{m \in M} \gamma_{Y|s}(t) = \frac{i_Y(t) - (p_l i_Y(t_l) + p_r i_Y(t_r))}{i_Y(t)}. \tag{4}$$

At the first stage we can select more than one predictor, namely we can order the predictors with respect to the values of $\gamma_{Y|X_m}(t)$ so that we can rank the predictors with respect to their predictability power. The selected predictors are used to generate the set of splits used at the second stage.

Notice also that the numerator of (4) is equivalent to the decrease in impurity (1). As a result, the splitting rule in CART can be defined in terms of the dependency index γ instead of the decrease in impurity. Indeed, if we consider in place of S the set Q of all possible splitting variables generated by all predictors then we could use directly (4) as a splitting rule in CART. In this way, we provide at least two new interpretations of CART splitting rules: one in terms of the predictability τ index, the other in terms of the conditional entropy index. Using this result, Mola and Siciliano (1997a) have recently introduced a fast splitting algorithm that is related to the two-stage criterion but that allows to find the same solution of CART splitting criterion.

2. The behavior of splitting criteria

In this section we discuss the behavior of splitting criteria focalizing our attention on some aspects concerning their performance with respect to the structure of the dependency in the data, the problem to choose among several splitting rules, the role played by the type of predictors and the stability of the classification rule. For sake of brevity, we present a simulation study and some applications on a real data set in order to contribute to the discussion on the above mentioned aspects with some methodological issues.

2.1 The dependence data structure

Although there are methodological differences in the splitting procedures proposed in literature we believe that empirical differences in results strongly depend on the structure of the data set analyzed. Our experience allows us to say that when passing from a data set without structure of dependency to a data set where some of the predictors have high predictability power on the response variable different tree procedures might converge to the same result.

As an example, we present some results of a simulation study when comparing the CART splitting procedure, using either the Gini index or the entropy index as a splitting rule, with the two-stage splitting procedure, using either the predictability τ index or the conditional entropy index as a splitting rule.

We have considered a 4×2 factorial design, with I (number of categories for each categorical predictor) taking values $3, 4, 5, 6$ and J (response classes) taking values $2, 3$. For each combination of the factorial design we have generated 1000 matrices of dimensions 100×11 (where 100 is the sample size, and 11 is given by 1 response variable plus 10 predictors).

We have generated all the variables from a uniform distribution without imposing any dependency. Tables 1 and 2 describe the results using the Gini index of hetero-geneity and the entropy index respectively in order to find the best split at the root node. In particular, for each combination of the factorial design each cell of the table gives the percentage of times that the best split of CART splitting criterion is found by the two-stage splitting criterion through the first best predictor $X_{(1)}$, the second predictor $X_{(2)}$ and the third predictor $X_{(3)}$ in the order respectively (see section 1.2). For instance, for $I = 3$ and $J = 2$ in table 1 the best split according to CART has been found in predictor $X_{(1)}$ for 88% of times and in predictor $X_{(2)}$ for the remaining 12% of times.

		$J = 2$				$J = 3$		
I	$X_{(1)}$	$X_{(2)}$	$X_{(3)}$	overall	$X_{(1)}$	$X_{(2)}$	$X_{(3)}$	overall
3	88%	12%	0%	100%	79%	16%	4%	99%
4	88%	10%	2%	100%	72%	22%	4%	98%
5	81%	16%	2%	99%	75%	14%	5%	94%
6	72%	21%	7%	100%	76%	13%	3%	92%

Table 1: Percentage of times that the two-stage splitting criterion finds the best split of CART (using the Gini index of heterogeneity)

		$J = 2$				$J = 3$		
I	$X_{(1)}$	$X_{(2)}$	$X_{(3)}$	overall	$X_{(1)}$	$X_{(2)}$	$X_{(3)}$	overall
3	88%	12%	0%	100%	79%	18%	2%	99%
4	89%	9%	2%	100%	77%	15%	5%	97%
5	85%	12%	2%	99%	74%	20%	1%	95%
6	76%	17%	7%	100%	70%	11%	8%	89%

Table 2: Percentage of times that the two-stage splitting criterion finds the best split of CART (using the entropy index)

Both tables 1 and 2 show that whatever is the adopted splitting rule considering three best predictors in two-stage splitting criterion yields to find with a high percentage of times (see column "overall") the same best split of CART, especially when $J = 2$, although we have not imposed any dependency structure.

As soon as we have one of the predictors related to the response variable then the CART splitting criterion and the two-stage splitting criterion using only one best predictor give the same result. This has been the result of a further simulation study in which one of the predictors was generated according to a dependency structure (passing from a low level of dependency to a high level of dependency as measured with the dependency γ index).

2.2 The choice of the splitting rule

Another interesting point is the role played by the splitting rule in the adopted split-ting criterion, that is to verify whether the selection of the best split does depend on the choice of the splitting rule. This aspect is strictly related to the problem discussed in section 2.1, as we believe that if there exists a dependency of the response variable on some of the predictors then different splitting rules agree on the same best split.

On this purpose we consider the factorial design described in section 2.1 and we calculate the percentage of times that the Gini index of heterogeneity and the entropy index in the CART splitting procedure attain the same result: analogously, we calculate the percentage of times that the predictability τ index and the conditional entropy index in the two-stage splitting procedure yield to the same result. We describe these results in tables 3 and 4 where we notice a certain coherence in the results especially for $J = 2$.

I	$J = 2$ Gini-Entropy	$J = 3$ Gini-Entropy
3	99%	90%
4	98%	84%
5	97%	82%
6	93%	70%

Table 3: Percentage of times that the best split is the same using the Gini index and the entropy index in CART splitting criterion

I	$J = 2$ Gini-Entropy	$J = 3$ Gini-Entropy
3	99%	91%
4	99%	88%
5	97%	84%
6	97%	72%

Table 4: Percentage of times that the best split is the same using the τ index and the conditional entropy index in two-stage splitting criterion

2.3 The role played by the type of predictors

One of the innovative aspects of the CART splitting criterion has been the simultaneous treatment of numerical and categorical predictors. Neverthless we have very often verified that numerical predictors play a privileged role in the analysis in the sense that the best splits in the upper part of the tree are very often generated by numerical predictors rather than categorical ones.

As an example, we have considered a well known real data set such as the "low birth weight" (medical) data set from Hosmer and Lemeshow (1990). The response variable is a dummy variable that is equal to 1 if birth-weight is less than $2\,500\,g$ and 0 otherwise. For each mother, two numerical predictors were recorded, age of the mother in years (X_1) and the weight in pounds at the last menstrual period (X_2). Six categorical variables were also recorded: race (X_3, 1=white, 2=black, 3=other), smoking status during pregnancy (X_4, 0=non-smoker, 1=smoker), history of premature abort (X_5 0=no, 1=yes), history of hypertension (X_6, 0=no, 1=yes), presence of uterine irritability (X_7, 0=no, 1=yes), physician visits during last trimester (X_8, 0=none, 1=one or more).

We have analyzed this data set using the CART splitting procedure. The split sequence to grow the final binary tree is shown in table 5. Most of the splits have been generated by numerical predictors. We have repeated the analysis considering a categorization of the numerical predictors. The result is shown in table 6. It is interesting to notice how the categorization of numerical predictors modifies the tree structure and thus the misclassification rate in both nonterminal and terminal nodes (see both columns "cases" and "error rate").

		split sequence			terminal
t	cases %	error rate (%)	best pred.	split	label class
1	100.0	31.0	X_5	0 vs 1	
2	16.76	40.0	X_1	< 31.5	
4	15.08	33.0	X_2	< 156.5	
8	13.97	28.0			1
9	1.12	0.0			0
5	1.68	0.0			0
3	83.24	26.0	X_2	< 106.0	
6	15.64	46.0	X_1	< 22.5	
12	10.06	28.0			0
13	5.59	20.0			1
7	67.70	21.0	X_6	no	
14	63.69	18.0	X_7	no	
28	56.42	14.0	X_1	< 27	
29	7.26	46.0	X_2	< 122.5	
58	4.47	25.0			0
59	2.79	20.0			1
15	3.91	29.0			1

Table 5: Split sequence using numerical and categorical predictors

		split sequence			terminal
t	cases %	error rate (%)	best pred.	split	label class
1	100.0	31.0	X_5	0 vs 1	
2	16.76	40.0	X_2	1.2 vs 3.4	
4	4.47	38.0			0
5	12.29	32.0			1
3	83.24	26.0	X_6	0 vs 1	
6	78.77	23.0	X_2	1 vs 2.3.4	
12	51.40	16.0			0
13	27.37	37.0	X_2	2 vs 0.1	
26	13.41	46.0	X_7	0 vs 1	
52	10.61	37.0			0
53	2.79	20.0			1
27	13.97	28.0	X_8	1 vs 0	
54	6.70	17.0			0
55	7.26	38.0	X_1		
110	6.15	27.0			0
111	1.12	0.0			1
7	4.47	38.0			1

Table 6: Split sequence using categorized and categorical predictors

2.4 The stability of the classification rule

A skeptical researcher might consider "unstable" the results of tree procedures: this is in fact a crucial point that we discuss in this section.

Through some resampling methods we analyze the stability of classification tree procedures with the respect to the structure of the final binary tree and the related misclassification rates $(R(t))$. In particular, we have analyzed the behavior of the CART splitting procedure when the test sample is used to validate the classification rule (see Breiman *et al.*, 1984) . Considering the same data set described in section 2.3 we have repeated the analysis 1000 times taking 30% of cases randomly chosen in the test sample; again we have repeated the analysis 1000 times taking 20% of cases in the test sample; finally we have repeated the analysis 1000 times taking 10% of cases in the test sample. For each analysis we have considered final classification trees with 3 and 4 terminal nodes respectively and we have calculated the related misclassification rates. In figure 1 we show two series of boxplots in order to describe the distribution of the misclassification rate considering 3 terminal nodes (boxplots above) and 4 terminal nodes (boxplots below). As a result, the misclassification rate appears to be quite "unstable" when the test sample takes 30% of cases. In conclusion, when the learning sample is not too large the cross-validation is recommended since it can provide more stable validations of the classification rule.

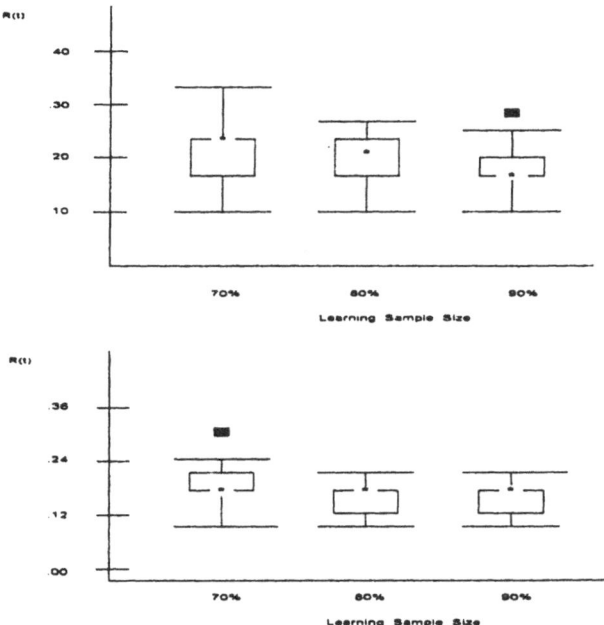

Figure 1: Boxplots of the misclassification error for different binary trees

3 Concluding remarks

In this paper we have discussed important aspects concerning the behavior of splitting criteria in classification trees. We have shown how the two–stage splitting criterion can be fruitfully used to select a number of predictors that generate with high confidence level the best split according to the CART criterion.

We have also verified that the structure of the binary tree is not influenced by the choice among alternative splitting rules, but rather by the type of predictors and their treatment in the splitting procedure.

There are several classification tree procedures proposed in literature and in recent years a lot of attention has been given to the specialized software for applying such procedures (i.e. CART. RECPAM). Furthermore. it is possible to find binary segmentation procedures also in statistical packages such as SPSS, S+, SPAD.S.

As a result, the number of utilizers of such procedures is increasing and "nonexpert researchers" might be willing to apply classification tree procedures for statistical analysis. It becomes then evident that a correct use of such methods requires a certain experience or at least the attention for some crucial aspects such as the simultaneous treatment of numerical and categorical predictors. the choice of the splitting rule, the method for validating the tree and so on. We believe that is worthwhile to discuss some of the problems and peculiar aspects of classification tree procedures as we have described in this paper; we hope that the present contribution provides a good step in this direction.

Acknowledgements: This research was supported for the first author by CNR research funds number 95.02041.CT10 and for the second author by CNR research funds number 92.1872 P

References

Breiman L., Friedman J.H., Olshen R.A., Stone C.J., (1984): *Classification and Regression Trees*. Belmont C.A. Wadsworth.

Celeux, G. and Lechevallier Y. (1982): Methodes de segmentation non parametriques, *Revue de statistique appliquees*. 4, 39–53.

Ciampi. A. and Thiffault, J. (1987): Recursive Partition and Amalgamation (RECPAM) for Censored Survival Data: Criteria for tree selector, *Statistical Software Newsletter*, **2**, vol. 14, 78–81.

Hosmer. D. W. and Lemeshow, S. (1990): *Applied Logistic Regression*, J. Wiley, New York.

Mingers. J. (1988): An empirical comparison of selection measures for decision tree induction. *Machine learning*, 3, 319–342.

Mola. F. (1993): *Aspetti metodologici e computazionali delle tecniche di segmentazione binaria. Un contributo basato su funzioni di predizione*. PhD dissertation, University of Naples.

Mola. F. and Siciliano, R. (1992): A Two–Stage Predictive Splitting Algorithm in Binary Segmentation, *Computational Statistics*, Dodge, Y. and Whittaker, J. (eds.), **1**, 179–184, (Compstat '92 Proceedings). Physica Verlag.

Mola. F. and Siciliano, R. (1994): Alternative Strategies and CATANOVA Testing in Two--Stage Binary Segmentation. *New Approaches in Classification and Data Analysis*, Diday. E. et al. (eds.).316–323, Springer Verlag.

Mola, F. and Siciliano, R. (1997a): A Fast Splitting Procedure for Classification Trees, *Statistics and Computing* (to appear).

Mola, F. and Siciliano, R. (1997b): Visualizing Data in Tree-Structured Classification, *Proceedings of IFCS-96: Data Science, Classification and Related Methods*, (Hayashi, C. *et al.*, eds.). Springer Verlag. Tokyo.

Mola, F., Klaschka, J. and Siciliano, R. (1996): Logistic Classification Trees, *COMPSTAT 96 Proceedings* (A. Prat, ed.), Physica Verlag.

Taylor, P.C. and Silverman, B.W. (1993): Block Diagrams and Splitting Criteria for Classification Trees. *Statistics and Computing*. 3, 147–161.

Fitting Pre-specified Blockmodels

Vladimir Batagelj[1], Anuška Ferligoj[2], and Patrick Doreian[3]

[1] University of Ljubljana, FMF, Dept. of Mathematics
Jadranska 19, 1000 Ljubljana, Slovenia

[2] University of Ljubljana, Faculty of Social Sciences
P.O. Box 47, 1109 Ljubljana, Slovenia

[3] University of Pittsburgh, Dept. of Sociology
Pittsburgh, PA 15260, USA

Summary: In this paper an optimization approach to blockmodeling for fitting an observed relation to a pre-specified blockmodel is used. The proposed deductive approach to blockmodeling is applied to a concrete example. Some further research directions are also suggested.

1. Introduction.

The goal of conventional blockmodeling is to reduce a large, potentially incoherent network to a smaller comprehensible structure that can be interpreted more readily. Blockmodeling, as an empirical procedure, is based on the idea that units in a network can be grouped according to the extent to which they are equivalent, under some *meaningful* definition of equivalence.

There are many *inductive* approaches for establishing blockmodels for a set of social relations defined over a set of social actors. Some form of equivalence is specified and clusterings are sought that are consistent with the specified equivalence. In all cases, the analyses respond to empirical information in order to establish the blockmodel. Another view of blockmodeling is *deductive* in the sense of starting with a blockmodel that is specified in terms of substance prior to an analysis. In this paper we present methods where a set of observed relations are fitted to a pre-specified blockmodel.

2. Basic Terms

Network: Let $E = \{x_1, x_2, \ldots, x_n\}$ be a finite set of *units*. The units are related by binary *relations* $R_t \subseteq E \times E$, $t = 1, \ldots, r$ which determine a *network* $\mathcal{N} = (E, R_1, R_2, \ldots, R_r)$. In the following we restrict our discussion to a single relation R described by a corresponding binary matrix $\mathbf{R} = [r_{ij}]_{n \times n}$ where

$$r_{ij} = \begin{cases} 1 & x_i R x_j \\ 0 & \text{otherwise} \end{cases}$$

In some applications r_{ij} can be a nonnegative real number expressing the strength of the relation R between units x_i and x_j.

Cluster, clustering: One of the main procedural goals of blockmodeling is to identify, in a given network, *clusters* (classes) of units that share structural characteristics defined in terms of R. The units within a cluster have the same or similar connection patterns to other units. They form a *clustering* $C = \{C_1, C_2, \ldots, C_k\}$ which is a partition of the set E: $\bigcup_i C_i = E$ and $i \neq j \Rightarrow C_i \cap C_j = \emptyset$. Each partition determines an equivalence relation (and vice versa). Let us denote by $\sim$ the relation determined

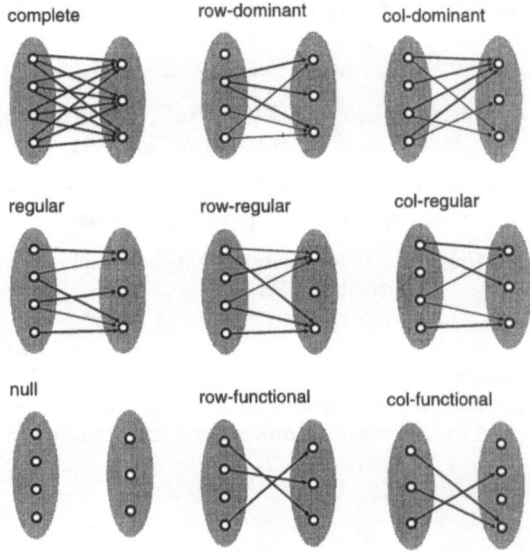

Figure 1: Types of connection between two sets; the left set is the ego-set.

by partition C.

Block: A clustering C partitions also the relation R into *blocks* $R(C_i, C_j) = R \cap C_i \times C_j$. Each such block is defined by units belonging to clusters C_i and C_j in terms of the arcs leading from cluster C_i to cluster C_j. If $i = j$, a block $R(C_i, C_i)$ is called a *diagonal* block.

Blockmodel: A *blockmodel* consists of structures obtained by identifying all units from the same cluster of the clustering C. For an exact definition of a blockmodel we have to be precise also about which blocks produce an arc in the *reduced graph* and which do not. The reduced graph can be presented by a matrix, called also *image matrix*.

Block Types: Several possible block types can be defined. In Figure 1 nine block types are presented (Batagelj, 1993). In the relational matrix below three types of blocks can be found:

$$
\begin{array}{cccc|cccc}
1 & 1 & 1 & 1 & 1 & 1 & 0 & 0 \\
1 & 1 & 1 & 1 & 0 & 1 & 0 & 1 \\
1 & 1 & 1 & 1 & 0 & 0 & 1 & 0 \\
1 & 1 & 1 & 1 & 1 & 0 & 0 & 0 \\
\hline
0 & 0 & 0 & 0 & 0 & 1 & 1 & 1 \\
0 & 0 & 0 & 0 & 1 & 0 & 1 & 1 \\
0 & 0 & 0 & 0 & 1 & 1 & 0 & 1 \\
0 & 0 & 0 & 0 & 1 & 1 & 1 & 0 \\
\end{array}
\implies
$$

complete	regular
null	complete

3. Blockmodeling - Formalization

A *blockmodel* is an ordered sextuple $\mathcal{M} = (U, K, \mathcal{T}, Q, \pi, \alpha)$ where:

- U is a set of *types* of units (images or representatives of classes);

- $K \subseteq U \times U$ is a set of *connections*;

- $\mathcal{T}$ is a set of predicates used to describe the types of connections between different classes (clusters, groups, types of units) in a network. We assume that nul $\in \mathcal{T}$. A mapping $\pi : K \to \mathcal{T} \setminus \{\text{nul}\}$ assigns predicates to connections;

- Q is a set of *averaging rules*. A mapping $\alpha : K \to Q$ determines rules for computing values of connections.

For $\mu : V \to U$ we define for $t \in U$ $C(t) = \mu^{-1}(t) = \{x \in V : \mu(x) = t\}$. Therefore $\mathcal{C}(\mu) = \{C(t) : t \in U\}$ is a partition (clustering) of the set of units V.

A (surjective) mapping $\mu : V \to U$ determines a blockmodel $\mathcal{M}$ of network $\mathcal{N}$ iff it satisfies the conditions:

$$\forall (t, w) \in K : \pi(t, w)(C(t), C(w))$$

and

$$\forall (t, w) \in U \times U \setminus K : \text{nul}(C(t), C(w)).$$

Note, $\mathcal{T} = \{\text{nul}, \text{com}\}$ implies a structural blockmodel (Lorrain and White, 1971); and, $\mathcal{T} = \{\text{nul}, \text{reg}\}$ implies a regular blockmodel (White and Reitz, 1983).

Let $\approx$ be an equivalence relation over V and $[x] = \{y \in V : x \approx y\}$. We say that $\approx$ is *compatible* with $\mathcal{T}$ over a network $\mathcal{N}$ iff

$$\forall x, y \in V \exists T \in \mathcal{T} : T([x], [y]).$$

It is easy to verify that the notion of compatibility for $\mathcal{T} = \{\text{nul}, \text{reg}\}$ reduces to the usual definition of regular equivalence (Borgatti and Everett, 1989). For a compatible equivalence $\approx$ the mapping $\mu : x \mapsto [x]$ determines a blockmodel.

4. Optimization

4.1 A Criterion Function

One of the possible ways of constructing a criterion function that directly reflects the considered equivalence is to measure the fit of a clustering to an ideal one with perfect relations within each cluster and between clusters according to the considered equivalence (Batagelj, Doreian, and Ferligoj, 1992; Batagelj, 1993; Doreian, Batagelj, and Ferligoj, 1994).

Given a set of types of connection $\mathcal{T}$ and a block $R(X, Y)$, $X, Y \subseteq V$. we can determine the strongest (according to the ordering of the set $\mathcal{T}$) type T which is satisfied by $R(X, Y)$. In this case we set

$$\pi(\mu(X), \mu(Y)) = T$$

We need to consider also the (many) cases where no type from $\mathcal{T}$ is satisfied. One approach is to introduce the set of *ideal blocks* for a given type $T \in \mathcal{T}$

$$\mathcal{B}(X, Y; T) = \{B \subseteq X \times Y : T(B)\}$$

and define the *deviation* $\delta(X, Y; T)$ of a block $R(X, Y)$ from the nearest ideal block. We can efficiently test whether the block $R(X, Y)$ is of the type T (see Table 1). On

Table 1: Characterizations of types of blocks

null	all 0 (except may be diagonal)
complete	all 1 (except may be diagonal)
row-dominant	$\exists$ all 1 row (except may be diagonal)
col-dominant	$\exists$ all 1 column (except may be diagonal)
regular	1-covered rows and 1-covered columns

the basis of these characterizations we can construct also the corresponding measures of deviation from the ideal realization. For the proposed types, all deviations are *sensitive*

$$\delta(X, Y; T) = 0 \Leftrightarrow T(R(X, Y)).$$

Therefore a block $R(X, Y)$ is of a type T exactly when the corresponding deviation $\delta(X, Y; T)$ is 0. In the deviation δ we can also incorporate the values of lines, ν, if the network has valued arcs.

Based on the deviation $\delta(X, Y; T)$ we introduce the *block-error* $\varepsilon(X, Y; T)$ of $R(X, Y)$ for type T. Two examples of block-errors are

$$\varepsilon_1(X, Y; T) = w(T)\delta(X, Y; T) \quad \text{and} \quad \varepsilon_2(X, Y; T) = \frac{w(T)}{n_r n_c}(1 + \delta(X, Y; T)),$$

where $w(T) > 0$ is a weight for type T. We extend the block-error to the set of feasible types $\mathcal{T}$ by defining

$$\varepsilon(X, Y; \mathcal{T}) = \min_{T \in \mathcal{T}} \varepsilon(X, Y; T) \quad \text{and} \quad \pi(\mu(X), \mu(Y)) = \text{argmin}_{T \in \mathcal{T}} \varepsilon(X, Y; T)$$

To make π well-defined, we order (priorities) the set $\mathcal{T}$ and select the first type from $\mathcal{T}$ which minimizes ε. We combine block-errors into a *total error* – a blockmodeling *criterion function*

$$P(\mu; \mathcal{T}) = \sum_{(t,w) \in U \times U} \varepsilon(C(t), C(w); \mathcal{T}).$$

The criterion functions based on block-errors ε_1 and ε_2 are denoted P_1 and P_2 respectively.

For the criterion function $P_1(\mu)$ we have $P_1(\mu) = 0 \Leftrightarrow \mu$ is an exact blockmodeling. Also for P_2, we obtain an exact blockmodeling μ iff the deviations of all blocks are 0.

The obtained optimization problem can be solved by local optimization. Once a partitioning μ and types of connection π are determined, we can also compute the values of connections by using the *averaging rules*.

4.2 Local Optimization

For solving the blockmodeling problem we use a local optimization procedure (a relocation algorithm):

Determine the initial clustering $\mathcal{C}$:
repeat:
       if in the neighborhood of the current clustering $\mathcal{C}$
       there exists a clustering $\mathcal{C}'$ such that $P(\mathcal{C}') < P(\mathcal{C})$
       then move to clustering $\mathcal{C}'$.

Table 2: Student Government Matrix

		m 1	p 2	m 3	m 4	m 5	m 6	m 7	m 8	a 9	a 10	a 11
minister 1	1	0	1	1	0	0	1	0	0	0	0	0
p.minister	2	0	0	0	0	0	0	0	1	0	0	0
minister 2	3	1	1	0	1	0	1	1	1	0	0	0
minister 3	4	0	0	0	0	0	0	1	1	0	0	0
minister 4	5	0	1	0	1	0	1	1	1	0	0	0
minister 5	6	0	1	0	1	1	0	1	1	0	0	0
minister 6	7	0	0	0	1	0	0	0	1	1	0	1
minister 7	8	0	1	0	1	0	0	1	0	0	0	1
adviser 1	9	0	0	0	1	0	0	1	1	0	0	1
adviser 2	10	1	0	1	1	1	0	0	0	0	0	0
adviser 3	11	0	0	0	0	0	1	0	1	1	0	0

The neighborhood in this local optimization procedure is determined by the following two transformations: *moving* a unit x_k from cluster C_p to cluster C_q (*transition*); and *interchanging* units x_u and x_v from different clusters C_p and C_q (*transposition*).

4.3 Benefits from Optimization Approach

There are several benefits of using the optimization approach to blockmodeling:

- *ordinary / inductive blockmodeling*: Given a network $\mathcal{N}$ and set of types of connection $\mathcal{T}$, determine $\mathcal{M}$, i.e., μ, π and α;

- *evaluation of the quality of a model, comparing different models, analyzing the evolution of a network* (Sampson (1968) data, Doreian and Mrvar, 1996): Given a network $\mathcal{N}$, a model $\mathcal{M}$, and blockmodeling μ, compute the corresponding criterion function;

- *model fitting / deductive blockmodeling*: Given a network $\mathcal{N}$, set of types $\mathcal{T}$, and a model $\mathcal{M}$, determine μ which minimizes the criterion function.

- we can fit the network to a partial model and analyze the residual afterward;

- we can also introduce different constraints on the model, for example: units x and y are of the same type; or, types of units x and y are not connected; ...

5. Example: Student Government

The example network consists of communication interactions among twelve members and advisors of the Student Government at the University in Ljubljana (Hlebec, 1993). The results of the measurement are not real interactions among actors but their cognitions about communication interactions.

Data were collected through face to face interviews that were conducted in May 1992. Communication flow among actors was identified by the following question: Of the members and advisors of the Student Government, whom do you (most often) talk with?

The content of the communication flow was limited to the matters of the Student Government. The time frame was also defined: the question was referred to a six

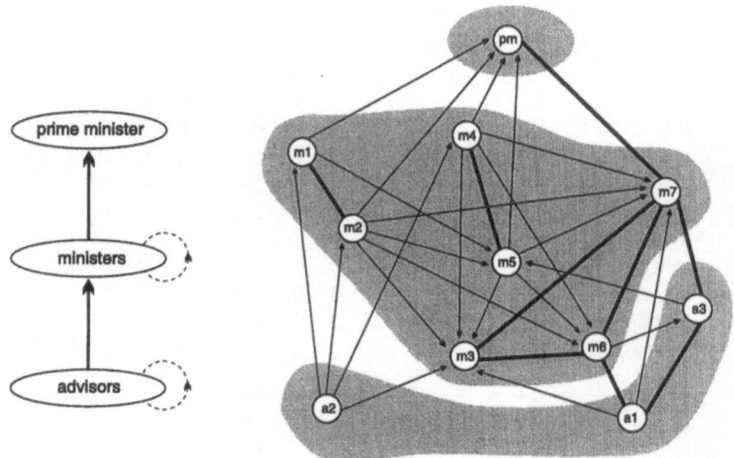

Figure 2: Student Government – hypothetical blockmodel

month period. One respondent refused to cooperate in the experiment. As he was not considered in the analysis, the network consists of eleven actors and is presented in Table 2. The computations were done with the program MODEL2 which is for PC available on WWW address http://vlado.fmf.uni-lj.si/pub/networks/.

The *hypothetical structure* of the Student Government network, a hierarchy where each advisor communicates at least to one minister and the ministers to the prime minister, is presented in Figure 2.

First, let us analyze the network *inductively* (without assuming the hypothetical hierarchical structure) assuming 3 clusters and regular equivalence. We obtained many regular solutions with 7 errors.

The basis for the *deductive* blockmodeling is the assumed hierarchical structure of the Student Government presented in Figure 2.

The *first pre-specified blockmodel* is defined by assuming 3 clusters, regular equivalence and the following pre-specified image matrix

	1	2	3
1	{ 0, reg }	{ 0 }	{ 0 }
2	{ reg }	{ 0, reg }	{ 0 }
3	{ 0 }	{ reg }	{ 0, reg }

The result is a subset of the set of inductive solutions with 7 errors. One of the solutions is the following:

$$C_1 = \{\{pm, m7, a3\}, \{m1, m2, m3, m4, m5, m6, a1\}, \{a2\}\}$$

The solution is presented also in Figure 3. The black dots on the arcs denote superflous arcs (errors) according to the ideal solution and the white dots the missing arcs.

We can constrain our pre-specified blockmodel further by *additional constraint* on

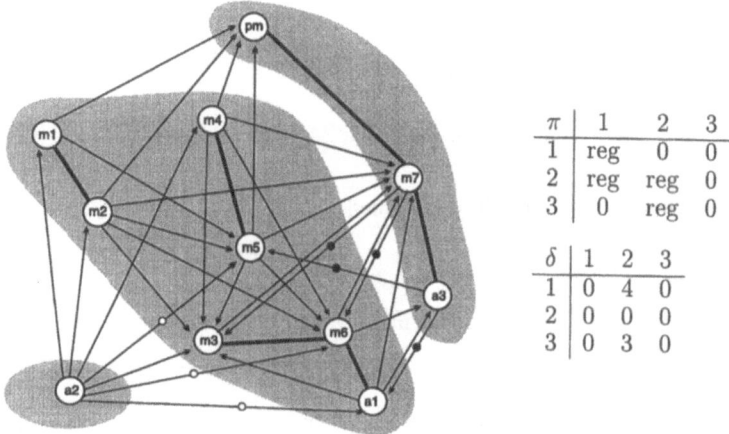

π	1	2	3
1	reg	0	0
2	reg	reg	0
3	0	reg	0

δ	1	2	3
1	0	4	0
2	0	0	0
3	0	3	0

Figure 3: Student Government – first deductive solution

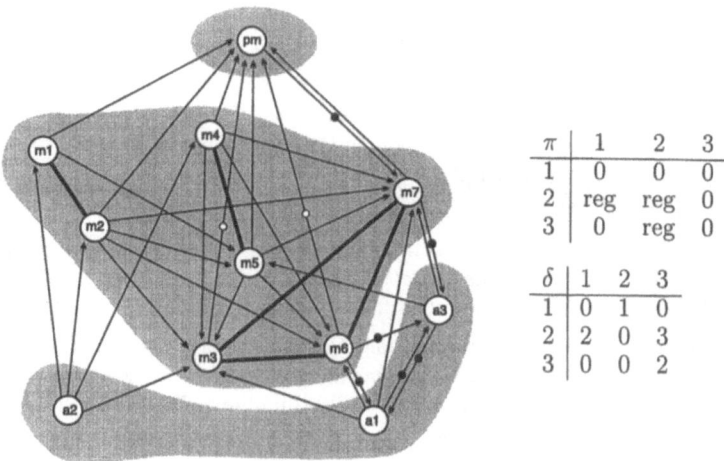

π	1	2	3
1	0	0	0
2	reg	reg	0
3	0	reg	0

δ	1	2	3
1	0	1	0
2	2	0	3
3	0	0	2

Figure 4: Student Government – second deductive solution

units in clusters: all advisers are in cluster 3. We obtained a single solution with 8 errors.

$$\mathcal{C}_2 = \{\{pm\}, \{m1, m2, m3, m4, m5, m6, m7\}, \{a1, a2, a3\}\}$$

The solution is also presented in Figure 4.

The results of model fitting show that the hypothetical hierachical model was obtained with a minimal increase of error compared to inductive solution (from 7 to 8). This indicates that it represents the network structure well.

6. Further Research

There are several possible directions for further research in the field of blockmodeling. At least two questions need an attention:

- to define additional block types which are more appropriate for describing specific network structures:

- to elaborate blockmodeling of valued networks.

References:

Batagelj, V. (1991): *STRAN - STRucture ANalysis*, Manual, Ljubljana.

Batagelj, V., Doreian, P. and Ferligoj, A. (1992): An optimizational approach to regular equivalence, *Social Networks*, **14**, 121–135.

Batagelj, V., Ferligoj, A. and Doreian, P. (1992): Direct and indirect methods for structural equivalence, *Social Networks*, **14**, 63–90.

Batagelj, V. (1993): Notes on block modelling. In *Abstracts and Short Versions of Papers*, 3rd European Conference on Social Network Analysis, München: DJI, 1-9. Extended version in print in *Social Networks* 1997.

Borgatti, S.P. and Everett, M.G. (1989): The class of all regular equivalences: Algebraic structure and computation, *Social Networks*, 11:65–88.

Doreian, P., Batagelj, V. and Ferligoj, A.(1994): Partitioning Networks on Generalized Concepts of Equivalence, *Journal of Mathematical Sociology*, **19**, 1, 1–27.

Doreian, P. and Mrvar, A. (1996): A Partitioning Approach to Structural Balance, *Social Networks*, **18**, 2, 149–168.

Ferligoj, A., Batagelj, V. and Doreian, P. (1994): On Connecting Network Analysis and Cluster Analysis, In *Contributions to Mathematical Psychology, Psychometrics, and Methodology* (G.H. Fischer and D. Laming, Eds.), New York: Springer.

Hlebec, V. (1993): Recall versus recognition: Comparison of two alternative procedures for collecting social network data, In *Developments in Statistics and Methodology* (A. Ferligoj and A. Kramberger, Eds.), Metodološki zvezki 9, Ljubljana: FDV, 121-128.

Lorrain, F. and White, H.C. (1971): Structural equivalence of individuals in social networks, *Journal of Mathematical Sociology*, 1. 49–80.

Sampson, S.F. (1968): *A Novitiate in a Period of Change: An Experimental and Case Study of Social Relationships*. PhD thesis, Cornell University.

White, D.R. and Reitz. K.P. (1983): Graph and semigroup homomorphisms on networks of relations, *Social Networks*, **5**, 193–234.

Robust impurity measures in decision trees

Tomàs Aluja-Banet, Eduard Nafria

Dept. of Statistics and Operational Research
Universitat Politcnica de Catalunya
c. Pau Gargallo. 5. 08028 Barcelona. Spain
E-mail: aluja@eio.upc.es

Summary: Tree-based methods are a statistical procedure for automatic learning from data, their main characteristic being the simplicity of the results obtained. Their virtue is also their defect since the tree growing process is very dependent on data; small fluctuations in data may cause a big change in the tree growing process. Our main objective was to define data diagnostics to prevent internal instability in the tree growing process before a particular split has been made. We present a general formulation for the impurity of a node, a function of the proximity between the individuals in the node and its representative. Then, we compute a stability measure of a split and hence we can define more robust splits. Also, we have studied the theoretical complexity of this algorithm and its applicability to large data sets.

1. Introduction

The objective of tree-based methods is to automatically detect which variables serve to explain the behaviour of a response variable, whether quantitative or categorical. They can be applied in the same context as other alternative methods, such as multiple regression, discriminant analysis, logistic regression or neural networks. Its main advantage is the simplicity of the results obtained and the possibility of automatic generation of decision rules. This property links this methodology with the AI techniques. Thus the main usage is decision making. Its strength is also its weakness, since the tree growing process is very dependent on data; a small fluctuation in data may cause a major change in the topology of the tree. This raises the problem of the stability of a tree. We distinguish internal stability from external stability in the same sense as that stated by Greenacre (1984). External stability refers to the sensitivity of the tree to independent random samples, and can be assessed by means of a test sample or cross-validation, whereas by internal stability we mean the influence exerted by each observation in the learning sample on the formed tree. Another problem relating to the tree methodology is the computational cost, due to the recursive nature of the algorithms and the large number of possible splits, which can be very costly for large data sets. For this reason we have studied the complexity of the algorithm in order to optimise it, proposing an efficient heuristic capable of coping with large data sets, with almost linear cost depending on the number of individuals and variables and the depth of the tree.

2. Tree growing methodology

Since the pioneering work of AID. Sonquist et al. (1964), tree growing methodology has consisted of splitting[1] each group of individuals (node) recursively into two groups. starting from the total sample n. according to a statistical criterion relating the condition for splitting to the response variable. Since then. a great deal of research have been done into the threshold-based criterion. Kass (1980) developed tree methodology for a categorical response variable using a Chi-square split criterion. and Celeux et al. (1982) proposed for the latter case a split criterion based on a distance between distribution functions. while Ciampi (1991) proposed instead the use of the deviance of a generalised linear model. Although the results obtained can be satisfactory in applied research. with an error rate of the same order of alternative methods. they do not escape the criticism of the optimality and goodness of the tree obtained. The CART approach, introduced by Breiman et al. (1984). was an attempt to solve these problems. Its main innovations consisted of:

1. Unification of the case of a categorical response variable (classification trees) with that of a quantitative response variable (regression trees) within a similar framework.
2. Use of an impurity index to measure the heterogeneity of a node.
3. Pruning from a maximal tree instead of using a stop criterion.
5. Giving right honest estimates of the misclassification error.

The impurity indices measure the heterogeneity of a node:

$$i(t) = \quad \mathcal{F}(p(j \mid t)) \quad \text{for a classification tree} \qquad (1)$$
$$i(t) = \quad \mathcal{F}(y_j \mid j \in t) \quad \text{for a regression tree}$$

where $p(j \mid t)$ is the probability of class j in node t and y_j is the value of the response. for an individual in node t.

Impurity indices should have a maximum value for classes with equal probability. a value of 0 for a pure node and should be a decreasing function through the splitting process:

$$i(t) \geq \alpha\, i(t_l) + (1-\alpha)\, i(t_r) \quad 0 \leq \alpha \leq 1$$

The most usual impurity indices are:

$$i(t) = \quad \sum_{j \neq k} p(j|t)p(k|t) \qquad \text{Gini index}$$
$$i(t) = \quad 1 - max_j p(j|t) \qquad \text{misclassification index}$$
$$i(t) = \quad -\sum_j p(j|t)log(p(j|t)) \qquad \text{entropy index}$$
$$i(t) = \quad \tfrac{w \cdot w}{4}(\sum_j |p(j|t_l) - p(j|t_r)|)^2 \quad \text{Twoing index}$$

And for a regression tree:

$$i(t) = \quad \tfrac{1}{n_t}\sum_{i \in t}(y_i - y_t)^2 \quad \text{variance index}$$
$$i(t) = \quad \tfrac{1}{n_t}\sum_{i \in t}|y_i - y_t| \quad \text{absolute deviation index}$$

Then. the split criterion consists of selecting the split which maximises the weighted reduction of impurity between the parent node and its offspring (left and right).

$$\Delta i(t) = i(t) - \frac{n_{tl}}{n_t}i(t_l) - \frac{n_{tr}}{n_t}i(t_r) \qquad (2)$$

---

[1] Here, we just consider trees made of binary splits

The problem of defining a right-sized tree is solved, instead of using a threshold, by growing a maximal tree (a tree with every terminal node pure or with say, five individuals or fewer) and from that tree, defining a nested sequence of optimum subtrees by successively removing non informative branches of the large tree, minimising an error complexity measure (pruning step). Then, the problem is transformed in choosing the subtree with minimum misclassification error. This is done by using a test sample or a tenfold cross-validation to obtain honest estimates of the misclassification error.

Although this approach solves many of the shortcomings for which its ancestors were criticised, the tree growing process is still very dependent on data. It is easy to see that every split depends on the presence of some outliers in the actual node leading to instable trees. Moreover, the choice of an impurity measurement is closely related to the stability of the tree. This is particularly true using for example the Gini index which attempts to favour small but very pure nodes rather than equal-sized but less pure ones. This is why the pruning process very often results in a severe reduction. In order to tackle this problem, we have studied the internal stability of a partition and then defined more robust impurity indices.

3. General formulation of impurity

Let us suppose that we have a node t with n_t individuals to classify according to k classes of a response variable[2], and let w_{it} be the weight of one individual in node t. To generalise the notion of impurity, we use a geometric approach where each class defines a point $(1, 0, 0, \ldots) \in R^k$. Then, we define the impurity as a function of the distances between each individual in the node and the representative of the node m_t, defined as the point of the convex polygon of R^k, which minimises $i(t)$ (Figure 1).

$$i(t) = \frac{\sum_{i \in t} w_{it} \delta(i, m_t)}{\sum_{i \in t} w_{it}} \tag{3}$$

where $\delta(i, m_t)$ is the distance of an individual i and m_t (obviously, all individuals in the same response class j share the same distance). This formula reduces for a categorical response variable with uniform weight to:

$$i(t) = \frac{\sum_{j=1}^{k} n_{jt} \delta(j, m_t)}{n_t} \tag{4}$$

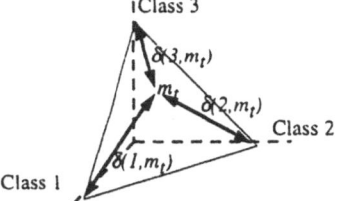

Fig. 1 Convex polygon of classes of a node with its representative

---

[2] We first present the case of a categorical response variable.

For a regression tree, the geometrical interpretation is easier since the values of the response variable are represented in the real line m_t being the point on the real line representing the node t, which minimises its impurity.

This formulation being very general, we can choose the distances $\delta(j, m_t)$ in a very general sense. In particular, we can use the L_2 norm. Then, it is easy to show that for a classification tree the representative of the node coincides with the multinomial vector of probabilities of classes, and the impurity index reduces to the well known Gini index; and that for a regression tree, the representative of the node is the mean of the response in a node and the impurity is the variance. On the other hand, in the case of an L_1 norm, for a classification tree the representative coincides with the class of maximum probability and the index reduces to twice the misclassification index. and for a regression tree the representative is the median of the response in the node and the impurity is the absolute deviation.

Furthermore, this formulation can allow for different misclassification costs. Let C be the matrix of misclassification costs. and c_{ji} the cost of misclassifying an individual in class j when it belongs to class i. Then c_i will represent the overall cost of misclassifying an individual of class i.

$$i(t) = \frac{\sum_{i \in t} c_{\cdot i} w_{it} \delta(i, m_t)}{\sum_{i \in t} c_{\cdot i} w_{it}} \tag{5}$$

We can see that, in fact, introducing the misclassification costs entails overweighting those response classes for which it is most dangerous to make a wrong assignment.

In any case, the reduction to impurity can be expressed as a function of the distances of the individuals to the representative of the parent node and the distance to the corresponding successor.

$$\Delta i(t) = \frac{\sum_{i \in t} w_{it} \delta(i, m_t) - \sum_{i \in t_l} w_{it_l} \delta(i, m_{t_l}) - \sum_{i \in t_r} w_{it_r} \delta(i, m_{t_r})}{\sum_{i \in t} w_{it}} \tag{6}$$

where t_l and t_r represent the left and right child nodes of node t.

4. Stability analysis

Thus, from Formula 6, it is easy to calculate the contribution of any individual to the reduction of impurity. It is simply the difference in the distances of this individual to the representative of the parent node and to the representative of its corresponding child node:

$$c_i = w_{it}(\delta(i, m_t) - \delta(i, m_{t_\cdot})) \tag{7}$$

Notice that the contribution to the reduction of impurity can be positive or negative. although on average this contribution coincides with the overall impurity reduction.

$$\bar{c}_{it} = \frac{\sum_{i \in t} c_{it}}{\sum_{i \in t} w_{it}} = \Delta i(t) \tag{8}$$

Then. the ratio for the average reduction of impurity $\frac{c_{it}}{\Delta i(t)}$ is an easy way to diagnose individuals with a strong influence in the split. Moreover. the distance between the representatives of the child nodes is an indicator of the stability of the present split.

It is clear that in classification trees with the L_2 metric, due to the quadratic form of the impurity index, instabilities occur when the splitting process leads to nodes with very few members of at least one response class j, whereas the most stable case is when the probability of classes are similar, that is, the representative of the node m_t is close to the centre of the convex polygon. In the L_1 metric, the locus of the representative of the node coincides with the class with maximum probability, thus the distance of the remaining classes to the representative is equal to 2, that is, each individual of the latter classes has the same influence.

In regression trees, instability may occur when dealing with nodes with some outlying values; a split attempting to accommodate for these outliers will reduce the impurity significantly and hence the distance between the representatives of both child nodes will be large. Of course, one way of achieving a robust split that is insensitive to the effect of outliers would be by using the L_1 norm, that is, using the absolute deviation as an impurity measure.

4.1 Function of the impurity reduction

For each predictor variable we can define a function of the impurity reduction relative to the impurity in the parent node, defined over all possible splits of this variable, defined as follows:

$$f(u) = 1 - \frac{\sum_{i \in t_l} w_{it_l} \delta(i, m_{t_l}) + \sum_{i \in t_r} w_{it_r} \delta(i, m_{t_r})}{\sum_{i \in t} w_{it} \delta(i, m_t)} \tag{9}$$

where u means a particular split of a variable. This function can be represented graphically provided that there is an ordering of the splits of the predictor variable. This is the case for continuous, ordinal predictors, and also for any type of predictor when the response variable is continuous or binary[3].

Then, a sharply peaked function is sign of an instable split, since a small change in the level of the predictor will imply a major change in the impurity reduction, whereas a smooth function would indicate more stable splits. Also, it is useful to plot the function of impurity reduction together with the distribution functions of response classes in the case of a classification tree. In Figure 2 we represent two functions of impurity reduction (in thick) with the empirical distribution function of the response classes (in thin). The first illustrates an instable case, whereas the second correspond to a more stable case, although it contains one inversion in the empirical distribution functions.

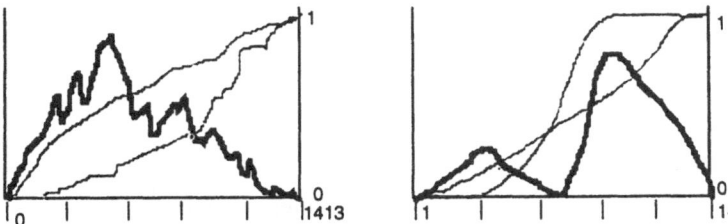

Fig. 2 Two examples of impurity reduction functions

---

[3]Then, an ordering of splits is induced by the response variable.

5. A split criterion based on a distance between distribution functions

For the case of a categorical response variable, it is natural to compute for every node the empirical distribution function of every response class F_j and compare them with the average distribution function F_t in node t. Then, the split point can be defined by a distance between these functions.

$$Max\ d_u = \sum_{j=1}^{k} |F_j(u) - F_t(u)| \qquad (10)$$

where $F_j(u)$ and $F_t(u)$ are the empirical distribution function of class j and the average distribution function of node t evaluated at split u. To use this split criterion there should be an ordering among the possible splits of the predictor variable.

It is easy to see that this distance coincides with the Smirnov distance for the case of two classes (Figure 3).

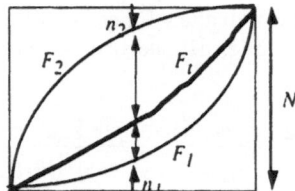

Fig. 3 Distance between the distribution functions of classes

$$max\ [F_1(u) - F_t(u)| + |F_2(u) - F_t(u)|] = max\ |F_1(u) - F_2(u)| \qquad (11)$$

Moreover, for this case, it also coincides with the misclassification index with uniform weight of classes.

$$max\ \Delta i(t) \quad \Rightarrow \quad min\ [n_l i(t_l) + n_r i(t_r)] = min\ [n_l(1 - \frac{n_1}{n_l}) + n_r(1 - \frac{n_2}{n_r})] =$$
$$= \quad max\ [n_1 + (n_2 - N)] = max\ (F_1(u) - F_2(u)) \qquad (12)$$

Furthermore, it coincides with the Celeux-Lechevallier (1982) index with uniform weighing of the response classes. In fact, the difference with this latter index consist of in the different weighing of the response classes.

6. Complexity

The search for an optimal tree is an NP-complete problem. Thus, we should use efficient heuristics. The most used heuristic consists of finding at each step the best split among the whole set of binary partitions. This solution leads to fairly good results obtained in a hierarchical fashion. However, the computational cost of this heuristic is very high, and it requires an efficient algorithm to guarantee a reasonable speed.

We have designed an algorithm of almost linear cost with the parameters of a tree. Lets us first define the parameters of a tree. These are the number of individuals n, the number of total splits s, and the maximum depth of the tree d. Obviously, the total number of splits depends on the number of variables and its type. See Table 1 for the number of splits according the type of variable. Let p be the total number of variables, which can be split according to type: $p = p_b + p_o + p_n + p_c$. For each variable we have s_j splits with $\sum_{j=1}^{p} s_j = s$ and for a maximum depth of d we have $l \leq 2^{d-1}$ nodes.

Type of variable	splits
Binary	1
Ordinal with k modalities	$k - 1$
Nominal with k modalities	$2^{k-1} - 1$
Continuous with m different values	$m - 1$

Table 1. Number of splits according to the type of variables

The complexity can be decomposed in the following steps:

1. Cost of a split for a given variable: $O(n_t) + C$. This cost depends only on the number of individuals in the node plus a constant.

2. Cost of all splits for a given variable: $\sum_{i=1}^{s_j} O(n_t) = O(n_t \cdot s_j)$. This cost depends on the number of splits of the variable, which according to its type can be optimised to the following results:
$$\begin{cases} O(n_t) & \text{(for a binary variable)} \\ O(n_t) + O(k) & \text{(for an ordinal variable)} \\ O(n_t) + O(2^{k-1}) & \text{(for a nominal variable)} \\ O(n_t) + O(n \cdot log(n)) & \text{(for a continuous variable)} \end{cases}$$

3. Cost of all splits for a node: $\sum_{j=1}^{p} O(n_t) = O(n_t \cdot s)$. This is, of course, simply the sum of the costs for all the active variables in a node, which, according to the above result reduces in most cases to: $O(n_t \cdot p)$.

4. Cost of all splits in every node: $\sum_{t=1}^{l \leq 2^{d-1}} O(n_t \cdot p) = O(p) \sum_{t=1}^{d-1} O(n) = O(p \cdot n \cdot d)$. Moreover, we have the cost of assigning every individual to its node: $O(l \cdot n)$.

Thus, the total cost can be written in the following way:

$$O(p_c \cdot n \cdot log(n)) + O(l \cdot n) + O(p \cdot n \cdot d) \tag{13}$$

A critical point is the total number of splits when $s \geq n$, then $O(s \cdot n \cdot d) \geq O(n^2 \cdot d)$. This is particularly dangerous for a nominal response variable with a large number of classes k; when $k \geq n_t$ the cost becomes quadratic or even exponential. See Mola et al. (1992) for the treatment of a multiple class response variable. Also, when l becomes large the cost increases quadratically.

This algorithm, named SAAD (*Segmentació Automàtica per Arbres de Decisió*), runs on a PC platform in a Windows environment and is able to cope with problems with up to 100.000 individuals and 100 variables. Here we present the time in seconds of two problems, one corresponding to a classification tree into 4 classes, with 9 explanatory variables (6 of them were categorical with a maximum of 8 categories each and 3 were continuous), and the other being a regression tree with 18 explanatory variables (half categorical and the other half continuous). For each problem we have varied the number of individuals considered and the depth of the tree produced, obtaining the results shown in Table 2. As can be seen, linearity is preserved approximately up to a depth of 8.

individuals	depth	classification tree	regression tree
1078	1	5"	9"
	3	7"	13"
	5	9"	14"
	8	20"	23"
	13	46"	60"
9862	1	21"	32"
	3	30"	50"
	5	41"	62"
	8	91"	111"
	13	391"	381"
37575	1	108"	155"
	3	126"	269"
	5	159"	352"
	8	376"	641"
	13	1544"	2847"

Table 2. Execution time in seconds for the SAAD segmentation algorithm on an HP 486/66

References

Aluja T., Nafria E. (1995). Generalised impurity measures and data diagnostics in decision trees. *Visualising Categorical Data*. Cologne.

Breiman L., Friedman J.H., Olshen R.A., and Stone C.J. (1984). *Classification and Regression Trees*. Waldsworth International Group, Belmont, California.

Celeux G., Lechevallier Y. (1982). Méthodes de Segementation non Paramétriques. *Revue de Statistique Appliquée*, XXX (4), 39-53.

Ciampi A. (1991). Generalized Regression Trees. *Computational Statistics and Data Analysis*, 12, 57-78. North Holland.

Greenacre M. (1984). *Theory and Application of Correspondence Analysis*. Academic Press.

Gueguen A., Nakache J.P. (1988). Méthode de discrimination basée sur la construction d'un arbre de décision binaire. *Revue de Statistique Appliquée*, XXXVI (1), 19-38.

Kass G.V. (1980). An Exploratory Technique for Investigating Large Quantities of Categorical Data. *Applied Statistics*, 29, n 2, pp. 119-127.

Mola F., Siciliano R. (1992). A two-stage predictive splitting algorithm in binary segmentation. *Computational Statistics*. vol. 1. Y. Dodge and J. Whittaker ed. Physica Verlag.

Sonquist J.A., Morgan J.N. (1964). *The Detection of Interaction Effects*. Ann Arbor: Institute for Social Research. University of Michigan.

Induction of Decision Trees
Based on the Rough Set Theory

Tu Bao Ho, Trong Dung Nguyen, Masayuki Kimura

Japan Advanced Institute of Science and Technology, Hokuriku
Tatsunokuchi, Ishikawa, 923-12 JAPAN

Summary: This paper aimed at two following objectives. One was the introduction of a new measure (R-measure) of dependency between groups of attributes in a data set, inspired by the notion of dependency of attribute in the rough set theory. The second was the application of this measure to the problem of attribute selection in decision tree induction, and an experimental comparative evaluation of decision tree systems using R-measure and other different attribute selection measures most of them are widely used in machine learning: gain-ratio, gini-index, d_N distance, relevance, χ^2.

1. Introduction

The goal of inductive classification learning is to learn a classifier from a training set that correctly predicts classes of unseen instances. Among approaches to inductive classification learning, decision trees is certainly the most active and applicable one. During the last decade, many top-down induction of decision tree (TDIDT) systems have been developed, most notably CART of Breiman et al. (1984), ID3 of Quinlan (1986) and its successor C4.5 (1993). There are two crucial problems of *variable selection* (choosing the "best" attribute to split a decision node) and *pruning* (avoiding overfitting) in TDIDT systems. Most heuristics for estimating multivalued attributes are information-based measures, such as Quinlan's information gain or gain-ratio (1986, 1993), Mantaras' normalized information gain (1991), etc., and statistics-based measures, such as Breiman's Gini-index (1984), χ^2 (Liu and White, 1994), Baim's relevance (1989), etc. An analysis of eleven measures for estimating the quality of the multi-valued attributes was given by Kononenko (1995).

Rough set theory, introduced by Pawlak in early 1980s (Pawlak, 1991), is a mathematical tool to deal with vagueness and uncertainty, in particular for the approximation of classifications. Although the rough set methodology of approximation sets has been successful in many real-life applications, there are still several theoretical problems to solve. For example, we will show that one of its fundamental notions - the measure of dependency of an attribute set Q on an attribute set P - is not always robust with noisy data and not enough sensitive with "partial" dependency between Q and P. Inspired by this measure for dependency of attributes, we introduce in this paper a new measure of attribute dependency, called R-measure. We show experimentally that R-measure can be applied with success to the problem of attribute selection in decision tree induction.

2. A measure for attribute dependency

2.1 Attribute dependency measure in rough set theory

The theory of rough sets was recognized as a fruitful theory for discovering relationship in data. Though closely related to statistics, its approach is entirely different: rough sets are based on equivalence relations describing partitions made of classes of indiscernible objects instead of employing probability to express data vagueness.

The starting point of the rough set theory is the assumption that our "view" on elements of the object set $\mathcal{O}$ depends on an equivalence relation $E \subseteq \mathcal{O} \times \mathcal{O}$. Two objects $o_1, o_2 \in \mathcal{O}$ are called to be *indiscernible* in E if $o_1 E o_2$. The *lower* and *upper* approximations of any $X \subseteq \mathcal{O}$ consisting all objects which surely and possibly belong to the X, respectively, regarding the relation E. These lower approximation $E_*(X)$ and upper approximation $E^*(X)$ are defined as

$$E_*(X) = \{o \in \mathcal{O} : [o]_E \subseteq X\} \tag{1}$$

$$E^*(X) = \{o \in \mathcal{O} : [o]_E \cap X \neq \emptyset\} \tag{2}$$

where $[o]_E$ denotes the equivalence class of objects indiscernible with o in the equivalence relation E.

Pawlak (1991) has pointed out that one of the most important and fundamental notions to the rough sets philosophy is the need to discovery redundancy and dependencies between attributes. A key concept in the rough set theory is the *measure of dependency* of a set of attributes Q on a set of attributes P, denoted by $\mu_P(Q)$ $(0 \leq \mu_P(Q) \leq 1)$, defined as

$$\mu_P(Q) = \frac{card(\bigcup_{[o]_Q} P_*([o]_Q))}{card(\mathcal{O})} \tag{3}$$

If $\mu_P(Q) = 1$ then Q is totally depends on P; if $0 < \mu_P(Q) < 1$ then Q is partially depends on P; if $\mu_P(Q) = 0$ then Q is independent of P. The measure of dependency is a basic notion in the rough set theory as based on it many other notions are defined, such as reducts and significance of attributes. The formula (3) can be rewritten as

$$\mu_P(Q) = \frac{card(\bigcup_{[o]_Q} P_*([o]_Q))}{card(\mathcal{O})} = \frac{card(\{o \in \mathcal{O} : [o]_P \subseteq [o]_Q\})}{card(\mathcal{O})} \tag{4}$$

An interpretation of the formula (4) is given through Table 1, slightly modified from the example of Pawlak et al. (1995), consisting of eight objects described by two descriptive attributes Temperature, Headache, and the class attribute Flu.

Table 1. Information table

	Temperature	Headache	Flu
e_1	normal	yes	no
e_2	high	yes	yes
e_3	very_high	yes	yes
e_4	normal	no	no
e_5	high	no	no
e_6	very_high	no	yes
e_7	high	no	no
e_8	very_high	yes	yes

Consider how the attribute Flu depends on the attribute Temperature. We express the causal relation between these attributes in the form of usual rules

If Temperature = normal then Flu = no
If Temperature = very_high then Flu = yes

The number of objects satisfied these rules is 5 out of 8. In the other words, the proportion of objects whose values on Flu are correctly predicted by values of Temperature is 5/8. This argument is analogous with the definition of degree of dependency, where each rule corresponds to an equivalent class w.r.t P which is included in an equivalent class w.r.t Q.

2.1 R-measure

From the above argument we can see that the formula (3) takes account only precise rules. However, in real-world data, rules with some probability may be found more often. For example, consider the dependency of Flu on Headache, the following probabilistic predicted rules can be see from the Table 1

> If Headache = yes then Flu = yes (3/4)
> If Headache = no then Flu = no (3/4)

These rules show that Flu depends somehow on Headache, but the formula (3), by its value 0 in this case, says that *Flu* is independent of *Headache*. Taking account of only precise rules causes some limitations of the above degree of dependency: not robust with noisy data and not enough sensitive with "partial" dependency. Let consider further probabilistic rules. Suppose that the value on Headache of a new object is known, and an agent wants to predict the value on Flu of this object. For example, if Headache = yes, then there are two possibilities: Flu = yes (3/4), or Flu = no (1/4). To minimize the probability of error, Flu = yes is certainly chosen as the value with the maximum likelihood of occurrence among all possibilities. Due to the risk of Flu = no, this prediction is uncertain and has an estimated accuracy of 3/4. Similarly, the value Flu = no will be predicted if Headache = no with the estimated accuracy is also 3/4. Denote by X the event that the prediction of the agent is true, we have

$$
\begin{aligned}
P(X) &\doteq P(\text{Headache} = \text{yes}) \times P(X \mid \text{Headache} = \text{yes}) \\
&\quad + P(\text{Headache} = \text{no}) \times P(X \mid \text{Headache} = \text{no}) \\
&= 1/2 \times 3/4 + 1/2 \times 3/4 = 3/4
\end{aligned}
$$

This value can be interpreted as the degree of dependency of Flu on Headache established by the above argument. This argument can be generalized and formulated for a measure of degree of dependency of an attribute set Q on an attribute set P

$$
\mu'_P(Q) = \frac{1}{card(O)} \sum_{[o]_P} max_{[o]_Q} card([o]_Q \cap [o]_P) \tag{5}
$$

The degree of dependency of *Flu* on *Temperature* calculated by (3) is 3/4. The main difference between $\mu_P(Q)$ and $\mu'_P(Q)$ is the latter measures the dependency of Q on P in maximizing the predicted membership of an instance in the family of equivalence classes generated by Q given its membership in the family of equivalence classes generated by P. We have obtained the following property (Ho and Nguyen, 1997).

Theorem. *For every set P and Q we have*

$$
\frac{max_{[o]_Q} card([o]_Q)}{card(O)} \leq \mu'_P(Q) \leq 1 \tag{6}
$$

From this theorem we can define that Q totally depends on P iff $\mu'_P(Q) = 1$; Q partially depends on P iff $max_{[o]_Q} card([o]_Q) / card(O) < \mu'_P(Q) < 1$; Q is independent of P iff $\mu'_P(Q) = max_{[o]_Q} card([o]_Q)/card(O)$. In practice, to emphasize rules those have the higher generalities we use the following formula, and call it R-measure

$$
\tilde{\mu}_P(Q) = \frac{1}{card(O)} \sum_{[o]_P} max_{[o]_Q} \frac{card([o]_Q \cap [o]_P)^2}{card([o]_P)} \tag{7}
$$

3. Decision tree induction and selection measures

In this paper R-measure is applied to the attribute selection problem in decision tree induction. R-measure is compared experimentally with five different attribute selection measures by a careful designed benchmark. In order to facilitate a common understanding of different attribute selection measures, we use the notations presented by Liu and White (1994) and Kononenko (1995). Suppose that we are dealing with a problem of learning a classifier with k classes C_i $(i = 1, k)$ from a set $\mathcal{O}$ of objects described by a set of attributes $\mathcal{A}$. We assume that all attributes are discrete each of which is with a finite number of possible values. Let $n_{..}$ denotes the total number of objects in $\mathcal{O}$, $n_{i.}$ the number of objects from class C_i, $n_{.j}$ the number of objects with the j-th value of the given attribute A, and n_{ij} the number of objects from class C_i and with the j-th value of A. Let further

$$p_{ij} = \frac{n_{ij}}{n_{..}}, \quad p_{i.} = \frac{n_{i.}}{n_{..}}, \quad p_{.j} = \frac{n_{.j}}{n_{..}}, \quad p_{i|j} = \frac{n_{ij}}{n_{.j}} \tag{8}$$

denote the approximation of the probabilities from the object set $\mathcal{O}$. Let

$$H_C = -\sum_i p_{i.} log p_{i.}, \quad H_A = -\sum_j p_{.j} log p_{.j}, \tag{9}$$

$$H_{CA} = -\sum_i \sum_j p_{ij} log p_{ij}, \quad H_{C|A} = H_{CA} - H_A \tag{10}$$

be the entropy of the classes, of the values of the given attribute, of the joint example class–attribute value, and of the class given the value of the attribute, respectively (all logarithms introduced here are of the base two).

The well-known decision tree algorithm C4.5 use the gain-ratio (Quinlan, 1993)

$$GainR = \frac{H_C + H_A - H_{CA}}{H_A} \tag{11}$$

The measure d_N (López de Mantaras, 1991) is based on the definition of distance between two partitions and can be rewritten as in (Kononenko, 1995)

$$d_N = 1 - \frac{H_C + H_A - H_{CA}}{H_{CA}} \tag{12}$$

The author has reported experiments with two data sets "hepatitis" and "breast cancer". Gini-index used in decision tree learning algorithm CART (Breiman et al., 1984) can be rewritten as

$$Gini = \sum_j p_{.j} \sum_i p_{i|j}^2 - \sum_i p_{i.}^2 \tag{13}$$

Baim (1988) introduced a selection measure called *relevance* and showed an experiment in craniostenosis syndrome identification

$$Relev = 1 - \frac{1}{1-k} \sum_j \sum_{i \neq i_m(j)} \frac{n_{ij}}{n_{i.}}, \quad i_m(j) = arg\ max_i\{\frac{n_{ij}}{n_{i.}}\} \tag{14}$$

Another statistics-based measure of interest is χ^2

$$\chi^2 = \sum_i \sum_j \frac{(e_{ij} - n_{ij})^2}{e_{ij}}, \quad e_{ij} = \frac{n_{.j} n_{i.}}{n_{..}} \tag{15}$$

Suppose that $P = \{A_1, A_2, ..., A_p\}$, and $Q = \{B_1, B_2, ..., B_q\}$. Let $n_{j_1 j_2 ... j_p}$ denotes the number of instances with the j_1-th, j_2-th,..., j_p-th values of attributes $A_1, A_2, ..., A_p$, respectively, and $n_{i_1 i_2 ... i_q | j_1 j_2 ... j_p}$ the number of instances with the i_1-th, i_2-th,..., i_q-th values of attributes $B_1, B_2, ..., B_q$ and with the j_1-th, j_2-th,..., j_p-th values of A_1, A_2, ...,A_p, simultaneously. We also denote by $p_{.j_1 j_2 ... j_p}$ and $p_{i_1 i_2 ... i_q | j_1 j_2 ... j_p}$ the approximations of these probabilities from the training set. We can rewrite (7) as

$$\tilde{\mu}_P(Q) = \sum_{j_1, ..., j_p} p_{.j_1 ... j_p} max_{i_1, ..., i_q} p^2_{i_1 ... i_q | j_1 ... j_p} \tag{16}$$

In a special case of (7) when Q stands for the class attribute and P stands for the given attribute A, $\tilde{\mu}_P(Q)$ can be used to measure the dependency of the class attribute on A, and can be written as

$$\tilde{\mu}_A = \sum_j p_{.j} max_i \{ p^2_{i|j} \} \tag{17}$$

4. An experimental comparative study

The studies of Minger (1989), White and Liu (1994) have shown how difficult it is to be thorough and fair when evaluating the accuracy of learning methods. Generally, it is not adequate to evaluate methods either with a small number of data sets or with only a single train-and-test experiment. In order to attain a reliable estimation of the mentioned attribute selection measures we designed experiments as follows

1. Implement programs for different measures in an unique scheme;
2. Use a large number of public data sets;
3. Use the stratified cross-validation for the evaluation.

Six mentioned attribute selection measures are implemented in six systems based on the common scheme of Concept Learning System (CLS) for decision tree induction. This scheme can be briefly described in the following steps

1. Choose the "best" attribute by given *selection measure*.
2. Extend tree by each attribute value.
3. Sort training examples to leaf nodes.
4. If training examples unambiguously classified Then Stop Else Goto steps 1.

In this paper, to be independent with pruning techniques, we considered the results of these programs on unpruned trees. The pruning technique of cost complexity (Breiman, 1984) is used and its experimental results are reported in (Ho and Nguyen, 1997). In this work we use the entropy-based technique (Dougherty, 1995) for discretization of numeric attributes.

Twelve data sets are taken from the UCI repository of machine learning databases. They included Wisconsin Breast Cancer, Congressional Votes, Audiology, Hayes-Roth, Heart Disease, Image Segmentation, Ionosphere, Lung Cancer, Promoters and Splice-Junction of sequences in molecular-biology, Solar Flare, Soybean Disease,

Satellite Image, Tic-Tac-Toe endgame, King-rook-vs-King-pawn. Their properties are listed in Table 2 with the following abbreviations: att, inst, and class stand for the number of attributes, instances, and classes; type stands for the type of attributes (symbolic or mixture of symbolic and numeric).

Table 2. Properties of chosen data sets

Data sets	att	inst	class	type
Vote	16	435	2	sym
Breast cancer	9	699	2	sym
Soybean-small	35	46	4	sym
Tic-tac-toe	9	862	9	sym
Lung cancer	56	32	3	sym
Hayes-roth	5	160	3	sym
Kr-vs-kp	36	3196	2	sym
Audiology	69	226	24	sym
Splice	60	3190	3	sym
Promoters	58	106	2	sym
Heart-disease	13	303	2	mix
Solar Flare	13	1389	7	mix

Traditionally, in machine learning data are usually divided into two sets of training and testing data. Training data are used to produce the classifier by a method and testing data are used to estimate the prediction accuracy of the method. A single train-and-test experiment is often used in machine learning for estimating performance of learning systems.

It is recognized that multiple train-and-test experiments can do better than single train-and-test experiments. Recent works showed that cross validation is a suitable for accuracy estimation, particularly the 10-fold stratified cross validation (Kohavi, 1995). However, cross validation is still not widely used in machine learning as it is computationally expensive.

Table 3. Error rates estimated for different measures

data set	GainR	Gini	χ^2	d_N	Relev	R-m
Vote	5.5	6.2	6.2	7.6	6.4	6.4
Breast cancer	6.7	5.9	6.0	7.7	6.6	6.1
Soybean-small	2.2	2.2	0.0	2.2	2.2	2.2
Tic-tac-toe	12.0	13.8	13.6	13.7	11.9	14.1
Lung cancer	23.3	10.0	10.0	36.6	20.0	13.3
Hayes-roth	25.0	25.0	26.3	24.4	25.6	23.7
Kr-vs-kp	0.5	0.3	0.3	3.2	0.3	0.5
Audiology	23.4	22.5	21.7	31.7	44.1	21.6
Splice	8.0	7.9	8.4	10.4	10.3	8.4
Heart-disease	26.7	29.1	27.7	31.7	30.1	27.8
Flare	28.4	28.7	28.1	29.8	29.3	27.5
Promoters	11.5	14.5	16.0	29.5	17.2	14.5
Average	13.3	12.8	12.7	17.6	15.7	12.8

To obtain a fair estimation of the performance of mentioned selection measures, we carried out a 10-fold cross validation in our experiments. The data set $\mathcal{O}$ is randomly divided into 10 mutually exclusive subsets $O_1, O_2, ..., O_{10}$ of approximately equal size. Each attribute selection measure is tested 10 times. Each time for $k \in \{1, 2, ..., 10\}$, a decision tree is generated on $\mathcal{O} \setminus O_k$ and tested on O_k. A *k-fold stratified cross validation* is a k-fold cross validation the folds are stratified so that they contain approximately the same proportions of labels as in the original dataset. The error rate of each measure is the average of its error rates after 10 running times.

Table 3 shows the estimated error rates of six attribute selection measures obtained in our experiments. Each line presents error rates of six measures for a data set in which the bold number indicates the measure with minimum error rate and the italic number indicates the measure with maximum error rate. The abbreviation R-m stands for R-measure.

Averages of error rates are given in the last line of Table 3. Though average of error

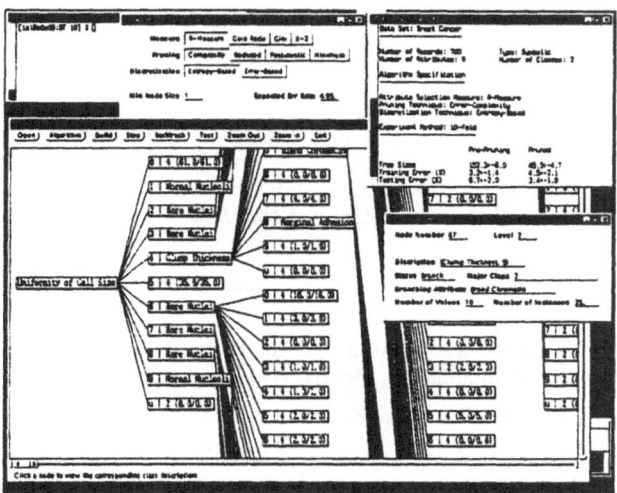

Figure 1: Main screen of the system

rates can not be used directly to evaluate measures, it may, however, provide a snapshot of measure comparison, and show the stability of these measures.

5. Discussions and conclusions

Some conclusions can be drawn from the above experimental results. First, we reconfirm that the quality of the decision tree construction is affected by the choice of attribute selection measures. Somehow different from experimental results in some previous works, in these experiments the error rates of Relevance and d_N distance were relatively higher than those of others, and they were relatively unstable (as showed the results with data sets Lung-cancer, Audiology, Splice, Promoters). Measures Gain-ratio, Gini-index, χ^2 and R-measure have error rates which are not really much different. In comparison with estimation of learning performance by a single train-and-test experiment, i.g. in Buntine and Niblett (1991), a multiple train-and-test experiment such as 10-fold stratified cross-validation in these experiments often gives higher error rates. Thus, the stratified cross validation technique can allow us to avoid an usual overoptimistic estimation of learning performance. An extension of these experiments has been made with more data sets and with noisy data on pruned trees. Careful evaluation in these conditions hopefully allows us to obtain more reliable and new results on R-measure (Ho and Nguyen, 1997).

In this paper we have introduced R-measure for the degree of dependency between two groups of attributes in a data set. R-measure is inspired by the same notion of attribute dependency in the rough set theory and it aims at overcoming some limitations of that notion. We have applied R-measure to the general scheme of decision tree induction and carried out carefully an experimental comparative study of R-measure with five attribute selection measures which are well-known in the machine learning literature.

References:

Baim, P.W. (1988): A method for attribute selection in inductive learning systems. *IEEE Trans. on PAMI*, **10**, 888-896.

Breiman, L., Friedman, J., Olshen, R., Stone, C. (1984): *Classification and Regression Trees*, Belmont, CA: Wadsworth.

Buntine, W., Niblett, T. (1991): A further comparison of splitting rules for decision-tree induction. *Machine Learning*, **8**, 75-85

Dougherty, J., Kohavi, R. and Sahami, M. (1995): Supervised and Unsupervised Discretization of Continuous Features. *Proceedings 12th International Conference on Machine Learning*, Morgan Kaufmann, 194-202.

Ho, T.B., Nguyen, T.D. (1997): An interactive-graphic system for decision tree induction (under review).

Kononenko, I. (1995): On biases in estimating multi-valued attributes. *Proc. 14th Inter. Joint. Conf. on Artificial Intelligence*, Montreal, Morgan Kaufmann, 1034-1040.

Kohavi, R (1995): A study of cross-validation and bootstrap for accuracy estimation and model selection. *Proc. Int. Joint Conf. on Artificial Intelligence IJCAI'95*, 1137-1143.

Liu, W.Z., White, A.P. (1994): The importance of attribute selection measures in decision tree induction. *Machine Learning*, **15**, 25-41.

López de Mantaras, R. (1991): A distance-based attribute selection measure for decision tree induction. *Machine Learning*, **6**, 81-92.

Mingers, J. (1989): An empirical comparison of selection measures for decision-tree induction. *Machine Learning*, **3**, 319-342.

Pawlak, Z. (1991): *Rough Sets: Theoretical Aspects of Reasoning About Data*, Kluwer Academic Publishers.

Pawlak, Z., Grzymala-Busse, J., Slowinski, R., Ziarko, W. (1995): Rough sets. *Communications of the ACM*, **38**, 89-95.

Quinlan, J. R. (1993): *C4.5: Programs for Machine Learning*, Morgan Kaufmann.

Wille, R. (1992): Concept lattice and conceptual knowledge systems. *Computers and Mathematics with Applications*, **23**, 493-515.

Visualizing Data in Tree–Structured Classification

Francesco Mola, Roberta Siciliano

Dipartimento di Matematica e Statistica
Università di Napoli Federico II
Via Cintia - Monte S.Angelo
80126 - Naples - Italy
e-mail: f.mola@dmsna.dms.unina.it
r.sic@dmsna.dms.unina.it

Summary: This paper provides a classification tree methodology to analyze and visualize data in multi–way cross–classifications of a categorical response variable and a high number of predictors observed in a large sample. The idea is to perform recursively a factorial method such as nonsymmetric correspondence analysis to every finer partitions of the given sample. Some new insights on the graphic displays of nonsymmetric correspondence analysis are considered for defining classification criteria based on factorial scores. As a result, we grow particular types of classification trees with two aims: 1) to enrich the interpretation of the dependence structure using predictability measures and graphic displays; 2) to obtain a classification rule for new cases of unknown class on the basis of a factorial model.

1. The proposed methodology

Consider a data matrix with a response categorical variable (Y) and M categorical predictors ($X_1 \ldots X_M$) observed on a sample of N cases. The response variable is used as criterion variable and its categories represent the *a-priori* classification of the observed cases. Objective of the proposed methodology is to analyze and visualize the dependency in such data matrix and, in the same time, to obtain a classification rule for new cases of unknown class.

The idea is to use recursively, on continuously finer partitions of the N cases, a factorial method such as nonsymmetric correspondence analysis (Lauro and D'Ambra, 1984; D'Ambra and Lauro, 1989; Lauro and Siciliano, 1989; Siciliano, Mooijaart and van der Heijden, 1993; Mola and Siciliano, 1995). The approach consists in selecting at each node the best set of predictor categories for visualizing the dependence structure by nonsymmetric correspondence analysis, successively, in partitioning cases on the basis of some properties concerning the factorial axes.

As a result, we define a tree–structured classification of the N cases where for passing from a coarser partition to a finer partition we use a factorial model and where a factorial representation is assigned to each partition.

In the following subsections we describe the main phases of the partitioning procedure to divide N cases into two subgroups. Then the overall procedure repeats for each subgroup until stopping according to a given rule.

1.1 Table selection by the predictability index τ

We consider the set of M contingency tables where each table cross–classifies the response categories of Y with the categories of each predictor. In this phase we select the best predictor X^* by maximizing the predictability index τ of Goodman and Kruskal, that is:

$$\max_{m \in M} \tau_{Y|X_m} = \tau_{Y|X^*} = \frac{\sum_i \sum_j (p_{ij}/p_{\cdot j} - p_{i\cdot})^2 p_{\cdot j}}{(1 - \sum_i p_{i\cdot}^2)} \tag{1}$$

where p_{ij} are the proportions of the selected contingency table that cross–classifies the response categories $i = 1, \dots, I$ of Y with the categories $j = 1, \dots, J$ of the best predictor X^*. The usual dot notation is used for summations. i.e., $\sum_j p_{ij} = p_{i\cdot}$.

1.2 Dependence analysis by a factorial model

In this phase we analyze the dependency in the selected table using the factorial model of nonsymmetric correspondence analysis:

$$\frac{p_{ij}}{p_{\cdot j}} = p_{i\cdot} + \sum_k \lambda_k r_{ik} c_{jk}, \tag{2}$$

for $k = 1, \dots, K^* = \min(I - 1, J - 1)$. and $\lambda_1 \geq \dots \geq \lambda_k^* \geq 0$. The row scores r_{ik} and the column scores c_{jk} satisfy the following centering and orthonormality conditions:

$$\sum_i r_{ik} = 0, \sum_j c_{jk} p_{\cdot j} = 0. \tag{3}$$

$$\sum_i r_{ik} r_{ik^*} = \delta_{kk^*}, \sum_j c_{jk} c_{jk^*} p_{\cdot j} = \delta_{kk^*}, \tag{4}$$

where δ_{kk^*} is the Kronecker's delta.

Nonsymmetric correspondence analysis decomposes into a number of dimensions the predictability index τ: $\tau_{Y|X^*}(1 - \sum_i p_{i\cdot}^2) = \sum_k \lambda_k^2$. The objective is however to decompose the observed table by using a number of factors K lower than K^*.

To each level of the partitioning procedure we assign a two–dimensional factorial representation of the dependence structure between the response categories and the selected predictor categories (see for example Lauro and Siciliano, 1989; Siciliano, Mooijaart and van der Heijden, 1993).

1.3 Partitioning criterion using factorial scores

Let $E = (\epsilon_1, \dots, \epsilon_N)$ be the set of the N observed cases. We denote by $G_{(2)} = (g_l, g_r)$ a partition of E into two subgroups called "left" and "right" such that a case ϵ_n can belong to only one of the two subgroups and all cases are classified.

The classification criterion is defined by using the weighted predictor scores of the best predictor X^* in the first dimension. i.e.. $c_{jk} p_{\cdot j}$ for $k = 1$, that sum up to zero due to the centering condition (3): we can say that the predictor categories having a negative score value behave in the opposite way with respect to the categories having a positive score value. Notice in fact that the centering condition (3) for the column scores holds since $\sum_j (\frac{p_{ij}}{p_{\cdot j}} - p_{i\cdot}) p_j = 0$ where $p_{i\cdot}$ is the centroid of the J conditional distributions. In this way we can discriminate between two subgroups of categories: the first subgroup includes the categories j such that $c_{j1} \leq 0$ and the second subgroup includes the categories j such that $c_{j1} > 0$.

To this discrimination it corresponds a partition $G_{(2)}$ of the set E of N cases into two subgroups: the "left" subgroup g_l includes the N_l cases that present one of the predictor categories with nonpositive score value, whereas the "right" subgroup g_r includes the N_r cases that present one of the predictor categories with positive score value.

In addition, we characterize this partitioning of both categories and cases with some predictability measures. Using the orthonormality conditions (4) we can prove that the following relation holds

$$\sum_k \lambda_k^2 = \sum_i \sum_k (r_{ik}\lambda_k)^2 = \sum_j p_{.j} \sum_k (c_{jk}\lambda_k)^2. \tag{5}$$

Since the index τ is proportional to $\sum_k \lambda_k^2$. (5) shows that we can further decompose such predictability over the row categories as well as over the column categories. In particular, we can identify which predictor categories contribute at most in predicting the response variable by defining the following predictability measure of category j: $pred(C_j) = (p_{.j}\sum_k (c_{jk}\lambda_k)^2)/(\sum_k \lambda_k^2)$, which sum up to one over the index j. When we consider the first factorial axis in the partitioning criterion this formulation simplifies to $pred(C_j) = c_{j1}^2 p_{.j}$. We can compare $pred(C_j)$ with the weight $p_{.j}$ of category j: we can say that category j is a *strong category* when $pred(C_j) \geq p_{.j}$ whereas category j is a *weak category* when $pred(C_j) < p_{.j}$. By definition, we can equivalently check the conditions $|c_{jk}| \geq 1$ and $|c_{jk}| < 1$ respectively.

1.4 Stopping rule and class assignment rule

For stopping the recursive partitioning procedure we consider a stopping rule based on the CATANOVA statistic defined in our context as $C_t = (N_t - 1)(I_t - 1)\tau_{Y|s_t^*}$, where s_t^* is the best split at node t, I_t is the number of unempty response classes at node t and N_t is the number of cases at node t (Mola and Siciliano, 1994); this statistic is asymptotically approximated as $\chi^2_{(I_t-1)}$ under the null hypothesis of independence. Once we define the best split we test the significance by applying the CATANOVA statistic. In case we reject the null hypothesis of independence we continue partitioning the cases; otherwise, we stop at the current node. When there are no groups that could be divided further then the partitioning procedure is completed.

In the end. we can assign the class label to each subgroup of the final partition. This can be done by considering the response category having the highest proportion of cases within the given subgroup.

2. Example

A real data set is analyzed with the proposed methodology. The set consists of 286 graduates of the Economy and Commerce Faculty of the University of Naples over the period 1986 − 1989, on which the following variables were observed:

- *Final score* with categories *Low (L). Medium-Low (ML). Medium-High (MH). High (H)*;

- *Sex* with categories *male, female*;

- *Origin* with categories *Naples. county, other counties*;

- *Age* with categories *-25. 26-30, 31-35. +36*;

- *Diploma* with categories *classical. scientific. technical. magistral. professional*;

- *Study plan* with categories *official, managerial, economics, quantitative, public. professional*;

- *Time needed to graduate* with categories *4 years, 5-6 years. +7 years*;

- *Thesis subject* with categories *economy. law. quantitative. history. management*.

The variable *Final Score* is the response variable. In table 1 we summarize the partitioning sequence to grow the binary tree shown in figure 1.

node	N	N1	N2	N3	N4	predictor	% τ	left	right
1	286	54	78	81	73	*Age*	94%	-25 years 26–30 years	31–35 years +36 years
2	239	33	59	74	73	*Diploma*	88%	Classical Magistral	Professional Scientific Tecnical
3	47	21	19	7	0	*Study Plan*	87%	Official Quantitative Public	Managerial Economics Professional
4	54	5	3	19	27	*Origin*	87%	Naples	Other Counties County
5	185	28	56	55	46	*Study Plan*	84%	Official Public Economics Quantitative	Managerial Professional
10	174	24	54	50	46	*Origin*	83%	Naples	Other counties County
20	81	7	18	28	28	*Thesis subject*	83%	Economy Quantitative Management	Law History

Tab. 1: Partitioning Sequence (strong categories are in bold)

For sake of brevity, we analyze only some partitions. The first table that is selected cross–classifies the response variable *Final Score* with the predictor *Age*. In figure 2 we present the two–dimensional factorial representation of nonsymmetric correspondence analysis applied to this table. The first factorial axis explains a very high percentage of τ index (94 %).

We notice an axial opposition between *old graduated* and *young graduated* as well as between *graduated with high score* and *graduated with low score*. Only the age category *26–30 years* is a weak category (very nearby the origin). These considerations enrich the interpretation of the partitioning at node 1 that divides the 286 cases into two subgroups of 239 and 47 cases respectively.

At node 2 the selected predictor is *Diploma*. Figure 3 shows the graphic display where the first factorial axis explains again a high percentage of τ index (88 %).

We notice an axial opposition between *classical diploma* and *professional diploma* that represent the strong predictor categories. Considering the simultaneous interpretation of predictor and response categories we underline that graduated with *high score* are better predicted by the category *classical diploma*, whereas the category *professional diploma* predicts more the graduated with *medium low score*. Instead, the category *magistral diploma* predicts more the graduated with *medium high score*. The partitioning at node 2 divides the 239 cases into two subgroups of 54 and 185 cases respectively.

The selected predictor at node 5 is *Study Plan*. In the graphic display of figure 4 the prediction explained by the first factorial axis is 84 %. The axial opposition is now provided by *official* and *public* against *managerial* and *professional* study plans. The other predictor categories, *economics* and *quantitative*, are weak categories (very nearby the origin). The partitioning divides the 185 cases at node 5 into two subgroups of 174 and 11 cases respectively.

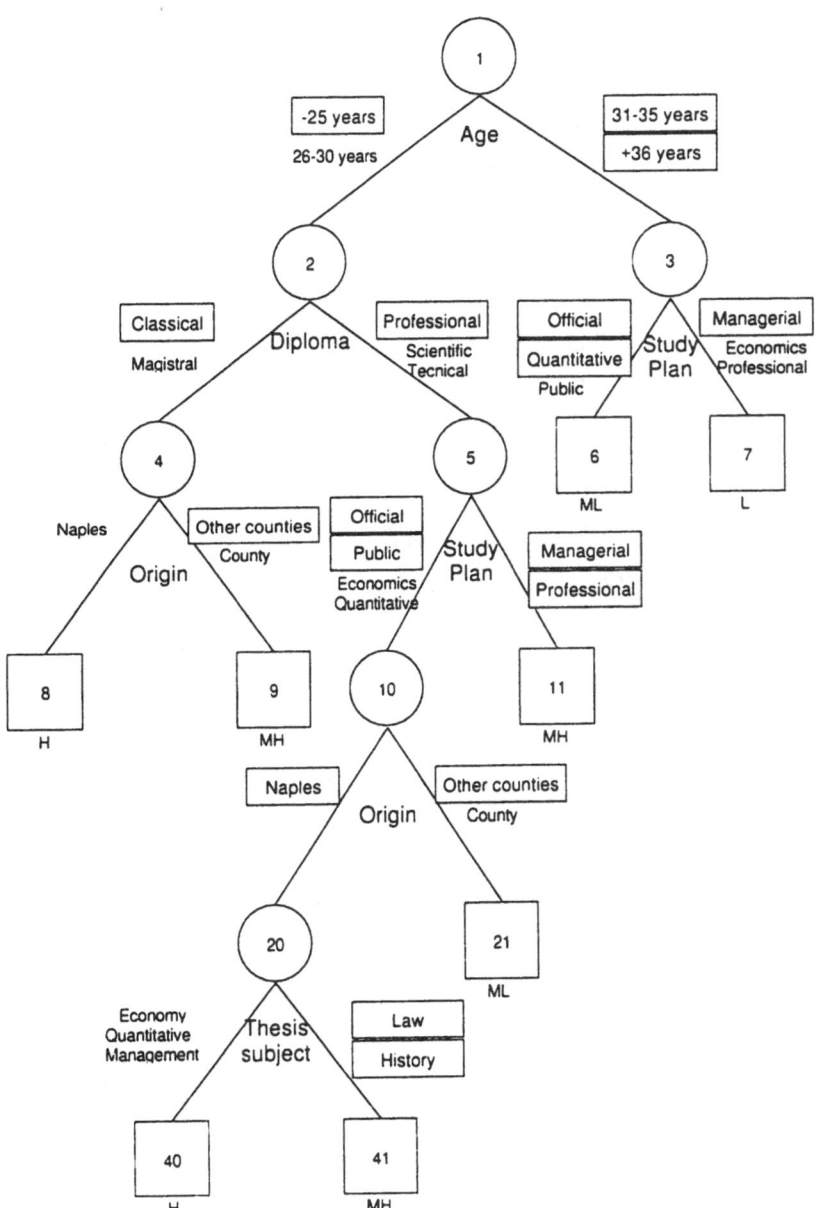

Figure 1: Final binary tree (strong categories are in a box)

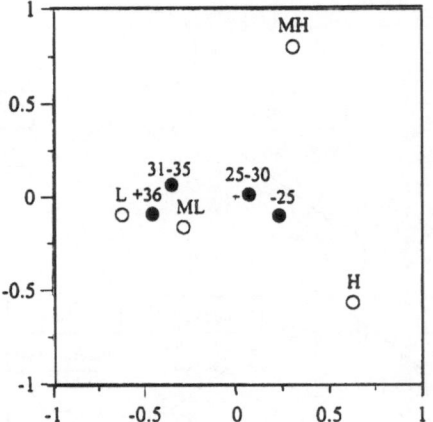

Figure 2. Factorial representation at node 1: Final Score vs Age

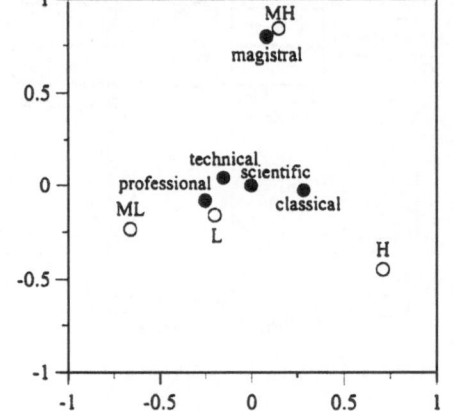

Figure 3. Factorial representation at node 2: Final Score vs Diploma

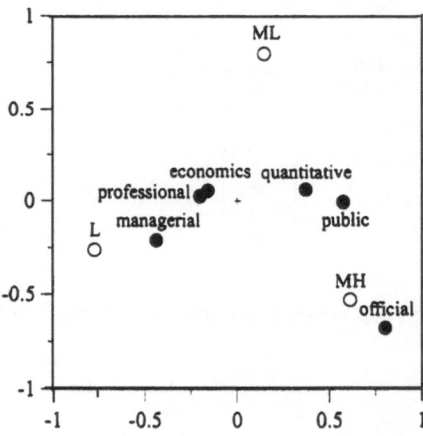

Figure 4. Factorial representation at node 5: Final Score vs Study Plan

Table 2 summarizes the terminal node information with the values of the CATANOVA statistic and the assigned class. The left side of table 3 shows the misclassification matrix obtained with the proposed tree-classification procedure whereas the right side of table 3 shows the misclassification matrix obtained with a standard classification tree procedure such as for example the CART methodology (Breiman, Friedman, Olshen and Stone, 1984). We can compare the two misclassification matrices of table 3. Notice that the proportions of misclassified cases are very similar in the two analyses and for some response categories the proposed method has classified better than CART.

node	cases	n1	n2	n3	n4	CATANOVA	d.f.	assigned class
6	17	2	8	7	0	4.6	2	2
7	30	19	11	0	0	3.29	1	1
8	29	0	0	7	22	3.96	1	4
9	25	5	3	12	5	9.96	3	3
11	11	4	2	5	0	6.11	2	3
21	93	17	36	22	18	11.03	3	2
40	53	5	10	14	24	7.72	3	4
41	28	2	8	14	4	4.00	3	3

Tab. 2: Terminal Node Information

		true class			
		L	ML	MH	H
predicted class	L	0.35	0.14	0.00	0.00
	ML	0.35	0.56	0.36	0.25
	MH	0.20	0.17	0.38	0.12
	H	0.10	0.13	0.26	0.63

the proposed method

		true class			
		L	ML	MH	H
predicted class	L	0.35	0.14	0.01	0.00
	ML	0.37	0.42	0.41	0.27
	MH	0.17	0.26	0.31	0.18
	H	0.11	0.18	0.27	0.55

CART

Tab. 3: Misclassification Matrices

3. Some extensions and relations

It is interesting to consider some extensions of the basic method; in the first phase it is possible to select more then one predictor. In fact we can extend the proposed method to the case of joint predictors that cross–classify two or more predictors.

In the partitioning phase we can consider a partition into three subgroups considering more levels of dependency (*strong dependency* with positive or negative score value and *weak dependency*), growing up not necessarily binary trees but ternary trees too (Siciliano and Mola, 1997). Moreover, the splitting criterion can be based on two factorial axes instead of one. We consider also the possibility to have fuzzy subgroups, that is the possibility to send the same case in two different subgroups (Mola and Siciliano, 1995).

Our methodology has some relations with two–stage binary segmentation procedures (Mola and Siciliano, 1992, 1994; Mola, Klaschka and Siciliano, 1996) although a factorial model is used as a splitting rule. Instead of the usual classification tree procedures (CART, Breiman, Friedman, Olshen and Stone, 1984; FACT, Loh and Vanichsetakul, 1988), the proposed method can also grow ternary classification trees where it is possible to distinguish between nodes where there is a strong dependency and nodes where there is a weak dependency.

As main results of applications on real data sets, the proposed methodology has shown to achieve several purposes:

- to build up a classification rule (binary tree),

- to predict the response classes (binary tree),

- to visualize the dependency structure in multi–way cross–classifications with many predictors and large sample size (recursive factorial representations),

- to detect outliers (ternary tree).

Acknowledgements: The Authors wish to thank Carlo Lauro and Jaromir Antoch for helpful comments on a previous version of this paper. This research was supported for the first author by CNR research funds number 92.1872 P and for the second author by CNR research funds number 95.02041.CT10.

References

Breiman L., Friedman J.H., Olshen R.A., Stone C.J., (1984): *Classification and Regression Trees*, Belmont C.A. Wadsworth.

D'Ambra, L. and Lauro, C. (1989): Nonsymmetrical Analysis of Three–way Contingency Tables, in *Multiway Data Analysis*, Coppi, R. and Bolasco, S. (eds.), North Holland, Amsterdam.

Lauro, N.C. and D'Ambra, L. (1984): l'Analyse non Symmetrique des Correspondances. *Data Analysis and Informatics III*. E. Diday et al. (eds.), 433-446. North Holland, Amsterdam.

Lauro, N.C. and Siciliano, R. (1989): Exploratory methods and modelling for contingency tables analysis: an integrated approach. *Statistica Applicata. Italian Journal of Applied Statistics*, 1, 5-32.

Loh, W. and Vanichsetakul, N. (1988): Tree–Structured Classification via Generalized Discriminant Analysis. *Journal of the American Statistical Association*, 83, 715–728.

Mola, F. and Siciliano, R. (1992): A Two–Stage Predictive Splitting Algorithm in Binary Segmentation. *Computational Statistics*, Dodge, Y. and Whittaker, J. (eds.), 1, 179–184, (Compstat '92 Proceedings). Physica Verlag.

Mola, F. and Siciliano, R. (1994): Alternative Strategies and CATANOVA Testing in Two–Stage Binary Segmentation. *New Approaches in Classification and Data Analysis*, Diday, E. et al. (eds.), 316–323, Springer Verlag.

Mola, F. and Siciliano, R. (1995): Nonsymmetric correspondence analysis for tree-structured classification, research internal report, conditionally accepted for *Applied Stochastic Models and Data Analysis*.

Mola, F., Klaschka, J. and Siciliano, R. (1996): Logistic Classification Trees, *Compstat 96 Proc.*, Prat, A. (ed.), Physica Verlag.

Siciliano, R. and Mola, F. (1997), Ternary classification trees: a factorial approach, *Visualization of categorical data* (Greenacre, M., Blasius, J., eds.), Academic Press, CA.

Siciliano, R., Mooijaart, A. and van der Heijden, P.G.M. (1993): A Probabilistic Model for Nonsymmetric Correspondence Analysis and Prediction in Contingency Tables. *Journal of the Italian Statistical Society*, 2, 1, 85-106.

Adaptive Cluster Analysis Techniques - Software and Applications

Hans-Joachim Mucha[1] , Rainer Siegmund-Schultze[1] and Karl Dübon[2]

[1] Weierstrass Institute for Applied Analysis and Stochastics (WIAS)
Mohrenstrasse 39
D-10117 Berlin, Germany

[2] Daimler-Benz AG, Deptartment Research and Technology
Postfach 2360
D-89013 Ulm, Germany

Summary: Well-known cluster analysis techniques like for instance the K-means method can be improved in almost every case by using adaptive distances. For this one has to estimate at least „appropriate" weights of variables, i.e. appropriate contributions to cluster analysis. Recently, adaptive classification techniques for two class models are under development. Here usually both the weights of variables and the weights (masses) of observations play an important role. For instance, observations that are harder to classify get increasingly larger weights. Quite successful applications of these techniques can be reported from the area of credit scoring systems for consumer loans or credit cards. The software *ClusCorr* (running under Microsoft EXCEL) perform classification, cluster analysis and multivariate graphics of (huge) high-dimensional data sets containing numerical values (quantitative or categorical) as well as non-numerical information.

1. Introduction

What can our statistical software do for you? For example, with the help of the EXCEL-Add-In *ClusCorr* one is able to look for classification rules in data sets like the following one:

```
k1,kB,kFF,k2,k1,k1,k2,k0,k0,k0,kD,k1,k2,k4,k8,k7,k1
k0,kS,kME,k0,k1,k1,k3,k0,k0,k0,kD,k0,k4,k1,k0,k7,k1
k0,kS,kFF,k1,k1,k1,k1,k0,k0,k0,kD,k0,k4,k3,k2,k9,k0
k0,kS,kMB,k2,k0,k2,k1,k0,k0,k1,kD,k2,k0,k3,k5,k9,k1
k1,kS,kAD,k1,k1,k1,k1,k0,k0,k1,kD,k7,k6,k2,k7,k5,k1
k0,kS,kFF,k1,k0,k1,k3,k0,k0,k0,kD,k0,k4,k2,k7,k9,k0
...
```

Moreover, one can look at multivariate graphics of such data sets in order to get a first view on the way of understanding the data at hand. Above, every line contains information (whatever this may be in reality) about an applicant for credit. At the end of each line the code „k1" characterises a good applicant, whereas a „k0" stands for a bad one (which is not able to pay the amount of credit back to the bank, telephone company, mail-order house, or department store). Depending on the kind of credit often additional numerical (quantitative, categorical, ordinal,...) information has to be taken into account in order to optimise the decision about a new applicant. In fact that is no problem for the classification technique described later on.

2. Adaptive clustering techniques

Some years ago new adaptive clustering techniques were developed (for details see, e.g., Mucha 1992, 1995). They appear to show some intelligence because of their ability to learn the appropriate (adaptive) distance measures. For instance, the squared weighted Euclidean distance

$$d_Q^2(\mathbf{x}_i, \mathbf{x}_l) = (\mathbf{x}_i - \mathbf{x}_l)' \mathbf{Q}(\mathbf{x}_i - \mathbf{x}_l) = \left\| \mathbf{x}_i - \mathbf{x}_l \right\|_Q^2 \tag{1}$$

or the weighted absolute distance

$$t_Q(\mathbf{x}_i, \mathbf{x}_l) = \sum_{j=1}^{J} q_j \left| x_{ij} - x_{ij} \right| \tag{2}$$

between two observations $\mathbf{x}_i$ and $\mathbf{x}_l$ are well-known dissimilarity measures for metric scaled data which are often used in cluster analysis. Here the number of variables of a data matrix $\mathbf{X}$ is denoted by J. In the simple case which we consider here, the metric $\mathbf{Q}$ is diagonal with non-negative diagonal elements q_j which are unknown usually and therefore have to be estimated during an adaptive clustering procedure. For example, the general variant of the well-known K-means clustering method (MacQueen 1967) takes into consideration both the weights of variables q_j introduced above and the non-negative weights (masses) of observations m_i in order to minimise the sum of the within-clusters variances

$$V_K = \sum_{k=1}^{K} \sum_{i=1}^{I} \partial_{ik} m_i d_Q^2(\mathbf{x}_i, \bar{\mathbf{x}}_k) \tag{3}$$

concerning a fixed number of clusters K. Here the number of observations of the data table $\mathbf{X}$ is denoted by I. The indicator function ∂_{ik} is equals 1 if the observation $\mathbf{x}_i$ comes from cluster k, or 0 otherwise. The vector $\bar{\mathbf{x}}_k$ contains the usual arithmetic mean values in the cluster k. On the one hand the weights can be chosen in conformity with data and model (usually in this case these weights are termed special weights). On the other hand the weights can be adaptive ones estimated in some ways in an iterative manner regarding some calibration constraints in (3). In any case, as a result the contributions of the variables to cluster analysis differ among one another. As a further result it is no longer necessary to provide a selection of variables.

The performance of adaptive clustering methods was investigated using extensive simulation studies (resampling as well as random error techniques). We used measures which are proposed by Hubert and Arabie (1985). The usefulness of adaptive techniques has been shown in practice in several applications. Because of weights q_j and w_i accompanied by various kinds of data transformations like rank transformation, clustering techniques based on (3) have a wide area of applications. For instance, one can look for clusters in contingency tables using K-means technique or Ward's hierarchical clustering method (Mucha and Klinke 1993; see Figure 1 below: a dendrogram which is drawn onto the plane of the first two factors of correspondence analysis). It should be mentioned already here that the dendrograms and other graphics of the software *ClusCorr* are linked by click with data values. Another example of a successful application of adaptive clustering based on (3) is the detection of groups in rank order data (Mucha 1992).

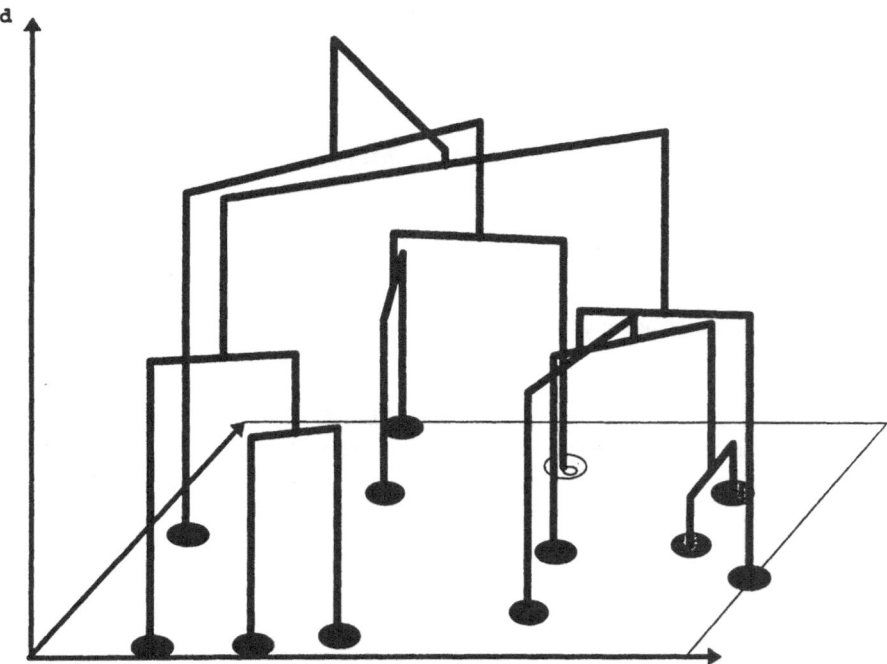

Fig. 1: Application in archaeology. The so-called plot-dendrogram combining both the output of a hierarchical cluster analysis (clusters of tools from stoneage) and a two-dimensional plot (for instance a scatterplot).

Moreover, adaptive distances are very important in order to obtain highly informative multivariate plots. In that way, both the interactive data analysis and the interpretation of clustering results become much easier.

A basic approach for a simultaneous classification and visualization of data is proposed by Bock (1987). Some methods are descibed by Bock (1996). These methods are based on a well-specified goodness-of fit criterion.

In the case of mixed data several distance measures can be used side by side, for example, in an additive fashion.

3. A distance approach in Credit scoring

If some information about classes is known beforehand, then one has a problem in supervised learning (see Michie et.al. 1994). This knowledge has to be taken into account in order to give categorical data a quantitative meaning (Nishisato 1994). Credit scoring is a new area of application of adaptive distances. Let us consider very briefly the first few steps of a simple distance approach for two class models. A training data set containing a class membership variable c is used to estimate the classification rule. The elements of the vector c can either take the values 0 (bad applicant) or 1 (good applicant). One can start with equal weights of observations, namely $m_i = 1$ for all $i = 1, 2, ..., I$.

First of all we recommend an optimal scaling of the data values with regard to the given class membership by using a kind of correspondence analysis technique (Nishisato

1980). Generally, we want to obtain a new variable **x** so as to make the values (scores) within the given classes ss_w as similar as possible and the scores between classes ss_b as different as possible. Let us look at a contingency table which can be obtained by crossing a categorical variable j (K_j categories, where $K_j \geq 2$ is considered) with the class membership variable (2 categories at least). That is (regarding some constraints in the frame of the dual scaling approach), the squared correlation ratio has to be maximised:

$$\eta^2 = \frac{ss_b}{ss_b + ss_w} \qquad \left(= \frac{ss_b}{ss_t} \right).$$

(4)

Because of the well-known variance decomposition

$$ss_t = ss_w + ss_b$$

(5)

the squared correlation ratio lies in the interval [0,1]. Considering the special case of two classes now the correspondence analysis can be used without the calculation of orthogonal eigenvectors (Mucha 1996a): a given category y_{kj} of a variable j is transformed into an optimally scaled one by

$$x_{kj} = \frac{p_{kj}^{(1)}}{p_{kj}^{(1)} + p_{kj}^{(0)}}.$$

(6)

Here $p_{kj}^{(1)}$ is an estimate of the probability of being a good applicant when coming from category k, whereas on the other side $p_{kj}^{(0)}$ is an estimate of the probability of being a bad applicant when coming from category k. Additionally, these estimates can take into consideration weights of the observations. For instance observations that are harder to classify get increasingly larger weights during the adaptive classification process. The transformation (6) is quite a simple one compared with, for example, another one given by Fahrmeir and Hamerle (1984), but it has important advantages (Mucha 1996a).

Without loss of generality the hypothetical worst case $\mathbf{x}_1 \equiv 0$ will be considered here (otherwise one can look at the best case model with $\mathbf{x}_1 \equiv 1$). Then the L_1-distance (2) between an observation (applicant) $\mathbf{x}_i$ and the worst case is simply

$$t^*(\mathbf{x}_i, \mathbf{x}_l) = \sum_{j=1}^{J} x_{ij}.$$

(7)

Instead of t^* here we prefer

$$t(\mathbf{x}_i, \mathbf{x}_l) = \frac{1}{J} \sum_{j=1}^{J} x_{ij}$$

(8)

because of independence of the number of variables J. Now the question arises, how can a suitable cut-off point be determined on the distance scale? The simplest way is to take the cut-off point which gives the minimum error rate for the training sample (Figure 2). Usually (without consideration of assumptions on distributions) the cut-off point is chosen by sampling techniques. For example, one can take the mean or median of a set of cut-off points (corresponding to minimum error rates obtained from many different samples which are drawn randomly from the training data).

Afterwards, as usual, we are able to classify the observations of the test sample using the scores (6) and the cut-off point of the training sample.

The obtained results of the first step (described above) can be improved in almost every case by altering the weights of observations (i.e. by changing the amount of influence of the observations). This leads to a local adaptive distance approach.

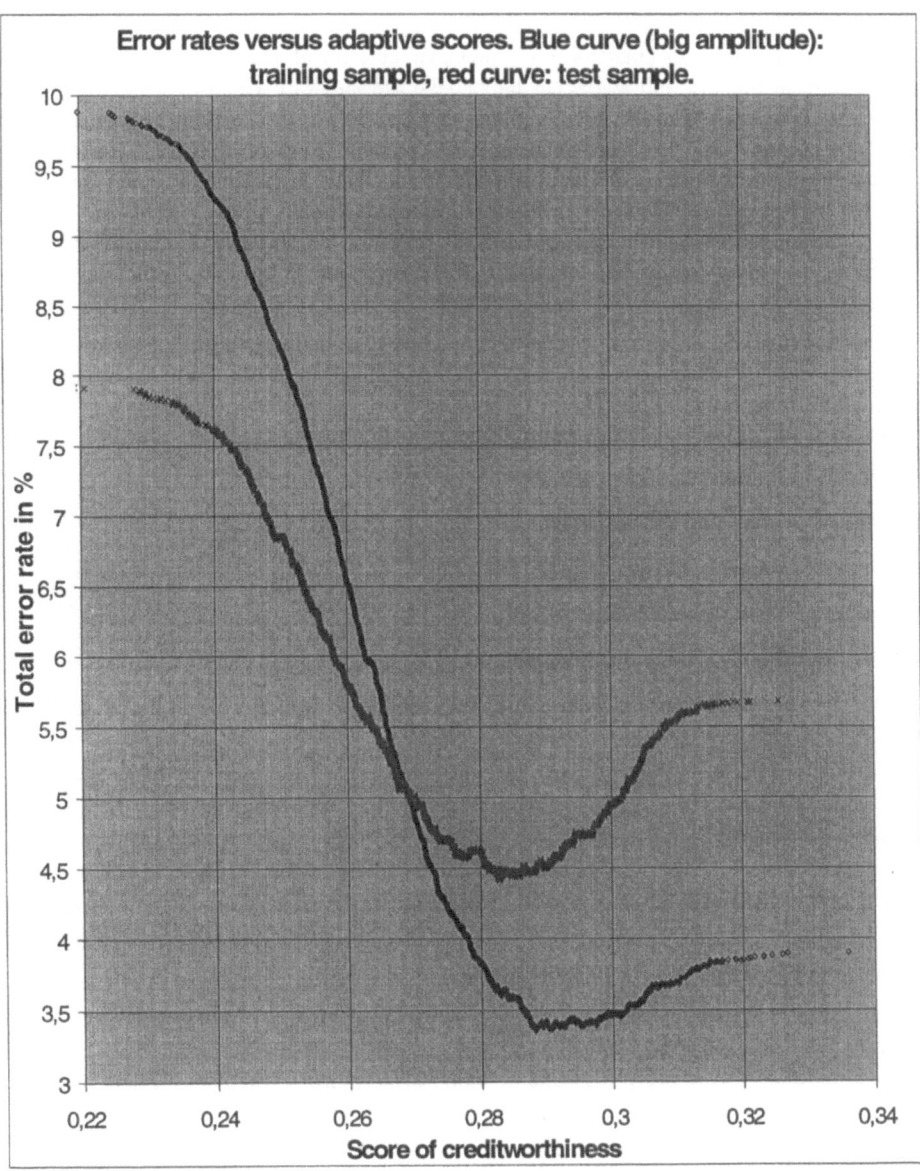

Fig. 2: Credit scoring. Example 1. Cut-off point on the distance axis versus error rate (dark curve: training sample, grey curve: test sample).

4. The software

The *XClust*-library (Mucha, 1995; Mucha and Klinke, 1993) of the interactive statistical computing environment *XploRe* contains well-known cluster analysis methods as well as new adaptive ones. Additionally, highly interactive and dynamic graphics of *XploRe* support both the search for and the interpretation of clusters. Everyone who is a little bit familiar with matrix notations can write new distance functions (for instance for mixed data) by using the macro language of *XploRe*.

In order to make cluster analysis techniques available for almost everyone (without any knowledge in algebra and statistical languages) the software *ClusCorr*, running under Microsoft Windows, is under development (Mucha 1996b). It is written in Visual Basic for Applications (VBA) to function as an Add-In under Microsoft EXCEL 5 (or higher). Hardware requirements are defined from this point. Have regard to the wide family of EXCEL users in mind, *ClusCorr* is designed on the one hand to have teachware properties including an extensive help system and decision support for choice of distance measures, clustering techniques, multivariate graphics, and so on, and on the other hand to perform the cluster analysis of huge data sets.

In consideration of the kind of results obtained one can distinguish roughly between hierarchical and partitioning techniques. The hierarchical methods form a sequence of nested partitions, a hierarchy (Figure 1). In *ClusCorr* (as well as in *XClust*) well-known hierarchical ascending clustering methods are available, for instance

- Ward's minimum variance method,
- Single Linkage (nearest neighbour),
- Complete Linkage (furthest neighbour),
- Average Linkage (group average), and
- Centroid Method (weighted centroid).

About twenty well-known distance measure can be selected out simply by a click: Euclid, L_1, Jaccard, Cosine,...

What is one to do if one has to carry out a hierarchical cluster analysis for a million of observations? We offer a special principal axes cutting algorithm in order to reduce a huge data set to, for instance, 250 „pre-clusters". This algorithm is quite fast because it takes 25 minutes of time (on a Pentium 200 MHz) to reduce one million objects to some few hundred pre-clusters. Afterwards one can carry out a hierarchical cluster analysis.

The (adaptive) K-means method is available in different variants: exchange method, gradient technique, minimum distance method,...

The stability of cluster analysis results can be investigated by simulation studies based on measures for comparing partitions. In that way one can check automatically whether the adaptive clustering performs better or not. Furthermore, in that way one can validate the number of clusters, or the user can assess the importance of the variables.

5. Credit scoring - some results

In general, credit scoring systems should be able to estimate approximately an applicant's creditworthiness. Usually, to this end one quantifies numerically the variables (attributes) and adds them up to form a so-called creditworthiness score (see above). On this basis a decision about a new applicant is derived.

Example 1: The training sample A (see above) consists of 16795 applicants, whereas the test sample B contains 14745 persons. Several statistical methods (see, for example, Quinlan 1993) are used which give the following total error rates:

Method	A	B
C4.5 (decision tree by Quinlan)	3,3%	4,62%
logistic regression	5,85%	5,03%
linear discriminant analysis	3,9%	5,67%
ClusCorr (local adaptive distances)	3,3%	4,47%

Example 2: Credit scoring data from Fahrmeir and Hamerle (1984): the sample consists of 1000 applicants, whereas 300 are bad clients and 700 are good ones. All 20 variables are used. The following total error rates were obtained by resubstitution (R) and cross-validation (C) (leaving-one-out-method):

Method	R	C
quadratic discriminant analysis	26,7% [2]	32,3% [2]
linear discriminant analysis	27,3% [2]	29,0% [2]
ClusCorr (local adaptive distances)	19,7%	22,9%

[2] see Fahrmeir and Hamerle (1984). The following average error rates (sum of error rates of each class divided by 2) are given by the authors: quadratic discriminant analysis: 24,3% (R), 31,2% (C); linear discriminant analysis: 27,0% (R), 28,9% (C).

6. References

Bock, H. H. (1987): On the interface between cluster analysis, principal component analysis, and multidimensional scaling, In: *Multivariate statistical modeling and data analysis*, Bozdogan, H. and Gupta A. K. (eds.), 17-34, D. Reidel, Dordrecht.

Bock, H. H. (1996): Simultaneous visualization and classification methods as an alternative to Kohonen's neural networks, In: *Classification and Multivariate Graphics: Models, Software and Applications*, Mucha, H.-J. and Bock, H. H. (eds.), Report No. 10 (ISSN 0956-8838), 15-24, Weierstrass Institute for Applied Analysis and Stochastics, Berlin.

Fahrmeir, L. and Hamerle, A. (1984): *Multivariate statistische Verfahren*, De Gruyter, Berlin.

Hubert, L. J. and Arabie, P. (1985): Comparing partitions, *Journal of Classification*, **2**, 193-218.

MacQueen, J. (1967): Some methods for classification and analysis of multivariate observations, In: *Proc. 5th Berkeley Symp. Math, Statist. Prob. 1965/66*, LeCam, L. and Neyman, J. (eds.), Vol. 1, 281-297, Univ. California Press, Berkeley.

Michie, D. M., Spiegelhalter, D. J. and Taylor, C. C. (eds.)(1994): *Machine Learning, Neural and Statistical Classification*, Ellis Horwood, Chichester.

Mucha, H.-J. (1992): *Clusteranalyse mit Mikrocomputern*, Akademie Verlag, Berlin.

Mucha, H.-J. (1995): XClust: clustering in an interactive way, In: *XploRe: an Interactive Statistical Computing Environment*, Härdle, W., Klinke, S., and Turlach, B. A. (eds.), 141-168, Series Statistics and Computing, Springer-Verlag, New York.

Mucha, H.-J. (1996a): Distance based credit scoring, In: *Classification and Multivariate Graphics: Models, Software and Applications*, Mucha, H.-J. and Bock, H. H. (eds.), Report No. 10 (ISSN 0956-8838), 69-76, Weierstrass Institute for Applied Analysis and Stochastics, Berlin.

Mucha, H.-J. (1996b): *ClusCorr*: cluster analysis and multivariate graphics under MS EXCEL, In: *Classification and Multivariate Graphics: Models, Software and Applications*, Mucha, H.-J. and Bock, H. H. (Eds.), Report No. 10 (ISSN 0956-8838), 97-105, Weierstrass Institute for Applied Analysis and Stochastics, Berlin.

Mucha, H.-J. and Klinke, S. (1993): Clustering techniques in the interactive statistical computing environment XploRe, *Discussion Paper 9318*. Institute de Statistique, Universite Catholique de Louvain, Louvain-la-Neuve.

Nishisato, S. (1980): *Analysis of Categorical Data: Dual Scaling and its Applications*, University of Toronto Press, Toronto.

Nishisato, S. (1994): *Elements of Dual Scaling: An Introduction to Practical Data Analysis*, Lawrence Erlbaum Associates, Publishers, Hillsdale.

Quinlan, J. R. (1993): *C4.5: Programs for Machine Learning*, CA: Morgan Kaufmann, San Mateo.

Part III

Classification and Discrimination

· Statistical Approaches for Classification Problems

· Discrimination and Pattern Analysis

The Most Random Partition of a Finite Set and Its Application to Classification

Masaaki Sibuya

Takachiho University
2-19-1 Ohmiya, Suginaki-ku, Tokyo 168, Japan

Summary: The notion of 'the most random partition of a finite set' is defined and proposed as a null model in classification. A Hamming-type distance between two independent and most random partitions is used for justifying its randomness, and is used for testing this null hypothesis. The probability distribution of the distance is studied for the latter purpose.

1. Validation of classification

In order to justify a classification procedure from the viewpoint of classical statistics, one should test the null hypothesis that a given or resulting classification is purely random, thus meaningless against the alternative that the classification is somehow structured or meaningful. Many models for the null hypothesis were proposed as reviewed by Gordon (1996) and Bock (1996) at this conference. Typically, a probabilistic null model assumes the uniform distribution of the observable variables on a domain, or a uniform random structure of a similarity or dissimilarity matrix.

The approach presented in this report is completely new and different from those which were previously proposed (see also Sibuya 1993 a,b). The new proposal is based on the notion of 'the most random partition of a finite set' which yields directly the null hypothesis of 'random partition without any regularity'. A Hamming-type distance between two independent and most random partitions plays the role of the conventional chi-square goodness of fit statistic.

2. Preliminaries

2.1 Random partitions

In this paper, a classification means a partition $M = \{m_1, m_2, \ldots\}$ of a finite set $\mathcal{N}_n = \{1, 2, \ldots, n\}$ of n objects or elements. The ordering of the classes or subsets is disregarded. Let the set of all partitions of $\mathcal{N}_n$ be denoted by $\mathcal{A}_n$. A discrete probability distribution P on $\mathcal{A}_n$ given by

$$P(M) \geq 0, \ M \in \mathcal{A}_n, \ \text{and} \ \sum_{M \in \mathcal{A}_n} P(M) = 1,$$

is called a random partition A. Typically it has a random number K of classes: $A = \{a_1, \ldots, a_K\}$. A special random partition, the 'most random partition', will be introduced later.

2.2 Distance between partitions

The discrepancy between two given partitions of $\mathcal{N}_n$ is measured by a distance $d(\cdot, \cdot)$, satisfying the axioms of the distance.

Definition (Rand, 1971). Let $L, M \in \mathcal{A}_n$ and $i, j \in \mathcal{N}_n, i \neq j$, and let

$$d_0(i, j; L, M) = \begin{cases} 0, & \text{if } (i,j) \text{ belong to the same class in } L \\ & \text{and the same class in } M, \\ 0, & \text{if } (i,j) \text{ belong to different classes in } L \\ & \text{and to different classes in } M, \\ 1, & \text{if } (i,j) \text{ belong to different classes in } L \\ & \text{and to the same class in } M, \\ 1, & \text{if } (i,j) \text{ belong to the same class in } L \\ & \text{and to different classes in } M. \end{cases}$$

Then the distance d between two partitions L and M is defined by

$$d(L, M) = \sum_{1 \leq i < j \leq n} d_0(i, j; L, M), \tag{1}$$

i.e., the number of inconsistent class assignments in L and M. We have $d(L, M) = 0$ if and only if $L = M$. Let $V = \{\mathcal{N}_n\}$ be a partition with the single class, and let $W = \{\{1\}, \dots, \{n\}\}$ be the partition with all singletons. Then

$$d(L, M) \leq d(V, W) = \binom{n}{2} \quad \text{for any } L, M \in \mathcal{A}_n. \tag{2}$$

The distance d is used also by Zahn (1964).

3. A single parameter family of random partitions

For an arbitrary partition $M = \{m_1, \dots, m_k\} \in \mathcal{A}_n$ define by s_j the number of classes of cardinality j, namely,

$$s_j := |\{i : |a_i| = j\}| \geq 0 \text{ for } 1 \leq j \leq n, \quad \sum_{j=1}^{n} s_j = k, \text{ and } \sum_{j=1}^{n} j s_j = n.$$

This defines the vector $S(M) = s = (s_1, \dots, s_n)$ which is a partition of the natural number n.

If a random partition $A = \{a_1, \dots, a_K\}$ on $\mathcal{A}_n$ with K classes has the probability distribution,

$$P(K = k \text{ and } A = \{m_1, \dots, m_k\}) = \frac{\alpha^k}{\alpha^{[n]}} \prod_{j=1}^{n} ((j-1)!)^{s_j} =: g(n; s),$$

$$\text{for } 1 \leq k \leq n, \ M = \{m_1, \dots, m_k\} \in \mathcal{A}_n, \tag{3}$$

we say that it has the distribution $\mathcal{P}(n, \alpha)$ and write $A \sim \mathcal{P}(n, \alpha)$. Here $s = S(M)$, $\alpha > 0$ is a real parameter and $\alpha^{[n]} = \alpha(\alpha + 1) \cdots (\alpha + n - 1)$. An illustrative interpretation of $\mathcal{P}(n, \alpha)$ is given in Section 5.

We can prove the following recursions of $g(n; s)$. g with a negative argument $s_j < 0, 1 \leq j \leq n$ is regarded as 0.
(i)

$$n g(n; s) = \rho_{n-1} s_1 g(n-1; s_1 - 1, s_2, \dots, s_{n-1})$$
$$+ \frac{1 - \rho_{n-1}}{n-1} \sum_{j=1}^{n-1} j(j+1) s_{j+1} g(n-1; s_1, \dots, s_{j-1}, s_j + 1, s_{j+1} - 1, s_{j+2}, \dots, s_{n-1}), \tag{4}$$

where $\rho_{n-1} = \alpha/(\alpha + n - 1)$.

(ii)

$$g(n-1; s_1, \ldots, s_{n-1}) = g(n; s_1 + 1, s_2, \ldots, s_{n-1}, 0)$$
$$+ \sum_{j=1}^{n-1} s_j g(n; s_1, \ldots, s_{j-1}, s_j - 1, s_{j+1} + 1, s_{j+2}, \ldots, s_n). \qquad (5)$$

This is a sort of dual of (4). It means that if an arbitrarily chosen element, say n, of A is deleted, the remaining random partition has the distribution $\mathcal{P}(n-1, \alpha)$.

(iii)

$$g(n; s) = \frac{\alpha(j-1)!}{(\alpha + n - j)^{[j]}} g(n - j; s_1, \ldots, s_{j-1}, s_j - 1, s_{j+1}, \ldots, s_{n-j}). \qquad (6)$$

This relation means that if an arbitrarily chosen element, say n, of A belongs to a class with j elements, and if the class is deleted from A, then the remaining partition has the distribution $\mathcal{P}(n-j, \alpha)$.

Any one of the three conditions with some part of the other two conditions characterizes the distribution $\mathcal{P}(n, \alpha)$, Sibuya and Yamato (1995).

4. The most random partition

Let A be any random partition on $\mathcal{A}_n$, and let $M_0 \in \mathcal{A}_n$ be a fixed partition. If

$$E(d(A, M_0)) \leq E(d(A, M)), \quad \text{for all } M \in \mathcal{A}_n,$$

then M_0 is called the 'center', a sort of mean, of the random partition A.

Proposition 1 *(Snijders. 1996) Suppose that $A \sim \mathcal{P}(n, \alpha)$.*
(i) If $\alpha < 1$ the center of A is V.
(ii) If $\alpha = 1$ $E(d(A, M))$ is a constant for all $M \in \mathcal{A}_n$. and any fixed partition M is the center of A.
(iii) If $\alpha > 1$ the center of A is W.

Proposition 2 *If A and B are any independent and identically distributed random partitions, then*

$$E(d(A, B)) \leq \frac{1}{2} \binom{n}{2}.$$

Equality holds if and only if the marginal distribution of partitions of any subset $A_0 = \{a_i, a_j\}$ with two elements of A is $\mathcal{P}(2, 1)$. The condition is satisfied if $A \sim \mathcal{P}(n, 1)$.

Definition $\mathcal{P}(n, 1)$ defines the 'most random partition' A on $\mathcal{A}_n$.

From the properties of $\mathcal{P}(n, \alpha)$ and from the propositions, $\mathcal{P}(n, 1)$ is qualified to be called the most random partition. The next section illustrates this definition.

The following table lists six random partitions A from $\mathcal{P}(n, 1)$ for $n = 26$ objects (letters):

1	2	3	4	5	6
az	abi	ajkwz	ag	af	ackqr
bilrs	cw	bcdegmt	bix	bcmr	b
cfkmp	dp	flv	cdjmt	diu	d
dhtw	elrtx	hos	eh	eo	e
eq	fhjnou	iu	fsz	gkpqtz	fho
g	gmy	nx	kno	hlv	gimntz
jv	k	py	l	jn	jpwx
nox	qsz	qr	pqrw	swxy	l
uy	v		u		sv
			v		u
			y		y

In this table. the elements in a class are alphabetically ordered, and the classes of a partition are lexicographically ordered. The latter order, which is independent of the ordering of elements in a class, is natural and automatic if the elements of the original set is linearly ordered. In this paper, the order of classes and elements are disregarded.

5. Geneses of the family and the most random partition

5.1 A clustering process

The random partition $\mathcal{P}(n, \alpha)$ is generated by the following model. Suppose that balls and urns are labeled $1, 2, \ldots$ (one ball and one urn for each label) and the balls are sequentially thrown at random into the urns as follows. Ball 1 is put into Urn 1. Ball 2 is put into Urn 1 with probability $1/(\alpha + 1)$, and into empty Urn 2 with probability $\alpha/(\alpha + 1)$, where $\alpha > 0$ is a constant. After n balls, suppose that Urn j has c_j balls, $j = 1, \ldots, k$, and $c_1 + \cdots + c_k = n$. Ball $n + 1$ is put into Urn j with probability $c_j/(\alpha + n)$ for $j = 1, \ldots, k$, and into a new empty Urn $k + 1$ with probability $\alpha/(\alpha + n)$. After n steps, the balls in an urn are regarded as a cluster, and we obtain an ordered random partition like the above table. Disregarding the urn numbers we obtain a random partition $A \sim \mathcal{P}(n, \alpha)$. For $\alpha = 1$, the assignment probabilities for Ball $n + 1$ are proportional to the class sizes $c_1, \ldots, c_k$ and 1. See Sibuya (1993a). This process is essentially the same as Hoppe's urn model (see, e.g.. Ewens 1990).

If the balls are indistinguishable we observe a random partition $|a_1| + |a_2| + \cdots + |a_K| = n$ of a number n. Again, the partitioned numbers are unordered (or ordered) provided that the urn numbers are disregarded (or considered). In the theory of population genetics, these cases are discussed in Hoppe's urn model, and the unordered and ordered partitions of number are known as Ewens' sampling formula, and Donnelly-Tavaré-Griffiths formula, respectively (see, e.g., Ewens 1990). The logical aspects of these random partitions are studied by Zabell (1992).

5.2 Partition of a random permutation into cycles

Let σ be a permutation of $\mathcal{N}_n$, i.e. a bijection of $\mathcal{N}_n$ onto itself. The sequences $(j, \sigma(j), \sigma(\sigma(j)), \ldots)$, $j \in \mathcal{N}_n$, form cycles and partition $\mathcal{N}_n$ into cycles. This is a natural way to get a partition of $\mathcal{N}_n$. Assume that all $n!$ permutations of $\mathcal{N}_n$ are equally probable, and the probability distribution on A_n induced by cycles of the random permutation σ is $\mathcal{P}(n, 1)$. The model appears frequently in applied probability, statistics and computer science. and was studied in Sibuya (1993b).

6. Applications

Let A, B be independent random partitions on $\mathcal{A}_n$ following $\mathcal{P}(n,1)$. Let F_n be the distribution function (d.f.):

$$F_n(x) = \Pr\{d(A, B) \le x\}, \quad 0 \le x \le \binom{n}{2}.$$

The d.f. F_n is used for measuring the performance of classification. Some properties of F_n are shown in the last section.

Chernoff's faces are useful for subjective classification of a multivariate dataset. Its performance depends on the design of comical faces, the allocation of variates to the face features, and the training of judges. The effects of these factors can be measured in terms of $F_n(d(A, B))$. A trial experiment and its analysis are reported by Harada and Sibuya (1991). The planning of the classification experiments of Section 1 can be examined in a similar way.

In numerical taxonomy, suppose that we have a 'standard' classification procedure, and some new candidate procedure. Sample data are generated from a mixture of some known population distributions. Each multivariate observation is known to belong to its true population, and observations from the same population are expected to be classified in the same subset. Consider the distance between the classifications A, B of sample data into the true populations and that obtained by a classification procedure. The distance $d(A, B)$ measures the difficulty of the classification of a mixed population if the classification is the 'standard' one. Otherwise, the distance measures the performance of a new candidate, to be compared with that of the standard one.

7. Distance between two independent most random partitions

For the computation of the probability function of $D_n = d(A, B)$ and its moments, for independent random partitions A and B of n objects following $\mathcal{P}(n, \alpha)$, the following facts are useful:

Proposition 3 *Let A be a random partition following $\mathcal{P}(n, \alpha)$ and $M \in \mathcal{A}_n$ be a fixed partition. The probability distribution of $d(A, M)$ is determined by $S(M)$.*

If $\alpha = 1$ the probability distribution of D_n is the same as that of $d(A, M)$, for any $M \in \mathcal{A}_n$. The choices $M = V$ or W are convenient.

Proposition 4 *The moment of D_n of degree r, for any n. can be calculated from the probability function of $D_{2r}, r = 1, 2, \ldots$*

From these, we calculate the lower moments:

$$E(D_n) = \binom{n}{2} \frac{2\alpha}{(\alpha + 1)^2},$$

$$E(D_n^2) = \binom{n}{2} \frac{2\alpha}{(\alpha + 1)^2} + \binom{n}{3} \frac{12\alpha(3\alpha + 2)}{(\alpha + 1)^2(\alpha + 2)^2} + \binom{n}{4} \frac{12\alpha(\alpha^3 + 9\alpha^2 + 21\alpha + 6)}{(\alpha + 1)^2(\alpha + 2)^2(\alpha + 3)^2},$$

$$Var\left(D_n \Big/ \binom{n}{2}\right) = \frac{4\alpha(4\alpha^4 + 3\alpha^3 - 3\alpha^2 - 3\alpha + 6)}{(\alpha + 1)^2(\alpha + 2)^2(\alpha + 3)^2} + O\left(\frac{1}{n}\right), \quad \text{for } n \to \infty.$$

The variance of $D_n/\binom{n}{2}$ at $\alpha = 1$ is neither maximum nor minimum, but a rather small value 0.008. The numerical value of the probability function of $\mathcal{P}(n,1)$ for $n = 2,\ldots,8$, and its moments up to the degree four are shown in Sibuya (1993b). The probability function of $\mathcal{P}(n,1)$ is multi-modal, and is, as well as its limit, strongly mesokurtic.

Acknowledgements

The author has benefited from valuable discussions with participants of the IFCS conference and with Prof. H. Yamato.

References:

Bock, H. H. (1996): Probabilistic aspects in classification, *IFCS'96, Kobe, March 1996*, Invited Lecture 5.

Ewens, W. J. (1990): Population genetics theory – the past and the future, In: *Mathematical and Statistical Developments of Evolutionary Theory*, Lessard, S. ed., NATO Adv. Sci. Inst. Ser. C-299, Kluwer, Dordrecht, 177–227.

Gordon, A. D. (1996): Cluster validation, *IFCS'96, Kobe, March 1996* , Invited Lecture 1.

Harada, M. and Sibuya, M. (1991): Effectiveness of the classification using Chernoff faces, *Japanese Journal of Applied Statistics*, **20**, 39–48 (in Japanese).

Rand, W. M. (1971): Objective criteria for the evaluation of clustering methods, *Journal of American Statistical Association*, **66**, 846–850.

Sibuya, M. (1992): Distance between random partitions of a finite set, In: *Distancia '92* , Joly, S. and Le Calvé, G. (eds.) 143–145, June 22-26, Rennes.

Sibuya, M. (1993a): A random clustering process, *Ann. Inst. Statist. Math.*, **45**, 459–465.

Sibuya, M. (1993b): Random partition of a finite set by cycles of permutation. *Japan Journal of Industrial and Applied Mathematics*, **10**, 69–84.

Sibuya, M. and Yamato, H. (1995): Characterization of some random partitions, *Japan Journal of Industrial and Applied Mathematics*, **12**, 237–263.

Snijders, T. A. B. (1996): Private communications.

Zabell, S. L. (1992): Predicting the unpredictable, *Syntheses*, **90**, 205–232.

Zahn, C. T., Jr. (1964): Approximating symmetric relations by equivalence relations. *J. Soc. Indust. Appl. Math.*, **12**, 840–847.

A Mixture Model To Classify Individual Profiles Of Repeated Measurements

Toshiro Tango

Division of Theoretical Epidemiology, The Institute of Public Health
4-6-1 Shirokanedai, Minato-ku, Tokyo 108, JAPAN
E-mail: tango@iph.go.jp

Summary : To evaluate the treatment effects in a randomized experiment, Tango(1989, *Japanese Journal of Applied Statistics*,18, 143-161) and Skene and White(1992, *Statistics in Medicine*,11, 2111-2122), independently , proposed similar mixture models in which the effects of treatment were characterized by the shape of common latent profiles and the mixing proportion of these profiles. This paper describes the difference between the two procedures and presents a generalized model to cope with improper longitudinal records. The usefulness of the proposed methods are illustrated with data from clinical trials.

1. Introduction

In clinical medicine, drugs are usually administered to control some response variable, X, reflecting the patient's disease state directly or indirectly, within a specified range. So, in many clinical trials, some response variable is scheduled to be observed at regular intervals for assessing changes from the baseline. Figure 1 shows the mean treatment profiles of (a) log-trasformed serum levels of glutamate pyruvate transaminase (GPT) for 124 patients with chronic hepatitis patients randmly assigned to recieve the new treatment A or the standard B in a double-blinded clinical trial, which are measured at baseline and 1 week interval thereafter up to 4 weeks and (b) its change from the baseline level for each of treatment groups. In this paper, only complete cases are shown and used for illustration purposes. In this clinical trial, the effects of treatment can be observed as "decrease" in levels of GPT as compared with baseline level and must be evaluated at the last observation time (4 weeks). In this kind of clinical trials, the difference between mean treatment profiles can generally be defined as the size of interaction term TREATMENT x TIME. Classical and still most frequently used procedures in medical literature will be to repeat Student's t-test or Wilcoxon's rank sum test at each time point for the treatment difference in change from the baseline shown in Figure 1-(b). Repeated application of two-tailed t-test resulted in no significant differences between the two groups at all the time points. Since the test results are the same regardless of the time point, we tend to conclude "no difference". However, such multiple comparisons inflate the over-all siginificance level and often show siginificant differences at some points but not at other points, which will generate confusion and may lead to the post hoc selection of the most highly significant difference.

To avoid this problem, the following two kinds of procedures are well known : 1) Univariate repeated measure ANOVA where the degrees of freedom associated with F test for repeated factors are reduced by one of two procedures, Greenhouse-Geisser method and Huynh-Feldt method , and 2)maximum likelihood based ANOVA which can allow for more general within-subjects covariance structure, missing values and irregularly spaced data. But all are assuming within-group covariance structure to be homogeneous between treatment groups, which seems to be a difficult assumption to justify in clinical trials. The results of applying these procedures using BMDP

Table 1: Comparison of existing procedures to test for interaction term TREAT-MENT x TIME

Methods	P-values	-2log(L)
ANOVA with d.f. reduction		
Greenhouse-Geisser	0.623	
Huyng-Feldt	0.627	
ANOVA with specified Cov. Structure		
Compound symmetry	0.793	862
First order autoregressive	0.802	528
General autoregressive	0.873	472
Unstructured	0.879	429
Random-effects growth curve		
2nd degree polynomial	0.753	552
3rd degree polynomial	0.870	490

5V including random coefficient growth curve models are summarized in Table 1, in which P-value for the interaction term TREATMENT x TIME in each model is shown. All the results indicate non-siginificant regardless of goodness-of-fit of models.

Recently the use of summary statistics for the analysis of repeated measurements have received a growing interest instead of using complicated models described above in a randomized clinical trials. This seems to be mainly due to ease of interpretation and communication. Most types of summary statistics, say slope, change from the baseline, etc, can be expressed in the form of a weighted linear combination of repeated measurements. Of course, these summary statistics are also devised for detecting differences in mean treatment profiles between groups. Amongst others, Frison and Pocock(1992) recommended analysis of covariance for the post treatment mean adjusted for pretreatment baseline values. Therefore, let us apply ANOCOVA for the last measurement (at 4 weeks) adjusted for the baseline valued. The resultant $F = 0.148$ with $(1, 121)$ degrees of freedom and $p = 0.701$, indicating nonsignificant.

Based upon all the results discussed above, we might tend to conclude that the effects of new treatment A is not statistically different from the standard B. However, what should be taken into account before applying these procedures is to consider the implication of mean treatment profiles of the response variable. If each group is really homogeneous, namely, SUBJECT x TIME interaction within each group is negligible, then the mean treatment profile could reasonably represent the pattern of the treatment effects over time in each group. However, if SUBJECT x TIME interaction is not negligible, the problem is not so simple. As procedures which are capable of dealing with heterogeneity among patients' profile, random-effects or random coefficient models have been proposed. But, they are still based upon the "homogeneous" assumption. To this problem, Tango(1989) proposed a mixture model for the analysis of repeated measurements in a randomized clinical trial assuming that 1) each treatment group consists of a mixture of several distinct latent profiles of changes from the baseline level, common to all the treatment groups, 2)a low-degree polynomial can represent the "mean profile" for each of these latent profiles and 3)the effects of treatment is characterized by the shape of the latent profiles and the mixing propotions of these latent profiles. Skene and White(1992) , probably without knowing my paper since it has been written in Japanese, also proposed a similar mixture model but it is unsuitable to data with "improper records" and undesirable since it estimates unrealistic ragged profiles.

The purpose of this paper is to make clear the difference between the two models, to present a generalized formulation of my model to cope with improper records and describe how these procedures are useful and essentially important to analyze data and interprete the results in some sorts of randomized clinical trials.

2.Model

Suppose that a randomized clinical trials specify the following protocol:

1. Patients are randomly assigned to receive one of G treatments, with N_i patients on the ith treatment group.

2. The response variable X is measured $T + 1$ times at baseline and at equally spaced intervals, where T is at most 4 to 6.

3. The effects of treatment is evaluated at the last measurement time.

But, in practice, the occurrence of missing values and measurements at irregularily spaced intervals is inevitable. Further, recent tendency of "intent-to-treat" requires that all the patients registered should be included in the analysis regardless of the degree of completeness of records. Both Tango and Skene and White formulated the model only for complete data. Therefore, we shall here generalize the model to allow for incomplete records of patients. Let $X_{ij}(t_{ijk})$ denote the measurement made at the time $t_{ijk}(k = 0, 1, ..., U_{ij} \leq T)$ of the jth subject $(j = 1, 2, ..., N_i)$ in the ith treatment group $(i = 1, 2, ..., G)$ and $t_{ij0} = 0$. Thus $X_{ij}(0)$ indicates the baseline level. Without loss of generality, the "improvement" induced by a treatment is defined by the "decrease" in levels of the response variable X as compared with the baseline level. Let $Y_{ij}(t_{ijk}) = X_{ij}(t_{ijk}) - X_{ij}(0)$, the change from the baseline level and assume the existence of M latent profiles common to all the treatment groups. Then, under the condition that the jth patient of the ith group follows the mth latent profile $(m = 0, 1, 2, ..., M - 1)$, it can be assumed

$$Y_{ij}(t) = \mu_m(t) + \epsilon_{ijt}, \quad \epsilon_{ijt} \sim N(0, \sigma^2) \tag{1}$$

where mth latent profile and ϵ_{ijt} are, conditional on m, mutually independent. With regard to the mean profile $\mu_m(t)$, Tango proposed a smooth function of time by a low degree polynomial. For example, when $M = 3$, we have

$$\mu(t) = \begin{cases} \mu_0(t) = 0 & \text{if the subject belongs to "unchanged"}, \\ \mu_1(t) = \sum_{k=1}^{R} \beta_{1k} t^k (< 0) & \text{if the subject belongs to "improved"}, \\ \mu_2(t) = \sum_{k=1}^{R} \beta_{2k} t^k (> 0) & \text{if the subject belongs to "worsened"}, \end{cases} \tag{2}$$

where R is the degree of polynomial common to all the profiles except for the unchanged profile. On the other hand, Skene and White proposed a profile vector $\mu_m = (\mu_{m1}, ..., \mu_{mT})$. This kind of parameterization is seemingly more flexible in representing profiles but tends to estimate undesirably ragged profiles. Further, it cannot allow for incomplete data with missing values or measured at irregularily spaced intervals, which are also pointed out in the discussion section of their paper.

In this model, the response pattern for the jth subject of the ith group, $\mathbf{Y}_{ij} = (Y_{ij}(t_{ij1}), ..., Y_{ij}(t_{ijU_{i_j}}))'$, $j = 1, ..., N_i$, have the following mixture density

$$g_i(\mathbf{Y}_{ij}|\theta) = \sum_{m=0}^{M-1} p_{im} f_m(\mathbf{Y}_{ij}) \tag{3}$$

where p_{im} implies the mixing proportion of the mth latent profile in the ith treatment group and $f_m(.)$ denotes the density function of the mth latent profile and are given by

$$f_m(\mathbf{Y}_{ij}) = \left(\frac{1}{2\pi\sigma}\right)^{U_{ij}/2} \cdot exp\left(-\frac{1}{2\sigma^2}\sum_{k=1}^{U_{ij}}(Y_{ij}(t_{ijk}) - \mu_m(t_{ijk}))^2\right), \quad m = 0, ..., M - 1$$

$$(4)$$

where m=0 means "unchanged" profile. The log-likelihood for the parameters of $\theta = (p_{im}, \beta_{mk}, \sigma^2), i = 1, ..., G; m = 0, 1, ..., M - 1; k = 1, ..., R$ is

$$L = \sum_{i=1}^{G}\sum_{j=1}^{N_i} g_i(\mathbf{Y}_{ij}|\theta). \tag{5}$$

and the comparison of treatment effects might be reduced to the test of the following null hypothesis:

$$H_0 : p_{1m} = p_{2m} = ... = p_{Gm}, \quad m = 0, 1, ..., M - 1 \tag{6}$$

Furthermore, this model provides us with a criterion of classification of subjects into one of the components based upon the maximum of the posterior probability that the jth subject of the ith group comes from the mth profile :

$$\hat{Q}_{ij}(m) = p_{im}f_m(\mathbf{Y}_{ij})/g_i(\mathbf{Y}_{ij}|\hat{\theta}). \tag{7}$$

where $\hat{\theta}$ is the maximum likelihood estimate(MLE).

3.EM algorithm

To obtain the maximum likelihood estimates (MLE) of parameters of θ of the mixture distribution, a Newton-Raphson algorithm or the EM algorithm have been used. Both I and Skene and White recommended the use of EM algorithm especially for its easy implementation. Its solution, however, heavily depends on the initial values and then careful examination is neccessary to assure that the global, rather than a local, maximum was attained. The EM algorithm is briefly outlined below:

1. **Step 0** : Give starting values of the posterior probability $\hat{Q}_{ij}(m)$.

2. **Step 1[M-step]**:Given the $\hat{Q}_{ij}(m)$, the parameters p_{im} are easily given by

$$\hat{p}_{im} = \sum_{j=1}^{N_i} \hat{Q}_{ij}(m)/N_i \tag{8}$$

$\hat{\beta}_{mk}$ are obtained by using a weighted least squares procedure minimizing

$$\sum_{i=1}^{G}\sum_{j=1}^{N_i}\sum_{m=0}^{M-1} \hat{Q}_{ij}(m) \sum_{k=1}^{U_{ij}}(Y_{ij}(t_{ijk}) - \mu_m(t_{ijk}))^2,$$

and $\hat{\sigma}^2$ is obtained as

$$\hat{\sigma}^2 = \sum_{i=1}^{G}\sum_{j=1}^{N_i}\sum_{m=0}^{M-1} \hat{Q}_{ij}(m) \sum_{k=1}^{U_{ij}}(Y_{ij}(t_{ijk}) - \hat{\mu}_m(t_{ijk}))^2 / \sum_{i=1}^{G}\sum_{j=1}^{N_i} U_{ij}, \tag{9}$$

3. **Step 2[E-step]** :Calculate the posterior probability $\hat{Q}_{ij}(m)$ based on the estimates $\hat{\theta}$ obtained in the step 1.

4. **Step 3**:Check to see if $\hat{\theta}$ has converged; if not, repeat M-step and E-step.

When we construct a particular alternative hypothesis, we need to apply some constraints to the p_{im}'s, say $p_{11} = p_{21}$. In this case, expression (9) must be changed. But this kind of extra work can easily be handled in GLIM or S-PLUS.

4.Number of latent profiles

As is well known, the classical asymptotic theory of distributions of likelihood ratio tests cannot be applied for testing the number of components in mixture models because it does not satisfy the regularity condition. Self and Liang(1987) discussed its theoretical issues and gave several examples in which the distribution can be expressed as mixtures of chi-squared distribution. But, its application to mixture models are difficult. Everitt(1981), Thode, Finch and Mendell(1988) conducted a simulation study to find the percentage points of likelihood ratio test. McLachlan(1987) applied bootstrapping methods. But, most of their works were focussed on the test of a single normal density versus a mixture of two normal densities in the univariate case. Skene and White examined the possibility of the use of empirical semi-variogram plot as an exploratory tool. But it is in nature subjective and is not so easy to determine the number of profiles based on such plots. So, its use cannot be recommended especially for confirmatory clinical trials. Basically, this problem is identical to that of choosing the number of clusters in cluster analysis. So far, the challenges to this kind of problem have never been successful in obtaining the clear solution statistically. All such procedures are exploratory. Therefore, especially in a confirmatory trials such as phase III clinical trials, the number of latent profiles should not be selected statistically but be carefully discussed before starting clinical trials and be fixed in the protocol. Given the number of latent profiles, the problem becomes simple and clear and usual likelihood ratio tests can be applied to compare the goodness-of-fit of nested models (hypotheses) and to estimate optimal θ and R.

5. Examples

We shall consider again the data of GPT shown in Figure 1. Empirically the distribution of GPT in healthy subjects can be approximated by log-normal, then let $X_{ij}(t_{ijk})$ denotes here the transformed value log(GPT), natural logarithm. Further assume M=5 since several other endpoints in this trial are to be evaluated by 5 ordered categories for each patient. As an criterion which gives initial values of $Q_{ij}(m)$, we may use the 5 ordered categories based on the value of $S_{ij} = \sum_{k=1}^{U_{ij}} Y_{ij}(t_{ijk})$, where $U_{ij} = M - 1$ for all the patients in this complete set of data. Several other initial values were examined to assure whether the result derived below is optimal. The main results are summarized in Table 2.

Based on the likelihood ratio tests, one alternative hypothesis H_1: $p_{11} = p_{21}, R = 3$ was selected as the most appropriate model. Compared with models for the null hypothesis H_0 for each of three kinds of mean profiles, R=2, 3, and Skene and White's vector μ_m, fitting the model with constraints $p_{11} = p_{21}$ gave a significant decrease in deviance of 8.9 on 3 d.f., 9.6 on 3 d.f. and 8.5 on 3 d.f., respectively, regardless of the goodness-of-fit of models. Among others, a cubic polynomial effected the highest decrease with $p = 0.022$. Goodness-of-fit of these models were also investigated by observing each patient's response profile in relation to the estimated 95% region of

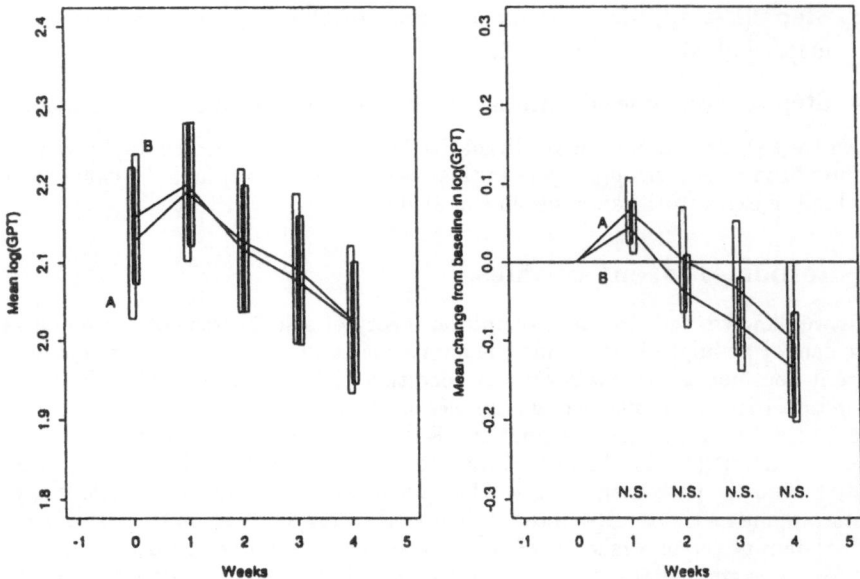

Figure 1. The mean treatment profiles and *mean* $\pm 2SD$ at each time point, of (a) log(GPT) and (b) its change from the baseline level for each of new treatment A and standard treatment B. The difference in change from the baseline was not significant ($p > 0.05$ by two-tailed Student's t test) at each time point.

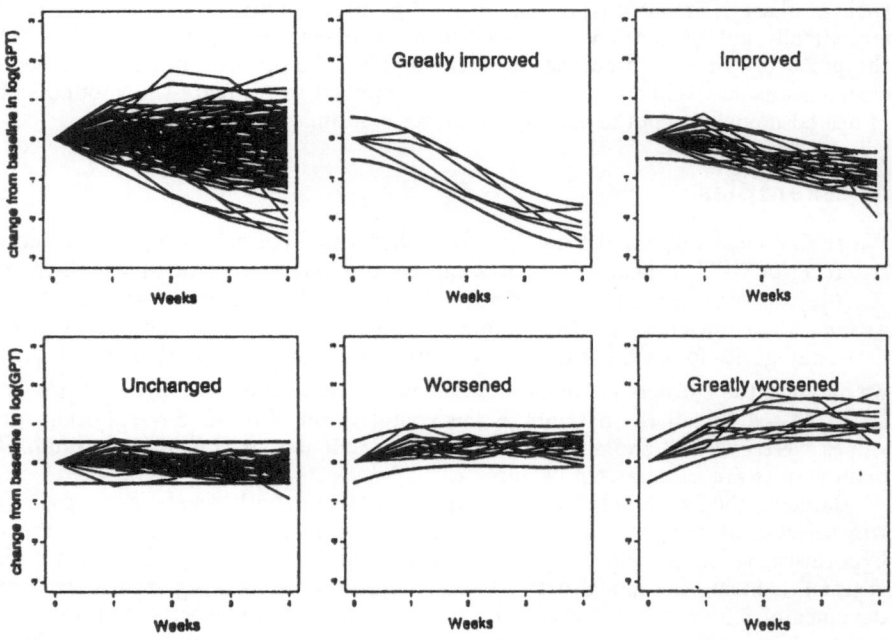

Figure 2. (a)Individual profiles for all the patients. (b)-(f) Estimated 95% region of profiles, $\hat{\mu}_m(t) \pm 2\hat{\sigma}, m = 0, 1, ..., 4$ and individual profiles classified into the corresponding region regardless of the treatment group.

Table 2: -2log L for each of mixture models assuming $M = 5$. Degrees of freedom are shown in parentheses.

Hypothesis	form of latent profile		
	quadratic polynomial	cubic polynomial	Skene and White
H_0	402.7 (13)	389.1 (17)	384.7 (21)
$K_1(p_{11} = p_{21})$	393.8 (16)	380.5 (20)	376.2 (24)
$H_1(p_{1m} \neq p_{2m})$	393.6 (17)	380.3 (21)	376.0 (25)
Difference: H_0 vs. K_1	8.9 (3)	9.6 (3)	8.5 (3)
p-value	0.031	0.022	0.037

profiles, $\hat{\mu}_m(t) \pm 2\hat{\sigma}$, of the mth latent profile into which each patient was classifed according to the maximum of estimated $\hat{Q}_{ij}(m)$. In Figure 2, the estimated 95% region of latent profiles for the optimal model with $p_{11} = p_{21}$ and $R = 3$ are illustrated together with individual profiles classified into the corresponding profile regardless of treatment groups. Table 3 presents the classification of each patient into one of those 5 profiles. As would be expected, χ^2 test based on this table yielded P-value = 0.023 very close to that of the likelihood ratio test. These results are summarized using estimated mixing proportions $\hat{p}_{mk}$'s as follows: Compared with the standard treatment B, the treatment A has

1. the same proportion of "greatly improved" (4.1%),

2. higher proportions of "improved" (32.7% vs 24.7%), but also higher "worsened" (24.5% vs 16.7%) and "greatly worsened" (11.8% vs 2.9%)

3. a lower proportion of "unchanged" (26.7% vs 51.6%).

Therefore, based on estimated proportions and latent profiles, we can not say that treatment A is better than B, but we can say that the effects are significantly different. These characterizations of the efficacy of treatments seem to be medically important espeicaly for finding the key factors on baseline factors to discriminate responders from non-responders, but cannot be recognized by observing the mean treatment profiles over time. Figure 2 suggests how the mean treatment profiles shown in Figure 1 is misleading. Example illustrated here is not exceptional but rather typical one. Tango(1989) illustrated the method with another two sets of data from randomized clinical trials.

Table 3: Classification of patients by the optimal mixture model with $M = 5, R = 3$ and constraint $p_{11} = p_{21}$.

Group	Greatly improved	improved	unchanged	worsened	Greatly worsened	total
A	3	20	17	14	8	62
B	2	13	34	11	2	62

6. Discussion

It is well recognized that some unknown prognostic factors could have larger effects on the response variable than treatment under study. Therefore, if they really exist, random allocation of patients into one of treatment groups help these unknown prognostic factors also to distribute equally likely between groups. Namely, several distinct latent profiles common to all the groups could be explained by these prognostic factors and the mixture model provide useful data to investigate and identify these unknowns in the next stage of research.

On the other hand, it seems to me that recent literature concerning to the analysis of repeated measurements has been concentrated too much on modelling the within-group covariance structure assuming homogeneity between groups which seems to be unrealistic especially for clinical trials. As Skene and White has pointed out, the observed autocorelation could be a consequence of under-specifying the mean structure for subjects of each treatment group. Therefore, before applying such statistically flexible but clinically unrealistic models, more attention should be placed on the validity of these assumptions and on the reasons why we take observations over time.

ACKNOWLEDGEMENTS

The author is indebted to the Japanese Foundation for Multidisciplinary Treatment of Cancer. This study was supported in part by Grant-in-Aid for Scientific Research (Grant No. 05302064) from the Ministry of Education, Science and Culture of Japan.

References:

Crowder,M.J. and Hand,D.J. (1990): *Analysis of Repeated Measures*, Chapman and Hall.

Dempster,A.P., Laird,N.M. and Rubin,D.B.(1977): Maximum likelihood from incomplete data via the EM algorithm, *Journal of the Royal Statistical Society, Series B*,**39**,1-22.

Diggle,P., Liang,K.Y. and Zeger,S.L. (1994): *Analysis of Longitudinal Data*, Oxford Science Publication.

Everitt,B.S.(1981): A Monte Carlo investigation of the likelihood ratio test for the number of components in a mixture of normal distributions, *Multi. Behav. Res.*,**16**, 171-180.

Frison,L. and Pocock,S.J.(1992): Repeated measures in clinical trials: analysis using mean summary statistics and its implications for design, *Statistics in Medicine*, **11**, 1685-1704.

McLachlan,G.J. (1987): On bootstrapping the likelihood ratio test statistics for the number of components in a normal mixture, *Applied Statistics*,**36**, 318-324.

Self, S.G. and Liang,K.Y.(1987): Asymptotic properties of maximum likelihood estimates and likelihood ratio tests under nonstandard conditions, *Journal of American Statistical Association*,**82**, 605-610.

Skene,A.M. and White,S.A. (1992): A latent class model for repeated measurments experiments, *Statistics in Medicine*,**11**,2111-2122.

Tango,T. (1989): Mixture models for the analysis of repeated measurements in clinical trials,*Japanese Journal of Applied Statistics*,**18**, 143-161.

Thode,Jr.,H.C, Finch,S.J. and Mendell,N.R.(1988): Simulated percentage points for the null distribution of the likelihood ratio test for a mixture of two normals, *Biometrics*,**44**, 1195-1201.

Titterrington,D.M., Smith,A.F.M. and Makov, U.F. (1985): *Statistical Analysis of Finite Mixture Distributions*. New York. Wiley and Sons.

Irregularly Spaced AR (ISAR) Models

Jeffrey S.C. Pai[1], Wolfgang Polasek[2] and Hideo Kozumi[3]

[1] Faculty of Management, University of Manitoba
181 Freedman Crescent, Winnipeg, Manitoba, R3T 5V4, Canada

[2] Institute of Statistics and Econometrics, University of Basel
Holbeinstrasse 12, CH-4051 Basel, Switzerland

[3] Faculty of Economics and Business Administration, Hokkaido University
Kita 9 Nishi 7 Kita-ku, Sapporo 060, Japan

Summary: High frequency data in finance are time series which are often measured at unequally or irregularly spaced time intervals. This paper suggests a modeling approach by so-called AR response surfaces where the AR coefficients are declining functions in continuous lag time. The irregularly spaced ISAR models contain the usual AR models as a special case if the time series is equally spaced. We illustrate our methodology with two examples.

1. Introduction

For some years now the set of available data from financial markets has increased rapidly. So far only a small subset of the information available has been used. In the 70'ies, most of the empirical studies were based on yearly, quarterly, or monthly data. This data could typically be modeled by random walks or linear models such as ARIMA-models (Box and Jenkins, 1976). In the 80'ies, the study of weekly and daily financial data lead to non-linear models such as ARCH models (Engle, 1982). Recently, empirical studies through analyzing intra-daily data are gaining new insights into the behavior of financial markets (see Guillaume et al. 1994).

For the Foreign Exchange (FX) market, Müller et al. (1990) include the daily heteroskedasticity of the volatility while Goodhart and Figliuoli (1991) discovered negative first order autocorrelation at one minute intervals. In fact, daily data are computed on the basis of the average of five intra-daily quoted prices of the largest banks around a particular time. The spot intra-daily FX data are observed as an irregularly spaced time series (see Olsen & Associates, 1993). The standard methods of data analysis, on the contrary, are based upon equally spaced data. Typically, a certain fixed interval is chosen, and some averaging procedure for all the transactions within the intervals is done in order to apply standard methods of analysis. Several problems arise if the data from irregularly spaced time series are converted to regularly spaced time series. The appropriate interval will depend upon the transaction frequency. These intervals vary with different markets. There could be a problem of missing data if the length of the interval is too short. On the other hand, information is lost if the length of the interval is too long.

Data from financial markets exhibit high correlation between the ticking frequencies and the volatility of the time series. It is widely believed that the durations between transactions may carry information about the volatility.

In Section 2, we describe a new class of AR models for irregularly spaced time series, the ISAR(p) models, as well as the ordinary least squares result. Section 3 illustrates our methodology with two examples. Section 4 concludes.

2. The ISAR(p) models

Consider the time series model with p general lag response functions $\Phi_1[\cdot], \ldots, \Phi_p[\cdot]$ observed at n irregular time points

$$Y_{t_i} = \sum_{j=1}^{p} \Phi_j[\Delta^j t_i] Y_{t_{i-j}} + \epsilon_i, \quad i = 1, \ldots, n, \tag{1}$$

where Y_{t_i} is the observed time series at time t_i, and Δt_i is the time span between two adjacent observations, $\Delta^j t_i$ is the distance between observations which are j "ticks" apart: $\Delta^j t_i = t_i - t_{i-j}$. For the residual process we assume $\epsilon_i \sim N(0, \gamma_0^2)$, where γ_0^2 is the white noise variance. Note that the parameter functions $\Phi_j[\Delta^j t_i]$ are functions of $\Delta^j t_i$ as well. These functions will pick up the effect from the previous observations and noises. Possible parametrization for the lag response functions, the decay of Φ-functions are:
–Constant functions (in $\Delta^j t_i$, or in time)

$$\Phi_j[\Delta^j t_i] = \phi_{a_j}. \tag{2}$$

This special case is the usual AR models for equally spaced time series.
–Exponential functions

$$\Phi_j[\Delta^j t_i] = \phi_{a_j} + \phi_{b_j} e^{-\Delta^j t_i}. \tag{3}$$

–Reciprocal functions

$$\Phi_j[\Delta^j t_i] = \phi_{a_j} + \phi_{b_j}/\Delta^j t_i. \tag{4}$$

For different ϕ_a and ϕ_b parameters, we will have different decay functions (in absolute value). The stationarity condition for the irregularly spaced model in (1) depends on the parameter space of ϕ's as well as the distribution of Δt_i. This may be obtained by taking the expectation of $\Phi_j[\Delta^j t_i]$ with respect to Δt_i and assuming Y_{t_i} are observed at regularly spaced intervals. For example, assume
(S1) Δt_i are independent and identically distributed as Gamma random variables with parameters α and β,
(S2) Δt_i and ϵ_i are independent.
From the exponential function we have

$$\phi_j = E(\Phi_j[\Delta^j t_i]) = \phi_{a_j} + \phi_{b_j}(\beta + 1)^{-\alpha j}, \quad \alpha > 1. \tag{5}$$

For the process to be stationary, the roots of $\phi(z) = 0$ must lie outside the unit circle.

Consider the linear equation $Y = X\Phi + \epsilon$ where the parameter vector is $\Phi = (\phi_{a_1}, \phi_{b_1}, \cdots, \phi_{a_p}, \phi_{b_p})'$, and the dependent variable vector is $Y = (Y_1, \ldots, Y_n)'$. The $n \times 2p$ regression matrix X is built up by lagged unweighted and weighted dependent variables, where the weights depend on the elapsed duration time. For each lag j the first regressor component is

$$X_{i,j \cdot 2-1} = Y_{t_{i-j}},$$

and the second regressor component is

$$X_{i,j \cdot 2} = Y_{t_{i-j}}/e^{\Delta^j t_i}, \text{(for exponential function)},$$

$$X_{i,j \cdot 2} = Y_{t_{i-j}}/\Delta^j t_i, \text{(for reciprocal function)},$$

for $i = 1, \cdots, n$ and $j = 1, \cdots, p$. Note that the constant model has a $n \times p$ regression matrix with the first regressor component. Thus, an unrestricted estimate of the decay parameters of the exponential or reciprocal model is given by the ordinary least squares estimate of Φ:

$$\hat{\Phi} = (X'X)^{-1}X'Y.$$

3. Illustrative example

We present two examples to illustrate our methodology. The first example is based on a simulated ISAR(1) model and the second example is a high frequency exchange rate of the FX market.

Example 1: The first data set consists of three time series each with length 1000 simulated from ISAR(1) with
(i) constant function ($\phi = 0.3$),
(ii) exponential function ($\phi_a = 0.3, \phi_b = 0.3$),
(iii) reciprocal function ($\phi_a = 0.3, \phi_b = 0.3$),
respectively. Δ_i's are sampled from a Gamma distribution with parameter $\alpha = 2$ and $\beta = 0.5$ (giving the mean ticking frequency of 1) and ϵ_i's are sampled from a Normal distribution with mean 0 and variance 1. The ordinary least squares estimates together with the AIC values from three different functions are shown in Table 1. Our approach is successful in two aspects: First, the OLS estimates are all very close to the values we sampled from ($\phi = 0.3$):
(i) constant function ($\hat{\phi} = 0.28$),
(ii) exponential function ($\hat{\phi}_a = 0.31, \hat{\phi}_b = 0.23$),
(iii) reciprocal function ($\hat{\phi}_a = 0.29, \hat{\phi}_b = 0.30$).

Table 1: Simulated ISAR(1) model: ordinary least squares estimates

True parameters		Constant model*)	Exponential model*)	Reciprocal model*)
Const.	ϕ=0.3	$\hat{\phi}$=0.28(.02)	$\hat{\phi}_a$=0.32(.04)	$\hat{\phi}_a$=0.28(.02)
			$\hat{\phi}_b$=0.10(.08)	$\hat{\phi}_b$=0.00(.01)
	AIC	8471.64	8472.15	8473.64
Expon.	ϕ_a=0.3	$\hat{\phi}$=0.41(.02)	$\hat{\phi}_a$=0.31(.04)	$\hat{\phi}_a$=0.38(.02)
	ϕ_b=0.3		$\hat{\phi}_b$=0.23(.08)	$\hat{\phi}_b$=0.02(.01)
	AIC	8479.27	8472.25	8474.22
Recip.	ϕ_a=0.3	$\hat{\phi}$=0.86(.01)	$\hat{\phi}_a$=0.00(.01)	$\hat{\phi}_a$=0.29(.01)
	ϕ_b=0.3		$\hat{\phi}_b$=1.77(.01)	$\hat{\phi}_b$=0.30(.01)
	AIC	30952.96	16502.87	8469.70

*)Standard errors are in parentheses

Second, we are able to select the correct model based on AIC. More interesting results can be seen from Table 1. The estimated parameter ($\hat{\phi} = 0.41$) modeled by the constant function with data sampled from the exponential function is very closed to 0.42 which is value in (5) with $\alpha = 2$ and $\beta = 0.5$. Similarly, from the reciprocal function we have

$$\phi_1 = E(\Phi_1[\Delta t_i]) = \phi_{a_1} + \frac{\phi_{b_1}}{(\alpha - 1)\beta}, \quad \alpha > 1. \tag{6}$$

This procedure will give 0.9 which is very close to 0.86 from Table 1.

Example 2: The data consists of the FX rate quotes for the DEM-JPY exchange rate distributed by Olsen and Associates (1993). We first introduce the data definitions based on Guillaume et al. (1994). The logarithmic price Z is defined as the log of the geometric mean of the bid and the ask prices, P_{bid} and P_{ask},

$$Z = \log\left(\sqrt{P_{bid} \cdot P_{ask}}\right) = \frac{\log(P_{bid}) + \log(P_{ask})}{2}. \tag{7}$$

The change of price at time t_i, Y_{t_i}, is defined as

$$Y_{t_i} \equiv \frac{Z_{t_i} - Z_{t_{i-1}}}{\Delta t}, \tag{8}$$

where Δt is some fixed time interval. The change of the logarithmic price is often referred to as "FX return".

For the study of the irregularly spaced time series, we adopt the same data definitions in Guillaume et al. except (8). We define the return as duration dependent difference of the logarithmic price Z_{t_i}:

$$Y_{t_i} \equiv \frac{Z_{t_i} - Z_{t_{i-1}}}{t_i - t_{i-1}}. \tag{9}$$

Figure 1: Irregularly spaced time series plot of FX return in hours

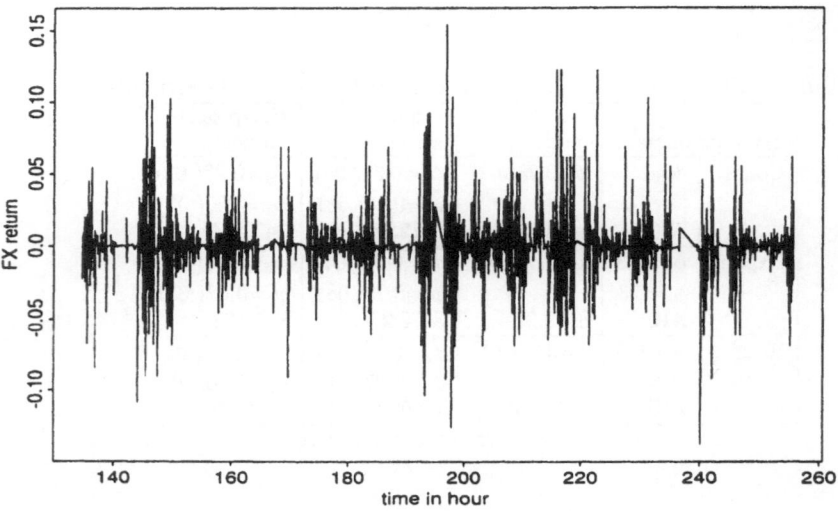

Differencing the time series means taking the difference of adjacent points and divide it by the space between. We define as the difference of an irregularly spaced time series the sequence of slopes between adjacent points. Instead of interpolating data for a fixed time interval Δt, we define Y_{t_i} in a natural way as shown in (9). The difference of the logarithmic series gives the growth rates, therefore the logarithmic price change is simply the average of the growth rates of the bid and the ask time series.

Figure 1 shows the plot of the return in hours for the first week in June, 1993 with sample size $n=3000$. Table 2 shows the ordinary least squares estimators by fitting ISAR(p) models. The negative first order autocorrelation is consistent with the study by Goodhart et al. (1991).

The OLS estimates on Table 2 for the ISAR(p) model shows a clear negative estimate which is significant up to order 2. But the higher order effects do not reduce the residual variance substantially. This result is also confirmed by the ML estimation method. A similar picture can be seen for the ISAR(p) model with exponential or reciprocal decay functions. All parameter estimates are negative where the limiting decay parameter are not significant. The slope decay parameter is significant up to order 2, but the residual variance is again rather constant.

Table 2: Ordinary least squares estimates for the DEM/JPY exchange rate

p	ϕ_1	ϕ_2	ϕ_3	γ_0
	parameter estimates (standard error)			residual variance
CONSTANT function				
1	-.163(.018)			.0195
2	-.170(.018)	-.047(.018)		.0195
3	-.171(.018)	-.049(.019)	-.016(.018)	.0195
EXPONENTIAL function				
1	.964(.314) -1.163(.323)			.0195
2	1.008(.316) -1.217(.325)	.262(.228) -.328(.241)		.0195
3	1.007(.316) -1.217(.325)	.274(.232) -.343(.245)	.048(.189) -.070(.205)	.0195
RECIPROCAL function				
1	.038(.029) -.0022(.0003)			.0193
2	.038(.029) -.0024(.0003)	.036(.030) -.0024(.0007)		.0192
3	.037(.029) -.0024(.0003)	.037(.030) -.0026(.0007)	.021(.032) -.0018(.0013)	.0192

4. Conclusions

This paper has demonstrated how we can estimate irregularly spaced AR models by reciprocal or exponential ISAR(p) models where the attributes "reciprocal" and "exponential" refer to the form of the response function of the AR model over the lag interval. Based on the setup for ISAR(p) models from the previous section, we can include an autoregressive conditional heteroskedasticity component into our irregularly spaced models (see Pai et al. 1995). Also, the modeling process is flexible enough to incorporate an ISAR model for the ticking process, i.e., the irregularly observed time spacing process as well.

References:

Box, G.E.P. and Jenkins, G.M. (1976): Time Series Analysis: Forecasting and Control. Holden Day: San Francisco.

Engle, R.F., (1982): Autoregressive conditional heteroskedasticity with estimates of the variance of U.K. inflation. *Econometrica*, 50, 987-1008.

Goodhart, C.A.E. and Figliuoli, L. (1991): Every minute counts in financial markets. *Journal of International Money and Finance*, 10, 23-52.

Guillaume, D.M., Dacorogna, M.M., Davé, R.R., Müller, U.A., Olsen, R.B. and Pictet O.V. (1994): From the bird's eye to the microscope: A survey of new stylized facts of the intra-daily foreign exchange. Olsen and Associates.

Müller, U.A., Dacorogna, M.M., Olsen, R.B., Pictet, O.V., Schwarz, M., and Morgenegg, C. (1990): Statistical study of foreign exchange rates, empirical evidence of a price change scaling law, and intraday analysis. *Journal of Banking and Finance*, 14, 1189-1208.

Olsen and Associates (1993): Data distribution for HFDF - 1.

Pai, J.S.C., Polasek, W. and Kozumi, H. (1995): Irregularly spaced AR and ARCH models. WWZ-Discussion Paper Nr. 9509, University of Basel.

Two Types of Partial Least Squares Method in Linear Discriminant Analysis

Hyun Bin Kim[1] and Yutaka Tanaka[2]

[1] System Engineering Research Institute,
P.O.BOX 1, Yusung-Gu, Taejon, 305-600, Korea

[2] Department of Environmental and Mathematical Sciences, Okayama University,
Tsushima, Okayama 700, Japan

Summary : Partial least squares linear discriminant function (PLSD) is a new discriminant function proposed by Kim and Tanaka (1995a). PLSD uses the idea of partial least squares (PLS) method, which was originally developed in multiple regression analysis, in discriminant analysis. In this paper, two types of PLSD are investigated and evaluated in a simulation study. In the first type named PLSDA(all), a common pooled within-group covariance matrix of all groups is used in modeling PLSD to discriminate all pairs of groups. In the second type named PLSDT(two), pooled within-group covariance matrices based on the related two groups are used in modeling PLSD to discriminate pairs of groups. As the results of the simulation study PLSDA has the better performance than PLSDT in all situations when the covariance matrices are equal in all groups, while PLSDT is better than PLSDA in well conditioned situations when the covariance matrices are different among the groups.

1. Introduction

Partial least squares regression (PLSR), which was originally developed by Wold (1975) in the field of chemometrics, is a regression method which intends to reduce the effect of multicollinearity by the reduction of dimensionality of the explanatory variables like principal components regression (PCR). The performance of PLSR has been investigated by some authors including Frank and Friedman (1993) and Kim and Tanaka (1994) in their simulation studies. It is known that PLSR has worked well also in many practical problems of chemical fields.

The basic idea of PLS is closely related to the conjugate gradient method for solving linear equations or for calculating inverse matrices in numerical analysis (see, e.g., Wold et al. 1984). Taking this aspect into consideration, we can apply the algorithm of PLS to other statistical methods in which we need to calculate the inverse of the covariance matrix. In discriminant analysis the inverse of the covariance matrix is needed. The direct application of ordinary linear discriminant functions can not be successful for the so-called ill-conditioned or multicollinear data set.

Kim and Tanaka (1995a) proposed a new linear discriminant function using partial least squares method (PLSD) and compared the performance of PLSD with that of the ordinary linear discriminant function (LDF) by applying to two real data sets, i.e., Fisher's iris data and Yoshimura's arc pattern data (see, Yoshimura et al., 1993) and by a Monte Carlo simulation study (Kim and Tanaka (1995a, 1996)). The results of these studies suggest that there are no great differences between the performances of PLSD and LDF in case of no multicollinearity and that the performance of PLSD is remarkably better than that of LDF in case of high degree of multicollinearity or poorly conditioned situations.

In this paper we consider two types of PLSD. In the first type, a common pooled within-group covariance matrix is calculated using the observations in all groups. In the second type, each one of within-group covariance matrices is calculated using the

261

observations in only related two groups. We abbreviate the former PLSDA(all) and the latter PLSDT(two). These two types of PLSD are compared through a simulation study.

In sections 2, 3 and 4 the algorithms of PLSR, the ordinary LDF and the proposed PLSD are briefly reviewed, respectively. In section 5 two types of PLSD are described and compared through a simulation study. Finally, section 6 provides a short discussion on PLSD.

2. Wold's PLSR algorithm

PLSR (Wold 1975) is a method in which a covariance vector w_1 between y and X is calculated at first, and it is used as a coefficient vector to calculate t_1, a linear combination of the columns of X. Simple OLS regression is then conducted to predict each of X and y using t_1 as the explanatory variable, and the residuals X_1 and y_1 from these regressions are computed. Next, another covariance vector w_2 based on these residuals and t_2, a linear combination of X_1, are calculated and simple regressions on t_2 to each of X_1 and y_1 are conducted. The unfitted parts by the regressions on t_1 are then complemented by the regressions on t_2. In the same way, covariance vectors $(w_3, w_4, \cdots)$ and linear combinations $(t_3, t_4, \cdots)$ are calculated and simple regressions are conducted until good fitting is obtained. The algorithm of PLSR proposed by Wold is described as follows :

1 Initialization (Centering) :

$$X_0 \leftarrow (X - 1\bar{x}^t), \quad y_0 \leftarrow (y - \bar{y}1)$$

2 For $k = 1, 2, \cdots$ to K :

    2.1 $w_k = X_{k-1}^t y_{k-1}$

    2.2 $t_k = X_{k-1} w_k$

    2.3 $p_k = X_{k-1}^t t_k / t_k^t t_k$ $(= X^t t_k / t_k^t t_k)$

    2.4 $q_k = y_{k-1}^t t_k / t_k^t t_k$ $(= y^t t_k / t_k^t t_k)$

    2.5 $X_k = X_{k-1} - t_k p_k^t$

    2.6 $y_k = y_{k-1} - t_k q_k$

3 Calculation of regression coefficients :

$$\hat{\beta}_K = W_K (W_K^t X^t X W_K)^{-1} W_K^t X^t y, \qquad W_K = (w_1, \cdots, w_K)$$

4 Prediction equation :

$$\hat{y} = (\bar{y} - \hat{\beta}_K^t \bar{x}) + \hat{\beta}_K^t x,$$

where n and p indicate the numbers of observations and explanatory variables, respectively, X is an $n \times p$ matrix of explanatory variables, y is an $n \times 1$ vector of response variable, 1 is a unit vector with 1 in its all elements, $\bar{x}$ is the mean vector of X, $\bar{y}$ is the mean of y, K ($\leq$ rank of X) is the number of components employed in the model. PLS becomes equivalent to OLS if it uses all possible components. It is important how to determine the number of components K in applying the PLSR.

3. LDF

Suppose that there exist g groups $\pi_1, \pi_2, \cdots, \pi_g$ and that an observation x from group π_i follows a p-variate normal distribution $N(\mu_i, \Sigma)$ with a common covariance matrix

Σ. Also suppose that the costs $c(j \mid i)$ due to misclassifying an observation from π_i to π_j for $i, j = 1, 2, \cdots, g$ are the same for all pairs (i, j). Then the LDF for the i-th group which minimizes the total cost is expressed as

$$\delta_i(x) = \boldsymbol{\mu_i}^t \, \boldsymbol{\Sigma}^{-1} x - \frac{1}{2} \boldsymbol{\mu_i}^t \, \boldsymbol{\Sigma}^{-1} \boldsymbol{\mu_i} + \ln(q_i), \tag{1}$$

where q_i is the prior probability of drawing an observation from group π_i. In sample version $\boldsymbol{\mu_i}$, $\boldsymbol{\Sigma}$ and q_i are replaced by the estimated $\bar{x}_i$, $\widehat{\boldsymbol{\Sigma}}$ and $\hat{q}_i$, respectively. Applying the LDF in sample version to, say, the i-th and j-th groups, we obtain LDF_{ij} for those groups defined by

$$\mathrm{LDF}_{ij}(x) = (\bar{x}_i - \bar{x}_j)^t \, \widehat{\boldsymbol{\Sigma}}^{-1} x - \frac{1}{2}(\bar{x}_i - \bar{x}_j)^t \, \widehat{\boldsymbol{\Sigma}}^{-1}(\bar{x}_i + \bar{x}_j) + \ln(\hat{q}_i) - \ln(\hat{q}_j), \tag{2}$$

with the classification rule as

$$\text{assign } x \text{ to } \pi_i \text{ if } \mathrm{LDF}_{ij}(x) \geq 0 \quad \text{or} \quad \pi_j \text{ if } \mathrm{LDF}_{ij}(x) < 0.$$

4. Classification by PLS method

4.1 Regression coefficient vector of PLSR

The regression coefficient vector of PLSR can be described as follows

$$\hat{\beta}_{PLSR} = \boldsymbol{W}_K (\boldsymbol{W}_K^t \boldsymbol{X}^t \boldsymbol{X} \boldsymbol{W}_K)^{-1} \boldsymbol{W}_K^t \boldsymbol{X}^t \boldsymbol{y}$$

$$= \boldsymbol{W}_K (\boldsymbol{W}_K^t \boldsymbol{X}^t \boldsymbol{X} \boldsymbol{W}_K)^{-1} \boldsymbol{W}_K^t (\boldsymbol{X}^t \boldsymbol{X})(\boldsymbol{X}^t \boldsymbol{X})^{-1} \boldsymbol{X}^t \boldsymbol{y}$$

$$= \boldsymbol{H}_K \hat{\beta}_{OLS},$$

where

$$\boldsymbol{H}_K = \boldsymbol{W}_K (\boldsymbol{W}_K^t \boldsymbol{X}^t \boldsymbol{X} \boldsymbol{W}_K)^{-1} \boldsymbol{W}_K^t \boldsymbol{X}^t \boldsymbol{X}. \tag{3}$$

Let $\mathcal{U}(\boldsymbol{W}_K)$ be the linear subspace spanned by the columns of matrix $\boldsymbol{W}_K$ and the inner product in $\mathcal{U}$ be defined by $(a, b) = a^t \boldsymbol{X}^t \boldsymbol{X} b$. Then $\boldsymbol{H}_K$ is the explicit expression for an orthogonal projector onto the subspace $\mathcal{U}$ with K dimensions (Rao 1973). Consequently the coefficient $\hat{\beta}_{PLSR}$ is obtained by projecting the ordinary regression coefficient vector $\hat{\beta}_{OLS}$ onto the subspace $\mathcal{U}$ to stabilize the estimator by reducing the dimensions.

In the principal components regression (PCR), the regression coefficient vector is stabilized by reducing the dimension of the eigenspace of $\boldsymbol{X}^t \boldsymbol{X}$, which is calculated from only $\boldsymbol{X}$. But in PLSR the regression coefficient vector is stabilized by reducing the dimension of the subspace spanned by successively derived covariance vectors between $\boldsymbol{X}$ and $\boldsymbol{y}$. We are of the opinion that the above difference between PCR and PLSR makes PLSR to have a slightly better performance than PCR. The comparison between PLSR and PCR was reported by Frank and Friedman (1993), Kim and Tanaka (1994) among others.

4.2 PLSD

PLSD was proposed by Kim and Tanaka (1995a) for the purpose of reducing the effects of multicollinearity. The proposed PLSD is obtained by replacing $\widehat{\boldsymbol{\Sigma}}^{-1}$ in LDF_{ij} for the i-th and j-th groups, by $\ddot{\boldsymbol{H}}_{ijK}$ defined as

$$\ddot{\boldsymbol{H}}_{ijK} = \boldsymbol{W}_{ijK} (\boldsymbol{W}_{ijK}^t \widehat{\boldsymbol{\Sigma}} \boldsymbol{W}_{ijK})^{-1} \boldsymbol{W}_{ijK}^t. \tag{4}$$

Namely,

$$\mathrm{PLSD}_{ij}(x) = (\bar{x}_i - \bar{x}_j)^t \, \ddot{H}_{ijK} x - \frac{1}{2}(\bar{x}_i - \bar{x}_j)^t \, \ddot{H}_{ijK}(\bar{x}_i + \bar{x}_j) + \ln(\hat{q}_i) - \ln(\hat{q}_j). \quad (5)$$

Here W_{ijK} consists of K covariance vectors obtained by applying PLSR to the data of the i-th and j-th groups, where a dummy variable with the values $-n_j/(n_i + n_j)$ and $n_i/(n_i + n_j)$ is used as the response variable to indicate which group an observation belongs to. The basic idea behind this is that LDF is mathematically equivalent to the OLS regression of binary responses.

In case of more than three groups, $_gC_2$ PLSD_{ij}s are calculated and an observation x is assigned to the i-th or j-th group based on PLSD_{ij} for all possible pairs. Then, x is classified into the group to which x is most often assigned. The proposed method coped very well with the real data with more than three groups such as Fisher's iris data with three groups (Kim and Tanaka, 1995a) and Yoshimura et. al.'s arc pattern data with twenty groups (Kim and Tanaka, 1996).

The classification obtained by PLSD_{ij} becomes equivalent to that by LDF if all possible components are used in PLSD_{ij}, by the same reason that PLSR becomes OLS regression if full components are employed. This fact suggests that PLSD has at least equal and possibly better performance than LDF if the number of components are properly determined. It is important, therefore, how to choose the number of components K like in case of PLSR.

As a criterion to choose K, the correct discrimination rate (CDR), CDR $= 100 \times \sum_{i=1}^{g} c_i/n_i$, is computed using the cross-validation method, where c_i is the numbers of correctly classified observations in the i-th group. We search for a PLSD model with the maximum value of cross-validated CDR. In the case where two or more PLSD models have the same maximum value of CDR, we choose the PLSD model with the least number of components as a tentative rule.

It is shown that PLSD has the better performance than LDF in Kim and Tanaka (1996), where two methods are compared through a simulation study and a real data.

5. PLSDA and PLSDT

As discussed in section 1, two types of PLSD, named PLSDA(All) and PLSDT(Two), are studied. In PLSDA the pooled within-group covariance matrix of all groups is used for $\widehat{\Sigma}_{all}$ in eq. (5) and in PLSDT the pooled within-group covariance matrix of the related two i-th and j-th groups is used for $\widehat{\Sigma}_{ij}$ in eq. (5).

$$\ddot{H}_{ijK} \text{ of PLSDA } = W_{ijK}(W_{ijK}^t \widehat{\Sigma}_{all} W_{ijK})^{-1} W_{ijK}^t. \quad (6)$$

$$\ddot{H}_{ijK} \text{ of PLDST } = W_{ijK}(W_{ijK}^t \widehat{\Sigma}_{ij} W_{ijK})^{-1} W_{ijK}^t. \quad (7)$$

Naturally, it is expected that PLSDA fits the case where the covariance matrices are equal for all groups and that PLSDT is suitable for the case where groups have unequal covariance matrices.

5.1 Design of simulation study

This section shows a summary of the design of Monte Carlo experiments for evaluating the performance of two types of PLSD. Twenty four different situations were set up in this simulation study. The number of groups was fixed at three and the distances among the centroids of the three groups were fixed with ratios 3:4:5. The situations were classified by the training-sample sizes (balanced data, $n_1 = 20, n_2 = 20, n_3 = 20$; unbalanced data, $n_1 = 10, n_2 = 20, n_3 = 30$), the structure of the (population) covariance matrices (equal covariances, $\Sigma_1 = \Sigma_2 = \Sigma_3$; unequal covariances, $\Sigma_1 \neq$

$\Sigma_2 \neq \Sigma_3$) and the numbers of variables ($p = 5, 15, 25, 35, 45, 55$). Here artificial data of three groups in case of equal covariances were generated so that they had the same (population) condition number κ, defined by $\lambda_{max}/\lambda_{min}$ of eigenvalues of Σ, with the value $\kappa = 900.0$. In case of unequal covariance type the condition numbers were assigned as $\kappa_1 = 900.0$, $\kappa_2 = 400.0$ and $\kappa_3 = 1.0$. For each of 24 (= 2 (sample size) $\times$ 2 (covariance matrix) $\times$ 6 (variable number)) situations, 50 (= N_t) repetitions were performed.

The criterion to evaluate the performance was the average value of CDRs based on 50 repetitions in which CDR was evaluated with the independently generated test-samples of the sizes of 50 times as large as the training-samples.

$$ \text{CDR} = \frac{1}{N_t} \sum_{i=1}^{N_t} \text{CDR}_i. $$

Artificial data with the preassigned degrees of multicollinearity in this experiment were generated by the method proposed by Kim and Tanaka (1995b).

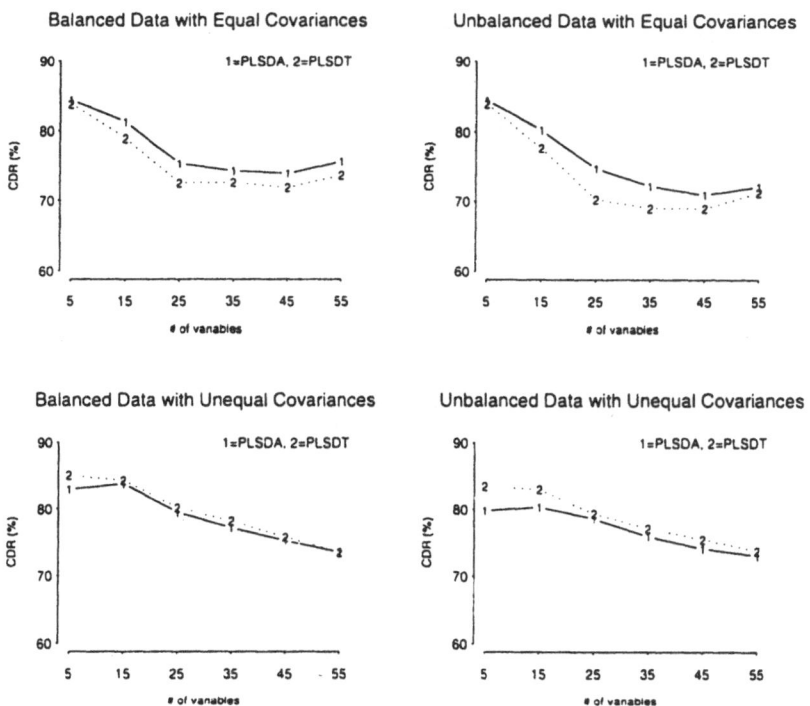

Figure 1: Comparisons between PLSDA and PLSDT

5.2 Comparison between PLSDA and PLSDT

Results of the experiment are given in Figure 1. Note that, since the training-sample size is fixed at 60 for all groups, the degrees of freedom of the sample covariance matrices decrease and, therefore, the degree of ill-conditioning increases as the number of variables increases.

As expected, PLSDA shows remarkably better performance than PLSDT at all situations in case of equal covariances. In case of unequal covariances, there are great differences between CDRs of PLSDA and PLSDT at $p = 5$, 15 (well conditioned situations), but no great differences at $p = 25$, 35, 45, 55 (poorly conditioned situations).

We can explain the results in such a way that the advantage of PLSDT in case of unequal covariances is not large enough to overcome the disadvantage of the smaller degrees of freedom of the covariances compared to the case of PLSDA. That is, the precision of the estimated values of Σ_{ij} is worse in PLSDT than in PLSDA, because PLSDT always has smaller sample size than PLSDA for estmating the covariance matrices. We think there is the relationship between sample size and dimension such that the more the number of dimension is in explanatory data, the better result is able to be obtained if sample size is in well conditioned, but no good if ill-conditioned or multicollinear data set because of the unstable covariance matrix. So, we can give one suggestion that the technique of dimension reduction is useful in order to avoid the problem of singularity or multicollinearity as a result of ill-conditioned or multicollinear data set.

Data	Cov.	5	15	25	35	45	55
Bal.	Equal	2.16/2.11	2.69/3.29	3.42/4.69	3.56/4.00	4.10/5.38	3.75/4.83
Bal.	Uneq.	5.41/5.30	3.11/3.28	4.05/4.34	3.92/3.23	4.49/3.56	3.72/3.28
Unbal.	Equal	1.65/2.58	3.48/3.83	3.82/5.53	4.23/5.77	4.95/4.91	3.96/3.78
Unbal.	Uneq.	5.41/4.79	4.05/3.90	4.01/4.72	3.82/3.85	3.98/3.35	3.64/4.08

TABLE 1. Standard deviation of PLSDA and PLSDT at each situation. Left value in each cell is for PLSDA, right for PLSDT.

6. Discussion

In this paper two types of partial least squares linear discriminant function (PLSD) are investigated through a simulation study. In the first type, which is abbreviated as PLSDA, the pooled within-group covariance matrix of all groups is used as an estimate for the common covariance matrix in the discriminant function, while in the second type, which is abbreviated as PLSDT, the pooled within-group covariance matrix of only the i-th and j-th groups is used for discriminating these two groups. From the results of the simulation study we can conclude :

(1) When the covariance matrices are common in all groups, PLSDA has better performance than PLSDT as is expected.

(2) When the covariance matrices are different, it is expected that PLSDT has better performance than PLSDA. However, we can say so only in well-conditioned situations, not in poorly or ill-conditioned situations.

As discussed by Flury et al. (1994), it is natural to use principal components in discriminant analysis (PCD) to reduce the dimensionality of explanatory variables as in regression analysis. So, we have a plan to compare PLSD with PCD. Moreover, it is our open and future task to compare PLSD with other kinds of discriminant functions such as Friedman's (1989) regularized discriminant function(RDF) using shrinkage estimators (see, e.g. James and Stein 1961; Efron and Morris 1976; Stigler 1990), which intend to reduce the effects of multicollinearity and can be applied in poorly conditioned or ill-conditioned situations (see, e.g, Titterington 1985; O'Sullivan 1986).

References :

Efron, B., and Morris, C. (1976), Multivariate Empirical Bayes and Estimation of Covariance Matrices, The Annals of Statistics, Vol. 4, pp.22-32.

Flury, B., Schmid, M. J. and Narayanan, A. (1994), Error Rates in Quadratic Discrimination with Constraints on the Covariance Matrices, Journal of Classification, Vol. 11, pp.101-120.

Frank, I. E. and Friedman, J. H. (1993), A Statistical View of Some Chemometrics Regression Tools, Technometrics, Vol. 35, No. 2, pp.109-148.

Friedman, J. H. (1989), Regularized Discriminant Analysis, Journal of the American Statistical Association, Vol. 84, pp.165-175.

James, W., and Stein, C. (1961), Estimation with Quadratic Loss, Proceedings of the Fourth Berkeley Symposium on Mathematical Statistics and Probability, Vol. 1, pp.361-379, Berkeley: University of California Press.

Kim, H. B., and Tanaka, Y. (1994), A Numerical Study of Partial Least Squares Regression with an Emphasis on the Comparison with Principal Component Regression, Proceedings of the Eighth Japan and Korea Joint Conference of Statistics, pp.83-88, Okayama, Japan.

Kim, H. B., and Tanaka, Y. (1995a), Linear Discriminant Function Using Partial Least Squares Method, Proceedings of International Conference on Statistical Methods and Statistical Computing for Quality and Productivity Improvement (ICSQP '95), Vol. 2, pp.875-881, Seoul, Korea.

Kim, H. B., and Tanaka, Y. (1995b), Generating Artificial Data with Preassigned Degree of Multicollinearity by Using Singular Value Decomposition, The Journal of Japanese Society of Computational Statistics, Vol. 8., pp.1-8.

Kim, H. B., and Tanaka, Y. (1996), Application of Partial Least Squares Linear Discriminant Function to Writer Identification in Pattern Recognition, Journal of the Faculty of Environmental Science and Technology Okayama University, Vol. 1. pp.65-76.

O'Sullivan, F. (1986), A Statistical Perspective on Ill-Posed Inverse Problems, Statistical Science, Vol. 1, pp.502-527.

Rao, C. R. (1973), Linear Statistical Inference and Its Applications, 2nd Edition, John Wiley & Sons, Inc., New York.

Stigler, S. M. (1990), The 1988 Neyman Memorial Lecture: A Galtonian Perspective on Shrinkage Estimators, Statistical Science, Vol. 5, pp.147-155.

Titterington, D. M. (1985), Common Structure of Smoothing Techniques in Statistics, International Statistical Review, Vol. 53, pp.141-170.

Wold, H. (1975), Soft Modeling by Latent Variables; the Non-linear Iterative Partial Least Squares Approach, In Perspectives in Probability and Statistics, Papers in Honou of M. S. Bartlett, Edited J. Gani, Academic Press, Inc., London.

Wold, S., Wold, H., Dunn, W. J., and Ruhe, A. (1984), The Collinearity Problem in Linear Regression. The Partial Least Squares (PLS) Approach to Generalized Inverse, SIAM Journal on Scientific and Statistical Computing, Vol. 5, pp.735-743.

Yoshimura, M., Yoshimura, I. and Kim, H. B. (1993), A Text-Independent Off-Line Writer Identification Method for Japanese and Korean Sentences, IEICE Trans. Inf. & Syst., Vol. E76-D, No 4, pp.454-461.

Resampling Methods for Error Rate Estimation in Discriminant Analysis

Masayuki Honda [1] and Sadanori Konishi [2]

[1] Division of Medical Informatics
Chiba University Hospital
1-8-1 Inohana, Chuou-ku
Chiba 260. Japan

[2] Graduate School of Mathematics
Kyushu University
6-10-1 Hakozaki, Higashi-ku
Fukuoka 812, Japan

Summary: The performance of resampling methods, like the bootstrap or cross-validation methods, was investigated for estimating the error rates in linear and quadratic discriminant analyses. A Monte Carlo experiment was carried out under the assumption that two population distributions were characterized by a mixture of two multivariate normal distributions. Simulation results indicated that the bootstrap method gave good performance in the case of the linear discriminant function, but it was a little biased when the quadratic discriminant function was used. Cross-validation method was superior in regard to the unbiasedness, and the 0.632 bootstrap estimator outperformed in regard to the mean square error. The methods for error rate estimation were also examined through the analysis of real data in medical diagnosis.

1. Introduction

The main aim in discriminant analysis is to allocate a future observation to one of a finite number of distinct groups or populations on the basis of several characteristics of the observation. It is assumed here that an individual with observation on the p-dimensional random vector is allocated into one of two p-variate populations, and that allocation is carried out on the basis of Fisher's linear discriminant function or the quadratic discriminant function.

In practice it is important to estimate the error rates in allocating a randomly selected future observation. For the problem of estimating the actual error rates (conditional error rates), Efron (1979, 1983) proposed nonparametric bootstrap methods, such as the bootstrap bias-corrected apparent error rate and 'the 0.632 estimator'. Ganeshanandam and Krzanowski (1990) investigated several methods including the 0.632 estimator for error-rate estimation in Fisher's linear discriminant function. Konishi and Honda (1990) examined the parametric and nonparametric methods for estimating the error rates in linear discriminant analysis under normal and nonnormal populations.

Very little work has been done on evaluating error rate estimation procedures both in the linear and the quadratic discriminant functions simultaneously. In this paper we investigate several estimation methods for error rates in Fisher's linear discriminant function and the quadratic discriminant function, when the population distribution is assumed to be a mixture of two multivariate normal distributions. We examine the performance of the estimation methods through Monte Carlo simulations, in which evaluation in nonnormal situations and the quadratic discriminant function are especially respected. Section 2 describes the formation of two-group discriminant problem. Section 3 presents the methods of error rates estimation. Simulation results for evaluating the several estimation methods are summarized in Section 4. Application of the methods to a real data in the medical problems is executed in Section 5.

2. Two-group discriminant analysis

Suppose that the allocation of an individual x into one of two populations Π_i ($i = 1, 2$), i-th population having a p -variate probability distribution $F_i(x)$, is carried out on the basis of Fisher's linear discriminant function (LDF)

$$h(x|X_n) = \{x - \frac{1}{2}(\bar{x}_1 + \bar{x}_2)\}'S^{-1}(\bar{x}_1 - \bar{x}_2) \tag{1}$$

or the quadratic discriminant function (QDF)

$$2q(x|X_n) = \log|S_2| - \log|S_1| + (\bar{x}_2 - x)'S_2^{-1}(\bar{x}_2 - x) - (\bar{x}_1 - x)'S_1^{-1}(\bar{x}_1 - x) \tag{2}$$

where $\bar{x}_1, \bar{x}_2, S_1, S_2$ and S are, respectively, the sample means, the sample covariance matrices and the pooled sample covariance matrix based on the training sample $X_n = \{x_\alpha^{(i)} : \alpha = 1, 2, \cdots, N_i, i = 1, 2\}$.

A future observation x_0 is allocated to Π_1 or Π_2 according as x_0 belongs to the region of discrimination $\hat{R}_1$ or $\hat{R}_2$ given by

$$\hat{R}_1 = \hat{R}_1(X_n) = \{x; h(x|X_n) > 0\} \text{ or } = \{x; q(x|X_n) > 0\}, \tag{3}$$
$$\hat{R}_2 = \hat{R}_2(X_n) = \{x; h(x|X_n) \le 0\} \text{ or } = \{x; q(x|X_n) \le 0\}, \tag{4}$$

under the assumption of equality of *a priori* probabilities and costs of misallocation. Then the actual error rates are given by

$$e_i(F_i; X_n) = \int_{\hat{R}_j} dF_i(x), \quad (i \ne j; \ i, j = 1, 2). \tag{5}$$

By replacing the unknown distribution F_i by the empirical distribution function, $\hat{F}_i$, with equal weight of $1/N_i$ assigned to each point of the training sample $\{x_1^{(i)}, \cdots, x_{N_i}^{(i)}\}$, we obtain the apparent error rate (AP method)

$$\hat{e}_i(\hat{F}_i; X_n) = \int_{\hat{R}_j} d\hat{F}_i(x) = \int I(x|\hat{R}_j)d\hat{F}_i = \frac{1}{N_i}\sum_{\alpha=1}^{N_i} I(x_\alpha^{(i)}|\hat{R}_j), \tag{6}$$

where $I(x|A)$ is the indicator function defined by

$$I(x|A) = \begin{cases} 1 & if \quad x \in A \\ 0 & if \quad x \notin A \end{cases}. \tag{7}$$

Usually the apparent error rate provides an optimistic assessment of the actual error rate, and hence much attention has been given to the development of the estimation procedures.

In the next section we present several methods for estimation of error rate in linear and quadratic discriminant analyses.

3. Methods for error rate estimation

3.1 Bootstrap bias-corrected apparent error rates

The bootstrap methods introduced by Efron (1979) provide an approach to estimate the bias of the apparent error rate numerically.

Suppose that the unknown probability distribution $F_i(x)$ is estimated by the empirical distribution function $\hat{F}_i(x)$ based on the training sample $\{x_\alpha^{(i)} : \alpha = 1, 2, \cdots, N_i\}$ $(i = 1, 2)$. We call the random sample $\{x_\alpha^{(i)*} : \alpha = 1, 2, \cdots, N_i\}$ $(i = 1, 2)$ taken from each $\hat{F}_i$ as the bootstrap sample. The bootstrap analogues of the actual error rate in (5) and the apparent error rate in (6) are, respectively,

$$\hat{e}_i(\hat{F}_i; X_n^*) = \int_{\hat{R}_j^-} d\hat{F}_i(x) = \frac{1}{N_i} \sum_{\alpha=1}^{N_i} I(x_\alpha^{(i)} | \hat{R}_j^*), \tag{8}$$

$$\hat{e}_i(\hat{F}_i^*; X_n^*) = \int_{\hat{R}_j^-} d\hat{F}_i^*(x) = \frac{1}{N_i} \sum_{\alpha=1}^{N_i} I(x_\alpha^{(i)*} | \hat{R}_j^*) \tag{9}$$

where X_n^* denotes the bootstrap sample of size $(N_1 + N_2)$, the discrimination regions $\hat{R}_i^*(i = 1, 2)$ is constructed based on the bootstrap sample and $\hat{F}_i^*$ is the empirical distribution function of the bootstrap sample $\{x_\alpha^{(i)*} : \alpha = 1, 2, \cdots, N_i\}$.

The bias of the apparent error rate is approximated by averaging $\{\hat{e}_i(\hat{F}_i; X_n^*) - \hat{e}_i(\hat{F}_i^*; X_n^*)\}$ over a large number of repeated bootstrap samples, say, $\hat{b}_i$ for $i = 1, 2$. Then we have the bias-corrected apparent error rate $\hat{e}_i(\hat{F}_i; X_n) + \hat{b}_i$ called BS method.

3.2 The 0.632 bootstrap estimator

Efron (1983) proposed the 0.632 bootstrap estimator given by

$$0.368 \, \hat{e}(\hat{F}_1, \hat{F}_2; X_n) + 0.632 \, \hat{e}_b, \tag{10}$$

where $\hat{e}(\hat{F}_1, \hat{F}_2; X_n)$ denotes the total apparent error rate over the discriminant rule $(n = N_1 + N_2)$ and

$$\hat{e}_b = \frac{1}{n_b} \sum_{j=1}^{B} \left\{ \sum_{\alpha=1}^{N_1} \delta_{\alpha j} I\left(x_\alpha^{(1)} | \hat{R}_2(X_n^*(j))\right) + \sum_{\alpha=1}^{N_2} \delta_{\alpha j} I\left(x_\alpha^{(2)} | \hat{R}_1(X_n^*(j))\right) \right\}. \tag{11}$$

Here $X_n^*(j)$ denotes the j-th bootstrap sample of size n for $j = 1, \cdots, B$ and $\delta_{\alpha j}$ equals 1 when $x_\alpha^{(i)}$ does not belong to $X_n^*(j)$, otherwise zero, and $n_b = \sum_{j=1}^{B}(\sum_{\alpha=1}^{N_1} \delta_{\alpha j} + \sum_{\alpha=1}^{N_2} \delta_{\alpha j})$. We call this estimator 632 method.

The 0.632 estimator is considered to be a summation of the apparent error rate and some kind of cross-validation with appropriate weights. Fitzmaurice et al. (1991) investigated the performance of the 0.632 estimator by a Monte Carlo simulation and showed that its performance depended on the true error rate.

3.3 Cross-validation method

In cross-validation or the leaving-one-out method, an individual is removed from the training samples and LDF $h(x|X_n)$ or QDF $q(x|X_n)$ is constructed with the remaining data. Then check whether or not the removed individual is allocated to the correct population. This is done for each individual in turn in the samples. The proportion of incorrect allocations gives the required estimate. We refer to this estimator by CV-method.

3.4 Asymptotic results for linear discriminant analysis in normal samples

Suppose that the population distributions $F_i(x)(i = 1, 2)$ are p-variate normal distribution with a common covariance matrix. McLachlan (1974) gave an asymptotic unbiased estimator of the actual error rate, say, QM method in the form

$$\Phi\left(-\frac{1}{2}D\right) + \left[\frac{p-1}{N_i D} + \frac{D}{32N}\left\{4(4p-1) - D^2\right\}\right]\phi\left(-\frac{1}{2}D\right), \tag{12}$$

where $D^2 = (\overline{x}_1 - \overline{x}_2)'S^{-1}(\overline{x}_1 - \overline{x}_2)$, $N = N_1 + N_2 - 2$ and $\Phi(x)$ and $\phi(x)$ are, respectively, a standard normal distribution function and its density function.

4. Monte Carlo study

Monte Carlo experiments are carried out under the assumption that two population distributions are characterized by a mixture of two multivariate normal distributions. To put the problem into canonical form (see Ashikaga and Chang (1981)), we make a linear transformation so that

$$
\begin{aligned}
\Pi_1 : \quad f_1(x) &= (1-\varepsilon)n_p(\mathbf{0}, I_p) + \varepsilon n_p(\nu, \sigma_1^2 I_p) \\
\Pi_2 : \quad f_2(x) &= (1-\varepsilon)n_p(\mu, I_p) + \varepsilon n_p(\mu + \nu, \sigma_2^2 I_p)
\end{aligned}
\tag{13}
$$

where ε is a rate of contamination ($0 \le \varepsilon \le 1$), and $\mu = (m, 0, \cdots, 0)'$, $\nu = (\nu_1, \nu_2, 0, \cdots, 0)'$, I_p is the identity matrix of order p, and $n_p(\mu, \Sigma)$ denotes a p-variate normal density function with mean vector μ and covariance matrix Σ. Then the population means and covariance matrices are given by

$$
\begin{aligned}
\Pi_1 : \quad \mu_1 &= \varepsilon\nu, \quad \Sigma_1 = \left\{1 + \varepsilon(\sigma_1^2 - 1)\right\}I_p + \varepsilon(1 - \varepsilon)\nu\nu', \\
\Pi_2 : \quad \mu_2 &= \mu + \varepsilon\nu, \quad \Sigma_2 = \left\{1 + \varepsilon(\sigma_2^2 - 1)\right\}I_p + \varepsilon(1 - \varepsilon)\nu\nu'.
\end{aligned}
\tag{14}
$$

When ε equals 0 or 1, population distributions are specified normal and otherwise nonnormal. Covariance matrices are unequal in the case of $\varepsilon \ne 0$ and $\sigma_1^2 \ne \sigma_2^2$. We investigate the performance of several estimation methods not only in normal and/or equal covariance situations, but also in nonnormal and/or unequal covariance situations.

In the Monte Carlo simulation, random samples were generated from a population for different combinations of parameters and sample sizes, and 200 bootstrap replications were taken for each trial. For ν, we took the value $\nu = (1, 1, 0, \cdots, 0)'$. Simulation entries in Table 1, 2, 3 and 4 are estimated by averaging over 200 repeated Monte Carlo trials. As a criterion for evaluating the methods, we took the expected actual error rate, $E[\{e_1(F_1; X_n) + e_2(F_2; X_n)\}/2]$, say TV. Values of TV were calculated by 10,000 generated samples as future observations for each trial. The means and the mean square errors (MSE)($\times 10^5$) are computed and listed in Tables 1 and 2. The dimension (p) in Table 2 is twice of that in Table 1, being fixed sample sizes. Difference between TV and the error rate estimate based on each method and mean square error are respectively focused in Tables 3 and 4, in which two populations are close together ($m = 1$). The degree of deviation from normality is assessed by the multivariate skewness $\beta_{1,p}$ and kurtosis ($\beta_{2,p}$) proposed by Mardia (1970). The values of $\beta_{1,p}$ and $\beta_{2,p}$ given in Table 5 for a mixture of two multivariate normal distributions in (13) were calculated, using the formulae obtained by Konishi and Honda (1990).

Several findings as a summary of the simulation study are given in the followings.

- Under the assumption of multivariate normality (i.e. $\beta_{1,p} = 0, \beta_{2,p} = p(p+2)$) and equal covariance matrices, all methods besides BS method in QDF provide estimates with small biases, when two populations are not close together (see the cases that $\varepsilon = 0$ in Tables 1 and 2).

- Apparent error rates (AP method) clearly underestimate the actual error rates (TV) in the case of normal and nonnormal situations. The difference between TV and AP in QDF is larger than one in LDF.

- Bootstrap method (BS method) gives good performance in LDF, but it is a little biased in QDF. High dimension gives worse influence to BS method when QDF is used, because AP method has more severely underestimated in the case of $p = 8$ than in the case of $p = 4$ (see Tables 1 and 2).

- 632 method performs fairly well with regard to the mean square errors in the case of normal and nonnormal situations. But it slightly underestimates the actual error rate in high dimensional case (Table 2) of nonnormal situations in QDF and the case of closer populations (Table 3).

- Cross Validation method (CV) has a little bias to the actual error rates but has larger mean square errors than 632 method in LDF. When QDF is used, CV performs well with regard to the unbiasedness.

- Parametric estimator (QM method) gives overestimated results in the cases further from normality and equality of covariance. But it totally performs very well in the case of $\varepsilon = 0$.

- With total consideration, CV method is superior to other methods in regard to the unbiasedness, and 632 method is far superior in regard to the mean square error.

Table 1. Comparison of methods for estimation of error rates
$(N_1 = N_2 = 30, p = 4, \sigma_1^2 = 3.0, \sigma_2^2 = 9.0, m = 3)$

ε		LDF						QDF			
		TV	AP	BS	CV	632	QM	TV	AP	BS	CV	632
0.0	Mean	.076	.058	.073	.076	.075	.074	.084	.051	.077	.083	.084
	MSE		135	125	120	95	70		196	127	133	98
0.1	Mean	.107	.086	.107	.107	.107	.107	.121	.076	.112	.123	.119
	MSE		182	170	176	138	183		336	177	222	134
0.3	Mean	.153	.124	.148	.149	.148	.152	.181	.120	.166	.180	.170
	MSE		287	238	258	197	301		544	253	269	192
0.5	Mean	.198	.169	.199	.202	.200	.231	.229	.172	.224	.234	.222
	MSE		317	269	258	216	344		543	275	298	214
0.7	Mean	.234	.197	.231	.236	.231	.255	.243	.177	.233	.246	.234
	MSE		399	313	300	261	287		661	285	303	219
0.9	Mean	.265	.228	.262	.263	.259	.278	.224	.154	.209	.223	.216
	MSE		446	370	371	302	297		780	368	315	265
1.0	Mean	.273	.240	.277	.279	.270	.288	.202	.134	.190	.204	.200
	MSE		360	303	311	245	273		669	264	275	199

Table 2. Comparison of methods for estimation of error rates
$(N_1 = N_2 = 30, p = 8, \sigma_1^2 = 3.0, \sigma_2^2 = 9.0, m = 3)$

ε		LDF						QDF				
		TV	AP	BS	CV	632	QM	TV	AP	BS	CV	632
0.0	Mean	.089	.053	.085	.096	.091	.083	.125	.033	.099	.133	.132
	MSE		218	131	153	107	98		909	176	224	114
0.1	Mean	.121	.074	.113	.121	.117	.120	.165	.049	.125	.168	.159
	MSE		342	173	178	128	152		1445	317	270	134
0.3	Mean	.173	.116	.165	.175	.169	.192	.239	.098	.188	.245	.214
	MSE		530	284	311	224	311		2208	568	496	272
0.5	Mean	.219	.157	.216	.228	.218	.244	.288	.139	.239	.299	.260
	MSE		634	326	316	259	359		2446	557	407	289
0.7	Mean	.255	.180	.243	.256	.246	.270	.284	.135	.238	.299	.267
	MSE		845	382	365	288	357		2417	512	452	261
0.9	Mean	.284	.205	.273	.288	.272	.291	.236	.088	.189	.247	.234
	MSE		922	415	484	329	365		2374	450	362	195
1.0	Mean	.294	.212	.282	.291	.279	.296	.199	.050	.151	.202	.204
	MSE		1002	431	455	328	357		2340	416	375	172

Table 3. Difference between TV and each estimate: (TV - each estimate)
$(N_1 = N_2 = 30, p = 4, \sigma_1^2 = 3.0, \sigma_2^2 = 9.0, m = 1)$

ε	LDF					QDF			
	AP	BS	CV	632	QM	AP	BS	CV	632
0.0	.049	.007	.003	.009	.007	.117	.031	.003	.027
0.1	.049	.002	-.004	.004	-.014	.112	.029	.005	.032
0.3	.064	.014	.007	.015	-.007	.101	.024	.005	.031
0.5	.063	.009	-.003	.013	-.007	.088	.014	.000	.021
0.7	.078	.030	.002	.024	-.004	.088	.016	-.002	.012
0.9	.074	.014	.000	.021	-.006	.089	.015	-.006	.004
1.0	.082	.022	.006	.028	-.005	.093	.021	.001	.009

Table 4. Mean square error
$(N_1 = N_2 = 30, p = 4, \sigma_1^2 = 3.0, \sigma_2^2 = 9.0, m = 1)$

ε	LDF					QDF			
	AP	BS	CV	632	QM	AP	BS	CV	632
0.0	577	416	396	337	297	1560	455	460	342
0.1	595	457	509	364	407	1587	502	486	385
0.3	781	480	484	363	343	1312	401	417	346
0.5	723	434	536	339	415	1013	329	416	290
0.7	1015	571	730	456	550	1011	352	416	279
0.9	891	457	557	327	345	1051	353	408	262
1.0	999	481	603	398	429	1094	347	342	353

Table 5. Multivariate skewness ($\beta_{1,p}$) and kurtosis ($\beta_{2,p}$)

	$p=4$				$p=8$			
	Π_1		Π_2		Π_1		Π_2	
ε	$\beta_{1,p}$	$\beta_{2,p}$	$\beta_{1,p}$	$\beta_{2,p}$	$\beta_{1,p}$	$\beta_{2,p}$	$\beta_{1,p}$	$\beta_{2,p}$
0.0	0.00	24.0	0.00	24.0	0.00	80.0	0.00	80.0
0.1	0.64	31.8	2.82	68.6	1.04	103.0	4.76	225.6
0.3	1.11	32.3	2.12	51.6	1.93	107.3	3.66	172.6
0.5	0.74	29.5	0.95	38.7	1.34	99.3	1.65	130.1
0.7	0.30	26.7	0.30	30.8	0.56	90.4	0.52	103.8
0.9	0.04	24.7	0.03	25.8	0.07	83.0	0.05	86.5
1.0	0.00	24.0	0.00	24.0	0.00	80.0	0.00	80.0

In practical situations, there often occur nonnormal populations and close together of two populations. When ε varies 0.1 through 0.5, populations seem to be nonnormal along the measure of nonnormality given in Table 5. In such cases we would like to recommend 632 and BS method in linear discriminant analysis and CV and 632 method in quadratic discriminant analysis.

Results of our simulation study also indicate that QDF is superior to LDF for two normal populations with unequal covariance matrices provided that the sample sizes are sufficiently large. QDF has poor performance with high-dimension of p relative to the sample sizes. Sample size is a critical factor in choosing between LDF and QDF with normal data, as shown by Marks and Dunn (1974), Wahl and Kronmal (1977).

5. Application to medical data

We applied the methods for error rate estimation to the problem of medical diagnosis. Source data under consideration are written in Andrews and Herzberg (1985). The two groups consist of 44 cases without crystals of calcium oxalate in urine and 33 cases with crystals. The data set consists of six physical characteristics of urine, namely,

x_1 : specific gravity x_2 : pH x_3 : osmolarity
x_4 : conductivity x_5 : urea concentration x_6 : calcium concentration

The estimates of the actual error rate based on each method are

(LDF) AP: 0.189 BS: 0.227 CV: 0.246 632: 0.215 QM: 0.236
(QDF) AP: 0.186 BS: 0.243 CV: 0.288 632: 0.241

in which 200 bootstrap replications were taken for BS and 632 methods.

The total consideration referring to the results concludes that the estimate of the actual error rate is about 22 % in the linear discriminant analysis and about 28 % in the quadratic discriminant analysis. We recommend the simultaneous use of several estimators with caution to data characteristics for estimation of the actual error rates.

Acknowledgment

The authors would like to thank referees for their helpful comments and suggestions.

References

Andrews, D. F. and Herzberg, A. M. (1985): Data: A Collection of problems from many fields for the student and research worker. Springer-Verlag, New York.

Ashikaga, T. and Chang, P. C. (1981): Robustness of Fisher's linear discriminant function under two-component mixed normal models. *J. Amer. Statist. Assoc.* **76**,676-680.

Efron, B. (1979): Bootstrap methods: Another look at the jackknife. *Ann. Statist.* **7**, 1-26.

Efron, B. (1983): Estimating the error rate of a prediction rule : Improvement on cross-validation. *J. Amer. Statist. Assoc.* **78**, 316-331.

Fitzmaurice, G. M., Krzanowski, W. J. and Hand, D. J. (1991): A Monte Carlo study of the 632 bootstrap estimator of error rate. *J. of Classification* **8**, 239-250.

Ganeshanandam, S and Krzanowski, W. J. (1990): Error-rate estimation in two-group discriminant analysis using the linear discriminant function. *J. Statist. Comput. Simul.* **36**, 157-175.

Konishi, S. and Honda, M. (1990): Comparison of procedures for estimation of error rates in discriminant analysis under nonnormal populations. *J. Statist. Comput. Simul.* **36**, 105-115.

Mardia, K. V. (1970): Measures of multivariate skewness and kurtosis with applications. *Biometrika* **57**, 519-530.

Marks, S and Dunn, O. J. (1974): Discriminant functions when covariance matrices are unequal. *J. Amer. Statist. Assoc.* **69**, 555-559.

McLachlan, G. J. (1974): An asymptotic unbiased technique for estimating the error rates in discriminant analysis. *Biometrics* **30**, 239-249.

Wahl, P and Kronmal, R. (1977): Discriminant functions when covariances are unequal and sample sizes are moderate. *Biometrics* **33**, 479-484.

A Short Overview of the Methods for Spatial Data Analysis

Masaharu Tanemura

The Institute of Statistical Mathematics
4-6-7, Minami-Azabu, Minato-ku
Tokyo 106, Japan

Summary: Some methods of spatial data analysis are given in a manner of short overview. Here the recent development of this field is also included. At first, it is shown that spatial indices based on quadrat counts or nearest neighbour distances, which have been devised mostly by ecologists, are still useful for preliminary analysis of spatial data. Then, it is discussed that distance functions such as nearest neighbour distribution and K function are useful to the diagnostic analysis of spatial data. Further, it is shown that the maximum likelihood procedures for estimating and fitting pair interaction potential models are very useful for a wide class of spatial patterns. Finally, it is pointed out that Markov chain Monte Carlo (MCMC) methods are powerful tools for spatial data analysis and for other fields.

1. Spatial Data Analysis ?

Suppose we have a set of data of spatially distributed objects. Let us assume the objects be plants or animals and let us call each object an 'individual'. There might be various types of spatial data. Among such types, we assume a mapped spatial point pattern is given and let the coordinates of N individuals in an area S be $X \equiv (x_1, x_2, \cdots, x_N)$. This is considered as the most informative.

Then our purpose of spatial data analysis would be such that: which type of spatial pattern the observed data is classified; how to model the observed pattern; what is the estimated value of parameters for the model; and how to predict the pattern which will appear in future.

The field of statistics which investigates such problems as above is called 'spatial statistics'. In this short overview, we will summarize a part of its status by putting emphasis on how the recent development of spatial statistics is related to the early studies of spatial data analysis in other fields of science, such as ecology.

Due to the limited space of this article, subjects considered here are mainly confined to those the present author had concerned and some references cited here are omitted from the list of references. For the full list and for more informations, readers can refer, for instance, Hasegawa and Tanemura (1986) and Cressie (1993).

2. Obtaining Spatial Index

It will be quite natural to ask what the type of the observed point pattern is and how can it be represented by a certain quantity.

As a preliminary study of spatial data analysis, it is usual to investigate whether the observed pattern has an indication of departure from complete spatial randomness (or homogeneous Poisson point process). More concretely, we put as a null hypothesis H_0: the homogeneous Poisson point process, and make statistical test against H_0. As alternatives, we consider two types of patterns, i.e., 'regular' and 'clustered'.

In the field study, the spatial data are often sampled according to the following methods:

1. Quadrat method : Counts of individuals in contiguous quadrats are obtained. Let s_i be the counts in the i-th quadrat $(i = 1, 2, \cdots, q)$, where q is the number of quadrats.

2. Nearest neighbour distance method : Usually the two types of nearest neighbour distances are measured; the distance between a randomly sampled point to its nearest individual (r_1); and the distance between a randomly sampled individual to its nearest individual (r_2).

For these types of spatial data, many spatial indices had been considered by mainly ecologists in order to represent the degree of aggregation of individuals.

2.1 Spatial indices for quadrat counts

As regards quadrat counts, we cite here only three indices. David and Moore (1954) presented an index $I = V/m - 1$, where m and V are, respectively, sample mean and variance of s_i's. Morisita (1959) introduced the index $I_\delta = q \sum_{i=1}^{q} s_i(s_i - 1)/(N(N - 1))$. This is often called Morisita's I_δ. Lloyd (1967) presented an index $\overset{*}{m} = m + V/m - 1$. The index $\overset{*}{m}$ is called a mean crowdedness.

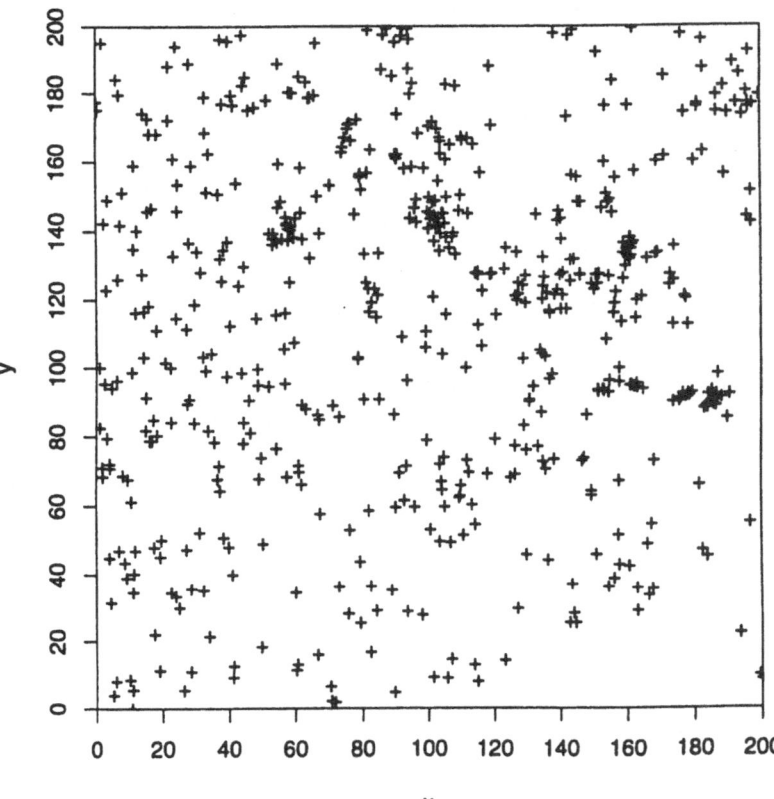

Fig. 1: A pattern of 584 trees of longleaf pines (*Pinus palustris*) taken from Cressie (1993).

Let us show an example of values of these indices. Figure 1 is the stands of longleaf pines (*Pinus palustris*) (Cressie, 1993). For this data, the values of the above indices are, respectively, $I = 1.335$, $I_\delta = 1.930$ and $\overset{\cdot}{m} = 2.765$. These values indicate that the stands of longleaf in Fig. 1 is a clustered pattern.

2.2 Spatial indices for nearest neighbour distances

As regards nearest neighbour distances, we show three indices here, too. Hopkins and Skellam (1954) gave an index $A = \sum r_1^2 / \sum r_2^2$, Clark and Evans (1954) proposed the index $R = 2\sqrt{\lambda} \sum r_2/n$ and Besag and Gleaves (1973) presented the index $T = 2 \sum r_1^2 / \sum r_3^2$. Here, n is the number of samples of r_2, λ the intensity of Poisson point process and r_3 is the so-called T-square samples devised by Besag and Gleaves.

For the data of Fig. 1, the values of indices given above are: $A = 2.049$, $R = 0.849$ and $T = 1.646$. These values again indicate the clustered nature of the longleaf pines data.

3. Nearest Neighbour Distance Distribution

For the data of nearest neighbour distances, we can further consider their distributions in order to get more detailed informations from the data.

For that purpose, let $\{r_{1,i}; \ i = 1, 2, \cdots, n_1\}$ be the nearest neighbour distance data between point and individual, and let $\{r_{2,i}; \ i = 1, 2, \cdots, n_2\}$ be the data between individual and individual. Here, n_1 and n_2 are the number of respective samples.

3.1 Empirical distributions for r_1 and r_2

Then, we obtain the empirical distributions for $\{r_1\}$ and $\{r_2\}$ as:

$$\hat{p}(r) = \sum_{i=1}^{n_1} I(r_{1,i} \leq r)/n_1, \quad \text{and} \quad \hat{q}(r) = \sum_{i=1}^{n_2} I(r_{2,i} \leq r)/n_2,$$

where $I(B)$ is the indicator function, such that $I(B) = 1$, if B is true; $= 0$, if B is false.

It is interesting to note, for the Poisson point process with intensity λ, $p(r) = q(r) = 1 - \exp(-\lambda \pi r^2)$ hold in the limit of $n_1, n_2 \to \infty$.

3.2 Exact empirical distribution for r_1

It is further interesting to point out that an 'exact' empirical distribution $p(r)$ for r_1 is obtained if the mapped point pattern is given. See Okabe and Miki (1981) and Tanemura (1983). Here, the term 'exact' is meant by the distribution in the limit of $n_1 \to \infty$.

This can be seen by considering the following relations:

$$
\begin{aligned}
p(r) &= \Pr\{r_1 \leq r\} \\
&= \Pr\{\text{a random point lies within distance } r \text{ from a certain individual}\} \\
&= \sum_{i=1}^{N} \Pr\{\text{a random point lies within a disc with center } x_i \text{ and with radius } r\} \\
&= |S|^{-1}\{\text{area of union of discs with radius } r \text{ and with center at every individual}\}.
\end{aligned}
$$

This is illustrated in Fig. 2. In Fig. 2, the radius of each shaded disc is r. In order to compute the area of the union of discs, it will be most easy to use Voronoi tessellations as shown in Fig. 2.

It is also important to note that we can get an exact empirical 'density' $f(r) = dp(r)/dr$ for r_1 in the following manner:

$$f(r) \;=\; |S|^{-1}\{\text{peripheral length of union of discs with radius } r \text{ and with center at every individual}\}.$$

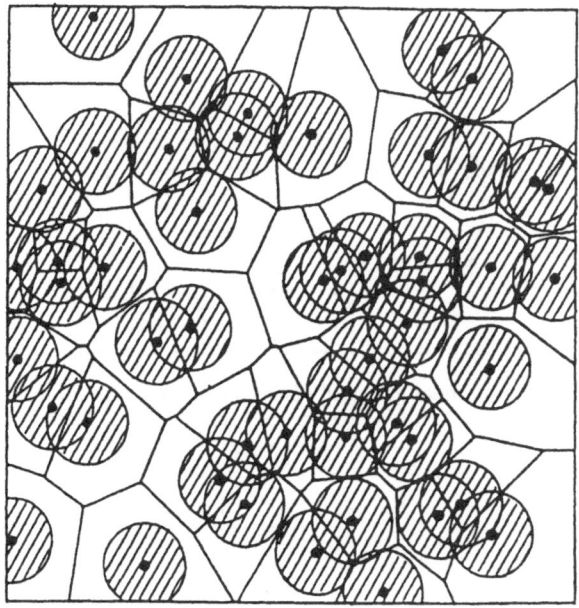

Fig. 2: Illustration for computing exact empirical distribution $p(r)$ and its density $f(r)$.

In Fig. 2, the value of $f(r)$ is obtained by computing the total peripheral length of union discs. This is again not difficult if we consider the Voronoi tessellation as in Fig. 2.

Let us show an example. In Fig. 3, a pattern of nests of gray gulls (*Larus modestus*) is given. In Fig. 4(a), the exact empirical distribution estimated for the gray gull data of Fig. 3 is shown as the curve with crosses. We have done computer simulations of Poisson point process for the same number of individuals in the same size of area. In this figure, the envelopes of $p(r)$'s obtained from 19 simulations are represented as curves. As a comparison, we give in Fig. 4(b) similar curves for the empirical distribution $q(r)$ for the same data sets as in Fig. 4(a). It is obvious that the range of envelopes in Fig. 4(a) is narrower than in Fig. 4(b). This indicates the power of test against Poisson model is bigger for $p(r)$ than for $q(r)$. Actually, we can see from Fig. 4(a) that the curve of $p(r)$ for gray gull data systematically deviates from the envelopes for the Poisson model. This indicates the rejection of the Poisson model. The figures such as Figs. 4(a) and (b) indicates their usefulness as diagnostic analyses.

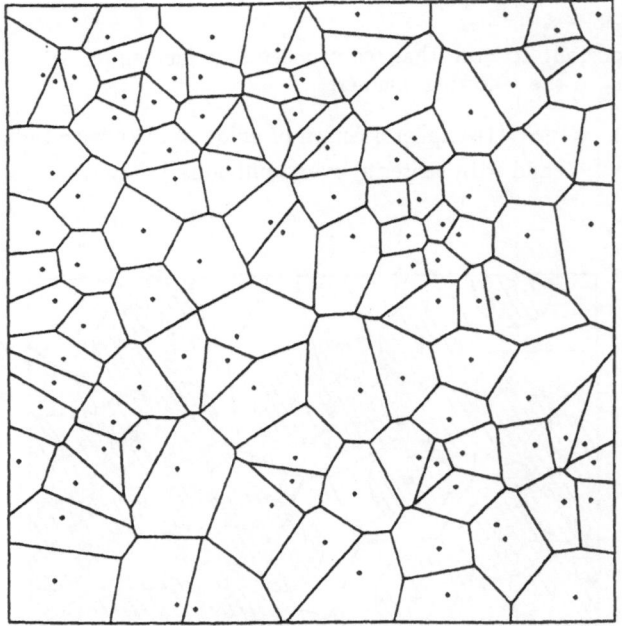

Fig. 3: Pattern of 110 nests of gray gull (*Larus modestus*) with its Voronoi tessellation.

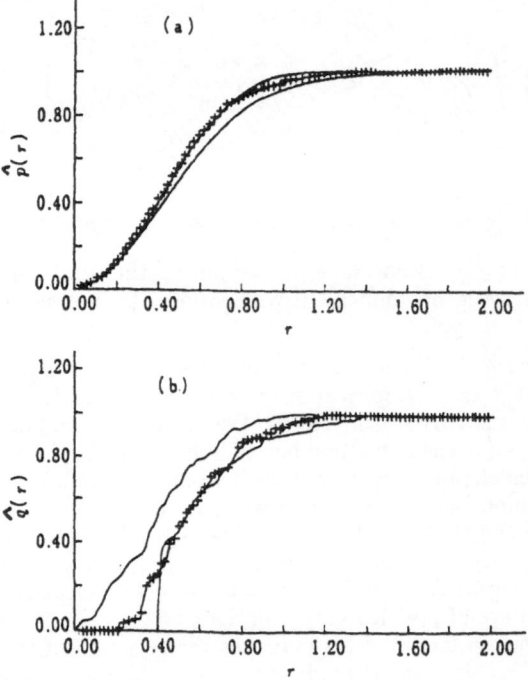

Fig. 4: Empirical distributions of nearest neighbour distance. (a) exact empirical distribution $p(r)$ for r_1. (b) empirical distribution $q(r)$ for r_2.

4. Second Moment Measures

From the above example, we see that the number of nearest neighbour distance data between individual and individual is at most of order N, i.e., the total number of individual. In contrast to this, if we consider the second neighbours, the third neighbours,..., we can get more detailed informations. If all of distances between individuals are considered, the number of data amounts to be of order N^2.

The distribution of distances between all pair of individuals is already known as the 'pair correlation function' in the field of statistical physics, and this function is experimentally observed by X-ray diffractions for crystals or liquids.

In spatial statistics, the distribution is called the 'second moment measure'. It is often called a K function and is defined by:

$$K(r) = \lambda^{-1}E(\text{number of extra individuals within distance } r$$
$$\text{from an arbitrary individual}\},$$

and its empirical distribution is given by

$$\hat{K}(r) = \hat{\lambda}^{-1}\sum_{i<j}^{N} I(|x_i - x_j| \leq r)/N,$$

where λ is the intensity of the observed process.

The derivative of $K(r)$ is sometimes convenient. It is called a 'radial distribution function' and is given by $g(r) = (1/2\pi r)dK(r)/dr$. A so-called L function $L(r) = \sqrt{K(r)/\pi}$ is also useful. For Poisson point processes, these functions have the forms: $K(r) = \pi r^2$, $g(r) = 1$, $L(r) = r$. It is important to note that these functions are also useful for the diagnostic analysis.

5. Spatial Point Process Models and Model Fitting

In order to proceed the analysis of spatial data further, we should take into account models which may include mechanisms of pattern formation. Among such models, we consider here the Gibbs point process model because of its generality.

5.1 Gibbs point process model

Assume the observed mapped point pattern $X = (x_1, x_2, \cdots, x_N)$ is a random sample from a Gibbs canonical distribution which is characterized by a pair interaction potential function $\Phi(|x_i - x_j|)$:

$$f(X) = Z^{-1}\exp\{-U_N(X)\}, \quad \text{where } U_N(X) = \sum\sum_{i<j}^{N}\Phi(|x_i - x_j|),$$

and

$$Z = Z(\Phi; N, S) = \int_{S^N} \exp\{-U_N(X)\}dx_1 dx_2 \cdots dx_N.$$

Here, $U_N(X)$ is called a 'total potential energy' and $Z(\Phi; N, S)$ a partition function or a normalizing constant.

5.2 Maximum likelihood estimation of pair interaction potential function

By parametrizing $\Phi(r)$ as $\{\Phi_\theta(r); \theta \in \Theta\}$ in a certain parameter space Θ, we can get a log-likelihood function from the above Gibbs canonical distribution as:

$$L(\Phi_\theta; X) = -\sum_{i<j}^{N}\Phi_\theta(|x_i - x_j|) - \log \bar{Z}(\Phi_\theta; N, S).$$

where $\bar{Z} = Z/|S|^N$.

For Poisson model, it holds $\Phi_\theta(r) \equiv 0$ and $Z(0; N, S) = |S|^N$.

In order to perform the likelihood procedure, we encounter a serious difficulty of obtaining $L(\Phi_\theta; X)$ and $Z(\Phi_\theta; N, S)$ for a general $\Phi_\theta(r)$ as a function of θ. This is due to the high multiplicity of integration in the normalizing constant Z.

To overcome this difficulty, efforts for devising methods for obtaining approximate log-likelihoods have been done. Some of such methods are the cluster expansion, the virial expansion and the polynomial approximation through computer experiments (Ogata and Tanemura, 1981, 1984, 1989). For the computer experiments of Gibbs point process, the so-called Markov chain Monte Carlo method is most suitable. This is discussed more in the next section.

As the interaction potential models, it would be desirable that they can cover a wide class of patterns of individuals, including regular and clustered patterns which respectively correspond to repulsive and attractive interactions. For that purpose, we are necessary to consider potential models with several parameters, even in different class of potential family. In order to select a suitable model among competing models, the information criterion AIC is useful:

AIC = −2 (maximum Log-likelihood) +2 (number of adjusted parameters).

The model which has minimal AIC value is most suitable. By using this criterion, the Poisson model is always considered as a candidate, since AIC $\equiv$ 0 for this model.

Here, we will show some examples. Let us consider the so-called 'soft-core potential models' $\Phi_{\sigma,n}(r) = (\sigma/r)^n$, $\sigma > 0, n > 2$ (Ogata and Tanemura, 1984, 1989). This family of potential can cover a certain class of point patterns with repulsive interactions. Table 1 shows some of results of fitting the soft-core models (Ogata and Tanemura, 1989).

Data	$\hat{\sigma}$	$\hat{n}$	$\hat{\tau}$
Pines	0.14 meters	∞	0.04
Gulls	2.26 meters	6.9	0.06
Iowa	16.8 miles	6.2	0.37
Balls	1.10 millimeters	5.8	0.42

Table 1: Results of fitting soft-core models to real data.

Here, $\tau = N\sigma^2/A$ represents a crampedness of a pattern. The data for 'Gulls' corresponds to Fig. 3.

6. Markov Chain Monte Carlo Methods and Spatial Data Analysis

Finally, we mention the computational methods which we have already used to simulate Gibbs point processes. These methods are developing recently due to their usefulness to other fields as well as to spatial statistics (see, for example, Geyer, 1992; Besag et al., 1995).

Let $u(X)$ be a certain probability density of $X \in \mathcal{X}$. Then, the expectation of a function $g(X)$:

$$E_u[g(X)] = \sum_{X \in \mathcal{X}} g(X)u(X),$$

is obtained by $(1/t)\sum_{i=1}^{t} g(X_i)$ after generating a Markov chain $(X_1, X_2, \cdots, X_t, \cdots)$ without knowing the normalizing constant of the density u. It is because $X_t \to X \sim u(X)$ and $(1/t)\sum_{i=1}^{t} g(X_i) \to E_u[g(X)]$ as $t \to \infty$, almost surely.

Let $q(x, x')$ be an arbitrary transition probability such that, if $X_t = x$, a sample x' drawn from $q(x, x')$ is considered as a proposed possible value for X_{t+1}. Then, the essence of the Markov chain Monte Carlo (MCMC) method would be to choose the transition probability $p(X, X')$ in such a way:

$$p(X, X') = \begin{cases} q(X, X')\alpha(X, X') & \text{if } X' \neq X \\ 1 - \sum_{X''} q(X, X'')\alpha(X, X'') & \text{if } X' = X, \end{cases}$$

where $\alpha(x, x')$ is the acceptance probability such that we accept x' for X_{t+1} with probability α: otherwise we reject it and set $X_{t+1} = x$.

In the Metropolis-Hastings method,

$$\alpha(X, X') = \begin{cases} \min\{u(X')q(X', X)/u(X)q(X, X'), 1\} & \text{if } u(X)q(X, X') > 0, \\ 1 & \text{if } u(X)q(X, X') = 0 \end{cases}$$

is chosen. Then, $u(X)p(X, X') = u(X')p(X', X)$ holds. This property is called a 'detailed balance'. The property of 'ergodicity' $u(X)p^{(n)}(X, X') \to u(X')$ as $n \to \infty$ also holds. Thus, from the above choice of $\alpha(X, X')$, it is obvious that a Markov chain can be constructed without knowing the normalizing constant as pointed out above. According to mainly to this property, the Markov chain Monte Carlo (abbreviated as MCMC) methods are becoming important tools, besides for spatial data analysis, for image analysis and for other fields, which are somehow related to Bayesian inferences.

References:

Besag, J., Green, P., Higdon, D. and Mengersen, K. (1995): Bayesian computation and stochastic systems, *Statistical Science*, **10**, 3–66.

Cressie, N.A.C. (1993): *Statistics for Spatial Data*. Revised edition, John Wiley & Sons, New York.

Geyer, C.J. (1992): Practical Markov chain Monte Carlo, *Statistical Science*, **7**, 473–483.

Hasegawa, M. and Tanemura, M. (1986): *Ecology of Territories — Statistics of Ecological Models and Spatial Patterns*, Tokai University Press (in Japanese).

Morisita, M. (1959): Measuring of the dispersion and analysis of distribution patterns, *Memoires of the Faculty of Science, Kyushu University, Series E. Biology*, **2**, 215–235.

Ogata, Y. and Tanemura, M. (1981): Estimation of interaction potentials of spatial point patterns through the maximum likelihood procedure, *Annals of the Institute of Statistical Mathematics*, **33B**, 315–338.

Ogata, Y. and Tanemura, M. (1984): Likelihood analysis of spatial point patterns. *Journal of the Royal Statistical Society, Series B*, **46**, 496–518.

Ogata, Y. and Tanemura, M. (1989): Likelihood estimation of soft-core interaction potentials for Gibbsian point patterns, *Annals of the Institute of Statistical Mathematics*. **41**. 583–600.

Okabe, A. and Miki, K. (1981): A statistical method for the analysis of a point distribution in relation to networks and structural points, and its empirical application. *DP-3*, Department of Urban Engineering, University of Tokyo.

Tanemura, M. (1983): Statistics of spatial patterns — Territorial patterns and their formation mechanisms. *Sûri Kagaku* (Mathematical Science), **246**. 25–32 (in Japanese).

Choice of Multiple Representative Signatures for On-line Signature Verification Using a Clustering Procedure

Isao Yoshimura [1], Mitsu Yoshimura[2] and Shin-ichi Matsuda [3]

[1] Faculty of Engineering, Science University of Tokyo
1–3 Kagurazaka, Shinjuku-ku, Tokyo 162, Japan

[2] Chubu University
Matsumoto-cho, Kasugai-shi, Aichi 487, Japan

[3] Nanzan University
Showa-ku, Naogya-shi Aichi 466, Japan

Summary: The task of signature verification is to judge whether the writer of a signature is truly the declared person or not, by referring to a previously provided signature database. The verification is completed when the observed dissimilarity between the questioned signature and a set of signatures, which were previously written by the declared person and included in the database, is less than a given threshold. This paper proved, through a verification experiment using a signature database supplied by CADIX Co. Ltd., that the use of multiple representatives, which represent respective clusters constructed by a clustering procedure, is effective. According to the experiment, the resulting error rate decreased from about 6% to about 1% on average by increasing the number of representatives from one to three.

1. Introduction

This paper deals with an automatic signature verification for the identification of an individual based on on-line information, where the problem is to verify an input signature to be a realization of the declared autograph by comparing it with a set of authentic signatures previously provided.

From a statistical viewpoint, this problem can be formulated as one of judging whether a questioned sample belongs to the declared population of authentic signatures based on a set of samples from this population. This will be referred to as the reference sample in the following. When we can assume a suitable distribution of the population, the problem is reduced to one of estimating the population parameters based on the reference sample and of judging whether the questioned sample is located near the central part of the estimated distribution.

In real situations, however, the population is not so homogeneous that we can assume a simple distribution. It seems better for us to use signatures themselves to represent the population without assuming any distribution. Then there occurs a problem of how to construct the representative signature from the reference sample.

In our experience, people often write their signatures in two or three different forms, (examples of which are shown in Fig. 1) although most of them are similar, as are shown in Fig. 2. From this observation we considered an idea that we can construct a signature verification system with a good performance by using two or three signatures as the representatives. Fortunately, the first two authors of this paper have had some experience of signature verification; they devised a signature verification system and, at the same time, provided a database of on-line signatures. They had the idea to cluster the reference sample in two or three categories and to choose one representative from each cluster. They subsequently tried to realize this idea in a signature verification system.

The purpose of this paper is to explain the devised system and to show its effectiveness through an experiment.

Sincerely yours,

Blaire Mossman

(Mrs.) Blaire Mossman
Managing Editor

Sincerely yours,

Blaire V. Mossman

(Mrs.) Blaire V. Mossman
Managing Editor

Figure 1   An example of two different forms of signatures written by the same person

Figure 2   Examples of a set of on-line signatures written by the same person
Sampling points are traced by straight lines along the time.

2. Devised system

Our system gets any signature as a set of time series $\{(x(t), y(t), p(t)); t = 1, 2, \ldots, T\}$ of pen position (x, y) and writing pressure p, where T is the number of sampling points automatically fixed by the sampling time inherent in the input device. We retain in the system, in advance, a set of authentic signatures as a database to be referred to in verification. After a stage of preprocessing, such as the normalization of size and sampling points, the system classifies these reference signatures into a certain number (three in this paper) of clusters by using a complete linkage method (Cf. Yanagisawa and Ohsumi (1979).) and chooses one representative from each cluster, where the distance for clustering is set as the dissimilarity measure explained in the next section.

When the system receives a questioned signature it measures the dissimilarity between the questioned signature and each representative and uses their minimum as the dissimilarity between the questioned signature and the declared autograph. If the dissimilarity is less than a given threshold the questioned signature is judged as genuine; otherwise it is judged as forged. The threshold is defined for each autograph as the mean value of dissimilarity measure among authentic signatures multiplied by an optional coefficient, C, which is specified by the system manager. Note that although there are some optional parts such as the method of normalization or the choice of weights in the system, they are properly fixed in this paper because they do not seriously affect the conclusion induced from the experiment.

3. Dissimilarity measure

Although there are various proposals concerning the dissimilarity measure (See Yoshimura and Yoshimura (1996)), we adopted a measure similar to that of Sato and Kogure (1982), whose characteristic feature is the adoption of a dynamic programming matching method for adjusting the time scale of two signatures. Note that this adjustment is very important to measure the dissimilarity, because anyone often takes irregular rests of pen movement during the course of writings and these meaningless time duration is better to be omitted.

The adjustment in this sense can be realized by determining a matching of two time coordinates, say ξ and η, for the two signatures. The matching thus determined is referred to as the "warping function" in this paper, which is visually illustrated by a thick line on Fig. 3.

Practically, the system first considers a temporal dissimilarity measure $D^{*}(z_A, z_B)$ between two signatures z_A and z_B as

$$
\begin{aligned}
D^{*}(z_A, z_B) = & \sum_{s=1}^{n(A,B)} [w_p \delta(s, s-1) + 1 - \delta(s, s-1)] \\
& \times [w_1\{(x_A(\xi_s) - y_B(\eta_s))^2 + (y_A(\xi_s) - y_B(\eta_s))^2\} \\
& + w_2\{p_A(\xi_s) - p_B(\eta_s)\}^2 \\
& + w_3\{(d_{xA}(\xi_s) - d_{xB}(\eta_s))^2 + (d_{yA}(\xi_s) - d_{yB}(\eta_s))^2\}],
\end{aligned} \tag{1}
$$

where the subscripts A and B correspond to two signatures respectively, (ξ_s, η_s) is a coordinate of a warping function $\tau = \{(\xi_s, \eta_s); s = 1, 2, \ldots, n(A, B)\}$ which represents an adjustment in the time domain, $n(A, B)$ is the number of sampling points, w_1, w_2, and w_3 are properly chosen constants, δ is an indicator constant related to the s-th

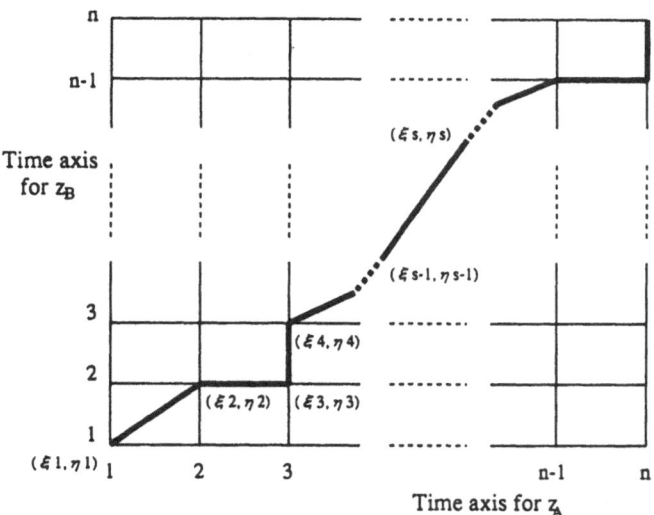

Warping function

Figure 3   Illustration of the warping function τ and indicator constant δ.

τ is a path of coordinates (ξ_s, η_s) on grids running from the lower left $(1,1)$ to the upper right (T,T). $\delta(s, s-1) = 1$ if the span from $(s-1)$ th grid to s-th grid is horizontal or vertical and $= 0$ otherwise. In the above figure, $(\xi_1, \eta_1) = (1,1), (\xi_2, \eta_2) = (2,2), (\xi_3, \eta_3) = (3,2), (\xi_4, \eta_4) = (3,3), \dots; \delta(2,1) = 0, \delta(3,2) = 1, \dots$

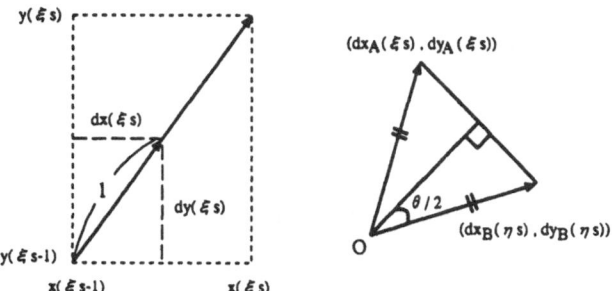

Figure 4   Illustration of the direction vector and related variables

$(d_x(\xi_s), d_y(\xi_s))$ is the unit vector with the same direction as $(x(\xi_s) - x(\xi_{s-1}), y(\xi_s) - y(\xi_{s-1}))$. When the angle of two unit vectors $(d_{xA}(\xi_s), d_{yA}(\xi_s))$ and $(d_{xB}(\eta_s), d_{yB}(\eta_s))$ is θ then $((d_{xA}(\xi_s) - (d_{xB}(\eta_s))^2 + ((d_{yA}(\xi_s) - (d_{yB}(\eta_s))^2 = 4 \times \sin^2(\theta/2)$ and therefore, the last term of the equation (1) reflects the direction of the pen-movement.

point on the warping function, and $(d_x(\xi_s), d_y(\xi_s))$ is the unit vector with the same direction as $(x(\xi_s) - x(\xi_{s-1}), y(\xi_s) - y(\xi_{s-1}))$ (See Figs. 3 and 4), i.e., as follows:

$$d_{xA}(\xi_s) = \frac{x_A(\xi_s) - x_A(\xi_{s-1})}{\sqrt{(x_A(\xi_s) - x_A(\xi_{s-1}))^2 + (y_A(\xi_s) - y_A(\xi_{s-1}))^2}} \tag{2}$$

$$d_{yA}(\xi_s) = \frac{y_A(\xi_s) - y_A(\xi_{s-1})}{\sqrt{(x_A(\xi_s) - x_A(\xi_{s-1}))^2 + (y_A(\xi_s) - y_A(\xi_{s-1}))^2}} \tag{3}$$

Next it calculates, using a DP matching algorithm, the dissimilarity measure $D(z_A, z_B)$ as

$$D(z_A, z_B) = min D^*(z_A, z_B), \tag{4}$$

where the minimum is taken over the possible change of the warping function τ under a certain restriction, the detail of which is explained in Yoshimura et al. (1991) and Yoshimura and Yoshimura (1992).

In the proposed method, although the dissimilarity of questioned signature to the representative signatures are measured, there are some five optional parameters $\delta, w_p, w_1, w_2,$ and w_3 in the definition of this dissimilarity measure. Among them, δ and w_p are introduced to impose a penalty for warping the time scale. As our proposal, δ was set as one when $\xi_s = \xi_{s-1}$ or $\eta_s = \eta_{s-1}$, and as zero otherwise, additionally w_p was set as two. These setting implies that the additional increment in a coordinate at the s-th step is not the same as the one in the other coordinate, then the dissimilarity value is increased depending on the degree of distortion of the time scale. Other constants represent the weights on the three variables of $x, y,$ and p with respect to the difference between the two signatures. The optimum values of them are to be determined adaptively to the signature data. In the experiment below, they were set as $(w_1, w_2, w_3, w_p) = (0.5, 0.4, 0.1, 2.0)$ based on a preliminary experiment.

4. Experiment

An experiment to examine the effectiveness of our proposal was performed based on a database of on-line signatures provided by CADIX Co. Ltd. The database is composed of 2203 signatures, including both authentic and forged signatures, for 28 autographs (See Table 1). These contain 10 in Roman letters by non-Japanese, 14 in Japanese letters by Japanese and 4 in Roman letters by Japanese; "Roman letters" implies English alphabet such as a, b etc. Examples of signatures in the database are shown in Fig. 5. In the experiment, ten authentic signatures for each autograph and five forged signatures were used as references and the remaining signatures were used to evaluate the performance of our system.

In the experiment, the number of clusters were set in three cases of one to three in order to evaluate the effectiveness of the increase of representative signatures.

The most difficult obstacle to achieving a good performance of the system was the determination of optional constants inherent in the system, such as the weights in the dissimilarity measure and the threshold coefficient C. We utilized the reference sample to optimize these constants. It was important that forged signatures were included in the reference because, by comparing the dissimilarities within authentic signatures with those between authentic and forged, we could reasonably evaluate the suitability of various values of C.

The system is composed of two parts: a preprocessing part and verification part. In the preprocessing part various normalizations and standardizations such as the reduction of sampling points, are possible which may affect the achieved error rates. In this paper, however, we fixed the normalization and standardization in one form because we could make sure, through preliminary experiments, that the effect of such preprocessing did not influence the relative capability of the three cases of the number of representative signatures.

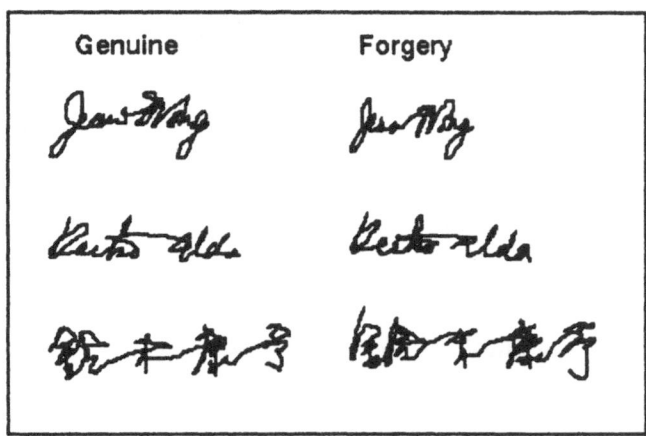

Figure 5   Examples of signatures in CADIX database

Table 1 Number of signatures in the CADIX database

Case RN: Roman sig. by non-Japanese

No.	Gn.	Forg. (repetitions)
1	30	54 (2,4,5,5,5,5,5,5,6,12)
2	25	64 (4,5,5,5,5,5,5,6,6,6,12)
3	51	53 (5,5,5,5,5,5,5,5,6,7)
4	25	51 (5,5,5,5,5,5,5,5,5,6)
5	25	51 (5,5,5,5,5,5,5,5,5,6)
6	33	56 (5,5,5,5,5,5,5,5,5,6)
7	30	65 (5,5,5,5,5,5,5,5,5,5,5,5,5)
8	28	60 (5,5,5,5,5,5,5,5,5,5,5)
9	25	64 (4,4,5,5,5,5,5,5,5,5,5,6)
10	35	40 (5,5,10,10,10)

Case RJ: Roman sig. by Japanese

No.	Gn.	Forg. (repetitions)
1	27	62 (1,2,3,4,5,5,5,5,5,5,5,6,6)
2	42	62 (5,5,5,5,5,5,5,5,5,5,6,6)
3	31	55 (4,5,5,5,5,5,5,5,5,6)
4	25	64 (5,5,5,5,5,5,5,5,5,5,7,7)

Case JJ: Japanese sig. by Japanese

No.	Gn.	Forg. (repetitions)
1	27	57 (4,5,5,5,5,6,6,6,6,9)
2	25	53 (5,5,5,5,5,5,5,5,8)
3	25	16 (2,6,8)
4	20	52 (3,4,5,5,5,5,5,5,5,5,5)
5	24	53 (5,5,5,5,5,5,5,5,6,7)
6	25	21 (2,7,12)
7	25	9 (9)
8	28	53 (5,5,5,5,5,5,56,6,6)
9	26	67 (4,5,5,5,5,5,5,5,5,6,7)
10	26	56 (4,5,5,5,5,5,5,5,6,11)
11	28	53 (5,5,5,5,5,5,5,6,6,6)
12	26	50 (5,5,5,5,5,5,5,5,5,5)
13	34	50 (5,5,5,5,5,5,5,5,5,5)
14	26	15 (2,3,,5,5)

5. Result

The effectiveness of the use of multiple representatives is clearly shown in Table 2, which is a part of the results of the experiment with w_1, w_2, w_3 and w_p fixed as 0.5, 0.4, 0.1, and 2.0, respectively. The error rates in Table 2 are an average of those for foreigners and Japanese. For any choice of the threshold coefficient C, at least two representatives are necessary to get good performance of the system in the verification, while more than three do not yield any improvement.

Table 2 Error rates in %

Coef.	# of representatives		
C	1	2	3
1.4	5.22	1.51	1.46
1.5	6.28	1.27	1.11
1.6	8.07	1.83	1.21
1.7	10.31	2.89	1.73

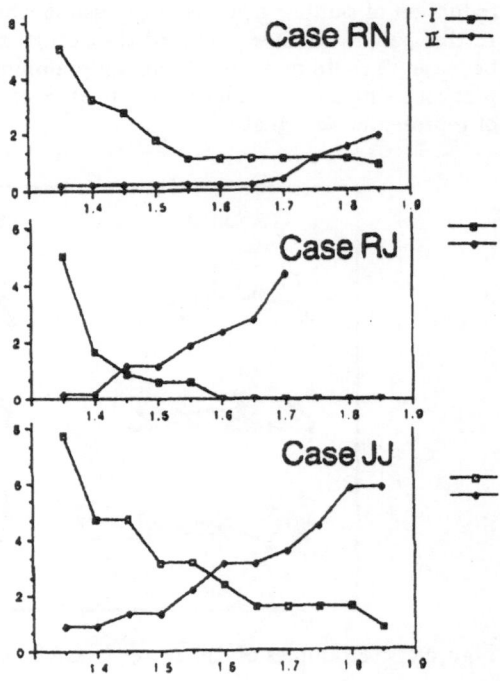

Figure 6   Type I and II errors in % against the threshold coefficient C

RN: Roman signatures by non-Japanese.
RJ: Roman signatures by Japanese.
JJ: Japanese signatures by Japanese.

6. Discussions

The use of multiple representatives of authentic signatures based on a clustering procedure was proved, through an experiment using a signature database, to be effective in decreasing error rates in on-line signature verification. The authors think that it is due to the existence of clusters even in a set of authentic signatures written by one person. It also implies that various type of signatures should be included in the database for signature verification.

The achieved error rates are an average of two types of errors and each error varies depending on the threshold or threshold coefficient C in our setting of the threshold. If it moves to greater values, type I errors decrease whereas type II errors increase. The trade-off relation is obvious by looking at Fig.6. The determination of the reasonable value of C is the responsibility of the system manager, while C = 1.5 is a generally good coefficient in our experience.

An example of clustering and of representative signatures obtained from each cluster is shown in Fig. 7. While the result of clustering is not visually clear average error rates decreased, which may imply that the automatic verification is better than visual verification. Although the experiment was limited to on-line verification, similar conclusions will be obtained for off-line verification, too.

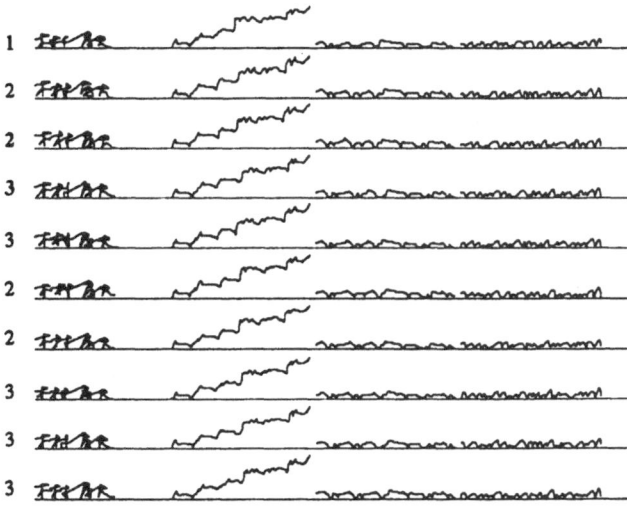

Figure 7   Example of realized clusters and representatives with time series of (x, y, p)

The number placed at the head of signature implies the cluster it belongs to when three clusters are constructed. The bold figure denotes the representative signature. The three traced curves to time in abscissa imply the time series of $x(t), y(t), p(t)$ in this order from left to right.

References:

Sato, Y. and Kogure, K. (1982): On-line signature verification based on shape, motion, and writing pressure, *Proc. 6th Int. Conf. Pattern Recog.*, 823–826.

Yanagisawa, Y. and Ohsumi, N. (1979): Evaluation procedure for estimating the number of clusters in hierarchical clustering system, *The Japanese Journal of Applied Statistics*, **8**, 51–71.

Yoshimura, I. et al. (1991): On-line signature verification incorporating he direction of pen movement, *Trans IEICE Japan*, **E74**, 2083–2092.

Yoshimura, I. and Yoshimura, M. (1992): On-line signature verification incorporating he direction of pen movement– An experimental examination of the effectiveness, In: *From Pixels to Features*, Impedove, S. and Simon, J. C. (eds.), 353–361, North Holland, Amsterdam.

Yoshimura, M. and Yoshimura, I. (1996): The-state-of-the-art and issues to be addressed in writer recognition, *Technical Report of IEICE*, **PRMU96-48**, 81-90 (in Japanese).

Part IV

Related Approaches for Classification

- Fuzzy and Probabilistic Modeling Methods

- Spatial Clustering and Neural Networks

- Symbolic and Conceptual Data Analysis

Algorithms for L_1 and L_p Fuzzy c-Means and Their Convergence

Sadaaki Miyamoto[1] and Yudi Agusta[2]

[1] Institute of Information Sciences and Electronics
University of Tsukuba, Ibaraki 305, Japan

[2] Program Research, Development and Documentation Division
Central Bureau of Statitics, Indonesia

Summary: Algorithms for L_1 and L_p based fuzzy c-means are proposed. These algorithms calculate cluster centers in the general alternating algorithm of the fuzzy c-means. The algorithm for the L_1 space is based on a simple linear search on nodes of step functions derived from derivatives of components of the objective function for the fuzzy c-means, whereas the algorithm for the L_p spaces use binary search on the nodes and then the interval to which the cluster center belong. Termination of the algorithms based on different criteria for the convergence is discussed. The algorithm for the L_1 space is proved to be convergent after a finite number of iterations. A numerical example is shown.

1. Introduction

The L_1 and L_p spaces have sometimes been referred to in studies of data analysis such as the regression analysis in the L_1 space (Bloomfield and Steiger, 1983), although these spaces have not extensively been applied to cluster analysis yet. Recently, many researchers have studied fuzzy clustering, which uses membership degrees in the unit interval interpreted as fuzzy classification. The method of fuzzy c-means, abbreviated as FCM, is most well-known among different techniques in fuzzy clustering (Bezdek, 1981).

Fuzzy c-means clustering based on the L_1 or L_p space has recently been considered by Jajuga (1991) and by Bobrowski and Bezdek (1991). These two studies have shown difficulties in solving fuzzy c-means in the L_p spaces. The general fuzzy c-means algorithm is an iterative procedure in which the step of determining grades and that of determining cluster centers are repeated. A unified formula can be used for the determination of grades for different distances, whereas the calculation of cluster centers strongly depends on a selected distance. Cluster centers in the Euclidean space are derived by a simple formula of weighted average, whereas the calculation of cluster centers seems to require much computation for the L_1 and L_p spaces, however.

In this paper we propose efficient algorithms for the FCM in the L_p spaces. The results are divided into those for the L_1 space and for the L_p spaces, which implies that stronger results are obtained for the L_1 case.

In the case of the L_1 space, the fact that each coordinate of a cluster center is the minimizing element of a piecewise affine function is utilized, whereas a binary search is considered for the L_p spaces.

Moreover we prove several theorems of convergence of the algorithms. The proofs of the convergence theorems for the L_1 case use the finiteness of the search region for the cluster centers and the uniqueness of the minimum of strictly convex functions with respect to U.

A numerical example is given to show that the algorithm actually works well on a large set of data with a small computation time.

2. Fuzzy c-means based on L_p spaces

The problem herein is that n objects, each of which is represented by an h-dimensional real vector $x_k = (x_{k1}, ..., x_{kh}) \in \mathbf{R}^h$, $k = 1, ..., n$, should be divided into c fuzzy clusters. Namely, the grade u_{ik}, $1 \leq i \leq c$, $1 \leq k \leq n$, by which the object k belongs to the cluster i should be determined. For each object k, the grade of membership should satisfy the condition of the fuzzy partition:

$$M = \{ (u_{ik}) : \sum_{i=1}^{c} u_{ik} = 1,\ 1 \leq k \leq n;\quad 0 \leq u_{ik} \leq 1,\ 1 \leq i \leq c,\ 1 \leq k \leq n \}.$$

The formulation by Bezdek (1981) is the optimization of the objective function

$$J(U, v) = \sum_{i=1}^{c} \sum_{k=1}^{n} (u_{ik})^m d(x_k, v_i)$$

in which $d(x, v)$ is a measure of dissimilarity between x and v, m is a real parameter such that $m > 1$, v_i is the center of the fuzzy cluster i, and $U = (u_{ik})$ and $v = (v_1, ..., v_c) \in \mathbf{R}^{ch}$.

Here the dissimilarity d is assumed to be the pth power of the L_p $(p \geq 1)$ norm:
$d(x_k, v_i) = \|x_k - v_i\|_p^p = \sum_{j=1}^{h} |x_{kj} - v_{ij}|^p$, where $v_i = (v_{i1}, ..., v_{ih})$. Namely, the objective function is

$$J(U, v) = \sum_{i=1}^{c} \sum_{k=1}^{n} (u_{ik})^m \|x_k - v_i\|_p^p. \tag{1}$$

It is well-known that the direct optimization of J by (U, v): $\min\limits_{U \in M, v \in \mathbf{R}^{ch}} J(U, v)$ is difficult. A two stage iteration algorithm is therefore used.

A General Algorithm of FCM (cf. Bezdek, 1981):

(a) Initialize $U^{(0)}$; Set $s = 0$.

(b) Calculate cluster centers $v^{(s)} = (v_1^{(s)}, ..., v_c^{(s)})$ that minimize $J(U^{(s)}, \cdot)$:

$$J(U^{(s)}, v^{(s)}) = \min_{v \in \mathbf{R}^{ch}} J(U^{(s)}, v).$$

(c) Update U: calculate $U^{(s+1)}$ that minimizes $J(\cdot, v^{(s)})$:

$$J(U^{(s+1)}, v^{(s)}) = \min_{U \in M} J(U, v^{(s)}).$$

(d) Check convergence using a given $\epsilon > 0$: If the convergence criterion is satisfied, stop; otherwise $s = s + 1$ and go to (b).

The convergence criterion in general should be one of the following (I–III).

(I)
$$|J(U^{(s+1)}, v^{(s)}) - J(U^{(s)}, v^{(s-1)})| \leq \epsilon.$$

(II)

$$\|v^{(s+1)} - v^{(s)}\| \leq \epsilon.$$

(III) Using a suitable matrix norm,

$$\|U^{(s+1)} - U^{(s)}\| \leq \epsilon.$$

Remark: The above term of convergence simply implies that the algorithm will eventually terminate; the convergence does not guarantee that the obtained solution is the correct one, nor the solution is the global optimum. In general it is difficult to derive an efficient algorithm that guarantees the true optimal solution.

In general, calculation of $U^{(s+1)}$ does not depend on a particular choice of a norm. It is well-known that u_{ik} is easily derived by using the Lagrange multiplier. Namely, for x_k such that $x_k \neq v_i$, $i = 1, ..., c$, and $m > 1$,

$$u_{ik} = \frac{1}{\sum_{j=1}^{c} \left(\frac{d(x_k, v_i)}{d(x_k, v_j)} \right)^{\frac{1}{m-1}}}. \tag{2}$$

On the other hand, calculation of cluster centers is not simple for the L_p spaces, and therefore the problem to be solved is the minimization with respect to v in the step (b) of the above FCM. The algorithms for the step (b) that we propose here are based on the following ideas.

(i) Each component of a cluster center can independently be calculated from other coordinates, by decomposing the function $J(U, v)$ to be optimized with respect to v into a sum of hc functions $F_{ij}(w) = \sum_{k=1}^{n} (u_{ik})^m |x_{kj} - w|^p$, $i = 1, .., c$, $j = 1, .., h$:

$$J(U, v) = \sum_{i=1}^{c} \sum_{j=1}^{h} F_{ij}(v_{ij}),$$

where each F_{ij} depends solely on the jth coordinate of a cluster center. Notice that U is a parameter in this subproblem. Thus, concerning the search of cluster centers, we can limit ourselves to the minimization of $F_{ij}(w)$.

(ii) The function F_{ij} is convex with repect to each coordinate of a cluster center, and in particular, it is a piecewise affine function in the L_1 case.

(iii) Seeing the properties (i) and (ii), we can use one-dimensional search for the minimization of $F_{ij}(w)$. For the L_1 space, however, a more efficient algorithm can be derived using the piecewise affine property: the coordinate is calculated by a linear search on the derivative of the function $F_{ij}(w)$, which is remarkably simple.

3. Algorithms for calculating cluster centers

3.1 Algorithm for the L_1 space

Let us first consider the L_1 case. Since the function $F_{ij}(w)$ is piecewise affine for the L_1 space, the solution

$$\min_{w \in \mathbf{R}} F_{ij}(w) \tag{3}$$

can be limited to the coordinates of the data: $\{x_{1j}, ..., x_{nj}\}$.

Two ideas are used in the following algorithm: ordering of $\{x_{kj}\}$ and derivative of

F_{ij}. We assume that when $\{x_{1j}, ..., x_{nj}\}$ is ordered, first subscripts are changed using a permutation function $q_j(k)$, $k = 1, ..., n$, that is, $x_{q_j(1)j} \leq x_{q_j(2)j} \leq \cdots \leq x_{q_j(n)j}$. Using $\{x_{q_j(k)j}\}$,

$$F_{ij}(w) = \sum_{k=1}^{n} (u_{iq_j(k)})^m |w - x_{q_j(k)j}|.$$

Although $F_{ij}(w)$ is not differentiable on $\mathbf{R}$, we extend the derivative of $F_{ij}(w)$ on $\{x_{q_j(k)j}\}$:

$$dF_{ij}^{-}(w) = \sum_{k=1}^{n} (u_{iq_j(k)})^m sign^{+}(w - x_{q_j(k)j})$$

where

$$sign^{+}(z) = \begin{cases} 1 & (z \geq 0), \\ -1 & (z < 0). \end{cases}$$

Thus, $dF_{ij}^{-}(w)$ is a step function which is right continuous and monotone nondecreasing in view of its convexity and piecewise affine property. Now, it is easy to see that the minimizing element for (3) is one of $x_{q_j(k)j}$ at which $dF_{ij}^{+}(w)$ changes its sign. More precisely, $x_{q_j(t)j}$ is the optimal solution of (3) if and only if $dF_{ij}^{-}(w) < 0$ for $w < x_{q_j(t)j}$ and $dF_{ij}^{+}(w) \geq 0$ for $w \geq x_{q_j(t)j}$.

Let $w = x_{q_j(r)j}$, then

$$dF_{ij}^{+}(x_{q_j(r)j}) = \sum_{k=1}^{r} (u_{iq_j(k)})^m - \sum_{k=r+1}^{n} (u_{iq_j(k)})^m.$$

These observations lead us to the next algorithm.

```
begin
    S := - ∑_{k=1}^{n} (u_{ik})^m;
    r := 0;
    while ( S < 0 ) do begin
        r := r + 1;
        S := S + 2(u_{iq_j(r)})^m
    end;
    output v_{ij} = x_{q_j(r)j} as the j-th coordinate of
    the cluster center v_i
end.
```

It is easy to see that this algorithm correctly calculates one coordinate of the cluster center.

Remark: A similar idea to the above algorithm can be found in the L_1 regression. See Bloomfield and Steiger (1983).

3.2 Algorithm for the L_p spaces

For the L_p $(p > 1)$ spaces, the function $F_{ij}(w)$ is differentiable. For calculating the solution $\bar{w}_{ij}$: $\dfrac{dF_{ij}}{dw}(\bar{w}_{ij}) = 0$, the following algorithm is natural.

(i) Search $x_{q_j(r)j}$ and $x_{q_j(r-1)j}$ such that $\dfrac{dF_{ij}}{dw}(x_{q_j(r)j}) \leq 0$ and $\dfrac{dF_{ij}}{dw}(x_{q_j(r-1)j}) \geq 0$.

(ii) Search the solution $\bar{w}_{ij}$ in the interval $[x_{q_j(r)j}, x_{q_j(r+1)j}]$ using, e.g., binary search.

4. Convergence of the algorithms

Different results are obtained for the foregoing convergence criteria (I–III). First, it should be remarked that during the iterations, the value of the objective function is monotone nonincreasing:

$$J(U^{(s)}, v^{(s)}) \geq J(U^{(s+1)}, v^{(s)}) \geq J(U^{(s+1)}, v^{(s-1)}), \quad s = 0, 1, 2, \ldots, \tag{4}$$

since the FCM algorithm is the alternative optimization with respect to U and v.

Then, an obvious result for the convergence of the above algorithms using (I) can be stated as follows.

Theorem 1. For arbitrary given $\epsilon > 0$ and the convergence criterion (I) used in FCM, the L_1 and L_p algorithms converge.

(Proof) The proof is obvious, seeing that the sequence $a(s) = J(U^{(s)}, v^{(s)})$ is monotone nonincreasing and bounded from below by the obvious bound $a(s) \geq 0$. Hence the basic theorem of convergence of monotone and bounded sequences is applied. (QED)

Theorem 2. For $\epsilon = 0$ and the convergence criterion (I), the L_1 algorithm converges after a finite number of iterations in FCM.

(Proof) Consider the set

$$\Pi = \{y = (y_1, \ldots, y_h) \mid y_j \in \{x_{1j}, \ldots, x_{nj}\}, \quad j = 1, \ldots, h \}.$$

Let Π^c be the Cartesian product of Π: $\Pi^c = \overbrace{\Pi \times \Pi \times \ldots \times \Pi}^{c}$, and $|\Pi^c|$ be the number of elements in Π^c. This set is finite and we can easily see that if $J(U^{(s)}, v^{(s-1)}) > J(U^{(s)}, v^{(s)})$ then $v^{(s-1)}$ and $v^{(s)}$ are different points of Π^c. Thus, within $|\Pi^c| + 1$ times of iterations in FCM, $J(U^{(s)}, v^{(s-1)}) = J(U^{(s)}, v^{(s)})$ occurs and then the algorithm for the L_1 space terminates with $\epsilon = 0$. (QED)

In most cases of iterative calculations, the convergence is checked using the solutions such as (II) or (III) instead of the function (I). Unfortunately, it is difficult to prove a theoretical property for the L_p space algorithm using (II) or (III). In contrast, however, the L_1 algorithm can further be analyzed.

In general $J(U^{(s)}, v^{(s-1)}) = J(U^{(s)}, v^{(s)})$ does not imply $v^{(s-1)} = v^{(s)}$, since the solution for (3) may not be unique. It is, however, easy to modify the L_1 algorithm so that

$$\text{if } J(U^{(s)}, v^{(s-1)}) = J(U^{(s)}, v^{(s)}) \text{ then } v^{(s-1)} = v^{(s)} \tag{5}$$

holds. The modification is obvious:

> Check if $\hat{w} = v_{ij}^{(s-1)}$, the (i, j) component of the previous solution $v^{(s-1)}$, satisfies $dF_{ij}^-(w) = 0$. When $dF_{ij}^+(\hat{w}) = 0$, i.e., the previous solution is still optimal, use the previous solution as $v_{ij}^{(s)} = \hat{w}$, the (i, j) component of the new $v^{(s)}$.

Theorem 3. For $\epsilon = 0$ and the convergence criterion (II), the L_1 algorithm with the above modification converges after a finite number of iterations in FCM.

(Proof) Since we have modified the algorithm so that (5) is satisfied, the conclusion follows from the observation stated in the proof of theorem 2: within $|\Pi^c| + 1$ times of iterations, $J(U^{(s)}, v^{(s-1)}) = J(U^{(s)}, v^{(s)})$ occurs. (QED)

Finally, we should be careful about reduction of the number of clusters. Namely, when more than one clusters have the same cluster centers, these centers actually indicate a unique cluster, and hence the number of clusters should accordingly be reduced. Specifically, if

$$x_k = v_{i_1} = v_{i_2} = ... = v_{i_\gamma}, \qquad \gamma > 1,$$

the clusters $i_1, i_2, ..., i_\gamma$ should be replaced by a unique cluster, say i_1, and the other cluster numbers $i_2, ..., i_\gamma$ will not be used thereafter. After this reduction we can set $u_{i_1 k} = 1$, since $x_k = v_{i_1}$.

Theorem 4. For $\epsilon = 0$ and the convergence criterion (III), the L_1 algorithm with the above two modifications converges after a finite number of iterations in FCM.

(Proof) By the above consideration, the matrix U is uniquely determined, since (1) is a strictly convex function with respect to u_{ik} with an arbitrarily fixed v for $x_k \neq v_i$ for all i. When $x_k = v_i$, the previous modification uniquely determines the corresponding part of U. Thus, if $v^{(s-1)} = v^{(s)}$, we have $U^{(s)} = U^{(s+1)}$. (QED)

5. A numerical example

In this example we assume that $m = 2$, the number of clusters $c = 2$, the dimension $p = 2$, and the number of points $n = 10,000$. A region is considered and data points are scattered over the region. Namely, two squares $ABCD$ and $EFGH$ of unit size with the intersecting square $PEQC$ are considered, as shown in Figure 1. The square $PEQC$ has edge length a ($0 < a < 1$). Data points have been scattered over the area surrounded by $ABQFGHPDA$ using the uniformly distributed random numbers.

The initial value for the grade $u_{1k}^{(0)}$ for each x_k has been generated by the pseudo random numbers uniformly distributed over $[0,1]$; $u_{2k}^{(0)} = 1 - u_{1k}^{(0)}$ to form a fuzzy partition. Ten trials with different initial grades $U^{(0)}$ have been carried out. The criterion (III) has been used for the convergence test in all cases.

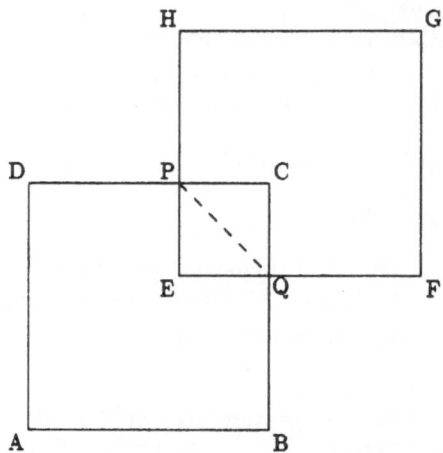

Fig. 1: Two overlapped squares.

Fuzzy clusters have been transformed into crisp clusters using the α - cut of $\alpha = 0.5$. Then, a measure of *misclassification* has been introduced for a quantitative evaluation of the results. Namely, when a data point that is in the left and lower side of the broken line segment PQ in Figure 1 is classified into the same class as the north east cluster, i.e., the one to which data in the area surrounded by $PQFGHP$ belong, the former data point is called *misclassified*. In the same way, when a data point that is in the right and upper side of the broken segment PQ in Figure 1 is classified into the same class as the south west cluster, i.e., the one to which data in the area surrounded by $QPDABQ$ belong, the data point is also called *misclassified*.

Table 1 shows the number of successes, the average number of misclassified data, the maximum number of iterations, and the average CPU time (sec) throughout the ten trials for three values of the parameter a: $a = 0.1, 0.2, 0.3$. Moreover this table compares results by the L_1 c-means, Euclidean c-means, and the L_p c-means with $p = 3$.

The number seven, for example, of successes means that seven trials out of the ten have produced good results, while the other three have led to unacceptable classifications of large numbers of misclassified data. The average number of misclassifications has been calculated from the successful trials: if the seven trials are successful, the data of the other three trials are not used for the calculation. The CPU time is for one cycle of calculating $v^{(s)}$ and $U^{(s+1)}$ in the main loop of FCM. The total CPU time needed until the convergence is, for example, $0.758 \times 11 \cong 8.34$ for $a = 0.1$ by L_1 FCM.

Comparison of the statistics given in Table 1 leads to the following observations.

(a) The computation for one cycle by L_1 FCM is faster than Euclidean FCM, whereas the L_p algorithm is far slower than the other two.

(b) The number of iterations by L_1 FCM is less than that by Euclidean FCM in every case; the L_p algorithm requires more iterations than Euclidean FCM.

(c) For the numbers of misclassifications, we do not find a remarkable difference among these three algorithms.

(d) L_1 FCM has failed 4 times in all 30 trials, while Euclidean and L_p FCM algorithms have succeeded in all trials.

We have analyzed the cases of the failure of the L_1 method, and found that the iteration stopped at $s = 1$, i.e., after $(v^{(1)}, U^{(2)})$ had been calculated, or at $s = 2$. Thus, the failure to produce an appropriate result occurred when the iteration terminated too early. A simple technique for improving the algorithm is to incorporate an empirical rule into the L_1 algorithm, whereby if an early termination is detected, the calculation starts again with renewed initial membership values.

This failure has not been caused by the present algorithm, since the L_1 method exactly calculates the optimal solution for the cluster center, without any approximation. In other words, one cannot theoretically expect a better result by replacing the present L_1 algorithm by any other procedure for calculating the cluster centers. (This does not mean, however, that we are unable to improve the algorithm by using heuristic or *ad hoc* rules.)

6. Conclusions

We have presented two algorithms, i.e., the L_1 algorithm based on a linear search on

Table 1: The number of successes out of ten trials, the average number of misclassifications, and the maximum number of iterations, and CPU time for $a = 0.1$, 0.2, 0.3 by L_1, L_2 (Euclidean), and L_p FCM ($p = 3$).

L_1 FCM				
a	successes	misclassifications	iterations(max)	CPU time(sec)
0.1	7	5.4	11	0.758
0.2	9	10.8	13	0.752
0.3	10	19.8	13	0.755
Euclidean FCM				
a	successes	misclassifications	iterations(max)	CPU time(sec)
0.1	10	3.3	14	0.813
0.2	10	6.0	15	0.812
0.3	10	24.4	14	0.812
L_p FCM ($p = 3$)				
a	successes	misclassifications	iterations(max)	CPU time(sec)
0.1	10	3.90	17	36.465
0.2	10	13.90	18	34.366
0.3	10	25.10	18	35.053

the nodes of step functions, and the L_p algorithm using the binary search. Moreover we have shown theorems of convergence under three stopping criteria. The numerical example has shown that the L_1 algorithm is as efficient as the Euclidean algorithm. The L_p algorithm has required 40 times more processing time than the other algorithms, since it uses an iterative procedure in calculating a cluster center.

For the L_p ($p > 1$) algorithm, improvements are still possible, whereas further improvement cannot be expected for the L_1 case, since the present algorithm is already simple enough and has theoretically good properties as shown in the theorems of convergence.

Further studies on the L_1 and L_p clustering include: (i) to find applications when the L_1 and L_p spaces are more appropriate than the Euclidean space; (ii) analysis of data obtained from real applications using these two algorithms; (iii) development of a system of softwares for these methods.

Acknowledgment:

This research has partly been supported by TARA (Tsukuba Advanced Research Alliance), University of Tsukuba.

References:

Bezdek, J.C. (1981): *Pattern Recognition with Fuzzy Objective Function Algorithms*. Plenum.

Bloomfield, P. and Steiger, W.L. (1983): *Least Absolute Deviations: Theory, Applications, and Algorithms*, Birkhäuser.

Bobrowski, L. and Bezdek, J.C. (1991): c-means clustering with the ℓ_1 and ℓ_∞ norms. *IEEE Transactions on Systems, Man, and Cybern.*, **21**, 3, 545–554.

Jajuga, K. (1991): L_1-norm based fuzzy clustering. *Fuzzy Sets and Systems*, **39**, 43–50.

General Approach to the Construction
of Measures of Fuzziness of Fuzzy K-Partitions

Slavka Bodjanova

Department of Mathematics,
Texas A&M University-Kingsville,
Kingsville, TX 78363, U.S.A.

Summary: One of the most important characterizations of fuzzy partitions is the amount of their fuzziness. This paper proposes an axiomatic framework for measures of fuzziness of nonhierarchical fuzzy partitions. Mathematical conditions for measures of fuzziness are discussed and a way of measuring fuzziness in terms of dissimilarity between a fuzzy partition and its complement is presented.

1. Introduction

Research on the theory of fuzzy sets and on the broad variety of applications of fuzzy sets has been growing steadily since the inception of the theory in the mid sixties by Lotfi Zadeh (1965). In the area of classification, an impressive number of papers has been published on fuzzy clustering algorithms, fuzzy pattern recognition, etc. As a result of fuzzy clustering of a finite set of objects, each object may be assigned to the multiple clusters with some degree of certainty. The amount of fuzziness of the resulting partition is an important characterization of the structure of data. If the fuzziness is low, it means that clusters are reasonably separable. On the other hand, if the fuzziness is large, the fuzzy cluster separability is low and either the partition does not reflect the real structure well or there is not a clear structure present in data.

Bezdek (1981) introduced the partition entropy as a measure of fuzziness of nonhierarchical partitions which is a complete formal analogy to Shannon's entropy and generalization of nonprobabilistic entropy of fuzzy sets introduced by de Luca and Termini (1972). Backer (1987) proposed to measure fuzziness of fuzzy partitions in terms of fuzziness of fuzzy clusters. There are some other characteristics of fuzzy partitions which could be interpreted as measures of fuzziness, but the comprehensive theory of measures of fuzziness of fuzzy partitions is still missing.

The aim of this paper is to develop an axiomatic framework for measures of fuzziness of nonhierarchical fuzzy partitions and to show a more general way of constructing these measures.

In the first part of our paper we briefly review the matrix characterization of k-partitions and the sharpness of fuzzy k-partitions.

De Luca and Termini (1972) formulated three essential requirements that adequately capture the intuitive comprehension of fuzziness of fuzzy set. Generalization of these requirements for fuzzy partitions is the cornerstone of our definition of measure of fuzziness of a fuzzy k-partition introduced in the second part of our paper. Empotz (1981) studied the mathematical background of measures of fuzziness of fuzzy sets. We discuss conditions under which some nonnegative real functions on the interval $[0, 1]$ could be used for constructing measures of fuzziness of fuzzy partitions.

Fuzziness of fuzzy sets is often measured in terms of lack of distinction between the set and its complement, or by a metric distance between its membership grade function and the characteristic function of the nearest crisp set (Klir and Yuan (1995)).

In the last part of our paper we show how this idea can be used in the evaluation of the amount of fuzziness of fuzzy partition.

2. Fuzzy k-partitions

2.1 Matrix characterization

Let $X = \{x_1, x_2, \ldots, x_n\}$ be a given set of objects. Fix the integer k, $2 \leq k < n$ and denote by V_{kn} the usual vector space of real $k \times n$ matrices. Bezdek (1981) proposed the following matrix characterization of partitions of X into k clusters.

Fuzzy k-partition space associated with X:

$$P_{fk} = \{U \in V_{kn}; u_{ij} \in [0,1]; \sum_i u_{ij} = 1 \text{ for all } j; \sum_j u_{ij} > 0 \text{ for all } i\}. \tag{1}$$

Here u_{ij} is the grade of membership of object $x_j \in X$ in fuzzy cluster u_i.

Hard k-partition space associated with X:

$$P_k = \{U \in V_{kn}; u_{ij} \in \{0,1\}; \sum_i u_{ij} = 1 \text{ for all } j; \sum_j u_{ij} > 0 \text{ for all } i\}. \tag{2}$$

If we relax the condition $\sum_j u_{ij} > 0$ we will get degenerate partitions.

Degenerate fuzzy k-partition space associated with X:

$$P_{fko} = \{U \in V_{kn}; u_{ij} \in [0,1]; \sum_i u_{ij} = 1 \text{ for all } j\}. \tag{3}$$

Degenerate hard k-partition space associated with X:

$$P_{ko} = \{U \in V_{kn}; u_{ij} \in \{0,1\}; \sum_i u_{ij} = 1 \text{ for all } j\}. \tag{4}$$

It is obvious that $P_k \subset P_{fk} \subset P_{fko}$ and $P_k \subset P_{ko} \subset P_{fko}$.

2.2 Sharpness

Partitions from P_{ko} are certain, i.e. they have zero amount of uncertainty. On the other hand, the partiton $\bar{U} = [\frac{1}{k}] \in P_{fko}$ is maximally uncertain, maximally fuzzy. If a partition $U \in P_{fko}$ is moving from $\bar{U}$ to a partition $V \in P_{ko}$ its amount of fuzziness decreases, we say that the partition U becomes " sharper". We propose the following definition of sharpness of fuzzy k-partitions.

Definition 1 *Let $U, V \in P_{fko}$. We say that U is sharper than V, denoted by $U \prec V$, if and only if for all i, j:*

$$u_{ij} \leq v_{ij} \qquad \text{for } v_{ij} \leq \frac{1}{k}, \tag{5}$$

$$u_{ij} \geq v_{ij} \qquad \text{for } v_{ij} \geq \frac{1}{k}. \tag{6}$$

Note:

For $k = 2$ we get the relation sharpness defined by De Luca and Termini (1972) for fuzzy sets.

Property 1 *Relation ≺ satisfies the following properties:*

1. P_{fko} *is partially ordered by* ≺.

2. *For all* $U \in P_{fko} : U \prec \bar{U} = [\frac{1}{k}]$.

3. *Let* $U \in P_{fko}$ *such that for all* j *card* $\{u_{ij} : 0 < u_{ij} < \frac{1}{k}\} = 0$. *Let* $V \prec U$. *Then* $V = U$.

4. *Let* $U \in P_{fko} - P_{ko}, U \neq \bar{U}$. *Then there exists* $V \in P_{ko}$ *such that* $V \prec U$, *if and only if for all* j *card* $\{u_{ij} : u_{ij} > \frac{1}{k}\} = 1$.

Definition 2 *Function* $F : P_{fko} \rightarrow R$ *is called a function preserving sharpness of fuzzy partitions if and only if for all* $U, V \in P_{fko}$ *such that* $U \prec V$ *we get* $F(U) \leq F(V)$.

Example 1:
Function $F_1 : P_{fko} \rightarrow R$ defined by

$$F_1(U) = 1 - \sum_i \sum_j u_{ij}^2 / n \qquad (7)$$

is a sharpness preserving function.

Example 2:
Function $F_2 : P_{fko} \rightarrow R$ defined by

$$F_2 = \sum_{j=1}^{n} (\sum_{i=2}^{k} u_{ij} u_{i-1,j}) / u_{mj}, \qquad (8)$$

where $u_{mj} = \min_{i, u_{ij} \geq \frac{1}{k}} \{u_{ij}\}$, is not a sharpness preserving function.

For example:
Let $U, V, W \in P_{f4o}$ be partitions of $X = \{x_1, x_2, x_3, x_4, x_5\}$
given by matrices

$$U = \begin{pmatrix} 0.1 & 1 & 0 & 0 & 0 \\ 0.1 & 0 & 0 & 0 & 0 \\ 0.3 & 0 & 1 & 1 & 0 \\ 0.5 & 0 & 0 & 0 & 1 \end{pmatrix} \quad V = \begin{pmatrix} 0.05 & 1 & 0 & 0 & 0 \\ 0.10 & 0 & 0 & 0 & 0 \\ 0.30 & 0 & 1 & 1 & 0 \\ 0.55 & 0 & 0 & 0 & 1 \end{pmatrix}$$

$$W = \begin{pmatrix} 0.0 & 1 & 0 & 0 & 0 \\ 0.0 & 0 & 0 & 0 & 0 \\ 0.4 & 0 & 1 & 1 & 0 \\ 0.6 & 0 & 0 & 0 & 1 \end{pmatrix}$$

Obviously $W \prec V \prec U$, but $F_2(W) = 0.6 \leq F_2(V) = 0.66$,
and $F_2(V) = 0.66 \geq F_2(U) = 0.63$.

3. Fuzziness of fuzzy k-partitions

3.1 Measures of fuzziness

Several measures of uncertainty of fuzzy partitions have been used as objective functions in fuzzy clustering algorithms or as measures of the quality of a fuzzy classification. The most frequent is the partition entropy introduced by Bezdek (1981) as follows:

$$H(U,k) = -\frac{1}{n} \sum_i \sum_j u_{ij} \log_a u_{ij}, \tag{9}$$

where $a \in (1,\infty)$ and $u_{ij} \log_a u_{ij} = 0$ for $u_{ij} = 0$.

Let us denote by $F(X)$ the set of all fuzzy sets defined on X. De Luca and Termini (1972) raised an interesting question of assigning to any fuzzy set $u \in F(X)$ some measure of its fuzziness which would express the degree to which the boundary of u is not sharp. They proposed as a measure of fuzziness a function $\xi : F(X) \to R^+$ which satisfies the following properties:

1. $\xi(u) = 0$ if and only if $u(x) \in \{0,1\}$ for all $x \in X$,

2. $\xi(u) = \max_{v \in F(X)} \xi(v)$ if and only if $u(x) = \frac{1}{2}$ for all $x \in X$,

3. if $u, v \in F(X)$ such that $u(x) \le v(x) \le \frac{1}{2}$ and $u(x) \ge v(x) \ge \frac{1}{2}$ for all $x \in X$, then $\xi(u) \le \xi(v)$.

We propose the following definition of a measure of fuzziness of fuzzy k- partitions:

Definition 3 *Function $\phi : P_{fko} \to R^+$ satisfying the following properties*

1. *ϕ is a sharpness preserving function, i.e. if $U \prec V$ then $\phi(U) \le \phi(V)$,*

2. *$\phi(U) = 0$ if and only if $U \in P_{ko}$,*

3. *$\phi(U) = \max_{V \in P_{fko}} \phi(V)$ if and only if $U = \bar{U}$,*

is a measure of fuzziness of partitions from P_{fko}.

We will say that ϕ is a normalized measure of fuzziness if $\phi(\bar{U}) = 1$.

Note:

Partition entropy where $a = \frac{1}{k}$ is a normalized measure of fuzziness of $U \in P_{fko}$.

Example 3:

The following functions satisfy the properties of Definition 3:

$$\phi(U) = 1 - \frac{1}{n} \sum_i \sum_j u_{ij}^2 \tag{10}$$

$$\phi(U) = 1 - (1 - \max_j(\min_i u_{ij})) \tag{11}$$

$$\phi(U) = 1 - \frac{k}{2n(k-1)} \sum_i \sum_j |u_{ij} - \frac{1}{k}| \tag{12}$$

3.2 Fuzziness measured by real functions

Measures of fuzziness (9), (10), (12) have the following general form:

$$\phi(U) = \alpha \sum_i \sum_j f(u_{ij}) + \beta, \tag{13}$$

where f is a real function on $[0,1]$.

Two questions arise:

1. What are the properties of function f defined on $[0,1]$ which guarantee that function $\phi : P_{fko} \to R$ defined by

$$\phi(U) = \alpha \sum_i \sum_j f(u_{ij}) + \beta, \tag{14}$$

is a measure of fuzziness of fuzzy k-partitions ?

2. How to find constants α and β in order to obtain a normalized measure of fuzziness of fuzzy partitions ?

We answer these questions in the following remarks:

Property 2 *Let f be a real continuous function defined on $[0,1]$ and let*

$$A = \inf_{0 \leq x < y \leq \frac{1}{k}} \frac{f(y)-f(x)}{y-x} \geq \sup_{\frac{1}{k} \leq r < s \leq 1} \frac{f(s)-f(r)}{s-r} = B.$$

Then for $U, V \in P_{fko}$ such that $U \prec V$

$$\sum_i \sum_j f(u_{ij}) \leq \sum_i \sum_j f(v_{ij}). \tag{15}$$

Property 3 *Let f be a real continuous function defined on $[0,1]$ and let*

$$C = \sup_{0 \leq x < y \leq \frac{1}{k}} \frac{f(y)-f(x)}{y-x} \leq \inf_{\frac{1}{k} \leq r < s \leq 1} \frac{f(s)-f(r)}{s-r} = D \ .$$

Then for $U, V \in P_{fko}$ such that $U \prec V$

$$\sum_i \sum_j f(u_{ij}) \geq \sum_i \sum_j f(v_{ij}). \tag{16}$$

As consequence of Property 3 we have:

1. Let $f : [0,1] \to R$ be any nonconstant continuous function increasing on $[0, \frac{1}{k}]$ and decreasing on $[\frac{1}{k}, 1]$. Then there exist constants α, β such that

$$\phi(U) = \alpha \sum_i \sum_j f(u_{ij}) + \beta \tag{17}$$

is a function preserving sharpness of fuzzy k-partitions.

2. Let $f : [0,1] \to R$ be any convex or any concave nonlinear function. Then there exist constants α, β such that

$$\phi(U) = \alpha \sum_i \sum_j f(u_{ij}) + \beta \tag{18}$$

is a function preserving sharpness of fuzzy k-partitions.

Property 4 *Let* $U \in P_{fko} - P_{ko}, U \neq \bar{U}$. *Let* $f : [0,1] \to R$ *be a continuous function satisfying one of the two next conditions i), ii) and the condition iii):*

i) $A = \inf_{0 \le x < y \le \frac{1}{k}} \frac{f(y) - f(x)}{y - x} \ge \sup_{\frac{1}{k} \le r < s \le 1} \frac{f(s) - f(r)}{s - r} = B,$

ii) $C = \sup_{0 \le x < y \le \frac{1}{k}} \frac{f(y) - f(x)}{y - x} \le \inf_{\frac{1}{k} \le r < s \le 1} \frac{f(s) - f(r)}{s - r} = D,$

iii) $(A - B)^2 + (C - D)^2 \neq 0.$

Then

$$\phi(U) = \alpha \sum_i \sum_j f(u_{ij}) + \beta \tag{19}$$

where

$$\alpha = \frac{1}{n[k.f(\frac{1}{k}) - (k-1).f(0) - f(1)]} \tag{20}$$

and

$$\beta = [-f(1) - (k-1).f(0)]\alpha.n \tag{21}$$

is a normalized measure of fuzziness of fuzzy k-partitions.

Example 5:
Examples of functions which could be used in (19):

$$f(t) = t^2 \text{ for } t \in [0,1]$$

$$f(t) = \{ \begin{array}{ll} t & \text{if } t \in [0, \frac{1}{k}] \\ \frac{1}{1-k}(t-1) & \text{if } t \in (\frac{1}{k}, 1] \end{array}$$

$$f(t) = \{ \begin{array}{ll} 0 & \text{if } t = 0 \\ -t \log t & \text{if } t \in (0,1] \end{array}$$

$$f(t) = 1 - |t - \frac{1}{k}| \text{ for } t \in [0,1]$$

For $k = 4$ we can use

$$f(t) = \{ \begin{array}{ll} 5t + 2 & \text{if } t \in [0, 0.1] \\ 2.5 & \text{if } t \in (0.1, 0.2] \\ -10t + 4.5 & \text{if } t \in (0.2, \frac{1}{k}] \\ 10t - 0.5 & \text{if } t \in (\frac{1}{k}, 1] \end{array}$$

4. Dissimilarity and fuzziness of fuzzy partitions

4.1 Measures of dissimilarity

Definition 4 *Function $D : P_{fko} \times P_{fko} \to R^+$ is a measure of dissimilarity between two partitions from P_{fko} if it satisfies the following properties:*

$$D(U,V) = 0 \text{ iff } U = V, \tag{22}$$

$$D(U,V) = D(V,U). \tag{23}$$

Definition 5 *A dissimilarity measure D is called a dissimilarity measure preserving sharpness of fuzzy partitions if for $U, V \in P_{fko}$ such that $U \prec V$ we get $D(U, \bar{U}) \geq D(V, \bar{U})$.*

Property 5 *Let $d : [0,1] \times [0,1] \to R$ be a distance function. Let $U \in P_{fko}$. Then $D(U,V) = \sum_i \sum_j d(u_{ij}, v_{ij})$ is a sharpness preserving dissimilarity function which can be used in construction of measures of fuzziness of fuzzy k-partitions.*

Example 6:

Let U, V be two fuzzy k-partitions of $X = \{x_1, ..., x_n\}$. Let $l^p(u(x_j), v(x_j))$ be the distance function of Minkowski class defined by

$$l^p(u(x_j), v(x_j)) = \left(\sum_i |u_{ij} - v_{ij}|^p \right)^{\frac{1}{p}}, \ 1 \leq p \leq \infty \tag{24}$$

Then

$$D_p(U,V) = \sum_j l^p(u(x_j), v(x_j)) \tag{25}$$

is measure of dissimilarity preserving sharpness of fuzzy partitions.

Example 7:

Measure

$$D(U,V) = \sum_j \left(1 - \frac{\sum_i u_{ij} v_{ij}}{(\sum_i u_{ij}^2 \sum_i v_{ij}^2)^{\frac{1}{2}}} \right) \tag{26}$$

is a dissimilarity measure preserving sharpness.

Example 8:

Measure

$$D(U,V) = \sum_{j=1}^n \frac{\sum_{i=2}^k \|u_{ij} - u_{i-1,j}| - |v_{ij} - v_{i-1,j}\|}{u_{mj} v_{mj}} \tag{27}$$

where $u_{mj} = \min_{i, u_{ij} \geq \frac{1}{k}} \{u_{ij}\}$, and $v_{mj} = \min_{i, v_{ij} \geq \frac{1}{k}} \{v_{ij}\}$, is a dissimilarity measure which does not preserve sharpness of fuzzy partitions.

For example:

Let $U, V, W \in P_{f4o}$ be partitions given in Example 2. Obviously $W \prec V \prec U$, but $D(W, \bar{U}) = 18 \leq D(V, \bar{U}) = 18.67$, and $D(V, \bar{U}) = 18.67 \geq D(U, \bar{U}) = 12.53$.

4.2 Dissimilarity between a fuzzy partition and its complement

One way of measuring the amount of fuzziness of a fuzzy partition is to view the fuzziness in terms of lack of distinction between the partition and its complement. The less the partition differs from its complement the fuzzier it is. We use the definition of complement introduced by Bodjanova (1994).

Definition 6 *Let* $U \in P_{fko}$. *The complement of* U *is a fuzzy partition* $\sim U \in P_{fko}$ *defined by*

$$\sim u_{ij} = \frac{u_{ij} - \frac{\lambda_j}{k}}{1 - \lambda_j} \tag{28}$$

where

$$\lambda_j = \begin{cases} \frac{k(\max_i u_{ij} - \min_i u_{ij})}{1 - k \min_i u_{ij}} & \text{if } \min_i u_{ij} < \frac{1}{k} \\ 0 & \text{if } \min_i u_{ij} = \frac{1}{k}. \end{cases} \tag{29}$$

Note:

Let $U \in P_{fko}$, where $k = 2$. Then $\sim u_{ij} = 1 - u_{ij}$ for all i, j, which is Zadeh's definition of complementation of fuzzy sets.

Property 6 *Let* $U, V \in P_{fko}$. *Then*

1. $\sim (\sim U) = U$,

2. *if* $V \prec U$ *then* $\sim U \prec \sim V$,

3. $U =\sim U$ *iff* $U = \bar{U}$, *therefore* $\bar{U}$ *is the unique equilibrium of* P_{fko}.

Example 9:

Let us consider the fuzzy partition $V \in P_{f3o}$ given by the matrix

$$V = \begin{pmatrix} 0.9 & 0.3 & 0 & 0.4 \\ 0.1 & 0.2 & 1 & 0.2 \\ 0.0 & 0.5 & 0 & 0.4 \end{pmatrix}$$

Complement of V:

$$\sim V = \begin{pmatrix} 0.00 & 0.36 & 0.50 & 0.20 \\ 0.47 & 0.44 & 0.00 & 0.60 \\ 0.53 & 0.20 & 0.50 & 0.20 \end{pmatrix}$$

Property 7 *Let* $U \in P_{fko}$ *and* $D : P_{fko} \times P_{fko} \to R$ *be a sharpness preserving dissimilarity measure such that if* $V_1, V_2 \in P_{ko}$, *and* $W \in P_{fko} - P_{ko}$ *then*

$$D(V_1, \sim V_1) = D(V_2, \sim V_2) > D(W, \sim W). \tag{30}$$

Then there exist constants $\alpha \neq 0$ *and* $\beta \in R$ *such that*

$$\phi(U) = \alpha D(U, \sim U) + \beta \tag{31}$$

is a measure of fuzziness of fuzzy partitions from P_{fko}.

It is obvious, that the fuzzier a partition $U \in P_{fko}$, the more similar to the equilibrium of P_{fko}. Therefore, another way of measuring the amount of fuzziness of a fuzzy partition is to evaluate its dissimilarity with $\bar{U}$.

Property 8 *Let $U \in P_{fko}$ and $D : P_{fko} \times P_{fko} \rightarrow R$ be a sharpness preserving dissimilarity measure such that if $V_1, V_2 \in P_{ko}$, and $W \in P_{fko} - P_{ko}$ then*

$$D(V_1, \bar{U}) = D(V_2, \bar{U}) > D(W, \bar{U}). \tag{32}$$

Then there exist constants $\alpha \neq 0$ and $\beta \in R$ such that

$$\phi(U) = \alpha D(U, \bar{U}) + \beta \tag{33}$$

is a measure of fuzziness of fuzzy partitions from P_{fko}.

5. Conclusion

We have proposed a definition of a measure of fuzziness of fuzzy partitions. Our definition is a generalization of the measure of fuzziness of fuzzy sets. We identified classes of real functions which could be used for constructing measures of fuzziness of fuzzy partitions. We also explained how the fuzziness of fuzzy partitions can be evaluated by the dissimilarity between a fuzzy partition and its complement. Since fuzziness is one of the most important characterizations of fuzzy partitions, more theoretical work needs to be conducted in this area.

References:

Backer, E. (1987): *Cluster analysis by optimal decomposition of induced fuzzy sets* , Delftse Universitaire Pres, Delft.

Bezdek, J.C. (1981): *Pattern recognition with fuzzy objective function algorithms*, Plenum Press, New York.

Bodjanova, S. (1994): Complement of fuzzy k-partitions, *Fuzzy Sets and Systems* , **62**, 175–184.

De Luca, A. and Termini, S. (1972) : A definition of a nonprobabilistic entropy in the setting of fuzzy sets theory, *Information and Control*, **20**, 4, 301–312.

Empotz, H. (1981) : Nonprobabilistic entropies and indetermination measures in the setting of fuzzy set theory, *Fuzzy Sets and Systems*, **5**, 307–317.

Klir, J.G. and Youan, B. (1995): *Fuzzy sets and fuzzy logic: theory and applicaitons*, Prentice Hall, Englewood Cliffs.

Zadeh, L.A. (1965): Fuzzy sets, *Information and Control*, **8**, 3, 338–353.

Additive Clustering Model and Its Generalization

Mika Sato[1], Yoshiharu Sato[2]

[1] Institute of Policy and Planning Science, University of Tsukuba
Tenodai 1-1-1, Tsukuba 305, Japan
e-mail : mika@shako.sk.tsukuba.ac.jp

[2] Department of Information and Management Science, Hokkaido University
Kita 13, Nishi 8, Kita-ku, Sapporo 060, Japan
e-mail : ysato@huie.hokudai.ac.jp

Summary: ADCLUS (ADditive CLUStering) is known as a clustering model which is designated for the purpose of finding the structure of the similarity data. The aim of this paper is to generalize this model from several points of view. The first point of view is to extend the degree of belongingness of the objects to the continuous value in the interval [0,1], namely to an additive fuzzy clustering model, because the combinatorial optimization is inevitable in the algorithm for ADCLUS. The second point of view is to generalize the model for an asymmetric similarity data. And the third point of view is that we introduce the aggregation operator in the model to represent the degree of simultaneous belongingness of the objects to each cluster.

1. Introduction

The concept of ADCLUS (ADditive CLUStering model) was proposed by Shepard and Arabie (1979). This model is intended to find the structure of a similarity relation between the pair of objects by clusters. The model is defined by the following:

$$s_{ij} = \sum_{k=1}^{K} w_k p_{ik} p_{jk} + \varepsilon_{ij}, \tag{1}$$

where s_{ij} $(0 \leq s_{ij} \leq 1;\ i,j = 1,2,\cdots,n)$ is the observed similarity between objects i and j, K is the number of clusters, and w_k is a weight representing the salience of the property corresponding to cluster k. If object i has the property of cluster k, then $p_{ik} = 1$, otherwise it is 0. In this model, the similarity is represented by the sum of weights of clusters $k_1, k_2, \cdots, k_m$, to which both objects i and j belong. That is,

$$\hat{s}_{ij} = w_{k_1} + w_{k_2} + \cdots + w_{k_m}.$$

This shows that if objects i and j belong to cluster k_ℓ, then the degree of contribution to the similarities is w_{k_ℓ}. Moreover, if the pair of objects shares some common properties, the grades which the pair of objects contributes to the similarities $w_{k_1}, \cdots, w_{k_m}$ are additive.

In fuzzy clustering, a fuzzy cluster is defined to be a fuzzy subset on a set of objects and the fuzzy grade of each object represents the degree of belongingness. The degree of belongingness of object i to cluster k is denoted by u_{ik}. $(0 \leq u_{ik} \leq 1)$. To avoid conditions in which objects do not belong to any clusters, we assume that

$$u_{ik} \geq 0, \ \sum_{k=1}^{K} u_{ik} = 1. \tag{2}$$

A pioneering work for applying the concept of fuzzy sets to a cluster analysis was made by E. Ruspini (1969). Since the fuzzy c-means clustering algorithm was proposed by J.C.

Bezdek (1987) and J.C. Dunn (1973), several methods of fuzzy clustering have rapidly developed and many applications have been suggested (Dave, *et al.* (1992), Hall, *et al.* (1992)).

In order to construct a fuzzy clustering model, we have to define a function $\rho(u_{ik}, u_{jk})$ which represents the grade of belongingness of objects i and j to cluster k. Generally, $\rho(x, y)$ is a function from $[0, 1] \times [0, 1]$ to $[0, 1]$. By using this function ρ, the clustering model (1) is extended as follows (M. Sato and Y. Sato (1994a, 1994b)):

$$s_{ij} = \sum_{k=1}^{K} \rho(u_{ik}, u_{jk}) + \varepsilon_{ij}. \tag{3}$$

Namely, the similarity s_{ij} is represented by the addition of the functions $\rho(u_{ik}, u_{jk})$, where ε_{ij} is an error. T-norms (K. Menger (1942), M. Mizumoto (1989)) are well known as a class of concrete functions $\rho(x, y)$.

In the practical applications, the similarity data s_{ij} is not always symmetric — for instance, the mobility data, the input-output data, the perceptual confusion data and so on. Recently, clustering techniques based on such asymmetric similarity have generated tremendous interest among a number of researchers.

In conventional clustering methods for asymmetric similarity, A.D. Gordon (1987) proposed a method using only the symmetric part of the data. This method is based on the idea in which the asymmetry of the given similarity data can be regarded as errors of the symmetric similarity data, that is,

$$\tilde{S} = \frac{1}{2}(S + S'),$$

where S is a similarity matrix and S' is the transposed matrix of S. As for the data, $\tilde{S}$ was used.

L. Hubert (1973) proposed a method to select a maximum element in the corresponding elements, that is,

$$\bar{s}_{ij} = \max(s_{ij}, s_{ji}), \quad \tilde{s}_{ij} = \min(s_{ij}, s_{ji}).$$

R.E. Tarjan (1983) proposed an algorithm to hierarchically decompose a directed graph with weighted edges which is used for clustering of asymmetric similarity.

We have proposed a model under which the cluster is constructed with similar objects, but the similarity between clusters is not symmetric. (see M. Sato and Y. Sato (1995a, 1995b, 1995c).

In the above conventional clustering algorithm, the concept of similarity between a pair of objects is eventually reduced to symmetry. Therefore, we will propose a new concept of asymmetric aggregation operators in order to represent the asymmetric relationship between a pair of objects. Introducing these asymmetric aggregation operators into the above fuzzy clustering model (3), a new model is proposed in order to obtain clusters in which objects are not only similar to each other but also asymmetrically related. The validity of this model is shown by the numerical example and some features of the aggregation operators.

2. Generalized Clustering Model

Suppose that K fuzzy clusters exist on a set of n objects, that is, the partition matrix $U = (u_{ik})$ is given. u_{ik} is a fuzzy grade (the degree of belongingness) in which object i belongs to cluster k. This condition is assumed:

$$u_{ik} \geq 0, \quad \sum_{k=1}^{K} u_{ik} = 1. \tag{4}$$

Let $\rho(u_{ik}, u_{jl})$ be a function which denotes the degree of simultaneous belongingness of the pair of objects i and j to clusters k and l, namely, a degree of sharing common properties. Then a general model for the similarity s_{ij} is defined as follows:

$$s_{ij} = \gamma(\rho_{ij}) + \varepsilon_{ij}, \tag{5}$$

$$\rho_{ij} = (\rho(u_{i1}, u_{j1}), \cdots, \rho(u_{i1}, u_{jK}), \cdots, \rho(u_{iK}, u_{j1}), \cdots, \rho(u_{iK}, u_{jK})).$$

To state simply, we assume that if all of $\rho(u_{ik} u_{jl})$ are multiplied by α, then the similarity is also multiplied by α. Therefore, the function γ itself must satisfy the condition "positively homogeneous of degree 1 in the ρ", that is,

$$\alpha\gamma(\rho_{ij}) = \gamma(\alpha\rho_{ij}), \qquad \alpha > 0.$$

We consider the following function as a typical function of γ:

$$s_{ij} = \sum_{k=1}^{K} \rho(u_{ik}, u_{jk}) + \varepsilon_{ij}.$$

The ρ is assumed to satisfy the following conditions:

1. $0 \le \rho(u_{ik}, u_{jl}) \le 1$, $\rho(u_{ik}, 0) = 0$, $\rho(u_{ik}, 1) = u_{ik}$
2. $\rho(u_{ik}, u_{jl}) \le \rho(u_{sk}, u_{tl})$ whenever $u_{ik} \le u_{sk}, u_{jl} \le u_{tl}$
3. $\rho(u_{ik}, u_{jl}) = \rho(u_{jl}, u_{ik})$

We can consider the T-norms as a function which satisfies the above conditions. (Weber, 1983).

The structure of an observed similarity is usually unknown and complicated, so various clustering models are required to identify the latent structure of the similarity data. Therefore, we define the general class of clustering models so as to represent many different structures of the similarity data. Moreover, the robustness of this model is confirmed by simulation.

3. Model for Asymmetric Similarity Data

3.1 Similarity between Clusters

If the observed similarity data is asymmetric, then the proposed additive fuzzy clustering models in the foregoing sections are not available. We then extend Model (5) as follows:

$$s_{ij} = \sum_{k=1}^{K} \sum_{l=1}^{K} w_{kl} \rho(u_{ik}, u_{jl}) + \varepsilon_{ij}.$$

In this model, the weight w_{kl} is considered to be a quantity which shows the asymmetric similarity between the pair of clusters. That is, we assume that the asymmetry of the similarity between the objects is caused by the asymmetry of the similarity between the clusters.

3.2 Asymmetric Aggregation Operator

In Model (5), if the obtained similarity is symmetric, then the model is represented by symmetric function ρ, and by a symmetric similarity between clusters.

On the other hand, if the similarity is asymmetric, two ways exist. The first is a way which represents the asymmetry between objects by the asymmetry between the clusters in the foregoing section. The second is a way using the new approach, that is the asymmetric aggregation operators. In this case, we have to create new aggregation operators which satisfy the following conditions: Boundary conditions, Monotonicity, and Asymmetry.

Suppose. $f(x)$ is a generator function of t-norms, and $o(x)$ is a continuous monotone decreasing function satisfying

$$\varrho : [0,1] \rightarrow [0,\infty], \quad o(1) = 1.$$

Then we define the following $\gamma(x, y)$ as asymmetric aggregation operators:

$$\gamma(x, y) = f^{[-1]}(f(x) + o(x)f(y)).$$

Using the generator function of the Hamacher product (R Fullér (1991)), i.e. $f(x) = \dfrac{1-x}{x}$ and the monotone decreasing function $o(x) = \dfrac{1}{x^m}$ $(m > 0)$, the asymmetric aggregation operator is defined as

$$\gamma(x, y) = \frac{x^m y}{1 - y + x^{(m-1)}y},$$

which is shown in Figure 1. In Figure 2, the dotted curve shows the intersecting curve of the surface shown in Figure 1 and the plane $x = y$, and the solid curve is the intersection with $x + y = 1$. >From the solid curve, we find the asymmetry of the proposed aggregation operator. Figure 3 shows the asymmetric aggregation operator defined as

$$\gamma(x, y) = \frac{xy}{y + x(2 - x)^m(1 - y)},$$

where the generator function is the Hamacher product, i.e. $f(x) = \dfrac{1-x}{x}$ and the monotone decreasing function $\varphi(x) = (2 - x)^m$. Figure 4 shows intersecting curves with $x = y$ and $x + y = 1$. In the case of the generator function of the Algebraic product, i.e. $f(x) = -\log x$ and the monotone decreasing function $o(x) = 2 - x^m$ (shown in Figure 5), the asymmetric aggregation operator is defined as

$$\gamma(x, y) = xy^{(2-x^m)}.$$

Intersecting curves are shown in Figure 6.

Generally, $\gamma(x, y)$ satisfies the inequality

$$\gamma(x, y) \le \rho(x, y) \tag{6}$$

by $f(x) + f(y) \le f(x) + o(x)f(y)$, because $o(x) \ge 1$. Since the following inequality:

$$\sum_{k=1}^{K} \gamma(u_{ik}, u_{jk}) \le \sum_{k=1}^{K} \rho(u_{ik}, u_{jk}) \le 1$$

is satisfied by (4) and (6), we assume the condition $0 \le s_{ij} \le 1$.

Using the asymmetric aggregation operators $\gamma(x, y)$, we define the model for asymmetric similarity data as follows:

$$s_{ij} = \sum_{k=1}^{K} \gamma(u_{ik}, u_{jk}) + \varepsilon_{ij}, \tag{7}$$

where $\gamma(u_{ik}, u_{jk}) \ne \gamma(u_{jk}, u_{ik})$.

316

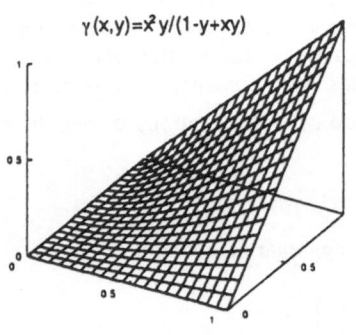

$\gamma(x,y)=x^2y/(1-y+xy)$

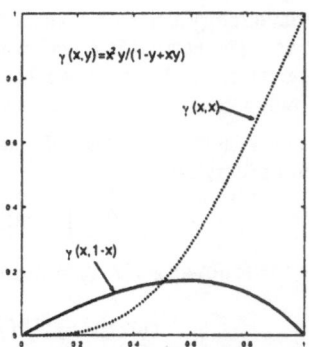

$\gamma(x,y)=x^2y/(1-y+xy)$

$\gamma(x,x)$

$\gamma(x,1-x)$

Fig. 1: Asymmetric Aggregation Operator
($m = 2$)

Fig. 2: Intersecting Curves

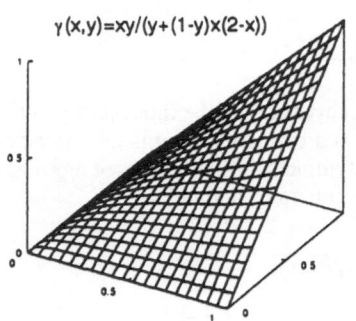

$\gamma(x,y)=xy/(y+(1-y)\times(2-x))$

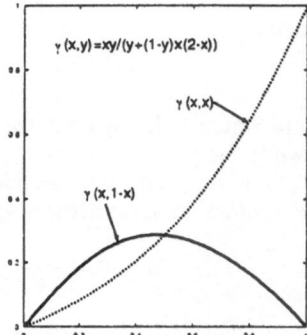

$\gamma(x,y)=xy/(y+(1-y)\times(2-x))$

$\gamma(x,x)$

$\gamma(x,1-x)$

Fig. 3: Asymmetric Aggregation Operator
($m = 1$)

Fig. 4: Intersecting Curves

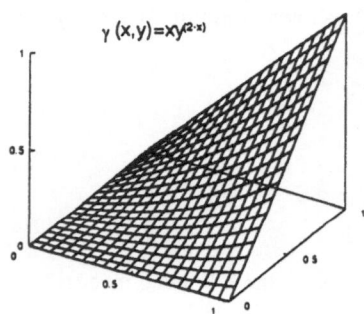

$\gamma(x,y)=xy^{(2-x)}$

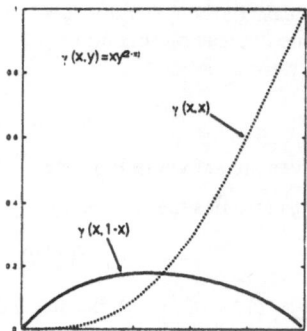

$\gamma(x,y)=xy^{(2-x)}$

$\gamma(x,x)$

$\gamma(x,1-x)$

Fig. 5: Asymmetric Aggregation Operator
($m = 1$)

Fig. 6: Intersecting Curves

4. Numerical Example

To demonstrate the applications of Model (7), we will use data which shows telephone traffic from one prefecture to another. In the optimization algorithm used in this example, 20 sets of initial values are given by using uniform pseudorandom numbers in the interval $[0, \frac{\pi}{2}]$, and in the end, we select the best result. The number of clusters is determined based on the value of fitness. By increasing the number of clusters, the value of fitness decreases, but even if the number of clusters is greater than 4, there is no severe decrease of the fitness. >From the principle of parsimony, it should be considered that the number of clusters is determined to be 4.

The results of the analysis using the asymmetric aggregation operator defined in Figure 1 are shown in Table 1 and Figure 7. In Figure 7, monotone gradation in this figure shows the degree of belongingness of a prefecture to a cluster. The darker the shade, the larger the degree of belonging. As for the results, we find that geographical distance is closely connected with the telephone communication. Moreover, the results show that large cities become influx points.

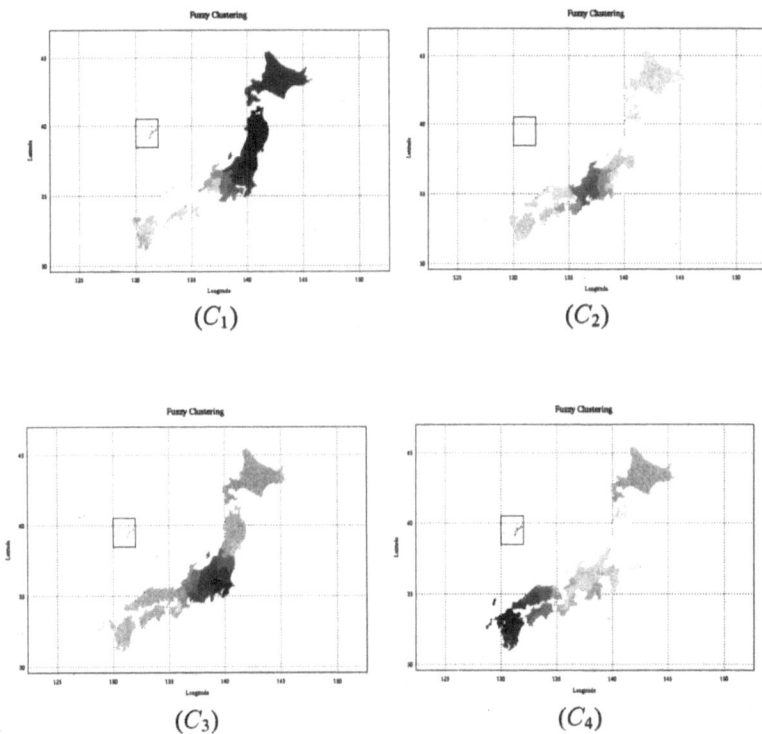

(C_1)　　(C_2)

(C_3)　　(C_4)

Fig. 7: Fuzzy Clustering

Tab. 1: Clustering using Asymmetric Aggregation Operator

No. of Prefectures	C_1	C_2	C_3	C_4
Hokkaido	.3269	.1702	.1953	.3076
Aomori	.5469	.0992	.0999	.2540
Iwate	.6248	.0574	.0684	.2494
Miyagi	.7782	.0000	.0000	.2218
Akita	.5956	.0584	.0785	.2675
Yamagata	.6063	.0417	.0739	.2781
Fukushima	.4882	.0532	.0737	.3849
Ibaragi	.2093	.1022	.0974	.5911
Tochigi	.2484	.0915	.0908	.5693
Gunma	.2173	.1145	.1106	.5576
Saitama	.1862	.0003	.0056	.8079
Chiba	.0951	.0847	.0985	.7218
Tokyo	.0000	.0000	.0000	1.0000
Kanagawa	.0537	.0591	.1071	.7800
Niigata	.2979	.0999	.1762	.4260
Toyama	.2277	.1223	.3414	.3085
Ishikawa	.2063	.1060	.3934	.2942
Fukui	.1993	.1294	.4218	.2495
Yamanashi	.1880	.1665	.1612	.4843
Nagano	.1657	.1167	.2292	.4883
Gifu	.1261	.1177	.4326	.3236
Shizuoka	.1454	.1373	.2527	.4645
Aichi	.0000	.0000	.5544	.4456
Miye	.1286	.1217	.4716	.2780
Shiga	.1054	.1091	.5867	.1988
Kyoto	.0988	.0234	.6947	.1831
Osaka	.0000	.0000	1.0000	.0000
Hyogo	.0000	.1597	.6592	.1811
Nara	.1203	.1152	.6121	.1525
Wakayama	.1715	.1532	.4937	.1816
Tottori	.0808	.2584	.4700	.1908
Shimane	.0790	.3073	.4226	.1911
Okayama	.0660	.2810	.4652	.1878
Hiroshima	.0000	.3785	.4287	.1928
Yamaguchi	.0530	.4795	.2784	.1891
Tokushima	.1416	.2270	.4428	.1887
Kagawa	.0756	.2431	.4756	.2057
Ehime	.1033	.2811	.4153	.2003
Kochi	.1537	.2519	.3947	.1997
Fukuoka	.0000	.9098	.0000	.0902
Saga	.1106	.6236	.1127	.1531
Nagasaki	.0975	.5546	.1616	.1863
Kumamoto	.0738	.5996	.1456	.1810
Oita	.0945	.5480	.1741	.1834
Miyazaki	.1076	.5098	.1843	.1984
Kagoshima	.1063	.4796	.2045	.2096
Okinawa	.2347	.3276	.2050	.2327

Goodness of fit = 0.0099

Acknowledgement

A part of this work was supported by a grant for Scientific Research from the Ministry of Education, Science and Culture of Japan.

References:

Bezdek, J.C. (1987). *Pattern Recognition with Fuzzy Objective Function Algorithms*. Plenum Press.

Dave, R.N. and Bhaswan, K. (1992). Adaptive Fuzzy c-Shells Clustering and Detection of Ellipses. *IEEE Transactions on Neural Networks*, 3, 643-662.

Dunn, J.C. (1973). A Fuzzy Relative of the ISODATA Process and Its Use in Detecting Compact Well-Separated Clusters. *Jour. Cybernetics*, 3, 3, 32-57.

Fullér, R. (1991). On Hamacher-sum of Triangular Fuzzy Numbers. *Fuzzy Sets and Systems*, 42, 205-212.

Gordon, A.D. (1987). A Review of Hierarchical Classification. *Journal of the Royal Statistical Society*, Series A, 150 ,119-137.

Hall, L.O. and Bensaid, A.M. *et al.*. (1992). A Comparison of Neural Network and Fuzzy Clustering Techniques in Segmenting Magnetic Resonance Images of the Brain. *IEEE Transactions on Neural Networks*, 3, 672-682.

Hubert, L. (1973). Min and Max Hierarchical Clustering Using Asymmetric Similarity Measures. *Psychometrika*, 38, 63-72.

Menger, K. (1942). Statistical Metrics. *Mathematics*, 28, 535-537.

Mizumoto, M. (1989). Pictorial Representations of Fuzzy Connectives, Part I: Cases of t-Norms, t-Conorms And Averaging Operators. *Fuzzy Sets and Systems*, 31, 217-242.

Ruspini, E.H. (1969). A New Approach to Clustering, *Inform. Control.*, 15, 1, 22-32.

Sato, M. and Sato, Y. (1994a). An Additive Fuzzy Clustering Model. *Japanese Journal of Fuzzy Theory and Systems*, 6, 2, 185-204.

Sato, M. and Sato, Y. (1994b). Structural Model of Similarity for Fuzzy Clustering. *Journal of the Japanese Society of Computational Statistics*, 7, 27-46.

Sato, M. and Sato, Y. (1995a). Extended Fuzzy Clustering Models for Asymmetric Similarity. *Fuzzy Logic and Soft Computing*, World Scientific, 228-237.

Sato, M. and Sato, Y. (1995b). A General Fuzzy Clustering Model Based On Aggregation Operators. *Behaviormetrika*, 22, 2, 115-128.

Sato, M. and Sato, Y. (1995c). On a General Fuzzy Additive Clustering Model. *International Journal of Intelligent Automation and Soft Computing*, 1, 4, 439-448.

Shepard, R.N. and Arabie, P. (1979). Additive Clustering: Representation of Similarities as Combinations of Discrete Overlapping Properties. *Psychological Review*, 86, 87-123.

Tarjan, R.E. (1983). An improved algorithm for hierarchical clustering using strong components. *Information Processing Letters*, 17, 37-41.

Weber, S. (1983). A General Concept of Fuzzy Connectives, Negations and Implications Based on T-Norms and T-Conorms. *Fuzzy Sets and Systems*, 11, 115-134.

A Proposal of an Extended Model of ADCLUS Model

Tadashi Imaizumi

Department of Management & Information Sciences
Tama University
4-1-1 Hijirogaoka, Tama-shi,Tokyo 206, Japan

Summary: This paper presents an extended model of ADCLUS model as a model for overlapping clustering. It is assumed that a degree of the belongings of each object to a cluster is represented as a binary random variable, and similarity between two objects is defined by a weighted expectation of a cross product of these variables. The problems on the visualization of the results is also discussed. An extension of the proposed model to a two-mode, two-way data is also described.

1. Introduction

When we want to analyze a proximity data set among objects, it is very important how to extract a hidden information(structure) of it by using several models and several methods. MDS(Multi-Dimensional Scaling) models and methods have been used for extracting it. We can extract a continuos structure of a data set using these model in which dissimilarity or similarity are related to the inter-point distance of the corresponding two points.

A combinatorial theoretic model will be proper ones when the structure of data set is assumed to be the descrete one though these MDS methods are also applicable(Arabie and Hubert, 1992). For example, this combinatorial theoretic model has been used to analyze the data set which represents the structure of a confusions between 16 consonant phonemes (Arabie and Carroll, 1980; Soli, Arabie and Carroll, 1986), that of a kinship relations(Carroll and Arabie, 1983).

As Shepard and Arabie (1979) first introduced ADCLUS model as a model for overlapping clustering of one-mode, two-way data, several models and several methods which based on this model are proposed. The basic ADCLUS model for one mode, two-way data matrix is expressed as

$$S_{ij} = \sum_{t=1}^{R} w_t p_{it} p_{jt} + c + e_{ij}. \tag{1}$$

An extended model for two-mode, two-way data matrix or three-way matrices is proposed as

$$S_{ij} = \sum_{t=1}^{R} w_t p_{it} q_{ji} + c + e_{ij}, \tag{2}$$

$$S_{ijk} = \sum_{t=1}^{R} w_{kt} p_{it} p_{jt} + c_k + e_{ijk}, \tag{3}$$

where p_{it} or q_{jt} is a binary variable whose value is 0 or 1, w_t or w_{kt} is a positive value. Arabie and Carroll(1980) proposed an alternative algorithm,MAPCLUS program, for

320

fitting ADCLUS model. An extended models for three-way matrices are named IN-DCLUS model. In these models, N objects are clustered into R possibly overlapping clusters. Similarity between two objects that are not members of a cluster is assumed to be zero.

There are four states (p_{it}, p_{jt}) of two objects which represents two objects belonging to the same cluster or not. The state $(1,1)$ represents two objects belonging to the same cluster, the state $(0,0)$ represents them not belonging, and the states $(1,0)$ or $(0,1)$ represents one object belonging to the cluster and the other one not. The property of ADCLUS model considering the state $(1,1)$ only will also lead the number of clusters being increased and the visualization of the resultant clusters being difficult. In the ADCLUS model, only the state $(1,1)$ contributes to the similarity and the other three states are ignored.

As this ADCLUS model will be rewritten as the matching model of two binary sequences $(p_{i1}, p_{i2}, \ldots, p_{it}, \ldots, p_{iR})$ and $(p_{j1}, p_{j2}, \ldots, p_{jt}, \ldots, p_{jR})$ which represent whether an object has the property t or not. The binary 0 or 1 are exchangeable for any cluster t, the state $(0,0)$ or $(1,1)$ will contribute the similarity between the corresponding two objects. From this point of view, we propose an extended model of ADCLUS model in this paper.

2. Model

We present an extended model for overlapping clustering by defining similarity between two objects that are not members of a cluster. Let X_{it} denote an binary random variable such that

$$X_{it} \sim Bin[p_{it}, p_{it}(1 - p_{it})], \tag{4}$$

We also assume

X_{it} and X_{js} (i $\neq$ j, t $\neq$ s) are independently distributed.

The similarity between two objects o_i and o_j, S_{ij} will be defined by

$$S_{ij} = \sum_{t=1}^{R} \{w_t X_{it} X_{jt} + u_t(1 - X_{it})(1 - X_{jt})\} + c, (i \neq j) \tag{5}$$

where w_t, u_t and c denote real value respectively. In this model, similarity between o_i and o_j for cluster t, S_{ijt} takes one of three values

$$S_{ijt} = \begin{cases} 0 & (X_{it}, X_{jt}) = (1, 0) \\ 0 & (X_{it}, X_{jt}) = (0, 1) \\ w_t & (X_{it}, X_{jt}) = (1, 1) \\ u_t & (X_{it}, X_{jt}) = (0, 0) \end{cases} \tag{6}$$

We also assume that the observed similarity $s_{ij}(i = 1, 2, \cdots, n; j = 1, 2, \cdots, n; j \neq i)$ are a realization of $E(S_{ij})$. We have

$$s_{ij} = E[X_i W X_j' + (J - X_i)U(J - X_j)'] + c, (i \neq j). \tag{7}$$

where $S = (s_{ij})$ is $N \times N$ a similarity matrix of N objects, $X = (X_k)$ is a $N \times R$ binary random matrix, J is $1 \times R$ matrix whose elements are all 1, W and U is $R \times R$ diagonal matrix of positive weights respectively. This model is expressed as

$$s_{ij} = trace(U) - E(X_i - X_j)U(X_i - X_j)' + E(X_i)(W - U)E(X_j)' + c, (i \neq j). \tag{8}$$

The expection $E(X_i - X_j)U(X_i - X_j)'$ is interpreted as the weighed squared distance between X_i and X_j, and indicates the distance property of the proposed model. As the term $|w_t - u_t|$ is decreasing to 0 for each cluster t, a visualization of estimated parameters will be possible. The following two models are special sub-model of the proposed model.

2.1 Case W=U

In this case, Two objects which are in the same state were equally weighted. A mean similarity s_{ij} is expressed as

$$s_{ij} = trace(U) - E(X_i - X_j)U(X_i - X_j)' + c. \tag{9}$$

A resultant clustering will be represented as a geometric configuration.

2.2 Case W=0 or U=0

The proposed model is same to the ADCLUS model. Since W and U are exchangeable, two objects which belongs to the same cluster are weighted.

$$s_{ij} = E(X_i)WE(X_j)' + c. \tag{10}$$

3. The Algorithm

The observed mean similarity s_{ij} is represented as

$$s_{ij} = \sum_{t=1}^{R}\{w_t p_{it} p_{jt} + u_t(1 - p_{it})(1 - p_{jt})\} + c + e_{ij}, \tag{11}$$

with the constraints $0 \leq p_{it} \leq 1$. To estimate these parameters, we use the ordinary least squares method with the constraints on the parameters. The estimation procedure for given the number of clusters R is an iterative process consisted of three steps.

3.1 Initial value of p_{it}

When there is no rational initial value of p_{it}, p_{it}^0, we generate the initial $P^0 = (p_{it}^0)$ by using uniform random number $(0,1)$,

$$P^0 \sim Uniform(0, 1). \tag{12}$$

3.2 The estimation of w_t, u_t and c

As there are constraints with w_t and u_t being positive, they are estimated by using the quadratic programming procedure with the active set method which minimizes,

$$\sum_{i=1}^{n} \sum_{j \neq i, j=1}^{n} \{s_{ij} - w_t p_{it} p_{jt} - u_t(1 - p_{it})(1 - p_{jt}) - c\}^2. \tag{13}$$

3.3 The estimation of p_{it}

For given w_t, u_t and c, we estimate $p_{it}(t = 1, 2, \cdots, R)$ using ALS procedure with other $p_{jt}(j = 1, 2, \cdots, i-1, i+1, \cdots, n)$ being fixed. Similarity of object o_i, $s_{ij}(j = 1, 2, \cdots, i-1, i+1, \cdots, n)$, will be expressed as follows:

$$s_{ij} = \sum_{t=1}^{R} u_t(1 - p_{jt}) + c + \sum_{t=1}^{R}\{(w_t + u_t)p_{jt} - u_t\}p_{it} + e_{ij} \tag{14}$$

with the constraints $0 \leq p_{it} \leq 1$. We use the quadratic programming prodecure with the active set method to obtain a feasible solution of $(p_{i1}, p_{i2}, \cdots, p_{iR})$.

These iterative processes 3.2 and 3.3 are repeated until some the convergence criteria are satisfied. This iterative process is simple one than that of MAPCLUS procedure since we estimate p_{it} instead of x_{it}, a realization of X_{it}.

4. How to represent the resultant configuration as a space structure

The resultant configuration represents the descrete structure of data set. We compare each objects on the uni-dimensional representations of R clusters since cluster analysis is not to embed objects as points in lower dimensional space but to find cluster structure in data. However,an geometric representation will be informative since we will be able to understand the relationship between clusters.

The geometric representation as space structure seems to be difficult. This representation will be done in case of $W = U$ or $W = 0$ or $U = 0$. Since

$$E(S_{ij}) = trace(U) - E(X_i - X_j)U(X_i - X_j)' + E(X_i)(W - U)E(X_j)' + c, \quad (15)$$

the quantity $E(X_i)(W - U)E(X_j)'$ of two object o_i and o_j will indicates the degree of the resultant configuration being not represented as the space structure. To represent the resultant configuration as space structure, we assume the model with $W = U$,

$$E(S_{ij}) = trace(U) - E(X_i - X_j)U(X_i - X_j)' + c, \quad (16)$$

as the target model to be represented. This configuration can be represented as a space structure. We define the shifting factor , F_i, of object o_i to the other objects by

$$F_i = E(X_i)(W - U) \sum_{i \neq j, j=1}^{n} E(X_j)'/2(n - 1). \quad (17)$$

When this quantity F_i is positive one, a weighted squared distance $E(X_i - X_j)U(X_i - X_j)'$ should be increased to represent as space structure. When F_i is negative one, the distance should be decreased. The quantity F_i is interpreted as a deviate from the point $E(X_i)U$ to embed as point. This suggests that the resultant configuration will be more informative with a vector,

$$F_i = \sum_{j \neq i, j=1}^{n} E(X_j)/(n - 1) \quad (18)$$

from the point $E(X_i)U$.

This visual representation of the resultant configuration as a space structure will be appropriate for the upto 3 clusters. When number of clusters R is 4 or more, we must decide what representation is more informative. We will be able to choose these clusters, for example, by computing the sum of cross product between any 2 clusters

$$cr_{st} = \sum_{i=1}^{N} p_{is}p_{jt}, (s \neq t), \quad (19)$$

and inspecting this $R \times R$ matrix or the eigen vectors of this matrix.

5 Applications

We applied the proposed model to the confusion matrix of Morse Code signals collected by Rothkopt(1957). This data was analyzed by Shepard using Kruskal's non-metric MDS procedure(Shepard, 1963). We analyzed the symmetrized similarity matrix s_{ij} whose (i,j) element is defined by

$$s_{ij} \overset{\text{def}}{=} (M_{ij} + M_{ji})/2. \tag{20}$$

where M_{ij} is (i, j) element of the confusion matrix. The analysis was done using the number of clusters from 5 to 1. Ratio of the minimized squared loss over trace(S) in 5 clusters to 1 cluster were 0.051, 0.069, 0.082, 0.109, and 0.183. By using elbow criterion and to compare result by MDS, we chose two-cluster result as a solution. The estimated w, u and c are shown Table 1. The estimated P is shown Table 2 and is illustrated in Figure 1,

Tab. 1: The estimated w, u and c in number of clusters = 2

	Cluster 1	Cluster 2	
w	57.820	66.198	
u	39.281	17.285	
c			-22.935

Table 1 shows that the state (1,1), two Morse code are a member of this cluster, and the state (0,0), they are not a member of this cluster, are contributed to similarity in cluster 1 since the ratio u_1/w_1 is about 0.679. In cluster 2, the state (1,1) only is contributed to similarity since the ratio u_2/w_2 is about 0.261 and it seems to be small.

Tab. 2: The estimated P in number of clusters = 2

Morse code	Clus.1	Clus. 2	Morse code	Clus. 1	Clus. 2
A .-	0.201	0.068	S ...	0.391	0.000
B -...	1.000	0.051	T -	0.154	0.099
C -.-.	0.873	0.398	U ..-	0.437	0.000
D -..	0.622	0.000	V ...-	0.675	0.000
E .	0.194	0.083	W .--	0.501	0.068
F ..-.	0.747	0.099	X -..-	1.000	0.151
G --.	0.415	0.224	Y -.--	0.872	0.512
H	0.521	0.000	Z --..	0.781	0.512
I ..	0.211	0.084	1 .----	0.375	0.873
J .---	0.643	0.544	2 ..---	0.565	0.744
K -.-	0.668	0.024	3 ...--	0.686	0.316
L .-..	0.943	0.105	4 ....-	0.690	0.032
M --	0.234	0.125	5	0.593	0.000
N -.	0.189	0.086	6 -....	0.929	0.209
O ---	0.393	0.292	7 --...	0.792	0.541
P .--.	0.808	0.565	8 ---..	0.503	0.817
Q --.-	0.818	0.676	9 ----.	0.233	0.855
R .-.	0.524	0.000	0 -----	0.202	0.692

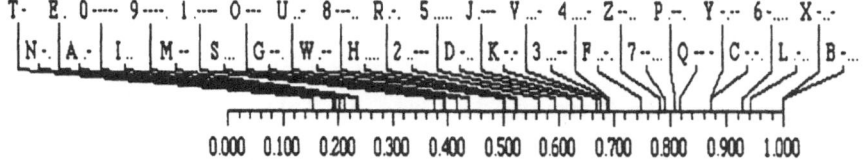

Cluster 1

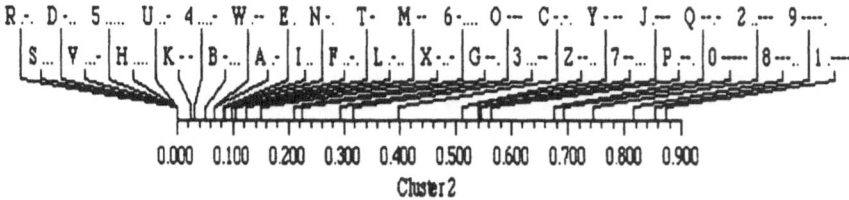

Cluster 2

Fig. 1: The estimated P.

From Table 2 and Figure 1, the cluster 1 seems to be interpreted as the degree of mixture of dash and dot. The cluster 2 is interpreted as dot-to-dash ratio.

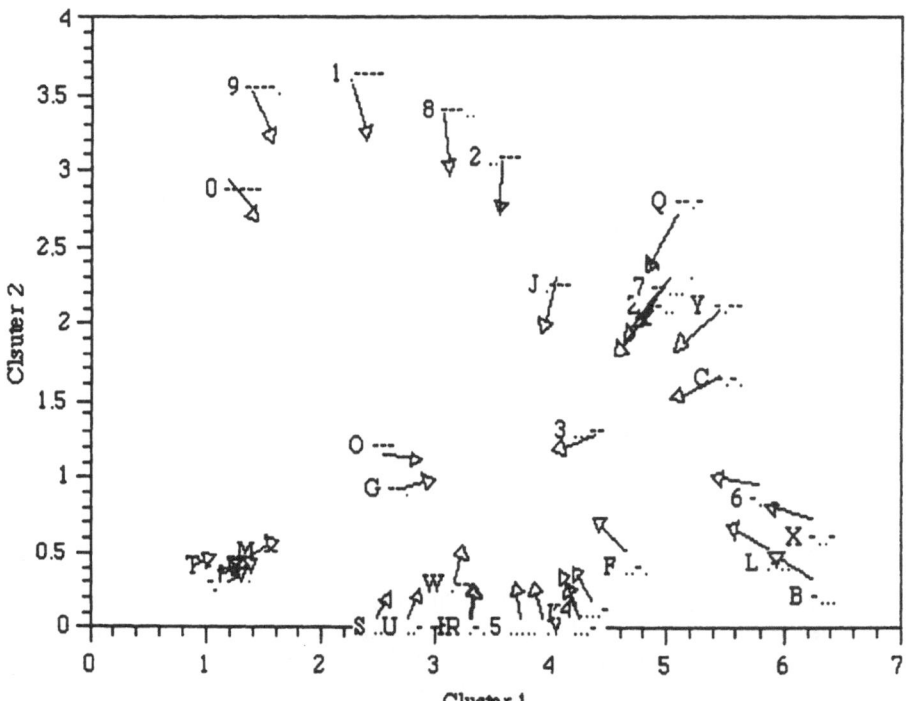

Fig. 2: A geometric representation with the shifting factor from each point.

In Figure 2, The estimated PU are illustrated as points in two-dimensional space.

Each vetcor from each point illustrates the shifting factor. each point was shifted to the direction of the centroid of other $n - 1$ points.

6. An extension to Two-mode Data Matrix

This model is applicable to the two-mode data matrix of similarity. Let Y_{jt} denote binary random variable for column object.

$$Y_{jt} \sim Bin\{q_{jt}, q_{jt}(1 - q_{jt})\}, \tag{21}$$

We also assume

$$Y_{jt} \text{ and } Y_{ks} \text{ (j} \neq \text{k, t} \neq \text{s) are independently distributed,}$$

and

$$X_{it} \text{ and } Y_{js} \text{ are also independently distributed,}$$

Then the model for two-mode data matrix of similarity will be defined as

$$S_{ij} = \sum_{t=1}^{R} w_t X_{it} Y_{jt} + u_t(1 - X_{it})(1 - Y_{jt}) + c \tag{22}$$

$$= E[X_i W Y_j' + (J - X_i)U(J - Y_j)'] + c, \tag{23}$$

where $S = (s_{ij})$ is a $N \times M$ similarity matrix, $X = (X_k)$ is a $N \times R$ binary random matrix ,$Y = (Y_j)$ is a $M \times R$ binary random matrix W and U is $R \times R$ diagonal matrix of positive weights.

The estimation procedure of p_{it} and q_{jt} are similar to that in section 3. For given w_t ,u_t, c, and $q_{jt}(j = 1, 2, \cdots, m; t = 1, 2, \cdots, R)$ we estimate $p_{it}(t = 1, 2, \cdots, R)$ using ALS procedure. Similarity of row object o_i, $s_{ij}(j = 1, 2, \cdots, i - 1, i + 1, \cdots, m)$, will be expressed as

$$s_{ij} = \sum_{t=1}^{R} u_t(1 - q_{jt}) + c + \sum_{t=1}^{R} \{(w_t + u_t)q_{jt} - u_t\}p_{it} + e_{ij} \tag{24}$$

with the constraints $0 \leq p_{it} \leq 1$. We use the quadratic programming prodecure with the active set method to obtain a feasible solution of $(p_{i1}, p_{i2}, \cdots, p_{iR})$. Then $q_{jt}(j = 1, 2, \cdots, m)$ will be updated using these updated $p_{it}(i = 1, 2, \cdots, n; t = 1, 2, \cdots, R)$ by similar procedure.

7 Discussions

We proposed the extended model of the ADCLUS model with random variables. By using this model, statistical inference on parameters will be applicable. The state $(0,0)$ of (X_{it}, X_{jt}) is also contributed to the similarity S_{ij}. As S_{ijt} takes one of three values, $0, w_t$, and u_t, the maximum distinct values of similarity matrix is $3 + 2^{t-1}$. The other hand, it is $2 + t - 1$ in ADCLUS model. The number of clusters needed will be less that that of ADCLUS model for analyzing a similarity matrix.

As the estimation procedure is simple one, the weighted least squares method will be applicable by using a consistent estimator of variance of S_{ij}.

The model discussed in section 6 is also applicable to the case of the preference data with some modification. It seems to be natural that each subject is correspond to each ideal object. Then we relate the degree of preference of subject s_i to object o_j with similarity between an idean object I_i and o_j. However,it seems to be difficult that the evaluated preference is a realization of the expection from statistical point of view. We will analyze the preference data with the latent class model in which the definition of the expection will make a sense.

When data matrix is one of preference data, we assume that there are G unobserved group and each subject belongs to one of these groups. Then we can apply this model by assuming that the preference to object o_j of subject s_i, who belongs group g $(g = 1, 2, \cdots, G)$, is related to the similarity between o_j and some ideal point g_k.

Let X_{gt} denote the binary random variable of group g in cluster t. We define the similarity between group g and object o_j by

$$S_{gj} = \sum_{t=1}^{R} w_{gt} X_{gt} Y_{jt} + u_{gt}(1 - X_{gt})(1 - Y_{jt}) + c_g. \tag{25}$$

Each group differently weights each property in this model. We assume that the preference to object o_j in group g is equall to S_{gj}. Each subject is assumed to be a sample from one of G groups. A clustering procedure will be rolled in the estimation procedure for allocating each subject to one of groups.

As the extension of the proposed model to analyze three-way data matrices, specially for analysis of individual differences, one model is same to INDCLUS model, and the other model is based on the latent class model in which each individual belongs to one of G groups.

8. References

Arabie, P. and Carroll, J. D.(1980): MAPCLUS: a mathematical programming approach to fitting the ADCLUS model *Psychometrika*, **45**,211-235.

Arabie, P. and Hubert, L. J.(1992): Combinatorial Data Analysis. *Annual Review of Psychology*,**43**,169-203.

Carroll, J. D. and Arabie, P.(1983):INDCLUS: An individual differences generalization of the ADCLUS model and the MAPCLUS algorithm *Psychometrika*, **48**,157-169.

Rothkopt, E. Z.(1957): A measurement of stimulus similarity ans errors in some paired-associate learning tasks. *Journal of Experimental Psychology*,**53** 94-101.

Soli, S. D., Arabie, P. and Carrol, J.D.(1986): Representation of discrete structure underlying observed confusions between consonant phonemes. *Journal of the Acoustical Society of America*, **79**,826-837

Shepard, R.N., and Arabie, P. (1979): Additive Clustering Representation of Similarities as Combinations of Descrete Overlapping Properties. *Psychological Review*,**86**,87-123.

Comparison of Pruning Algorithms
in Neural Networks

Yoshihiko Hamamoto, Toshinori Hase, Satoshi Nakai, and Shingo Tomita

Faculty of Engineering, Yamaguchi University
Ube, 755 Japan

Summary: In order to select the right-sized network, many pruning algorithms have been proposed. One may ask which of the pruning algorithms is best in terms of the generalization error of the resulting artificial neural network classifiers. In this paper, we compare the performance of four pruning algorithms in small training sample size situations. A comparative study with artificial and real data suggests that the weight-elimination method proposed by Weigend et al. is best.

1. Introduction

There are two fundamental problems in the design of artificial neural network (ANN) classifiers: finding training algorithms and selecting the right-sized network. Concerning training algorithms, the back-propagation algorithm (Rumelhart et al., 1986) has been widely used because of its simplicity. In the back-propagation learning, the following error function is minimized:

$$E = \sum_{k \in T}(t_k - o_k)^2 \tag{1}$$

where o_k is the output in the output layer for the k-th training sample, t_k is the corresponding target value, and T is the training set. However, BP has two serious disadvantages: the extremely long training times and the possibility of trapping in local minima. On the other hand, concerning right-sized network selection, its issue is that the network that is too large or too small can overfit or underfit the data, respectively. The use of the right-sized network leads to the improvement in the performance of the resulting ANN classifier. Hence, the problem of selecting the right-sized network is very important in neural networks. We will address only this problem.

In order to select the right-sized network, many pruning algorithms have been presented (Reed, 1993). Unfortunately, little is known about experimental comparison of the pruning algorithms in finite sample conditions. In this paper, we compare the performance of four pruning algorithms in terms of the generalization error of ANN classifiers in small training sample size situations. Our emphasis is on giving practical advice to designers and users of ANN classifiers.

2. ANN classifiers

We will consider ANN classifiers with one hidden layer. The units in the input layer correspond to the components of the feature vector to be classified. The hidden layer has m units. The units in the output layer are associated with pattern class labels. In the network discussed here, the inputs to the units in each successive layer are the outputs of the preceding layer. Initial weights were distributed uniformly in -0.5 to 0.5.

3. Pruning algorithms

In this section, we briefly describe four pruning algorithms. We will follow Reed (1993)'s notations.

A. *Karnin's method* (Karnin, 1990)

Karnin measures the sensitivity of the error function with respect to the removal of each connection and then prunes the weights with low sensitivity. The sensitivity of a weight w_{ij} is given as

$$S_{ij} = \sum_{n=0}^{N-1} [\Delta w_{ij}(n)]^2 \frac{w_{ij}^f}{\eta(w_{ij}^f - w_{ij}^i)} \tag{2}$$

where N is the number of training epochs, η is a learning rate, w^f is the final value of the weight after training, and w^i is the initial weight. Δw_{ij} in (2) can be calculated by the back-propagation algorithm.

B. *Optimal Brain Damage (OBD) method* (Le Cun et al., 1990)

When the weight vector w is perturbed, the change in the error is approximately given by

$$\delta E \cong \frac{1}{2} \sum_{i \in C} \frac{\partial^2 E}{\partial w_i^2} \delta w_i^2 \tag{3}$$

where the δw_i's are the components of δw and C is the set of all connections. The second derivatives can be calculated by a modified back-propagation algorithm. The saliency of the weight w_i is then

$$s_i = \frac{\partial^2 E}{\partial w_i^2} \cdot \frac{w_i^2}{2}. \tag{4}$$

Pruning is done iteratively: i.e., train to a reasonable error level, compute saliencies, delete low saliency weights, and resume training.

C. *Weight-Elimination (WE) method* (Weigend et al., 1991)

The following error function is minimized:

$$J = \sum_{k \in T} (t_k - o_k)^2 + \lambda \sum_{i \in C} \frac{w_i^2/w_0^2}{1 + w_i^2/w_0^2} \tag{5}$$

where λ is a parameter dynamically adjusted in training. The second term represents the complexity of the network as a function of the weight magnitudes relative to the constant w_0.

D. *Kruschke's method* (Kruschke, 1988)

Units are redundant when their weight vectors are nearly parallel or antiparallel and they compete with others that have similar directions. The gains g are adjusted according to

$$\Delta g_i^s = -\gamma \sum_{j \neq i} \cos^2 \angle(w_i^s, w_j^s) \cdot g_j^s \tag{6}$$

where γ is a small positive constant, $\angle(\cdot, \cdot)$ denotes the angle between vectors and the superscript s indexes the layer. If unit i has weights parallel to those of unit j, then the gain of each will decrease in proportion to the gain of the other and the one with the smaller gain will be driven to zero faster. Once a gain becomes zero, it remains zero. A unit with zero gain has a constant output. So, the unit can be removed.

4. Experimental Results

Jain and Chandrasekaran (1982) suggest that the size of training samples per class should be at least five to ten times the dimensionality. However, in practice, the number of available samples is limited. Hamamoto et al. (1996a) point out that the evaluation of ANN classifiers in the small-sample, high-dimensional setting is very important. In our experiments, thus, the ratio of the training sample size to the dimensionality is small. On the other hand, a large test sample size was used to accurately evaluate a classifier. Note that the estimated generalization error is a random variable, because it is a function of training and test samples. Thus, it is preferable to repeat the experiments several times independently.

To highlight the difference in the performance of four pruning algorithms, the following experiments were conducted.

4.1 Experiment 1

We briefly describe the Ness data set (Van Ness, 1980), which is used in this experiment. This data set has been used in order to evaluate the performance of classifiers such as the nearest neighbor, Parzen, linear, quadratic and neural network classifiers (Van Ness, 1980; Hamamoto et al., 1996a). The available samples were independently generated from n-dimensional Gaussian distributions $N(\mu_k, \Sigma_k)$ with the following parameters:

$$\mu_1 = [0, 0, \cdots, 0]^T, \quad \mu_2 = [\Delta/2, 0, \cdots, 0, \Delta/2]^T$$

$$\Sigma_1 = I_n, \quad \Sigma_2 = \begin{bmatrix} I_{n/2} & O \\ O & \frac{1}{2}I_{n/2} \end{bmatrix}$$

where Δ is the Mahalanobis distance between class ω_1 and class ω_2, μ_1 is the n-dimensional zero vector, and I_n is the $n \times n$ identity matrix. The true Bayes error can be controlled by the values of Δ and n. That is, the degree of overlapping between two distributions can be controlled by the values of Δ and n. For that reason, we used this data set.

The experimental condition is summarized as follows:

Data	:	Ness data set
Values of Δ	:	2, 4, 6
Dimensionality	:	2, 10, 20
No. of training samples	:	10 per class
No. of test samples	:	100 per class
Hidden unit size	:	100
No. of trials	:	100

Figs. 1–3 provide the mean of the estimated generalization error. For comparison, the generalization error of BP is also presented. It is well known that when a fixed number of training samples is used to design a classifier, the error of the classifier tends to increase as the dimensionality n gets large. That is, as n increases, the generalization problem becomes severe. In our limited experiment, the WE method works well regardless of the true Bayes error and the dimensionality, even in practical situations where the training sample size is relatively small for the dimensionality.

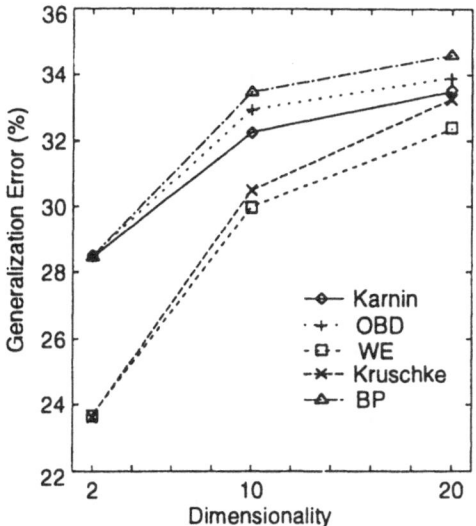

Fig. 1: Comparison of pruning algorithms in terms of the generalization error
(Ness data set with $\Delta = 2$).

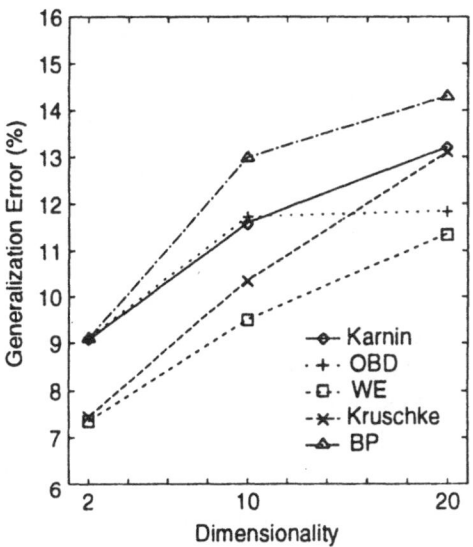

Fig. 2: Comparison of pruning algorithms in terms of the generalization error
(Ness data set with $\Delta = 4$).

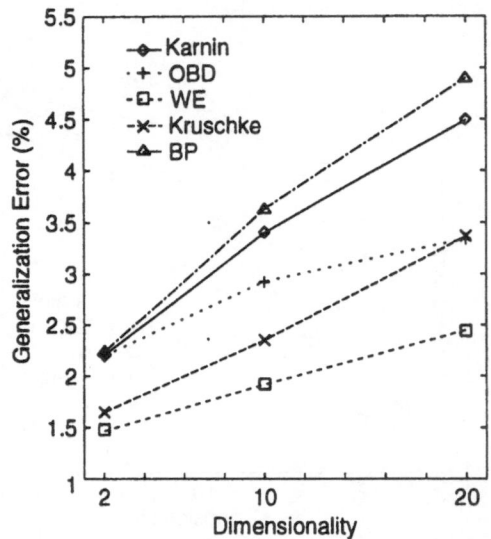

Fig. 3: Comparison of pruning algorithms in terms of the generalization error
(Ness data set with $\Delta = 6$).

4.2 Experiment 2

Next, we compare four pruning algorithms on a real data set. In this data set, each class represents one of 10 handwritten numerals. This data set contains 1400 128-dimensional feature vectors per class. In feature extraction, Gabor filters (Gabor, 1946) were applied to a character image. The outputs of Gabor filters produce a 128-dimensional feature vector. Gabor filters tend to detect line and edge segments, which seem to be good discriminating features. We call this the Gabor data set. For additional details refer to (Hamamoto et al., 1996b). We need to assure the independence between training and test sets. Thus, the following handwritten numeral character experiment was performed:

(1) Divide 1400 samples into the training set of size 100 and the test set of size 1300. Note that the two sets are mutually exclusive.

(2) Design an ANN classifier with 256 hidden units by using a pruning algorithm with the above training set.

(3) Estimate the generalization error of the ANN classifier by using the test set.

(4) Repeat steps (1)–(3) 5 times independently.

(5) Compute the average of the generalization error and its standard deviation.

Results are shown in Tab. 1. The performance of the classifiers trained only on 25 training samples per class, which are randomly selected out of 100 training samples per class, is also presented. It should be pointed out that as the training sample size decreases, the generalization problem becomes severe. Again, the WE method performs better than other pruning algorithms.

Tab. 1: Comparison of pruning algorithms in terms of the generalization error[%] in the 128-dimensional feature space. The first line in Table is the mean of the 5 trials. The second line is the standard deviation.

Training sample size per class	Karnin	OBD	WE	Kruschke	BP
25	8.77	8.57	7.65	7.72	8.58
	0.81	0.77	0.58	0.66	0.72
100	4.98	4.59	4.28	4.92	4.65
	0.61	0.28	0.17	0.26	0.33

5. Conclusions

We have compared four pruning algorithms for ANN classifier design, in small training sample size situations. The generalization error of resulting ANN classifiers was estimated on artificial and real data. Experimental results show that the WE method outperforms other pruning algorithms. Therefore, we believe that the WE method is best for ANN classifier design.

References:

Gabor, D. (1946): Theory of communication, J. Inst. Elect. Engr., 93, 429–459.

Hamamoto, Y. et al. (1996a): On the behavior of artificial neural network classifiers in high-dimensional spaces, *IEEE Trans. Pattern Analysis and Machine Intelligence*, 18, 5, 571–574.

Hamamoto, Y. et al. (1996b): Recognition of handwritten numerals using Gabor features, In *Proc. of 13th Int. Conf. Pattern Recognition*, Vienna, in press.

Jain, A. K. and Chandrasekaran, B. (1982): Dimensionality and sample size considerations in pattern recognition practice, In *Handbook of Statistics*, Vol.2, P. R. Krishnaiah and L. N. Kanal, Eds., North-Holland, 835–855.

Karnin, E. D. (1990): A simple procedure for pruning back-propagation trained neural networks, *IEEE Trans. Neural Networks*, 1, 2, 239–242.

Kruschke, J. K. (1988): Creating local and distributed bottlenecks in hidden layers of back-propagation networks, In *Proc. 1988 Connectionist Models Summer School*, 120–126.

Le Cun, Y. et al. (1990): Optimal brain damage, In *Advances in Neural Information Processing* (2), Denver, 598–605.

Reed, R. (1993): Pruning algorithms — A survey, *IEEE Trans. Neural Networks*, 4, 5, 740–747.

Rumelhart, D. E. et al. (1986): Learning internal representations by error propagation, In D. E. Rumelhart & J. L. McClelland (Eds.), *Parallel Distributed Processing: Explorations in the Microstructure of Cognition, Vol.1 : Foundations.*, MIT Press.

Van Ness, J. (1980): On the dominance of non-parametric Bayes rule discriminant algorithms in high dimensions, *Pattern Recognition*, 12, 355–368.

Weigend, A. S. et al. (1991): Generalization by weight-elimination with application to forecasting, In *Advances in Neural Information Processing* (3), 875–882.

Classification Method by Using the Associative Memories in Cellular Neural Networks

Akihiro Kanagawa, Hiroaki Kawabata, and Hiromitsu Takahashi

Faculty of Computer Science and System Engineering
Okayama Prefectural University

Soja, Okayama 719-11, Japan

Summary : This paper deals with a classification problem, such as medical diag-nosis, which classes are defined by categorical forms. Classification should be done by careful and synthetical judgement for a lot of characteristic values taking each individual variations into account. We use the associative memory function of the cellular neural networks to classify by means of remembering one category from among the preregistered categories.

1. Introduction

Supposing there are K objects, and each object has n kinds of characteristic values $q_1, q_2, \cdots, q_j, \cdots, q_n$. The problem to classify these objects into given m classes $\{C_1, C_2, \cdots, C_i, \cdots, C_m\}$ based on their characteristic values has been discussed for a long time. Classification problem discussed here is a sort of the diagnosis problem which includes some of individual variations. Classic methods using the multivariate normal distribution theory including the discriminant analysis are difficult to be applied to these problems. To cope with these, Pawlak(1984) proposed the concept of rough sets, which can be employed to discuss the consistency of the classification given by human experts with the observed attribute values of each sample. Shigenaga *et al.* (1993) modified the Pawlak's method to reduce more attributes by considering the given classification. In the case of rough sets, choice of descriptive function or fundamental sets is difficult; while in the case of fuzzy if-then rules, it is so troublesome to determine a set of membership functions or logical structure. In addition, if there are some lacking or missing data, these methods easily have a bad effect upon their classification results. This paper aims to apply the associative memories of cellular neural networks to this classification problem. The associative memory is a function of nueral networks that the nueral netwok recall a pattern from patterns embedded in advance. Further we propose a CNN whose each cell has three output values to enhance its capability of association.

2. Three valued cellular neural networks (TVCNN)

Cellular neural network (hereafter CNN) is composed of some simple analog circuits called 'cell', which is arranged checkered (see Fig. 1). CNN is regarded as a sort of Hopfield network from the view of connecting. Whereas Hopfield network has full conection, CNN has limited connection. Namely each cell is connected with the adjacent cells, and varies its state by a dynamics of differential equations under the influence of the adjacent cells. A differential equation of a cell which is located i - row·and j - column is:

$$\dot{x}_{ij} = -x_{ij} + P_{ij} * y_{ij} + S_{ij} * u_{ij} + I_{ij} , \tag{1}$$

where x_{ij} and u_{ij} denote the state variable and control variable respectively, I_{ij} is the threshold value, and P_{ij} and S_{ij} are template matices. $P_{ij} * y_{ij}$ implies the sum of the influence terms from the adjacent cells.

$$P_{ij} * y_{ij} = \begin{vmatrix} P_{ij(-r,-r)} \cdots P_{ij(-r,0)} \cdots P_{ij(-r,r)} \\ \vdots \quad \ddots \quad \vdots \quad \ddots \quad \vdots \\ P_{ij(0,-r)} \cdots P_{ij(-r,-r)} \cdots P_{ij(0,r)} \\ \vdots \quad \ddots \quad \vdots \quad \ddots \quad \vdots \\ P_{ij(r,-r)} \cdots P_{ij(r,0)} \cdots P_{ij(r,r)} \end{vmatrix} * y_{ij} = \sum_{k=-r}^{r} \sum_{l=-r}^{r} P_{ij(k,l)} y_{i+k,j+l} . \tag{2}$$

To make it simplified, we reform Eq.(1) by the vector notation :

$$\left. \begin{aligned} \dot{x} &= -x + T y + I (x) \\ y &= \text{sat} (x) , \end{aligned} \right\} \tag{3}$$

where

$n = N \times M$

$y \in D^n \underline{\text{def}} \{ x \in \mathfrak{R}^n : -1 \le x_i \le 1, \ i = 1, \cdots, n \}$

$T = [T_{ij}] \in \mathfrak{R}^{n \times n}$

$I = (I_1, \cdots, I_n)^T \in \mathfrak{R}^n$

$\text{sat} (x) = [\text{sat} (x_1), \cdots, \text{sat} (x_n)]^T \in \mathfrak{R}^n.$

Original CNN has binary state output, and it is suit for coding as black-and-white picture. But the binary state outputs put fetters upon the approach of higher recognizing problem such as classification or diagnosis based on association. The reason is that it is essential to grasp these problems based on more than two valued logic such as low-middle-high or grade A - grade B - grade C - grade D. For example, in a group examination, personal condition of health is measured by each ingredient of the blood test. There is an ideal state (healthy condition) given by a certain range for each ingredient, and from where degrees of separation toward either side are comprehensively judged.
In this study, for the purpose of extending method of these associative memory functions of CNN, we propose a cellular neural network which has three valued outputs of the cell neurons. Here we devise the output function to apply it to classification problems.

$$y = \text{sat}(x) = \frac{1}{2} (\ |x + \frac{3}{2}| - |x + \frac{1}{2}| + |x - \frac{1}{2}| - |x - \frac{3}{2}|) . \quad \text{(See Fig.2)} \tag{4}$$

For $B^n \underline{\text{def}} \{ x \in \mathfrak{R}^n \ | \ |x| = 1 \ or \ x = 0 ; i = 1, \cdots, m\}$, we can define

$$C(\alpha) = \{ x \in \mathfrak{R}^n | \quad \begin{aligned} \alpha_i &= 1 \Rightarrow x_i > 1.5, \\ \alpha_i &= 0 \Rightarrow |x_i| < 0.5, \quad \} \\ \alpha_i &= -1 \Rightarrow x_i < -1.5. \end{aligned} \tag{5}$$

Then we have sat $(x) = \alpha$ for $x \in C (\alpha)$ because $\alpha = \pm 1$ for all i.

Therefore the control equation of x for $x \in C (\alpha)$ is

$$\dot{x} = -x + T\alpha + I . \tag{6}$$

As $\beta = T\alpha + I$ is constant, this equation has the apparent equilibrium point of $x_e = \beta$. If $\beta \in C(\alpha)$, this equilibrium point is also asymptotically stable since all eigen values of equation (5) are -1. From this, Liu and Michel discussed the designing method of

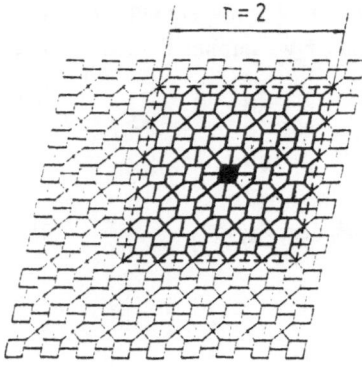

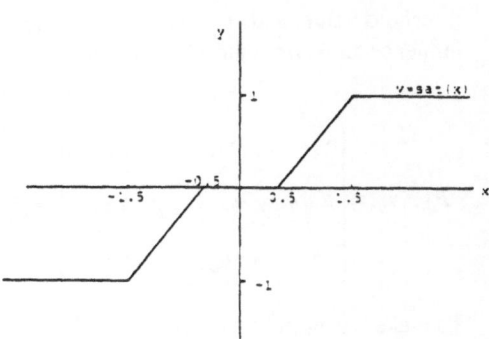

Fig.1: CNN with r-neighbor   Fig.2 : Output function with three levels

templates which can recollect the image pattern, and called it the associative memories. In the same manner we can easily make output functions with multiple levels.

3. Design procedure using a singular value decomposition

Now we introduce a design procedure for template matrix and bias vector given by Liu and Michel(1993). Suppose that we are given m vectors $\alpha_1, \alpha_2, \cdots , \alpha_m$ in B^n which are to be stored as memory vectors for CNN. First we choose a real number $c > 1$ and m vectors $\beta_1, \beta_2, \cdots, \beta_m$, such that $\beta_i = c\,\alpha_i$ ($i = 1, \cdots, m$). Then, our problem is to determine the template matrix T and the threshold vector I which simultaneously satisfy the following equations :

$$\left.\begin{array}{c} -\beta_1 + T\alpha_1 + I = 0 \\ -\beta_2 + T\alpha_2 + I = 0 \\ \vdots \\ -\beta_m + T\alpha_m + I = 0. \end{array}\right\} \tag{7}$$

Where each memory vector β_i corresponds to one of the equilibrium steady states of the vector differential equation system in CNN. Here we set the following matrices :

$$\left.\begin{array}{l} A = (\ \alpha_1 - \alpha_q,\ \alpha_2 - \alpha_q,\ \cdots, \alpha_{q-1} - \alpha_q\ ) \\ B = (\ \beta_1 - \beta_q,\ \beta_2 - \beta_q,\ \cdots,\ \beta_{q-1} - \beta_q\ ). \end{array}\right\} \tag{8}$$

Then the problem is equivalent to determine the matrix T and the vector I which satisfy :

$$\left.\begin{array}{c} B = T A \\ I = \beta_q - T\alpha_q. \end{array}\right\} \tag{9}$$

Now we focus the i-j cell in TVCNN. Then the next condition is needed to be satisfied :

$$b_k = t_k A \quad ; k = n(i-1) + j, \tag{10}$$

where b_k and t_k are the k th row vectors of the matrices B and T respectively. If we take the elements out of the matrix A and vectors b_k and t_k which belong to the r- neighbor cells of kth cell, we obtain the following equation :

$$b'_k = t'_k A', \tag{11}$$

where b'_k, t'_k and A' are the vectors and matrix in which the coupling coefficients having no influence on the kth cell are removed. Since A' is not a square matrix, we apply the singular value decomposition to the matrix A'. Then we obtain the relation :

$$t'_k = b'_k V_{k1} [\lambda]^{-1/2} U^T_{k1} + w_k U^T_{k2}, \tag{12}$$

where U_{k1}, U_{k2}, V_{k1} and V_{k2} are the unit orthogonal matrices which satisfy :

$$A' = [\, U_{k1} \;\; U_{k2}\,]\,[\,\lambda\,]^{1/2} \begin{vmatrix} V^T_{k1} \\ V^T_{k2} \end{vmatrix}, \tag{13}$$

where $[\lambda]$ is a diagonal matrix with the non-zero singular values of the matrix $[A']^T A'$. Thus we can obtain the desired matrix T and the vector I by calculating the above t'_k in each cell.

4. Formulation into Classification / Diagnosis Problem

The three valued CNN enables one to express three kinds of aspects, such as low - middle - high or small - moderate - large and so on. Next, we introduce the following classification procedure. Each data is expressed by a bit map pattern, and the TVCNN associates one of patterns registered beforehand. Namely, the TVCNN comes to classify oneself. The procedure is shown below :

(1) Make an $n = M \times N$ CNN, and allocate $q_1, q_2, \cdots, q_j, \cdots, q_n$ to each cell.

(2) Determine the number of output levels of each cell. From 2 to 4 levels are

recommended. In this paper we use 3 levels output functions for all cells.

(3) Make scaling functions $f_1, f_2, \cdots, f_j, \cdots, f_n$ such that

$$\left[\, \min_i q_j, \max_i q_j \,\right] \xrightarrow{f_j} [-1, 1] \text{ for each } j.$$

(4) Make $M \times N$ lattice patterns representing m classes $\{C_1, C_2, \cdots, C_i, \cdots, C_m\}$.

(5) Embed these patterns in MVCNN. Concretely, determine the template matrix T and the threshold vector I by means mentioned in § 3.

(6) Express respective data by $M \times N$ lattice patterns using the scaling functions

$f_1, f_2, \cdots, f_n$.

(7) Operate the TVCNN with the data pattern as an initial state, and observe a reached pattern.

(8) Classify the object to a class represented by a reached pattern.

5. Example of the classification method

In order to demonstrate this method, we take up an application to the automatic diagnosis problem of liver troubles. Diagnosis of the liver disease is very difficult since it is necessary to judge synthetically for a lot of medical inspection items taking their individual

338

q_1	UA	uric acid
q_2	BUN	blood urea nitrogen
q_3	LDH	lactate dehydrogenase
q_4	PLt	plate let
q_5	ALb	albumin
q_6	γ-GTP	γ-glutamyl transpeptitase
q_7	GOT	glutamic oxaloacetic transaminase
q_8	LAP	leucine aminopeptitase
q_9	TBil	total bilirubin
q_{10}	GOT/GPT	the ratio of GOT to GPT
q_{11}	GPT	glutamic pyruvic transaminase
q_{12}	AFP	alpha-1 fetoprotein
q_{13}	DBil	direct bilirubin
q_{14}	ChE	cholinesterase
q_{15}	ALP	alkaline phosphatase
q_{16}	AFP	alpha-1 fetoprotein

UA	BUN	LDH	PLt
ALb	γ GTP	GOT	LAP
TBil	GOT/GPT	GPT	AFP
DBil	ChE	ALP	AFP

Tab. 1 : Medical inspection indeces   Fig.3 : Allocation of each cell

variations into account. So even the medical specialist occasionary makes a wrong diagnosis when he must examine a large number of patient data. On the contrary, in case of diagnonsis of diabetes, it is sufficient to know only information of FPG (blood sugar).

(1) We select 14 medical inspection indices shown in Table 1, and make a 4×4 CNN, whose each cell is allocated a medical inspection index shown in Fig.3. The neiborhood r is 4. In order to see robustness against the missing or lacking data, we allocated q_8 - cell nevertheless we have no data for LAP.

(2) We adopt three valued output function for all cells because most of inspection indeces are grasped as three stages, for example, γ-GTP roughly has the following three levels : NORMAL (0 - 50), LIGHT EXCESS (50 - 100) and HEAVY EXCESS (100 -). For another example ChE has the levels: SHORTAGE (0 - 200) , NORMAL (200 - 400) and EXCESS (400 -). Both SHORTAGE and EXCESS are regarded as extraordinary.

(3) We make the scaling functions by consulting some technical books of medical science. It is shown in Table 2.

f_1	UA	$(q_1-5)/6$
f_2	BUN	$(q_2-14)/12$
f_3	LDH	$(q_3-175)/120$
f_4	PLt	$(q_4-25)/30$
f_5	ALb	$(q_5-4.2)/1.6$
f_6	γ-GTP	$q_6/100-1$
f_7	GOT	$q_7/80-1$
f_8	LAP	0 (lacking data)
f_9	TBil	$(q_9-5)/8$
f_{10}	GOT/GPT	$\ln(q_7/q_{11})$
f_{11}	GPT	$q_{11}/100-1$
f_{12}	AFP	$(q_{12}-210)/380$
f_{13}	DBil	$(q_{13}-70)/40$
f_{14}	ChE	$(q_{14}-375)/150$
f_{15}	ALP	$(q_{15}-60)/80$
f_{16}	AFP	$=f_{12}$

Tab. 2 : The scaling functions

(4) We take up four liver diseases. Registered patterns shown in Fig. 4 imply 'Healthy person', 'Hepatoma', 'Chronic hepatitis' and 'Liver cirrhosis' , respectively. In these pictures, 1 corresponds to a black pixel, 0 corresponds to a neutral gray pixel and −1 corresponds to a white pixel .

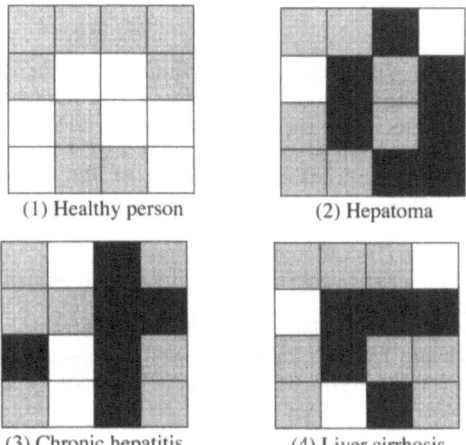

(1) Healthy person   (2) Hepatoma

(3) Chronic hepatitis   (4) Liver cirrhosis

Fig.4: Registered patterns representing four liver diseases

(5) Fig. 5 shows an evolution process of assoiation for the MVCNN. The initial state represents an actual patient's data. Resultingly, the MVCNN associates a pattern of 'Liver cirrhosis' with the initial pattern. Actually, this patient is diagnosed liver cirrhosis by a close medical examination.

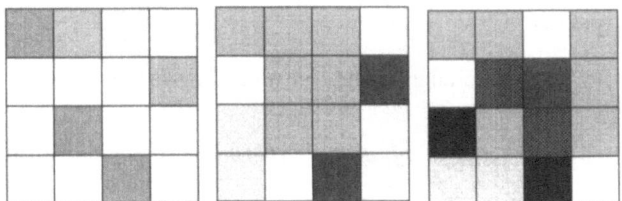

Fig.5: Evolution process of TVCNN

Statistical inference results should be shown. We are offered forty actual patient's data, which consist of ten healthy person data, ten hepatoma data, ten chronic hepatics data and ten liver cirrhosis data from Kawasaki medical college. Diagnostic results by TVCNN system is shown in Table 3, where 2 (#4) implies that TVCNN diagnosed two hepatoma cases wrongly as two liver cirrhosis cases. In Table 3, "irregular convergent" is that the TVCNN did not reach a pattern among the preregistered images; besides "diagnostic sensitivity" is the right diagnosis rate in the past literature (Shigenaga et al. (1993)). They used a concept of rough set and fuzzy if - then rules.

	right diagnosis	wrong diagnosis	irregular convergent	diagnostic sensitivity
(1) healthy person (10)	10	0	0	100.0 %
(2) hepatoma (10)	5	2 (#4)	3	58.8 %
(3) chronic hepatitis (10)	6	1(#1)	3	61.5 %
(4) liver cirrhosis (10)	4	0	6	64.7 %

Tab.3 : Infereance results and comparison of Shigenaga's method

5. Conclusion

We propose a CNN whose each cell has three output values. It is easily extended to one whose each cell has multiple output values We call this CNN multiple-valued cellular neural network (MVCNN). Associative memory function of MVCNN enables one to express several kinds of aspects. We give an application of TVCNN to a diagnosis ploblem of liver troubles. The classification power of the proposed method is demonstrated eqivalent or somewhat inferior in comparison with other method using the fuzzy if -then rules. But it should be emphasized our method is :
(1) Data of LAP is lacking, so it is allocated neutral grey,
(2) Scaling functions and disease patterns are made by us (amateur) without consulting the medical specialists,
(3) Sample size is rather small. Shigenaga *et al.* [8] used 500 sample data.
Thus our diagnosis system using TVCNN has much room for improvement. So it cannot be concluded from Table 3 that the classification power of TVCNN is somewhat inferior to that of fuzzy if-then rules. Generally, fuzzy expert system is effective to these problems. It is, however, complicated and troblesome to make a great number of fuzzy if-then rules when one should make an expert system. On the contrary, the diagnosis system using the TVCNN is designable by simple procedure.
Optimization of diagnosis system using TVCNN (MVCNN) is an important subject in the future study.

References:

Fisher, R.A.(1936) : The use of Multiple Mesurements in Taxonomic Problems, *Annals of Eugenics*, 7, pp.175-188.

Liu, D and Michel,A.N.(1993): Cellular Neural Networks for Associative Memories, *IEEE Trans. Circuits Syst. II*, 40, pp. 119 -121.

Pawlak, Z (1984): Rough Classification,*Intern. J. of Man Machine Studies*, 30, pp. 457-473.

Shigenaga, T., Ishibuchi,H. and Tanaka, H.(1993): Fuzzy Inference of Expert System Based on Rough Sets and Its Application to Classification Problems," *J. of Japan Society for Fuzzy Theory and Systems*, 5, 2, pp. 358-366.

Application of Kohonen maps to the
GPS Stochastic Tomography of the Ionosphere

M.Hernández-Pajares, J.M.Juan and J.Sanz

Research Group of Astronomy and Space Geodesy
Universitat Politècnica de Catalunya
Campus Nord Mod. C3,B4
c/. Gran Capità s/n., 08034 Barcelona, Spain
e-mail: matmhp@mat.upc.es

Summary: The adaptative classification of the rays received from a constellation of geo-detic satellites (GPS) by a set of ground receivers is performed using neural networks. This strategy allows to improve the reliability of reconstructing the Ionospheric electron density from GPS data. As an example, we present the evolution of the radially integrated electron density (Total Electron Content, TEC) during the day 18th October 1995, coinciding with an important geomagnetic storm. Also the problems in the vertical reconstruction of the electron density are discussed, including the data coming from one Low Earth Orbiter GPS receiver: the GPS/MET. Finally is proposed as main conclusion a new strategy to estimate the ionospheric electron distribution, using GPS data, at different scales —the 2-D distribution (TEC) at Global scale and the 3-D distribution (Electron density) at Regional scale—.

1. Introduction

about the Problem:

As it is well known, the Ionosphere is the part of the Earth Atmosphere containing free ions and causes a frequency-dependent delay in the propagated EM-signals, being proportional to the columnar density of electrons (TEC) (see for instance Davies 1990, page 73).

This is a *distorting* physical effect for Space Geodesy and Satellite Telecommunications activities, that can be used in a *positive sense* to estimate the global 3-D distribution of the free-electrons in the Atmosphere, from dual frequency delay observations, i.e. for the *Stochastic Tomography of the Ionosphere.*

about the Data:

To achieve this objective, we need during a certain time interval, a high *sampling rate* of the Atmosphere, with so many rays in so many orientations as possible. Nowadays, the unique system that provides so many observations, continuously and on a planetary scale, is the *Global Positioning System* (GPS). Its space segment contains a constellation with more than 24 satellites emitting continuously carrier and code phases in two frequencies L1 ($\simeq$1.6 GHz) and L2 ($\simeq$1.2 GHz) (see for instance more information in Seeber 1993, pages 209-349). In the GPS user segment, it is possible to get *few hours later* the public domain data gathered from a global network of permanent receivers, such as International GPS Service for Geodynamics (IGS, Zumberge et al. 1994), with more than 100 stations worldwide distributed, mainly concentrated in the Northern Hemisphere, in North America and Europe. Also the *Low Earth Orbiters* containing GPS receivers (LEO), are becoming usual. Hajj et al. (1994) conclude than the GPS/MET LEO observations, are important to resolve the vertical structure of the Ionosphere.

about the Model and Goals:

But the amount of data implied (more than $1.5\,10^6$ delays/day collected in the IGS network) jointly with its inhomogeneous distribution, makes it difficult to solve the problem, so that new algorithms and strategies must be considered to perform the tomography of the Ionosphere.

In this paper we mainly discuss the different data analysis problems encountered in the estimation of the ionospheric electron distribution using mainly ground data from the IGS network, and the solutions adopted, emphasizing three points:

1. The adaptative clustering of the rays, using the Kohonen neural network algorithm. This technique will be applied for bidimensional modeling (TEC) of the overall Ionosphere, i.e. at Global scale, generating cells adapted in size to the variable sparsity of the data.

2. The problems coming from the bad geometry to solve the vertical structure of the Ionosphere using data coming solely from ground receivers are also discussed. We will adopt the model using regular cells instead of the adaptative one, due to the better performance related with the lower discretization error in presence of high correlations.

3. The inclusion of a set of GPS/MET data, consisting on *orthogonal* rays to the ground data rays, improve the estimation of the ionospheric vertical structure.

2. The Model

The Scenario:

We have the following situation:

- From each ground station (see figure 1) we simultaneously measure with a certain sampling rate (i.e. 1 time/30 sec) the ionospheric delays experienced by the rays received from the visible satellites (i.e. 4-8).

- This rays cross different parts of the *nearby* Ionosphere to the respective station.

- Between observation epochs the Earth rotates, and the part of the sounded Ionosphere has changed.

- We assume that the Ionosphere is *stationary* in a Sun-fixed reference system[1].

- Then, we have chosen an Geocentric Equatorial pseudo-Inertial reference system (GEI), where the X-Axis points towards the Vernal Equinox and the Z-Axis points towards the Geographic North Pole; the XY plane is the celestial equator. In the GEI the Sun is only *moving* 1 degree/day. The associated spherical coordinates are the right ascension α (*azimuthal* angle) and the declination δ (angle referred to the equator).

The General Model:

Not taking into account the *bending-effect* of the ray, we have to solve the following

---

[1]This is not true in *second order*, due to the magnetic field effects and the variability of geomagnetic conditions (see for example Sanz el al.. 1996)

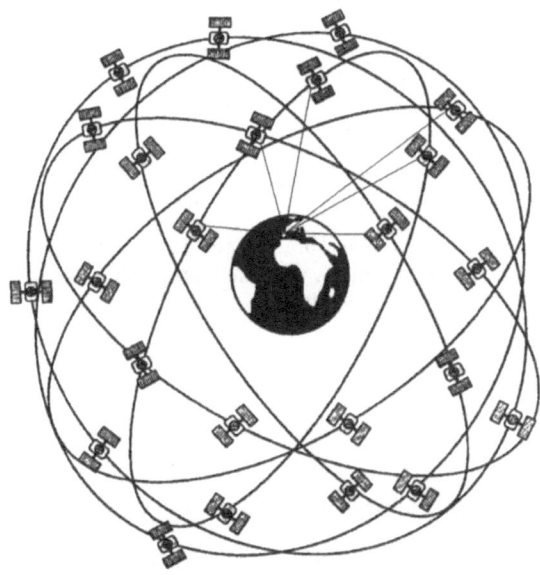

Figure 1: Layout of the GPS full constellation (24 satellites) orbiting around the Earth. The GPS rays corresponding to the observations of a given station at a given time are also represented.

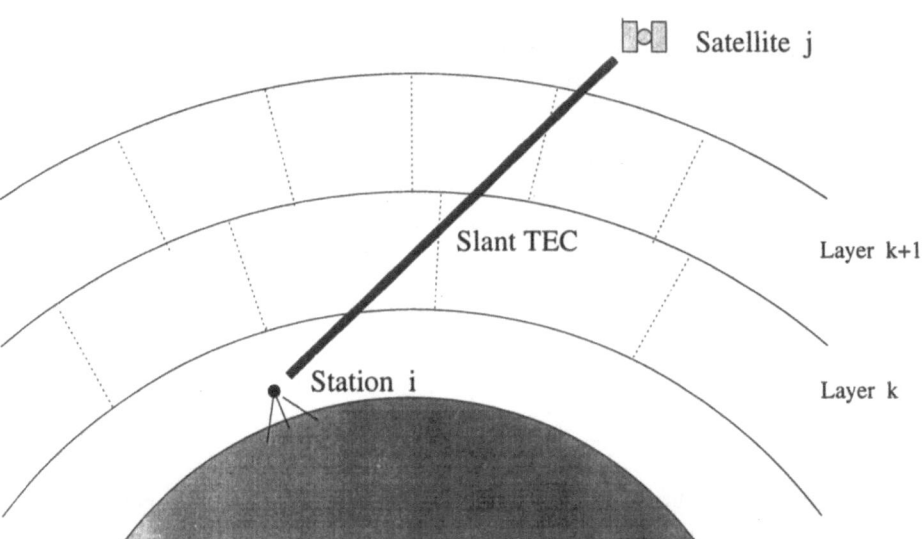

Figure 2: An scheme with one pair-station satellite

integral equation:

$$I_t(\vec{r}_i, \vec{r}^j) = \int_{\vec{r}_i}^{\vec{r}^j} N(\vec{r})\, ds + D_i + D^j \qquad (1)$$

where:

- $N(\vec{r})$ is the electron density at position $\vec{r}$, which is a point that belong to the ray between station i and satellite j at distance s of the station.

- $I_t(\vec{r}_i, \vec{r}^j)$ is the *ionospheric combination* corresponding to the ray from the satellite j for the station i at time t, obtained preprocessing the GPS observations (see for instance Sardon et al. 1994).

- $\vec{r}_i$, $\vec{r}^j$ are the position vectors of the station i and satellite j at the observation time.

- D_i and D^j are the instrumental delays associated to the station i and satellite j.

- The integral path is assumed to be extended over the linear ray.

In order to estimate the density N from equation 1, we can expand it within a certain basis functions,

$$N(\vec{r}) = \sum_l \lambda_l g_l(\vec{r}) \qquad (2)$$

in such a way that:

$$I_t(\vec{r}_i, \vec{r}^j) = \sum_l \lambda_l \int_{\vec{r}_i}^{\vec{r}^j} g_l(\vec{r})\, ds + D_i + D^j \qquad (3)$$

Our final purpose is to get an estimation of N, from the data I, knowing $\vec{r}_i$, $\vec{r}^j$. The instrumental delays D_i, D^j are also unknowns. Then, the general model proposed consists on a certain number of geocentric spherical shells, covering the Ionosphere sampled by the GPS rays. These shells define the floor and ceil of each layer, which is going to be partitioned within pixels.

For each layer k (see figure 2) we define the pixels $\{P_{k,l}\}_{\forall l}$ as those given by a set of centers $\{w_{k,l}\}$ with the minimum distance criterium:

$$P_{k,l}(\vec{r}) = \begin{cases} 1 & \text{if } \|\vec{r} - \vec{w}_{k,l}\| <= \|\vec{r} - \vec{w}_{k,l'}\| \ \forall l' \\ 0 & \text{otherwise} \end{cases} \qquad (4)$$

Then we can write applying to equation 3:

$$I_t(\vec{r}_i, \vec{r}^j) = \sum_k \sum_l N_{k,l} \int_{\vec{r}_i}^{\vec{r}^j} P_{k,l}(\vec{r})\, ds + D_i + D^j \qquad (5)$$

where the unknowns λ have been reinterpreted as the mean electronic density in the cell $N_{k,l}$. From the last equations, we get:

$$I_t(\vec{r}_i, \vec{r}^j) = \sum_k \sum_l N_{k,l} \Delta s_{k,l} + D_i + D^j \qquad (6)$$

where $N_{k,l}$, $\Delta s_{k,l}$ are the mean density, and fraction of the ray length respectively, corresponding to the cell l of layer k.

	Regular grid	Adaptative cells
Cells poor populated?	Yes	No
Discretization error?	Lower in general	Higher in general
CPU time	Fast	Slow ($\simeq 3$ times slower)
Kalman filtering	Easy	Non-immediate

Table 1: Comparison of some features of the *regular* and em adaptative grid to get cells for each considered layer of the Ionosphere

Finally we have in equation 6 a linear overdetermined system, that will be solved by Least Squares taking into account techniques such the Singular Value Decomposition following Hajj et al. 1994.

Adaptative pixels versus Regular pixels:

If we choose an pixel basis function to expand the electron density, two approaches –among others– are possible to define the pixels or cells:

1. To consider a *regular grid*

2. To get *adaptative* cells in such a way that an approximate equal number of rays per cell let be assured.

Some advantages and disadvantages are summarized in Table 1.

The main reason to choose in this paper an adaptative basis is that it is a more robust approach adapting the cell size to the data density, avoiding is this manner the existence of void or *few visited* cells. The centers are obtained applying the unsupervised classifier known as the *Konohen neural network* or Self Organizing Map (Kohonen, 1990) to the crossing points of each ray with the mean shell of the given layer. The resulting cells are adapted in size allowing to guarantee a minimum number of crossing points per cell (see appendix A for a detailed explanation of the algorithm, and Murtagh & Hernández-Pajares 1995 for an assessment).

Taking into account the main problems of poor populated cells, i.e. cells crossed by few rays, and the discretization error, we propose the following combined strategy to estimate the electron density in equations 6:

Firstly, we estimate a 2-D Global Model of the overall Ionosphere (using all the IGS data). To do so we solve equation 6, with only 1 layer, avoiding in this manner the high correlations between layers as is commented in Hajj et al 1994. Then it is possible to use adaptative basis of functions to avoid the problem of poor populated cells (mainly in the South Hemisphere du to the lack of stations). In fact the low correlations of the 2-D model makes not critical the greater discretization error of the adaptative grid model regarding on the regular grid model.

Secondly, we solve equation 6 with several layers, performing a 3-D Regional model of the Ionosphere, in a region crossed by a high number of rays to avoid the problem of poor populated cells. Due to the high correlations between layers that appear in the estimation, it is useful to constrain the instrumental delays with the values obtained in the first run, and to use a regular gridding. The regular cells model presents lower discretization error, that is a critical aspect when we have high correlations.

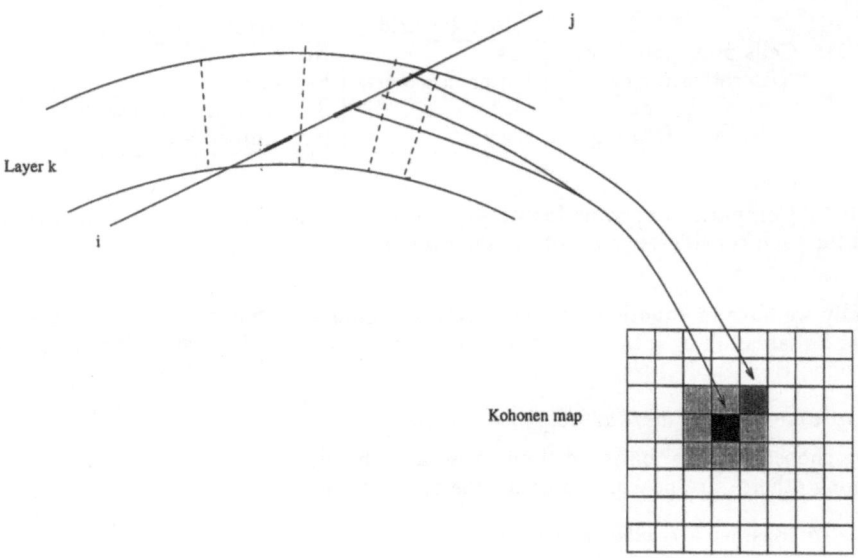

Figure 3: Scheme of the strategy adopted to classify the *sub-rays* to the corresponding cells, with a minimum computation load –explanations in the text–.

Some practical aspects of the problem:

In order to implement the adaptative strategy, we have to deal with the problem of the non-regular boundaries between the cells. Indeed, to solve Eq. 6, we have to compute for each ray and for a given layer, the cells crossed and the ray path fraction for each one. One procedure is to *digitize* the ray within *sub-rays* of length L, counting how many of them are contained in the crossed cells (see figure 3).

This approach implies to get the cell to which each sub-ray belongs to, and this can be an expensive operation, due to the amount of sub-rays that we have. One way to avoid such computation load, is to use the physical continuity of the ray, knowing the neighboring between the cells. The Kohonen neural network at the same time that constructs cells with an approximate equal number of elements, gives a bidimensional topological map of the cells (see appendix A).

The strategy adopted is:

1. **Computing the cells:** From the crossing points of all the rays with the mean spherical shell of the layer, the centers of the cells are obtained. Simultaneously we get the cell membership of all the rays and the 2-D Kohonen map of the centers, which maintains their 3-D neighboring relationships. The cells are defined in the usual way by the *nearest center* criterium.

2. **Computing the crossing fractions of the rays:** For a given ray, we know now to which cell the center sub-ray belongs to (step 1), and hence where it is placed in the Kohonen map. For a enough small length and taking into account the continuity of the ray, the next upper (lower) sub-ray must belong to the same or to a neighbor cell (see figure 3); Then only a small subset of centers must be explored, neighbors in the Kohonen map to the center of the last subray (see again figure 3).

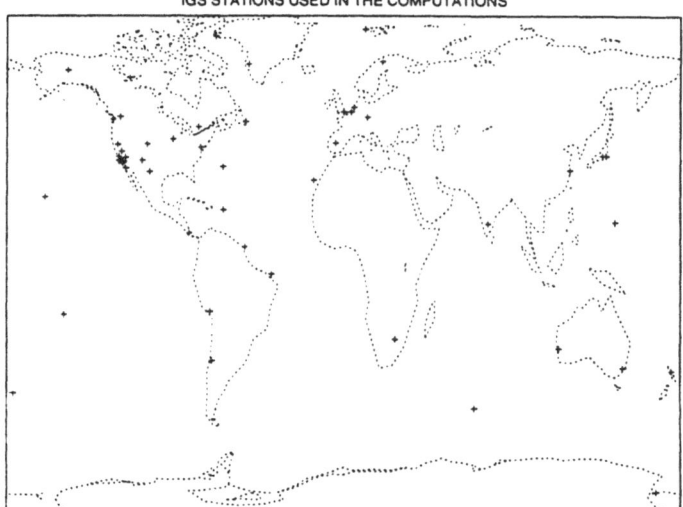

IGS STATIONS USED IN THE COMPUTATIONS

Figure 4:

3. Computations and Results

The Data

Some features of the data sets used —which have been provided by the IGS via anonymous ftp— are summarized in Table 2. Also the stations distribution is plotted in figure 4.

Estimated Models

We have computed basically two different models using Least Squares:

2-D Global Ionosphere; we use the Self Organizing Map algorithm to generate the adaptative cells in order to compute the TEC for the Data Set G (subsets 2-7h, 7-12h, 12-17h and 17-22h, see figure 5)[2]. We have solved the equation 6 for one unique layer between 300 and 400 km, taking 400 cells (self-organized along a 20x20 Kohonen map) and a subray length of 5 km. The cells presents sizes ranging from few squared-degrees to 10-100 times greater, depending on the small or large sparsity of the data. In figure 5 appears an important result: the detection of the TEC increase due to the start of the geomagnetic storm in the last time interval (from 17 to 22 h UT).

3-D Regional Ionosphere; we describe the electron density with 5 spherical layers in equation 6, with height boundaries at 50, 200, 350, 500, 650 and 800 km. The regular gridding consists on cells of 5x5⁰. We have performed two computations: one with the data set RGROUND (only ground data, see table 2) and other using also one GPS/MET occultation (data set RGROUND+RMET). In both cases we have reemplazed the real observations (ionospheric delays) by those coming

---

[2]During this period one geomagnetic storm happened, with a high variability of the ionospheric electron distribution (see for instance Web document at http://bolero.gsfc.nasa.gov/gov/ solart/cloud/cloud.html)

	Global		Regional
	Data set G	Data set RGROUND	Data set RMET
GPS Receivers	60	31	1(GPS/MET occultation 0207)
Time Interval	2-22h UT	18-23h UT	20h30m-20h31m UT approx.
Number of subsets	4	1	1
Right ascension range of the rays at 330 km height	0 to 360^0	185 to 230^0	185 to 230^0 approx.
Declination range of the rays at 330 km height	-90 to 90^0	20 to 60^0	20 to 60^0 approx.
Elevation mask	0^0	0^0	none (occultation geometry)
Number of rays	$\simeq$200000	6581	3347

Table 2: Description of the Data Sets considered in the computations, all of them corresponding to the 18th October 1995, with a geomagnetic storm and P-code not encrypted (Antispoofing Off)

from a very simple model: the Ionosphere as a one spherical layer with constant density of 10^{12} e/m^2, with boundaries at 240 and 400 km height. Nevertheless the geometry (the rays) is the real one. The results (figure 6) are quite significative showing the important improvement in the estimation when we add to the ground data, the observations coming from one orbital GPS receiver such the GPS/MET.

5. Conclusions

We present in this paper an study of a difficult problem: To reconstruct the 3-D Ionospheric electron distribution from GPS data.

The use of the neural network in this work makes it possible to overcome the problem of the non-homogeneous sampling, dividing each layer into a partition of cells or clusters with a similar number of rays. The existence of the topological relationships between the neighbors centers in the Kohonen map, is taken into account to reduce the computation load. In order to solve it, we have consider a new approach than consist on a basis of adaptative pixels (*Kohonen adaptative pixel basis*) that are defined from a certain number of centers , obtained with the Kohonen artificial neural network.

As the main conclusion, a new modeling of the Ionosphere is proposed within two steps: firstly a bidimensional Global model, describing the TEC and instrumental delays by mean of adaptative cells; and secondly a tridimensional regional model with regular cells for the electron density within a region with a high number of observations (Northern hemisphere). We ought to include Low Earth Orbital GPS receivers, data such GPS/MET, to improve the vertical resolution, and constraining the instrumental delays with the values obtained in step 1. The cells in this case are chosen regular in angular size, in order to diminish the discretization error in the description of electron density, in a model with high correlations between layers.

This new strategy is supported by the main results obtained for the day 18th October 1995:

- the detection in the global model of the increase of electron content during the geomagnetic storm start.

- the capability to estimate the 3-D structure of the ionosphere with a large number of regional data, specially when we include orbital GPS occultation data from

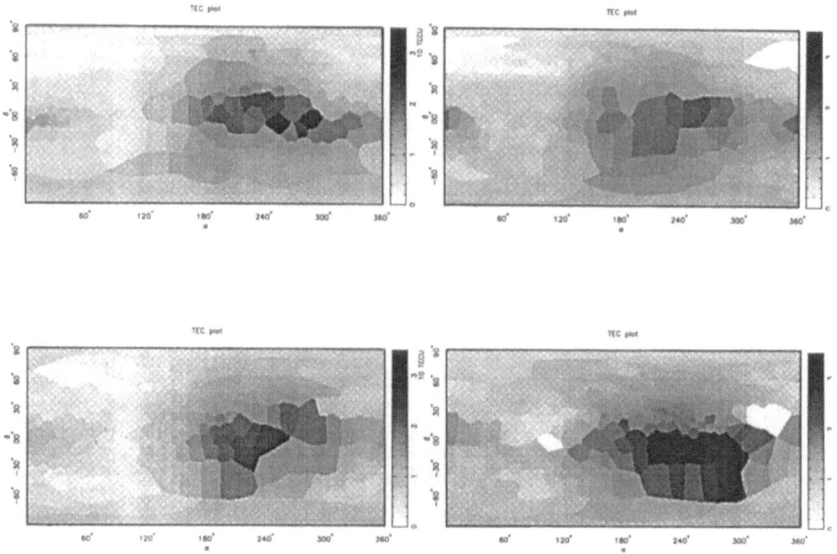

Figure 5: Global model of the TEC for the day 18th October 1995, for the data subsets 2-7h, 7-12h, 12-17h and 17-22h respectively

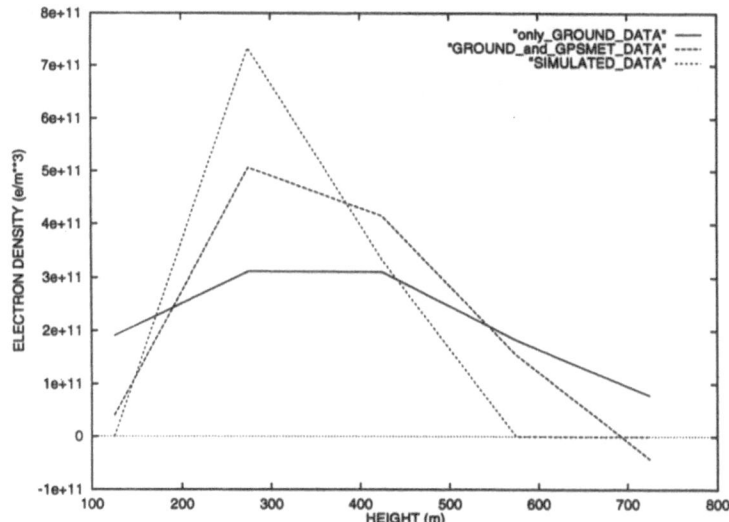

Figure 6: Comparison of the electron density vertical profile at $\alpha = 207.5^0$ and $\delta = 37.50^0$, estimated from the data sets RGROUND and RGROUND+RMET (day 18th October 1995) with the real rays but with simulated ionospheric delays with a very simple model.

orbital receivers. This last point confirms the study of Hajj et al. about the importance of including orbital data to reconstruct the Ionosphere with GPS.

6. Acknowledgments

We would like to acknowledge to the International GPS Service for Geodynamics and to the University Corporation for Atmospheric Research for the availability ot the GPS data used in this research. This work has been partially supported with funds from the Spanish government projects PB94-1205 and PB94-0905 (DGICYT).

References:

Davies K. (1990): Ionospheric Radio. IEE ElectroMagnetic Waves Series 31, Peter Perigrinus Ltd., London.

Hajj G.A., Ibañez-Meier R., Kursinski E.R., Romans L.J. (1994): Imaging the Ionosphere with the Global Positioning System. *Imaging Systems and Technology*, Vol.5, 174-184.

Kohonen T. (1990): The self-organizing map. *Proceedings of the IEEE*, Vol.78, pages 1464-1480.

Murtagh F., Hernández-Pajares M. (1995): The Kohonen self-organizing map method: an assessment. *Journal of Classification*, Vol.12, 165-190.

Press W.H., Flannery B.P., Teukolsky S.A., Vetterling W.T. (1986): Numerical Recipes. The Art of Scientific Computing. Cambridge Univ. Press, Cambridge.

Sanz J., Juan J.M., Hernández-Pajares M., Madrigal A.M. (1996): GPS Ionosphere imaging during the "October-18th 1995" magnetic cloud. European Geophysical Society meeting, The Hague, May 1996.

Sardón E., Rius. A., Zarraoa N. (1994): Estimation of the transmitter and receiver differential biases and the ionospheric total electron content from Global Positioning System observations. *Radio Science*, Vol.29, No.3, pages 577-586.

Seeber G (1993): Satellite Geodesy. Walter de Gruyter, Berlin.

Zumberge J., Neilan R., Beutler G., Gurtner W., 1994. The International GPS Service for Geodynamics-Benefits to Users. ION GPS-94, Salt Lake City, Utah.

APPENDICES

A The Self Organizing Feature Map algorithm

One of the competitive neural network algorithms is the *Self-Organizing Map* (SOMA, also named *Kohonen network*). It has the special property of creating spatially organized *representatives* of the *centroids* (weights of the output neurons) found in the input vectors. The resulting maps resemble real neural structures that appear in the cortices of developed animal brains. The SOMA has also been successful in various pattern recognition tasks involving noisy signals such as speech recognition (see a summarized review in Kohonen 1990).

The basic aim of this neural network is to find a smaller set $\{\vec{w}_1, \ldots, \vec{w}_c\}$ of c centroids that provides a good approximation of the original set S of n objects — *input space* —, with m attributes, encoded as *vectors* $\vec{x} \in S$. Intuitively, this should mean that for each $\vec{x} \in S$ the distance $\| \vec{x} - w_{\vec{f}(x)} \|$ between x and the closest centroid $w_{\vec{f}(x)}$ should be small. Simultaneously the algorithm arranges the centroids so that the associated map $f(.)$ from S to A,

$$
\begin{array}{ccc}
f : S \subset \mathbb{R}^n & \longrightarrow & A \subset \mathbb{N}^2 \\
\vec{w}_l & \longrightarrow & f(\vec{w}_l) = (i_l, j_l)
\end{array}
\tag{7}
$$

reflects the topology of the set S in a least distorting way, where A is the *representation space*, a 2-dimensional set of indices known as the *Self-Organizing Feature Map* or *Kohonen Map*. Proximity in A means similarity between the global properties of the associated groups of objects.

From a detailed point of view, the *Kohonen network* is composed of a set of c nodes or neurons. The algorithm scheme is:

1. We initialize at random the weights of the c nodes of the grid with small values: $\{\vec{w}_1(0), \ldots, \vec{w}_c(0)\}$. After training, every neuron $l \in \{1, \ldots c\}$ will represent a group of objects (stars) with similar features, and the weight vector $\vec{w}_l$ will approximate to the centroid of these associated objects.

2. For each of the n training vectors of the overall database, $\vec{x}^p$:

    (a) We find the node k whose weight $\vec{w}_k$ best approaches $\vec{x}^p$ (d can represent the Euclidean distance): $d(\vec{w}_k, \vec{x}^p) \leq d(\vec{w}_l, \vec{x}^p), \forall l \in \{1, \ldots, c\}$.

    (b) We update the weight of the winning node k and its neighbours, $N_k(t)$, approaching the training vector as closely as possible:

$$
\vec{w}_l(t) = \left\{ \begin{array}{ll} \vec{w}_l(t-1) + \alpha(t) H(t) \left(\vec{x}^p - \vec{w}_l(t-1) \right) & l \in N_k(t) \\ \vec{w}_l(t-1) & l \notin N_k(t) \end{array} \right. \forall l \in \{1, \ldots, c\}
\tag{8}
$$

    where:

    - $\alpha(t)$ is the learning rate, a suitable monotonic decreasing sequence of scalar-valued gain coefficients, $0 < \alpha(t) < 1$.
    - The radius R_t of the activated neighbourhood, $N_k(t)$, is a monotonically decreasing function of the iteration t.
    - $H(t) = \exp \left(\frac{-d^2}{4R_t^2} \right)$ is a function that represents in Eq. 8 the decay of the activation depending on the distance d in A between the winner unit k and the considered unit l, $d = \sqrt{(i_l - i_k)^2 + (j_l - j_k)^2}$.

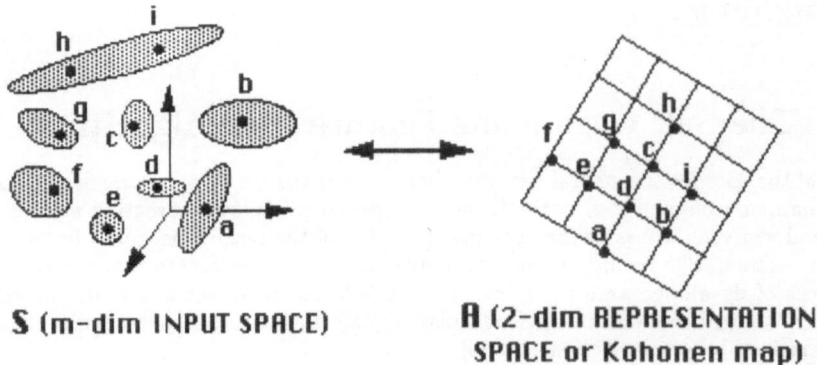

S (m-dim INPUT SPACE)   **A (2-dim REPRESENTATION SPACE or Kohonen map)**

Figure 7: Ordering induced by the Kohonen network: after training the data, the centroids which are close within the *representation space* A will also be close within the input space S.

Updating neighbours' weights instead of just that of the winning node assures the ordering of the net, in such a way that centroids which are close in representation space A (dimension 2) are updated to become similar in input space S of dimension m (see Fig. 7).

3. Process 2 is repeated for the overall database until good final training is obtained.

The final point density function of $\{\vec{w}_1, \ldots, \vec{w}_c\}$ is an approximation of the continuous probability density function of the vectorial input variable $\vec{g}(\vec{x})$ (Kohonen 1990, p. 1466).

Capacities, Credibilities in Analysis of Probabilistic Objects by Histograms and Lattices

Edwin Diday[1],   Richard Emilion[2]

[1] *CEREMADE - Université Paris 9*

INRIA Rocquencourt, 78135 le Chesnay Cedex, France.

e-mail : Edwin.Diday@inria.fr

[2] *Centre de Calcul Informatique*

Université de Dakar, Sénégal.

Summary: Capacities and credibilities appear in the modelling of objects described by random variables with probability distributions. The general aim is to extend standard data analysis to such objects. Only histograms and lattices are investigated in this more general setting. Sub-additive ergodic theorems are used as a tool in the histogram analysis of these objects.

1. Introduction

In this paper, "first order objects" are described by rows and columns of a data table (which generalizes standard data tables) in which each cell represents a random variable and contains its probability distribution. For instance, if a row is associated to an animal and the weight to a column, then the cell corresponding to this row and this column contains the probability distribution of the weight of this animal. Such a row defines what we call a "probabilistic object". Note that standard data tables are a simple particular case of our model : take distributions concentrated on a single point.

A "second order object" is defined by a class of first order objects. For example, in order to describe the colour of a class of individuals in which some are yellow and others are red, we say that "the colour of a member of the class is yellow or red". In this way we "generalize" the description of the individuals by giving the description (yellow or red) to the colour of the class. In other words the colour yellow enter in the description of the class if "at least one of its members is yellow". We extend this notion of generalization to the case of individuals described by probabilistic objects, in describing the class by the probability that at least one member of the class be yellow. We call this probability the "capacity" of the class to be yellow since it can be shown that it satisfies the mathematical properties of a "capacity" in the sense of Choquet (1954).

We may also be interested in a specialization of the class. Instead of obtaining a "generalized" second order object we obtain a "specialized" second order object. It is described by all the properties which appear in the description of all the individuals of the class. For example, the colour of a specialized second order object is yellow if all the members of the class that it represents are yellow. When the individuals are described by probabilistic objects the specialized class is described by "the probability that all the members of a class are yellow". This probability is called the "credibility" of the colour yellow for this class since it can be shown that it satisfies the mathematical properties of a "credibility" (or "belief") in the sense of Schafer (1976). When the n random variables associated to each member of the class are independent, the capacity and credibility may respectively be calculated by a t-conorm and t-norm as defined by Schweizer and Sklar (1983).

A "concept" is usually defined by an intent and extent. It may be modelized by a "symbolic object" also defined by an intent and an extent, where the intent is the description of a second order object (in terms of capacities or credibilities, in this paper) and the extent is defined by the set of individuals (described by probabilistic objects, in this paper) which satisfy this description as well as possible. The general aim of this paper is to extract such (concepts) from a set of probabilistic objects versus an extension of standard data analysis methods to such objects.

2. Probabilistic objects

2.1 Basic model:

Several real situations can be modelled as follows :
Let C be a set whose elements c are called objects and let $\mathcal{C}$ denote a σ-algebra on C. Let (Ω, F, μ) be a measure space and I a set of indices.
A description of the object c is a family $(X_{c,i})_{i \in I}$, of measurable maps defined on Ω and taking values in a measurable space $(O_i, \mathcal{O}_i)$. If μ is a probability $\mathcal{P}$ then the objects will be called probabilistic objects, described by the random variables $X_{c,i}$ with laws denoted by $\tilde{X}_{c,i}$. By definition the distribution law $\tilde{X}_{c,i}$ is a probability on $\mathcal{O}_i$ such that $\tilde{X}_{c,i}(O) = \mathcal{P}\{X_{c,i}^{-1}(O)\}$ for any $O \in \mathcal{O}_i$.

2.2 Examples :

– Let C be a set of computer processors which are working in various environments i, $i \in I$. The time of execution of a task $\omega \in \Omega$ for the processor c under condition i is not deterministic, therefore it will be given by $X_{c,i}(\omega) \in O_i$ where $X_{c,i}$ is a r.v.
– Let C be a set of individuals who are submitted to different tests of type i. The random result of the individual c to a test $\omega \in \Omega$ of type i will be $X_{c,i}(\omega)$.
– Let C be a set of specimen (for example, insects or plants from a same species) given with different descriptors i, $i \in I$ (for example, the size, the age, etc...). Due to the variability of these descriptor's values on a same specimen when time varies, the random value of a descriptor i of the specimen c at time $\omega \in \Omega$ will be $X_{c,i}(\omega)$.

3. Capacities and credibilities

3.1 Capacities

Let $i \in I$, $A \in \mathcal{C}$ and $O \in O_i$. We will say that the system A is *not able* to reach the objective O if, for all $c \in A$, $\mu(X_{c,i} \in O) = 0$. Hence, we are tempted to evaluate the *ability* or the *capacity* of A to reach O, by the number $\mu(\bigcup_{c \in A}(X_{c,i} \in O))$ (for countable A) or by the number $\sup_{c \in A}(\mu(X_{c,i} \in O))$. In the above examples we will then get the capacity for a set of processors to complete a task before t seconds, the capacity of a set of individuals to succeed, etc... It turns out that these definitions agree with the capacities as defined by Choquet (1954) :

Definition :

A capacity on $(C, \mathcal{C})$ is a map κ from $\mathcal{C}$ to $\Re^+$ such that

i) $\kappa(\emptyset) = 0$ 　　　　　　　　　　ii) $\kappa(A_1 \bigcup A_2) \leq \kappa(A_1) + \kappa(A_2)$

iii) $A \subseteq B \Longrightarrow \kappa(A) \leq \kappa(B)$ 　　iv) $\kappa(\lim \uparrow A_n) = \lim \uparrow \kappa(A_n)$

The capacity κ is a strong capacity if, in addition, we have:

ii') $\kappa(A_1 \bigcup A_2) + \kappa(A_1 \bigcap A_2) \leq \kappa(A_1) + \kappa(A_2)$

The capacity κ is a capacity of order ∞ if, in addition we have

$\kappa(A_1 \bigcup A_2 ... \bigcup A_n) \leq \sum_i (\kappa A_i) - \sum_{i<j} \kappa(A_i \bigcap A_j) + ... + (-1)^{n+1} \kappa(\bigcap_j A_j)$.

Proposition 1 :

Let $\kappa(A, O) = \sup\limits_{B finite \subseteq A} (\mu\{ \bigcup\limits_{c \in B} (X_{c,i})^{-1}(O) \})$; then the map $A \longrightarrow \kappa(A, O)$ (resp.

$O \longrightarrow \kappa(A, O))$ is a capacity of order ∞ on $\mathcal{C}$ (resp. on $\mathcal{O}_i$).

3.2 Credibilities

What happens if in section 3.1, we replace union by intersection ? Actually, passing to complementary, we get in a natural way credibilities instead of capacities.

Definition :

A credibility on $(C, \mathcal{C})$ is a map β from $\mathcal{C}$ to $\Re^+$ such that

i) $\beta(\emptyset) = 0$

ii) $\beta(A_1 \bigcap A_2 ... \bigcap A_n) \leq \sum\limits_{j=1}^{n} \beta(A_j) - \sum\limits_{i<j} \beta(A_i \bigcup A_j) + ... + (-1)^{n+1} \beta(\bigcup\limits_{j=1}^{n} A_j)$

Proposition 2 :

Let $\beta(O) = \mathcal{P}\{ \bigcap\limits_{c \in A} (X_{c,i}(O))^{-1} \}$ for any fixed countable $A \in \mathcal{C}$; then the map $O \longrightarrow$

$\beta(O)$ is a credibility on $\mathcal{O}_i$.

3.3 T-norms and t-conorms

We will now get some new capacities by using some special maps introduced by Schweizer and Sklar (1983).

Definition

A t-norm T is a map from $[0,1]^2$ to $[0.1]$. $(u, v) \longrightarrow u \, T \, v$ such as for all t, u, v, w in $[0,1]$

i) $u \, T \, (v \, T \, w) = (u \, T \, v) \, T \, w$ 　　　　(Associativity)

ii) $t \, T \, u \leq v \, T \, w$ whenever $t \leq v, u \leq w$ 　(Increasing)

iii) $u \, T \, v = v \, T \, u$ 　　　　　　　　(Commutativity)

iv) $u \, T \, 1 = u$ 　　　　　　　　　　(Identity law)

Examples

$u \, T_1 \, v = u \times v$ and its dual $u \, T_1^* \, v = u + v - u \times v$,

$u \, T_2 \, v = min(u, v)$ and its dual $u \, T_2^* \, v = mar(u, v)$,

$u \, T_3 \, v = mar(u + v - 1, 0)$ and its dual $u \, T_3^* \, v = min(u + v, 1)$

$u \, T_4 \, v = 1 - min\{((1-u)^p + (1-v)^p)\}^{\frac{1}{p}}$ and its dual $u \, T_4^* \, v = min((u^p + v^p)^{\frac{1}{p}}); p > 0$.

$u \, T_5 \, v = log_\alpha(1 + \frac{(\alpha^u - 1)(\alpha^v - 1)}{(\alpha - 1)})$ $(0 < \alpha < \infty)$

and its dual $u \, T_5^* \, v = 1 - log_\alpha(1 + \frac{(\alpha^{1-u} - 1)(\alpha^{1-v} - 1)}{(\alpha - 1)})$

where $log_\alpha(u)$ denotes logarithm to the base $\alpha > 0$.

4. Capacities and credibilities histograms

Let $X_{c,i}$ be a real-valued r.v. and a sample $X_{c,i}^1, X_{c,i}^2, ..., X_{c,i}^n$ of n independent r.v.

distributed as a r.v $X_{c,i}$. In this case, each $X_{c,i}^j$ is defined on Ω^n. For example, let $\omega = (\omega_1, ..., \omega_n) \in \Omega^n$ the result of the individual c to the test w_j of type i is $X_{c,i}^j(\omega)$. Consider the probability measure defined by :
$\mathcal{P}_n(\omega)([s,t]) = ($ number of $X_{c,i}^1(\omega), X_{c,i}^2(\omega), ..., X_{c,i}^n(\omega)$ between s and $t)/n$.
Histograms of frequencies can be derived from this measure : given k real numbers $s_1 < s_2 ... < s_k$, we represent the function $\dfrac{H_{s_j, s_{j+1}}}{(s_{j+1} - s_j)}$ on $[s_j, s_{j+1}[$ where $H_{s_j, s_{j+1}}$ denotes $\mathcal{P}_n(\omega)([s_j, s_{j+1}[)$.
A consequence of the large numbers strong law is that $\mu_n([s,t[)$ is a good approximation of $\mathcal{P}(X_{c,i} \in [s,t])$ if n is large enough. Moreover, as an application of Lebesgue differentiation theorem the preceding function is a good approximation of the density of $X_{c,i}$, (if its distribution has a density) when the steps $s_{j+1} - s_j$ are small enough. Now, starting with two (or more) r.v. $X_{c,i}, X_{d,i}$ as above, and putting $A = \{c, d\}$, it is natural to compute the capacity $\kappa_{A,n}(\omega) = \mathcal{P}_{c,n}(\omega) * \mathcal{P}_{d,n}(\omega)$ and the credibility by $\beta_{A,n} = \mathcal{P}_{c,n}(\omega) \, T \, \mathcal{P}_{d,n}(\omega)$, where T is a t-norm and $*$ a t-conorm. (We have omitted the index i). Note that in case of independent r.v. $X_{c,i}$ we can take $u \, T \, v = u \times v$ and $u * v = u + v - uv$.
Put $\kappa_{s,t} = \lim\limits_{n \to \infty} \kappa_{A,n}(\omega)[s,t[$ and $\beta_{s,t} = \lim\limits_{n \to \infty} \beta_{A,n}(\omega)[s,t[$. Due to the continuity of t-norms and t-conorms, these capacities and credibilities converge as n tends to ∞. Since we have $\kappa_{s,u} \le \kappa_{s,t} + \kappa_{t,u}$ and $\beta_{s,u} \ge \beta_{s,t} + \beta_{t,u}$ for $(s \le t \le u)$, we get two new type of histograms : *histogram of capacity which are sub-additive and a histogram of credibility which are super-additive*, while usual frequencies histograms are additive.

Theorem : If $\kappa_{0,t}$ and $\beta_{0,t}$ are $\mathcal{O}(|t|)$ then $\lim\limits_{t \to s} \dfrac{\kappa_{s,t}}{(t-s)}$ and $\lim\limits_{t \to s} \dfrac{\beta_{s,t}}{(t-s)}$ exist for almost all s, as t tends to s. The limit function f is the smallest (resp. the greatest) positive function such as $\kappa_{s,t} \le \int_{[s,t[} f$ (resp. $\beta_{s,t} \ge \int_{[s,t[} f$).

Proof :
Let $(T_t)_{t \ge 0}$ be the tanslations semigroup operating on mesurable positive functions f, that is $T_t(f)(x) = f(x + t)$. Let $F_t(x) = \kappa_{x,x+t}$ so that

$$F_{t+s}(x) = \kappa_{x,x+t+s} \le \kappa_{x,x+t} + \kappa_{x+t,x+t+s} = F_t(x) + T_t F_s(x).$$

Since T_t preserves the integral of integrable functions, the subadditive local ergodic theorem of Akcoglu-Krengel (1987, 1981) yields the convergence. The assertion concerning the limit follows from Feyel's (1982) proof of the local ergodic theorem. Moreover $F_t = G_t + H_t$ with G_t additive and H_t subadditive such that $\lim\limits_{t \to 0} \dfrac{H_t}{t} = 0$, and $G_t(x) = \int_{[x,x+t[} f + S_t$ with S_t singular, f being necessarily the limit.

5. Lattices

5.1 Lattices of a set of probabilistic objects

For any fixed i let $V_i \in O_i$, where V_i may be defined, for instance, by a percentile, and consider the subset $(X_{c,i})^{-1}(V_i) = \{\omega \in \Omega / X_{c,i}(\omega) \in V_i\}$. Denoting by $F_{V_i,c}$ the characteristic function of this set, the family $d_c = \{F_{V_i,c}, i \in I\}$ is a partial description of the object c depending on the choice of the values sets V_i.
Consider the complete lattice generated by d_c, $c \in C$ with respect to the order $F_{V_i,c} \le F_{V_i,d}$ for all i. This order corresponds of course to the inclusion order of the sets $\{X_{c,i}^{-1}(V_i)\}$.

A natural valuation of the lattice is obtained by using capacities and credibilities : in the join semi-lattice of all the unions of $\{(X_{c,i}^{-1})(V_i)\}_{c \in B}$, the formula $\kappa(B, V_i) = \mathcal{P}\{\bigcup_{c \in B}(X_{c,i})^{-1}(V_i)\}$ for B countable is the order ∞ capacity of section 3.1. Similarly in the met semi-lattice of all intersections $\{(X_{c,i})^{-1}(V_i)\}_{c \in B}$, $\beta(V_i) = \mathcal{P}\{\bigcap_{c \in B}(X_{c,i})^{-1}(V_i)\}$ defines a credibility function on $(O_i, \mathcal{O}_i)$ when B is fixed.

5.2 Symbolic objects

A symbolic object is a mathematical model of a "concept" (see Diday (1995) for more details). It is defined by an intent (i.e. the set of properties satisfied by a set of individuals) an a way of computing its extent (i.e. the set of units which satisfy the intent). A symbolic object a is more general (resp. more *specific*) than a symbolic object b if the extent of a contains (resp. is contained in) the extent of b.

An interesting problem is to associat to each node of the lattices defined above, a symbolic object in such a way that the intent and the extent are compatible with the subset associated to this node. For instance, considering the lattice defined in (5.1), the symbolic object, denoted by a, associated to an upper semi-lattice node which is associated to the subset $\bigcup_{c \in A}(X_{c,i})^{-1}(V_i)$, may be defined as follows :

The intent restricted to each V_i is defined by $\kappa(A, V_i) = \mathcal{P}\{\bigcup_{c \in A}(X_{c,i})^{-1}(V_i)\}$ and denoted by : $a = \{((\kappa(A, V_i)V_i)/i \in I\}$.

The extent of a is defined by a mapping denoted by "cap" from C to $[0, 1]$ such that: $cap(d) = \prod_{i \in I} \mathcal{P}(\bigcup_{c \in A}((X_{c,i})^{-1}(V_i)) \bigcap (X_{d,i})^{-1}(V_i))/\mathcal{P}\{(X_{d,i})^{-1}(V_i)\}$.

This mapping may be considered as a membership function, which indicates the degree to which d belongs to extent of a. Note that $cap(d) = 1$ if d belongs to A.

Conclusion:

This work shows a way of extending standard Data Analysis to objects defined by random variables with probability distributions. We show that capacities and credibilities constitute natural way of describing classes of such objects by generalization or specialization and of calculating the intent and the extent of the associated concepts, modelled by symbolic objects. In Diday and Emilion (1996), lattices, hierarchies, pyramids, principal components analysis and decision trees are investigated in this more general approach.

Acknowledgements

We would like to thank E. U. INTRS (93-725) program that supported partially this researsh.

♯ References

Choquet, G.(1954) Theory of capacities. Ann. Instit. Fourier.

Diday, E.(1995) Probabilistic, possibilist and belief objects for knowledge analysis. Annals of Operations Research $55, 227 - 276$.

Diday E., Emilion, R. (1996) Capacities and credibilities in analysis of probabilistic objects. Proceedings of OSDA'95. Springer Verlag.

Schafer G. (1976) A mathematical theory of evidence. Princeton University Press.

Schweizer B., Sklar B. (1983) Probabilistic metric spaces. Elsevier North-Holland, New-York.

Symbolic Pattern Classifiers Based on
the Cartesian System Model

Manabu Ichino and Hiroyuki Yaguchi

Tokyo Denki University
Hatoyama, Saitama 350-03, Japan
E-mail: ichino@k.dendai.ac.jp

Summary: As symbolic pattern classifiers, this paper presents region oriented methods based on the *Cartesian system model* which is a mathematical model to treat *symbolic data*. Our region oriented methods are able to use locally effective information to discriminate between pattern classes. This fact may achieve, at least superficially, a perfect discrimination of the pattern classes under a finite design set. Therefore, we have to take a ballance between the *separability* between classes and the *generality* of class desciptions. We describe this viewpoint theoretically and experimentally in order to assert the importance of *feature selection* which is essentially important in any pattern classification problem. We present also an example based on *symbolic data* in order to illustrate the usefulness of our approach.

1. Introduction

Traditional approaches to pattern classification (e.g. Bow (1992)) may be divided into two categories as follows.
1) *Boundary oriented approach* :
The purpose in this category is fo find the equations of decision boundaries. Linear classifiers and Bayes classifiers are examples for this category.
2) *Similarity based approach* : The purpose in this category is to find standard patterns for pattern classes and to use an appropriate similarity measure between the standard patterns and new patterns to be classified. Nearest neighbor rules and various matching methods are examples for this category.

As the third category of classification methods, several authors developed *region oriented approaches*. Stoffel (1974) used the prime events to describe class regions for binary feature variables. The prime events for a class cover only training samples for the class in the feature space. Michalski (1980) developed a very general approach to pattern classification based on his mathematical model so called the variable valued logic system. In his approach, various feature types can be used simultaneously to describe sample patterns, and *feature selection* is performed in the process to find class regions. According to the recent terminology, we may use the term *symbolic data* (Diday (1988)) for this general type of sample patterns. Ichino (1979, 1981) used *hyperrectangles* to describe pattern classes in the *feature space*. This approach can treat ordinal and binary feature variables simultaneously. However, a further generalization is necessary to treat *symbolic data* (Ichino(1986, 1988, 1993, 1995)).

The purpose of this paper is to present *symbolic pattern classifiers* based on region oriented approaches. In particular, we point out the importance of *feature selection* under a limited number of design samples, since we may be disturbed to achieve a proper classification ability by the *pretended simplicity* appearing in classification problems. In Section 2, we describe the *Cartesian System Model* (*CSM*) as the mathematical model to treat *symbolic data*. The *CSM* is represented as $(U^{(d)}, \boxplus, \boxtimes)$, where $U^{(d)}$ is the *feature space* in which each sample pattern is represented by a mixture of various feature types, $\boxplus$ is the *Cartesian join operator* which generates a generalized description from given descriptions in the *feature space*, and $\boxtimes$ is the *Cartesian meet operator* which extracts a common description from given descriptions in the *feature space*. In Section 3, we define graphs so called the *relative neighborhood graph* (*RNG*) and the *mutual neighborhood graph* (*MNG*). The *MNG* for a pattern class yields the *interclass structure* against the other pattern class. On the other hand, the *RNG* for a pattern class indicates the *intraclass structure* for the class. Under the assumption that the sample size of each pattern class is finite, the *MNG* and the *RNG* approach complete graphs, when we increase the number of features used to describe sample patterns. The completeness of the *MNG* means that a *perfect separability* between pattern classes is achieved, while the completeness of the *RNG* means that the class description has a *minimum generality* even for the given design set. Then, based on the properties of the *RNG* and the *MNG*, we restate the *Pretended simplicity theorem* (Ichino(1993)) by an improved way in Section 4. We describe our *symbolic pattern classifiers* and we compare our approach to other well known approaches, the *ID3* (Quinlan(1986)) and the *backpropagation neural network* (Rumelhart(1986)), by using an example of *symbolic data* in Section 5. Section 6 is a summary.

2. The Cartesian system model

2.1 Description of sample patterns

Each sample pattern is represented by the Cartesian product set:

$$E = E_1 \times E_2 \times \cdots \times E_d, \tag{1}$$

where E_k is the feature value taken by the feature X_k. We can treat the following five feature types.
1) *continuous quantitative feature* (e.g. height, weight, etc.)
2) *discrete quantitative feature* (e.g. the number of family members, etc.)
3) *ordinal qualitative feature* (e.g. academic career etc.; appropriate numerical coding is assumed)
4) *nominal qualitative feature* (e.g. sex, blood type, etc.)
5) *tree structured feature* (see Fig. 1, where terminal values are taken as feature values)
Feature types 1), 2), and 3) are permitted to take interval values of the form [a,b], and feature types 4) and 5) are permitted to take finite sets as feature values. The Cartesian product (1) described in terms of features 1) $\sim$ 5) is called an *event*. It should be noted that a sample pattern is an *event*. Let U_k be the domain of the feature X_k. Let U_k be a finite interval when the feature type is 1), 2), or 3), and be a finite set when the feature type is 4) or 5). Then, the *feature space* is given by the product set

$$U^{(d)} = U_1 \times U_2 \times \cdots \times U_d. \tag{2}$$

2.2 The Cartesian join operator

The *Cartesian join* $A \boxplus B$ of a pair of *events* A and B in the *feature space* $U^{(d)}$ is defined by:

$$A \boxplus B = (A_1 \boxplus B_1) \times (A_2 \boxplus B_2) \times \cdots \times (A_d \boxplus B_d), \tag{3}$$

where $A_k \boxplus B_k$ is the *Cartesian join* of feature values A_k and B_k for feature X_k and is defined as follows.
1) When X_k is a quantitative or an ordinal qualitative feature, $A_k \boxplus B_k$ is a closed interval given by:

$$A_k \boxplus B_k = [min(A_{kL}, B_{kL}), max(A_{kU}, B_{kU})], \tag{4}$$

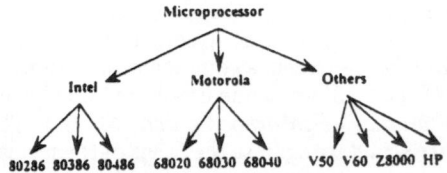

Figure 1: A tree sructured feature.

where A_{kL} and A_{kU} are the minimum value and the maximum value of the interval A_k, respectively; and $min(A_{kL}, B_{kL})$ and $max(A_{kU}, B_{kU})$ are the operators which select the minimum and the maximum values from the sets $\{A_{kL}, B_{kL}\}$ and $\{A_{kU}, B_{kU}\}$, respectively.

2) When X_k is a nominal feature, $A_k \boxplus B_k$ is the union:

$$A_k \boxplus B_k = A_k \cup B_k. \tag{5}$$

3) When X_k is a tree structured feature, let $N(A_k)$ be the nearest parent node which is common to all terminal values included in A_k. Then, if $N(A_k) = N(B_k)$,

$$A_k \boxplus B_k = A_k \cup B_k; \tag{6}$$

and if $N(A_k) \neq N(B_k)$,

$$A_k \boxplus B_k = \text{the set of all terminal values branched from the node } N(A_k \cup B_k), \tag{7}$$

where for each feature value A_k, we assume that

$$A_k \boxplus A_k = A_k. \tag{8}$$

2.3 The Cartesian meet operator

The *Cartesian meet* $A \boxtimes B$ of A and B in the *feature space* $U^{(d)}$ is defined by:

$$A \boxtimes B = (A_1 \boxtimes B_1) \times (A_2 \boxtimes B_2) \times \cdots \times (A_d \boxtimes B_d), \tag{9}$$

where $A_k \boxtimes B_k$ is the *Cartesian meet* of the k-th feature values A_k and B_k, and is defined by the intersection:

$$A_k \boxtimes B_k = A_k \cap B_k. \tag{10}$$

When the intersection (10) takes the empty value ϕ at least one feature, *events A and B have no common part.* We denote this fact by

$$A \boxtimes B = \Phi, \tag{11}$$

and we call that "*A and B are completely distiguishable.*" Fig. 2(a) and Fig. 2(b) illustrate the *Cartesian join* and the *Cartesian meet* in the Euclidean plane, respectively.

We call the triple $(U^{(d)}, \boxplus, \boxtimes)$ as the *Cartesian System Model (CSM)*. [1]

---

[1]We used the name *Cartesian Space Model* initially. However, according to the suggestion by Prof. E. Diday, we renamed to prevent misunderstanding about the model.

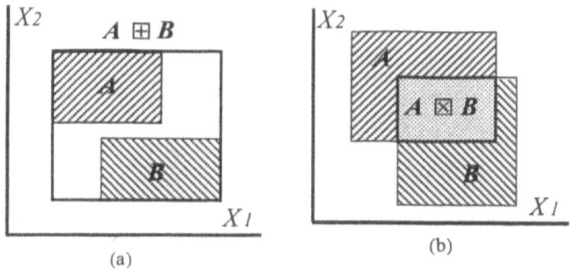

Figure 2: The *Cartesian join* and the *Cartesian meet* in the Euclidean plane.

3. The graph concepts

3.1 Sample patterns

Let ω_1 and ω_2 be two pattern classes, and let their sets of sample patterns be

$$\omega_k = \{E_{k1}, E_{k2}, ..., E_{kN_k}\}, k = 1, 2, \tag{12}$$

where sample pattern E_{kj}, $j = 1, 2, ..., N_k$, is represented by

$$E_{kj} = E_{kj1} \times E_{kj2} \times \cdots \times E_{kjd}. \tag{13}$$

We assume that each sample in ω_1 is *completely distinguishable* from any sample in ω_2:

$$E_{1p} \boxtimes E_{2q} = \Phi, p = 1, 2, ..., N_1, q = 1, 2, ..., N_2. \tag{14}$$

3.2 The relative neighborhood graph

The *relative neighborhood graph* (*RNG*) by Ichino and Sklansky(1985) yields relative relationships between samples in a class, and is generalized under the CSM as follows. Two samples E_{ip} and E_{iq} in ω_i are *relative neighbors* if

$$E_{ik} \boxtimes (E_{ip} \boxplus E_{iq}) = \Phi, k = 1, 2, ..., N_i, (p, q \neq k). \tag{15}$$

If two samples are *relative neighbors*, the *Cartesian join* of these samples never includes other samples. In this context, samples which are *relative neighbors* are isolated and singular from other samples. The *relative neighborhood graph*, written $RNG(\omega_i)$, is a graph constructed by joining all pairs of sample patterns, E_{ip}, $E_{iq} \in \omega_i$, which are *relative neighbors*. Fig. 3 (a) illustrates the RNG in the Euclidean plane, where we omit all edges which represent the fact that each sample is itself *relative neighbor*.

3.3 The Mutual neighborhood graph and the silhouette

The *mutual neighborhood graph* (*MNG*) (Ichino (1986)) yields the information about interclass structure, and is defined as follows. Two samples E_{1p} and E_{1q} in ω_1 are *mutual neighbors* against ω_2 if

$$E_{2k} \boxtimes (E_{1p} \boxplus E_{1q}) = \Phi, k = 1, 2, ..., N_2. \tag{16}$$

362

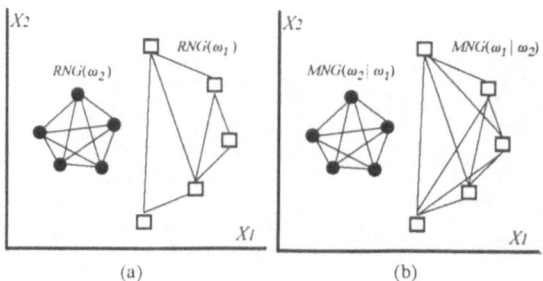

(a)   (b)

Figure 3: The *RNG* and the *MNG* in the Euclidean plane.

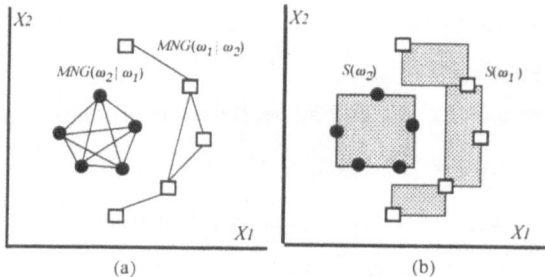

(a)   (b)

Figure 4: The *MNG* and the *silhouette* in the Euclidean plane.

The *mutual neighborhood graph (MNG)* of ω_1 against ω_2, written $MNG(\omega_1|\,\omega_2)$, is a graph constructed by joining all pairs of sample patterns, E_{1p}, $E_{1q} \in \omega_1$, which are *mutual neighbors* against ω_2.

Fig. 3(b) and Fig. 4(a) illustrate the *MNG* in the Euclidean plane, where we omit all edges which represent the fact that each sample pattern is itself *mutual neighbor*. When two pattern classes are well separated (Fig. 3(b)), the *MNG* becomes a complete graph or a near complete graph (e.g., $MNG(\omega_1|\omega_2)$ and $MNG(\omega_2|\omega_1)$ in Fig. 3(b)). On the other hand, the number of edges of the *MNG* decreases according to the closeness of the two pattern classes (e.g., $MNG(\omega_1|\omega_2)$ in Fig. 4(a)). The shaded regions in Fig. 4(b) are called the *silhouettes* of the pattern classes. The *silhouette* $S(\omega_i)$ approximates the region for pattern class ω_i, and is defined by

$$S(\omega_i) = \cup_p \cup_q (E_{ip} \boxplus E_{iq}), \tag{17}$$

where the union is taken for all *mutual neighbor* pairs in ω_i against the other class.

It should be noted that the *MNG* and the *silhouette* of a pattern class are the descriptions of the class from the view point of *relativity* against the other class. However, the *MNG* is an abstract mathematical description, while the *silhouette* is an actual description in the *feature space*.

4. The Pretended simplicity theorem

The *Pretended simplicity theorem* (Ichino (1993)) was introduced in order to assert the importance of *feature selection* in the design of *symbolic pattern classifiers*. We restate this theorem here in an improved way.

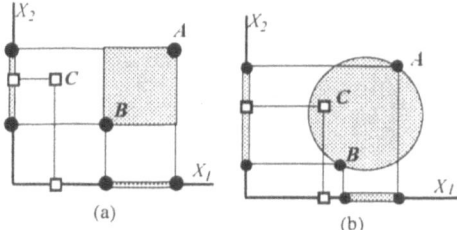

Figure 5: Illustration for the invariability for *join regions*.

4.1 As the boy so the man theorem

We start from a simple example of three samples in the Euclidean plane (see Fig. 5(a)). In this example. the *Cartesian join* of two samples A and B excludes sample C along feature X_1. The property of this exclusion is invariable by the addition of feature X_2. In general, once the exclusion of a sample from the given *Cartesian join region* is achieved by a set of features, called simply as a *feature set*, it is invariable for the addtion of new features to the *feature set*. This invariability is a special property for our *Cartesian join*. In fact, if we use the *circle of influence* as the *join region* of two sample patterns. our invariability is no longer obtained (see Fig. 5(b)).

Now we assume again the data sets in (12). Let two samples E_{1i}, E_{1j} in ω_1 be *mutual neighbos* against ω_2. Then for each E_{2k} in ω_2, it is required the existence of at least one feature X_p such that

$$E_{2kp} \boxtimes (E_{1ip} \boxplus E_{1jp}) = \phi. \tag{18}$$

The condition in (18), however, is not related to any other sample in ω_2 and is not related to any other feature. In other words, once the property that the *Cartesian join* of two samples E_{1i} and E_{1j} are *completely distinguishable* from sample E_{2k} is obtained for a *feature set*. the property is invariable for the addition of new features to the *feature set*. It should be noted that this invariablity is also true for the *relative neighbors* by confining ourselves to a single class.

Theorem 1 (*As the boy so the man theorem*)
Once the properties of the *relative neighbors* and the *mutual neighbors* are obtained for a *feature set*, the properties are invariable for the addition of new features to the *feature set*.

4.2 Generality and separability

We introduce two additional terms *generality* and *separability* for the description by the *Cartesian join*. Let F be a *feature set*. Let E_{1i} and E_{1j} be two samples of class ω_1, and let α_{ij} be the number of samples, in class ω_1, which are included in the *Cartesian join* of E_{1i} and E_{1j} under the *feature set* F. Then we define the *generality* of samples E_{1i} and E_{1j} under the *feature set* F as follows:

$$Gen(i, j|F) = \alpha_{ij}/(N_1 - 2), \tag{19}$$

where N_1 is the number of samples given for class ω_1. On the other hand, let β_{ij} be the number of samples, in class ω_2, which are included in the *Cartesian join* of E_{1i} and E_{1j} under the *feature set* F. Then, we define the *separability* of the *Cartesian join* of E_{1i} and E_{1j} from the other class ω_2 under the *feature set* F as follows:

$$Sep(i, j|F) = 1 - \beta_{ij}/N_2, \tag{20}$$

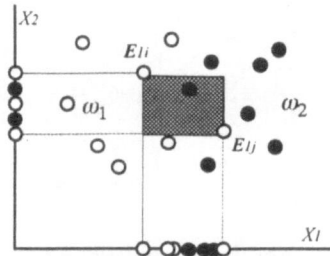

Figure 6: Illustration for the *generality* and the *separability*.

where N_2 is the number of samples given for class ω_2. It is clear that

$$0 \leq Gen(i, j|F), Sep(i, j|F) \leq 1, \tag{21}$$

and the *generality* and the *separability* become maximum when $Gen(i, j|F) = 1$ and $Sep(i, j|F) = 1$, respectively.

We illustrate the *generality* and the *separability* by using a two dimensional example in Fig. 6. In this figure, we have 8 samples for class ω_1 and 7 samples for class ω_2. In the *feature space* by X_1, the *Cartesian join* of samples E_{1i} and E_{1j} includes 2 ω_1 samples and 3 ω_2 samples. Hence, we have

$$Gen(i, j|X_1) = 2/(8-2) = 1/3, Sep(i, j|X_1) = 1 - 3/7 = 4/7 \tag{22}$$

In the *feature space* by X_2, we have

$$Gen(i, j|X_2) = 1/(8-2) = 1/6, Sep(i, j|X_2) = 1 - 2/7 = 5/7. \tag{23}$$

On the other hand, in the two dimensional *feature space* by X_1 and X_2, we have

$$Gen(i, j|X_1, X_2) = 0, Sep(i, j|X_1, X_2) = 6/7. \tag{24}$$

From this example, it may be clear that the *generality* is monotonically decreased and the *separability* is monotonically increased by the increasion of features to describe sample patterns. We shoulld point out that this monotonic property of the *generality* and the *separability* is based on the *As the boy so the man theorem*.

4.3 Pretended simplicilty theorem

Now we assume that the sample size of each pattern class is finite, and that, for each pair of samples E_{1i} and E_{1j} in ω_1, there exist features by which the Cartesian join $E_{1i} \boxplus E_{1j}$ is completely distinguishable from any other samle E_{1k} in ω_1 and from any sample E_{2k} in ω_2. Then, we have the following theorem.

Theorem 2 (*Pretended simplicity theorem*)
By adding features appropriately to the *feature set* F:
1) The *generality* $Gen(i, j|F)$ becomes *zero* and the *separability* $Sep(i, j|F)$ becomes *one* for each pair of samples E_{1i} and E_{1j} in ω_1;
2) The $RNG(\omega_1)$ and the $MNG(\omega_1 \mid \omega_2)$ approach complete graphs; and
3) The *silhouette* $S(\omega_1)$ has a *perfect separability* from class ω_2, but it has a *minimum generality* as a description of class ω_1.

The properties 1) and thus 2) are direct conclusions from the *As the boy so the man theorem*. Then the property 3) is derived from 1) and 2).

This theorem asserts that: 1) The *silhouette* $S(\omega_1)$ becomes a connected single cluster, and it never includes any sample in class ω_2, since all sample pairs in ω_1 are *mutual neighbors*; but 2) The *silhoutte* $S(\omega_1)$ yields a very sparse description for class ω_1, and it yields only a very poor covering ability even for other design samples in the same class ω_1, since all sample pairs of class ω_1 are also *relative neighbors*. Therefore, the simplicity for the interclass structure obtained here is superficial and is a "pretended simplicity", and thus the *selection of globally effective features* is absolutely important in order to achieve a realistic classification performance.

4.4 Example 1

We generate $2N$ d-dimensional Gaussian samples, where d features are mutually independent identically distributed with the *zero mean* and the *unit variance*. We devide $2N$ samples randomly into two sets of N samples. These two sets are used as the design sets for pattern classes ω_1 and ω_2. Therefore, two pattern classes are completely overlapped in the d-dimensional *feature space*. Fig. 7(a) illustrates the distributions in a three dimensional *feature space*. The *Pretended simplicity theorem* asserts that if we fix N and increase d, the $MNG(\omega_1|\omega_2)$ $(MNG(\omega_2|\omega_1))$ and $RNG(\omega_1)$ $(RNG(\omega_2))$ approach complete graphs, namely their number of edges approach the maximum number $_NC_2$. Fig. 7(b) summarizes our experimental results. For example, when $N = 500$, the numbers of edges of MNGs and RNGs increase by the addition of features and approach the maximum number $_NC_2 =124750$ at arround 11 features. This is a remarkable fact, since we can separate our mixed up pattern classes by using only a small number of *very* locally effective features. The *silhouettes* $S(\omega_1)$ and $S(\omega_2)$ may be mutually overlapped, but they never include any sample from their counter pattern class. Therefore, we achieved a *perfect separability* between classes in terms of our design sets, although it is *pretended simplicity* from the view point of the given interclass structure. Furthermore, the *silhouette* $S(\omega_k)$ includes all samples of the class ω_k, but each *Cartesian join region* of $S(\omega_k)$ which is spanned by a pair of samples of ω_k never includes other samples of ω_k except the pair of samples. Therefore, the *silhouette* $S(\omega_k)$ has a *minimum generality* in the description of the class ω_k. In fact, for a new sample pattern independent from the design sets, the *silhouttes* $S(\omega_1)$ and $S(\omega_2)$ have exactly the same possiblity to cover the pattern.

This example asserts again that we have to *select* only sufficiently effective features in order to take a ballance between the *separability* between pattern classes and the *generality* of class descriptions.

5. Symbolic pattern classlifiers

5.1 Region oriented methods

For pattern class ω_i, $i = 1, 2$, we assume the data sets in (12). Let R_{ij}, $j = 1, 2, ..., M_i$. $i = 1, 2$, be the *events* such that

$$E_{ij} \subseteq \bigcup_{k=1}^{M_i} R_{ik}, j = 1, 2, ..., N_i, i = 1, 2 \tag{25}$$

$$E_{ij} \not\subseteq R_{pq}, j = 1, 2, ..., N_i, q = 1, 2, ..., M_i, (i \neq p), \tag{26}$$

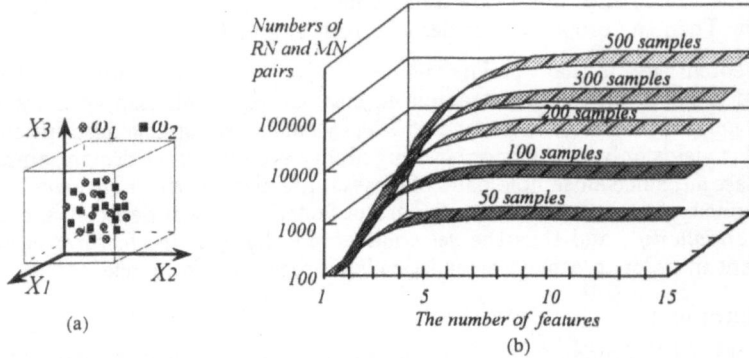

Figure 7: Example 1.

where $M_t < N_t$, $i = 1, 2$, in general. Then, we can use the following decision rule to classify a given pattern sample E.

1) E is determined to come from class ω_i if there exists an R_{ik} for which $E \subseteq R_{ik}$ and if $E \not\subseteq R_{pq}$ for all q where $p \neq i$.

2) It is rejected as type-I reject if it is covered by *events* of ω_1 and ω_2 simultaneously.

3) It is rejected as type-II reject if no *events* cover it.

Now we can state our basic problem as: "Generate an appropriate set of *events* which satisfy (25) and (26)". We can assert the following theorem.

Theorem 3 (*Existence thorem* (Ichino (1988)))

If the given training sets ω_1 and ω_2 are mutually *completely distinguishable*, there exist *events* which satisfy (25) and (26).

This theorem may be clear by assuming that $R_{ik} = E_{ik}, k = 1, 2, ..., N_i, i = 1, 2$. However, realistic covering ability of R_{ij} for new patterns will be achieved by the *events* which are expanded from given sample patterns.

As an approach, we may use the *silhouette* $S(\omega_i)$ in (17) to describe the region for each pattern class ω_i. We point out a principle of the *relativity* that *silhouettes* can be relatively described in the *feature space* according to the mutual separability of the pattern classes (see Fig. 4). Therefore, if we can find a set of minimum number of sample patterns which span the *silhouettes*, we may obtain a realistic symbolic classifier.

As a different approach. Ichino (1986, 1988) presented an algorithm which approximates the *silhouette* of a class by a lesser number of *events*. This algorithm generates *events* so that the *events* cover the sample patterns which yield the densely connected portions of the *mutual neighborhood graph* (see Fig. 8).

In the above decision rule, a given new sample is rejected to assign class name, when the sample is not included in any *event*. In this case, we can suggest the nearest pattern class ω_i by using *membership grade* of sample $E = E_1 \times E_2 \times \cdots \times E_d$ from the event $R_{ik} = R_{ik1} \times R_{ik2} \times \cdots \times R_{ikd}$ for class ω_i defined by

$$MG(E \mid R_{ik}) = \sum_{p=1}^{d} \frac{\mid R_{ikp} \mid}{d \mid R_{ikp} \boxplus E_p \mid}, k = 1, 2, ..., M_i \qquad (27)$$

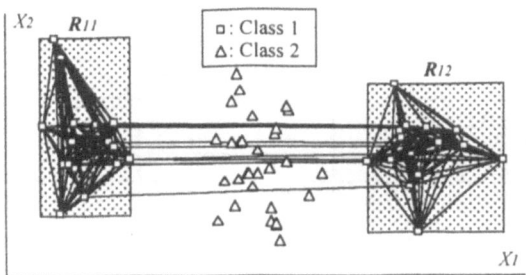

Figure 8: The MNG and *events*.

where $| \star |$ is the length of the interval $\star$ when the $p - th$ feature is continuous quantitative, and is the number of possible feature values included in $\star$ when the $p - th$ feature is discrete quantitative, ordinal qualitative, or structural. It is clear that this membership function takes the value in the unit interval $[0, 1]$. Thus, the function may be regarded as a *fuzzy membership function*.

5.2 Example 2

As an example of symbolic pattern classification problem, we treat here the data of "TOYOTA" and "NISSAN" car models in 1992. Each sample (car model) is described by 23 quantitative features and 3 qualitative features. We prepared 181 samples as the design set. The experiments were performed in the following way.

Step 1: We applied the *furthest neighbor method (complete linkage method)* for hierarchical clustering based on the *generalized Euclidean distance* by Ichino and Yaguchi (1994) defined for a pair of samples $A = A_1 \times A_2 \times \cdots \times A_d$ and $B = B_1 \times B_2 \times \cdots \times B_d$ by

$$d(A, B) = [\sum_{k=1}^{d} \psi(A_k, B_k)^2]^{1/2}, \tag{28}$$

$$\psi(A_k, B_k) = \frac{| A_k \boxplus B_k | - | A_k \boxtimes B_k | + 0.5(2 | A_k \boxtimes B_k | - | A_k | - | B_k |)}{| U_k |}, \tag{29}$$

where $| \star |$ is the same in (27).

We found *five* clusters (pattern classes) which were well correspondent to usually used concepts of "luxury cars", "sport cars", "leisure vehicles", etc..

Step 2: We found *events* in (25) and (26) for each pattern class by using the method in Ichino (1986, 1988), where our multiclass problem was treated as a set of dualclass problems in the sense that "class ω_i versus other classes except class ω_i". Each pattern class was described by one or two *event*(s). Then, for each *event*, we found a minimum set of features by which the *event* is separated from other pattern classes by using a *modified zero-one integer programming* (Ichino (1986, 1988)). The total number of features was reduced from 26 to 16. The selected features were *1) Weights, 2) Width, 3) Length, 4) Height, 5) Wheel base, 6) Front tread, 7) Rear tread, 8) Minimum turning radius, 9) Maximum power, 10) Rev/Max power, 11) Max torque, 12) Rev/Max torque, 13) Engine stroke, 14) Cylinder layout, 15) Final gear ratio, and 16) 10-mode mileage*, where *14*-th feature is a tree structured feature and others are quantitative features (some of them are interval valued features).

No.	Company	Model	Correct Answer	Decisions		
				Neural Network	ID3	Proposed system
1	HONDA	NSX	2	2	2,3,5	2
2	HONDA	Legend	1	1	1,2,3,4,5	1
3	HONDA	Prelude	3	3	2,3,5	3
4	HONDA	Accord Wagon	4	3	2,3,4,5	4
5	HONDA	Accord	3	3	2,3,4,5	3
6	HONDA	Integra	3	3	2,3,5	3
7	HONDA	Civic	3	3	2,3,5	3
8	MAZDA	Sentia	1	2	1,2,3,4,5	1
9	MAZDA	Eunos Cosmo	2	4	-	2
10	MAZDA	Efini RX-7	2	3	-	2
11	MAZDA	Efini MPV	4	1	-	4
12	MAZDA	Familia	3	3	-	3
13	MAZDA	Revue	5	2	-	5
14	MAZDA	Carol	5	5	-	5
15	SUBARU	Legacy wagon	4	4	-	4
16	SUBARU	Vivio	5	5	-	5

Table 1: Results of Example 2.

Step 3: We classified 1992 car models of "HONDA", "MAZDA", and "SUBARU" as the test data, by using the membership function defined in (27). Our classification results were shown in Table 1.

We applied also well known the *ID3* (Quinlan(1986)) and the *backpropagation neural network* (Rumelhart(1986)). In the case of the neural network system, almost test samples were classified correctly, however, some test samples were incorrectly classified. We performed *training* of the network 200 times, where we excluded *14*-th feature from the limitation of the use of qualitative fatures. However, it was still insufficient for classes 2, 4, and 5.

On the other hand, in the case of the *ID3* , all car models of "MAZDA" and "SUBARU" were rejected. Each car model of "HONDA" was assigned to several classes which include correct one. This is because that the *ID3* evaluate a single feature at a time in the learning process to generate a decision tree.

In our system, all test samples were classified correctly. In the region oriented approach, each pattern is included or not in the predetermined *events* of the class. This will yield the reliable classification results for the *events* properly generated. However, many patterns appearing in the future may be rejected as type-II reject because that they are not included in any prepared *events*. In order to overcome this drawback we introduced a *fuzzy memebership function* in (27).

6. Concluding remarks

This paper presented region oriented methods for *symbolic pattern classifiers* based on the *Cartesian system model*. Our methods use the *mutual neighborhood graph (MNG)* as a tool to understand the interclass structure. This graph is able to pick up very local discrimination information to describe class regions. This property requires to take a ballance between the *separability* of classes and the *generality* of class descriptions. In order to assert this viewpoint we presented the *Pretended simplicity theorem*, and we pointed out the importance of *feature selection* in symbolic pattern classification. We compared our approach to well known the *ID3* and the *backpropagation neural network* based on the symbolic data of car models.

Acknowledgment

The authors thank Professor Edwin Diday for his helpful discussions. The authors wish to thank also the referees for their suggestions leading to improvements in this paper.

References

Bow, S. T. (1992): *Pattern Recognition and Image Preprocessing*, Mercel Dekker.

Diday, E. (1988): The symbolic approach in clustering. *In Classification and Related Methods of Data Analysis,*, Bock, H. H. (ed.), Elsevier.

Stoffel, J. C. (1974): A classifier design technique for discrete pattern recognition problems. *IEEE Trans. Compt., C-23, pp. 428-441.*

Michalski, R. S. (1980): Pattern recognition as rule-guided inductive inference. *IEEE Trans. Pattern Anal. and Mach. Intell. PAMI-2, pp. 349-361.*

Quinlan, J.R. (1986): Introduction of Decision Tree, *Machine Learning, 1, pp. 81-106.*

Rumelhart, D.E.R. and McClelland (1986): Parallel Distributed Processing, MIT Press.

Ichino, M. (1979): A nonparametric multiclass pattern classifier. *IEEE Trans. Syst. Man, Cybern. 9, pp.345-352.*

Ichino, M. (1981): Nonparametric feature selection method based on local interclass structure, *IEEE Trans. on Syst. Man. Cybern. 11. pp. 289-296.*

Ichino, M and Sklansky, J. (1985): The relative neighborhood graph for mixed feature variables, *Pattern Recognition, 18, 2. pp. 161-167.*

Ichino, M. (1986): Pattern classification based on the Cartesian join system: A general tool for feature selection, *In Proc. IEEE Int. Conf. on SMC (Atlanta).*

Ichino, M. (1988): A general pattern classification method for mixed feature problems. *Trans IEICE Japan J-71-D, PP. 92-101* (in Japanese).

Ichino, M. (1993): Feature selection for symbolic data classification. *In New Approaches in Classification and Data Analysis,* Diday, E. et al. (ed.), Springer-Verlag.

Ichino, M and Yaguchi, H. (1995): Generalized Minkowski metrics for mixed feture-type data analysis, *IEEE Trans. Syst. Man. Cybern. 24. 4. pp. 698-708.*

Ichino, M., Yaguchi, H. and Diday, E. (1995): A fuzzy symbolic pattern classifier. *OSDA '95. Paris.*

Yaguchi, H., Ichino, M. and Diday, E. (1995): A knowledge acquisition system based on the Cartesian space model. *OSDA '95. Paris.*

Extension based proximities between constrained Boolean symbolic objects

Francisco de A. T. de Carvalho[1]

[1] Departamento de Estatistica - CCEN / UFPE
Av. Prof. Luiz Freire, s/n - Cidade Universitaria
50.740-540 Recife - PE BRASIL
Fax: ++55 +81 2718422 and E-mail:fatc@di.ufpe.br

Summary: In conventional exploratory data analysis each variable takes a single value. In real life applications, the data will be more general spreading from single values to interval or set of values and including constraints between variables. Such data set are identified as Boolean symbolic data. The purpose of this paper is to present two extension based approaches to calculate proximities between constrained Boolean symbolic objects. Both approaches compares a pair of these objects at the level of the whole set of variables by functions based on the description potential of its join, union and conjunctions. The first comparison function is inspired on a function proposed by Ichino and Yaguchi (1994) while the others are based on the proximity indices related to arrays of binary variables.

1. Introduction.

Constrained Boolean symbolic objects (Diday (1991)) are better adapted than usual objects of data analysis to describe classes of individuals taking into account simultaneously variability, as a disjunction of values on a variable, and logical dependencies between variables. For example, if an expert wishes to describe the fruits produced by a village, by the fact that "the weight is between 300 and 400 and the colour is white or red and if the colour is white then the weight is lower than 350", it is not possible to put this kind of information on a usual data table where rows represent villages and columns descriptors of the fruits. Instead, this description may be represented by a constrained Boolean symbolic object as $a_j = $ [weight = [300, 400]] $\wedge$ [colour = {white, red}] $\wedge$ [[colour = {white}] $\Rightarrow$ [weight = [300, 350]]] where a_j, which represents the jth village, is a mapping defined on the set of fruits such that for a given fruit ω, $a_j(\omega) = true$ iff the weight of ω belongs to the interval [300, 400], its colour is white or red and if it is white then its weight is less than 350.

2. The Boolean symbolic objects.

Let Ω be a set of individuals and $\omega \in \Omega$ an individual. A variable is a function $y_i : \Omega \longrightarrow O_i$ where O_i is the set of values that y_i may take. A variable may take no values, a single value or several values (a discrete set or an interval) for a symbolic object. Let $\mathcal{Y} = \{y_1, \ldots, y_p\}$ be the set of p variables, defined on Ω and taking their values on $O_1, \ldots, O_p$, respectively. Let $V = \{V_1, \ldots, V_p\}$ where $V_i \subseteq O_i$, $i \in \{1, \ldots, p\}$.

2.1 No constrained Boolean symbolic objects.

A Boolean symbolic object is a logical conjunction of properties. Formally, a Boolean symbolic object a is defined by the function $a_{yy} : \Omega \longrightarrow \{true, false\}$ such that $a_{yy}(\omega) = $ true iff $\forall i \in \{1, \ldots, p\}, y_i(\omega) \in V_i$. The **intention** of a, denoted as $a = [y_1 \in V_1] \wedge \ldots \wedge [y_p \in V_p] = \wedge_{i=1}^{p}[y_i \in V_i]$, states that *variable y_1 takes its values in V_1 and ... and variable y_p takes its values in V_p*. The **extension** of a is defined by $| a |_\Omega = \{\omega \in \Omega / y_i(\omega) \in V_i, \forall i \in \{1, \ldots, p\}\}$.

Example 1: Let $a = [\text{colour} \in \{\text{red,blue}\}] \wedge [\text{height} \in [0, 15]]$. An individual $\omega \in \Omega$ is such that $a(\omega) = \text{true}$ iff its colour is red *or* blue **and** its height is between 0 and 15.

2.2 Constrained Boolean symbolic objects.

To represent actual knowledge, the description of a class of individuals by a Boolean symbolic object must take into consideration different kinds of constraints given by *logical dependencies* between variables. We distinguish two kinds of dependencies: conditional and logical correlation dependencies.

2.2.1 Conditional dependence.

A variable y_i may become inapplicable if another variable y_j takes its values on a subset S_j of its observation set O_j. As an example, we can not describe the colour of the hat of a mushroom which has no hat. We note $y_i = \text{NA}$ to indicate that variable y_i became inapplicable and y_i should be considered as non-existent or meaningless. In any case NA should not be considered as a variable value, and it is just for notation conveniences that we denote $y_i = \text{NA}$.

In the case of single dependence, the conditional dependence is expressed by the rules $r_1 : [[y_j \in S_j] \Rightarrow [y_i = \text{NA}]]$ and $r_2 : [[y_i = \text{NA}] \Rightarrow [y_j \in S_j]]$. The rule r_1 expresses the fact that if y_j takes its values in S_j then y_i becomes inapplicable while rule r_2 expresses the fact that y_i becomes inapplicable only, and only if y_j takes its values in S_j (Vignes (1991)).

2.2.2 Logical correlation dependence.

The set $S_j \subseteq O_j$, where O_j is the observation set of y_j, may be on correspondence with the set $S_i \subseteq O_i$, where O_i is the observation set of y_i. In the case of single dependence, the logical correlation dependence is expressed by the rule $r_1 : [[y_j \in S_j] \Rightarrow [y_i \in S_i]]$.

Example 2. Let $a = [\text{Height} \in [50, 150]] \wedge [\text{Weight} \in [25, 75]]$. The Fig. 1 shows the Cartesian representation of a, respectively, (a) under hypothesis of independence between the variables *Height* and *Weight*, (b) under hypothesis of conditional dependence between these variables expressed by the rules $r_1 : [[\text{Height} \in [100, 150]] \Rightarrow [\text{Weight} = \text{NA}]]$ and $r_2 : [[\text{Weight} = \text{NA}] \Rightarrow [\text{Height} \in [100, 150]]]$ and (c) under hypothesis of logical correlation dependence between these variables expressed by the rule $r_3 : [[\text{Height} \in [100, 150]] \Rightarrow [\text{Weight} \in [25, 50]]]$.

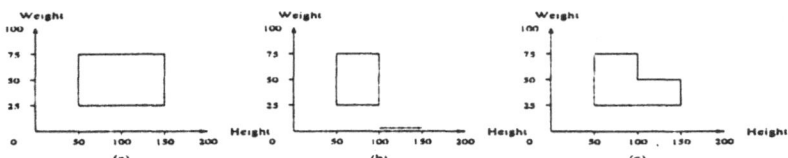

Fig. 1: Cartesian representation of a: (a) independence; (b) conditional dependence; (c) logical correlation dependence.

2.2.3 Multiple logical dependencies.

The logical dependencies may be visualised as a set of connected graphs. In each graph, the nodes are the variables. These graphs are directed (from the variable premise to the variable conclusion) and labelled CD become inapplicable in the case of conditional dependence) (conditional dependence) or LCD (logical correlation dependence). The figure 2 shows, for example, that there is a conditional dependence between variables y_1 and y_2 and that there is no logical dependence between variables y_2 and y_3. We consider now several important cases.

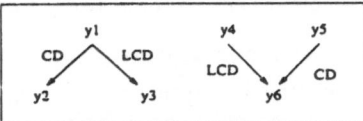

Fig. 2: Graphs of logical dependencies between variables.

Non Applicability propagation. Suppose that a variable y_i may become inapplicable by a variable y_j which on the same time may become inapplicable by another variable y_k:

$$y_k \xrightarrow{\text{CD}} y_j \xrightarrow{\text{CD}} y_i.$$

In this case, where the variable y_j may also become inapplicable, the conditional dependence between the variables y_i and y_j is expressed by the rules $r_1 : [[y_j \in S_j] \Rightarrow [y_i = \text{NA}]]$, $r_2 : [[y_i = \text{NA}] \Rightarrow [y_j \in S_j]]$ and $r_3 : [[y_j = \text{NA}] \Rightarrow [y_i = \text{NA}]]$. The rule r_3 expresses the Non Applicability propagation which is a kind of inheritance induced by the conditional dependence.

Relaxation. Suppose now a variable y_i has a logical correlation dependence with a variable y_j which may become inapplicable by another variable y_k:

$$y_k \xrightarrow{\text{CD}} y_j \xrightarrow{\text{LCD}} y_i.$$

In this case, the logical correlation dependence between the variables y_i and y_j is expressed by the rules $r_1 : [[y_j \in S_j] \Rightarrow [y_i \in S_i]]$ and $r_2 : [[y_j = \text{NA}] \Rightarrow [y_i \in O_i]]$. The rule r_2 expresses the fact that when y_j is not applicable y_i is not affected by the logical dependence (relaxation).

Subordination. Suppose a variable y_i has a logical correlation dependence with a variable y_j and on the same time may become inapplicable by another variable y_k:

$$y_j \xrightarrow{\text{LCD}} y_i \xleftarrow{\text{CD}} y_k.$$

In that case, the logical correlation between the variables y_i and y_j is expressed by the rule $r_1 : [[y_j \in S_j] \Rightarrow [y_i \in S_i \cup \{\text{NA}\}]]$. The rule r_1 expresses the fact that $y_j \in S_j$ implies $y_i \in S_i$ only if y_i is applicable. The logical correlation dependence is subordinate to conditional dependence.

A single variable may become inapplicable by multiple variables. Suppose a variable y_i may become inapplicable by variables $y_{j1} \ldots y_{jm}$. In that case, the m conditional dependencies are expressed by $m + 1$ rules $r_1 : [[y_{j1} \in S_{j1}] \Rightarrow [y_i = \text{NA}]], \ldots, r_m : [[y_{jm} \in S_{jm}] \Rightarrow [y_i = \text{NA}]]$ and $r_{m+1} : [[y_i = \text{NA}] \Rightarrow [y_{j1} \in S_{j1}] \vee \ldots \vee [y_{jm} \in S_{jm}]]$.

Multiple variables may become inapplicable by a single variable. Suppose the variables $y_{j1} \ldots y_{jm}$ may become inapplicable by a variable y_i. In that case, the m conditional dependencies are expressed by m pairs of rules $r_{11} : [[y_{j1} \in S_{j1}] \Rightarrow [y_i = \text{NA}]]$ and $r_{12} : [[y_i = \text{NA}] \Rightarrow [y_{j1} \in S_{j1}]], \ldots r_{m1} : [[y_{jm} \in S_{jm}] \Rightarrow [y_i = \text{NA}]]$ and $r_{m2} : [[y_i = \text{NA}] \Rightarrow [y_{jm} \in S_{jm}]]$.

We say that a description of an individual (or a group of individuals) by a Boolean symbolic object is coherent if it does not contradict the set of rules which expresses the logical dependencies between the variables.

2.3 Join, Union, Disjunction, Conjunction and Equality.

Let $a = \wedge_{i=1}^{p}[y_i \in A_i]$ and $b = \wedge_{i=1}^{p}[y_i \in B_i]$, where $A_i = [l_i^a, u_i^a]$ and $B_i = [l_i^b, u_i^b]$ if y_i is a quantitative or a ordinal qualitative variable. The join (Ichino and Yaguchi (1994)) between these two Boolean symbolic objects is defined as $a \oplus_s b = \wedge_{i=1}^{p}[y_i \in A_i \oplus B_i]$, where $A_i \oplus B_i = [\min(l_i^a, l_i^b), \max(u_i^a, u_i^b)]$ if y_i is a quantitative or a ordinal qualitative variable, and $A_i \oplus B_i = A_i \cup B_i$ if y_i is a nominal qualitative variable. The union is defined as $a \cup_s b = \wedge_{i=1}^{p}[y_i \in A_i \cup B_i]$, the disjunction is defined as $a \vee_s b = \{\wedge_{i=1}^{p}[y_i \in A_i]\} \vee \{\wedge_{i=1}^{p}[y_i \in B_i]\}$, the conjunction is defined as $a \wedge_s b = \wedge_{i=1}^{p}[y_i \in A_i \cap B_i]$ and we say that $a =_s b$ iff $\forall i, A_i = B_i$. The figure 3 illustrate these operations.

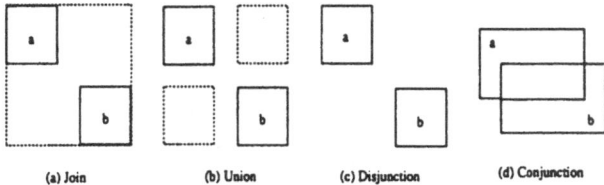

(a) Join   (b) Union   (c) Disjunction   (d) Conjunction

Fig. 3: Symbolic operations.

2.4 Description potential of a Boolean symbolic object.

Let $a = \wedge_{i=1}^{p}[y_i \in A_i]$ be a Boolean symbolic object We define the *Description potential* of a, denoted $\pi(a)$, as the volume of the part of the Cartesian product $A_1 \times \ldots \times A_p$ formed only by the descriptions of individuals given by a which are *coherent*.

2.4.1 No constrained Boolean symbolic objects

If we suppose there are no logical dependencies between variables in the knowledge base, the Boolean symbolic object a is coherent and also all individuals descriptions given by the Cartesian product $A_1 \times \ldots \times A_p$. In this case $\pi(a)$ is calculated by the following expression:

$$\pi(a) = \prod_{i=1}^{p} \mu(A_i) \tag{1}$$

where

$$\mu(A_i) = \begin{cases} \text{cardinal}(A_i), & \text{if } y_i \text{ is qualitative or discrete quantitative} \\ \text{range}(A_i), & \text{if } y_i \text{ is quantitative continuous} \end{cases} \tag{2}$$

range(A_i) being the sum of absolute value of the difference between the upper bound and the lower bound of each interval, where A_i is a set of real intervals.

Proposition 1 *If $\{a_1, \ldots, a_n\}$ is a set of Boolean symbolic objects, where $a_j = \wedge_{i=1}^{p}[y_i \in A_{ij}]$ with $j \in \{2, \ldots, n\}$, then $\pi(a_1 \vee \ldots \vee a_n) = \sum_{j=1}^{n} \pi(a_j) - \sum_{j<k} \pi(a_j \wedge a_k) + \sum_{j<k<l} \pi(a_j \wedge a_k \wedge a_l) + \ldots + (-1)^{n-1}\pi(a_1 \wedge \ldots \wedge a_n)$.*

2.4.2 Constrained Boolean symbolic object.

Suppose now there are logical dependencies between variables in the knowledge base. Let $a = \wedge_{i=1}^{p}[y_i \in A_i]$ be a constrained Boolean symbolic object, where $NA \in A_i$ if y_i may become inapplicable, and let $\{r_1, \ldots, r_t\}$ be the set of rules expressing the dependencies between the variables. The description potential of a will be now calculated as the difference between the volume of the Cartesian product $A_1 \times \ldots \times A_p$ and the part of this volume which is formed by the individual descriptions which are not coherent, i. e.,

$$\pi(a) = \prod_{i=1}^{p} \mu(A_i) - \pi(a \wedge (\neg(r_1 \wedge \ldots \wedge r_t))) \tag{3}$$

We have

$$\pi(a \wedge (\neg(r_1 \wedge \ldots \wedge r_t))) = \pi((a \wedge \neg r_1) \vee \ldots \vee (a \wedge \neg r_t))$$

and therefore, according to proposition 1,

$$\begin{aligned} \pi(a \wedge (\neg(r_1 \wedge \ldots \wedge r_t))) &= \sum_{j=1}^{t} \pi(a \wedge \neg r_j) - \sum_{j<k} \pi((a \wedge \neg r_j) \wedge \neg r_k) + \ldots \\ &+ (-1)^{t-1} \pi((a \wedge \neg r_1) \wedge \neg r_2) \ldots) \wedge \neg r_{t-1}) \wedge \neg r_t) \end{aligned} \tag{4}$$

The complexity of the calculation of the description potential of a constrained Boolean symbolic object is exponential on the number of rules and linear on the number of variables to each connected graph of dependencies.

3. Extension based proximity indices.

Proximities measures play an important role in exploratory data analysis. Thus, clustering and ordination may have as input data a matrix of proximities between objects. Usually, to calculate the proximity between a pair of objects, we use a comparison function, in order to compare the objects at the level of each variable, and an aggregation function, in order to aggregate each comparison to obtain a global measure of proximity. The Minkowsky metric is a classical example of a proximity measure which uses a comparison and an aggregation functions.

In symbolic data analysis, approaches which use a comparison and an aggregation function to calculate the proximities between Boolean symbolic objects were proposed. We can find some approaches concerning the measurement of the proximity between no constrained Boolean symbolic objects in Ichino and Yaguchi (1994) and Gowda and Diday (1991). A first approach to calculate the proximity between constrained Boolean symbolic objects can be find in De Carvalho (1994).

In this paper, we present two extension based approaches to calculate the proximity between no constrained and constrained Boolean symbolic objects. Both approaches use only a comparison function based on the description potential of the join or union and disjunction of these objects. The description potential of a Boolean symbolic object is related to its extension.

Let $a = \wedge_{i=1}^{p}[y_i \in A_i]$ and $b = \wedge_{i=1}^{p}[y_i \in B_i]$ and let $\mathcal{Y} = \{y_1, \ldots, y_p\}$. We define $a \wedge_{\mathcal{Y}} b = \wedge_{y_i \in \mathcal{Y}'}[y_i = A_i \cap B_i]$, where $\mathcal{Y}' = \{y_i \in \mathcal{Y}/V_i^j \cap V_i^k \neq \emptyset\}$.

The first type of comparison functions is inspired on the comparison function proposed by Ichino and Yaguchi (1994):

$$d_1(a, b) = \pi(a \text{ op } b) - \pi(a \wedge b) + \gamma(2\pi(a \wedge b) - \pi(a) - \pi(b)) \tag{5}$$

where op is the join operator or the union operator and $0 \leq \gamma \leq 0.5$.

Two normalised versions of equation 5 are:

$$d_2(a, b) = \frac{\pi(a \text{ op } b) - \pi(a \wedge b) + \gamma(2\pi(a \wedge b) - \pi(a) - \pi(b))}{\pi(a^{\Omega})} \tag{6}$$

where $a^\Omega = \wedge_{i=1}^{p}[y_i \in O_i]$, $0 \leq \gamma \leq 0.5$, and

$$d_3(a,b) = \frac{\pi(a \text{ op } b) - \pi(a \wedge b) + \gamma(2\pi(a \wedge b) - \pi(a) - \pi(b))}{\pi(a \text{ op } b)} \qquad (7)$$

where $0 < \gamma \leq 0.5$.

Proposition 2 *Let (a, b) be a pair of no constrained Boolean symbolic objects.*

a) d_1 *and* d_2 *are semi metrics, i. e., the triangular inequality does not hold, and* $d_1 \sim d_2$. d_2 *is normalised between 0 and 1;*

b) d_3 *is a metric and it is normalised between 0 and 1;*

c) *the quasi order defined as* $(a, b) \preceq_d (c, d) \Leftrightarrow d_1(a, b) \leq d_1(c, d)$ *is not affected by a change of scale of continuous variables by a linear transformation;*

d) d_2 *and* d_3 *are not affected by a change of scale of continuous variables by a linear transformation.*

The second type of comparison functions are based on the proximity indices related to arrays of binary variables:

$$d_4(a,b) = \frac{\pi(a \text{ op } b) - \pi(a \wedge_{y'} b)}{\pi(a \text{ op } b) + \gamma\, \pi(a \wedge_{y'} b)} \qquad (8)$$

$$d_5(a,b) = \frac{1}{2}\left[\frac{[\pi(a \text{ op } b) - \pi(a))}{\pi(a \text{ op } b) + \pi(a \wedge_{y'} b) - \pi(a)} + \frac{(\pi(a \text{ op } b) - \pi(b))}{\pi(a \text{ op } b) + \pi(a \wedge_{y'} b) - \pi(b)}\right] \qquad (9)$$

$$d_6(a,b) = 1 - \frac{\pi(a \wedge_{y'} b)}{\sqrt{(\pi(a \text{ op } b) + \pi(a \wedge_{y'} b) - \pi(a))(\pi(a \text{ op } b) + \pi(a \wedge_{y'} b) - \pi(b))}} \qquad (10)$$

where *op* is again the join operator or the union operator and $\gamma \in \{-0.5, 0, , 1\}$.

These comparison functions are normalised between 0 and 1 and they are valid only to discrete variables because in the case of continuous variables it is possible to have $\pi(a \text{ op } b) < \pi(a \wedge_{y'} b)$. In the case of nominal qualitative variables, there is no difference between join and union operators and we may use indifferently one of them in equations 8 to 10.

Proposition 3 *Let (a, b) be a pair of no constrained Boolean symbolic objects.*

a) d_4 *is a metric if* $\gamma \in \{-0.5, 0\}$;

b) d_4 *($\gamma = 1$), d_5 and d_6 are semi metrics;*

c) d_4, d_5 *and* d_6 *are not affected by a change of scale of continuous variables by a linear transformation;*

d) d_4 *($\gamma = -0.5$)* $\sim d_4$ *($\gamma = 0$)* $\sim d_4$ *($\gamma = 1$).*

4. Case studies.

We present two examples in order to illustrate the usefulness of our dissimilarity indices.

4.1 Fats and Oils.

The chemical data base presented by Ichino and Yaguchi (1994) describes 8 items of fats and oils by 5 variables (four continuous which take interval as values and one discrete which take finite sets as values). Each item is described by a no constrained Boolean symbolic object. It is known that each of item pairs (linseed oil, perilla oil), (cottonseed oil, sesame oil), (camellia oil, olive oil) and (beef tallow, hog fat) has similar properties. Ichino and Yaguchi studied this data base by using one of its proximity measures:

$$d_7(a,b) = \frac{1}{p}\left[\sum_{i=1}^{p}\left\{\frac{\mu(A_i \text{ op } B_i) - \mu(A_i \cap B_i) + \gamma\left(2\mu(A_i \cap B_i) - \mu(A_i) - \mu(B_i)\right)}{\mu(O_i)}\right\}^2\right]^{\frac{1}{2}}$$

This data base, which includes continuous variables, cannot be studied by the comparison functions of equations 8, 9 and 10. Fig. 4 shows the dendrograms obtained by the complete linkage method by using d_7 (Ichino and Yaguchi function, Fig. 4a to Fig. 4d), d_2 (equation 6, Fig. 4e to Fig. 4h) and d_3 (equation 7, Fig. 4i to Fig. 4l). It is not necessary to show the results obtained by d_1 (equation 5) because this proximity measure is equivalent to d_2 (see proposition 1).

It seems that parameter γ has no influence on the proximity indices d_2 and d_3. To γ fixed in d_2, d_3 and d_7, the join operator furnishes better results than union operator. This is because in the case of continuous variables position is important. With join operator, proximity measure d_3 is able to obtain the five groups of fats and oils indicated by experts (Fig. 4l). This is not the case by using Ichino and Yaguchi index (Fig. 4d).

4.2 Freshwater insect orders.

The biological knowledge base (Vignes, 1991) concerns freshwater insect orders. It includes items described by 12 nominal qualitative variables (which one takes finite sets as values), where the number of modalities is between 2 and 6, and there are 3 different pairs of variables presenting conditional dependencies. Both types of comparison functions (equations 5 to 10) may be used to study this knowledge base.

Fig. 5 shows the dendrograms obtained by the complete linkage method by using d_3 ($\gamma = 0.5$) as proximity measure (a) under hypothesis of logical independence between variables (each insect order is described by a no constrained Boolean symbolic object) and (b) under hypothesis of conditional dependencies between variables (each insect order is described by a constrained Boolean symbolic objects). In this figure, I means imago, L means larva, N means naiad and P means pupa.

In both cases all the orders larva are grouped together but only under hypothesis of conditional dependencies the orders naiad and imago are in the same subgroup. It seems that we have got better results when we describe each item of this knowledge base by a constrained Boolean symbolic object.

5. Conclusions.

Two approaches to calculate proximities (dissimilarities) between constrained Boolean symbolic objects are presented. These approaches use comparisons functions which are based on the description potential of its join or union and conjunction and are inspired from Ichino and Yaguchi (1994) functions and from the proximity indices of binary variables. Classical properties concerning proximity indices such as type, equivalence and metric properties are presented. Simulations with a chemical data and with a biological knowledge base seems corroborate the approaches.

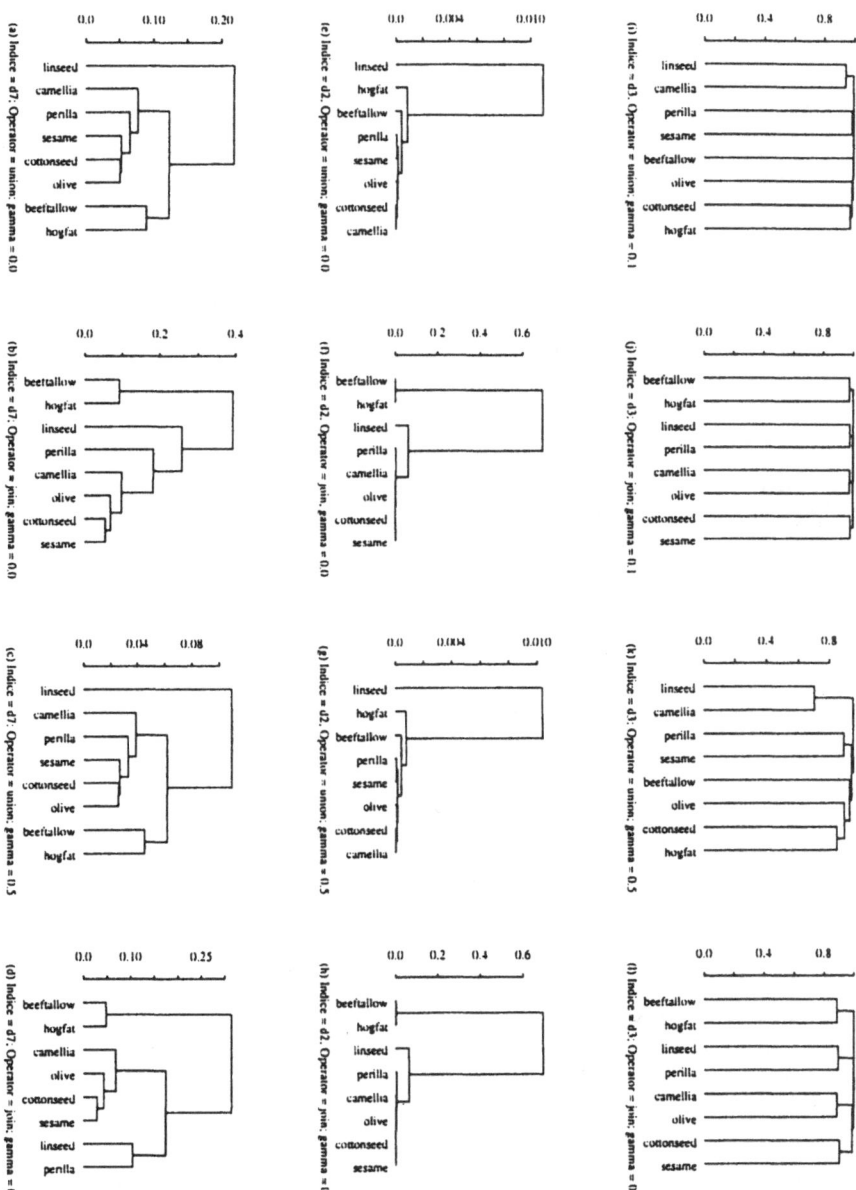

Fig. 4: Dendrograms by the complete linkage method (Fats and Oils).

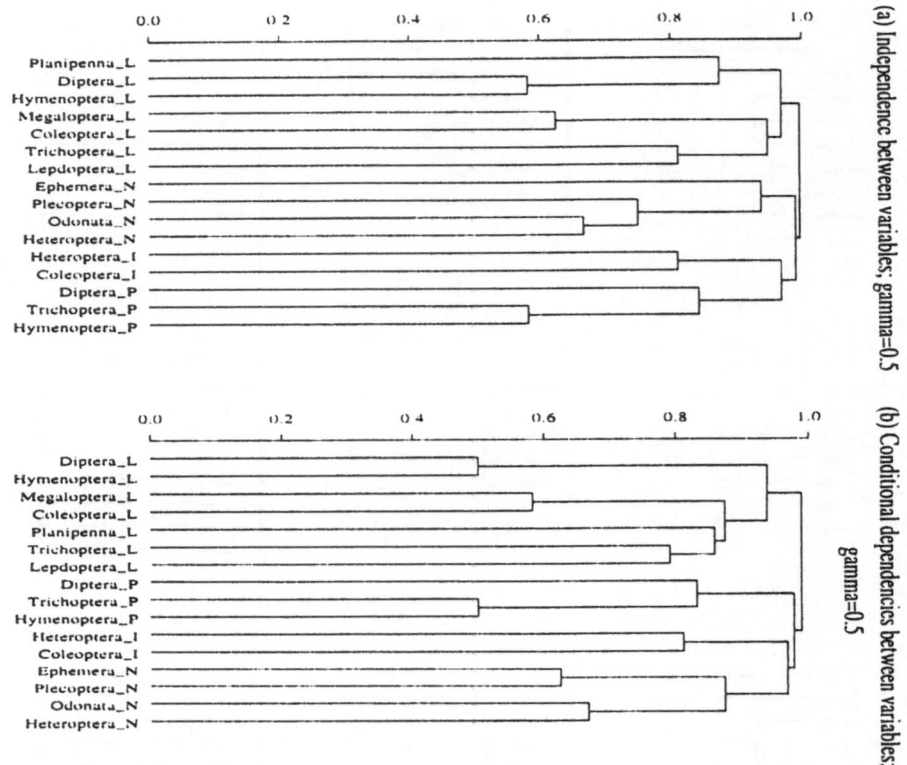

Fig. 5: Dendrograms by the complete linkage method (Freshwater insect orders).

References:

De Carvalho. F.A.T. (1994): Proximity Coefficients between Boolean symbolic objects. *In New Approaches in Classification and Data Analysis*. Diday, E. et al. (eds.), 387–394, Springer-Verlag, Heidelberg.

Diday, E. (1991): Des objets de l'analyse de données à ceux de l'analyse de connaissances. *In Induction symbolique et numérique à partir de donnés*. Diday, E. and Kodratoff. Y. (eds.), 9–75, Cepadue Editions. Toulouse.

Gowda, K.C. and Diday. E. (1991): Symbolic clustering using a new dissimilarity measure. *Pattern Recognition*, **24**, 6, 567–578.

Ichino, M. and Yaguchi. H. (1994): Generalised Minkowsky Metrics for Mixed Features Type Data Analysis. *IEEE Transactions on System. Man and Cybernetics*. **24**, 4, 698–708.

Vignes, R. (1991): Caractérisation automatique de groupes biologiques. *Thèse de Doctorat*. Université Paris-VI Pierre et Marie Curie. Paris.

Towards A Normal Symbolic Form

Marc Csernel[1] , Francisco de A.T. de Carvalho[2]

[1] Université Paris Dauphine 75016 Paris France.&
INRIA-Rocquencourt
Domaine de Voluceau
Le Chesnay Cedex France
E-mail : Marc.Csernel@inria.fr

[2] Departamento de Estatistica - CCEN/UPFE
Av Luiz Freire, s/n - Cidade Universitaria
50.740 - 540 - Recife - PE BRASIL
E-mail fatc@di.ufpe.br Fax ++55 +81 2710359

Summary: Boolean Symbolic Objects were introduced by Diday (1988) and since that time a large number of applications have been defined, using these objects, but relatively few of them take constraints on the variables into account. Even in this case, when the graph of dependencies becomes too large, the computational time becomes huge because dependencies are treated in a combinatorial way. We present a method inspired by the technique used in relational data bases (Codd 1972) leading to a decomposition of symbolic objects into a Normal Symbolic Form which allows an easier calculation, however huge the graph of dependencies rules may be. We will apply our method to distance computation following a method due to De Carvalho and inspired by Ichino (1994) but the normal form we present in this paper could be used for other purposes. In our first trials we obtained a 90% reduction of the computational time. In the present text we will only deal with nominal boolean Symbolic Objects, but the method could be used with other kinds of symbolic objects.

1. Introduction

Constrained Boolean Symbolic Objects defined by Diday (1991) are better adapted than usual objects of data analysis to describe classes of individuals such as populations or species, being able to take into account variability. They are expressed by a logical disjunction of elements called elementary events, and each of these elementary events represents a set of values associated with a variable. Each boolean symbolic object describes a volume which is a subset of the Cartesian product of the description variable.

A symbolic object can be constrained by different kinds of rules which express logical dependencies between the variables. These rules reduce the description space described by symbolic objects, they interfere greatly on computation of distance between them.

We shall use for distance computation a comparison function based on the description potential (De Carvalho (1994)) of each object. We define the description potential as the part of the volume described by a symbolic object which is coherent. i.e. where all the values are satisfying all the dependencies rules.

Until now the different methods used to compute distances between symbolic objects were taking rules into account by computing the incoherent part of each object or each computed element. This computation can become huge when the dependencies graph is deep and has to be repeated for each pair of objects each time you choose a different distance indice.

To avoid this kind of problem, we were induced to propose a representation of symbolic objects where only the coherent part of an object is represented. We recall that a boolean symbolic object (if no dependence rule applies) describes a subset of a Cartesian product, this is just the definition of a relation.

Since long people dealing with data bases are familiar with relations: they use a relational model. E. Codd has introduced some normal forms to structure more efficiently the data base relational schema, particularly the third one which concerns the case where it exists functional dependencies between the variables. Normal forms are used in relational data bases to offer a better factorization of data, thus providing a simpler and easier way to update data.

All this has induced us to introduce a normalization of boolean symbolic objects, inspired by the Codd's third normal form, which allows a representation of the only coherent part of a symbolic object. By reference to the relational normalization we call it Normal Symbolic Form(NSF).

We shall expose in section 2 a mathematical definition of boolean symbolic object and examine different possible kinds of dependencies rules. In the section 3 we shall see rules influence on distance computation. In the section 4 we shall see the definition of Codd's third normal form, the definition of the NSF and an example of decomposition it induces. In section 5 we shall examine the principle of the decomposition process, and in the section 6 we shall see how to use a NSF description of symbolic objects to perform some usual calculus needed by distance computation and then conclude in section 7.

2. Constrained Boolean Symbolic Objects

Let Ω be a set of elementary objects generally called "individuals", described by p variables y_i where $i \in \{1...p\}$. Let O_i be a set of observations where the variable y_i takes its values. The set $O = O_1 \mathrm{x} O_2 \mathrm{x}...O_p$ is then called the description space.

An elementary event e_i, denoted by the symbolic expression

$$e_i = [y_i = V_i]$$

where $(i \in \{1, p\}, V_i \subseteq O_i)$, expresses that "the variable y_i takes its values in V_i".

A Boolean symbolic object a, is a conjunction of elementary events of the form:

$$a = \bigwedge_{i \in \{1...p\}} e_i = [y_i = V_i]$$

where the different $V_i \subseteq O_i$, represent the intention of a set of individuals $C \subset \Omega$. It can be interpreted as follows: the values of y_i for any individual $c \in C$ are in V_i.

Example : a = [colour = {blue, red}] $\wedge$ [size = {small, medium}]

means that the colour of a is red or blue and the size small or medium.

We define the symbolic union denoted $\bigcup_s$:

$$a_1 \bigcup_s a_2 = \bigwedge_{i \in \{1...p\}} (e_{i1} \bigcup e_{i2}) \text{ with } (e_{i1} \bigcup e_{i2}) = [y_i = V_{i1} \bigcup V_{i2}]$$

Constrained boolean symbolic objects are usual symbolic objects associated with a Domain-Knowledge which constraints the description given by the objects. Such constraints are expressed as logical dependencies between the variables. We take into account two kinds of dependencies. The first one, called conditional dependencies is of the form if A = a1 then B has No Sense, as in :

if wings = Absent then wings_colour = No_Sense. (r1)

The other form, called logical correlation dependencies, is of the form if A = a1 then B = b1, as in :

$$if \ wings\_colour = red \ then \ Thorax\_colour = blue. \ (r2)$$
$$if \ Thorax\_colour = blue \ then \ Thorax\_size = big \ . \ (r3)$$

The term No_Sense means that the variable does not exist, hence it's value is not applicable.

The difference between these two kinds of rules appears at different levels:

- In case of single dependencies:

- the conditional dependence rule reduces the description space, the logical correlation does not.

- the No_Sense value can be taken only when a conditional dependence rule is active.

- In case of multiple dependencies the way each rule is propagated among the dependencies graph is different, for example:

- If we consider r2 and r3 together, then *wings_colour = red* implies that *Thorax_size = big*. This is the *usual propagation*.

- If we consider r1 and r2 together we can see that when *wings = absent* we can say nothing about *wings_colour* which is not relevant and by consequence, we can not consider any particular value of *Thorax_colour*. This is the *rule relaxation*.

- if we consider the following rules together :
  if y1 = a1 then y3 = No_Sense; if y2 = b1 then y3 = c1;
  The second rule implies that the variable y3 has a meaning, hence that the first rule premise is not true. The logical correlation is subordinated to the conditional dependence.

We shall call premise variable the variable associated with the premise and conclusion variable the variable associated with the conclusion.

3. Dependency rules influence on distance computation

Some approaches to calculating the proximity between constrained Boolean objects have been proposed for instance by De Carvahlo (1994). The use of dependencies rules could be considered as creating "holes" in the Cartesian product of the variables which must be taken into account for an accurate distance computation.

So far, existing algorithms have not solved this problem in a satisfactory way, especially when the constraints are dependent on one another. To accomplish distance computation we use a comparison measure based on the Description Potential of each object i.e. the *COHERENT* part of the volume described by each object. We can see on the following example how a rule can interfere on the comparison between two objects. If we have the two following objects:

$a1 = [y1 = [5, 30]] \wedge [y2 = [10, 30]];$ $a2 = [y1 = [15, 30]] \wedge [y2 = [15, 35]];$

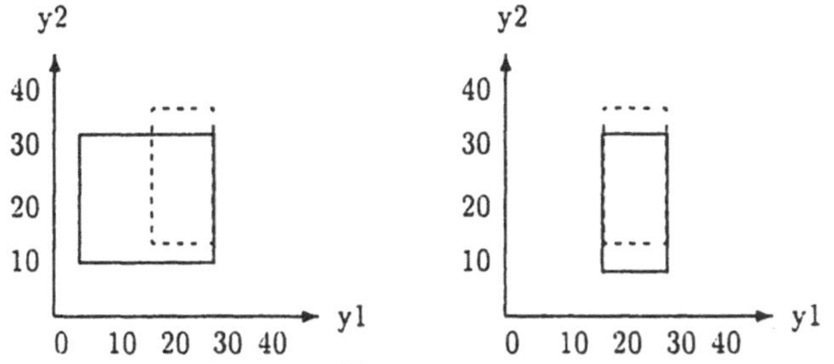

Figure 1

the left side of the Figure 1 shows a rectangle in plain line which area represents the description potential of $a1$, and a rectangle in dot line which area represents the description potential of $a2$. While there is no dependence rule, the description potential is here equivalent to the Cartesian product given by the description of respectively $a1$ and $a2$.

In the right side of the Figure 1 we represent the description potential of $a1$ and $a2$ considering the following dependence rule:

$$\text{if } y1 \in [5, 15[\text{ then } y2 \text{ has No Sense};$$

Examining Figure 1 it seems that $a1$ and $a2$ are more similar when the dependence rule is considered. We believe that the distance must take into account this evidence.

4. The Normal Symbolic Form

In a way symbolic Boolean objects are very close to the relations used in data bases (i.e. a subset of a Cartesian product). People using relational data bases have been long familiar with the decomposition in normal forms introduced by Codd (1972). We will focus on the third normal form which states roughly " a relation is in third normal form if there are no functional dependencies between a non-key attribute S and other non-key attributes $S_1, ..., S_n$ ".

Attributes have the same meaning as variable,and relations are presented as arrays where each column represents an attribute (or variable). Each line represents a tuple of values (an individual) and each line can be identified by a key. A key is an attribute (or a set of attributes) which can be used to identify a tuple in a relation. Codd's definition of a functional dependency says that "an attribute Y is functionally dependent of an attribute X if each X-value is associated with precisely one Y-value".

Usually it is necessary to decompose a relation if you want it to follow the third normal form.

The third normal form is used in relational data bases to offer a better factorization of data, thus providing a simpler and easier way to update data and a reduction of the space amount necessary.

Considering Boolean symbolic objects as a relation, if no dependencies rules are considered, suggests the possibility of a normal form.

We were induced to think that it can be useful to split an array of symbolic objects in two parts when it exists dependencies between some variables.

If some logical correlation exists between a premise variable X and some conclusion variables $Y_1, Y_2..., Y_n$, we will have then to split, as for Codd's normal form, our original table in two, the second table containing all values associated with $X: Y_1, Y_2, ..., Y_n$. We will refer the different tables, as original, main and secondary.

Simultaneously, we will represent the premise variable X in a flat way. By 'flat way' we mean that the values corresponding to a premise must appear one by one and not set by set as it occurs in the usual representation. This will allow to represent in an independent way premise and conclusion values and allow to **not** represent the inconsistent values. We will refer, to the elements of secondary tables by their line number.

Remark that for each object, we will have to refer to as many lines in the secondary table that there was values in the premise variable of the the initial table.

We define the Normal Symbolic Form (NSF) as follows: **" if no dependency occurs between the variables, or, if the dependencies occur between the first variable V1 and the others, then V1 must have a single value"**. One can remark that the variable order has no importance on the description but it is more convenient to have the premise variable as the first one.

Most of the time a symbolic object has to be decomposed to follow the Normal Symbolic Form (NSF), as we can see in the following example,

	wings	wings_colour	Thorax_colour	Thorax_size
a1	{present,absent}	{blue,red}	{blue,yellow}	{big,small}
a2	{present,absent}	{green,red}	{blue,red}	{small}

Table 1 original table

The previous array represents two boolean symbolic objects called a1 and a2, the dependencies rules r1,r2 are associated with the definition.

$$if\ wings = Absent\ then\ wings\_colour = No\_Sense.\ (r1)$$
$$if\ wings\_colour = red\ then\ Thorax\_colour = blue.\ (r2)$$

The description of the objects a1 and a2 representing two different(imaginary) insect species is obviously not NSF, because the description of wings in a1 has two values and there is a dependency between wings and wings_colour (r1). There is also a dependency between wings_colour (r2) and Thorax_colour and Wings_colour is not the first variable. Then the description has to be transformed in the sequence of the three following tables to be NSF. In these tables the upper left corner contains the table name, a new kind of column appears where the values are integers referring to a line in another table with the same name as the column. The first table has no name, it refers to the initial table.

	wings ...	Thorax_size
a1	{ 1, 3}	{big,small}
a2	{2,4}	{small}

main table

wings ...	wings	colour
1	absent	4
2	absent	5
3	present	{ 1, 2 }
4	present	{ 1, 3 }

secondary table 1

colour	wings_colour	Thorax_colour
1	{ red }	{blue }
2	{ blue }	{ blue, yellow }
3	{ green }	{blue, red }
4	NS	{ blue, yellow }
5	NS	{ blue, red }

secondary table 2

We have now three tables instead of a single one, but only the valid parts of the objects are represented: now, the tables include the rules.

5. How to Transform a set of Description in a NSF Form

The aim of the NSF is to provide a description of a symbolic object where only the part of the object satisfying dependencies rules is represented, so the description potential of the object can be calculated directly.

As we mentioned it before, we will have to split the original table into a main table and some secondary ones. We will follow the dependencies graph to proceed this task. Generally each premise variable will generate a new table containing the premise variable and all the conclusion variables depending from the premise.

At first, we must precise we will only consider the case where the graph between the variables induced by dependencies forms a tree or a set of trees i.e. no variable can be the conclusion of more that one premise variable.

The transformation process can be decomposed in two different phases:
1) Definition of the new tables   2) Fill the new tables

The secondary tables definition follows the variables dependencies graph. For each non terminal node N of the graph, we build a new table Tv composed of the variable V associated with N as the first variable, and the variables associated with each of the sons of N as other variables. The variable V will be replaced in its original table by a reference variable Rv which will contain lines number of the new table.

The table filling is a little more complicated and the lack of space does not allow us to describe it in full detail. It is decomposed in two processes the first one consists in a construction process, the second in a factorization one.

The first process induces a table growing, the second one a table reduction. With the real examples we did proceed, the reduction factor was more important than the growing factor, and we did obtain a reduction in size of the secondary tables greater than 30% .

For commodity reason we will expose the construction process under an algorithm form.

```
for each symbolic object
  { for each variable V
      if (V is not a premise nor a conclusion)
          put value of V in the main table;
      else if (V is a premise)
          put the references provided by GetrRef(V)
  }
GetrRef(V)
{ for each value Val of V // the premise variable
  // build a new line in Tv
  for each other variables Vc in Tv
      { restrict the values of Vc according to Val and the rule
      if (Vc is not a premise)
          put the corresponding values in Tv
      else
          put the references provided GetrRef(Vc) }
  return the list of lines builded
}
```

Once the construction is done, we need to factorize. For each newly builded line L, if L' = L is already present in the table we change the reference for a reference to L' and L is suppressed.

6. An application to distance computation

At present distance computation algorithms must generate (in the worst case) all possible combinations of variables, and then verify which ones are valid. In that case Mp combinations must be generated (M is the average number of modalities, p the number of variables) and verified. The NFS avoids this huge amount of verification.

Our distance measure uses, for the comparison process of the distance calculus, the volume of the description potential of the union of two objects. For nominal symbolic objects we will use a distance due to De Carvalho (to be published) inspired by Ichino (1994).

$$d(a_1, a_2) = \frac{potential(a_1 \bigcup_s a_2) - 0.5(potential(a_1) + potential(a_2))}{potential(a_1 \bigcup_s a_2)}$$

We will first show how to compute, using a NSF representation, the potential of a symbolic object, second, how to compute the potential of a symbolic union. We will illustrate the method, using the two objects a_1, a_2 described in our previous example.

7.1 Computing the potential of an object

Each line of a secondary table describes a coherent hyper-volume, and all the lines contributing to the representation of the same object describe hyper-volumes which do not intersect (by construction). So one has to sum up the potential described by each line of a secondary table.

On the following example, the potential of each line of the secondary table 2 has to be computed first, then the lines of the secondary table 1, and at last the potential of each object described in the main table.

For example line 3 of the secondary table 1, refers to the lines 1 and 2 of the secondary table 2. The potential is the sum of the potential described by these two lines: $1+2 = 3$. In the same way the potential of a_1 is obtained by multiplying the sum of the potentials of lines 1 and 3 of the the secondary table 1 $(2 +3)$ by the potential due to the variable Thorax_size (2) giving the result 10.

	wings ...	Thorax_size	pot
a1	{1,3}	{big,small}	10
a2	{2,4}	{small}	5

main table

	wings ...	wings	colour	pot
1		absent	4	2
2		absent	5	2
3		present	{1,2}	3
4		present	{1,3}	3

secondary table 1

colour	wings_colour	Thorax_colour	pot
1	red	{blue }	1
2	blue	{ blue, yellow }	2
3	green	{blue, red }	2
4	NS	{ blue, yellow }	2
5	NS	{ blue, red }	2

secondary table 2

7.2 Computing the potential of the symbolic union

The computation of the potential of the union of a_1 and a_2 must be split in two parts: the first one concerning the lines where the premise is verified the second concerning the lines where the premise is not verified.

For each part one needs to compute the union of two lines l_1 and l_2, l_1 participating to the description of a_1, l_2 participating to the description of a_2. These values must be multiplied by the number of lines of the part participating to the union and we obtain the potential related to the part. The potential of the union is obtained by summing up the potential computed for each part.

We will note $potU1(1,2)$ the union of lines 1 and 2 of the secondary table 1.
We will show now, on the previous example, how to compute
potential$(a_1 \cup_s a_2) = potU(\{big,small\})* potU1(\{1,3\},\{2,4\})$

For the secondary table 1:
$potU1(\{1,3\},\{2,4\}) = potU1(1,2)$ (premise verified) +
                    $potU1(3,4)$ (premise not verified)
$potU1(1,2) = 1*potU2(4,5)$ $potU1(3,4) = 1*potU2(\{1,2\},\{1,3\})$
For the secondary table 2:
$potU2(\{1,2\},\{1,3\}) = potU2(1,1)$ (premise verified) +
                    $potU2(2,3)$ (premise not verified)
$potU2(4,5) = pot(\{blue,red,yellow\}) = 3$ $potU2(1,1) = 1$
$potU2(2,3) = pot(\{blue,green\}) * pot(\{blue,red,yellow\}) = 2*3 = 6$

This is expressed by the following tables:

	wings ...	Thorax_s..	pot (U_s)
a1	{1,3}	{big,small}	2*(7+3)
a2	{2,4}	{small}	= 20

main table

wings..	wings	colour	pot (U_s)
1	absent	4	pot(1,2) =
2	absent	5	1*3 =3
3	present	{1,2}	pot(3,4)=
4	present	{1,3}	1+6 =7

secondary table 1

colour	wings_colour	Thorax_colour	pot
1	red	{blue }	1
2	blue	{blue, yellow}	2
3	green	{blue, red}	2
4	NS	{blue, yellow}	2
5	NS	{blue, red}	2

secondary table 2

7. Conclusion

The decomposition of symbolic objects following the Normal Symbolic Form induces the easiest way to take dependencies rules into account, as shown by our first application on distance computation. Including the construction of the NFS we have obtained on our first trials a reduction of about 90 % of the computational time. This encourages us to carry on our work.

We need to test it with a larger set of examples, to estimate better the amelioration it can provide. This will lead us to make a more formal approach of the complexity of the different computation phases needed in a distance processing with and without NSF. Because NFS is a normal form, we hope it will induce in the future, a better and easier interfacing of large sets of symbolic objects with data bases.

References:

Codd, E.F. (1972) : Further Normalization of the Data Base Relational Model. *In Data Base Systems, Courant Computer Science Symposia Series*, Vol 6. Englewood Cliffs, N.J. . Prentice-Hall

De Carvalho, F.A.T. (1994): Proximity Coefficients between Boolean symbolic objects. *In: E. Diday et al (eds.): New Approaches in Classification and Data Analysis.* Springer-Verlag, 387-394.

De Carvalho, F.A.T (to be published) : Extension based proximities between Boolean symbolic objects *In : Proceeding of Fifth Conference of the International Federation of Classification societies*

Diday, E. (1991): Des objets de l'analyse de données à ceux de l'analyse de connaissances. *In: Y. Kodratoff and E. Diday (eds.): Induction symbolique et numérique à partir de données.* Cepadue Editions,Toulouse, 9-75.

Diday, E. (1988) : The Symbolic Approach In Clustering. *In H.H. Bock (eds.): Classification and Related Methods of Data Analysis.* North Holland 673-683

Ichino, M. , Yaguchi, H (1994) : Generalized Minkowski Metrics for Mixed Feature-Type Data Analysis. *IEEE Transaction on Systems, Man, and Cybernetics* **24**,4,698-708.

The SODAS Project : a Software for Symbolic Data Analysis

Georges Hébrail

ELECTRICITE DE FRANCE - Research Center
1, Av. du Général de Gaulle
92141 CLAMART CEDEX - FRANCE
E_mail: Georges.Hebrail@der.edfgdf.fr

Summary: This paper presents an ESPRIT European project, whose goal is to develop a prototype software for symbolic data analysis. Symbolic data analysis is an extension of standard methods of data analysis (such as clustering, discrimination, or factorial analysis) to more complex data structures, called symbolic objects. After a short presentation of the model of symbolic objects, the different parts of the software are briefly described.

1. Introduction

Standard statistical data analysis methods, such as clustering, discrimination, or factorial analysis, apply to data which are basically structured as arrays. Each row represents an individual and each column for a particular row contains a *single* value, which is the value of a variable describing individuals. In many real world applications, for some variables, an individual may be described by sets of values, intervals of values, or probability distributions of values. Moreover, some a priori knowledge of the user may be associated with data, such as taxonomies in variable domains.

More complex data structures - called *symbolic objects* - have been proposed by Pr Diday in the last decade (see Diday (1991)). These data structures capture the complexity described above, but remain manageable regarding to computations performed in statistical data analysis methods. Beyond these data structures, some extensions of standard methods have been studied and evaluated to apply to symbolic objects. The extensions include clustering, discrimination, and factorial analysis.

But these new methods remain difficult to use in real applications for two main reasons (see Hébrail (1995)): there is no available software to do so (there are only disparate pieces of software in various universities), and it is difficult to manage data objects with a more complex structure than simple arrays. The goal of the SODAS project is to develop a prototype software to solve these problems and make these methods available to more users.

The SODAS project (for Symbolic Official Data Analysis System) is an European project within the DOSIS Programme (Development Of Statistical Information Systems), organized by EUROSTAT, the Statistical Office of the European Communities in Luxembourg. It gathers several partners, including national official statistics offices, industrials and universities; various European countries are represented. An important part of the project is devoted to benchmarks of real world applications, which will be

387

used to specify and test the software. These benchmarks are mainly provided by the national official statistics offices involved in the project.

In this communication, after a short presentation of the model of symbolic objects, we describe the main contents of this project, and especially the different parts of the software.

2. Symbolic objects

As mentioned before, standard methods of statistical data analysis accept as their input INDIVIDUALS by VARIABLES arrays. Each cell of such arrays contains the value taken by an individual for a variable. This value is said to be *atomic* in the sense that it is not a list or a set of values. For instance, if individuals are people, and if variables are AGE and SOCIO-PROFESSIONAL CATEGORY (SPC), the AGE cell for a person contains one value (the age of the person) and the SPC cell contains one value (the SPC of the person).

Symbolic objects introduced by Pr Diday extend the classical data structure to INDIVIDUALS by VARIABLES arrays where the value taken by an individual on a variable may be *non-atomic*: possibly a set of values, intervals of values, or a probability distribution. For instance, if individuals represent groups of people, and if variables are still AGE and SPC, a cell of this new array may contain, for each individual (i.e. each group of people), the interval of people age values in the group for the AGE variable, and the list of SPC of people of the group for the SPC variable.

We recall below, in an informal way, the basic data structures defined by Pr DIDAY. Additional structures have already been defined (see Diday (1991)), but are not presented here. The benchmarks of the project will be used to define the final list of data structures supported by the software.

Boolean elementary events

Boolean elementary events are expressions of the form:

SPC = { Worker, Employee }
    meaning that SPC takes one of the two values Worker or Employee.
AGE = { [25,27], [48,55] }
    meaning that AGE is between 25 and 27, or between 48 and 55.

Probabilistic elementary events

Probabilistic elementary events are expressions of the form:

SPC = { Worker(0.8), Employee(0.1), Farmer(0.1) }
    meaning that SPC is Worker in 80% of cases, Employee in 10% of cases, or
    Farmer in 10% of cases.
AGE = { [26,30] (0.7), [31,35] (0.1), [36,40] (0.1), [41,45] (0.1) }
    describing a probability distribution of ages on a sub-population (for instance a
    district).

Assertions

Assertions are conjunctions of boolean and/or probabilistic elementary events, for instance:

Group 125 = [AGE = { 34, 29, 2, 1 }] ∧ [SPC = { Employee, Worker }]
District 92 = [AGE = { [25,30](0.2), [31,35](0.23), ... }]
∧ [SPC = { Executive manager(0.6), Worker(0.2), ...}]

The model of symbolic objects also enables the user to associate with the data some a priori knowledge (i.e. metadata), which is then used by methods applied to symbolic objects. This a priori knowledge can be defined by different means. We present below three ways to do so: mother-daughter variables, rules, and taxonomies in variable domains.

Mother-daughter variables

Mother-daughter variables offer the possibility of defining variables which are not applicable to all people, but only to people verifying some properties. For instance:

SPC is applicable only if AGE > 18

Expression of a priori knowledge with rules

Rules offer the possibility of defining a priori knowledge, in the form of a restriction of possible combinations of values for the different variables, for instance:

If AGE > 60 Then SPC = Retired

Variables with taxonomic domain

Variables with taxonomic domain offer the possibility of defining a taxonomy within the values taken by a variable. Figure 1 gives an example of such a taxonomy.

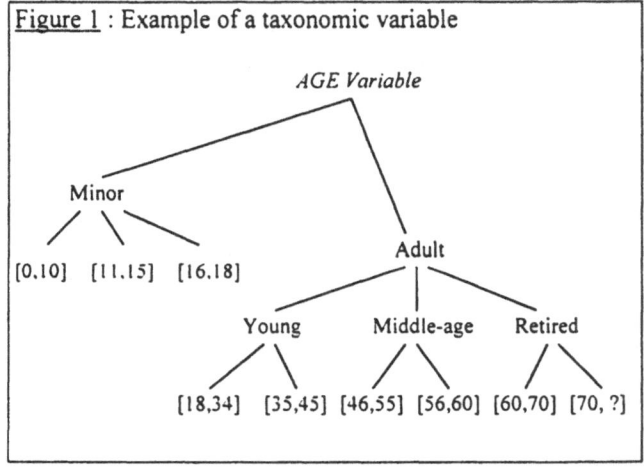

Figure 1 : Example of a taxonomic variable

As a summary, symbolic objects can represent individuals which are groups of elements of an underlying population, featuring variation within these groups. These objects can also describe uncertainty on data (with probabilistic objects) and metadata (with mother/daughter variables, taxonomies in variable domains, and rules).

3. Construction, management and manipulation of symbolic objects

A *symbolic object manager* will be developed as the kernel of the system. A normalized language will be defined to describe symbolic objects and a parser will enable the user to acquire symbolic objects from ascii files. Once objects are acquired, the system will store them physically in an ad'hoc structure and allow queries and updates to this symbolic object 'database'. Objects will also be accessible easily from programs by the means of some basic access functions.

As in many applications data are stored in relational databases, an interface will be developed to help the user to build symbolic objects from the contents of relational databases. Several operators will enable the user to build assertions from an underlying population stored in the database, as well as to extract a priori knowledge such as taxonomies or mother/daughter variables when they are available in the database.

4. Methods of symbolic data analysis

Several methods will be developed to apply to symbolic objects. All these methods will be extensions of standard methods to symbolic objects. There will be three classes of methods in the system: clustering, discriminant analysis, and factorial analysis methods.

4.1 Clustering methods

Symbolic clustering methods build clusters which are represented by symbolic objects. They can be applied either to standard data or to symbolic objects. In addition to that, these methods provide the following features:

- the clustering criterion favours interpretation of clusters instead of a complete optimization of a distance criterion.
- metadata associated with symbolic objects (mother/daughter variables, taxonomies, ...) are used in the clustering process.

The methods which will be developed include partitioning algorithms (see Chavent (1995)) and hierarchical clustering (see Brito (1995)). Hierarchical clustering will produce either disjoint or overlapping clusters (pyramids).

4.2 Discriminant analysis methods

Two standard methods of discriminant analysis will be extended to the case of individuals defined by symbolic objects.

The first extension will concern decision tree construction. This extension will take into account mother/daughter variables in addition to the possibility of discriminating between individuals defined by boolean assertions (see Lebbe and Vignes (1991)).

The second extension is the extension of non-parametric Bayesian discriminant analysis to symbolic objects defined by boolean and probabilistic assertions. In particular, it will be possible to discriminate between individuals described by variables which are probability distributions (see Granville and Rasson (1995)).

4.3 Factorial analysis

The method developed in the project will extend standard factorial analysis to individuals defined by symbolic objects featuring numerical variables of interval type. This method can be useful in the case of numerical data with uncertainty (see Chouakria *et al.* (1995)).

5. Link with standard data analysis

Two different approaches can be considered to use jointly standard and symbolic data analysis:

- the transformation of symbolic objects into classical data arrays, followed by the application of standard methods of data analysis,
- the use of the symbolic object formalism to describe groups of individuals found by standard methods.

These two approaches will be available in SODAS through the following features:

- an interface to call standard methods from SODAS,
- a tool for building disjunctive arrays of data from symbolic objects,
- a tool for building distance matrices from symbolic objects (see Ichino (1994), Carvalho (1996)),
- a tool for creating symbolic interpretations by symbolic objects to standard clustering and factor analysis methods (see Tong *et al.* (1996)).

6. User interface

A large part of the project will be devoted to the consideration of user's needs. As a working package of the project, a users' group will be created and animated.

This users' group will :

- gather different benchmarks from national official statistics offices and from industrial partners,
- list the symbolic object structures which are necessary in these benchmarks,
- check if the developed methods solve real problems and meet user's needs,
- test the software on real world applications.

From another point of view, the software will include a user-friendly interface to help the end-user to visualize *graphically* symbolic objects.

Within the budget of this project, it will not be possible to develop a fancy homogenized interface to all the methods. But some guidelines will be edited to homogenize interfaces of different methods of the software.

Finally, a scientific reference manual will be edited for the software. This book will contain a unified presentation of symbolic objects and methods applicable to them in the SODAS prototype software.

7. Partners

The project gathers 18 partners from various European countries. While THOMSON-CSF is the pilot of the project, Pr. Diday will be the scientific manager. The CISIA French company will be responsible for the development of the kernel of the software. For more information about this project and the distribution of tasks between partners, see SODAS Project (1995).

The main partners of the project are: Thomson-CSF (France), Université de Dauphine (France), Facultés Universitaires Notre Dame de la Paix (Belgium), Instituto Nacional de Estadistica (Portugal), and University of Athens (Greece).

The associated partners are: CISIA (France), Centre de recherche public STADE (Luxembourg), Central Statistical Office (England), Universita degli Studi dei Bari (Italy), Universita Federico II - Napoli (Italy), Electricité de France - Research center (France), EUSTAT (Spain), INRIA (France), Universidade de Lisboa (Portugal), Institute for Statistics - RWTH (Germany), Service des études et de la statistique du ministère de la région wallone (Belgium), and Universidad complutense de Madrid (Spain).

8. References

Brito P. (1995): « Symbolic objects: order structure and pyramidal clustering », in *Annals of Operations Research*, N°55, pp.277-297.

Carvalho F.A.T. (1996): « Extension based proximities between constrained boolean symbolic objects », in *Proceedings of the IFCS'96 Conference*, Kobe, March 96.

Chavent M. (1995): « Choix de base pour un partitionnement d'objets symboliques », in *Actes des 3èmes Rencontres de la Société Francophone de Classification (SFC-95)*, Namur, Sept.95.

Chouakria A., Cazes P., Diday E. (1995): « Extension de l'analyse factorielle des correspondances multiples à des données de type intervalle et de type ensemble » in *Actes des 3èmes Rencontres de la Société Francophone de Classification (SFC-95)*, Namur, Sept.95.

Diday E. (1991): « Des objets de l'analyse des données à ceux de l'analyse des connaissances », in *Induction Symbolique et Numérique à partir de Données*, Editeurs Y.Kodratoff et E.Diday, Cépaduès-Editions.

Granville V., Rasson J.P. (1995): « Multivariate discriminant analysis and maximum penalized likelihood density estimation », *Journal of Royal Statistical Society*, Series B, 57, pp. 501-517.

Hébrail G. (1995): « L'analyse de données symboliques : état de l'art et perspectives », *EDF-DER Research report*, n°HI-23/95-018.

Ichino M. (1994): « Generalised Minkowski metrics for mixed features type data analysis », in *IEEE Transactions on System, Man, and Cybernetics*, **24**, 4, pp. 698-708.

Lebbe J., Vignes R. (1991): « Génération de graphes d'identification à partir de descriptions de concepts », in *Induction Symbolique et Numérique à partir de Données*, Editeurs Y.Kodratoff et E.Diday, Cépaduès Editions.

Tong H.T.T., Summa M., Périnel E., Ferraris J. (1996): « Generating symbolic descriptions for classes », in *Proceedings of the IFCS'96 Conference*, Kobe, March 96.

SODAS Project (1995): Answer to the DOSIS Call for Proposals.

Classification Structures for Cognitive Maps

Stephen C. Hirtle and Guoray Cai

School of Information Sciences
University of Pittsburgh
Pittsburgh, PA 15260 USA
sch,gcai@sis.pitt.edu

Summary: The ability to create and manipulate meaningful data structures of cognitive spaces remains a problem for designers of geographic information systems. Methods to represent the inherent hierarchical structure in cognitive spaces are discussed. Several alternative scaling techniques for developing hierarchical and overlapping representations, including ordered trees, ultrametric trees, and semi-lattices, are presented and discussed. To demonstrate the differences among these three representation schemes, each of three techniques is applied to two small datasets collected on the recall of capitals or countries in Europe. The methods discussed here were chosen to illustrate the limitations of a strict, hierarchical representation and because they have been used in the past to model cognitive spaces.

1. Introduction

The ability to create and manipulate meaningful data structures of cognitive spaces remains a problem for designers of geographic information systems. The ability to present and to interpret spatial data in a method that is consistent with the internal cognitive map of the user would lead to systems that are more flexible and will provide greater functionality in terms of cognitive spatial tasks (Hirtle and Heidorn, 1993; Medyckyj-Scott and Blades, 1992).

A common conclusion that has emerged from the research on the structure of cognitive mapping is that spatial memory is organized hierarchically, which results in processing biases and errors in judgments (Couclelis, et al. 1987; Golledge, 1992; Hirtle and Jonides, 1985; McNamara, et al., 1989; Stevens and Coupe, 1978). However, as Hirtle (1995) argued recently, the claim that mental representations are inherently hierarchical is often made without providing an explicit alternative. For example, the first author and his colleagues have argued that their data is consistent with a "partially hierarchical model" (McNamara, et al., 1989) and have warned against the conclusion that only structure in a cognitive map is of a hierarchical nature (Hirtle and Jonides, 1985). While such qualifications are intriguing, they are often stated without proposing an explicit alternative. In this paper, several alternative scaling techniques for developing hierarchical and overlapping representations, including ordered trees, ultrametric trees, and semi- lattices, are considered.

2. Hierarchies

A strict hierarchy if often assumed for representing spatial concepts. For example, Stevens and Coupe (1978) showed how people consistently misjudged certain directions, such as assuming that Reno, Nevada is north and east of San Diego, California, when in fact Reno is north and west of San Diego. To account for such effects, Stevens and Coupe (1978) presented a nested, propositional model, with San Diego as part of California, Reno as part of Nevada, and California to the west of Nevada. Here, the reasoning processes occur on a hierarchical tree structure, which contains cities nested within states. Thus, a hierarchy is assumed to be formally equivalent to a rooted tree, in a graph-theoretic form. A hierarchy can be defined formally as a collection of sets such that for any two sets in the collection, either one set is contained in the other or the two sets are disjoint (Alexander, 1965).

Many real-world phenomena can be represented by a tree, such as cities within states, states within countries, countries within continents, and so on. However, Hirtle (1995) argued that most attempts to force a hierarchy onto anything other than artificial examples usually fail. Gary, Indiana, in terms of influences, transportation, and even time zones, is more closely associated with Chicago than with the rest of Indiana. Lake Tahoe represents a single geographical "neighborhood" that lies in both California and Nevada. Such examples might be considered noise in the data to be ignored. However, in discussing the structure of cities, Alexander (1965) has argued that a natural city is by nature not hierarchical, but contains overlapping clusters that are better represented in semi-lattice. Hirtle (1995) explored this hypothesis by examining a small subsample of a larger dataset. Here, we expand on this analysis by including the entire dataset and examining alternative distance metrics. We begin by constrasting two partially hierarchical structures, ordered trees and semi-lattices.

3. Ordered Trees

A technique that has proven useful for uncovering hierarchical structure in cognitive maps has been that of the ordered tree algorithm for free-recall data (Hirtle and Jonides, 1985; McNamara, et al., 1989). An ordered tree is a rooted tree where the children of a node, at any level, may be ordered, as a unidirectional or bidirectional node, or unordered, as a nondirectional node. Ordered trees, as discussed here, were first introduced by Reitman and Rueter (1980) and differ from two other uses in the literature of the term. Aho, et al. (1974) define an ordered tree as one in which all children are strictly ordered from left to right. In a third use of the term, Barthelemy, et al. (1986) define an ordered tree as a rooted tree where the nodes are ordered by the height of the nodes. In this paper, the discussion is restricted to the first use of the term, as defined by Reitman and Rueter (1980).

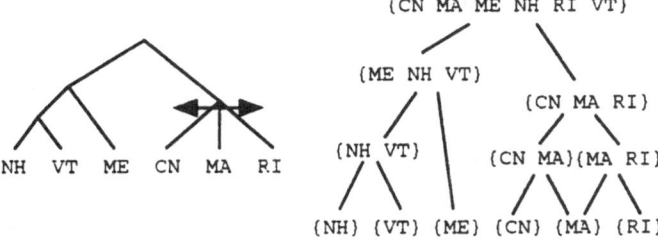

Fig. 1: Ordered dendrogram and set inclusion diagram for ordered tree.

An ordered tree is built by examining the regularities in a set of recalls over a fixed set of items. In fact, an ordered tree is a generalization that allows for some overlapping structure. As an example, the collection of sets {NH VT}, {ME NH VT}, {CN MA}, {MA RI}, {CN MA RI}, and {CN MA ME NH RI VT} can not be represented by a tree, since the sets {CN MA} and {MA RI} are overlapping and violate the definition of a hierarchy, given above. However, this collection can be represented by the ordered tree as seen in Figure 1.

One might be tempted to conclude that an ordered tree is simply a variant of non-binary tree. However, this is not the case. Note that in the previous example, a non-binary tree could be constructed by the removal of the overlappin sets {CN MA} and {MA RI}. However by the inclusion of these two sets, along with the explicit exclusion of the set {CN RI}, the collection of sets can no longer be represented by a strict hierarchy. Furthermore, many cognitive and real- world relations are best seen as exactly this type of ordered structure.

4. Semi-lattices

A semi-lattice is a generalization of an ordered tree. It is defined formally as a collection of sets, such that for any two overlapping sets in the collection, the intersection of the sets is also in the collection (Alexander, 1965). Therefore, if the sets {A B C D E F} and {B C E G H} are in the collection, then the set {B C E} must be in the collection, as well. As an example, consider the collection of sets {NH VT}, {CN MA}, {ME NH VT}, {CN MA VT}, {CN MA RI}, and {CN MA ME NH RI VT}. Such a collection cannot be represented as either a tree or an ordered tree, but can be represented as a semi-lattice. The sets {ME NH VT}, {CN MA VT} and {CN MA RI} are overlapping and thus violate the definition of an ordered tree, given above. However, this collection can be represented by the graph structure shown in Figure 2.

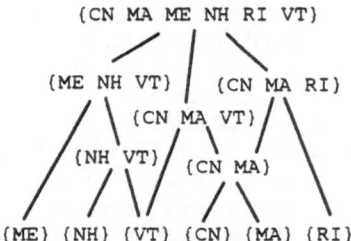

Fig. 2: Set inclusion diagram for semi-lattice structure.

Alexander (1965) notes that planned, or what he calls "artificial," cities often are designed using a strict tree structure. Two such examples are Columbia, Maryland, where clusters of exactly five neighborhoods combine to form villages and a 1943 plan of Greater London by Abercrombie and Forshaw argues for a "large number of communities, each separated from all adjacent communities." Each community is further subdivided into neighborhoods, each with their own shops and schools. Alexander (1965) goes on to argue that a natural, living city, despite the ill-advised wishes of the urban planner, does not conform to the hierarchical structure of a tree. Rather, aa post office, a local school, a social club, or water authority all serve areas of different sizes and scope. Thus, the resulting structure is better conceptualized to be that of a semi- lattice.

5. Mapping Data to Structures

5.1 Data and Trees

To demonstrate further the differences among these three representation schemes, we turn to two small datasets collected on the recall for countries in Europe. During the academic year of 1994-1995, students on two different campuses in two different European countries were asked to make an ordered list, from memory, of either all the countries in Europe, or all the capitals in Europe. No other instructions were given to the subjects. To equate the two samples, the capitals were converted into the country name for those receiving the capital task. It is further acknowledged that the capital task was harder and that exclusions might occur, not from forgetting the country, but because the subject does not know or is unsure of the name of the capital. However, these two datasets are considered only to highlight the differences between the representations discussed in this paper, and not to generalize about specific regional understanding of European geography. Furthermore, the purpose of this exercise was to explore the possible clustering that exists among countries and not the rather trivial set inclusion principle of aggregating a capital to its host country.

A group of 18 subjects in Norway, who were asked to recall countries of Europe, produced a total of 44 distinct entries. The entire set of recalls can be seen using a path-graph visualization developed by Hirtle (1991) in Figure 3. Here, the line width is proportional to the number of the times two countries were recalled simultaneously. By visually focusing on the thicker connections, several clear clusters, such as the Scandinavian countries, begin to emerge, as seen in Figure 3.

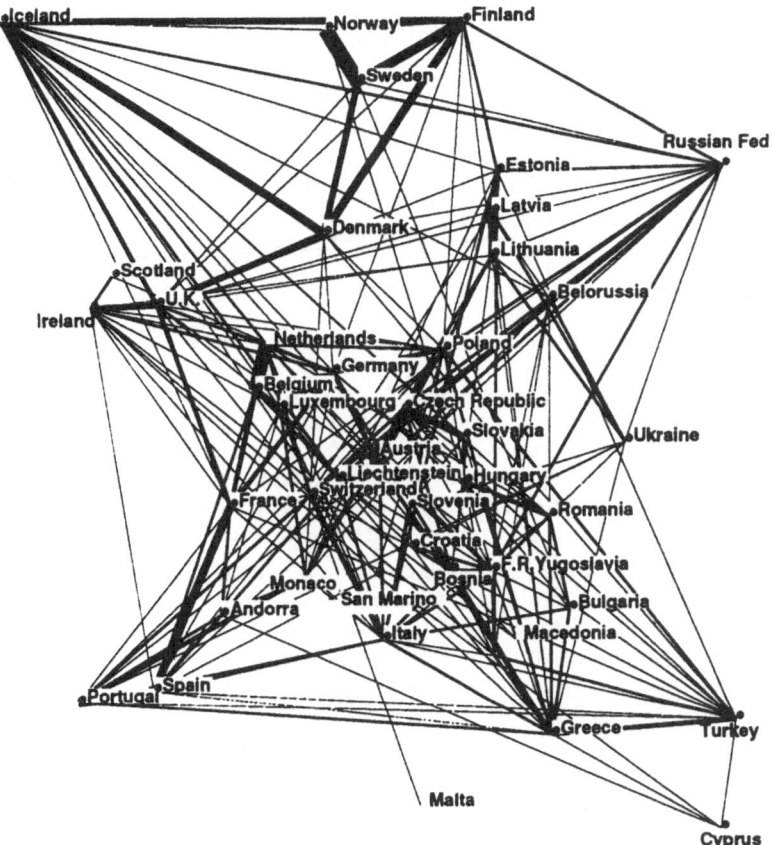

Fig. 3: Path graph of European countries generated by the Norwegian subjects

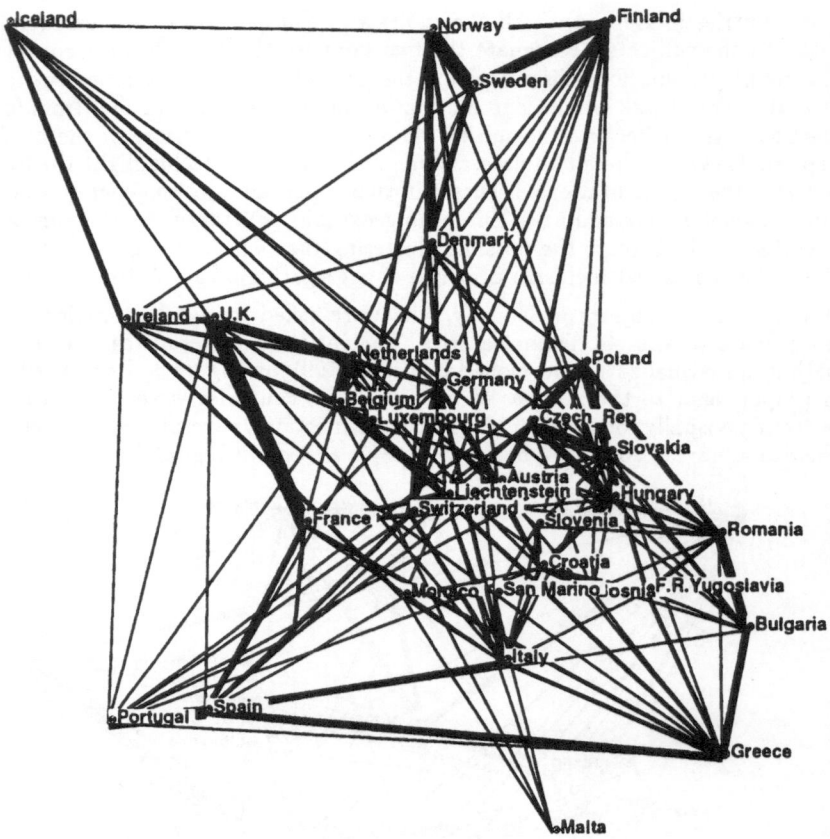

Fig. 4: Path graph of country names for the European capitals generated by the Austrian subjects

A group of 12 subjects in Austria, who were asked to list all the capital cities of Europe, produced a total of 32 distinct entries. The capitals were converted to the country names, and resulting complete path-graph, shown in Figure 4.

The ordered lists from each of the datasets were clustered into a strict hierarchical tree, using an average-link clustering algorithm (UPGMA). This was done using two different measures of distance, city-block and a log-based distance. The city-block metric is equivalent to stating that the distance between any two countries is proportional to the total number of intervening items between them across all the ordered lists. However, as items are further separated on the list, the actual numerical difference becomes less important. Therefore, we replicated the analysis with the logarithm of the difference. Furthermore, four countries were dropped from the analysis, due to a lack of data for calculating pairwise distances. For simplicity, only the later distance analysis is reported here. The resulting tree is shown in Figure 5.

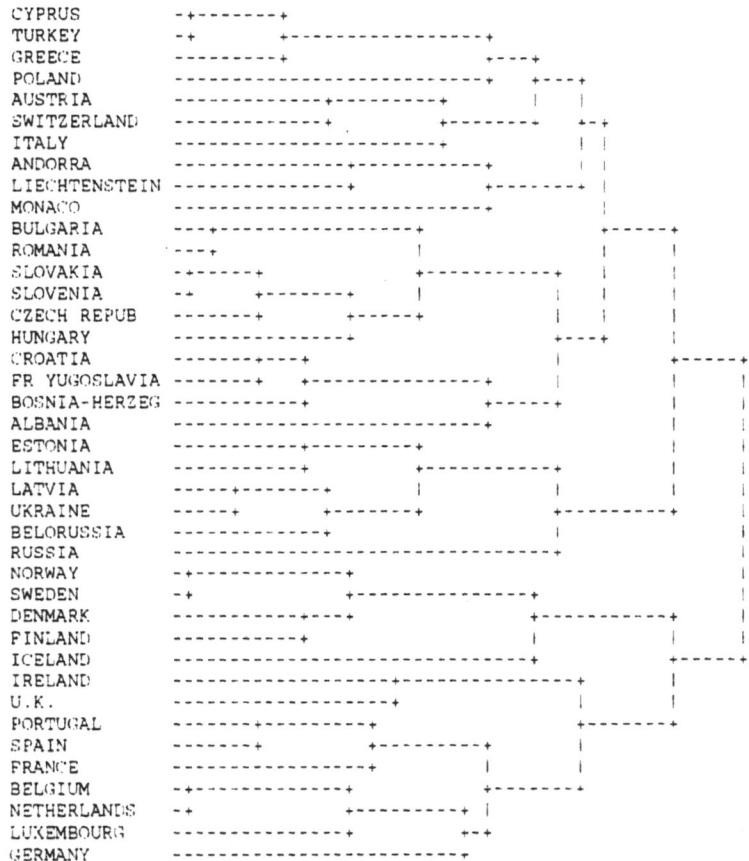

Fig. 5: Hierarchical tree of European countries generated by the Norwegian subjects

A group of 12 subjects in Austria, who were asked to list all the capital cities of Europe, produced a total of 32 distinct entries. The ordered recalls of these 32 countries were clustered into a strict hierarchical tree, also using the average-link clustering algorithm with the city-block distance and log-based distance. The resulting tree for log-based distances is shown in Figure 6.

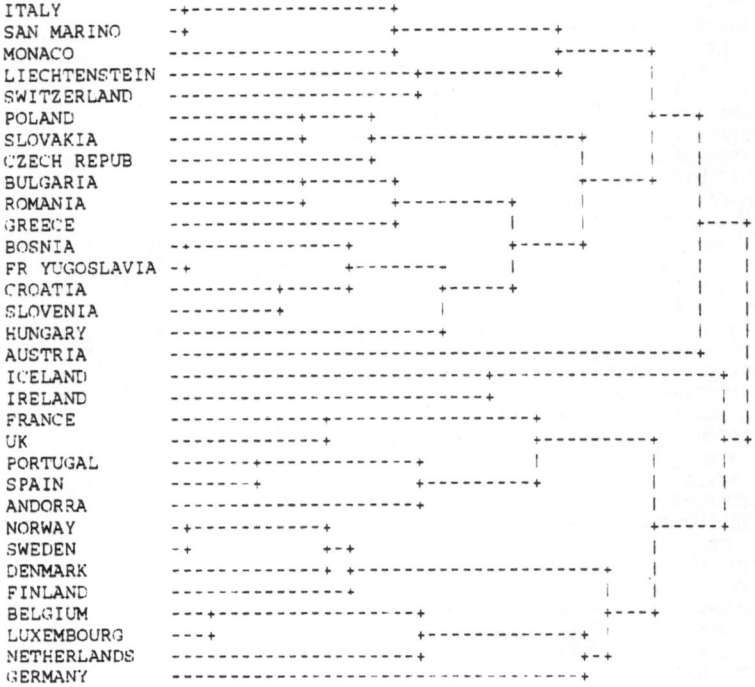

Fig. 6: Hierarchical tree of the country names for the European capitals generated by the Austrian subjects

In examining these trees, the limitation of a hierarchical tree becomes obvious. Each country is placed uniquely in a single cluster within the tree, by the very definition of a tree. The multiple relationships, which Alexander (1965) argued convincingly in favor of, are not able to be incorporated in the representational structure.

5.2 Ordered trees

An ordered tree might allow some overlapping relationships to emerge. Unfortunately, an immediate application of the existing ordered tree algorithm of Reitman and Rueter (1980) is not possible. The algorithm was developed to account for the strong representational structures within a single subject for a domain of interest and not to build an average structure across many subjects. Thus, the algorithm is deterministic and produces clusters that exist across all recall patterns. Within the Norwegian sample, there was not a single cluster that was common to all subjects, whereas in the Austrian sample, only the single cluster of Norway Sweden existed for all the subjects.

However, by examining subgroups of subjects within each sample, one can identify small groups of subjects with common strategies, for which one can calculate non-trivial ordered trees. Figure 7 shows one tree from a subset of the Norwegian subjects and Figure 8 shows trees from two subsets of the Austrian subjects. It is interesting to note the predominance of the home country, as expected, in each sample. In addition, the two ordered trees in Figure 8, from the Austrian sample, indicate two very different strategies, one that is geographically oriented (Figure 8a), and another that is ordered by prominence (Figure 8b). The former strategy resulted in Austria clustered with Switzerland and Liechtenstein, whereas the latter strategy resulted in Austria being followed by France and United Kingdom.

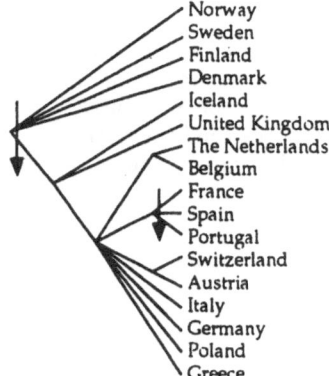

Fig. 7: An example of an ordered tree for a subgroup of Norwegian subjects

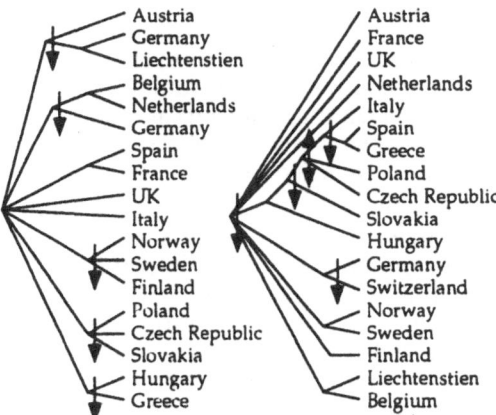

Fig. 8: An example of ordered trees for two subgroups of Austrian subjects

Two benefits arise from the ordered tree over the strict hierarchical tree. First, any order that might exist within a cluster is preserved. This can be seen by dominance of the home countries of Norway and Austria within their respective ordered trees in Figures 7 and 8. The ordered clusters also provide examples of implicit overlapping internal clusters. For example, in Figure 7, the ordered cluster of France, Spain, Portugal is created by the two underlying, overlapping clusters of France, Spain and Spain, Portugal both of which have strong surface validity, while the excluded relationship France, Portugal has much weaker surface validity. A strict hierarchical model would imply that every pair of items in a cluster would be associated at the same level.

5.3 Semi-lattices

The final representational scheme of a semi-lattice lacks any direct method to produce, which may account for why the representation of a semi-lattice for cognitive maps has not been considered to the extent of the previous two representations. One solution would be to use the MAPCLUS algorithm to fit the ADCLUS model (Shepard and Arabie, 1979) of overlapping clusters. These clusters could then provide a seed set of potential clusters to build a semi-lattice upon. An initial application of the MAPCLUS algorithm to the data from the Norwegian subjects resulted in four clusters, with only the Scandinavian cluster being distinct from the others. The data from the Austrian subjects also resulted in four overlapping clusters. One cluster consisted of northern European countries, including Scandinavia and the British Isles. The second consisted of eastern European countries. The third cluster consisted of prominent central European countries and the final of less prominent countries. While such an analysis is promising, it is clear that any implementation of semi-lattice models will require the additional development of appropriate algorithms.

6. Conclusions and summary

In summary, a tree structure is one realization for a hierarchical structure for the representation of space. It is easily constructed and understood, but it is also a rigid structure that does not allow for overlap. Ordered trees provide an extension that allows for some degree of overlap, whereas a semi-lattice is an even richer structure that appears to be consistent with many aspects of cognitive space (Alexander, 1965). There are many other possibilities for representing spatial clusters, including additive trees (Sattath and Tversky, 1977), pseudo-hierarchies or pyramids (Diday, 1986), extended trees (Carroll and Corter, 1995), and hybrid scaling models (Carroll and Pruzansky, 1980). A survey and review of the mathematical properties of many of these representations can be found in Van Cutsem (1994). The methods discussed here were chosen to illustrate the limitations of a strict, hierarchical representation and because they have been used in the past to model cognitive spaces.

As spatial information systems develop and evolve, the importance of considering alternative structures to strict hierarchical trees and the necessity of being explicit about the nature of the assumed representational structure will only increase. Spatial information systems, multimedia systems, and large information systems including the World Wide Web require a user to navigate through complex and often poorly differentiated spaces (Kim and Hirtle, 1995). The ability to generate a meaningful mutli-level structure should ease the cognitive burden imposed by the navigational task and allow users to focus on the informational task instead. Finally, it is important to note that a consistent theme behind all of the representations discussed is that of a highly structured representation. To replace use of hierarchical trees with unstructured representation, such as an undifferentiated network, would be a serious mistake. Rather, the goal of future research should be to clarify the exact nature of the underlying, structured representation of cognitive spaces.

7. Acknowledgments

This paper was prepared, in part, while the first author was on sabbatical at the Department of Computer Science, Molde College, in Molde, Norway. Their support is gratefully appreciated. The authors wish to thank Adrijana Car, Kai Olsen, and Phipps Arabie for their comments concerning the issues presented in this paper.

References:

Aho, A. V., et al. (1974): *The design and analysis of computer programs*, Addison-Wesley, Reading, MA.

Alexander (1965): A city is not a tree, *Design*, 46-55.

Barthelemy, J. P., et al. (1986): On the use of ordered sets in problems and consensus of classification, *Journal of Classification*, **3**, 187-224.

Carroll, J. D. and Corter, J. E. (1995): A graph-theoretic method for organizing overlapping clusters into trees, multiple trees, or extended trees, *Journal of Classification*, in press.

Carroll, J. D. and Pruzansky, S. (1980): Discrete and hybrid scaling models. In: *Similarity and Choice*, Lantermann, E. D. and Feger, H.. (eds.), Hans Huber, Bern.

Couclelis, H., et al. (1987): Exploring the anchor-point hypothesis of spatial cognition, *Journal of Environmental Psychology*, **7**, 99-122.

Diday, E. (1986): Orders and overlapping clusters in pyramids, In: *Multidimensional data analysis* de Leeuw, J., et al. (eds.), 201-234, DSWO Press, Leiden.

Golledge, R. G. (1992): Place recognition and wayfinding: Making sense of space, *Geoforum*, **23**, 199-214.

Hirtle, S. C. (1991): Knowledge representations of spatial relations. In: *Mathematical psychology: Current developments*, Doignon, J.-P. and Falmagne, J.-C. (eds.), 233-250, Springer-Verlag, New York.

Hirtle, S. C. (1995) Representational structures for cognitive space: Trees, ordered trees, and semi-lattices, In: *Spatial information theory: A theoretical basis for GIS*, Frank, A. V. and Kuhn, W. (eds.), Springer- Verlag, Berlin.

Hirtle, S. C. and Heidorn, P. B. (1993): The structure of cognitive maps: Representations and processes. In: , *Behavior and environment: Psychological and geographical approaches*, Garling, T. and Golledge, R. G. (eds.), 170-192, North-Holland, Amsterdam.

Hirtle, S. C. and Jonides, J. (1985): Evidence of hierarchies in cognitive maps, *Memory and Cognition*, **3**, 208-217.

Kim, H., and Hirtle, S. C. (1995). Spatial metaphors and disorientation in hypertext browsing. *Behaviour and Information Technology*, **14**, 239- 250.

McNamara, T. P., et al. (1989): Subjective hierarchies in spatial memory, *Journal of Experimental Psychology: Learning, Memory, and Cognition*, **15**, 211- 227.

Medyckyj-Scott, D. J. and Blades, M. (1992): Human spatial cognition, *Geoforum*, **2**, 215-226.

Reitman, J. S. and Rueter, H. R. (1980): Organization revealed by recall orders and confirmed by pauses, *Cognitive Psychology*, **12**, 554-581.

Sattath, S. and Tversky, A. (1977): Additive similarity trees. *Psychometrika*, **42**, 319-345.

Shepard, R. N. and Arabie, P. (1979): Additive clustering: Representation of similarities as combinations of discrete overlapping properties, *Psychological Review*, **86**, 87-123.

Stevens, A. and Coupe, P. (1978): Distortions in judged spatial relations, *Cognitive Psychology*, **10**, 422-437.

Van Cutsem, B. (Ed.) (1994): *Classification and dissimilarity analysis*, Lecture Notes in Statistics, No. 93, Springer-Verlag, New York.

Unsupervised Concept Learning
Using Rough Concept Analysis

Tu Bao Ho [1]

Japan Advanced Institute of Science and Technology
Tatsunokuchi, Ishikawa, 923-12 JAPAN

Summary: Formal concept analysis (Wille, 1982) offers an algebraic tool for representing and analyzing formal concepts, and the rough set theory (Pawlak, 1982) offers an alternative tool to deal with vagueness and uncertainty. Rough concept analysis (Kent, 1994) is an attempt to synthesize common features of these two theories. In this work we develop a method for unsupervised concept learning in the framework of rough concept analysis that aims at finding and using concepts with their approximations.

1. Introduction

The problem of finding not only hierarchical clusters of unlabelled objects but also their 'good' conceptual descriptions was addressed early in data analysis, e.g., Diday and Simon (1976). This problem is referred to as *unsupervised concept learning* in machine learning which can be defined as

- *Given* a set of unlabelled object descriptions;
- *Find* a hierarchical clustering that determines useful object subsets *(clustering)*;
- *Find* intensional definitions for these subsets of objects *(characterization)*.

Unsupervised concept learning techniques depend strongly on how concepts are understood and represented. The notion of concepts under the classical view and the generality relation was mathematically formulated in the theory of formal concept analysis by Wille and his colleagues during last fifteen years (Wille, 1992). Recently, several concept learning methods have been developed in this framework, e.g., those of Godin and Missaoui (1994), Carpineto and Romano (1996) that incrementally learn all possible concepts, or method OSHAM (Ho, 1995) that extracts a part of the hypothesis space in the form of concept hierarchies.

The theory of rough sets is a mathematical tool to deal with vagueness and uncertainty in interpreting given data (Pawlak, 1991). Recently, by combining common features between the rough set theory and formal concept analysis, Kent (1994) introduced a theory of *rough concept analysis* as a framework for representing and learning approximations of concepts. In this paper we developed unsupervised conceptual clustering method A-OSHAM, an extension of OSHAM, for inducing concept hierarchies with concept approximations by using rough concept analysis.

2. Formal concept analysis and rough concept analysis

2.1 Formal concept analysis

Basis of the most widely understanding of a concept is the function of collecting individuals into a group with certain common properties. One distinguishes these common properties as the *intent* of the concept that determines its *extent* which are

---

[1] Also with the Institute of Information Technology, Hanoi, Vietnam

objects sharing these properties and accepted as members of the concept. Formal concept analysis models formal concepts from a *formal context* which is a triple (O, A, R) where O is a set of objects, A is a set of attributes and R is a binary relation between O and A, i.e., $R \subseteq O \times A$. Notice that for the simplicity, formal concept analysis is always described with Boolean data. However, its notions can be extended for multi-valued data. In general, oRa is understood as "object o has attribute a" in Boolean domain, and can be extended to nominal or discrete numeric domains as "object o has attribute a with some value v". Data from continuous domains can be discretized into discrete data to be used with the framework. A *formal concept* of a given formal context is a pair of extent/intent (X, S) where the extent X contains *precisely* those objects sharing all attributes in the intent S, and vice-versa, the intent S contains *precisely* those attributes shared by all objects in the extent X. The relation between extent and intent of concepts can be described by two operators λ and ρ

$$S = \lambda(X) = \{a \in A \mid \forall o \in X : oRa\} = \bigcap_{o \in X} oR \subseteq A \tag{1}$$

$$X = \rho(S) = \{o \in O \mid \forall a \in S : oRa\} = \bigcap_{a \in S} Ra \subseteq O \tag{2}$$

objects	abb.	nw	lw	ll	nc	2lg	1lg	mo	lb	sk
Leech	Le	○	○					○		
Bream	Br	○	○					○	○	
Frog	Fr	○	○	○				○	○	
Dog	Dg	○		○				○	○	○
Spike-Weed	SW	○	○		○		○			
Reed	Rd	○	○	○	○		○			
Bean	Bn	○		○	○	○				
Maize	Mz	○		○	○		○			

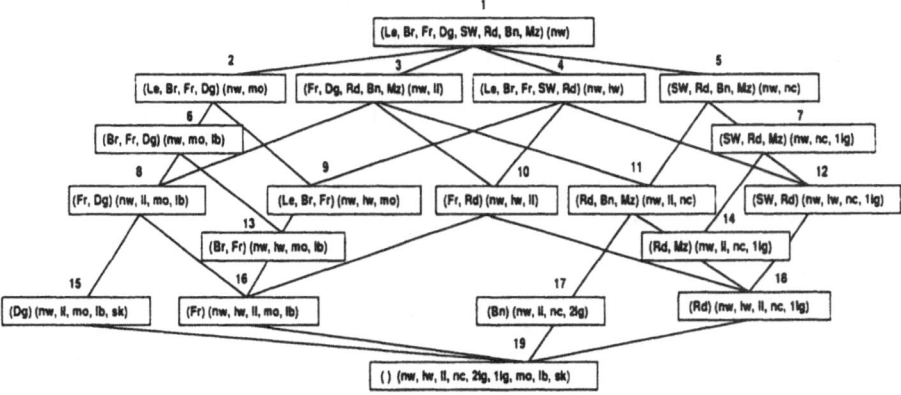

Figure 1: Formal context and concept lattice of living organisms

A concept C_1 is called a *superconcept* of a concept C_2 (C_2 is a *subconcept* of C_1) if the extent of C_2 is included in the extent of C_1. The basic theorem in this theory is that formal concepts of a formal context is a complete lattice[2] with respect to the superconcept-subconcept relation, called *concept lattice* and denoted by $\mathcal{L}(O, A, R)$.

Example: Figure 1 shows a Wille's illustration of a formal context of eight living organisms whose names are heading the rows (with corresponding abbreviations)

---

[2] A lattice $\mathcal{L}$ is complete when each of its subset X has a least upper bound and a greatest lower bound in $\mathcal{L}$.

{Leech (Le), Bream (Br), Frog (Fr), Dog (Dg), Spike-Weed (SW), Bean (Bn), Maize (Mz), Reed e(Rd)}. The living organisms are described by attributes {needs water (nw), lives in water (lw), lives on land (ll), needs chlorophyll (nc), 2 leaf germination (2lg), 1 leaf germination (1lg), is motile (mo), has limbs (lb), suckles young (sk)}. Furthermore, the $\circ$ indicates when an object has an attribute, i.e., which living organism has which attribute. The concept lattice $\mathcal{L}(O, A, R)$ consists of 19 formal concepts $C_1, C_2, ..., C_{19}$ (denoted in the figure only by their indexes), for example $C_9 = (\{Le, Br, Fr\}, \{nw, lw, mo\})$.

2.2 Rough sets and rough concept analysis

There have been different methods of approximating concepts, e.g., those employ the Bayesian decision theory or the well-known fuzzy set theory which characterize approximately concepts by a membership function with a range between 0 and 1. Rough set theory can be considered as an alternative way for approximating concepts. The starting point of this theory is the assumption that our "view" on elements of a set of objects O depends on some equivalence relation E on O. An *approximation space* is a pair (O, E) consisting of O and an equivalence relation $Ea \subseteq O \times O$. The key notion of the rough set theory is the *lower* and *upper approximations* of any subset $X \subseteq O$ which consist of all objects *surely* and *possibly* belonging to X, respectively. The lower approximation $E_*(X)$ and the upper approximation $E^*(X)$ are defined by

$$E_*(X) = \{o \in O : [o]_E \subseteq X\} \qquad E^*(X) = \{o \in O : [o]_E \cap X \neq \emptyset\} \qquad (3)$$

where $[o]_E$ denotes the equivalence class of objects indiscernible with o with respect to the equivalence relation E. Kent (1994) has pointed out common features between the theories of rough sets and formal concept analysis, and formulated the *rough concept analysis*. Saying that a given formal context (O, A, R) is not obtained completely and precisely means that the relation R is incomplete and imprecise. Let (O, E) be any approximation space on objects O, we wish to approximate R in terms of E. The lower approximation R_{*E} and the upper approximation R^{*E} of R w.r.t. E can be defined element-wise as

$$R_{*E}a = E_*(Ra) = \{o \in O \mid [o]_E \subseteq Ra\} \qquad (4)$$

$$R^{*E}a = E^*(Ra) = \{o \in O \mid [o]_E \cap Ra \neq \emptyset\} \qquad (5)$$

The formal context (O, A, R) can be then roughly approximated by two lower and upper formal contexts (O, A, R_{*E}) and (O, A, R^{*E}). These approximate contexts can be intuitively viewed as "truncated" and "filled up" contexts with respect to the equivalence relation E, as illustrated in Figure 2 and 3. Two formal context (O, A, R) and (O, A, R') are *E-roughly equal* if they have the same lower and upper formal contexts. i.e., $R_{*E} \equiv R'_{*E}$ and $R^{*E} \equiv R'^{*E}$. A *rough formal context* in (O, E) is a collection of formal contexts of object set O and attribute set A which have the same lower and upper formal contexts (roughly equal formal contexts).

The rough extent of an attribute subset $S \subseteq A$ w.r.t. R_{*E} and R^{*E} are defined as

$$\rho(S_{*E}) = \bigcap_{a \in S} R_{*E}a \qquad \rho(S^{*E}) = \bigcap_{a \in S} R^{*E}a \qquad (6)$$

Now, any formal concept $(X, S) \in \mathcal{L}(O, A, R)$ can be approximated by R_{*E} and R^{*E}. The *lower* and *upper E-approximation* of (X, S) are defined as

$$(X, S)_{*E} = (\rho(S_{*E}), \lambda\rho(S_{*E})) \in \mathcal{L}(O, A, R_{*E}) \qquad (7)$$

$$(X, S)^{*E} = (\rho(S^{*E}), \lambda\rho(S^{*E})) \in \mathcal{L}(O, A, R^{*E}) \qquad (8)$$

A *rough concept* of a formal concept (O, A, R) in (O, E) is the collection of concepts which have the same lower and upper E-approximations (roughly equal concepts).

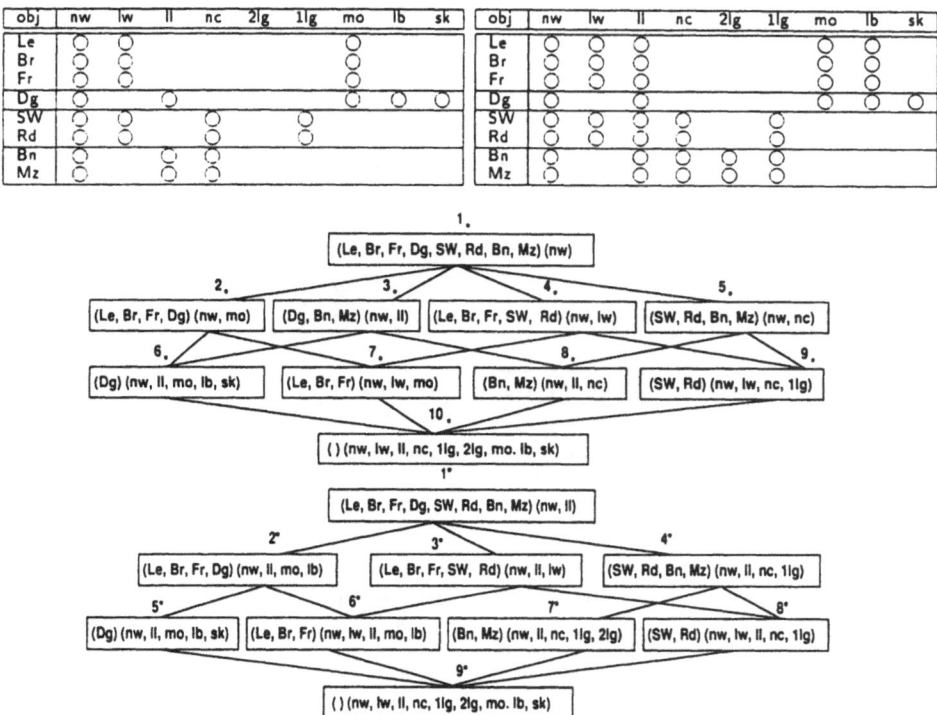

Figure 2: Lower and upper approximations of living organisms w.r.t. (lw, nc)

3. Learning in the framework of rough concept analysis

3.1 Inducing concept hierarchies with concept approximations

Note that approximate contexts of (O, A, R) in (O, E) vary according to the equivalence relation E. Figure 2 (modified from Kent, 1994) illustrates the lower and upper approximate contexts (left and right tables) as well as the lower and upper approximate concept lattices $\mathcal{L}(O, A, R_{*E})$ (up) and $\mathcal{L}(O, A, R^{*E})$ (down) of $\mathcal{L}(O, A, R)$ where the indiscernible relation E (denoted by E_1) is determined with respect to two features lives in water (lw) and needs chlorophyll (nc). These approximate concept lattices consist of 10 lower approximations and 9 upper approximations, denoted by $C_{1*}^1, C_{2*}^1, ..., C_{10*}^1$ and $C_{1*}^1, C_{2*}^1, ..., C_{9*}^1$, respectively. Figure 3 illustrates those of $\mathcal{L}(O, A, R)$ (similar positions) where the indiscernible relation E (denoted by E_2) is determined with respect to two features lives on land (ll) and is motile (mo). These approximate concept lattices also consist of 10 and 9 approximations, denoted by $\{C_{1*}^2, C_{2*}^2, ..., C_{10*}^2\}$ and $\{C_{1*}^2, C_{2*}^2, ..., C_{9*}^2\}$, respectively.

We can determine the maps of concepts from $\mathcal{L}(O, A, R)$ to their approximations in $\mathcal{L}(O, A, R_{*E_1})$ and $\mathcal{L}(O, A, R^{*E_1})$ and to $\mathcal{L}(O, A, R_{*E_2})$ and $\mathcal{L}(O, A, R^{*E_2})$.

Example: Some assignments from these maps

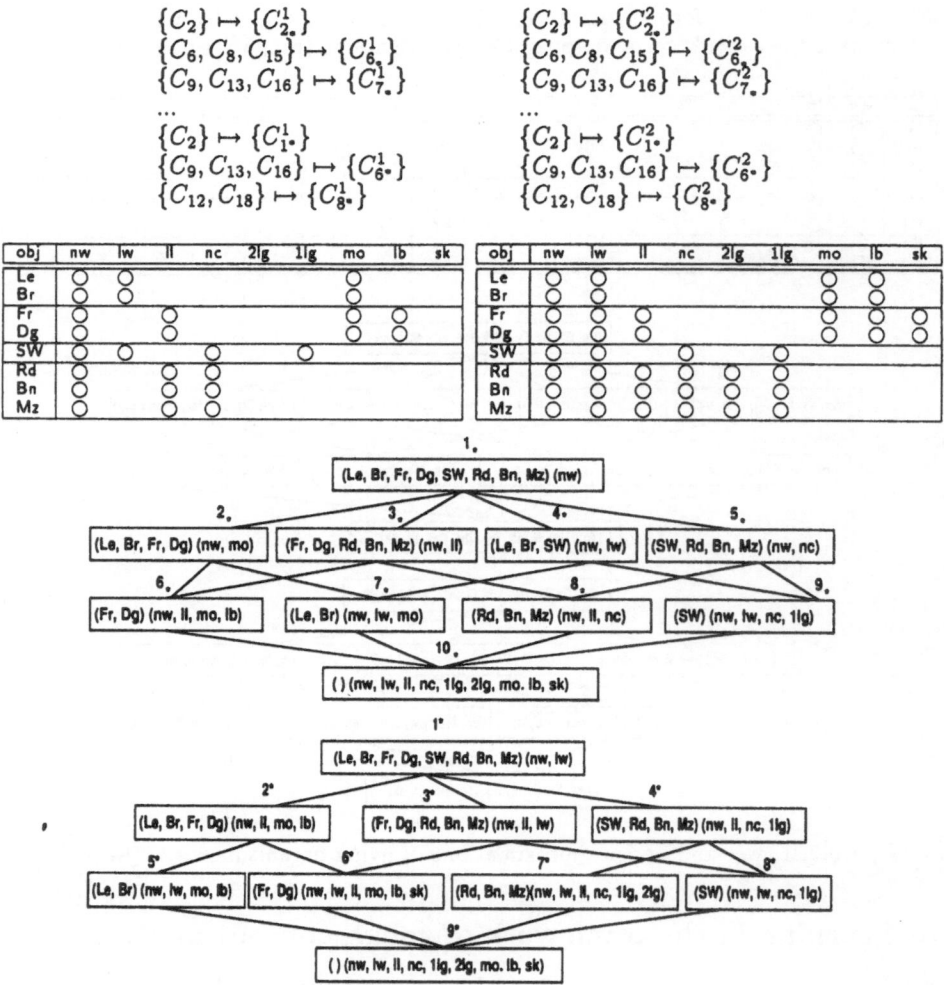

Figure 3: Lower and upper approximations of living organisms w.r.t. (ll, mo)

Consider concept $C_{13} = (\{Br, Fr\}, \{nw, lw, mo, lb\})$ in $\mathcal{L}(O, A, R)$. It has lower approximations $C_{7\bullet}^1 = (\{Le, Br, Fr\}, \{nw, lw, mo\})$ and $C_{2\bullet}^2 = (\{Le, Br, Fr, Dg\}, \{nw, mo\})$ in $\mathcal{L}(O, A, R_{\bullet E_1})$ and $\mathcal{L}(O, A, R_{\bullet E_2})$, and upper approximations $C_{6\bullet}^1 = (\{Le, Br, Fr\}, \{nw, lw, ll, mo, lb\})$ and $C_{2\bullet}^2 = (\{Le, Br, Fr, Dg\}, \{nw, ll, mo, lb\})$ in $\mathcal{L}(O, A, R^{\bullet E_1})$ and $\mathcal{L}(O, A, R^{\bullet E_2})$. Although some concepts in $\mathcal{L}(O, A, R)$ have similar indexes in different approximate lattices, their approximations may have different intent and extent.

OSHAM (Ho, 1995) is an unsupervised concept learning method that induces a concept hierarchy H from the concept lattice $\mathcal{L}(O, A, R)$. Inspired by this algorithm, we propose algorithm A-OSHAM for learning approximate concepts in the framework of rough concept analysis. Essentially, A-OSHAM induces a concept hierarchy in which each induced concept is associated with a pair of its lower and upper approximations. The search for the concept hierarchy is carried out through the hypothesis space of concept lattice $\mathcal{L}(O, A, R)$. The basis of this search is a generate-and-test operator to split a concept C into subconcepts at a lower level of H. Associated lower and upper approximations are computed from the approximate contexts generated corresponding to the heuristic of inducing the concept. Starting from the root concept with the whole set of training instances, A-OSHAM induces the concept hierarchy H recursively in a top-down direction as described in Table 1. Procedure for computing the lower and upper approximations in 1.(c) will be given in Table 2. Procedures for doing 1.(a), 1.(b), 1.(d), 1.(e) and 1.(f) are the same as for OSHAM (Ho, 1995).

Table 1: Algorithm A-OSHAM (C_k, H)

Input	concept hierarchy H and an existing splittable concept C_k.
Result	H formed gradually.
Top-level	call A-OSHAM(root concept, $\emptyset$).
Variables	α is a given threshold.

1. Suppose that $C_{k_1}, ..., C_{k_n}$ are subconcepts of $C_k = (X_k, S_k)$ found so far. While C_k is still splittable, find a new subconcept $C_{k_{n+1}} = (X_{k_{n+1}}, S_{k_{n+1}})$ of C_k and its approximations by doing:

    (a) Find attribute a^* so that $\bigcup_{i=1}^{n} X_{k_i} \cup \rho(\{a^*\})$ is the largest cover of X_k.

    (b) Find the largest attribute set S containing a^* satisfying $\lambda\rho(S) = S$.

    (c) Form subconcept $C_{k_{n+1}}$ with $\rho(S_{k_{n+1}}) = S$ and $X_{k_{n+1}} = \rho(S)$.

    (d) Find a lower approximation and an upper approximation of $C_{k_{n+1}}$ with respect to the chosen equivalence relation E.

    (e) From intersecting subconcepts corresponding to intersections of $\rho(S_{k_{n+1}})$ with extents of existing concepts on H excluding its superconcepts, and find their approximations.

2. Let $X_k^r = X_k \setminus \bigcup_{i=1}^{n+1} X_{k_i}$. If one of the following conditions holds then C_k is considered unsplittable:

    (a) There exist not any attribute set $S \subseteq S_k$ satisfying $\lambda\rho(S) = S$ in X_k.

    (b) $card(X_k^r) \leq \alpha$.

3. Apply A-OSHAM(C_{k_i}, H) to each C_{k_i} formed in the step 1.

A-OSHAM forms concepts at different levels of generality in the hierarchy each level corresponds to a partition of O. A-OSHAM generates concepts with their approximations recursively and gradually, once a level of the hierarchy is formed the procedure is repeated for each class.

There are many possible lower and upper approximations of a concept according to the the family $\mathcal{F}$ of equivalence relations E on O. There will be at least two ways of approximating concepts in the framework of the rough concept analysis: (1) to compute all possible approximations for each induced concept, and (2) to compute one plausible lower and upper approximation for each induced concept.

We investigate in this paper the case of one plausible lower and upper approximation for each induced concept. In each attempt of splitting a concept at the level n, A-OSHAM finds consequently its subconcepts at the level $n + 1$ by the specialized hypotheses with maximum coverage. Suppose that we want to find approximations $(X_k, S_k)_{\bullet E}$ and $(X_k, S_k)^{\bullet E}$ of the induced concept $C_k = (X_k, S_k)$. As S_k is chosen among the hypotheses specialized from the intent of its superconcept with the maximum coverage, it is reasonable to approximate C_k by the one-step further specialized hypothesis generated by the same way. Suppose that P is the partition of the superconcept extent with respect to S_k, we first generate a refinement of P with respect to

$S_k \cup \{a^*\}$ where the conjunction $S_k \cup \{a^*\}$ forms the largest cover of the C_k's super-concept extent, then we approximate C_k with equivalence classes of this refinement according to (8) and (9).

Table 2: Procedure to find approximations

---

1. Find attribute $a^* \in A \setminus S_k$ so that $S_k \cup \{a^*\}$ is the largest cover of the extent of the superconcept of C_k.

2. Find equivalence classes of the superconcept extent of C_k according to the equivalence relation E formed by adding a^* to S_k.

3. Find the lower and upper approximations of C_k with respect to E, i.e., $(X_k, S_k)_{*E}$ and $(X_k, S_k)^{*E}$.

Example: Figure 4 illustrates the concept hierarchy obtained by A-OSHAM from the living organisms context described in Figure 1, with the parameter $\alpha = 2$. The indexes are kept from the search space (Figure 1).

Example: Find the approximations of concept $C_4 = (\{\text{Le, Br, Fr, Sw, Rd}\}, \{\text{nw, lw}\})$. The equivalence relation to generate the approximate context is with respect to $\{\text{nw, lw, mo}\}$), the lower and upper approximations are $(\{\text{Le, Br, Fr}\}, \{\text{nw, lw, mo}\})$ and $(\{\text{Le, Br, Fr, Sw, Rd}\}, \{\text{nw, lw}\})$, respectively.

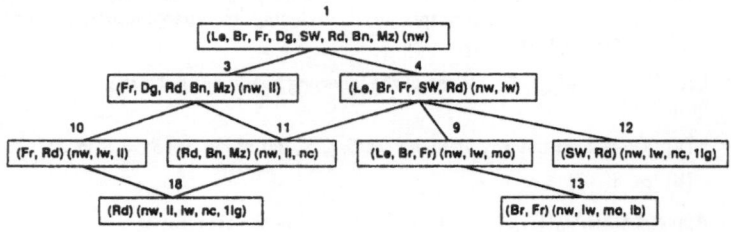

Figure 4: Concept hierarchy induced by A-OSHAM

3.2 Using concept approximations in predicting unknown instances

Concept approximations will have no meaning without some associated interpreter to exploit them in deciding to which induced concepts an unknown instance belongs. Traditionally, there are three types of outcomes when matching logically the induced concept hierarchy H with an unknown instance o:

1. *single-match:* only one concept on H that matches o;
2. *multiple-match:* many concepts on H that match o;
3. *no-match:* no concept on H that matches o.

Most conceptual clustering methods deal with the cases of no-match and multiple-match by employing a probabilistic estimation. Michalski (1990) developed the measure of fit for no-match cases and estimate of probability for multiple-match cases. Recently, we have developed an interpretation of unsupervised induction results that combines case-based reasoning with logical matching (Ho and Luong, 1997). Concept approximations in the framework of rough concept analysis offer an alternative solution to this problem.

The basis of this solution is a refinement of three common match outcomes by the concept approximations. For a concept $(X, S) \in H$, the approximations of its extent X_{*E} and X^{*E} are interpreted as the sets consisting of all objects surely and possibly, respectively, belong to X with respect to E. We have $X_{*E} \subseteq X \subseteq X^{*E}$ and dually $S^{*E} \subseteq S \subseteq S_{*E}$ for every E. We say that an object o *strongly matches* a concept (X, S) with respect to E if it matches the lower approximation S_{*E}. We say that an object o *matches* a concept (X, S) with respect to E if it matches the concept intent S. We say that an object o *weakly matches* a concept (X, S) with respect to E if it only matches the S^{*E} but does not match S. By the order $S^{*E} \subseteq S \subseteq S_{*E}$ we have the corresponding order of weakly match, match and strong match when matching an object o with (X, S). Three common match outcomes now can be refined as follows

1. *No match*: No concepts weakly match o;
2. *Single weak match*: Only one concept weakly matches o;
3. *Multiple weak match*: Many concepts weakly match o but no concepts match o;
4. *Single match but not strong*: One concept matches o but not strongly;
5. *Single strong match*: Only one concept strongly matches o;
6. *Multiple match with one strong match*: Many concepts match o and only one concept strongly matches o;
7. *Multiple match but no strong match*: Many concepts match o and no concepts strongly match o;
8. *Multiple strong match*: Many concepts match o and more than one concept strongly matches o.

The match outcomes 1–3 are the refinement of the exact no-match, the match outcomes 4–5 are the refinement of the exact single-match, and the match outcomes 6–8 are the refinement of the exact multiple-match. The notion of *degree of match satisfaction* can also be refined regarding these matching outcomes. In fact, cases 2 and 6 offer immediately a decision about the membership of the unknown object to the concept being considered; cases 3, 7, 8 require the same consideration as the usual multiple-match case.

References:

Carpineto, C., Romano G. (1996): A Lattice Conceptual Clustering System and its Application to Browsing Retrieval, *Machine Learning*, **10**, 95–122.

Diday, E. and Simon, J.C. (1976): Clustering analysis. In *Digital Pattern Recognition*, K.S. Fu (Ed.), 47–94, Springer-Verlag.

Godin, R. and Missaoui, R. (1994): An incremental concept formation approach for learning from databases. *Theoretical Computer Science*, **133**, 387–419.

Ho, T.B. (1995): An approach to concept formation based on formal concept analysis. *IEICE Trans. Information and Systems*, **E78-D**, 553–559.

Ho, T.B., Luong, C.M. (1997): Using Case-Based Reasoning in Interpreting Unsupervised Inductive Learning Results, *Inter. Joint Conf. on Artificial Intelligence*, Nagoya (in press).

Kent, R.E. (1994): Rough concept analysis. Proc. *Rough Sets, Fuzzy Sets and Knowledge Discovery*, 248–255, Springer-Verlag.

Michalski, R.S. (1990): Learning flexible concepts: Fundamental ideas and a method based on two-tiered representation. In *Machine Learning: An Artificial Intelligence Approach, Vol. III*. Michalski R.S. and Kodratoff Y. (Eds.), Morgan Kaufmann.

Pawlak, Z. (1991): *Rough Sets: Theoretical Aspects of Reasoning About Data*, Kluwer Academic Publishers.

Wille, R. (1992): Concept lattice and conceptual knowledge systems. *Computers and Mathematics with Applications*, **23**, 493–515.

Implicative statistical analysis

R. Gras[1,2], H. Briand[3],
P. Peter[3] and J. Philippe[3,4],
[1]IRMAR - Campus de Beaulieu, F-35042 Rennes Cedex, France
[2]IRESTE - CP 3003, F-44087 Nantes Cedex 03, France
[3]IRIN - Equipe SIC - 2 rue de la Houssinière
F-44072 Nantes Cedex 03, France
[4]PERFORMANSE - 3 rue Racine, F-44000 Nantes, France

Summary: Implicative analysis, due to a problem of evaluation in education, allows us to treat a table crossing subjects or objects and variables according to a non-symmetrical point of view. In term of method of data analysis, it structures the set of variables, leads to tree and hierarchical structures and leads to the calculation of the objects contribution to the structure of the variables. Furthermore, it appears to be an effective tool in artificial intelligence to explain a base of rules in a set of knowledge. An example of the treatment of a big corpus of human behaviours is presented. The results, given by the method, have been validated a posteriori by the expert (psychologist).

1 The theory

Every researcher interested by the relations between variables (for example psychologist, specialist of methods, didactic specialist, ...) questions himself as follows : "Let a and b be two binary variables, can I affirm that the observation of a leads to the observation of b ?". In fact, this non-symmetrical point of view on the couple (a, b), à contrary to methods of similarity analysis, expresses itself by the question : "Is it right that if a then b ?". Generally, the strict answer is not possible and the researcher must content himself with a quasi-implication. We propose, with the statistical implication, a concept and a method which allow to measure the degree of validity of an implicative proposition between (binary or not) variables. Furthermore, this method of data analysis allows to represent the partial order (or pre-order) which structures a set of variables.

1.1 Theorical aspects of the binary case

This is the generic situation of the binary case (Gras 1979, Lerman et al. 1981). We cross a set E of objects and a set V of variables. We now want to give a statistical meaning to a quasi-implication $a \Rightarrow b$ (logical implications are exceptional). We note A (respectively B) the subset of E where the variable a (respectively b) takes the value 1 (or true). Measuring the quasi-inclusion of a into b is similar to measuring this reduced form of implication. Intuitively and qualitatively, we can say that $a \Rightarrow b$ is admissible if the number of counter-examples (objects of $A \cap \overline{B}$), verifying $a \wedge \overline{b}$ in E is improbably small in comparaison with the number of objects expected in an absence-of-a-link hypothesis between a and b (or A and B).

The quality of the implication is measured with the implication intensity. The approach developed for elaborating the implication intensity is inspired by I.C. Lerman's theorical considerations for designing his similarity indexes (Lerman 1981). We associate A (and B respectively) with a random subset X (and respectively Y) of E which have the same cardinal. We then compare the cardinal of $A \cap \overline{B}$ to that of $X \cap \overline{Y}$ in an absence-of-a-link hypothesis. If the cardinal of $A \cap \overline{B}$ is improbably small in comparaison with the cardinal of $X \cap \overline{Y}$, the quasi-implication $a \Rightarrow b$ will be accepted; otherwise it will be refused. It has been demonstrated (Lerman et al. 1981), that the random variable $card(X \cap \overline{Y})$ follows a hypergeometrical law and, under certain conditions, follows a Poisson's law of parameter $card(A).card(\overline{B})/card(E)$. The implication intensity is defined by the function : $\varphi(a, \overline{b}) = 1 - Pr(card(X \cap \overline{Y}) \leq card(A) \cap \overline{B})) = 1 - \sum_{i=0}^{card(A \cap \overline{B})} \frac{\lambda^i}{i!} e^{-\lambda}$ with $\lambda = card(A).card(\overline{B})/card(E)$. We can say that the quasi-implication $a \Rightarrow b$ is admissible at the level of confidence α if and only if $\varphi(a, \overline{b}) \geq 1 - \alpha$.

For example, we have 100 students ($card(E) = 100$) who can have 2 behaviours a and b with : $card(A) = 6$, $card(B) = 15$ and $card(A \cap \overline{B}) = 1$

We can observe that the number of students (here 1) refuting the implication $a \Rightarrow b$ is improbably small in an absence-of-a-link hypothesis. In fact, $\varphi(a, \overline{b}) = 0.965$ that is to say a level of confidence equal to 96.5 per cent for the implication because the probability that $card(X \cap \overline{Y}) < 1$ is equal to 0.035.

In (Larher 1991), the notion of statistical implication is extended to modal (or qualitative) variables and to numerical (or quantitative) variables.

1.2 Implication graph

A great interest of the statistical implication consists to study together all the variables on a given population. We can associate at each couple (a, b) of variables, a measure of their implacation intensity. This will be represented by a valued oriented edge. When the cardinals of A and B are equals, there are two oriented edges ($a \rightarrow b$ and $b \rightarrow a$). If we fix a condition of transitivity of the implication (generally 0.5), it is possible to generate a transitive graph. For example, if we have a set of five variables, whose the implication intensities greater than 0.5, are given in the following table :

$\Rightarrow$	a	b	c	d	e
a		0.97			0.73
b					
c	0.82	0.975			0.82
d	0.78				0.92
e					

we obtain the following graph :

414

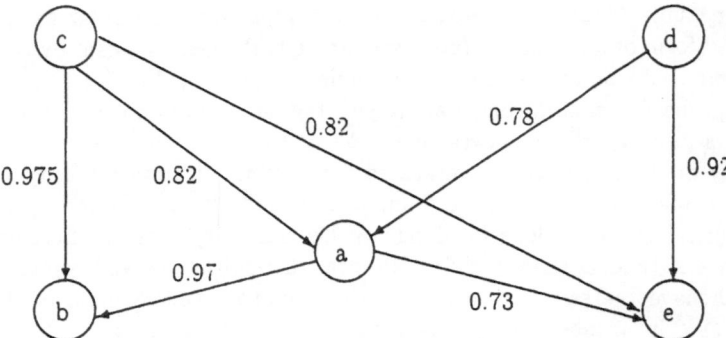

A. Larher (Larher 1991) has proved that the order between the intensities respects the order between the cardinals. So, for each pair of variables, we only keep the maximal intensity of the two couples defined by this pair. We can also proved (Gras and Larher 1992) the relation existing between the linear correlation coefficient and the statistical implication and the relation between the χ^2 of independance and the statistical implication.

1.3   Implication between classes of variables

When we consider the above graph, we are faced with some questions :
- can we aggregate c and e ?
- can we aggregate d and a ?
- can we have an implication from a class to another ?

This last notion have a (semantic) meaning only when the considered classes have a good cohesion. This cohesion must be measured. This measure is founded on the implication intensity. For example, if we consider the class (a, b, c), we observe : $\varphi(a, \bar{b}) = 0.97$, $\varphi(b, \bar{c}) = 0.95$ and $\varphi(a, \bar{c}) = 0.92$. We can say thatthe oriented class from a to c has a good cohesion. It would not be the case if the implication intensities were respectively equal to 0.82, 0.38 and 0.48. We then define the cohesion of a class like a notion which is opposed to the entropy (of the class). Then we can write (Gras and Larher 1992) :

Let be $p = max(\varphi(a, \bar{b}), \varphi(b, \bar{a}))$, the entropy (Shannon's definition) En of a class (a, b) equals : $En = -plog_2(p) - (1 - p)log_2(1 - p)$.
Let $coh(a, b)$ be the implicative cohesion of the class (a, b). It is defined as follows for a class with two elements :

$$\begin{cases} coh(a, b) = 1 \; if \; p = 1 \\ coh(a, b) = \sqrt{1 - En^2} \; if \; p \geq 0.5 \\ coh(a, b) = 0 \; if \; p < 0.5 \end{cases}$$

The cohesion $coh(a, b)$ is an increasing function of the implication intensity between a and b. The cohesion of a general class C is the geometrical average of the cohesion of the elements of C taken two by two.

$$coh(C) = \left(\prod_{\substack{i \in \{1,2,\ldots,r-1\} \\ j \in \{2,\ldots,r\}, j > i}} coh(a_i, a_j) \right)^{\frac{2}{r(r-1)}} \quad with \ C = (a_1, a_2, \ldots, a_r)$$

For the previous example, we find $coh(c, b) = 0.98$, $coh(d, e) = 0.91$ and $coh(a, (c, b)) = 0.89$.

The implication between classes must integrate their cohesion. For a class $A = \{a_1, \ldots, a_r\}$ and a class $B = \{b_1, \ldots, b_s\}$, we define the implication from class A to class B as follows :

$$\psi(A, B) = \{(sup(\varphi(a_i, \bar{b}_j)))^{r \cdot s} \cdot \sqrt{C(A) \cdot C(B)}, \ i \in \{1, \ldots, r\}, \ j \in \{1, \ldots, s\}\}$$

The implication between A and B increases according to their cohesion and decreases with their cardinal. As is the case with two variables, we retain the implication $A \Rightarrow B$ if $\psi(A, B) > \psi(B, A)$ or the implication $B \Rightarrow A$ if $\psi(B, A) > \psi(A, B)$.
We obtain, for the example $(a, (b, c)) \Rightarrow (d, e)$, an implication intensity equal to 0.27.

Using the classical algorithm of hierarchical clustering, we can build a hierarchy. However, here we refuse the aggregation of two classes if the cohesion of the resulting class is equal to zero. For the example, we obtain :

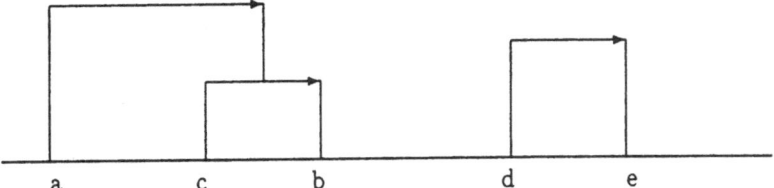

An appropriateness between the partition and the numerical index of the implicative clustering happens at some nodes of the hierarchical tree. These nodes are said significant and are studied in (Ratsimba-Rajohn 1992) and in (Gras and Ratsimba-Rajohn 1996). We find also in these papers a statistical tool which attributes at a given class of the hierarchy, the objects and sets of objects which contribute the more to this class. A software, CHIC (Ag Almouloud 1992), is available. Now we shall present an application of this method.

1.4 Use in learning algorithms

H. Briand's bibliographies and the linking of the initial type of problems as well as this concerning artificial intelligence allow a critical reflexion on the classical tools for the processing of these types of problems. The statistical implication, which has been newly defined, is resistant to noise, converges with the size of the sample, eliminates trivial rules and can be used with an incremental algorithm.
Furthermore, by expanding the application of the concepts of statistical implications, we have as is done in similarity analysis, defined the notion of a significant node and of a significant level.
Finally, we have defined the notion of "supplementary variables", as in factorial

analysis. These variables can help to explain the meaning of the classes which objects are responsible for their formation.

2 An expiremental knowledge discovery system Fiable1

2.1 A set of examples in human resources

We worked on populations in which each person has been represented by behaviour features. These behaviour features are used by our system (the knowledge discovery system Fiable1) to discover rules used by a Decision Aided System DIALECHO of the society PerformanSe at Nantes, France. In this context a person may be characterized by ten behaviour features : agressive (P), anxious (N), self-confidence (EST), need of affiliation (AFL), achievement (ACH), power (LED), intellectual dynamism (CLV), professional conscience (CON), receptivity (REC), extraversion (E). These ten features are defined by the expert. Each one may be valuate by one of these symbols : average (0), negative (-), positive (+). Example : N+ : the person is very anxious ; N- : the person is not anxious ; N0 : the person has an average degree of anxiety.

The objective of this experience is to provide to the expert the different relations between the different behaviour features in a population. Two populations have been examined : commercials of the firm (60 persons), workers of the same firm (40 persons).

With the agreement of the expert, the number of examples (Cab), the conditionnal probabilities, probability to have b true if a is true (Pab), and the forces of implications (Pfab) are the same in the discovery process on the two populations.

2.2 The relations between concepts : results and use by the expert

Relation : Commercial
Condition of discovery : Cab $\geq$ 5 and Pab $\geq$ 0.8 and Pfab $\geq$ 0.95
This condition, defined by the expert, represents knowledge discovery heuristics in the two populations.

Rules without counter-examples
IF E = E+ AND N = N- THEN EST = EST+ : 5 100 96 (Cab = 5, Pab=1, Pfab =0.96)
E = E+ : Extraversion is positive
N = N- : Anxiety is negative
EST = EST+ : Self-confidence is positive

There are 5 examples which verify this rule whose conditonnal probability is equal to 1, and whose intensity of implication is equal to 0.96. We notice that the condition of discovery is valid, this rule has no counter-example (it is a "logical" rule). This rule means that if the commercial has positive extraversion, and negative anxiety then his self-confidence is positive.

IF E = E0 AND P = P0 AND EST = EST0 THEN AFL = AFL0 : 6 100 96
IF E = E0 AND N = N0 AND EST = EST0 THEN ACH = ACH0 : 8 100 95

....

Rules with counter-examples

....

IF E = E0 AND EST = EST+ THEN REC = REC0 : 7 87 96

....

With the help of the expert psychologist, we will comment some rules discovered. Their analysis permits us to confirm and to ameliore the expertise. It permits us to valid the adequation of the theorical model of the construction to measured phenomena. We know that the psychometry assigns dimensions to qualitative objects that are not measurable.

An example which confirms psychometric theory
IF P = P+ AND CLV = CLV- THEN REC = REC- : 13 81 98

If a person (commercial) is aggressive and his intellectual dynamism is not very important then he does not listen other persons (no reception from another person). The coverage rate is equal to 0.22 (number of examples that satisfy the rule (13) / number of all examples (60)), and there are two counter-examples.

Some rules which proove that the same behaviour may be deduced by different premisses
IF EST = EST0 AND AFL = AFL- AND REC = REC- THEN P = P+ : 10 100 98
IF LED = LED+ AND REC = REC- THEN P = P+ : 12 100 99

If a person is animated by power motivation and he does not listen other persons then he is aggressive. Rate of coverage is equal to 0.2 and there is no counter-example.

2.3 Explanation of the concepts from the items

The learning set is composed of 864 examples and constitutes a medium sample from a students population given by the CIO (centre of informations and careers advising). The goal of this study is to explain the ten concepts of base. Each concept is characterized by three conceptual classes (cf 2.1.). The concepts are determined from a set of items. The results are used by the expert for validate the set of items in comparaison with concepts to determine. This completes the more classical process used in psychometry. The discovered rules are of the form :

IF item$_i$ = (1,2) AND $\cdots$ AND item$_j$ = (1,2) THEN concept = concept(-,0,+).

We present here the explanation from extraversion from the items. The expert can re-opens "item$_{41}$" which seems to determine E0 whatever the answer (rules 1 and 7).

1. IF item41 = 1 AND item47 = 1 AND item65 = 1 THEN E = E0 151 81 100

2. IF item15 = 2 AND item41 = 1 AND item65 = 1 THEN E = E+ 154 82 100

3. IF item4 = 2 AND item7 = 1 AND item43 = 1 THEN E = E0 151 81 99

4. IF item10 = 1 AND item43 = 1 THEN E = E- 90 82 100

5. IF item10 = 1 AND item30 = 1 AND item66 = 1 THEN E = E0 57 82 100

6. IF item10 = 1 AND item41 = 2 AND item43 = 2 THEN E = E- 31 83 100

7. IF item41 = 2 AND item47 = 1 AND item65 = 1 THEN E = E0 44 84 100

8. IF item4 = 2 AND item7 = 1 AND item62 = 1 THEN E = E+ 57 81 100

9. IF item7 = 1 AND item42 = 1 AND item53 = 1 THEN E = E0 43 81 96

10. IF item3 = 1 AND item47 = 1 THEN E = E- 17 85 99

11. IF item7 = 2 AND item42 = 1 THEN E = E0 14 82 95

The overall results are summarized in the next table where we can notice, among others, an excellent distribution for each concept of the number of explicative rules and of the number of covered examples in each class of definition of the concept.

	N1			N2			N3		
	-	0	+	-	0	+	-	0	+
E	142	462	220	3	6	2	138	460	211
P	193	386	245	3	6	4	186	374	242
N	186	438	200	5	6	3	185	428	190
ACH	178	478	168	4	6	2	173	474	157
CLV	205	242	377	7	12	10	194	239	374
CON	226	418	180	6	13	6	222	407	180
EST	163	495	166	2	4	2	154	494	162
LED	148	482	194	3	7	4	143	476	188
AFL	213	300	311	6	9	5	204	194	305
REC	220	384	220	6	14	7	205	382	218

where N1 is the number of examples which are members of the concept,
where N2 is the number of rules discovered by concept and
where N3 is the number of examples covered by the rules of each concept.

2.4   Conclusion

After the extraction process, the database contains many rules, many of them are not interesting, accurate and useful enough with respect to the end user's objectives. The quality of each generated rule has to be verified. So we need probabilistic tests to check if the rule actually describes some regularity in the data. So, the evaluation has to determine the usefulness of extracted rules, and decides which to save in the database. If a rule is no valid, by using the indexes, and more precisely the intensity of the implication, then it will be considered not interesting, and will not be saved in the database.

In the context of the collaboration between Knowledge and Information System team at IRIN and the firm PerformanSe SA, studies of other populations are underway in sportive and education domains.

3. References

Ag Almouloud, S. (1992) : L'ordinateur, outil d'aide à l'apprentissage de la démonstration et traitement de données didactiques. *Thèse de Doctorat de l'Université de Rennes I.*

Briand, H. et al. (1995): Mesure statistique de la robustesse d'une implication pour l'apprentissage symbolique. *Prépublication IRMAR 10-1995, Rennes*

Gras, R. (1979): Contribution à l'étude expérimentale et à l'analyse de certaines acquisitions cognitives et de certains objectifs didactiques. *Thèse d'Etat, Université de Rennes I.*

Gras, R. and Ratsimba-Rajohn H. (1996): Analyse non symétrique de données par l'implication statistique. *RAIRO - Recherche opérationnelle, 30-3 AFCET, Paris.*

Gras and al.(1996) : structuration sets with implication intensity, Proceeding of the International Conference on Ordinal and symbolic Data Analysis - *OSDA 95, E. Diday, Y. Chevallier, O. Opitz, Eds, Springer, Paris.*

Larher, A. (1991): Implication statistique et applications à l'analyse de démarches de preuve mathémathique. *Thèse de Doctorat de l'Université de Rennes I.*

Lerman, I.C. et al. (1981) : Evaluation et élaboration d'un indice d'implication pour des données binaires I et II. *Mathématiques et Sciences Humaines 74 p 5-35 et 75 p 5-47.*

Ratsimba-Rajohn, H. (1992) : Contribution à l'étude de la hiérarchie implicative. Application à l'analyse de la gestion didactique des phénomènes d'obstention et de contradictions. *Thèse de Doctorat de l'Université de Rennes I.*

Totohasina, A. (1992) : Méthode implicative en analyse des données et application à l'analyse de conception d'étudiants sur la notion de probabilité conditionnelle. *Thèse de Doctorat de l'Université de Rennes I.*

Part V

Correspondence Analysis, Quantification Methods, and Multidimensional Scaling

---

- Correspondence Analysis and Its Application

- Classification of Textual Data

- Multidimensional Scaling

Correspondence Analysis, Discrimination, and Neural Networks

Ludovic Lebart

Centre National de la Recherche Scientifique
Ecole Nationale Supérieure des Télécommunications
46 rue Barrault, 75013, Paris, France.

Summary: Correspondence Analysis of contingency tables (CA) is closely related to a particular Supervised Multilayer Perceptron (MLP) or can be described as an Unsupervised MLP as well. The unsupervised MLP model is also linked to various types of stochastic approximation algorithms that mimic the cognition process involved in reading and comprehending a data table.

1. CA: a tool at the junction of many different methods

Correspondence Analysis of contingency tables (CA), independently discovered by various authors, can be presented from nearly as many points of views. It can be viewed, for example, as a particular case of both Linear Discriminant Analysis (LDA) (performed on dummy variables) and Singular Value Decomposition (SVD) (performed after a proper scaling of the original data). After the seminal papers of Guttman (1941), Hayashi (1956) and Benzécri (1969a), various presentations of CA can be found in the available literature (see, for instance, Lebart et al. (1984), Greenacre (1984), Gifi (1990), Benzécri (1992), Gower and Hand (1996)).

In the context of neural networks - cf. the recent reviews of this fast-growing field by Cheng and Titterington (1994), Murtagh (1994), Ripley (1994) - Correspondence Analysis is also at the meeting point of many different techniques.

It can be described as a particular *Supervised Multilayer Perceptron* (MLP, section 2) (in that case, the input and the output layers are respectively the rows and the columns of the contingency table) or as an *Unsupervised Multilayer Perceptron* (UMLP, section 3) (in such a case the input layer, and the output layer as well, could be the rows, whereas the observations - also named examples, or elements of the training set - could be the columns of the table).

In both situations, the networks make use of the identity function as a transfer function. More general transfer functions might lead to interesting non-linear extensions of the method.

CA can also be obtained from *Linear Adaptive Networks* (section 4), a series of methods closely related to stochastic approximation algorithms.

2. A particular supervised Multilayer Perceptron

2.1 Reminder about the Multilayer Perceptron

Equivalence between Linear Discriminant Analysis and supervised Multilayer Perceptron (when transfer functions are identity functions) has been proved by Gallinari et al. (1988) and generalized to the case of more general models (such as non-linear discriminant analysis) by Asoh and Otsu (1989).

424

A general framework (see, e.g., Baldi and Hornik (1989)) can deal simultaneously with the supervised and the unsupervised cases.

Let $\mathbf{X}$ be the (n, q) matrix whose n rows contain the n observations of an *input* q-vector, and let $\mathbf{Y}$ be the (n, p) matrix containing (as rows) the n observations of an *output* p-vector.

$\mathbf{A}$ designates the (q, r) matrix of weights (a_{jm}) (see fig.1) before the hidden layer, and $\mathbf{B}$ the (r, p) matrix of weights (b_{mk}) following it (r ≤ p and r ≤ q).

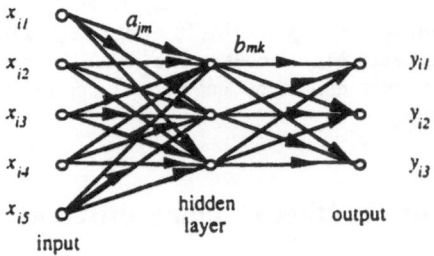

Fig. 1: Perceptron with one hidden layer (i-th observation)

A perceptron with a unique hidden layer is a model of the form:

$$y_{ik} = \Psi\left\{ \sum_{m=1}^{c} b_{mk} \; \Phi\left(\sum_{j=1}^{p} a_{jm} x_{ij} + c_m \right) + d_k \right\} + e_{ik} \qquad (1)$$

In the case of identity transfer functions (Φ and Ψ) and null constant terms, the model collapses to the simpler form:

$$y_{ik} = \left\{ \sum_{m=1}^{c} b_{mk} \left(\sum_{j=1}^{p} a_{jm} x_{ij} \right) \right\} = \sum_{j=1}^{p} \left(\sum_{m=1}^{c} b_{mk} a_{jm} \right) x_{ij} + e_{ik} \qquad (2)$$

2.2 Estimating the parameters

The *np* equations (2) are summarized by:

$$\mathbf{Y} = \mathbf{XAB} + \mathbf{E}. \qquad (3)$$

Denoting by $\mathbf{M}^T$ the transpose of matrix $\mathbf{M}$, the loss function to be minimized can be written:

$$f = trace \; \mathbf{E}^T\mathbf{E} = trace \; (\mathbf{Y} - \mathbf{XAB})^T (\mathbf{Y} - \mathbf{XAB}),$$

under the constraint:

$$\mathbf{BB}^T = \mathbf{I}_r \quad (\mathbf{I}_r \text{ is the identity (r, r) matrix)}.$$

This last constraint is introduced to remedy the indeterminacy of the model, since for any non-singular (r, r) matrix $\mathbf{H}$, $\mathbf{AH}$ and $\mathbf{H}^{-1}\mathbf{B}$ are solutions of the minimization problem as well as $\mathbf{A}$ and $\mathbf{B}$.

A and B could be estimated through a back-propagation algorithm, complemented with an orthonormalization of the rows of B at each step.

Since we are dealing here with the simpler case of identity transfer functions, we will focus on a direct analytical solution.

The minimization of f leads to equations (4) and (5):

$$BY^TX = A^TX^TX, \tag{4}$$

$$Y^TXA = B^TL \tag{5}$$

(L is an (r, r) matrix of Lagrange multipliers).

Equations (4) and (5), together with the previous constraint, lead to the following equation:

$$MB^T = B^TL,$$

the matrix M being defined as:

$$M = Y^TX(X^TX)^{-1}X^TY \tag{6}$$

We get a new expression for the criterion f:

$$f = trace\ Y^TY - trace\ L.$$

Minimizing f is then equivalent to maximizing $trace\ L$.

We can easily deduce from the preceding relationships and from this new criterion that L is a diagonal matrix containing the r largest eigenvalues of M as diagonal elements, the r rows of B being the corresponding unit eigenvectors.

We can then derive the value of A:

$$A = (X^TX)^{-1}X^TYB^T$$

This formula provides a generalization of that obtained in the simultaneous multiple regression, since (3) can be written:

$$Y = XW + E, \quad (with\ W = AB).$$

This generalization concerns the new situation where the matrix of coefficients W undergoes a constraint of rank.

Note that:

$$P = X(X^TX)^{-1}X^T$$

is the (idempotent) projector onto the subspace spanned by the columns of X.

Hence :

$$M = (PY)^T (PY).$$

Thus, the Multilayer Perceptron performs a *projected principal axes analysis* of Y, the projection being performed onto the space spanned by the columns of X. This analysis is also a *projected Principal Component Analysis*, if Y is centered columnwise.

2.3 The case of binary disjunctive data

When Y and X are binary disjunctive tables (dummy variables describing two partitions of the n observations into p and q classes), the matrix C defined as:

$$C = Y^TX$$

is the (p, q) contingency table crossing the two partitions.

The matrix D_q (resp. D_p) such that:

$$D_q = X^TX \quad (resp.\ D_p = Y^TY)$$

is the diagonal matrix whose q (resp. p) diagonal elements are the counts of the q classes (resp. p classes).

This particular Multilayer Perceptron, whose training entails the diagonalization of the matrix $\mathbf{M}$:

$$\mathbf{M} = \mathbf{CD_q^{-1}C},$$

performs a *Non Symmetrical Correspondence Analysis* (Lauro and D'Ambra (1984)) of the contingency table $\mathbf{C}$.

A classical Correspondence Analysis would imply a diagonalization of the matrix $\mathbf{M^*}$ such that :

$$\mathbf{M^*} = \mathbf{D_p^{-1}CD_q^{-1}C}$$

Note that $\mathbf{M^*}$ involves symmetrically the two sets (p columns of $\mathbf{X}$ on the one hand, q columns of $\mathbf{Y}$ on the other).

The Multilayer Perceptron will coincide with *Correspondence Analysis* if $\mathbf{D_p}$ is a scalar matrix (all the p classes have the same number of elements) or if the output matrix $\mathbf{Y}$ has been properly re-scaled during a preliminary step into $\hat{\mathbf{Y}}$ according to the following formula:

$$\hat{\mathbf{Y}} = \mathbf{YD_p^{-1/2}}$$

The new matrix to be diagonalized :

$$\mathbf{M_s} = \mathbf{D_p^{-1/2}CD_q^{-1}C\,D_p^{-1/2}}$$

has the same eigenvalues as $\mathbf{M^*}$, and has eigenvectors that can be easily derived from those of $\mathbf{M^*}$.

3. An unsupervised Multilayer Perceptron

In auto-associative neural networks, the output $\mathbf{Y}$ coincides with the input $\mathbf{X}$. The common value of $\mathbf{X}$ and $\mathbf{Y}$ is denoted by $\mathbf{Z}$.

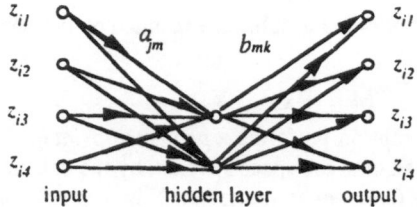

Fig. 2: Auto association strangulated network

It is an apparently trivial situation. In fact, these networks are of great interest if the hidden layer is narrower than the others, thus realizing a compression of the input signal (fig. 2).

Bourlard and Kamp (1988), Baldi and Hornik (1989) have stressed the link between SVD - and consequently Principal Component Analysis (PCA) - and these particular networks. The proof is straightforward if we replace both $\mathbf{Y}$ and $\mathbf{X}$ by $\mathbf{Z}$ in the formulas obtained in the previous section.

In this context, the matrix **M** given by the equation (6) is nothing but the product-moment matrix Z^TZ.

In this setting, the equivalence with Correspondence Analysis is obtained if **Z** is derived from a contingency table **K** according to the transformation (with usual notations):

$$z_{ij} = \frac{k_{ij} - k_{i.}k_{.j}}{\sqrt{k_{i.}k_{.j}}} \tag{7}$$

Note that the nature and the size of the input data involved in the two approaches of section 2 and 3 are radically different.

The network of section 2 is "fed" by n individual observations. It learns how to predict the output category corresponding to observation i, from the knowledge of its input category.

The network of section 3 is fed simultaneously by q observations of p categories (rows of **Z**) or equivalently by p observations of q categories (columns of **Z**). It learns how to summarize the input information.

Note that section 3 deals with properties common to Principal Component Analysis and Correspondence Analysis.

4. A Linear Adaptive Network

4.1 Brief review of some computational techniques involved in CA

Several distinct computational algorithms could be involved in Correspondence Analysis: Reciprocal averaging, iterated power, QR and QL algorithms, Jacobi method and its generalizations, Lanczos method, as well as other classical numerical procedure for SVD, (see, for example, Parlett (1980)).

The use of Back-Propagation method and other techniques usually associated with Multilayer Perceptron provides new numerical approaches and a better insight into the method. The unsupervised MLP model is also closely related to various types of stochastic approximation algorithms that could roughly outline the cognition process involved in perusing a data table. These algorithms are able to tackle huge data sets like those encountered in Automatic Information Retrieval.

Benzécri (1969b), Krasulina (1970) have proposed independently stochastic approximation algorithms for determining the largest eigenvalues of the expectation of a random matrix. Lebart (1974) has given a numerical proof of the convergence of Benzécri algorithm, and shown its interest in the case of sparse data matrices, such as those involved in Multiple Correspondence Analysis. Oja and Karhunen (1981) have proposed similar algorithms, adding new proofs and developments, reinforced by the results of Kushner and Clark (1978). The first mention of neural networks can be found in Oja (1982), who has proposed since then a wide variety of algorithms (see: Oja (1992)).

4.2 Basics of stochastic approximation algorithms

From our point of view, the basic idea is as follows:

X being the (n,p) matrix of properly re-scaled data, the product moment matrix X^TX can be written as a sum of n terms A_i.

$$X^TX = \sum_{i=1}^{i=n} A_i$$

with:

$$\mathbf{A}_i = \mathbf{x}_i \mathbf{x}_i^T , \quad (\mathbf{x}_i \text{ being the } i^{\text{th}} \text{ column of } \mathbf{X}^T)$$

The classical *iterated power algorithm* can then be performed using this decomposition, (cf. Wold (1966)) taking advantage of the possible sparsity of the data matrix $\mathbf{X}$.

Starting from a random vector $\mathbf{u}_0$, the step k of this algorithm, after setting $\mathbf{u}_k = \mathbf{0}$, consists of n assignments such as:

$$\textit{for } i = 1 \textit{ to } i = n, \quad do: \quad \mathbf{u}_k \leftarrow \mathbf{u}_k + \mathbf{A}_i \mathbf{u}_{k-1} \tag{8}$$

The vector $\mathbf{u}_k$ remains unchanged during the whole step k.

We can try to improve the algorithm by modifying the estimate of $\mathbf{u}_k$ during each assignment, according to the process:

$$\textit{for } j = 1 \textit{ to } j = \infty, \quad do: \quad \mathbf{u}_j \leftarrow \mathbf{u}_{j-1} + \gamma(j) \mathbf{A}_{i(j)} \mathbf{u}_{j-1} \tag{9}$$

where $\gamma(j)$ is a gain parameter.

During each step k, the index $i(j)$ of the matrix $\mathbf{A}$ takes values 1 to n.

At step $k : i(j) = j - (k-1)n$.

To ensure the convergence of $\mathbf{u}_j$ towards the largest eigenvector of $\mathbf{X}^T\mathbf{X}$, the series $\gamma(j)$ must diverge whereas the series $\gamma^2(j)$ must converge. The series $\gamma(j)$ could be chosen among series closely related to the *harmonic series* such as: $\gamma(j) = a/(b+j)$.

In fact, during step k, the iterated power algorithm (algorithm (8)) involves the operator:

$$\sum_i \mathbf{A}_i \tag{10}$$

whereas the stochastic approximation algorithm (algorithm (9)) replaces the operator (10) with the operator:

$$\prod_j \left(I + \gamma(j) \mathbf{A}_{i(j)} \right) \tag{11}$$

4.3 Stochastic approximation *versus* iterated power

Actually, if the terms of the series(j) are small enough, the two operators defined by (10) and (11) have similar unit eigenvectors. However, if the terms of the series $\gamma(j)$ are not too small, the operator (11) may have more separated eigenvalues, inducing a faster convergence of algorithm (9). Therefore, there is a trade-off between two options: fast convergence towards approximate eigenvectors, or slower convergence towards the exact values.

After several steps, because of the decrease in the values of $\gamma(j)$, operator (10) is definitely superior to operator (11).

In this sense, algorithm (9) can be considered as a mere technique of acceleration of the algorithm (8) (Lebart (1982)).

Unlike the algorithm (8), (9) depends on the order of the $\mathbf{A}_i$ within the sequence $(\mathbf{A}_1, \mathbf{A}_2, ..., \mathbf{A}_i, ..., \mathbf{A}_n)$. It can be shown that the speed of convergence can be improved if two consecutive sequences are read in reverse order (Lebart (1974)).

Both linear adaptive networks corresponding to algorithms (8) and (9) can produce simultaneously several eigenvectors, provided that orthonormalizations are carried out with a frequency that depends on the available precision. It is by no mean necessary to orthonormalize the estimates of eigenvectors at each reading (i.e. for each value of the index j when using the algorithm (9)).

It must be stressed that stochastic approximation algorithms such as algorithm (9) converge very slowly, their convergence being based on the divergence of the harmonic series. Iterated power algorithms (8) (whose firts steps could be speeded up by using stochastic approximation (9)) perform well if they confine themselve to finding a s-dimensional space V_s containing the t first eigenvectors (with: $t << s$). Then, the t dominant eigenvectors (and their corresponding eigenvalues) can be efficiently computed through a classical diagonalization algorithm applied to the (s, s) product-moment matrix obtained after projection onto the subspace V_s.

5. References

Asoh, H. and Otsu, N. (1989): Nonlinear Data Analysis and Multilayer Perceptrons. *IEEE, IJCNN*-89, **2**, 411-415.

Baldi, P. and Hornik, K. (1989): Neural networks and principal component analysis : learning from examples without local minima. *Neural Networks*, **2**, 52-58.

Benzécri.J.-P. (1969a): Statistical analysis as a tool to make patterns emerge from clouds. In : *Methodology of Pattern Recognition*, S.Watanabe, (ed.) Academic Press, 35-74.

Benzécri, J.-P. (1969b): Approximation stochastique dans une algèbre normée non commutative. *Bull. Soc. Math. France*, **97**, 225-241.

Benzécri J.-P. (1992): *Correspondence Analysis Handbook*. Marcel Dekker, New York.

Bourlard, H. and Kamp, Y. (1988): Auto-association by Multilayers perceptrons and singular value decomposition. *Biological Cybernetics*, **59**, 291-294.

Cheng, B. and Titterington, D.M. (1994): Neural networks: a review from a statistical perspective. *Statistical Science*, **9**, 2-54.

Gallinari, P., Thiria, S. and Fogelman-Soulie, F. (1988): Multilayers perceptrons and data analysis, *International Conference on neural Networks*, IEEE,, **1**, 391-399.

Gifi A. (1990): *Non Linear Multivariate Analysis*, J. Wiley, Chichester.

Greenacre M. (1984): *Theory and Applications of Correspondence Analysis*. Academic Press, London.

Guttman L. (1941): The quantification of a class of attributes: a theory and method of a scale construction. In : *The prediction of personal adjustment*, Horst P., (ed.) 251 -264, SSCR New York.

Hayashi C.(1956): Theory and examples of quantification. (II) *Proc. of the Institute of Statist. Math.* **4** (2), 19-30.

Hornik, K. (1994): Neural networks : more than "statistics for amateurs". In : *COMPSTAT*, Dutter R., Grossmann W. (eds.), Physica Verlag, Heidelberg, 223-235.

Krasulina, T. P. (1970): Method of stochastic approximation in the determination of the largest eigenvalue of the mathematical expectation of random matrices. *Automation and Remote Control*, Feb., 215-221.

Kushner, H. and Clark, D. (1978): *Stochastic approximation methods for constrained and unconstrained systems*, Springer, New York.

Lauro, N. C and D'Ambra, L. (1984): L'Analyse non-symétrique des Correspondances. In : *Data Analysis and Informatics*, III, Diday et al. (eds.), North-Holland, 433-446.

Lebart, L. (1974): On the Benzécri's method for finding eigenvectors by stochastic approximation. *COMPSTAT, Proceedings in Computational. Statist.*, Physica verlag, Vienna, 202-211.

Lebart, L. (1982): Exploratory analysis of large sparse matrices with application to textual data. *COMPSTAT, Proceedings in Computational. Statist.*, Physica Verlag, Vienna, 67- 76.

Lebart L., Morineau A., Warwick K. (1984): *Multivariate Descriptive Statistical Analysis*. J. Wiley, New York.

Murtagh, F. (1994): Neural network and related massively parallel methods of statistics: a short overview. *International Statistical Review*, **62**, 275-288.

Oja, E. (1982): A simplified neuron model as a principal components analyzer. *J. of Math. Biology*, **15**, 267-273.

Oja, E. (1992): Principal components, minor components, and linear neural networks. *Neural Networks*, **5**, 927-935.

Oja, E. and Karhunen, J. (1981): *On stochastic approximation of the eigenvectors and eigenvalues of the expectation of a random matrix*. Report of the Helsinki University of Technology (Dept of Technical Physics). Otaniemi, Finland.

Parlett B. N. (1980): *The Symmetric Eigenvalue Problem*. Prentice Hall, Englewood Cliffs, N.J.

Ripley, B. D. (1994): Neural nerworks and related methods of classification. *J. R. Statist. Soc. B*, **56**, 3, 409-456.

Wold, H. (1966): Estimation of principal components and related models by iterative least squares, in : *Multivariate Analysis*, Krishnaiah et al. (eds), Academic Press, New York, 391- 420.

Exploratory data analysis for
Hayashi's quantification method III by graphics

Tsutomu Komazawa and Takahiro Tsuchiya

The Institute of Statistical Mathematics
4-6-7, Minami-Azabu, Minato-ku
Tokyo 106, Japan

Summary: This paper gives an illustration of exploratory data analysis with Hayashi's Quantification Method III using graphics (hereafter we abbreviate this method as HQM III). It is shown that artificial data sets with ordinal structures can be expressed on the surface of a torus, and it is also pointed out that the torus suggests the results of HQM III. Some applications of HQM III to medical data using the graphical configuration are reported.

1. Introduction

Hayashi's quantification method III was presented by Hayashi in 1956. Correspondence analysis was later developed by the school of Benzécri in France in 1973. Artificial data sets with one-dimensional structure have been proposed by Guttman(1950), Iwatsubo(1987) and Okamoto, among others. Iwatsubo(1987) also presented data with circular structure.

The purpose of this report is to illustrate a graphical representation of qualitative data with ordinal structures and the usage of such graphics for exploratory analysis. We show that it is possible to express on the surface of torus a family of artificial data sets with ordinal structures, which includes the Guttman data and the Iwatsubo data. Hayashi's quantification method III is used to illustrate these data sets by a graphical configuration in three-dimensional spaces.

2. Solution of Hayashi's Quantification Method III (HQMIII) for Item-category data

We summarize the solution of HQM III for item-category data. The data matrix is

$$D = \left\{ \delta_{i(jk)} \right\},$$

where $\delta_{i(jk)}$, $n_{(jk)}$ and $n_{(jk)(uv)}$ are given by

$$\delta_{i(jk)} = \begin{cases} 1 & \text{if subject } i \text{ chooses category } k \text{ of item } j, \\ 0 & \text{otherwise}, \end{cases}$$

$$n_{(jk)} = \sum_{i=1}^{n} \delta_{i(jk)}, \quad n_{(jk)(uv)} = \sum_{i=1}^{n} \delta_{i(jk)} \delta_{i(uv)}.$$

for $i = 1, 2, \ldots, n$; $j, u = 1, 2, \ldots, m$; $k = 1, 2, \ldots, \ell_j$; $v = 1, 2, \ldots, \ell_u$, that n is the number of subjects, m is the number of items, ℓ_j or ℓ_u is the number of categories of item j or item u.

The solution of this problem HQM III is given by

$$AX = \lambda X,$$

where

$$A = \left\{ a_{(jk)(uv)} \right\}, \quad X = \left\{ x_{(jk)} \right\},$$

$$a_{(jk)(uv)} = \frac{1}{m \cdot n_{(jk)}} \left\{ n_{(jk)(uv)} - \frac{n_{(jk)} \cdot n_{(uv)}}{n} \right\}.$$

The eigenvector X of matrix A is a qualitative solution of item-categories, and the eigenvalue is a square of correlation coefficient r^2 between subject and item-category.

3. Torus Model

The following artificial data are the representative data with a one-dimensional structure.

(1) Guttman's perfect scale data

(2) Iwatsubo's (or Komazawa's) circular data

Generally, the data matrix D is given by a square matrix $(n \times n)$, where n is the number of subjects, m is the number of item, ℓ is the number of categories of an item. But,

$$n = \ell \cdot m$$

Then this data matrix D is,

$$D = \{ \delta_{ij} \}$$

where,

(1) $i \leq n - m + 1$

$$\delta_{ij} = \left\{ \begin{array}{ll} 1 & 0 \leq j - i < m \\ 0 & \text{otherwise} \end{array} \right.$$

(2) $i > n - m + 1$

$$\delta_{ij} = \left\{ \begin{array}{ll} 0 & 1 \leq i - j \leq n - m \\ 1 & \text{otherwise} \end{array} \right.$$

Table 1.1 shows a few examples generated from a matrix with n, m and ℓ being 20, 10 and 2 respectively.

1) The entire 20×20 matrix in Table 1.1 presents a circular model.

2) The 11×20 top half of the matrix of Table 1.1 is an example of Guttman's perfect scale model.

3) The 10×10 matrix, surrounded by a dotted line, is an example of Guttman's unidimensional scale model.

Table 1.1

item	1 2 3 4 5 6 7 8 9 10	1 2 3 4 5 6 7 8 9 10
category	1 1 1 1 1 1 1 1 1 1	2 2 2 2 2 2 2 2 2 2
1	■■■■■■■■■■	
2	■■■■■■■■■	□
3	■■■■■■■■	□□
4	■■■■■■■	□□□
5	■■■■■■	□□□□
6	■■■■■	□□□□□
7	■■■■	□□□□□□
8	■■■	□□□□□□□
9	■■	□□□□□□□□
10	■	□□□□□□□□□
11		□□□□□□□□□□
12	■	□□□□□□□□□
13	■■	□□□□□□□□
14	■■■	□□□□□□□
15	■■■■	□□□□□□
16	■■■■■	□□□□□
17	■■■■■■	□□□□
18	■■■■■■■	□□□
19	■■■■■■■■	□□
20	■■■■■■■■■	□

Table 1.2

item	1	2	3	4	5	6	7	8	9	10
category	1 2	1 2	1 2	1 2	1 2	1 2	1 2	1 2	1 2	1 2
1	■	■	■	■	■	■	■	■	■	■
2	□	■	■	■	■	■	■	■	■	■
3	□	□	■	■	■	■	■	■	■	■
4	□	□	□	■	■	■	■	■	■	■
5	□	□	□	□	■	■	■	■	■	■
6	□	□	□	□	□	■	■	■	■	■
7	□	□	□	□	□	□	■	■	■	■
8	□	□	□	□	□	□	□	■	■	■
9	□	□	□	□	□	□	□	□	■	■
10	□	□	□	□	□	□	□	□	□	■
11	□	□	□	□	□	□	□	□	□	□
12	■	□	□	□	□	□	□	□	□	□
13	■	■	□	□	□	□	□	□	□	□
14	■	■	■	□	□	□	□	□	□	□
15	■	■	■	■	□	□	□	□	□	□
16	■	■	■	■	■	□	□	□	□	□
17	■	■	■	■	■	■	□	□	□	□
18	■	■	■	■	■	■	■	□	□	□
19	■	■	■	■	■	■	■	■	□	□
20	■	■	■	■	■	■	■	■	■	□

Table 1.2 arranges Table 1.1 in corresponding similarity order.

Those data matrices can be indicated on the surface of torus as in Fig. 1.

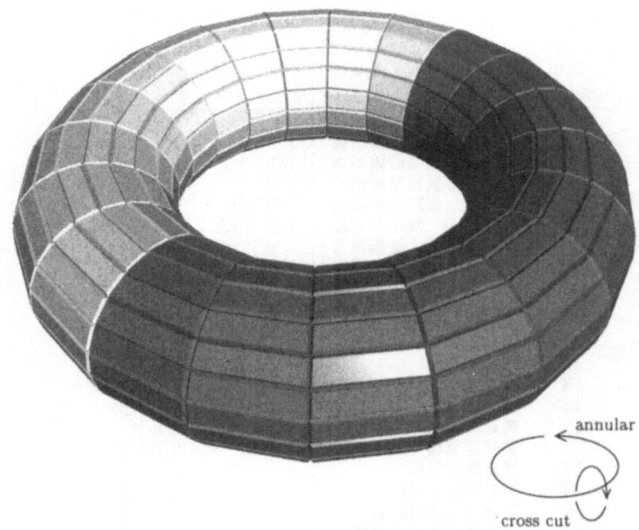

Fig. 1 Torus model

In this figure of torus, the column elements of item-categories of each model can be represented in the direction of cross cut, and the row elements of subjects in the annular direction. Using this graphical method, three models are plotted in three-dimensional space, together with the results obtained from Hayashi's quantication method Ⅲ, in Figures 2 ~ 4. Notice remarkable similarities between the two kinds of graphs in many respects.

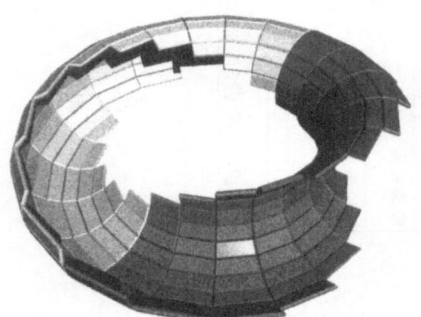

Fig. 2(a) Circular model

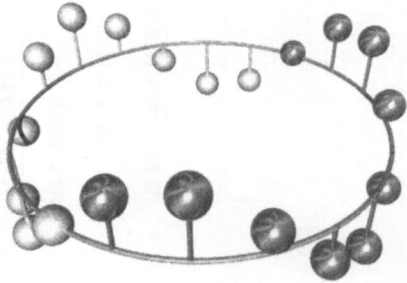

Fig. 2(b) Its configuration of HQM Ⅲ

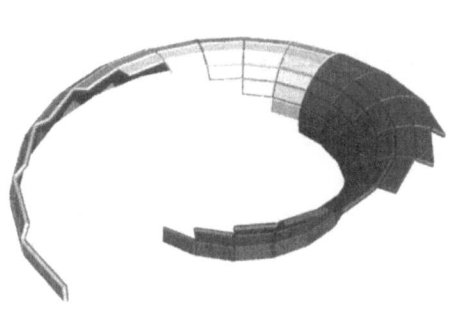

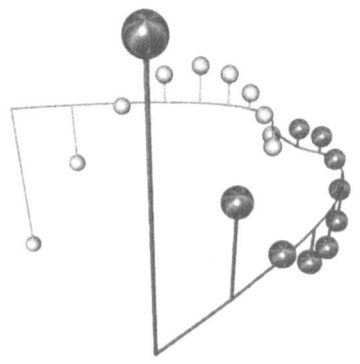

Fig. 3(a) Guttman's perfect
scale model

Fig. 3(b) Its configuration of HQM Ⅲ

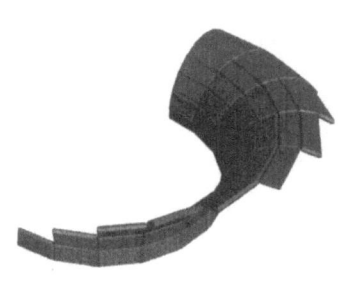

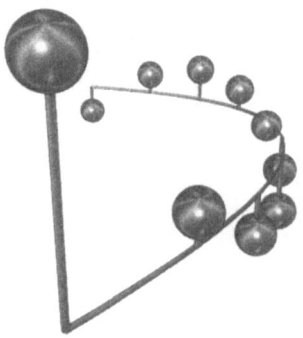

Fig. 4(a) Guttman's unidimensional
scale model

Fig. 4(b) Its configuration of HQM Ⅲ

4. Application of the method (HQMⅢ) to medical data

In this section, we show the results of HQM Ⅲ on an examination of heart functions. Let us consider a data set from *Treadmill excercise* study (Arai, 1987). This consists of data from 32 subjects ($n = 32$) on five variables ($m = 5$), and as shown in Table 2 the data set is augmented by item6, a classification variable which will be used later.

Table 2　　Data on Five Variables of the Heart Function from 32 Subjects

Items	1(HB)			2(SBP)				3(DBP)			4(MBP)				5(PP)			6(G)	
Categories	1	2	3	1	2	3	4	1	2	3	1	2	3	4	1	2	3	1	2
	−	+	++	−	±	+	++	−	+	++	−	±	+	++	−	+	++	Normal	Abnormal
Subjects																			
1		1			1			1			1				1			1	
2		1			1				1		1				1			1	
3		1		1				1			1				1			1	
4		1			1			1			1				1			1	
5	1				1				1		1				1			1	
6	1			1					1		1				1			1	
7		1			1			1			1				1			1	
8		1		1					1		1				1			1	
9	1			1					1		1				1			1	
10		1			1				1		1				1			1	
11		1			1				1		1				1			1	
12	1				1			1			1				1			1	
13	1				1				1		1				1			1	
14			1		1				1		1				1				1
15		1					1			1			1			1			1
16		1			1					1	1				1				1
17		1					1			1			1			1			1
18		1			1					1	1				1				1
19		1			1			1			1				1				1
20		1					1			1			1				1		1
21		1				1		1				1				1			1
22		1			1					1		1			1				1
23		1					1	1			1					1			1
24		1			1			1			1					1			1
25		1					1	1					1		1				1
26		1					1	1					1			1			1
27		1			1			1			1					1			1
28		1			1			1			1				1				1
29		1				1				1	1					1			1
30	1					1		1			1				1				1
31	1			1				1			1				1				1
32	1						1	1			1						1		1

[Notes]　1. HB　: Heart Beat　　　　　2. SBP　: Systolic Blood Pressure
　　　　　3. DBP : Diastolic Blood Pressure　　4. MBP : Mean Blood Pressure
　　　　　5. PP　: Pulse Pressure　　　　6. G　　: Group

Table 3 Data Rearranged in Order by HQM III

Item No.		1	2	4	3	5	2	4	3	1	2	1	4	5	3	2	5	4	Group	
Category No.		1	1	1	1	1	2	2	2	2	3	3	3	2	3	4	3	4	Normal	Abnormal
Subjects No.	Quantities	0.366	0.354	0.348	0.325	0.164	0.128	0.118	0.112	0.042	−0.178	−0.205	−0.243	−0.291	−0.293	−0.567	−0.637	−0.641		
6	1.346	1	1	1		1			1										1	
12	1.334	1		1	1	1	1												1	
13	1.120	1	1			1		1	1										1	
9	1.116		1		1	1		1		1									1	
8	1.022		1	1		1			1	1									1	
4	1.010			1	1	1	1			1									1	
3	1.006		1		1	1	1	1											1	
5	0.890	1				1	1	1	1										1	
11	0.796		1		1	1		1		1									1	
7	0.780				1	1	1	1		1									1	
1	0.566					1	1	1	1	1									1	
10	0.566					1	1	1	1	1									1	
19	0.566					1	1	1	1	1										1
14	0.318					1	1	1	1			1								1
28	0.318					1	1	1	1			1								1
2	0.163					1	1	1		1					1				1	
16	−0.085					1	1	1				1			1					1
18	−0.085					1	1	1				1			1					1
25	−0.103					1			1	1	1		1							1
24	−0.141						1	1	1			1		1						1
27	−0.141						1	1	1			1		1						1
23	−0.197							1	1	1	1			1						1
31	−0.202					1	1			1			1		1					1
22	−0.447					1	1					1	1		1					1
30	−0.506					1				1	1		1		1					1
21	−0.810								1		1	1	1	1						1
26	−1.198								1			1	1	1		1				1
29	−1.213										1	1	1	1	1					1
32	−1.697									1			1		1	1	1			1
15	−1.751									1				1	1	1		1		1
17	−1.999											1		1	1	1		1		1
20	−2.342											1			1	1	1	1		1

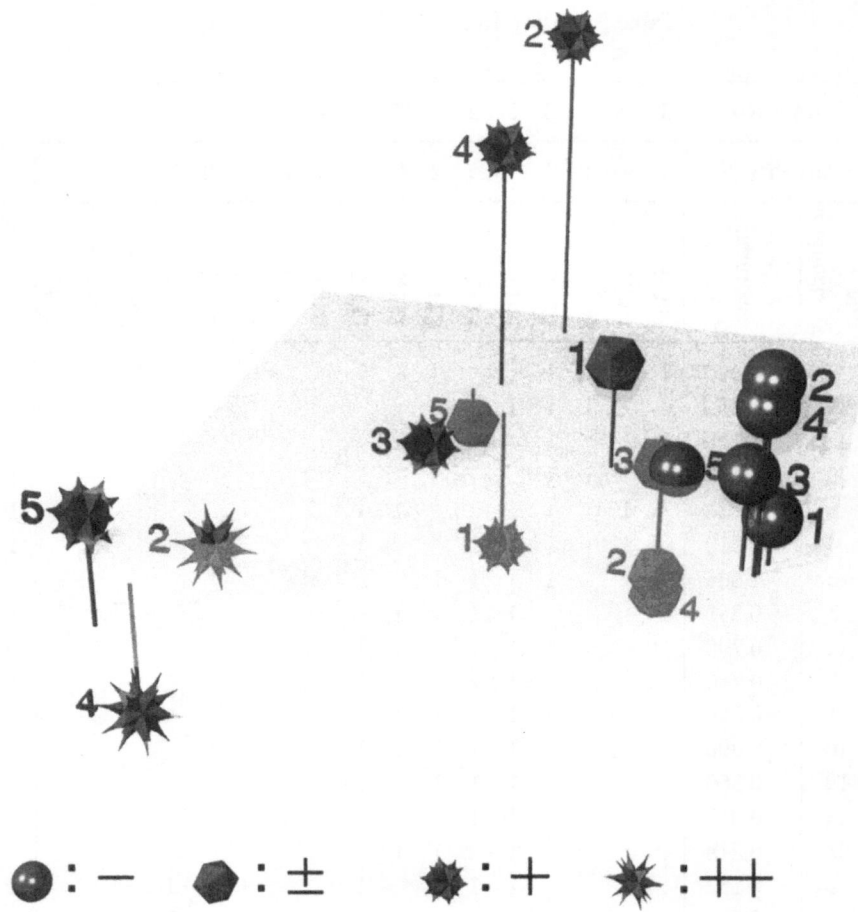

Fig. 5 Evaluation of Heart Function
(Examination Variables)

Figure 5 is a plot of the five variables with coordinates on the three-dimensional spaces, and Figure 6 shows a plot of subjects on the three-dimensional spaces. In these figures, we use the symbol 'ball and polygon star' to represent a category of an item and subjects for convenience. Smooth round balls indicate normal conditions or normal subjects, while their successive changes toward stars indicates increased abnormality as indicated on the scale show in Figure 5. Figure 5 and 6 reveal more precise in formation about the ordinal data structure than the torus representation.

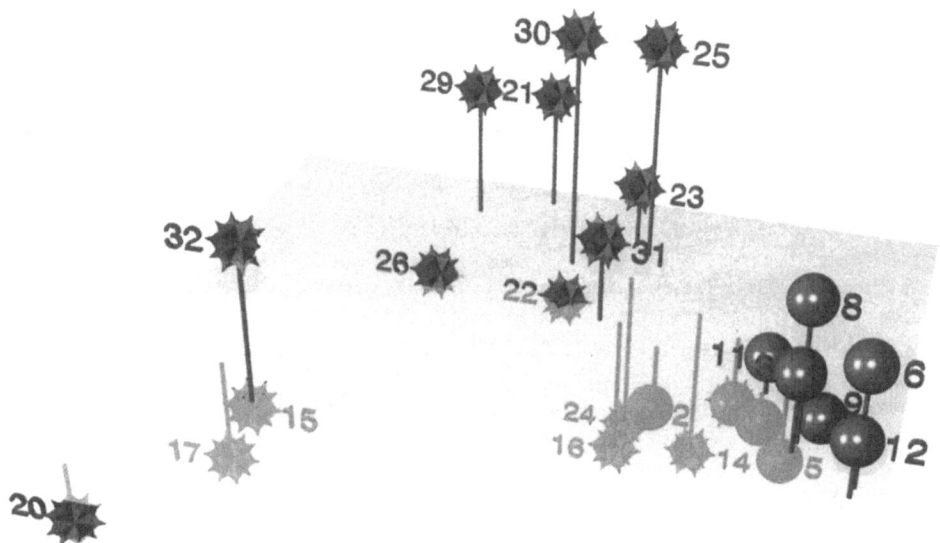

⬤ : Normal   ✴ : Abnormal

Fig. 6 Evaluation of Heart Function
(Subjects)

5. Conclusion

In this paper, we presented the use of the torus model and its comparison with HQM III. Artificial data by Guttman, Iwatsubo and Komazawa were used to investigate their properties by graphic data processing, specifically, the torus representation and graphical display based on HQM III. It was shown that graphical displays were helpful to clarify otherwise hidden properties of those data.

Acknowledgments

We would like to thank the editor and referees for a careful reading of the manuscript and for comment that improved it. Anonymous referee "A" was particularly helpful in detecting typographical errors and bring our attention to relevant references.

440

References:

Arai, C., Komazawa, T., et al. (1984): The effect of hypoxanthine riboside on integral value % for various hemodynamic parameters under upright treadmill exercise. *The Journal of Japanese College of Angiology*, **24**, 1, 75–81 (in Japanese) & 90 (abstract in English).

Guttman, L. (1950): The principal components of scale analysis. *Measurement and Prediction*, Stouffer, S. A. et al. (eds.), 312–361, Wiley, New York.

Hayashi, C. (1956): Theory and examples of quantification, Ⅱ. *Proc. of ISM*, **4**, 1, 19–30 (in Japanese).

Iwatsubo, S. (1987): *The Foundation of Quantification Method*. Asakura-shoten, Tokyo (in Japanese).

Komazawa, T. and Tsuchiya, T. (1995): Exploratory data analysis for Hayashi's quantification method Ⅲ by graphics: Its application of ordinal structure analysis to ninth nationwide survey of the Japanese national charactor. *Proc. of ISM*, **43**, 1, 161–176 (in Japanese).

Exploring Multidimensional Quantification Space

Shizuhiko Nishisato

The University of Toronto
OISE/UT, 252 Bloor Street West
Toronto, Ontario, Canada M5S 1V6

Summary: Dual scaling deals with two distinct types of categorical data, incidence data and dominance data. While perfect row-column association of an incidence data matrix does not mean that the data matrix can be fully explained by one solution, dominance data with perfect association can be explained by one solution. Considering a main role of quantification theory is to explain data in multidimensional space, the present study presents a non-technical look at some fundamental aspects of quantification space that are used for analysis of the two types of data.

1. Introduction

Consider principal component analysis (PCA) of n standardized variables. If the rank of the correlation matrix is 2, that is, $r(R) = 2$, then all the variables are positioned in two-dimensional space, and more specifically, on a circle of the radius of 1. In other words, each variable is located on a plane at the distance of 1 from the origin. Likewise, if $r(R) = 3$, all the variables are located on the surface of a sphere at the distance of 1 from the origin. One can extend the same statement to the general multidimensional case. Thus, no matter what rank the correlation matrix may have, one can infer that K dimensions are sufficient if the sum of squares of K coordinates of each variable is close to 1. Thus, with standardized continuous variables. both graphical display and the statistic "percentage accounted for" can tell us the dimensionality of data.

Suppose that we now consider PCA of deviation scores, that is, variables which are not standardized, but centered. This leads to principal component analysis of the variance-covariance matrix, V. Since the variance of each variable is different, even when $r(V) = 2$, variables are not positioned on a circle of any fixed radius. Although the shape of the distribution of variables in two-dimensional principal plane cannot tell us that $r(V) = 2$, the statistic of the percentage accounted for by each component tells us dimensionality and importance of each component. Thus, with non-standardized continuous variables, graphical display does not tell us the dimensionality of data, but the statistic "percentage accounted for" still does.

From the practioner's point of view, the real problem in choosing between R and V is not that of computational convenience, but rather that of which one is more meaningful out of the two. In the social sciences, many variates such as personality scores do not have any rational unit of measurement, and one tends to opt for standardizing variables for the purpose of comparability. Even in the natural sciences where there exists a rational unit of measurement, however, one may still face the choice between the use of available units (e.g., kilometers or miles) and standardization. There is no readily available guideline on which one to use. To make the task of choosing either

R or V more important and difficult than one may think, the results of PCA from the two can be vastly different, and there does not seem to be any mathematically traceable relationship between the two sets of PCA results (e.g., Nishisato and Yamauchi, 1974). This makes it almost impossible to consider a rational way of choosing one over the other.

When one looks at quantification methods such as Hayashi's quantification theory, correspondence analysis, homogeneity analysis and dual scaling, they are conceptually the same as PCA. In fact, Torgerson (1958) called these quantification methods PCA of categorical data. There are, however, a number of differences between them and PCA. On one hand, "PCA of categorical data" is regarded as singular value decomposition of standardized categorical variables, suggesting stronger resemblance to PCA of R than to PCA of V. On the other hand, there are many more numerical aspects that connect it to PCA of V rather than to PCA of R. If we consider such multiple-choice data in the response-pattern format (i.e., in the form of an indicator matrix) that can be explained by two components, we would realize almost immediately that the quantification results are more like PCA results associated with V than with R because a two-dimensional graph does not tell us that the data are two dimensional. In fact, it is the present author's view that the quantification method as we know by many different names is PCA of non-standardized categorical variables.

As French researchers in the area of correspondence analysis would say, graphical display of quantification results is almost indispensable for their interpretation. But, surprisingly, there seems to be little knowledge about the space used in quantification. For instance, under what circumstances can we infer from graphical display that the categorical data in hand are two-dimensional? To answer this question, it seems essential to know what multidimensional quantification space we are dealing with. For some reason or others, this fundamental question on the space has not been a topic of intensive investigation. The present study will be concerned with this problem, and some preliminary consideration will be presented.

2. Dual Scaling

As a method of quantification, dual scaling (Nishisato, 1980, 1994, 1996) will be considered to see what kind of multidimensional space it uses. Dual scaling is known for two aspects: (1) it handles a wide variety of categorical data, in particular *incidence data* (contingency tables, multiple-choice data, sorting data) and *dominance data* (rank-order data, paired comparison data, successive categories data), (2) it employs two distinct objectives, one for incidence data and the other for dominance data. The former is the familiar low-rank approximation to the input data, and the latter the low-rank approximation to the ranking of input data by the ranking of distances between each subject and a set of stimuli. The latter is, therefore, considered to offer a low-rank solution to the problem of multidimensional unfolding (Nishisato, 1994, 1996). Mathematical handling of the two major types of categorical data by dual scaling is based on singular value decomposition of appropriately transformed data matrices.

For *incidence data*, denote the data matrix by F, where the typical element f_{ij} is either 1 (presence) or 0 (absence), or the joint frequency of cells i and j, the diagonal matrices of row marginals and column marginals by D_r and D_c, respectively. Then singular value decomposition is carried out as follows,

$$D_r^{-\frac{1}{2}} F D_c^{-\frac{1}{2}} = V \Lambda W',$$

where $V'V = I$, $W'W = I$, and $\Lambda = diag(\lambda_j)$. Optimal weight matrices for rows, Y, and columns, X, are given by

$$Y = D_r^{\frac{1}{2}} V, X = D_c^{\frac{1}{2}} W \quad \text{so that} \quad F = f_t Y \Lambda X'$$

Optimal vectors are scaled as $x_k' D_c x_k = y_k' D_r y_k = f_t$, that is, the sum of all the elements in F.

For **dominance data**, responses collected are first transformed into the subject-by-stimulus matrix of dominance numbers E, where a typical element e_{ij} indicates the number of times Subject i judges Stimulus j higher (larger, more attractive) than other stimuli minus the number of times Subject i judges other stimuli higher than Stimulus j. If we indicate the number of subjects by N and the number of stimuli by n, the dominance matrix is Nxn, and the sum of elements of each row is zero. To define diagonal matrices D_r and D_c for the dominance matrix, each element is considered based on $n - 1$ comparisons. Thus, we can now specify the two diagonal matrices as follows: $D_r = n(n - 1)I$, and $D_c = N(n - 1)I$. Hence, $f_t = nN(n - 1)$. Then, with these newly defined diagonal matrices and dominance matrix E, we can carry out singular value decomposition of E to obtain optimal vectors y_k and x_k (Nishisato, 1978).

3. Multidimensional Quantification Space

It is interesting to know that probably due to the influence of literature in factor analysis and mental test theory most studies on quantification theory seem to have assumed more subjects than the number of stimuli. But, dual scaling is symmetric in its decomposition of data, and as such it is important to consider both cases: when the number of rows of a data matrix determines the dimensionality of the space, and when the number of columns determines it.

Generally speaking, one can make the following statement. When the rank of a data matrix is determined by its rows (e.g., five members of a family rank ten movies according to their order of preference; 100 students answer a multiple-choice personality questionnaire with 200 questions), each row variable (i.e., each subject) has sufficient space to reveal his or her unique pattern of responses or idiosyncratic responses; but each column variable (i.e., each stimulus) does not have enough room to be fully accounted for. Similarly, when the rank of a data matrix is determined by its columns, each column variable (stimulus) has enough space to be fully explained, and it is not the case with each row variable (subject). Thus, in the former case where the rank is determined by the rows, the contribution of each of those row variables to the quantification space can be expressed by a mathematical formula; in the latter case, the same applies to each of the column variables.

3.1. Multidimensional Analysis of Dominance Data

Since quantification of incidence data has been widely discussed in the literature of correspondence analysis, multiple correspondence analysis, homogeneity analysis and dual scaling, a number of aspects relevant to the theme of the current paper are relatively well known. Admittedly, however, they apply to the case in which there are more subjects (in the case of multiple correspondence analysis) than the rank of the data matrix. Therefore, the present paper will start with the quantification space pertaining to dual scaling of dominance data.

Table 1 shows the ranking of ten government services by 31 subjects, and Table 2 is the corresponding 31 × 10 dominance matrix. This dominance matrix can be fully explained by nine dual scaling solutions, and it is known that dual scaling offers a perfect solution to the problem of multidimensional unfolding (Nishisato. 1994, 1996). Let us first explain what the unfolding problem is. Coombs (1950) proposed a model for ranking judgement, in which a joint continuum for stimuli and subjects is postulated in such a way that a subject ranks first the stimulus most closely located to him or her on this continuum, second the second closest, and so on. The subject's

Table 1
Ranking of Ten Government Services
by 31 Subjects

1	7	9	10	2	6	3	8	5	4
6	10	9	5	3	1	7	2	4	8
9	8	4	3	5	6	10	2	1	7
2	10	5	6	3	1	4	8	7	9
2	10	6	7	4	1	5	3	9	8
1	3	5	6	7	8	2	4	10	9
7	10	1	6	5	3	8	4	2	9
2	10	6	7	4	1	5	3	9	8
2	10	5	8	4	1	6	3	7	9
2	10	5	9	8	7	4	1	3	6
9	10	7	6	5	1	4	2	3	8
6	10	7	4	2	1	3	9	8	5
1	10	3	9	6	4	5	2	7	8
8	6	5	3	10	7	9	2	1	4
8	10	9	6	4	1	3	2	5	7
3	5	10	4	6	9	8	2	1	7
1	10	8	9	3	5	2	6	7	4
5	4	9	3	10	8	7	2	1	6
2	10	6	7	8	1	5	4	3	9
1	4	2	10	9	7	6	3	5	8
2	10	5	7	3	1	4	6	8	9
6	3	9	4	10	8	7	2	1	5
6	9	10	4	8	7	5	2	1	3
5	2	1	9	10	4	8	6	3	7
2	10	6	7	9	1	3	4	5	8
7	10	9	5	2	6	3	1	4	8
8	7	10	3	5	9	4	2	1	6
3	8	6	7	5	10	9	2	4	1
2	10	7	9	4	1	5	3	6	8
2	10	9	1	4	7	5	3	6	8
4	10	9	7	5	1	3	2	6	8

Columns: (1) Public transit system, (2) Postal service, (3) Medical care (4) Sports/recreational facilities, (5) Police protection (6) Public Libraries, (7) Cleaning streets, (8) Restaurants (9) Theatres, (10) Overall planning and development position on the continuum is called his or her ideal point, and the ranking of each subject is interpreted as the ranking of stimuli on the continuum folded at each subject's ideal point. Thus, the problem of unfolding is that given a set of rank orders (i.e., folded continua) from subjects we wish to unfold the rank orders to recover the original single continuum, on which subjects and stimuli are jointly located. When a single continuum is replaced with multidimensional axes, unfolding becomes complicated, and the problem then is called that of multidimensional unfolding. A number of studies on the topic have been published (e.g., Coombs and Kao, 1960; Coombs, 1964; Schönemann, 1970; Schönemann and Wang, 1972; Carroll, 1972; Gold, 1973; Heiser, 1981; Greenacre and Browne, 1986). However, they have not made any reference to dual scaling, obviously being unaware of its relevance. Now we know that dual scaling always offers a perfect solution to the problem of multidimensional unfolding (Nishisato, 1994).

Table 2
Dominance Matrix E

9.	-3.	-7.	-9.	7.	-1.	5.	-5.	1.	3.
-1.	-9.	-7.	1.	5.	9.	-3.	7.	3.	-5.
-7.	-5.	3.	5.	1.	-1.	-9.	7.	9.	-3.
7.	-9.	1.	-1.	5.	9.	3.	-5.	-3.	-7.
7.	-9.	-1.	-3.	3.	9.	1.	5.	-7.	-5.
9.	5.	1.	-1.	-3.	-5.	7.	3.	-9.	-7.
-3.	-9.	9.	-1.	1.	5.	-5.	3.	7.	-7.
7.	-9.	-1.	-3.	3.	9.	1.	5.	-7.	-5.
7.	-9.	1.	-5.	3.	9.	-1.	5.	-3.	-7.
7.	-9.	1.	-7.	-5.	-3.	3.	9.	5.	-1.
-7.	-9.	-3.	-1.	1.	9.	3.	7.	5.	-5.
-1.	-9.	-3.	3.	7.	9.	5.	-7.	-5.	1.
9.	-9.	5.	-7.	-1.	3.	1.	7.	-3.	-5.
-5.	-1.	1.	5.	-9.	-3.	-7.	7.	9.	3.
-5.	-9.	-7.	-1.	3.	9.	5.	7.	1.	-3.
5.	1.	-9.	3.	-1.	-7.	-5.	7.	9.	-3.
9.	-9.	-5.	-7.	5.	1.	7.	-1.	-3.	3.
1.	3.	-7.	5.	-9.	-5.	-3.	7.	9.	-1.
7.	-9.	-1.	-3.	-5.	9.	1.	3.	5.	-7.
9.	3.	7.	-9.	-7.	-3.	-1.	5.	1.	-5.
7.	-9.	1.	-3.	5.	9.	3.	-1.	-5.	-7.
-1.	5.	-7.	3.	-9.	-5.	-3.	7.	9.	1.
-1.	-7.	-9.	3.	-5.	-3.	1.	7.	9.	5.
1.	7.	9.	-7.	-9.	3.	-5.	-1.	5.	-3.
7.	-9.	-1.	-3.	-7.	9.	5.	3.	1.	-5.
-3.	-9.	-7.	1.	7.	-1.	5.	9.	3.	-5.
-5.	-3.	-9.	5.	1.	-7.	3.	7.	9.	-1.
5.	-5.	-1.	-3.	1.	-9.	-7.	7.	3.	9.
7.	-9.	-3.	-7.	3.	9.	1.	5.	-1.	-5.
7.	-9.	-7.	9.	3.	-3.	1.	5.	-1.	-5.
3.	-9.	-7.	-3.	1.	9.	5.	7.	-1.	-5.

Suppose we consider the ranking of ten services by the first two subjects. As one can see easily, there is no constraints on each of the ten columns, while each row is constrained by the condition that each row sum of dominance numbers is zero. Thus the analysis of this 2×10 matrix yields at most two solutions (see Table 3). Similarly, if we analyze the ranking by the first three subjects, the 3×10 dominance matrix yields three dual scaling solutions (see Table 4). Since variates y_{ik} and x_{jk} do not span the same space, it is important that one of them is projected onto the space of

446

the other. For dual scaling to provide a perfect solution, we must project stimuli onto the space for subjects (Nishisato, 1994). In other words, we must plot y_{ik} and $\rho_k x_{jk}$. Figure 1 shows the results of dual scaling analysis of the 2×10 matrix: compute the distance between Subject 1 and each of the ten stimulus points, and rank order the distances from the closest (smallest distance) to the furthest, which reproduces the exact ranking of the ten services by Subject 1! Similarly, one can calculate the distances between Subject 2 and the stimuli, rank the distances, and see that the observed ranking is now again reproduced. For dual scaling of the 3×10 matrix, we need all the three solutions to recover the same ranking of the services by each of the three subjects from the plot of $(y_{ik}, \rho_k x_{jk})$.

Figure 1
A Perfect Two-dimensional Solution

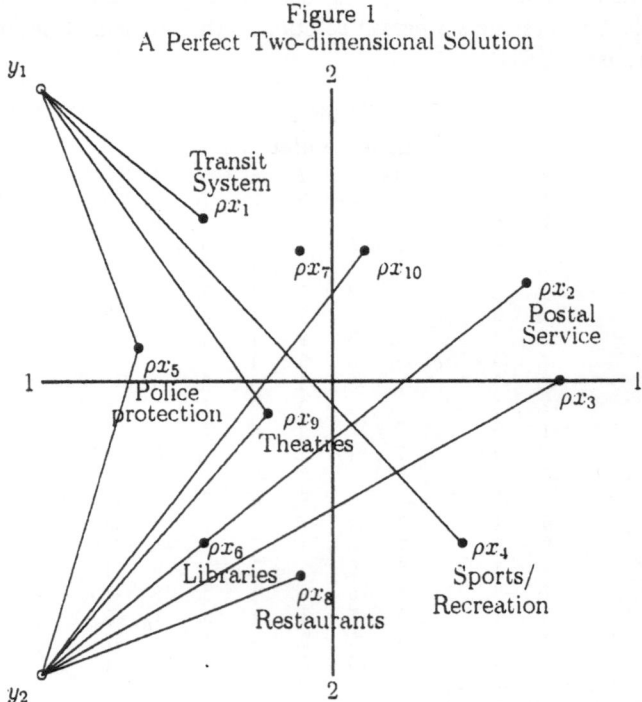

Suppose we increase the number of subjects to nine, and analyze the 9×10 dominance table. As expected, if we use all nine solutions, the ranking of distances between subjects and stimuli in 9-dimensional space will reproduce perfectly the ranking in the original data (Table 1). How about the case in which there are more subjects than nine? Would we fail to recover the ranking of the services by some subjects? A truly remarkable property of dual scaling is that no matter how many more subjects than the number of stimuli we may have the rankings of stimuli by all the subjects are reproduced in $(n-1)$-dimensional space, or N-dimensional space! This is a classical property of singular value decomposition. Table 5 shows the sums of squares of rank discrepancies between the input rankings and rankings approximated by one, two, three,...., and nine solutions. As you see in the table, the rank-9 approximation yields 0 discrepancy, indicating that it is a perfect solution to the problem of multidimensional unfolding.

Table 3

Normed Weights for 2 Subjects and
Projected Weights for 10 Stimuli (2 Solutions)

1	2	1	2
-0.9996	1.0005	-0.4442	0.5557
-1.0000	-0.9995	0.6668	0.3331
		0.7778	-0.0003
		0.4442	-0.5557
		-0.6666	0.1114
		-0.4447	-0.5554
		-0.1109	0.4445
		-0.1114	-0.6666
		-0.2223	-0.1110
		0.1113	0.4444

Table 4

Normed Weights for 3 Subjects and
Projected Weights for 10 Stimuli (3 Solutions)

1	2	3	1	2	3
0.9485	-1.0848	-0.9610	0.6677	-0.3027	-0.0398
-0.7304	-1.3498	0.8027	0.3699	0.5768	0.0608
-1.2517	-0.0342	-1.1967	-0.1956	0.6274	-0.0919
			-0.5750	0.3053	0.1285
			0.0643	-0.5325	-0.1448
			-0.2323	-0.4085	0.3475
			0.6740	-0.0395	0.1317
			-0.6895	-0.1580	0.0758
			-0.4633	-0.2016	-0.3453
			0.3797	0.1332	-0.1225

Thus, in dual scaling of dominance data, all we are interested in is the recovery of, or approximation to, the original rankings of stimuli by individual subjects. As such, the importance of shape of quantification space becomes secondary. The percentage of original rankings approximated by K solutions is sufficient for the investigator to know. Even so, however, it should be mentioned that multidimensional quantification space for dominance data is in some sense standardized since the total contribution of each subject to the total space is fixed and equal to $(n + 1)/[3(n - 1)]$, irrespective of N. One further characteristic of dominance data is that both D_r and D_c are scalar matrices.

3.2. Multidimensional Analysis of Incidence Data

All the complexities of multidimensional quantification space for incidence data seem to arise from the fact that D_r and D_c are not scalar matrices. Rather than reviewing relevant mathematical expressions for row and column contributions of a variety of incidence data to the total space, let us look at one numerical example. Consider multiple-choice data collected from seven subjects answering three multiple-choice questions with three response options per question. In this case, the number of solutions is determined by both rows and columns of the incidence matrix because $N - 1 = m - n = 6$. This is a special case, in which both row variables and column variables have space sufficient to accommodate their full contributions to the total space. Let us look at a numerical example (Table 6) and note the following regularities: (a) The sum of squared singular values (correlation ratios in dual scaling) is equal to the average number of options minus 1, that is, 2; (b) The sum of squares of quantified item scores is equal to $nN(m_j - 1)$, that is, $3 \times 7 \times (3 - 1) = 42$; (c) The sum of r_{jt}^2 over the total solutions (6) is equal to $m_j - 1$, which is, $3 - 1 = 2$; (d) The sum of normed scores of each subjects over the total solutions is equal to the total number of solutions; (e) The sum of squares of option weights over the total solutions is equal to $n(N - f_{jp})$, where f_{jp} is the number of subjects who chose option p of Item j, and; (f) The inter-subject squared distance in the total normed space is $2N$, which is 14 in the present example.

Table 5
Sum of Squares of Discrepancies between Observed and Reproduced Ranks

Subject	Number of Solutions								
	1	2	3	4	5	6	7	8	9
1	88	78	90	46	42	14	16	0	0
2	62	28	14	2	4	4	2	4	0
3	60	80	80	12	12	0	0	0	0
4	14	10	12	16	16	16	6	2	0
5	12	8	14	14	14	10	8	6	0
6	100	124	66	62	8	2	0	0	0
7	104	106	78	14	12	8	8	4	0
8	12	8	14	14	14	10	8	6	0
9	10	16	8	6	6	0	0	0	0
10	104	52	22	8	2	4	2	2	0
11	84	34	12	8	10	4	0	0	0
12	68	52	4	4	4	4	2	0	0
13	40	42	8	4	4	0	0	0	0
14	72	28	14	2	2	2	2	0	0
15	68	38	10	10	6	4	2	2	0
16	122	46	52	30	22	22	22	2	0
17	52	54	54	14	6	0	0	0	0
18	90	28	16	12	10	10	0	0	0
19	44	28	16	16	14	8	0	0	0
20	124	186	6	6	2	2	2	0	0
21	4	2	4	4	4	2	2	0	0
22	72	28	22	12	10	6	6	6	0
23	152	8	12	10	10	4	2	0	0
24	160	158	36	8	8	0	0	0	0
25	44	24	12	12	8	6	2	0	0
26	96	52	14	8	6	8	4	0	0
27	88	18	16	10	8	8	4	2	0
28	104	64	70	44	8	6	6	0	0
29	10	12	6	6	8	2	2	2	0
30	102	80	88	46	26	10	0	0	0
31	30	4	4	2	4	2	2	2	0

As stated earlier, these relations will be violated, depending whether the number of total solutions is determined by rows or columns. These cases, however, will not be discussed due to the limitation of the space.

Table 6

Summary Statistics on a Case Where
the Rank is Determined by Both Rows and Columns

	η^2	α	$\mathcal{E}(\%)$
1	0.8116	0.8839	40.5777
2	0.6071	0.6765	30.3575
3	0.2476	-0.5195	12.3794
4	0.1884	-1.1538	9.4206
5	0.1075	-3.1501	5.3762
6	0.0378	-11.7370	1.8886
Sum	2.0000	-15.0000	100.0000

Sum of Squares on Each Item.

Item 1	7.54	6.82	8.45	3.81	5.01	10.37	42.00
Item 2	7.54	6.82	8.45	3.81	5.01	10.37	42.00
Item 3	5.92	7.36	4.09	13.38	10.99	0.26	42.00

Squared Item-Total Correlation.

Item 1	0.8741	0.5915	0.2990	0.1026	0.0769	0.0560
Item 2	0.8741	0.5915	0.2990	0.1026	0.0769	0.0560
Item 3	0.6865	0.6385	0.1448	0.3601	0.1688	0.0014

Sums of Squared r_{jt}^2 over Dimensions

Item 1	2.00
Item 2	2.00
Item 3	2.00

Normed Score of Subjects, y_i

							SS
1	-0.9134	0.2923	-0.2465	-1.4013	-1.6140	0.6715	6.00
2	0.5128	-0.7759	1.7125	-0.1064	-0.5518	-1.3736	6.00
3	1.5501	1.2634	-0.7118	0.9488	-0.7685	-0.0591	6.00
4	-0.5128	-0.7759	-1.7125	-0.1064	0.5518	-1.3736	6.00
5	-1.5501	1.2634	0.7118	0.9488	0.7685	-0.0591	6.00
6	0.9134	0.2923	0.2465	-1.4013	1.6140	0.6715	6.00
7	0.0000	-1.5597	0.0000	1.1178	0.0000	1.5224	6.00

Normed Option Weights, x_{jp}

							SS
1	-1.1013	0.3336	-0.8355	-0.4292	-0.2986	-1.3056	12.00
2	0.2846	-1.4987	1.7208	1.1650	-0.8413	0.3830	15.00
3	1.3673	0.9983	-0.4675	-0.5212	1.2892	1.5754	15.00
4	-1.3673	0.9983	0.4675	-0.5212	-1.2892	1.5754	15.00
5	1.1013	0.3336	0.8355	-0.4292	0.2986	-1.3056	12.00
6	-0.2846	-1.4987	-1.7208	1.1650	0.8413	0.3830	15.00
7	1.7207	1.6214	-1.4305	2.1860	-2.3437	-0:3043	18.00
8	-1.7207	1.6214	1.4305	2.1860	2.3437	-0.3043	18.00
9	0.0000	-0.6486	0.0000	-0.8744	0.0000	0.1217	16.00

Inter-Subject Squared Distances in the Normed Space

0.000						
14.000	0.000					
14.000	14.000	0.000				
14.000	14.000	14.000	0.000			
14.000	14.000	14.000	14.000	0.000		
14.000	14.000	14.000	14.000	14.000	0.000	
14.000	14.000	14.000	14.000	14.000	14.000	0.000

4. Concluding Remarks

The current paper touched only on a surface of the topic. The two data types, incidence data and dominance data, are distinct as reflected on their respective objectives for quantification. From the view of multidimensional decomposition of data, however, the probably most important distinction lies in the fundamental premise of "what is multidimensionality for the two data types?" For incidence data, perfect association in the data (i.e., the correlation ratio of 1) *does not* mean that a single solution (component) can explain the data exhaustively. A simple example is a 10×10 contingency table of perfect row-column association (e.g., all non-zero entries are found only in the main diagonal). This data matrix yield nine perfect correlation ratios (i.e., 1), yet needing nine solutions to explain the data. In contrast, when association is perfect in dominance data (e.g., all the subjects rank the stimuli in the same way), one solution explains the data completely. This distinction creates a number of different characteristics between the data types, some of which are discussed in Nishisato (1993, 1994, 1996). From the graphical point of view as well as the interpretation point of view, the distinction between the two types should be well understood, but a number of further investigations into the differences are still needed before we are certain about our full understanding of the implications of the differences.

The last remark on both types of data is about the treatment of missing responses. Most imputation methods have the effect of increasing total information (i.e., the sum of squared singular values) in data. This is obviously undesirable, and should be regarded as fabrication of information. Thus, any method of imputation must be such that the total observed information be kept invariant. Such an example is rare (see dual scaling of rank order data, Nishisato, 1994), and the effects of missing responses on multdimensional space need to be further investigated.

5. References

Carroll, J.D. (1972). Individual differences and multidimensional scaling. In R.N. Shepard, A.K. Romney, and S.B. Nerlove (eds.), *Multidimensional Scaling: Theory and Applications in the Behavioral Sciences, Volume 1.* New York: Seminar Press.

Coombs, C.H. (1950). Psychological scaling without a unit of measurement. *Psychological Review,* 57, 148-158.

Coombs, C.H. (1964). *A Theory of Data.* New York: Wiley.

Coombs, C.H., and Kao, R.C. (1960). On a connection between factor analysis and multidimensional unfolding. *Psychometrika,* 25, 219-231.

Gold, E.M. (1973). Metric unfolding: Data requirements for unique solution and clarification of Schönemann's algorithm. *Psychometrika,* 38, 555-569.

Greenacre, M.J., and Browne, M.W. (1986). An efficient alternating least-squares algorithm to perform multidimensional unfolding. *Psychometrika,* 51, 241-250.

Heiser, W.J. (1981). *Unfolding analysis of proximity data.* Doctoral dissertation, Leiden University, The Netherlands.

Nishisato, S. (1978). Optimal scaling of paired comparison and rank-order data: An alternative to Guttman's formulation. *Psychometrika*, **43**, 263-271.

Nishisato, S. (1980). *Analysis of Categorical Data: Dual Scaling and Its Applications* . Toronto: University of Toronto Press.

Nishisato, S. (1993). On quantifying different types of categorical data. *Psychometrika*, **58**, 617-629.

Nishisato, S. (1994). *Elements of Dual Scaling: An Introduction to Practical Data Analysis*. Hillsdale, NJ: Lawrence Erlbaum Associates.

Nishisato, S. (1996). Gleaning in the field of dual scaling. *Psychometrika*, **61**, 559-599.

Nishisato, S., and Yamauchi, H. Principal components of deviation scores and standardized scores. *Japanese Psychological Research*, **16**, 162-170.

Schönemann, P.H. (1970). On metric multidimensional unfolding. *Psychometrika*, **35**, 167-176.

Schönemann, P.H., and Wang, M.M. (1972). An individual difference model for the multidimensional analysis of preference data. *Psychometrika*, **37**, 2 75-309.

Torgerson, W.S. (1958). *Theory and Methods of Scaling*. New York: Wiley.

Homogeneity Analysis for Partitioning Qualitative Variables

Takahiro Tsuchiya

The Institute of Statistical Mathematics
4-6-7, Minami-Azabu, Minato-ku
Tokyo 106, Japan

Summary: This paper proposes a method to construct multiple uni-dimensional scales by partitioning qualitative variables into mutually exclusive groups. The method is based on homogeneity analysis, and fuzzy c-means criterion is introduced for partitioning. Also, some goodness of fit indexes are proposed. Two artificial data sets and one real data set are analyzed as numerical examples. The results illustrate that the proposed method is more effective for partitioning qualitative variables compared with PCA with optimal scaling and Hayashi's Quantification Method III, or HOMALS.

1. Introduction

In the social and cultural sciences, uni-dimensional scaling is an important theme. A distinctive feature of scaling in those fields is that a data set often consists of qualitative variables. Multiple correspondence analysis, dual scaling and Hayashi's Quantification Method III (HQM III) are methods for analyzing the structure of qualitative variables. In case that all the variables are homogeneous (namely they have only one uni-dimensionality), the first solution of those methods can be used as a uni-dimensional scale. It is rare, however, that all the variables have a uni-dimensionality in practical situations. In order to construct uni-dimensional scales, selection or partitioning of variables is usually needed. HQM III is not necessarily appropriate for this purpose. To demonstrate it, let us consider the following example.

Tab. 1 shows an artificial data set 1, which consists of eight variables and 13 observations. The numerals in the table indicate category labels and n is an observation number. For example, observation 13 chose category 4 of variable 1. Although there are 8 variables in the data set, they should be partitioned into two groups. That is, an indicator matrix D_1 is obtained by using only variable 1 to 4 (Tab. 2). Zeros are omitted in the table. It is clear that D_1 has a Guttman's uni-dimensional structure. Another indicator matrix D_2 with uni-dimensional structure is also obtained by using variable 5 to 8 (Tab. 3). Because the orders of the rows are, however, different between D_1 and D_2, we should say that the data set in Tab. 1 have two uni-dimensional structures.

Tab. 1: Artificial Data Set 1

n	variable							
	1	2	3	4	5	6	7	8
1	1	1	1	1	2	1	1	1
2	2	1	1	1	2	2	2	1
3	2	2	1	1	3	3	2	2
4	2	2	2	1	2	2	1	1
5	2	2	2	2	3	3	3	3
6	3	2	2	2	2	2	2	2
7	3	3	2	2	1	1	1	1
8	3	3	3	2	3	2	2	2
9	3	3	3	3	4	4	4	3
10	4	3	3	3	4	4	3	3
11	4	4	3	3	3	3	3	2
12	4	4	4	3	4	4	4	4
13	4	4	4	4	4	3	3	3

452

Tab. 2: Indicator Matrix D_1

var.	1	2	3	4	1	2	3	4	1	2	3	4	1	2	3	4
cat.	1	1	1	1	2	2	2	2	3	3	3	3	4	4	4	4
1	1	1	1	1												
2		1	1	1	1											
3			1	1	1	1										
4				1	1	1	1									
5					1	1	1	1								
6						1	1	1	1							
7							1	1	1	1						
8								1	1	1	1					
9									1	1	1	1				
10										1	1	1	1			
11											1	1	1	1		
12												1	1	1	1	
13													1	1	1	1

Tab. 3: Indicator Matrix D_2

var.	5	6	7	8	5	6	7	8	5	6	7	8	5	6	7	8
cat.	1	1	1	1	2	2	2	2	3	3	3	3	4	4	4	4
7	1	1	1	1												
1		1	1	1	1											
4			1	1	1	1										
2				1	1	1	1									
6					1	1	1	1								
8						1	1	1	1							
3							1	1	1	1						
11								1	1	1	1					
5									1	1	1	1				
13										1	1	1	1			
10											1	1	1	1		
9												1	1	1	1	
12													1	1	1	1

Fig. 1 shows the first and the second axes of HQM III applied to this data set. In the figure, O indicates a category of variable 1 to 4, while ● indicates that of variable 5 to 8. It is well known that a horseshoe is obtained when a data set has one uni-dimensional structure. In Fig. 1, such a horseshoe appears and we might incorrectly conclude that the data set has only one uni-dimensional structure.

Fig. 2 shows the first and the third axes. Carefully observing the figure, we can see that there are two flows of categories and that they correspond to two groups of variables.

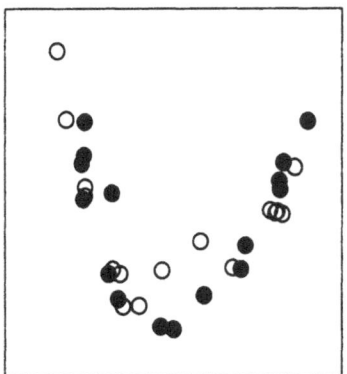

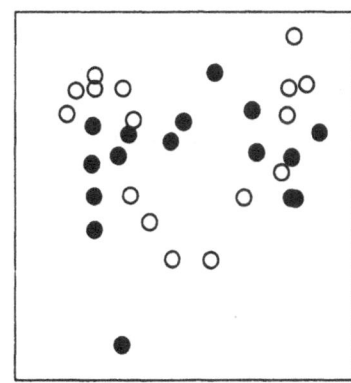

Fig. 1: The second axis versus the first   Fig. 2: The third axis versus the first

The above example illustrates that at least three-dimensional display of the result is needed for partitioning variables in order to construct multiple uni-dimensional scales. It is impractical, however, to plot all the categories when the number of variables increases because the figure will be illegible. In the example, difference between two groups of variables appeared on the third axis, but in case of other data sets, it is not certain which axis should be used.

Thus this paper proposes an easy method for partitioning qualitative variables to construct multiple uni-dimensional scales.

2. Method

Let $X_i(N \times C_i)$ be an indicator matrix of variable $i(i = 1, \ldots, I)$ where N is the number of observations and C_i is the number of categories of variable i. There is at most a single 1 in each row of X_i. Let $\Lambda_i(N \times N) = \mathrm{diag}(X_i 1)$ be a diagonal matrix whose nn-th diagonal element is the n-th element of $X_i 1$.

The method is developed based on homogeneity analysis (Gifi, 1990). It is well known that HQM III leads to the same equation as homogeneity analysis for analyzing $X_1, \ldots, X_I$. Homogeneity analysis is, however, more appropriate to explain the following method. In homogeneity analysis, sum of the squares of the distances between optimally transformed score vector, $X_i w_i$, and one common vector m is minimized under the constraint on m:

$$S(w_i, m) = \sum_{i=1}^{I} \|\Lambda_i(X_i w_i - m)\|^2 \longrightarrow \min, \tag{1}$$

$$\text{where} \quad \sum_{i=1}^{I} \|\Lambda_i m\|^2 = \sum_{i=1}^{I} \|\Lambda_i 1\|^2, \quad \sum_{i=1}^{I} 1' \Lambda_i m = 0.$$

If all the variables have only one uni-dimensional structure, it is possible to transform every score vector $X_i w_i$ close to each other. Since m in (1) is obtained as a mean vector of $X_1 w_1, \ldots, X_I w_I$, m can be a uni-dimensional score vector. In other words, a uni-dimensional score vector m has to be constructed from vectors which can be transformed near to each other. In case of the artificial data set 1, it is impossible to transform all the eight variables close. Hence, homogeneity analysis fails to construct uni-dimensional scales from data set 1. However, the first four variables can be transformed close to each other, and the same is true for the last four variables. If we prepare two score vectors, m_1 and m_2, and sum of the distances between m_1 and the first four variables and between m_2 and the last four are minimized, then both m_1 and m_2 can be uni-dimensional score vectors:

$$S(w_i, m_1, m_2) = \sum_{d=1}^{2} \sum_{i=1}^{8} \delta_{di} \|\Lambda_i(X_i w_i - m_d)\|^2,$$

where

$$\delta_{di} = \begin{cases} 1 & d = 1 \text{ and } 1 \le i \le 4, \quad d = 2 \text{ and } 5 \le i \le 8, \\ 0 & \text{otherwise.} \end{cases}$$

In practice, because we do not know in advance which variables have uni-dimensional structure, weight parameters summing up to 1 are introduced instead of δ_{di}:

$$S(a_{di}, w_{di}, m_d) = \sum_{d=1}^{D} \sum_{i=1}^{I} a_{di}^k \|\Lambda_i(X_i w_{di} - m_d)\|^2, \tag{2}$$

$$\text{where} \quad \sum_{i=1}^{I} a_{di}^k \|\Lambda_i m_d\|^2 = \sum_{i=1}^{I} a_{di}^k \|\Lambda_i 1\|^2, \quad \sum_{i=1}^{I} a_{di}^k m_d' \Lambda_i 1 = 0,$$

$$\sum_{d=1}^{D} a_{di} = 1, \quad a_{di} \ge 0.$$

We call D dimensionality because it indicates the number of uni-dimensional scales. k is given a priori in the region of $k > 1$. To explain the meaning of (2), let us first consider the case when $m_1, \ldots, m_D$ are obtained in some ways. For each i, a_{di} is constrained to $\sum_d a_{di} = 1$. Hence, in order to minimize the quantity S, a large value has to be assigned to a_{di} if corresponding $e_{di}^2 = ||\Lambda_i(X_i w_{di} - m_d)||^2$ is small. On the other hand, a small value must be assigned to a_{di} if e_{di}^2 is large. Thus a_{di} can be considered as an index of degree of correlation between $X_i w_{di}$ and m_d. Variable i is classified into the dimension d if a_{di} is the largest among $a_{1i}, \ldots, a_{Di}$. The variables which are classified into the same dimension d are close to each other because they are all in the neighborhood of one score vector m_d.

Next let us consider the case when all a_{di}'s are determined. In that case, m_d is obtained as a mean vector of $X_i w_{di}$ weighted by a_{di}. It implies that m_d is constructed from the vectors which are near to each other. Hence, we can say that m_d is a unidimensional score vector.

The same principle is used in fuzzy c-means clustering (Bezdek, 1981). $X_i w_{di}$ is, however, fixed in fuzzy c-means unlike (2). Also, $X_i w_{di}$ depends on dimension d in (2).

There are two reasons for not introducing k-means criterion (MacQueen, 1967) but fuzzy c-means. One is that fuzzy c-means includes k-means by letting $k \to 1$ in (2). The other is that the influence of variables which should be removed is decreased by letting $a_{di} = D^{-1}$. In practice, it is often the case that some variables have low correlation with the other variables. k-means criterion classifies such a variable into some dimension and m_d of that dimension can not be a uni-dimensional score.

The values of a_{di}, w_{di}, m_d, are obtained by means of alternating least squares method. Because of space limitations, the details of the algorithm are omitted.

An index

$$\lambda = 1 - \frac{D^{k-1}}{\sum_{i=1}^{I} 1' \Lambda_i 1} \min S, \tag{3}$$

which takes a value between 0 and 1, represents in a sense a goodness of fit to D dimensionality. Especially when $D = 1$, λ corresponds to the square of correlation coefficient in HQM III.

In case of $\Lambda_i = I$ for all i, such a goodness of fit index is obtained for each variable and each dimension. For each variable i,

$$\lambda_i = 1 - \frac{D^{k-1}}{N} \min S_i, \tag{4}$$

is an index where $S_i = \sum_{d=1}^{D} a_{di}^k ||X_i w_{di} - m_d||^2$. For each dimension d,

$$\lambda_d = 1 - \frac{D^k}{NI} \min S_d, \tag{5}$$

is an index where $S_d = \sum_{i=1}^{I} a_{di}^k ||X_i w_{di} - m_d||^2$. There is a following relation among λ, λ_i and λ_d.

$$\lambda = \frac{1}{I} \sum_{i=1}^{I} \lambda_i = \frac{1}{D} \sum_{d=1}^{D} \lambda_d \tag{6}$$

3. Example

Three data sets are analyzed. k is set to be 2 for all data sets.

3.1 Data Set 1

First, the artificial data set 1 of Tab. 1 is analyzed. Since an iterative algorithm often hits local minima, random starts are used for 100 times. The one giving the lowest S is considered as a global minimum.

Tab. 4 summarizes the value of D, frequency of global minimum, and λ. As the table indicates, the algorithm always converges to the same value when $D = 2$, but as D increases, it frequently falls into local minima. This result tells us that it is easy to partition 8 variables into 2 groups, and that partitioning into 3 or 4 groups is difficult. Further, λ of $D = 3$ is a little smaller than that of $D = 2$. Therefore, we can conclude that D should be 2.

Tab. 5 shows the values of a_{di} and goodness of fit indexes when $D = 2$. a_{di} of variable 1 to 4 is larger in dimension 1 than in dimension 2, while that of variable 5 to 8 is larger in dimension 2 than in dimension 1. That is, the first uni-dimensional scale can be constructed by variable 1 to 4 and the second can be constructed by variable 5 to 8. We should notice that this conclusion is the same as in the case when Fig. 2 was used in introduction.

Tab. 4: Summary of λ for each D

D	Frequency of global minimum	λ
1	100	.819
2	100	.879
3	23	.878
4	26	.896
5	12	.916

Tab. 5: a_{di} when $D = 2$

variable	dimension 1	dimension 2	λ_i
1	.875	.125	.873
2	.898	.102	.883
3	.853	.147	.881
4	.794	.206	.880
5	.189	.811	.878
6	.104	.896	.878
7	.132	.868	.885
8	.129	.871	.872
λ_d	.880	.878	.879

3.2 Data Set 2

Next, artificial data set 2 of Tab. 6 is analyzed. The purpose of this example is to compare the proposed method with principal component analysis model (PCA) or factor analysis model (FA). FA is usually used for partitioning quantitative variables. This example shows that PCA or FA for qualitative variables is not necessarily appropriate to partition qualitative variables.

The artificial data set 2 has also two uni-dimensional structures as data set 1; that is, variable 1 to 4 have one uni-dimensionality and variable 5 to 8 have another uni-dimensionality. Unlike

Tab. 6: Artificial Data Set 2

n	variable 1	2	3	4	5	6	7	8
1	1	1	1	1	1	1	3	3
2	2	1	1	1	1	1	1	3
3	2	2	1	1	3	3	3	3
4	2	2	2	1	1	3	3	3
5	2	2	2	2	4	1	1	1
6	3	2	2	2	4	4	1	1
7	3	3	2	2	1	1	1	1
8	3	3	3	2	4	4	4	4
9	3	3	3	3	2	2	4	4
10	4	3	3	3	2	2	2	4
11	4	4	3	3	4	4	4	1
12	4	4	4	3	2	2	2	2
13	4	4	4	4	2	4	4	4

data set 1, the order of categories of the last four variables is 3,1,4 and 2.

Tab. 7 is the result of principal components analysis with optimal scaling followed by VARIMAX rotation. The analysis was performed with the PRINQUAL procedure in SAS (A similar procedure is PRINCALS in SPSS.). The orders of categories are 1,2,3,4 for variable 1 to 4 and 3,1,4,2 for variable 5 to 8. Hence, the procedure found the proper orders. Variables are partitioned into two groups according to factor loadings. The first group consists of variable 4 to 8 and the second group consists of variable 1 to 3. This result does not represent the original data structure. This is because the assigned value for each category is optimal for applying PCA model but it is not optimal for partitioning of variables.

Tab. 7: Factor loadings for optimally transformed data set 2

variable	factor loading		category			
	1	2	1	2	3	4
1	.514	.814	1.29	1.87	3.05	4.00
2	.350	.931	1.06	1.92	3.08	3.96
3	.635	.735	1.02	1.88	3.24	3.72
4	.688	.684	0.92	2.07	3.18	3.30
5	.886	.411	1.16	3.85	0.36	2.65
6	.896	.380	1.63	4.09	0.56	3.02
7	.747	.614	1.93	4.13	0.88	3.59
8	.772	.569	2.55	4.32	1.03	3.84

Tab. 8 is a summary of a_{di} and λ_i, λ_d obtained by the proposed method. Variable 1 to 4 are classified into the first dimension and the other variables are classified into the second dimension. Hence, the proposed method succeeded in partitioning variables in order to construct uni-dimensional scales.

Tab. 8: a_{di} for data set 2

variable	dimension		λ_i
	1	2	
1	.766	.234	.895
2	.828	.173	.889
3	.634	.366	.912
4	.537	.463	.905
5	.184	.816	.896
6	.237	.763	.884
7	.404	.896	.906
8	.299	.702	.892
λ_d	.901	.894	.898

3.3 Kendall Data

The third example is taken from Kendall et al.(1983). The data consists of 15 variables on which 48 applicants for a post were judged. The original variables were scored on an 11-point scale, but Sato and Yanai (1985) recategorized each variable into three categories considering the category frequencies. Thus the recategorized data matrix is used in this paper.

Fig. 3 shows the first three axes of HQM III applied to the data set. There seems to be no distinctive structures in the figure.

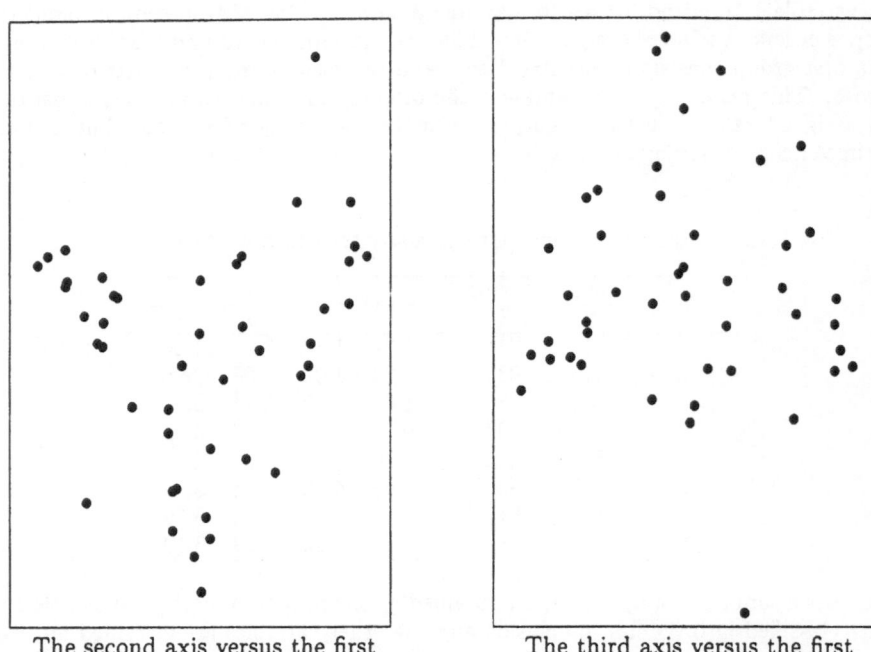

The second axis versus the first   The third axis versus the first

Fig. 3: Plots of categories of Kendall data

The proposed method was applied to the data set. The same as data set 1, the algorithm was randomly started for 100 times. Tab. 9 summarizes the value of D, frequency of global minimum, and λ. When $D = 3$, there are quite some local minima (34%), but the algorithm successfully reached a global minimum for 66 times. This indicates that we can not say it is too difficult to partition 15 variables into three groups. Tab. 10 is a summary of a_{di} and λ_i, λ_d when $D = 3$.

Tab. 9: Summary of λ for each D

D	Frequency of global minimum	λ
1	100	.469
2	100	.474
3	66	.492
4	6	.520

The first dimension consists of "Salesmanship", "Self-confidence", "Ambition" and so on. This dimension seems to represent "ambition to the job". The second dimension consists of "Potential", "Grasp", "Suitability" and "Experience". This dimension expresses "ability of the applicant". The third dimension consists of "Likeability",

"Honesty" and so on. This dimension means "external appearance" or "the first impression". These three dimensions are easy to understand, but it is difficult to obtain these dimensions from Fig. 3.

Tab. 10: a_{d_i} when $D = 3$

variable		dimension			λ_i
		1	2	3	
8	Salesmanship	.834	.116	.050	.872
5	Self-confidence	.679	.201	.120	.693
11	Ambition	.500	.366	.137	.693
10	Drive	.431	.410	.159	.655
6	Lucidity	.413	.395	.192	.493
14	Keenness to join the company	.351	.312	.337	.342
13	Potential	.144	.751	.106	.815
12	Grasp	.226	.640	.134	.698
15	Suitability	.338	.374	.288	.361
9	Experience	.327	.347	.326	.078
4	Likeability	.032	.042	.926	.916
7	Honesty	.257	.271	.473	.260
1	Form of letter of application	.306	.334	.360	.133
2	Appearance	.303	.346	.351	.302
3	Academic ability	.327	.332	.342	.064
	λ_d	.471	.465	.538	.492

4. Conclusion

In this paper a new method to construct multiple uni-dimensional scales by partitioning qualitative variables is proposed. As for the presented data sets, the results of the application of the model are successful.

References:

Bezdek, J.C. (1981): *Pattern Recognition with Fuzzy Objective Function Algorithm*. Plenum Press, New York.

Gifi, A. (1990): *Non Linear Multivariate Analysis*. John Wiley & Sons, Chichester.

Kendall, M.G. et al. (1983): *The Advanced Theory of Statistics, Volume, 3, 4th ed.* Charles Griffin.

MacQueen, J. (1967): Some methods for classification and analysis of multivariate observations. *Proceeding of the fifth Berkeley Symposium on Mathematical Statistics and Probability*, 1, 281–297.

Meulmann, J.J. (1996): Fitting a distance model to homogeneous subsets of variables: points of view analysis of categorical data. *Journal of Classification*, **13**, 249–267.

Sato, T. and Yanai, H. (1985): A method of simultaneous scaling of discrete variables. *Behaviormetrika*, **18**, 39–51.

Determining the Distance Index, II

Matevž Bren[1] and Vladimir Batagelj[2]

[1] University of Maribor, Faculty of Organizational Sciences,
Prešernova 11, 4000 Kranj, Slovenia

[2] University of Ljubljana, Department of Mathematics,
Jadranska 19, 1000 Ljubljana, Slovenia

Summary: We present the results of approximating the distance index by optimization. We do it for 16 standard dissimilarities on binary vectors and for some usual nonlinear transformations of these dissimilarities.

1. Preliminaries

1.1 Dissimilarities

A mapping $d: \mathcal{E} \times \mathcal{E} \to \mathbb{R}$ is a *dissimilarity measure* on the set of objects $\mathcal{E}$ iff it is

P1. symmetric: $d(x, y) = d(y, x)$ for all $x, y \in \mathcal{E}$,

P2. straight: $d(x, x) \leq d(x, y)$ for all $x, y \in \mathcal{E}$.

A dissimilarity measure d that is

D1. nonnegative: $d(x, y) \geq 0$ for all $x, y \in \mathcal{E}$ and

D2. vanishes on the diagonal: $d(x, x) = 0$ for all $x \in \mathcal{E}$

is a *dissimilarity* (on $\mathcal{E}$). The ordered pair $(\mathcal{E}, d)$ is a *dissimilarity space*. We denote with $\mathcal{D}_+$ the set of all dissimilarities on $\mathcal{E}$.

A dissimilarity d on $\mathcal{E}$ is said to be:

D3. *definite* iff $d(x, y) = 0 \Rightarrow x = y$;

D4. *even* iff $d(x, y) = 0 \Rightarrow$ for all $z \in \mathcal{E}$: $d(x, z) = d(y, z)$;

D5. a *semi-distance* iff the *triangle inequality*

$$\text{for all } x, y, z \in \mathcal{E}: \qquad d(x, y) \leq d(x, z) + d(y, z) \qquad \text{holds.}$$

1.2 Dissimilarity Spaces

A dissimilarity space $(\mathcal{E}, d)$ is a *semi-metric space* if and only if d is a semi-distance, and is a *metric space* if and only if d is also definite - d is a *distance*.

We denote the set of all semi-distances on $\mathcal{E}$ by $\mathcal{D}_\infty$.

In Joly and Le Calvé (1986) we can find the theorem:

- $d \in \mathcal{D}_\infty \Rightarrow d^\alpha \in \mathcal{D}_\infty$ *for all* α: $0 \leq \alpha \leq 1$;

- $d \in \mathcal{D}_+ \Rightarrow$ *there exists an unique positive number* $p \in \mathbb{R}$, *such that* $d^\alpha \in \mathcal{D}_\infty$ *for all* α: $\alpha \leq p$ *and* $d^\alpha \notin \mathcal{D}_\infty$ *for all* α: $\alpha > p$;

We call the threshold value $p = p(d)$ a *distance index* of the dissimilarity d. For details see Batagelj and Bren (1996).

2. Determining the Distance Index for Dissimilarities on Binary Vectors

In the first part of the paper we present:

- an analytical solution of this problem for two families of dissimilarities obtained from functions defined in Gower and Legendre (1986)

$$S_\theta = \frac{a+d}{a+d+\theta(b+c)} \quad \text{and} \quad T_\theta = \frac{a}{a+\theta(b+c)}$$

(where $\theta > 0$ to avoid negative values) that contain some well-known similarity measures (see Table 1: Kendall, Sokal-Michener; Rogers and Tanimoto; Jaccard; Dice, Czekanowski; Sokal and Sneath);

The quantities a, b, c and d have the usual meaning: for binary vectors $x, y \in \mathbb{B}^m$ we denote with $xy = \sum_{i=1}^m x_i y_i$ their scalar product, and with $\bar{x} = [1 - x_i]$ the complementary vector of x. We define counters: $a = xy$, $b = x\bar{y}$, $c = \bar{x}y$ and $d = \bar{x}\bar{y}$, where $a+b+c+d = m$. Using these counters several resemblance measures on binary vectors are defined (see Table 1). The use of symbol d for a dissimilarity measure and also for a counter might be confusing, but it's meaning is always clear from the context.

- and a computational approach to determining the distance index for other dissimilarity measures from Table 1.

In this part we study, using a computational approach, dissimilarity measures obtained by the following nonlinear transformations

$$D_1 = \frac{d}{1+d} \qquad D_2 = \frac{d}{1-d} \qquad D_3(t) = \frac{d}{1+t(1-d)}$$

$$D_4 = -\ln(1-d) \qquad D_5 = \frac{2}{\pi}\arctan d \qquad D_6 = 1 - |1 - 2d|$$

$$D_7 = 4d(1-d)$$

on dissimilarity measures from Table 1.

Since the triangle inequality implies evenness we consider only even (**D4**) dissimilarity measures.

For a dissimilarity measure that does not vanish on the diagonal the triangle inequality implies
$$d(x, x) \leq 2\,d(x, y) \qquad \text{for all } x, y \in \mathcal{E}.$$

But, since **P2** holds this is true for any dissimilarity measure.

For the indeterminate cases we use the definitions proposed in Batagelj and Bren (1995). To make this reading easier, the second column of Table 1 includes also the labels used there.

Table 1: Association coefficients

measure	s_i	definition	d_i	D2	D3	D4	D5
1. Russel and Rao (1940)	s_1	$\frac{a}{m}$	$1-s$	N	Y	Y	Y
2. Kendall, Sokal-Michener (1958)	s_2	$\frac{a+d}{m}$	$1-s$	Y	Y	Y	Y
3. Jaccard (1900)	s_6	$\frac{a}{a+b+c}$	$1-s$	Y	Y	Y	Y
4. Kulczynski (1927), T^{-1}	s_7	$\frac{a}{b+c}$	s^{-1}	Y	Y	Y	N
5. Kulczynski	s_{10}	$\frac{1}{2}\left(\frac{a}{a+b}+\frac{a}{a+c}\right)$	$1-s$	Y	Y	Y	N
6. Sokal & Sneath (1963), un_4	s_{11}	$\frac{1}{4}\left(\frac{a}{a+b}+\frac{a}{a+c}+\frac{d}{d+b}+\frac{d}{d+c}\right)$	$1-s$	Y	Y	N	N
7. Q_0	s_{12}	$\frac{bc}{ad}$	s	N	N	N	N
8. Yule (1927), Q	s_{14}	$\frac{ad-bc}{ad+bc}$	$\frac{1}{2}(1-s)$	N	N	N	N
9. $-bc-$	s_{15}	$\frac{4bc}{m^2}$	s	Y	N	N	N
10. Driver & Kroeber (1932), Ochiai (1957)	s_{16}	$\frac{a}{\sqrt{(a+b)(a+c)}}$	$1-s$	Y	Y	Y	N
11. Sokal & Sneath (1963), un_5	s_{17}	$\frac{ad}{\sqrt{(a+b)(a+c)(d+b)(d+c)}}$	$1-s$	Y	Y	Y	N
12. Pearson, ϕ	s_{18}	$\frac{ad-bc}{\sqrt{(a+b)(a+c)(d+b)(d+c)}}$	$\frac{1}{2}(1-s)$	Y	Y	Y	N
13. Baroni-Urbani, Buser (1976), S^{**}	s_{19}	$\frac{a+\sqrt{ad}}{a+b+c+\sqrt{ad}}$	$1-s$	Y	Y	Y	N
14. Braun-Blanquet (1932)	s_{20}	$\frac{a}{\max(a+b,a+c)}$	$1-s$	Y	Y	Y	Y
15. Simpson (1943)	s_{21}	$\frac{a}{\min(a+b,a+c)}$	$1-s$	Y	N	N	N
16. Michael (1920)	s_{22}	$\frac{4(ad-bc)}{(a+d)^2+(b+c)^2}$	$\frac{1}{2}(1-s)$	N	Y	Y	N

2.1 Approximating the Distance Index by Optimization

We can approach the problem of determining the distance index by solving numerically the corresponding optimization problem

$$p = \operatorname{argmax}\{\alpha : \forall i, j, k \in \mathbb{B}^m : d^\alpha(i,j) + d^\alpha(j,k) \geq d^\alpha(i,k)\}.$$

We use a local optimization procedure

```
initial (read, random) i, j, k
p := 1
while ∃(u, v, w) ∈ N(i, j, k) :
    d^p(u, v) + d^p(v, w) < d^p(u, w) do begin
    p :=Solve( α : d^α(u, v) + d^α(v, w) = d^α(u, w) )
    i := u; j := v; k := w
end
```

over a neighbourhood

$$N(i, j, k) = \{(i', j', k') : r \in 1..m \ \wedge \ (i'_r = \neg i_r \vee j'_r = \neg j_r \vee k'_r = \neg k_r)\}$$

From the local minima (i^*, j^*, k^*) obtained by this procedure we can usually guess a general pattern of 'extremal' triples from which we compute an upper bound for p_m:

$$\bar{p}_m = p(i^*, j^*, k^*), \qquad i^*, j^*, k^* \in \mathbb{B}^m.$$

We conjecture that for the obtained triples (i^*, j^*, k^*) the equality $p_m = \bar{p}_m$ often holds.

3. Results

In Table 2 there are values $\bar{p}_{20}$ for twelve even dissimilarity measures and their transformations computed with this local optimization procedure. The notation used:

Y   triangle inequality (**D5**) holds;
N-d   there exist such binary vectors that D is not defined;
N-e   **D5** doesn't hold because transform D is not even;
N-p   some values of transform D are not positive;
$p \to$ p tends to this value as dimension m increases.

In the following paragraphs we shall give explanation of the obtained results for most interesting cases.

Russel and Rao $d_1 = 1 - s_1 = 1 - \frac{a}{m} \in [0, 1]$

It does not vanish at the diagonal: for $0 := [0, \ldots, 0] \in \mathbb{B}^m$ we have $d_1(0, 0) = 1$. The triangle inequality holds with equality on vectors of the form

$$
\begin{aligned}
i^* &= \overbrace{[0, \ldots, 0}^{x}, \overbrace{1, \ldots, 1]}^{m-x} \\
j^* &= [1, \ldots, 1, 0, \ldots, 0] \\
k^* &= [1, \ldots, 1, 1, \ldots, 1] =: 1
\end{aligned}
$$

as $1 = d_1(i^*, j^*) = d_1(i^*, 1) + d_1(j^*, 1) = \frac{x}{m} + \frac{m-x}{m}$. So p is equal to 1.

The transform D_1 is a concave function on the interval $[0, 1]$. Hence strict triangle

Table 2: Results

measure		Linear	D_1	D_2	$D_3(2)$	D_4	D_5	D_6	D_7
Russel	s_1	Y	Y	N-d	0.5	N-d	Y	N-e	N-e
Kendall	s_2	Y	Y	N-d	0.5	N-d	Y	Y	Y
Jaccard	s_6	Y	Y	N-d	0.5	N-d	Y	N-e	N-e
T^{-1}	s_7	N-d	Y	N-d	N-d	N-d	0.7761	N-p	N-p
Kulczynski	s_{10}	0.3798 $p \to 0$	0.5283 $p \to 0$	N-d	0.2489	N-d	0.4375 $p \to 0$	N-e	N-e
	s_{11}	0.5004 $p \to 0$	0.6457	N-d	0.3847	N-d	0.5377	0.6186*	0.8385*
Driver	s_{16}	0.5535 $p \to 0.5$	0.7706	N-d	0.3286	N-d	0.6509	N-e	N-e
un_5	s_{17}	0.5946 $p \to 0.5645$	0.9319 $p \to 0.8755$ Y for $n \leq 10$	N-d	0.3412 $p \to 0.3286$	N-d	0.7257 $p \to 0.6938$		
ϕ	s_{18}	0.5696 $p \to 0.5645$	0.7383	N-d	0.4255	N-d	0.6067	0.6236	0.8965
$S^{\cdot\cdot}$	s_{19}	0.3785 $p \to 0$	0.5386 $p \to 0$	N-d	0.2460	N-d	0.4337	N-e	N-e
Braun	s_{20}	Y	Y	N-d	0.5	N-d	Y	N-e	N-e
Michael	s_{22}	0.2811	0.2979	N-d	0.2265	N-d	0.2855	0.2811	0.2985

inequality holds and it is an equality only for the "triples" $\mathbf{1}$, $\mathbf{1}$, i for all $i \in B^m$. Indeed, for the previous triple $i^*, j^*, \mathbf{1}$ we have

$$\frac{1}{2} = D_1(d_1(i^*, j^*)) \leq D_1(d_1(i^*, \mathbf{1})) + D_1(d_1(j^*, \mathbf{1})) = \frac{x}{m+x} + \frac{m-x}{2m-x}$$

The equality holds only when $x = m$; that is for $j^* = \mathbf{1}$.

The transform D_2 is undefined because $D_2(d_1(i^*, j^*)) = \frac{1}{1-1}$.

If we choose parameter $t = 2$ we get $D_3(2) = \frac{d_1}{3-2d_1}$, and for the previous triple

$$1 = D_3(2)(d_1(i^*, j^*)) > D_3(2)(d_1(i^*, \mathbf{1})) + D_3(2)(d_1(j^*, \mathbf{1})) = \frac{x}{3m-2x} + \frac{m-x}{m+2x}$$

The expression on the right side is a function of x ($0 \leq x \leq m$) with minimum at $x = \frac{m}{2}$. Therefore, for $x = \frac{m}{2}$, we get

$$1 = D_3(2)(d_1(i^*, j^*)) > D_3(2)(d_1(i^*, \mathbf{1})) + D_3(2)(d_1(j^*, \mathbf{1})) = \frac{1}{4} + \frac{1}{4}$$

and $\bar{p} = 0.5$.

The transform D_4 is undefined because $D_4(d_1(i^*, j^*)) = -\ln(1-1)$.

The transform D_5 is a concave function on the interval $[0, \infty]$. Hence strict triangle inequality holds.

The transformations D_6 and D_7 are not even: for the triple $i = [0, 1, 1, 0]$, $j = [0, 0, 0, 1]$ and $k = [0, 1, 1, 1]$, we have

pair	a	b	c	d	$d_1(-,-)$	D_6	D_7
ij	0	2	1	1	1	0	0
ik	2	0	1	1	$\frac{1}{2}$	1	1
jk	1	0	2	1	$\frac{3}{4}$	$\frac{1}{2}$	$\frac{3}{4}$

Sokal & Sneath $d_{17} = 1 - s_{17} = 1 - \frac{ad}{\sqrt{(a+b)(a+c)(d+b)(d+c)}} \in [0, 1]$

is a definite dissimilarity (see Table 1) but it does not obey the triangle inequality: for the triple $i^*, j^*, k^* \in \mathbb{B}^m$

$$i^* = [0, \ldots, 0, 0, 1]$$
$$j^* = [0, \ldots, 0, 1, 0]$$
$$k^* = [0, \ldots, 0, 1, 1]$$

we get $d_{17}(i^*, j^*) = 1$ and $d_{17}(i^*, k^*) = d_{17}(j^*, k^*) = 1 - \sqrt{\frac{m-2}{2(m-1)}}$. This is for dimension

$m = 2$:   also equal to 1 – triangle inequality (**D5**) holds;
$m = 3$:   equal to 1/2 – triangle equality holds;
$m = 4$:   equal to $1 - \frac{\sqrt{3}}{3} \doteq 0.42$ – **D5** doesn't hold;
$m \gg$:   equal to $1 - \frac{\sqrt{2}}{2} \doteq 0.29$ – **D5** doesn't hold.

So $\bar{p}_{20} = -\frac{\ln 2}{\ln(1-\sqrt{\frac{18}{38}})} \doteq 0.5946$ and when m tends to infinity $\bar{p} = -\frac{\ln 2}{\ln(1-\frac{\sqrt{2}}{2})} \doteq 0.5645$.

The transform D_1 is a concave function on the interval $[0, 1]$ hence the triangle inequality holds for larger dimensions. For the previous triple we get $D_1(d_{17}(i^*, j^*)) = 1/2$ and $D_1(d_{17}(i^*, k^*)) = D_1(d_{17}(j^*, k^*))$ is for dimension

$m = 2$:   also equal to $1/2$ – **D5** holds;
$m = 3$:   equal to $1/3$ – **D5** holds;
$m = 4$:   equal to $\frac{\sqrt{3}-1}{2\sqrt{3}-1} \doteq 0.30$ – **D5** holds;
$m = 10$:   equal to $1/4$ – triangle equality holds;
$m = 20$:   equal to $\frac{\sqrt{19}-3}{2\sqrt{19}-3} \doteq 0.24$ – **D5** doesn't hold;
$m \gg$:   equal to $\frac{\sqrt{2}-1}{2\sqrt{2}-1} \doteq 0.23$ – **D5** doesn't hold.

For $m > 10$ we obtain for the upper bound of p_m:

$$\bar{p}_{20} = -\frac{\ln 2}{\ln 2\frac{\sqrt{19}-3}{2\sqrt{19}-3}} \doteq 0.9319 \qquad \text{and for } m \gg \qquad \bar{p} = -\frac{\ln 2}{\ln 2\frac{\sqrt{2}-1}{2\sqrt{2}-1}} \doteq 0.8755$$

The transform D_2 is undefined because $D_2(d_{17}(i^*, j^*)) = \frac{1}{1-1}$.

For $t = 2$ the transform $D_3(2) = \frac{d_{17}}{3-2d_{17}}$. For the previous triple i^*, j^*, k^* we get $D_3(2)(d_{17}(i^*, j^*)) = 1$ and $D_3(2)(d_{17}(i^*, k^*)) = D_3(2)(d_{17}(j^*, k^*))$. This is for dimension

$m = 2$:   also equal to 1 – **D5** holds;
$m = 3$:   equal to $1/4$ – **D5** doesn't hold;
$m = 20$:   equal to $\frac{\sqrt{19}-3}{\sqrt{19}+6} \doteq 0.13$ – **D5** doesn't hold;
$m \gg$:   equal to $\frac{\sqrt{2}-1}{\sqrt{2}+2} \doteq 0.12$ – **D5** doesn't hold.

Now we can calculate the upper bound of p_m:

$$\bar{p}_{20} = -\frac{\ln 2}{\ln \frac{\sqrt{19}-3}{\sqrt{19}+6}} \doteq 0.3412 \qquad \text{and for } m \gg \qquad \bar{p} = -\frac{\ln 2}{\ln \frac{\sqrt{2}-1}{\sqrt{2}+2}} \doteq 0.3286$$

If we calculate $\bar{p}_{20}$ for different values of parameter t we get for $t = 1$: $\bar{p}_{20} = 0.4103$, for $t = 2$: $\bar{p}_{20} = 0.3412$ and for $t = 3$: $\bar{p}_{20} = 0.3033$.

The transform D_4 is undefined because $D_4(d_{17}(i^*, j^*)) = -\ln(1 - 1)$.

The transform D_5 is a concave function on the interval $[0, \infty]$. For the previous triple we get $D_5(d_{17}(i^*, j^*)) = 1/2$ and $D_5(d_{17}(i^*, k^*)) = D_5(d_{17}(j^*, k^*))$ is for dimension

$m = 2$:   also equal to $1/2$ – **D5** holds;
$m = 3$:   equal to $\frac{2}{\pi} \arctan \frac{1}{2} \doteq 0.29$ – **D5** holds;
$m = 4$:   equal to $\frac{2}{\pi} \arctan(1 - \frac{\sqrt{3}}{3}) \doteq 0.25$ – **D5** holds;
$m = 5$:   equal to $\frac{2}{\pi} \arctan(1 - \sqrt{\frac{3}{8}}) \doteq 0.23$ – **D5** doesn't hold;
$m = 20$:   equal to $\frac{2}{\pi} \arctan(1 - \frac{3\sqrt{19}}{19}) \doteq 0.19$ – **D5** doesn't hold;
$m \gg$:   equal to $\frac{2}{\pi} \arctan(1 - \frac{\sqrt{2}}{2}) \doteq 0.18$ – **D5** doesn't hold.

Hence for $m > 4$ we can calculate the upper bound of p_m:

$$\bar{p}_{20} \doteq 0.7257 \qquad \text{and for } m \gg \qquad \bar{p} \doteq 0.6938$$

Transformations D_6 and D_7 are not even: for the triple $i = [1, 1, 0, 0]$, $j = [1, 0, 1, 1]$ and $k = [1, 0, 1, 0]$, we get

pair	a	b	c	d	$d_{17}(\_,\_)$	D_6	D_7
ij	1	1	2	0	1	0	0
ik	1	1	1	1	$\frac{3}{4}$	$\frac{1}{2}$	$\frac{3}{4}$
jk	2	1	0	1	$1-\frac{\sqrt{3}}{3}\doteq 0.42$	$2(1-\frac{\sqrt{3}}{3})\doteq 0.84$	$\frac{4}{3}(\sqrt{3}-1)\doteq 0.98$

In the last two columns of Table 2 we can see that most of the considered dissimilarity measures are not even. This occurs because the transformations D_6 and D_7 map $1 \to 0$. Hence the transformed dissimilarity measure is even if and only if the original one has the property: for all pairs $i,j \in \mathbb{B}^m$ such that $d(i,j) = 1$

$$d(i,k) = d(j,k) \quad \text{or} \quad d(i,k) = 1 - d(j,k) \quad \text{for all } k \in \mathbb{B}^m.$$

Because the dissimilarity d_{11} has this property on $\mathbb{B}^m - \{0,1\}$ we put the * in it's row.

4. Conclusion

In the paper we present an approach to determining the distance index for a given dissimilarity on binary vectors and applied it to some well known dissimilarity measures and their transforms. The results obtained offer new information that can be used when selecting dissimilarity measure for applications.

We also expect that the proposed approach can be successfully applied to dissimilarities between other types of units.

5. References:

Batagelj, V., Bren, M. (1995): Comparing Resemblance Measures, *Journal of Classifications*, **12**, 1, 73–90.

Batagelj, V., Bren, M. (1996): Determining the Distance Index, In: *Ordinal and Symbolic Data Analysis:* proceedings of the International Conference on Ordinal and Symbolic Data Analysis – OSDA'95, Paris, June 20-24, 1995, Diday, E. at al. (eds), 238–252, Springer-Verlag: Berlin, Heidelberg, New York.

Critchley, F., Fichet, B. (1994): The Partial Order by Inclusion of the Principal Classes of Dissimilarity on Finite Set, and some of their Properties, In: *Classification and Dissimilarity Analysis*, Van Cutsem, B. (ed.), 5–67, Lecture Notes in Statistics, Springer Verlag: New York.

Gower, J.C., and Legendre, P. (1986): Metric and Euclidean properties of dissimilarity coefficients, *Journal of Classification*, **3**, 5–48.

Joly, S., Le Calvé, G. (1986): Etude des puissances d'une distance, *Statistique et Analyse des Données*, **11**, 3, 30–50.

Meta-data and Strategies of Textual Data Analysis: Problems and Instruments

Sergio Bolasco

Faculty of Economy
University of Rome "La Sapienza"
Via del Castro Laurenziano, 9 - 00161 Roma - Italy

Summary : In order to develop a proper multidimensional *content analysis*, we discuss some typical aspects of a pre-treatment of a textual data analysis. In particular: i) how to select the *peculiar* subset of the words in a text; ii) how to reduce the word ambiguity. Our proposal is to use both *frequency dictionaries* and *reference lexicons* as external lexical knowledge bases with respect to the corpus, by means of a comparison of ranking, inspired by Wegman's parallel coordinate method. The conditions of *iso-frequency* of unlernmatized forms as an indication of the need for lemmatization is considered. Finally in order to evaluate the opportunities of the choices (both disambiguations and fusions), we propose the reconstruction, by means of bootstrapping strategy, of some convex hulls - as word confidence areas - in a factorial plane. Some examples from a large corpus of parliamentary discourses are presented.

1. Introduction

In this paper we are concerned with the different phases of text pre-treatment necessitated by a *content analysis*, based on multidimensional statistical techniques. These phases have been modified in recent years by the growth in sizes of textual data corpora and their related vocabularies and by the increased availability of lexical resources.

As a consequence of this, some new problems arise. The first one is how to select the *fundamental core* of the corpus vocabulary, when it is composed of several thousands of elements. In other words how to identify the subset of the characteristic words within a text, regardless of their frequency, in order to optimize computing time and minimize interpretation problems. The second problem is how to reduce the ambiguity of language produced by the automatic treatment of a text. The main aspects of this are the choice of the unit of analysis and of lemmatization.

We also propose the validation of the lemmatization choices in terms of the stability of the word points on factorial planes in order to control the effects of this preliminary intervention.

To solve these problems, it is possible to use both external and internal information, concerning the corpus: i. e. both meta-data and data. Some examples of our proposals are applied to a very large corpus of parliamentary discourses on government programmes (called Tpg from now on). The size of the Tpg corpus (Tpg Program Discourses and Tpg Replies) is over 700.000 occurrences and the size of the Tpg vocabulary it is over 28.000 unlemmatized words, equivalent to 2500 pages of text.

2. - How to identify the fundamental core of the corpus vocabulary

Regarding the first problem, *frequency dictionaries* and *reference lexicons* play a crucial role as external lexical knowledge bases. The former can be represented as being several *models of language*. Just as a reminder a frequency dictionary is a vocabulary ranking by decreasing of headword frequency obtained by means of a very large corpus (at least one million occurences); this corpus is a representative sample of texts from some collections of the language. A reference lexicon is a complete inventory of the inflected forms or of any other collection of locutions or idiomatic expressions.

We can assume that every textual corpus (as discourse) is the reflection of an *idiom*, a *context* and a *situation* (i. e.: enunciation and historical period). So its vocabulary cannot but come out of these three components.

The *idiom* is identifiable through the base-dictionary of a given natural language. In Italian this base-dictionary is represented by a VdB of around 7000 most frequent words in everyday language (or the 2000 most frequent words in LIF, see Bortolini *et al.* 1971).

Some of the words of the corpus which belong to the VdB, in some cases, could be eliminated from the analysis inasmuch as they are necessary only to the construction of sentences (for instance the grammatical words).

Words such as support-verbs or idiomatic phrases can be clearly identified and their capture will contribute to the reduction of ambiguity. This capture is possible by means of a reference lexicon of locutions and phrasal verbs. For example, if we look at the Italian verb <andare> (to go, in English), we will see in tab. 1, from a reference lexicon, that there are over 200 different phrasal verbs that use this verb as support. Of these, of course, almost half do not exist or do not have an equivalent in English.

Tab. 1: Examples of idioms of the verb "andare" (*to go*) as phrasal verb

andar/bene/VAVV/V/DIGE/DCM541/ =	be/a/good/match/VDETAGGN/V
andare/a gli/estremi/VPN/V/DIGE/DCM693/ =	go/to/extremes/VPN/V
andare/a/fare/la/spesa/VPVDETN/V/DIGE/DCM980/ =	go/shopping/VAVV/V
andare/a/giornata/VDETN/V/DIGE/DCM721/ =	go/out/to/work/by he/day/VPVPN/V
andare/a/male/VPN/V/DIGE/DCM654/ =	go/bad/VAVV/V
andare/a/spasso/VPN/V/DIGE/CTS/ =	go/for a/walk/VPN/V
andare/a/zonzo/VPN/V/DIGE/DTA/ =	saunter/V/V
andare/avanti/VAVV/V/DIGE/DCM562/ =	progress/V/V
andare/direttamente/a lo/scopo/VAVVPN/V/DIGE/DCM661/ =	go/straight/to he/mark/VAVVPN/V
andare/fuori/VAVV/V/DIGE/DCM1026/ =	get/out/VAVV/V
andare/fuori/VAVV/V/DIGE/DCM1026/ =	go/out/VAVV/V
andare/fuori/VAVV/V/DIGE/DCM1026/ =	set/out/VAVV/V
andare/fuori/uso/VPN/V/DIGE/DCM1027/ =	wear/out/VAVV/V
andare/oltre i/limiti/VPN/V/DIGE/DCM827/ =	overstep/the/limits/VDETN/V
andare/per la/maggiore/VPAGG/V/DIGE/GV/ =	be/very/popular/VAVVAGG/V
andare/sotto il/nome/di/VPNP/V/DIGE/GV/ =	go/by the/name/of/VPNP/V

and so on, with over 200 different examples in Italian language and at least other 40 phrasal forms of "to go" in English

The context and the situation are characterized with the aid of a specialized frequency dictionary (political, scientific, or economic, etc.). In this event, the lexical inclusion percentage of the corpus vocabulary in the reference language model is a basic measure.

With regards to the Tpg, the chosen frequency dictionary is the lexicon of Press and Press Agencies Information (called Veli). This vocabulary is derived from a collection of over 10 million occurrences. On the assumption that the Veli vocabulary is the pertinent neutral model available of a formal language in social and political context, we can ask ourselves to what extent the Tpg corpus resembles it, or differs from it.

In this sense the *situation* can be identified by studying the original terms not included in this external knowledge base. In our case, the language of the situation is composed of the Tpg terms which does not belong to the Veli. This sub-set is interesting in itself.

On the contrary, the *context* can be identified through the *words in common* in the above two lexicons. Among these words, in general, the highly specific sectorial terms are measured by the largest diversities of use with respect to the chosen frequency dictionary.

In this way we are interested to identify one sub-set of characteristic words. The peculiarity or intrinsic specificity of this sub-set will be measured by calculating the diversities of use for each pair of words. As Lyne says (1985: 165): "The specific words are terms whose frequency differs characteristically from what is normal. The difference can be calculated from the theoretical frequency of a word in a given text, on the assumption that the latter is proportional to the length of the text." One possible measure *of specificity* could be the classical measure of *z* - like a normalized difference of the frequencies -

$$z = \frac{f_j - f^*}{\sqrt{f^*}}$$

where: the f_j is the relative number of occurrences in the corpus and f^* the correspondent in the frequency dictionary. Proposed by P. Guiraud in 1954, *z* usually is called *écart reduit*, and it is equivalent to the square root of the chi square.

It is possible to compare the coefficients of usage between the two vocabularies; where the latter is - for each headword - the frequency weighed with the measure of dispersion.

The above specificity measure can be either positive or negative. Using the Veli list as a yardstick, we can investigate the Tpg vocabulary. In fact as Lyne suggests (*ibidem*: 1985: 7): "The ranking favours those items which are most characteristic of our corpus, what we shall call, Positive .. Items. Conversely, towards the bottom of this list are found those items, Negative .. Items, which, although still occurring (in some instances frequently) in our corpus, are nevertheless least characteristic of it, since they occur relatively less frequently than in the reference dictionary"

Once the relative differences between the Tpg and the Veli vocabulary are measured in terms of *z*, it is possible to select and to visualize two comparative rankings of words in the above vocabularies. The threshold of selection can be the classical level of the absolute value of *z* (greater than or equal to 3). The set of these selected words can be visualized by using the method of "parallel coordinates" (Wegman, 1990). As known, Wegman's proposal consists in using the parallel coordinate representation as a high-dimensional data analysis tool. Wegman shows that this geometry has some interesting properties; in particular a statistical interpretation of the correlation can be given. For highly negatively correlated pairs, the dual line segments in parallel coordinates tend to cross near a single point between the two parallel axes. So the level of correlation can be visualized by means of the set of these segments (see Wegman's fig. 3, *ibidem*: 666).

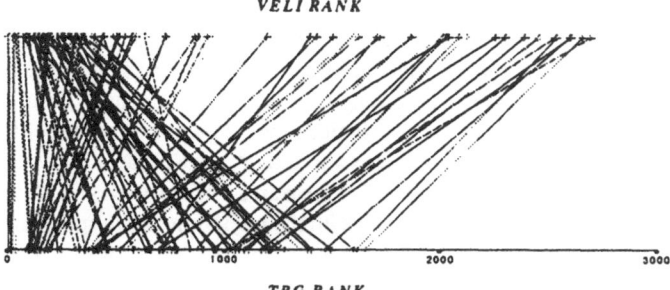

VELI RANK

TPG RANK

Examples of items, in the TPG, with Highest Positive   or   Negative Specificity

intendere = *to intend*	$z = + 35,1$	dire = *to say*	$z = - 28,2$
assicurare = *to assure*	$z = + 32,6$	fare = *to do*	$z = - 18,8$

Fig. 1a: *Comparison between TPG and VELI ranking of the 100 Verbs with the Highest Peculiarity in the TPG (either Positive or Negative intrinsic Specificity)*

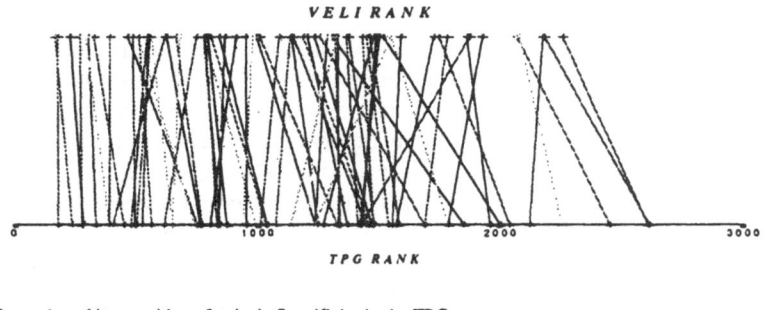

VELI RANK

TPG RANK

Examples of items without Intrinsic Specificity in the TPG

finanziare = *to finance*	$z = + 0.70$
valutare= *to evaluate*	$z = - 0,35$

Fig. 1b: *Comparison between TPG and VELI ranking examples of several Banal Verbs in the TPG (neither Positive or Negative Items)*

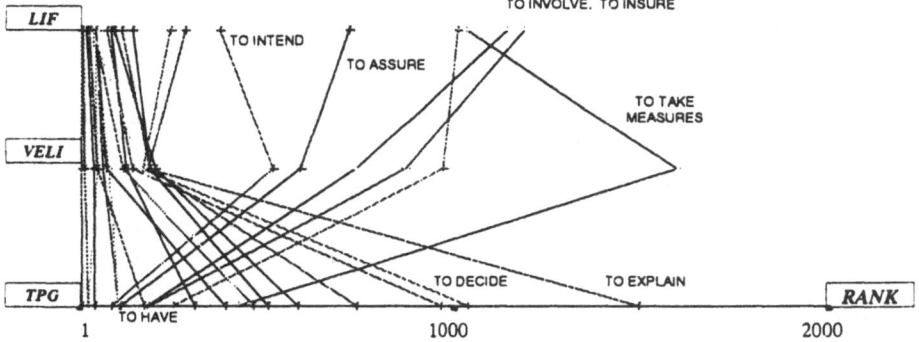

Fig. 1c: *LIF - VELI - TPG rank comparison. 15 most commonly used verbs in Italian and a selection of some highly peculiar Tpg verbs with positive or negative specificity*

Generally, only two dimensions are considered (fig. 1a, 1b), but it is possible to compare several (more than two) ranking lists from the related frequency dictionaries (fig. 1c).

Figures 1 illustrate the above selected verbs according to whether they occur more or less markedly in our Tpg corpus than in the Veli corpus. In fig. 1a we show the 50 verbs with the highest positive specificity, among these: <intendere>= to intend, <assicurare>= to assure, <impegnarsi>= to involve, <provvedere>= to take measures, <favorire>= to favour, <garantire>= to garantee; and also the other 50 verbs with the highest negative specificity in our Tpg. Among them, there are several most commonly used verbs like: <dire>= to say , <stare>= to stay, <fare>= to do, <vedere>= to see, <parlare>= to talk, <venire>= to come, but also <decidere>= to decide, <spiegare>= to explain, <andare>= to go. As you can see the criterion of negative specificity can clearly characterize certain words as "infrequent" words. In fact they are very relevant in their "rarity" (under-used or not so frequent) with respect to the chosen frequency dictionary, being consciously or unconsciously avoided by the writer or speaker. Also this selection of terms could be the subject of a study by itself.

In fig. 1b we show the group of words that are not specific, also called "banal", and could be discarded, because not so relevant as expressions of the context.

A further selection of items could be derived from the comparison of 3 ranking lists (Tpg - Veli - Lif). The figure 1c shows the first 15 most common verbs and some specific Tpg Verb, as Positive or Negative Items. From this illustration we can conclude that the most typical governmental verbs, among the Positive Items, are "to take measures" and "to intend". Conversely the most relevant among the negative ones, in comparison with Veli and Lif, are "to explain" and "to decide". Finally it is possibile to observe the situation of the same use, in the three dictionaries, of the verbs "to assure", "to involve", "to insure" as a set of high politic peculiarity due to their progressive ranking in the passage from the general language (Lif) to the sectorial one (Veli) up to the more specific one of government programs (Tpg).

3. How to solve problems of ambiguity

Regarding the two components, idiom and context, the corpus should be analysed at the level of *headwords (lemmas)* and therefore needs a lemmatization.

While with respect to the third component (situation) it is preferable to analyse the corpus in terms of *inflected forms* such as *graphical unlemmatized forms*, or, even better, through the choice of adequate units of analysis (like *lexias,* as linguists call them. The lexias is the minimal significant unit of meaning).

In general, if a whole sequence of words induces meaning (for example an idiomatic expression), it can be regarded as a single lexical item, and therefore as a single entry of vocabulary. If the frequency of the related forms composing the sequence is particularly high with respect to the chosen frequency dictionary, this reflects a highly peculiar terminology, and we can conclude that this segment is very representative and has an intrinsic specificity of its own in the corpus.

In all the above cases, the corpus vocabulary is both more precise and unambiguous. Moreover, it permits us to circumscribe the subsequent phases of lemmatization, that is disambiguation and fusion. A preliminary recognition of names, acronyms and polyforms shortens the lemmatization phase, especially from a semantic point of view. This requires the use of reference lexicons, such as a dictionary of locutions and of the principal support-verbs (Elia, 1995). The Institute of Linguistics at the University of Salerno has developed an integrated system of external lexical knowledge bases composed of the following inventories: one lexicon of over 110.000 simple entries - derived from a collection of 4 main dictionaries of the Italian language -, called DELAS; one lexicon of over 900.000 inflected simple forms, called DELAF; one lexicon of over 600.000 inflected polyforms, derived from 250.000 lexias, called DELAC. It is also available one dictionary of over 800.000 bilingual terms, called DEBIS. Elia's study show - for example - that in 13.790 simple forms there are 1.406 *polyrhematic* constructions (polyrhematic is a sequence of terms whose whole meaning is different from its elementary components), composed of 3.500 simple forms, equivalent to 25% of vocabulary. As we can see the density of polyrhematic forms is very high.

Therefore it could be very important to construct some *frequency dictionaries of polyforms*, in order to compare the corpus vocabulary of repeated segments (Salem, 1987) or, even better, of quasi-segments (Bécue, 1995), and select those sequences that are more significant. Up to now such frequency dictionaries are not available: an initial attempt to construct one is illustrated here in tab. 2, concerning the adverbial groups and other typical

Tab. 2: Example of Frequency Dictionary of Locutions derived from a collection of over 2 million occurrences (among a total of 250 locutions with occurrences > 30)

ITALIAN WORD	ENGLISH TRANSLATION	GEN Total	TPG Progr	TPG Repl	Other Corpora
DA PARTE	ON THE PART OF	855	227	368	260
IN MODO	IN THE WAY	853	309	288	256
IN ITALIA	IN ITALY	548	84	66	398
PER QUANTO RIGUARDA	WITH REGARDS TO	511	237	136	138
NON SOLO	NOT ONLY	477	176	119	182
IN PARTICOLARE	IN PARTICULAR	453	270	100	83
MA ANCHE	BUT ALSO	431	153	92	186
IN TERMINI	IN TERMS OF	429	92	94	243
DI FRONTE	IN FRONT OF	424	113	240	71
PER CUI	FOR WHICH	421	19	34	368
A LIVELLO	AT THE LEVEL	417	48	36	333
SI TRATTA	DEALS WITH	384	170	127	87
SUL PIANO	ON THE LEVEL OF	373	167	141	65
NELL'AMBITO	IN THE CONTEXT	368	149	132	87
NEI CONFRONTI	DEALING WITH	331	79	140	112
SEMPRE PIÙ	ALWAYS MORE	330	176	45	109
IN MATERIA	ON THE SUBJECT	321	143	160	18
NEL QUADRO	WITH THE REFERENCE TO	314	178	130	6
NEL SENSO	IN THE SENSE	297	27	35	235
IN CORSO	ON GOING	297	159	124	14
SULLA BASE	ON THE BASIS OF	277	153	102	22
PER QUANTO	IN AS FAR AS	273	61	37	175
NEL CAMPO	IN THE FIELD OF	273	107	76	90
PER ESEMPIO	FOR EXAMPLE	259	35	74	150
IN GRADO DI	ABLE TO	255	70	26	159
IN MANIERA	IN THE WAY	248	36	31	181
UNA VOLTA (da disambiguare)	ONCE, AT ONE TIME, ONCE UPON A TIME	248	35	48	165
AL FINE	IN ORDER TO	202	166	31	5

expressions. Preliminary matching with the corpus under study allows us to isolate the relevant parts of lexical items (either single or compound forms) and constitutes a valid system of text pre-categorization.

An additional possibility for this disambiguation emerges from the data. In every corpus it is possible to observe some equivalence of frequency - I call it *iso-frequency* - among the

474

inflected forms of the same adjectives or nouns. See in tab. 3 some examples of adjectives like economic, important and legislative.

Tab. 3: Examples of Iso-Frequency

--- *not ISO-FREQUENT NOUNS*			--- *ISO-FREQUENT ADJECTIVES*			
LEGGE (law)	(s) 622					
LEGGI (laws)	(p) 208	(DS = 0.33)	ECONOMICO	(ms)	315	(DS = 0,77)
			ECONOMICA	(fs)	461	
ECONOMIA (economy)	(s) 262		ECONOMICHE	(fp)	100	
ECONOMIE (economies)	(p) 35	(DS = 0.13)	ECONOMICI	(mp)	100	(DS = 1,00)
--- *ISO-FREQUENT NOUNS*						
			IMPORTANTE	(s)	117	
OBIETTIVO (purpose)	(s) 243		IMPORTANTI	(p)	116	(DS = 0,99)
OBIETTIVI (purposes)	(p) 286	(DS = 0,85)				
INTERESSE (interest)	(s) 193		LEGISLATIVO	(ms)	57	(DS = 0,84)
INTERESSI (interests)	(p) 178	(DS = 0,92)	LEGISLATIVA	(fs)	68	
			LEGISLATIVE	(fp)	53	
LIVELLI (levels)	(p) 110		LEGISLATIVI	mp	58	(DS = 0,91)
LIVELLO (level)	(s) 187-67=120	(DS = 0,91)				
	<a/livello> 48		LIBERA	(fs)	58	
	<al/livello> 19		LIBERO	(ms)	55	(DS = 0,95)
			LIBERE	(fp)	28	
FORZA (force)	(s) 105		LIBERI	(mp)	25	(DS = 0,89)
FORZE (forces)	(p) 259-166= 93	(DS = 0,88)				
	<forze politiche> 126		LOCALE (local)	(s)	80	
	<forze sociali> 40		LOCALI (local)	(p)	195-90 = 105	(DS = 0,89)
			<enti-locali> 90			

Legend: DS = occ A / occ B   with occ A < occ B

(s) singular   (p) plural   (ms) masculine and singular   (fs) feminine and singular
(mp) masculine and plural   (fp) feminine and plural

This iso-frequency can be the first clue to their equivalent use and meaning. On the contrary, in some cases, the *lack of iso-frequency* among the inflected forms of the same headword (Bolasco, 1993) suggests the need for disambiguation. In fact, this happens in presence of some compound forms, especially where the incidence of the occurrences of simple component forms is relevant. As you can see in words like <forza> (force) and <livello> (level). For example when we take away the frequency of the compound form of the word "level" (187) like "at (local) level" (48) and "at level of" (19), we return to the presence of iso-frequency (120) with the plural (110). As we will see later the differences among the inflected forms can be the clue to their different meanings. This should be verified by means of a bootstrapping approach.

4. Strategies for evaluating the lemmatization choices

For an optimal reconstruction of the main semantic axes of latent sense in a corpus we can use, as is well known, correspondence analysis (Lebart and Salem, 1994). Our objective, at this level, is to obtain stable representations. To assess the opportunities that both the disambiguations and fusions offer, we can test their significance by providing the factorial planes with confidence areas (Balbi, 1995). This assessment procedure is based on a bootstrapping strategy that generates a set of "word to subtext" frequency matrices. We assume Balbi's hypothesis which consists in generating a large number B of contingency tables by resampling, with replacement, into the original contingency table.

This set of bootstrapped matrices generates a three-way data structure; which could be analysed for example by means of a multiway technique, for constructing a reference matrix. A technique, such as STATIS, can be used, see Lavit (1988). In our example, in order to optimize computing time, the reference matrix is the average of these B matrices, due to the large dimensions of the original matrix (786 x 46) and of the number of bootstrapped matrices (B=200).

The stability of word points is graphically established by projecting them, as supplementary points, into the first factorial plane computed from a correspondence analysis of this reference matrix. Balbi proposes to use the non symmetrical correspondence analysis (ANSC). We have attempted this road but the results have not been comforting at level of interpretation. We believe that, in general, it is more opportune use the analysis of the simple correspondence analysis and only for special reasons the ANSC.The resulting clouds of points (for each word) constitute the empirical confidence areas, delimitated by a convex hull. The fig. 2 shows the convex hull regarding the word <way> and its locution <in/the/way>.

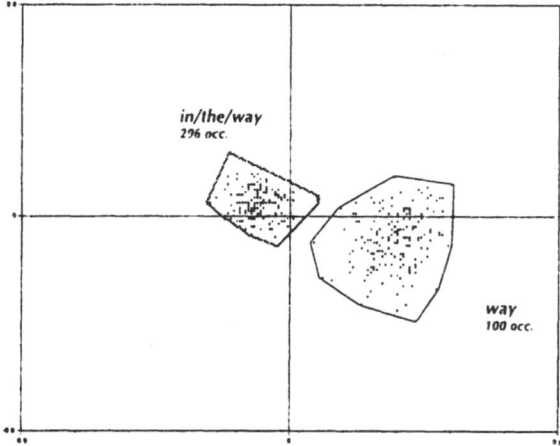

Fig. 2: Convex hulls of the locution IN THE WAY and of the word WAY

In practice, if two or more convex hulls do not overlap, disambiguation is absolutely necessary. See also, in fig. 3, the semantic disambiguation of the word <sviluppo> (development) in three different meanings: the first as "economic growth", the second as "progress" (in general political sense: social or civil), and the third as some "specific technological advance".

On the contrary, if the relative convex hull of different inflected forms or of some synonyms are (strongly) overlapping or included, their fusion is fruitful for the analysis.

476

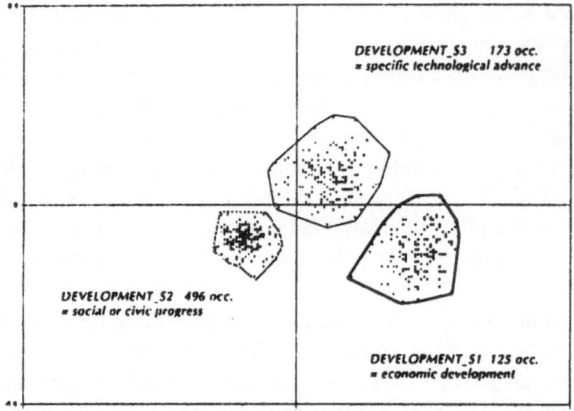

Fig. 3: Convex hulls of three different meanings of the word DEVELOPMENT (semantic disambiguation)

Let me give some examples concerning these situations. In fig. 4a "stato_verb" (equivalent to "been" in English) is clearly distant from "stato_noun" ("state"); but, conversely, the different unlemmatized forms (<stato/a/e/i>) of "stato_verb" (see fig. 4b) have their convex hulls completely overlapped and it is not important to distinguish them. Furthermore, if we look at the two meanings of "stato_noun" - they are further distinguished (fig. 4a). <Stato_s1> like state or nation and <stato_s2> like status or condition/situation (marital status, state of mind) have their relative convex hulls separated.

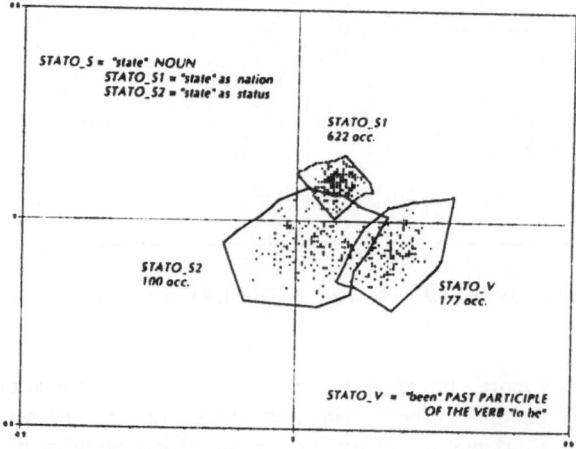

Fig. 4a: Convex hulls of the Italian word "STATO" after disambiguation

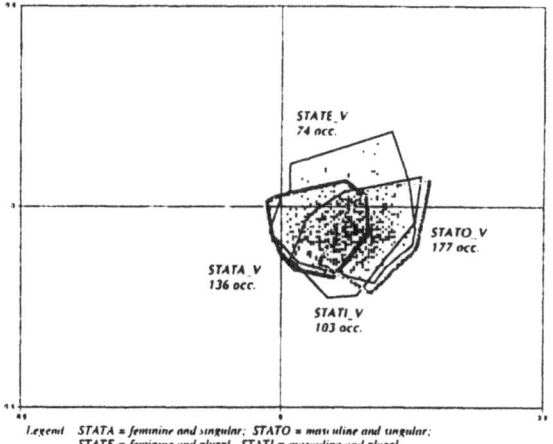

Fig. 4b: Convex hulls of Italian different infected forms of the past participle of the verb "to be"

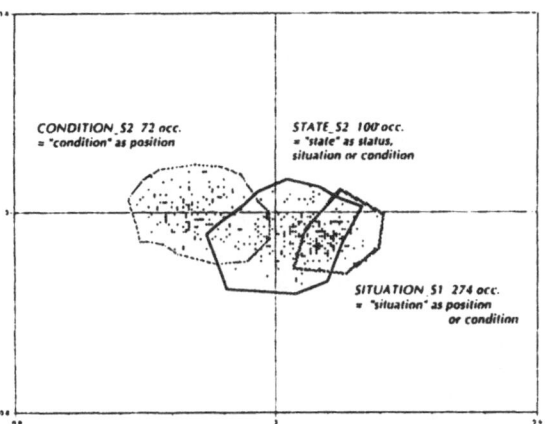

Fig 4c: Convex hulls of the synonym of "STATE" as "condition" or "situation"

In particular the latter does not overlap so much with the other synonyms such as "condition(_2)" and "situation(_1)", as you can see in fig. 4c. This shows how the use of these terms has changed over time in political discourse. Paying particular attention to fig. 4a, these words are always distant from state as the Italian State.

Now let me look at the significance and interpretation of convex hull sizes and positions, as shown in the following scheme:
1) a small convex hull, and therefore closeness of points, means high stability of representation but: a) when the points are located around the origin of the axes, it means evenness of these items in the various parts of the corpus, or b) when the points are in one particular quadrant of the plane, distant from the origin, it means the item is very characteristic and specific to some sub-set of the corpus. In this case, most of the time we obtain convex hulls not so small as above, because the factor scale of this region depends on the point distance from the origin (see the example of Politics in fig. 5);

478

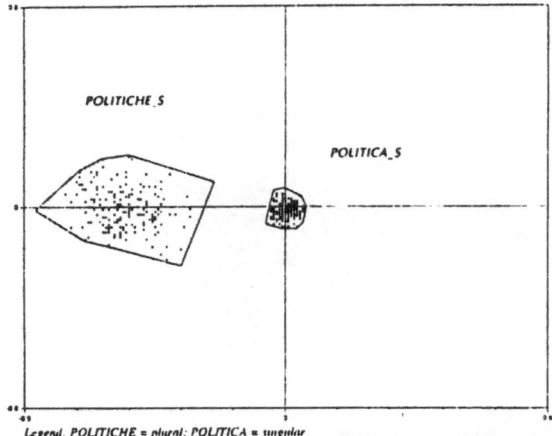

Legend: POLITICHE = plural; POLITICA = singular

Fig. 5: *Convex hulls of Italian inflected forms of the headword POLITICS*

2) a large convex hull, that is with a wide dispersion of points, means a not so strong stability of representation, and several different uses of this word in the corpus, but: a) if we do not have overlapping convex hulls, this means that the relative items have different meanings and that their fusion is not pertinent or, in other words, that their disambiguation is justified (see in fig. 4a the case of nation and status) or b) if, conversely, we have overlapping convex hulls, this means irrelevant disambiguation or justified fusion (*factual synonyms*).

In conclusion, having discussed how to identify the most significant part of the corpus and how to construct a more restricted and highly peculiar vocabulary composed of items with a high level of semantic quality, we can now finally proceed to an accurate and proper content multi-dimensional analysis, based on the above vocabulary, in which all the relevant units of analysis, which I have called "*textual forms*", are considered (Bolasco, 1993).
To this effect, such a vocabulary (see an example in tab. 4) will be composed of the items which are: 1) not banal with respect of some model of language (high intrinsic specificity or original terms); 2) significant as a minimal unit of meaning (*lexia*): either headwords (verbs and adjectives), or unlemmatized significant inflected forms (such as nouns in the plural with different meaning from the singular, i.e. forza/forze), or more frequent typical locutions and other idiomatic expressions (phrasal verbs and nominal groups).

References:

Balbi, S. (1995): Non symmetrical correspondence analysis of textual data and confidence regions for graphical forms. In: *JADT 1995 Analisi statistica dei dati testuali,* Bolasco, S. et al. (eds.), II, 5-12, CISU, Roma

Bécue, M. et Haeusler, L. (1995): Vers une post-codification automatique In: *JADT 1995 Analisi statistica dei dati testuali,* Bolasco, S. et al. (eds.), I, 35-42, CISU, Roma

Bolasco, S. (1993): *Choix de lemmatisation en vue de reconstructions syntagmatiques du texte par l'analyse des correspondances.* Proc. JADT 1993, 399-410, ENST-Telecom, Paris

Bolasco, S. (1994): *L'individuazione di forme testuali per lo studio statistico dei testi con tecniche di analisi multidimensionale*. Atti della XXXVII Riunione Scientifica della S.I.S., II, 95-103, CISU, Roma

Bortolini N., Tagliavini C., Zampolli A. (1971): *Lessico di frequenza della lingua italiana contemporanea*. Garzanti., Milano.

Dubois, J. et al. (1979): *Dizionario di Linguistica*, Bologna: Zanichelli

Elia, A. (1995): Per una disambiguazione semi-automatica di sintagmi composti: i dizionari elettronici lessico-grammaticali. In: *Ricerca Qualitativa e Computer*, Cipriani, R. e Bolasco, S. (eds.), 112-141, Franco Angeli, Milano

Cipriani, R. e Bolasco, S., eds. (1995): *Ricerca Qualitativa e Computer*. Franco Angeli, Milano

Lavit, Ch. (1988): *Analyse conjointe de tableaux quantitatifs*. Masson, Paris

Lebart, L. et Salem, A. (1994): *Statistique textuelle*. Dunod, Paris

Lyne A. A. (1985): *The vocabulary of french business correspondence*, Slatkine-Champion, Paris

Salem, A. (1987): *Pratique des segments répétés. Essai de statistique textuelle*. Klincksieck, Paris

Wegman, E. J. (1990): *Hyperdimensional Data Analysis Using Parallel Coordinates* JASA, **85**, 411, 664-675

Clustering of Texts using Semantic Graphs. Application to Open-ended Questions in Surveys

Monica Bécue Bertaut[1] and Ludovic Lebart[2]

[1] Universitat Politecnica de Catalunya.
Pau Gargallo,5
08028 Barcelona, Spain.

[2] Centre National de la Recherche Scientifique.
ENST, 46 rue Barrault 75013 Paris. France

Summary: A methodology for the automatic classification of short texts is proposed (leading cases are responses to open-ended questions in sample surveys, titles or abstracts of papers in documentary data bases). It aims to take into account a graph structure on the variables (elementary text units). This graph could be a semantic graph provided by an external source, or a co-occurrence graph, built from the corpus itself.

1. The basic problem

The starting point is to consider each text as described by its lexical profile, i.e. by a vector that contains the frequency of all the selected units in the text. The units could be words (or lemmas, or types), segments (sequences of words appearing with a certain frequency) or quasi-segments (segments allowing non-contiguous units). The corpus is represented by a (n,p) matrix X whose i-th row (i.e.: i-th respondent) contains vector whose p components are the frequencies of units (words).

However, an usual classification algorithm applied to the rows of X could lead to disappointing or misleading results.

1) the matrix X could be very sparse, many rows could have no element at all in common (in terms of responses to open-ended questions in sample surveys, this means that two answers may contain distinct words).

2) a wealth of meta-data is available and needs to be utilized (syntactic relationships, semantic networks, external corpus and lexicons, etc.).

To make more meaningful the distances between the lexical profiles of the texts we can add new variables obtained from a morpho-syntactic analyzer; it is then possible to tag the text units depending on their category. It is then possible to complement the p-vector associated to a response or a text with new components (Salem, 1995). It is also useful to take into account the available semantic information about units, information which can be stored in a "semantic weighted graph". From now on, we will focus on this latter issue and discuss the different ways of deriving such a graph from the data themselves.

Note that most algorithms of constrained classification (see for example Gordon (1996) for a recent review of hierarchical classification including this topic) involve a graph structure upon the set of *objects* to be classified. We are dealing here with a quite different situation: it is the set of *descriptors* (words) of the objects (texts) which is provided with a graph structure. This structure will be used to modify the distances between pairs of objects.

2. Semantic graph and contiguity analysis

There is no universal rule to establish that two words are semantically equivalent, despite the existence of synonymy dictionaries and thesaurus. In particular, in some sociological studies, it could be simplistic to consider that two different words (or expressions) have the same meaning for different categories of respondents. But it is clear that some units having a common extra-textual reference are used to designate the same "object".

Whereas the syntactic meta information can provide the user with new variables, the semantic information defined over the pairs of statistical units (words, lemmas) is described by a graph that can lead to a specific metric structure (see fig. 1).

a) The semantic graph can be constructed from an external source of information (a dictionary of synonyms, a thesaurus, for instance). In such a case, a preliminary lemmatization of the text must be performed.

b) It can be built up according to the associations observed in a separate (external) corpus.

c) Eventually, the semantic graph can also be extracted from the corpus itself. In this latter case, the similarity between two words (or other units) is derived from the proximity between their distributions (lexical profiles) within the corpus.

The vertices of this weighted undirected graph are the distinct units (words) j, (j=1,…, p). The edge (j, j') exists iff there is some non-zero similarity s(j, j') between j and j'. The weighted associated matrix $M= (m_{jj'})$, of order (p, p) associated to this graph, contains in the line j and column j' the weight s(j, j') of the edge (j, j'), or the value 0 if there is no edge between j and j'.

The repeated presence of a pair of words within a same sentence of a text is a relevant feature of a corpus. The words can help to disambiguate each other (see e.g.: Lewis and Croft, 1992). To take into account co-occurrence relationships allows one to use words in their most frequent contexts. In particular, we can also describe the most relevant co-occurrences by using a weighted undirected complete graph linking lexical units. Each pair of units are joined by an edge weighted by a co-occurrence intensity index.

At that stage, we find ourselves within the scope of a series of descriptive approaches working simultaneously with units and pairs of units (see for example Art and al., 1982), including contiguity analysis or local analysis (see: Lebart, 1969; Aluja and Lebart, 1984; Escofier, 1989; Cazes and Moreau, 1991).

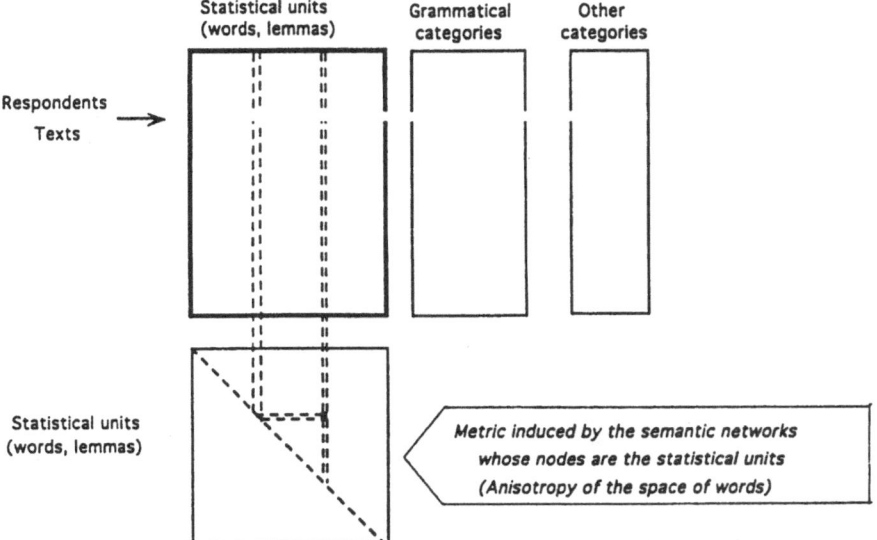

Fig. 1: How to use meta-data in text classification: metric, or new variables?

These visualization techniques are designed to modify the classical methods based on Singular Value Decomposition by taking into account a graph structure over the entries (row or/and column) of the data table. Visualizing the proximities using contiguity analysis is equivalent to performing a projection pursuit algorithm as described in Burtschy and Lebart (1991).

The classification can then be performed either :

1) by using as input data the principal coordinates issued from these contiguity analyses,

or :

2) by computing a new similarity index between texts. This new index is built from generalized lexical profiles (i.e.: original profiles complemented with weighted units that are neighbors (contiguous) in the semantic graph).

3. Different types of associated graphs

3.1 Case of an external semantic graph

A simple way to take into account the semantic neighbors leads to the transformation:

$$Y = X(I + \alpha M), \qquad (1)$$

where M is the matrix associated with the graph defined previously, and a a scalar allowing to calibrate the importance given to the semantic neighborhood.

This is equivalent to providing the p-dimensional space of words with a metric defined by the matrix:

$$Q = (I + \alpha M)^2$$

This leads immediately to a new similarity index that can be used for the classification of the rows. Due to the size of the involved data tables, the classification is often performed using distances computed with the r principal coordinates (a current order of magnitude is $r = 100$, whilst $p = 1500$).

The metric:

$$Q^* = (I - \alpha M)^{-2},$$

closely related to the previous one if α is small, has also some interesting properties.

3.2 Case of an internal co-occurrence graph: self learning.

A possible matrix M could be:

$$M = C - I$$

where C is the correlation matrix between words (allowing negative weights, which correspond to a negative co-occurrence intensity measure); if the columns of C have their variances equal to 1:

$$C = \frac{1}{n} X^T (I - \frac{1}{n} U) X, \text{ with } u_{ij} = 1 \text{ for all i and j}. \qquad (2)$$

Since:

$$Y = X (I + \alpha(C-I)),$$

the matrix S to be diagonalized when performing a Principal Component Analysis of Y, reads:

$$S = \frac{1}{n} Y^T (I - \frac{1}{n} U) Y \ = (1 - \alpha)^2 C + 2\alpha(1 - \alpha)C^2 + \alpha^2 C^3 \quad (3)$$

Therefore, the eigenvectors of S are the same as the eigenvectors of C. However, to an eigenvalue λ of C corresponds the eigenvalue μ of S such that:

$$\mu = (1 - \alpha)^2 \lambda + 2\alpha(1 - \alpha)\lambda^2 + \alpha^2 \lambda^3$$

The effect of the new metric is simply to re-weight the principal coordinates when recomputing the distances to perform the classification.

If $\alpha = 1$, for instance, we get λ^3 instead of λ, thus the relative importance of the first eigenvalues is strongly increased.

Such properties contribute to shed light on the prominent role played by the first principal axes, particularly in the techniques of *Latent Semantic Analysis* used in Automatic Information Retrieval: see Furnas and al. (1988).

Construction of the matrix M *through a hierarchical classification of words*

Another way of deriving a matrix M from the data themselves is to perform a hierarchical classification of words, and to cut the dendrogram at a low level of the index. It can either provides a graph associated with a partition, or a more general weighted graph if the nested set of partitions (corresponding to the lower values of the index) is taken into account.

The approaches of Salton and Mc Gill (1983), Celeux et al. (1991), Iwayama and Tokunaga (1995), in the framework of discriminant analysis (also named in this context "Text Categorization"), are similar to the case of a graph associated with a partition: to remedy the sparsity of the matrix X (leading generally to ill-conditionned problems of discrimination), words below a certain threshold of distance are aggregated.

However, in our context, it would be awkward to confine ourselves to semantic graphs having a partition structure, since the relation of semantic similarity is not transitive (consider for example the semantic chain: *fact - feature - aspect - appearance - illusion*).

It is more appropriate to assign to each pair of words (i, j) a value:

$$m_{ij} = f(d(i, j)), \quad \text{for} \ d(i, j) \leq t_0,$$

f(t) being a decreasing function of t, d(i, j) being the index corresponding to the smallest set containing both elements i and j, and:

$$m_{ij} = 0, \quad \text{for} \ d(i, j) > t_0.$$

We could use the initial distance instead of the ultrametric associated with the hierarchy, but it is convenient to obtain as a by-product the graphical representation induced by the ultrametric.

3.3 Internal co-occurrence graph after aggregation: another self learning procedure

The preceding case of self learning entails several serious problems. To derive semantic information from frequency distributions (see Harris (1954), Church and Hanks (1990)) becomes a fruitful operation only if the frequency profiles are defined by using a substantial amount of occurrences.

We define as *substitution relationship* between words the situation occurring in the following two circumstances:

    i) they have a similar context distribution

    ii) they don't appear (or seldom appear) in a same units string

More concretely, we may use a correspondence analysis (and/or a hierarchical clustering) of the (q, p) matrix $\mathbf{A}$, crossing aggregated responses and lexical units, to determine the distribution similarities.

Let us call $\mathbf{Z}$ the (n,q) binary matrix describing an *a priori* partition of the set of respondents (such that $z_{ij} = 1$ if respondent i belongs to category j, $z_{ij} = 0$ elsewhere).

We have in such a case: $\qquad\qquad \mathbf{A} = \mathbf{Z}^T\mathbf{X} \qquad$ ($\mathbf{Z}^T$ is the transpose of $\mathbf{Z}$)

Then, we could exclude those units that appear within a same string through a search for multi co-occurrences. We will see in the example below that the grouping of responses may have a too strong influence on the proximities between words. It appears to be more efficient to stack several contingency tables corresponding to various criteria (in a sample survey: *sex, age, region, status, occupation, education level,...*).

Note that the methods 3.2 and 3.3 do not provide semantic networks in the usual sense of the term. However, they have the advantage of being independent of the language, and of the context of the study as well. The "local network" obtained can be confronted to the external semantic network that can be available.

4. Example

This methodology has been applied to a corpus of 1563 answers to an open-ended question included in a sociological survey. We confine ourselves here to discussing the choice of the semantic graph that has been used to improve the classification of the respondents.

The following open-ended question was asked in a multinational survey conducted in seven countries (Japan, USA, United Kingdom, Germany, France, Italy and Netherlands) in the late nineteen eighties (Hayashi et al., 1992): *"What is the single most important thing in life for you? "* It was followed by the probe: *"What other things are very important to you?"*. Our illustrative example is limited to the American sample (Sample size: *1563*). Some aspects of this multinational survey concerning *general social attitudes* are described in: Sasaki and Susuki (1989).

Examples of answers to the first question were:

1 - *Family, being together as a family*

2 - *Mother, money, peace of mind, peace in the world*

Some words are connected to a single lemma (or dictionary word) (*be, is, are, being*). Also to be noted is the strong presence of function words (*a, and, for, of, the*). Note that the concept of a function word (sometimes referred to as empty word, or tool word or grammatical word in information retrieval) is widely used by text researchers. These words are obviously excluded from the semantic networks.

This corpus has a length of 13 999 occurrences and contains 1378 distinct words. Only the 126 words used at least 16 times are kept, the total length being then reduced to 10 752 occurrences.

Figure 2 shows three branches of the dendrogram of words obtained through a direct classification of the columns of $\mathbf{X}$ performed on the 15 first principal axes of a Correspondence Analysis of the sparse contingency table $\mathbf{X}$. We observe grouping of words according to their meanings (*home* and members of family, standard of living, together with function words (*on*) or pronouns (*my*), and repeated and isolated segments (*don't know*).

We note that some topics are characteristic of the produced clusters. These topics are less salient when a second dendrogram is computed from the 45 first principal axes.

If we cut both dendrograms at a level producing 30 classes, we obtain a ratio of variance (between classes variance divided by total variance) of 0.87 (case of 15 axes) and 0.64 (case of 45 axes). It is a further empirical proof of the ability of the first axes to gather structural features, ability already mentioned in section 3.2.

```
Index   Words

  .87   grandchildren  --------+                                                |
  .16   husband        ---+    |                                                |
 2.29   children       ---*----*---------+                                      |
  .25   kids           ---+              |                                      |
  .74   wife           ---*---+          |                                      |
  .25   home           ---+   |          |                                      |
15.61   my             ---*---*----------*------------------------*--------*--///-
                                                                                |
  .73   money          -------+                                                 |
  .14   on             --+    |                                                 |
  .29   live           --*+   |                                                 |
  .24   comfortably    ---*   |                                                 |
 6.90   enough         --*---*--------------------------------------------+     |
  .06   don't          --+                                                      |
-----   know           --*----------------------------------------------------///--
```

Fig. 2: Parts of the dendrogram obtained through a direct clustering of the columns of **X**

Fig.3 presents some similar branches of the dendrogram obtained through clustering of the columns of the aggregated table **A = Z^T X**. **Z** corresponds to a partition into 6 classes obtained by crossing the two nominal variables *sex*, and *age* (3 categories). The themes observed previously are now disseminated into various groups strongly influenced by the criteria of grouping.

```
Index   Words

 5.08   grandchildren  -----------------------------------+
 2.15   husband        ---------------+                   |
  .04   self           --+            |                   |
  .07   religion       --*            |                   |
  .28   peace          --*+           |                   |
  .10   mind           --+|           |                   |
  .80   in             --**--+        |                   |
 3.80   children       ------*--------*---------+         |
  .08   people         --+                      |         |
  .37   as             --*-+                     |         |
 1.49   health         ----*-------+            |         |
  .35   church         ----+       |            |         |
  .14   welfare        --+ |       |            |         |
  .04   our            --* |       |            |         |
-----   world          --*-*------*------------*------*---///---
```

Fig. 3: Parts of the dendrogram obtained through clustering of the columns of the aggregated table **A = Z^T X**.(**Z** corresponds to a partition into 6 classes according to *sex* and *age)*.

The category *"female-over 55 years"* being particular and homogeneous has strongly influenced the proximities between words. We do not find again *wife* (used by men), *kids* (used by younger people, mostly by men), *home* (used by younger people), etc..

Obviously, a composite partition (crossing or stacking various criteria) should be chosen instead of the nominal variable *"sex-age"* used to obtain higher frequencies of words. The quality of the partition of responses depends on the homogeneity of the classes and their interpretability. In such a context, only a group of experts could assess the obtained results. The use of a co-occurrence graph (with: $\alpha = 1$) issued from a direct classification of the columns of **X** enables a better classification of the responses that have a poor lexical profile, and leads to more meaningful grouping.

It is clear that we are dealing with experimental statistics: the value of the parameter α, the number of neighbours to be taken into account could probably vary according to the field of application. Reliable tools for assessing and comparing partitions are all the more needed.

5. References

Aluja Banet, T., Lebart, L. (1984): Local and Partial Principal Component Analysis and Correspondence Analysis. *COMPSTAT Proceedings*, 113-118, Physica Verlag, Vienna.

Art, D., Gnanadesikan, R., and Kettenring, J.R. (1982): Data Based Metrics for Cluster Analysis. *Utilitas Mathematica*, **21** A, 75-99.

Bécue, M. (1991): *Analisis de Datos Textuales. Metodos Estadisticos y Algoritmos.* CISIA, Paris.

Burtschy, B., Lebart, L. (1991): Contiguity analysis and projection pursuit, in : *Applied Stochastic Models and Data Analysis,* Gutierrez R. and Valderrama M.J., (eds), World scientific, Singapore, 117-128.

Cazes, P., Moreau, J. (1991): Analysis of a contingency table in which the rows and columns have a graph structure. in : *Symbolic and Numeric Data Analysis and Learning,* Diday E., and Lechevallier Y. (eds), 271-280, Novascience publisher, New York.

Celeux, G., Hebrail, G., Mkhadri, A., Suchard, M. (1991): Reduction of a large scale and ill-conditionned problem on textual data. in: *Applied Stochastic Model and Data Analysis,* Gutierrez R. and Valerrama N., J. (eds.), World Scientific, Singapore, 129-137.

Church, K. W., Hanks, P. (1990): Words association norms, mutual information and lexicography. *Computational Linguistics,* **16**, 22-29.

Escofier, B. (1989): Multiple correspondence analysis and neighboring relation *Data Analysis Learning Symbolic and Numeric knowledge,* Diday E. (eds), 55-62, Novascience publisher, New York.

Furnas, G. W. et al. (1988): Information retrieval using a singular value decomposition model of latent semantic structure. *Proceedings of the 14th ACM Conference on R. and D. in Information Retrieval,* 465-480.

Gordon, A.D. (1996): Hierarchical Classification. in: *Clustering and Classification.* P. Arabie, L. J. Hubert, G. De Soete (eds.) World Scientific, River Edge, NJ.

Harris, Z. S. (1954): Distributional Structure. *Word*, **2-3**, 146-162.

Hayashi, C., Suzuki, T., Sasaki, M. (1992): *Data Analysis for Social Comparative Research : International Perspective*. North-Holland, Amsterdam.

Iwayama, M., Tokunaga, T. (1995): Cluster-based text categorization: a comparison of category search strategies. in: *ACM / SIGIR'95*, (Fox E.A., Ingwersen P., Fidel R., eds), Seattle, WA, USA, 273-280.

Lebart, L. (1969): Analyse Statistique de la contiguité. *Publication de l'ISUP*, **28**, 81-112.

Lebart, L., Salem, A. (1994): *Statistique Textuelle*. Dunod, Paris.

Lebart, L., Salem, A., Berry, E. (1991): Recent development in the statistical processing of textual data, *Applied Stoch. Model and Data Analysis*, **7**, 47-62.

Lewis, D. D., Croft, W. (1990): Term clustering of syntactic phrases. *SIGIR- 90*,. 385-404.

Salem, A. (1995): Les unités lexicométriques. *Analisi Statistica dei Dati Testuali*, Bolasco et al. (eds), 19-27, CISU, Roma.

Salton, G., Mc Gill, M.J. (1983): *Introduction to Modern Information Retrieval*, International Student Edition.

Sasaki, M., Suzuki, T. (1989): New directions in the study of general social attitudes: trends and cross-national perspectives, *Behaviormetrika*, **26**, 9-30.

How to find the nearest by evaluating only few? Clustering techniques used to improve the efficiency of an Information Retrieval system based on Distributional Semantics

Martin Rajman and Arnon Rungsawang

ENST-Paris, Department of Computer Science
46 Rue Barrault, F-75634 Paris Cedex 13, France

Summary: The first objective of this contribution is to give a description of our textual information retrieval system based on distributional semantics. The central idea of the approach is to represent the retrievable units and the user queries in a unified way as projections in a vector space of pertinent terms. The projections are derived from a co-occurrence matrix computed on large reference (textual) corpora collecting the distributional semantic information. A similarity computation based on the cosine measure is then used to characterize the semantic proximity between queries and documents.

Retrieval effectiveness can be further improved by the use of relevance feedback techniques. A simple feedback method where document relevance is interactively integrated to the original query will also be presented and evaluated.

Although our first experiments lead to quite promising results, one major drawback of our IR system in its original form is that the satisfaction of a query requires the evaluation of the similarities between that query and *all* the documents in the textual base. Therefore, the second objective of this contribution is to investigate how clustering techniques can be applied to the textual database in order to retrieve the documents satisfying a query through a *partial* exploration of the base. A tentative solution based on hierarchical clustering will be suggested.

1. Introduction

Information Retrieval (IR) research is concerned with the analysis, the representation, and the searching of heterogeneous textual databases with wide varieties of vocabularies and unrestricted subject matters. Examples of such databases, the elements of which will be called hereafter *documents*, are databases containing newspaper articles, newswires, technical or scientific articles, magazines, encyclopedia entries and so on. Due to the enormous amount of information currently available on-line in the different computer networks (Internet, ...) and in the library environments, simple keyword search and browsing are not sufficient anymore. IR users need more sophisticated tools to help them to reach the relevant information.

Our retrieval model exploits co-occurrence properties of words to determine whether queries and texts are semantically related (*distributional semantics*). More precisely, documents and queries are represented in a unified way as projections of co-occurrence profile vectors in a multidimensional vector space of selected informative terms, in which the proximity is interpreted as semantic similarity. These co-occurrence profile vectors are derived from co-occurrence matrices, computed on large reference textual corpora. The cosine similarity measure is used to characterize the proximity and thus the relevance between user queries and documents in the textual database.

2. Distributional Semantics

Using distributional information for automatic extraction of general morphologic, syntactic or semantic properties of a given language has already been considered by several re-searchers (Schütze (1992), Gallant et al. (1992), Rungsawang & Rajman (1995)). Such properties correspond to observable regularities (frequency, distribution, co-frequency, ...) in large textual corpora. In our approach, we use a co-occurrence matrix $(cf_{ij})_{i,j}$ to fetch the semantic information by automatic co-occurrence computation on a large reference cor-pus of texts. The lines of such a matrix correspond to *all distinct terms* w_i found in the reference corpus and the columns correspond to selected informative terms t_j, called *perti-nent terms*, used to represent the meaning of all other terms in the textual database. Each element cf_{ij} records how often a term w_i *co-occurs* with a term t_j within some pre-defined textual units (e.g. sentences or paragraphs) in the reference corpus.

To build a co-occurrence matrix, several elements must be determined. First, the nature of the primary linguistic units which will be used as terms has to be defined. Tokens, as pro-duced by the simple stemmer (e.g. Stopper and Porter stemmer, (Frakes and Baeza-Yates 1992)), or words reduced to their radical forms (i.e. conjugated verbs to infinitives, nouns to singular forms) are frequently used. Words with their part-of-speech tags, for example, produced by a natural language tagger may also be considered.

Then, we need to determine the sets of terms and pertinent terms that define the rows and the columns of the co-occurrence matrix. To cover the maximum semantic informa-tion, all distinct terms (except perhaps the functional words) appearing in the reference corpus should be used. However, feasibility constraints have to be taken into consideration. Salton et al. (1975, 1976) indicate that terms which have the document frequency (i.e. the proportion of documents in which they appear) ranging between 1/100 and 1/10, possess good content discrimination in the document space and yield good retrieval effectiveness. Therefore, we decided to reduce the w-dimension of the matrix to terms appearing in at least 2 documents, and to use Salton's criterion for the pertinent terms in the t-dimension.

The third element to define is the textual unit in which the co-occurrences will be computed. Usually, sentences, paragraphs, fixed-size word windows or fixed-size character windows are chosen.

Once the co-occurrence matrix is built, a *distributional semantic hypothesis* is assumed pos-tulating a correlation between terms that co-occur in similar distributional environments. The semantics of a term w_i is then represented by its *co-occurrence profile vector*, the row corresponding to term w_i in the co-occurrence matrix. The geometric proximity between the co-occurrence profile vectors is interpreted as an indication of the semantic similarity between the corresponding terms, provided that the reference corpus is large enough to cover sufficient semantic information. The geometric proximity between these vectors is measured by the cosine value of the angle between them.

3. Document and Query Representation

In our model, the retrieval of document passages of varying size has been considered for the definition of document and query representation. Therefore, we consider that each docu-ment may be decomposed into several *retrievable text units* (RTUs). The RTU definition (i.e. document decomposition) is strongly related to the notion of a *text excerpt*, which can be defined as a varying-size piece of a document covering an amount of information that the user considers as a satisfactory answer to his/her information need. In addition to this, we do not formally distinguish between documents and queries. In our preliminary experiments, we simply define the RTUs as paragraphs and sections.

According to the distributional semantic hypothesis previously mentioned, we represent each RTU by the corresponding *indexing structure* (IS) vector. This IS vector is currently defined as the mean vector of the co-occurrence profile vectors associated with the terms contained in the RTU.

Since documents and queries are represented in a unified way in a common multi-dimensional

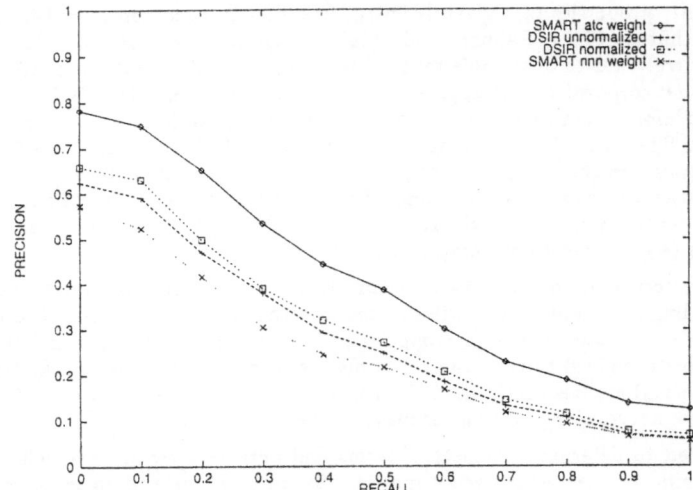

Figure 1: 11 points Recall-Precision curves (Cranfield standard test collection)

vector space, the document-query relevance can be defined on the basis of the proximity between the average IS vector representing the documents and the IS vector representing the query. We currently measure the proximity by the cosine of the angle between the IS vectors.

A benefit that we expect from our approach is that any two documents may have a high similarity score in the multi-dimensional space of pertinent terms even though they only have a few terms in common. For example, one document might contain the words "*corpus-based* linguistic *analysis*", whereas the other might contain words "*computational* linguistics". If the emphasized terms globally occur in the same distributional environment in the corpus that was taken as reference, the resulting IS vectors should correspond to similar directions in the multi-dimensional space of pertinent terms.

4. Preliminary Experiments

In the first phase of this ongoing research, we have implemented (in C and Perl languages on a SPARC workstation) a prototype corresponding to the system described in the previous section. With this prototype, we have conducted several experiments using the Cranfield standard test collection. The Cranfield collection is a test collection of 1400 documents and 225 queries in the field of aerodynamics which also contains, for each query, the list of the relevant documents. The collection is available in the SMART version 11.0 distribution [1], and has been used for several years to test many retrieval algorithms.

The 11 points Recall-Precision curves (Salton and McGill (1983)) comparing our system (denoted DSIR[2]) with the ones obtained with the standard version of the system SMART with term-frequency weights (denoted SMART nnn weight) and augmented inverse-document-frequency weights (denoted SMART atc weight) are given in figure 1.

---

[1] This contribution can be obtained from "ftp.cs.cornell.edu".

[2] Distributional Semantic Information Retrieval.

5. Experiments with Relevance Feedback

Relevance feedback is a process used to build improved queries especially through interaction with the user. It is based on the use of relevance information that the users can associate with previously retrieved documents. In a standard feedback process, the user indicates the relevance (or non-relevance) of either terms or documents to create new queries. In our experiments, IS vectors of previously retrieved (and user-evaluated) documents are merged with the original query vector to form the new one. More precisely, new feedback queries of rank k are made up by adding to the original query the IS vectors of all the relevant documents selected among the k first previously retrieved documents (ordered by decreasing similarity with the original query) and by subtracting from it the IS vector of the first non relevant document (among the k):

$$\vec{Q}^{new} = \vec{Q}^{old} + \beta \sum_{i=1}^{k} \delta_i \vec{IS}_i - \gamma \vec{IS}_{i1} \tag{1}$$

where $\vec{Q}^{new}$ is the new query vector, $\vec{Q}^{old}$ the original query vector. $\vec{IS}_i$ are the IS vectors of the k first previously retrieved documents. δ_i equals 1 when $\vec{IS}_i$ is the IS vector of a relevant document (and 0 otherwise), and $\vec{IS}_{i1}$ the IS vectors of the first non-relevant documents. β and γ are importance respectively given to relevant and non-relevant components. To eliminate the problem of residual (or ranking) effect (Hull (1993)), we created the feedback query vectors from one collection of documents and apply them to another one (see experimental setup below). Settings $\beta = 0.75$ and $\gamma = 0.25$ yield the best results in our experiments.

5.1 Document collection used in feedback experiments

The document base used in our feedback experiments was the OHSUMED collection, contributed by the Oregon Health Sciences University (Hersh (1994)). The OHSUMED collection contains all MEDLINE references from 270 journals, spanning from year 1987 through 1991. It also includes 106 queries with the associated relevance judgments. Document relevance was judged on a three-point scale: definitely relevant, possibly relevant, and not relevant. We have chosen to consider as relevant the documents judged as either definitely or possibly relevant. To cope with our limited computational resources, we only used a subset of the OHSUMED collection. The reduced collection was created as follows.

We extracted a first sub-collection, called the B1 sub-collection, consisting of the OHSUMED documents spanning from 1987 through 1989 (22040 documents, $\approx$ 24 M bytes) along with their relevant information. The second sub-collection, called the B2 sub-collection, contains the OHSUMED documents spanning from 1990 through 1991 (19950 documents, $\approx$ 24 M bytes) also with corresponding relevance judgment data.

For both sub-collections, we used the B1 sub-collection as reference corpus for co-occurrence matrix calculation. The document pre-processing consisted of removing the functional words by the Stopper algorithm, and normalizing the morphological variants to their radical forms by the Stemmer algorithm. We used the algorithms available from Frakes' contribution (Frakes and Baeza-Yates 1992)). Terms that only appeared in one document in the collection were removed. To limit the second dimension of the co-occurrence matrix, a simple frequency-based filtering technique was used to select pertinent attributes.

5.2 Our experimental setup

Our experimental set up follows the diagram shown in figure 2. Starting from the upper left portion of the diagram (arrows labeled with (1)), we first calculate the co-occurrence matrix (DSCOOCC program) using data from the B1 sub-collection. Then, the co-occurrence data and the B1 sub-collection are used to create (2) the B1 document IS vectors (DS-BASE program). Afterward, the DSSEARCH program retrieves (3) from the B1 collection

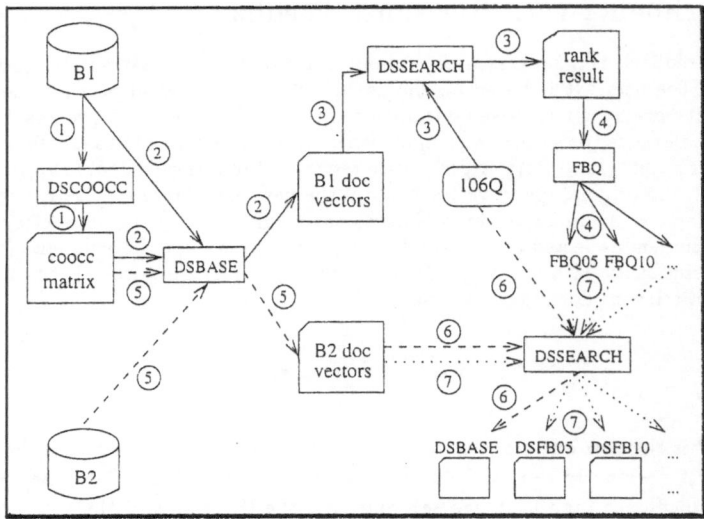

Figure 2: Feedback diagram.

the relevant documents corresponding to the original 106 queries, and present the result in a form of an ordered list, where the first document (rank 1) is the most similar to the query. Then (4), the FBQ program creates a new query based on the old one, the relevant documents (as evaluated by the user) and the first non-relevant document from rank 1 to rank 5 (FBQ05), rank 1 to rank 10 (FBQ10), etc.

Continuing at the lower portion of the diagram, the DSBASE program will create (5) the B2 document IS vectors, using previously calculated co-occurrence data and the B2 sub-collection. Then (6), the B2 document IS vectors and the original 106 queries are used by the DSSEARCH program to produce the DSBASE (result corresponding to the original queries) that will be used as a reference to evaluate the retrieval with feedback. Finally (7), the DSSEARCH program retrieves from the B2 document collection the relevant document sets (DSFB05, DSFB10, etc.) in response to the feedback queries FBQ05, FBQ10, etc.

5.3 The experimental evaluation and discussion

Our relevance feedback experiments were evaluated on the base of the standard 11 points recall-precision (RP) curves (Salton & McGill (1983)). We also conducted (on the B1 and B2 sub-collections) comparative retrieval experiments with the system SMART (version 11.0). The document-query weighting "ann-atn" was used for all SMART experiments (Hersh & Buckley(1994)).

The two curves shown in figure 3, and the first two columns in table 1 indicate the retrieval performance obtained by our system (DSIR) and the system SMART in the case of no relevance feedback on the B1 sub-collection.

The five curves given in figure 4, and the corresponding five columns in table 1 illustrate the results obtained by the system SMART (without relevance feedback) and our DSIR system (DSBASE: without relevance feedback, DSF05: feedback with documents under rank 5, etc.) on the B2 sub-collection. The curves (DSBASE vs. SMART) show the degradation of our system performance when no relevance feedback is used. DSIR seems not to retrieve relevant documents as well as SMART at lower recall level. However, its overall performance is still comparable to that of SMART (only 2.66% inferior). One possible

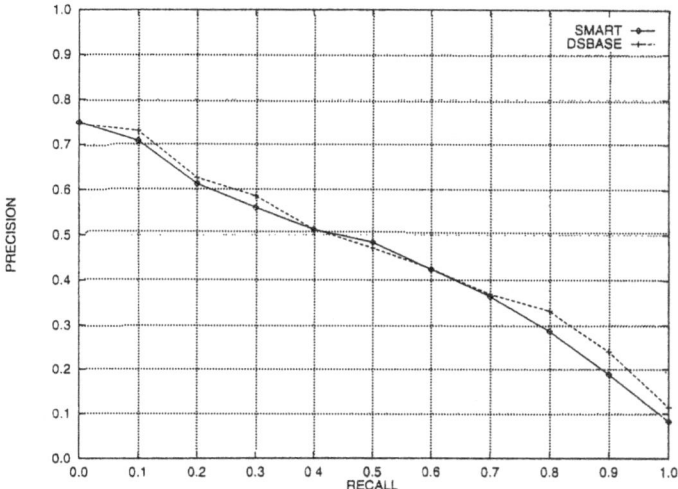

Figure 3: 11 points Recall-Precision curves derived from B1 sub-collection, without relevance feedback, using SMART and DSIR systems.

reason for this behavior could be the unsuitable use of co-occurrence data derived from B1 sub-collection. However, the result remains very encouraging especially when robustness considerations are taken into account, because our system can satisfyingly deal with many new documents ($\approx$ 20000 from the B2 sub-collection) without any parameter change, re-indexing a new co-occurrence computation. In addition to this, experiments conducted with the B2 sub-collection as reference corpus still confirm better results than SMART.

As far as feedback is concerned, the curves denoted DSFB05, DSFB10 and DSFB15 in figure 4 clearly indicate the improvement of our system's overall level of recall, and allow an interesting quantification of this improvement when various maximal (document) ranks are used for the feedback. The results in table 1 show the improvement in % of average precision change (compared with SMART). These results confirm the usefulness of relevance feedback as printed out in several previous references (Salton & Buckley (1990), Harman (1992), Allen (1995) and Buckley et al. (1995)).

6. Research Directions

The results of the experiments that we have conducted with our IR system on different standard test collections have convinced us of the feasibility of our approach.

However, as far as algorithmic efficiency is concerned, one major drawback of our system in its original form is that the satisfaction of a query requires the evaluation of the similarities between the query and *all* the documents in the textual base. Therefore, we are now investigating how clustering techniques can be used in association with the similarity measure in order to cluster the database in a way that allows the identification of the documents satisfying the query through a *partial* exploration of the base. We are currently working on a first tentative solution based on a very simple hierarchical partitioning process.

The process starts (step 0) with a unique initial class containing all the N documents of the textual database. At step n, each class c defined at step $n-1$ is divided into 2 subclasses corresponding respectively to elements with negative and positive coordinates in the first factorial dimension obtained by a factorial analysis performed on c. The partitioning pro-

Run / Recall - Precision	SMART B1	DSBASE B1	SMART B2	DSBASE B2	DSFB05 B2	DSFB10 B2	DSFB15 B2
0.00	0.7494	0.7460	0.7546	0.6900	0.7642	0.8238	0.8387
0.10	0.7095	0.7315	0.7260	0.6963	0.7629	0.7956	0.8160
0.20	0.6132	0.6267	0.6627	0.6276	0.6820	0.7234	0.7363
0.30	0.5611	0.5867	0.5888	0.5819	0.6172	0.6563	0.6790
0.40	0.5119	0.5124	0.5117	0.5111	0.5430	0.5767	0.5929
0.50	0.4839	0.4713	0.4660	0.4536	0.4978	0.5303	0.5493
0.60	0.4243	0.4254	0.3890	0.3982	0 4356	0.4504	0.4700
0.70	0.3639	0.3691	0.3258	0.3556	0.3808	0.4015	0.4076
0.80	0.2864	0.3317	0.2778	0.2738	0.2905	0.3045	0.3167
0.90	0.1879	0.2393	0.2090	0.1912	0.2046	0.2001	0.2075
1.00	0.0824	0.1143	0.1282	0.1283	0.1232	0.1196	0.1242
Average precision	0.4522	0.4686	0.4581	0.4459	0.4819	0.5075	0.5217
% Precision change		+3.6267		-2.6631	+5.1954	+10.7837	13.8834

Table 1: Average precision calculated over 11 recall points

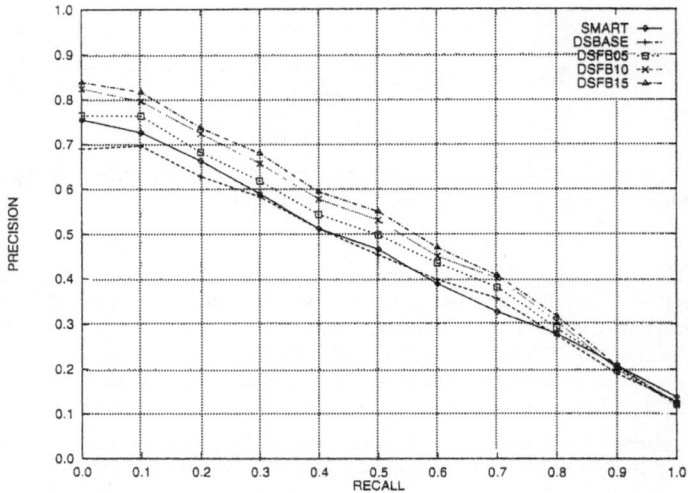

Figure 4: 11 points Recall-Precision curves derived from B2 sub-collection using SMART and DSIR with feedback query at rank 5, 10, 15.

cess is then iterated until all classes contain only one element.

By construction, the resulting set of classes corresponds to a binary tree T. In T, each non terminal node (i.e. class) is associated with a decision function based on the equation of the corresponding one-dimensional factorial vector space. Furthermore, consider, for any query q, the path $P(q)$, starting at the root of T, and in which, at each node, the branching decision is taken according to the result, for q, of the decision function associated with the node. Then $P(q)$ verifies the following interesting properties:

- $P(q)$ leads to a leaf of T that corresponds to a document representing a good approximation of the nearest document to q;

- the number of operations necessary to build $P(q)$ (i.e. to retrieve the document associated with the leaf) is at most $log_2(N)$.

We are currently integrating the partitioning process described above in our IR system in order to quantify, for the available test databases, the influence of such a method on the overall retrieval efficiency of the system. In addition to this, we are currently implementing a parallel version of our retrieval engine within the PVM programming environment (Geist et al. (1994)). This work will give us the means to conduct a realistic evaluation of the computational speed-up provided by the clustering based model.

References:

Allen, J. (1995): *Relevance Feedback with Too Much Data*, In Proceedings of the 18th Annual International ACM/SIGIR Conference on Research and Development in Information Retrieval, Seattle, USA.

Buckley, C. et al. (1995): *Automatic Query Expansion Using SMART: TREC3*, In the third Text REtrieval Conference (TREC-3), NIST Special Publication 500-225.

Frakes, W.B. and Baeza-Yates, R. (1992): *Information Retrieval: Data Structures & Algorithms*. Prentice Hall.

Gallant, S.I. et al. (1992): *HNC's MatchPlus System*. SIGIR FORUM, 16(2).

Geist, A. et al. (1994): *PVM: Parallel Virtual Machine, A Users' Guide and Tutorial for Networked Parallel Computing*, The MIT Press, Cambridge, England.

Harman, D. (1992): *Relevance Feedback Revisited.* In Proceedings of the 15th Annual International ACM/SIGIR Conference on Research and Development in Information Retrieval, Copehagen, Denmark.

Hersh, W. and Buckley C. (1994): *OHSUMED: An Interactive Retrieval Evaluation and New Large Test Collection for Research*, In Proceedings of the 17th Annual International ACM/SIGIR Conference on Research and Development in Information Retrieval, Dublin, Ireland.

Hull, D., (1993): *Using Statistical Testing in the Evaluation of Retrieval Experiments*, In Proceedings of the 16th Annual International ACM/SIGIR Conference on Research and Development in Information Retrieval, Pittsburgh, USA.

Rungsawang, A. and Rajman, M. (1995): Textual Information Retrieval Based on the Concept of the Distributional Semantics. *In Proceedings of the 3rd International Conference on Statistical Analysis of Textual Data*. Rome, Italy, December.

Schütze, H. (1992): Dimensions of Meaning. *In IEEE Proceedings of Supercomputing 92.*

Salton, G. and McGill, M.J. (1983): *Introduction to Modern Information Retrieval*. McGraw Hill.

Salton, G. et al. (1975): *A Theory of Term Importance in Automatic Text Analysis*. Journal of the American Society for Information Science.

Salton, G. et al. (1976): *Automatic Indexing Using Term Discrimination and Term Precision Measurement*. Information Processing & Management, 12.

Salton, G. and Buckley, C. (1990): *Improving Retrieval Performance by Relevance Feedback*. Journal of the American Society for Information Science, 41(4).

Fitting the CANDCLUS/MUMCLUS Models with Partitioning and Other Constraints

J. Douglas Carroll[1] and Anil Chaturvedi[2]

[1] Faculty of Management
Rutgers University
Management Education Center, Room 125
81 New Street
Newark, NJ 07102-1895
USA

[2] AT&T Bell Laboratories
Room 5C-133
600 Mountain Avenue
Murray Hill, NJ 07974
USA

Summary: The CANDCLUS (for CANonical Decompositon CLUStering) model and method is described for analysis of multiway data arrays in terms of multilinear models in which some ways (or modes) are modeled by continuous parameters defining spatial dimensions, other ways/modes by discrete parameters defining cluster or other categorical structures, and still others by mixtures of continuous *and* discrete parameters defining "hybrid" models in which spatial dimensional structure is combined with cluster-like categorical structure. A generalization of CANDCLUS, called MUMCLUS (for MUltiMode CLUStering), whose two-way special case corresponds to DeSarbo's GENNCLUS model, is also defined and discussed. Methods previously published for unconstrained fitting of the CANDCLUS/MUMCLUS family of models, based on a separability property observed by Chaturvedi, are extended to allow certain constraints on the discrete parameters—in particular a constraint that the cluster structure be a partition, and another that each entity in a particular mode may be a member of no more than C clusters. These constraints are implemented via an extended separability property (for vectors of discrete parameters, rather than for single parameters) which is defined. The possibility of fitting other constrained versions of these models within this general framework is discussed.

1. THE CANDCLUS MODEL

Canonical decomposition clustering (CANDCLUS) is a general multilinear model for multiway data arrays. We first state the CANDCLUS model in its most general form, for the N-way array $\mathbf{Y}$. It might be noted that this form of the CANDCLUS model is identical to that of the Carroll and Chang CANDECOMP (CANonical DECOMPosition) model (Carroll and Chang, 1970; Carroll and Pruzansky, 1984), whose most important application to date has been to provide the computational underpinnings of the INDSCAL approach to two-mode, three-way (individual differences) multidimensional scaling (MDS). The most important difference between CANDECOMP and CANDCLUS is that, while in the former, all parameters are assumed continuous, so that the models being assumed and fitted in the concomitant data analysis are all continuous spatial models (e.g., MDS models), in CANDCLUS, some or all of the dimensions for some or all ways/modes may be constrained to be discrete, typically binary (0-1) variables, which can be interpreted as class membership variables encoding whether a particular object (or other entity corresponding to a given level of a given mode/way) belongs (value = 1) or does not belong (value = 0) to a particular class or cluster. In CANDCLUS, the dimensions for various ways/modes may be continuous (spatial) or binary (clusterlike). Other possibilities include discretely valued dimensions with $k(> 2)$ distinct possible values for a particular dimension. Specifically, the CANDCLUS model for a general N-way array $\mathbf{Y}$ can be stated in the following form:

$$y_{i_1 i_2, \ldots, i_N} \sim \sum_{r=1}^{R} a_{i_1 r}^1 a_{i_2 r}^2 \ldots a_{i_N r}^N .$$
(1)

Define $\mathbf{A}_n$ as the parameter matrix of order $I_n \times R$ with elements $a_{i_n r}^n$ for the rth dimension and the nth way. The elements of matrices $\mathbf{A}_n (n = 1, \ldots, N)$ can take on any of the following values:

- Real values for all matrices $\mathbf{A}_n,$ $(n = 1, \ldots, N)$. This results in the R-dimensional CANDECOMP model of Carroll and Chang (1970; see also Carroll and Pruzansky, 1984).

- Discrete integer (or finite set of real number) values for all $\mathbf{A}_n$, $n = 1, \ldots, N$.

- Mixture of real parameters for some ways and discrete values for the rest. Other possibilities, such as those involving "hybrid" models, also exist—in a hybrid model for a particular way/mode, some dimensions are defined via continuous, and others via discrete parameters. See Carroll (1976), Carroll and Pruzansky (1980), De Soete and Carroll (1996), and Carroll and Arabie (in press) for discussion of hybrid models.

2. MUMCLUS (MULTIMODE CLUSTERING), GENNCLUS, GENERALIZED GENNCLUS, TUCKER'S THREE-MODE AND MULTIMODE FACTOR/COMPONENTS ANALYSIS, AND HYBRID MODELS OF GENERAL TUCKER FORM

2.1 DeSarbo's GENNCLUS Model (Two-way and Three-way Versions)

Given K nonsymmetric $I \times J$ matrices $\mathbf{S}_k$, $k = 1, \ldots, K$, the three-way GENNCLUS model is of the form

$$\mathbf{S}_k = \mathbf{A U}_k \mathbf{B} + \mathbf{C}_k + \text{ error},$$
(2)

where $\mathbf{A}$ is an $I \times R_a$ binary matrix, $\mathbf{B}$ is a $J \times R_b$ binary matrix, while $\mathbf{U}_k$ is a completely general $R_a \times R_b$ matrix.

In generalized GENNCLUS, $\mathbf{A}$ and $\mathbf{B}$ can each be continuous, discrete, or a mixture of continuous and discrete (i.e., hybrid) dimensions. $\mathbf{U}$ is generally assumed to have continuously valued entries.

Tucker's three-mode factor/components analysis (TMFA) model corresponds to the case in which both $\mathbf{A}$ and $\mathbf{B}$ (as well as $\mathbf{U}$) are entirely continuously valued. It should be noted that two-way GENNCLUS, as proposed by DeSarbo (1982), corresponds to the case where $K = 1$. While DeSarbo refers to a possible three-way generalization, he discusses only the two-way case.

2.2 N-Mode Factor/Components Analysis (NMFA), Multiway GENNCLUS, and Hybrid Models

Given an $I_1 \times I_2 \times \ldots \times I_N$ array, with general entry $y_{i_1 i_2} \ldots {}_{i_N}$, where $i_n = 1, 2, \ldots, I_n$ and $n = 1, 2, \ldots, N$. We fit it by a model of the general algebraic form

$$y_{i_1 i_2} \cdots {}_{i_N} \sim \sum_{t_1=1}^{T_1} \sum_{t_2=2}^{T_2} \cdots \sum_{t_N=1}^{T_N} a_{i_1 t_1}^1 a_{i_2 t_2}^2 \cdots a_{i_N t_N}^N u_{t_1 t_2} \cdots {}_{t_N} , \qquad (3)$$

where $\mathbf{A}_n$ is an $I_N \times T_N$ and $\mathbf{U}$ is a $T_1 \times T_2 \times \ldots \times T_N$ array (a "generalized" core array in Tucker's terminology). Each $\mathbf{A}_n$ may be continuous, discrete, or hybrid, while $\mathbf{U}$ will have continuously valued entries. It should be noted that the multiway/mode NMFA/GENNCLUS is the most general model, including all others described here as special cases. We call this highly general multilinear model "MUMCLUS," for MUlti-Mode CLUStering. (If all parameters are continuous, we call the resultant special case, which is a generalization of TMFA, "MUMSCAL.")

While both CANDCLUS and MUMCLUS, when one or more of the ways/modes entails binary (clustering) "dimensions," are, in general, overlapping clustering models/methods, it is possible as a special case that the resulting clustering will correspond to that in which the clusters are non-overlapping (mutually exclusive and collectively exhaustive)—i.e., comprise a partition of the objects or other entities corresponding to the mode/way in question. Assuming for the moment that the mode in question is treated completely cluster-wise, (not in terms of "hybrid" representation), the matrix (say $\mathbf{A}_1$) for that mode must then have the property that each row contains exactly one "1", all other entries in that row equaling 0.

Approaches for fitting either CANDCLUS or MUMCLUS models via either an OLS or LAD (least absolute deviation) criterion are described in Carroll and Chaturvedi (1995), for the general, unconstrained case. Slight modification of these methods would enable fitting via a WLS (weighted least squares) or weighted LAD (WLAD) criterion, or other even more general "additively decomposable" loss functions.

An additively decomposable loss function, L, is of the form:

$$L\left[\mathbf{Z}_n, \hat{\mathbf{Z}}_n\right] = \sum_{i_n=1}^{I_n} \sum_{j_n=1}^{J_n} \ell\left[z_{i_n j_n}^{(n)}, \hat{z}_{i_n j_n}^{(n)}\right] , \qquad (4)$$

where $J_n = I_1 I_2 \ldots I_{n-1} I_{n+1} \ldots I_N$, while j_n is an index ranging from 1 to J_n, varying systematically over the combinations of all values of all $N-1$ subscripts *excluding* i_n, while $\ell[z, \hat{z}]$ is a measure of discrepancy between the two scalar valued quantities z and $\hat{z}$; e.g., $(z - \hat{z})^2$ or $\mid z - \hat{z} \mid$, in the case of an OLS or LAD loss function, L, respectively.

As demonstrated for OLS fitting of the ADCLUS/INDCLUS model by Chaturvedi and Carroll (1994), and by Chaturvedi, Lakshmï-Ratan, and Carroll (1995) for the case of fitting ADCLUS/INDCLUS via a "least absolute deviation" (LAD) criterion (see, also, Carroll and Chaturvedi, 1995 for a general discussion of this in the context of the general unconstrained CANDCLUS/MUMCLUS models), very efficient general algorithms can be formulated for fitting these models via an OLS or LAD criterion, via what can be called a "one dimension at a time" *elementwise* approach.

That it is "one dimension at a time" simply means that, at each stage of an "outer iteration" process, only one dimension—whether continuous, or a discrete (usually binary) one (e.g., defining membership vs. non-membership in a cluster)—is estimated, *conditional* on fixed values of all the other $R-1$ dimensions (and/or clusters). This conditional estimation procedure is iterated over dimensions/clusters until convergence occurs. Within each of these "one dimension at a time" estimation steps another "*inner* estimation" process is used, in this case iterating over individual values of the (continuous or discrete) components of that dimension/cluster. Since this inner iteration process is, in fact, iterating over certain elements of the set of parameters matrices or arrays, we call this an *elementwise* procedure. At each of the most basic computational steps of the composite iterative process all elements of all parameter arrays are fixed, save the one which is currently being (re)estimated—*conditional* on the fixed values of the remaining parameters.

Concretely, in the stage in which conditional estimates are being made for dimension/cluster r for the general CANDCLUS model, as in the CANDECOMP algorithm (Carroll and Chang, 1970; Carroll and Pruzansky, 1984) when using the "one dimension at a time" approach, the CANDCLUS algorithm fixes the parameters for all ways *except* the nth, and *conditionally* (re)estimates the parameters for that nth way. In fact, if all parameters are continuous, and OLS estimation is done, the CANDCLUS algorithm is exactly equivalent to the CANDECOMP algorithm, implemented on the "one dimension at a time" basis. Thus a basic step in this overall algorithm entails conditional estimation of a vector of coordinates of just one (continuous or discrete) dimension, for just one way of the multiway data array, with all parameters for all other dimensions, and all other ways for the dimension currently being (re)estimated, held fixed at their current values.

Chaturvedi was the first to note that this adaptation of the one-dimension-at-a-time CANDECOMP algorithm to fitting clustering or other discrete models via an OLS (and later LAD) criterion could be greatly accelerated computationally based on a separability property resulting from the additively decomposable structure of these loss functions (see Chaturvedi and Carroll, 1994, Chaturvedi, Lakshmi-Ratan, and Carroll 1995 and Carroll and Chaturvedi, 1995 for details). This separability property enables optimization of the objective function over the entire vector of discretely valued parameter values via optimization for each discrete parameter separately. Thus, for example, in the "standard" case of binary valued parameters, this optimization, for a vector of I_n components (comprising the I_n values of the Rth dimension for the nth way) can be accomplished via $2I_n$ evaluations, rather than 2^{I_n} evaluations. (In the case of a K-ary discrete parameter, KI_n rather than K^{I_N} evaluations are required.) These discrete conditional estimation steps are implemented via what are called the "elementary discrete least squares or LAD procedures," respectively. Conditional OLS or LAD fitting of a vector of continuous parameters defining coordinates of a continuous (spatial) dimension can also be done via a sequence of 2^{I_N} (or K^{I_N}, in the K-ary case) quite simple OLS or LAD regression steps (called the "elementary continuous least squares or LAD procedures," respectively). Because of the one dimension at a time estimation in this case, it is quite straightforward to impose simple equality or inequality constraints (e.g., a nonnegativity constraint) on the continuous parameters, as is also discussed in the papers cited above.

It is straightforward to extend either the OLS or LAD estimation scheme for the unconstrained CANDCLUS model to weighted least squares (WLS) or weighted LAD (WLAD) estimation. It is also simple, in principle, to extend this to a criterion of fit based on minimizing an L_p—norm based loss function, for any $1 \leq p \leq \infty$. In fact, in certain circumstances this would also be sensible for an L_p norm for $0 \leq p < 1$, where the L_o norm is the limiting case as $p \to 0$, corresponding to the case of a "counting metric," appropriate for categorical data. An application of this "L_o loss function" to a clustering approach called "K-modes" has been discussed by Carroll, Chaturvedi, and Green (1994) and Chaturvedi, Green, and Carroll (1996). OLS and LAD fitting correspond, of course, to the L_2 (or Euclidean) and L_1 (or "city-block") norms, respectively.

The elementary discrete estimation procedures for any L_p-norm based loss function would be quite simple—merely entailing evaluating the loss function for each of the 2 (or K) values of each parameter, with all other parameters fixed, and choosing the parameter value with the lower (lowest) value of the loss function. There is, however, no closed form solution, in general, for an L_P norm based loss function for p other than 1 or 2 (and, also, for $p = 0$, for appropriate models and data types, where the solution will simply be the mode of certain values of an associated categorical variable, and for $p = \infty$, where the solution is the midrange of certain values).

In these cases, some form of line search algorithm or other unidimensional optimization procedure would have to be used to solve the elementary continuous estimation problem. These statements can be extended to *any* additively decomposable loss function of the form defined in equation (4); again the elementary discrete procedure would entail simple enumeration of the 2(or K) parameter values and choice of the one yielding the lower (lowest) value of the loss function, while the elementary continuous procedure would require either use of a unidimensional optimization method, or a closed form solution, if available. The separability property discussed above makes estimation of the CANDCLUS parameter based on any loss function of the form stated in equation (4) particularly efficient computationally for the discrete parameters, while reducing it to a unidimensional problem for the continuous ones (with a straightforward way of imposing such constraints as nonnegativity, if desired.)

Carroll and Chaturvedi (1995) also discuss an extension of this approach to estimation of the unconstrained MUMCLUS model, at least in the case of an OLS criterion of fit. LAD, or most other loss functions, would not be nearly as tractable for fitting MUMCLUS, however, since the estimation of the core array would be particularly difficult in this case (requiring, generally, use of some form of *multidimensional* optimization procedure, one not being simplifiable at all via a separability property such as that utilized in CANDCLUS). A WLS extension of MUMCLUS estimation would be quite straightforward, however, since estimation of the core array could be implemented in this case by use of a WLS multivariate regression procedure.

It turns out that a very straightforward modification of the CANDCLUS/MUMCLUS algorithm (for fitting $\mathbf{A}_1$ and/or matrices for any other modes)—namely fitting that matrix at each stage of the algorithm *row*wise (all "dimensions" and clusters being estimated simultaneously, rather than via the "one dimension at a time" *element*wise strategy discussed in the case of CANDCLUS/MUMCLUS)—enables fitting a solution constrained to have this partitioning form (for one or more of the ways/modes). In this approach, one simply seeks, for each row, to which column (cluster) the single "1" should be assigned to optimize the (OLS, LAD or other) loss function being minimized. The separability property, again, can be used—in this case it means that this decision can be made separately for each row of the matrix given that the other matrices are treated as fixed.

We describe the resulting approach for CANDCLUS, with partitioning constraints, below. To be specific, given current estimates of all other *matrices*, $\mathbf{A}_2, \mathbf{A}_3, \ldots \mathbf{A}_N$, (whether these are spatial/continuous, cluster-like/discrete, or hybrid), we define what is sometimes called the "column-wise Kronecker product" of $\mathbf{A}_2, \mathbf{A}_3, \ldots \mathbf{A}_N$, which we shall denote $\mathbf{A}_2 \otimes_c \mathbf{A}_3 \otimes_c \ldots \otimes_c \mathbf{A}_N$, and denote when appropriate as $\mathbf{Q}_1$—indexing $\mathbf{Q}$ by the index of the matrix, $\mathbf{A}_1$ in this case, which is *omitted* (see Carroll and Chang, 1970, and ten Berge and Kiers, 1996).

For two matrices, $\mathbf{A}_n$, and $\mathbf{A}_{n'}$, $I_n \times R$ and $I_{n'} \times R$ respectively (note, in particular, that both have a common column order, R) the columnwise Kronecker product is defined as :

$$\mathbf{A}_n \otimes_c \mathbf{A}_{n'} = \left(\mathbf{a}_1^{(n)} \otimes \mathbf{a}_1^{(n')}, \mathbf{a}_2^{(n)} \otimes \mathbf{a}_2^{(n')}, \ldots, \mathbf{a}_R^{(n)} \otimes \mathbf{a}_R^{(n')} \right) \tag{5}$$

where $\mathbf{a}_r^{(n)}$ is the rth column of matrix $\mathbf{A}_n$, and $\otimes$ is the ordinary Kronecker product (applied in the present case separately to corresponding columns $\mathbf{A}_n$, and $\mathbf{A}_{n'}$). Since $\mathbf{A}_n$ is $I_n \times R$, and $\mathbf{A}_{n'}$ is $I_{n'} \times R$, $\mathbf{A}_n \otimes_c \mathbf{A}_{n'}$ will be $I_n I_{n'} \times R$. We define such products as $\mathbf{A}_2 \otimes_c \mathbf{A}_3 \otimes_c \ldots \otimes_c \mathbf{A}_N$, recursively (as in the case of ordinary Kronecker products); thus, for example:

$$\mathbf{A}_2 \otimes_c \mathbf{A}_3 \otimes_c \ldots \otimes_c \mathbf{A}_N = (\mathbf{A}_2 \otimes_c \mathbf{A}_3 \otimes_c \ldots \otimes_c \mathbf{A}_{(N-1)}) \otimes_c \mathbf{A}_n \tag{6}$$

We also define a matrix $\mathbf{Z}_1$ $(I_1 \times I_2 I_3 \ldots I_N)$ as the matrix defined by concatenating the $N-1$ subscripts $i_2, i_3 \ldots i_N$ to form a row vector of $I_2 I_3 \ldots I_N$ components, for each $i_1 = 1, 2, \ldots I_1$, and then defining the $i_1 th$ row of $\mathbf{Z}_1$ as that $i_1 th$ row vector. (The order of the subscripts $i_2 i_3, \ldots, i_N$ is assumed to be identical to the order induced on these same susbscripts in the *columns* of the columnwise Kronecker product $\mathbf{Q}_1$.)

Having so defined $\mathbf{Z}_1$ and $\mathbf{Q}_1$ or, more generally, $\mathbf{Z}_n$ and $\mathbf{Q}_n$, for all $n = 1, 2, \ldots N$ (extending these definitions in the obvious way to the case of any one of the N matrices in the overall decomposition of $\mathbf{Y}$, say $\mathbf{A}_n$, denoting the corresponding matrices as $\mathbf{Z}_n$ and $\mathbf{Q}_n$, respectively, in this case with $\mathbf{Q}_n$ being defined in terms of current estimates of all $\mathbf{A}$ matrices except $\mathbf{A}_n$), we then can solve for a *new* estimate of $\mathbf{A}_n$, $\hat{\mathbf{A}}_n$, conditional on the current estimates of $\mathbf{A}_{n'}$ for all $n' \neq n$, by finding the constrained OLS, WLS, LAD, or other estimate of $\mathbf{A}_n$, by solving for an estimate, $\hat{\mathbf{A}}_n$, in the equation

$$\mathbf{Z}_n \cong \mathbf{A}_n \mathbf{Q}_n' , \tag{7}$$

(where $\cong$ can be taken as implying optimizing a fit criterion such as OLS, WLS, LAD, weighted LAD, or an objective function optimizing any other specified additively decomposable criterion of fit, of the additively decomposed form given in equation (4).

Given this general form of the loss function, whether OLS, WLS, LAD, WLAD or other, we can very simply minimize L by the use of a similar separability property as noted in the case of CANDCLUS/MUMCLUS. In this case, this separability of the overall loss function L is defined rowwise, for the entire matrix $\mathbf{A}_n$, since for a loss function of the additive form given in equation (5), given that $\hat{Z}_n = \mathbf{A}_n \mathbf{Q}_n'$ we have $\hat{z}_{i_n j_n}^{(n)} = \mathbf{a}_{i_n}^{(n)} (\mathbf{q}_{j_n}^{(n)})'$ so that

$$L = \sum_{i_n}^{I_n} \sum_{j_n}^{J_n} \ell \left[z_{i_n j_n}^n, \mathbf{a}_{i_n}^{(n)} (\mathbf{q}_{j_n}^{(n)})' \right] , \tag{8}$$

so that L is separable in each of the row vectors in the matrix $\mathbf{A}_n$, generalizing quite directly the elementwise separability property defined earlier (and used for *unconstrained* CANDCLUS/MUMCLUS) to a *rowwise* separability property.

In the case of a partition, the problem is particularly simple, of course, since each of these row vectors is constrained to have one and only one 1, with all other elements $= 0$. Thus, the optimal vector can be selected, at each stage of the overall algorithm, by simply computing the objective function being optimized (minimized) for each of the I_n unit vectors, and choosing the one optimizing (minimizing) the objective function—so this entails only I_n computations, rather than the 2^{I_n} computations that would be required for the unconstrained case.

Thus, the conditional minimization of L can be accomplished simply by sequentially minimizing

$$\ell[z_{i_n}^{(n)}, \hat{z}_{i_n}^{(n)}] = \ell[z_{i_n}^{(n)}, \mathbf{Q}_n(\hat{\mathbf{a}}_{i_n}^{(n)})'] = f_{i_n}[\hat{\mathbf{a}}_{i_n}^{(n)}] , \tag{9}$$

where $f_{i_n}[\hat{\mathbf{a}}_{i_n}^{(n)}]$ is a function of the vector $\hat{\mathbf{a}}_{i_n}^{(n)}$ *alone*, since all other variables are treated as constant, while f_{i_n} is minimized by a very simple exhaustive search. Given the constraints that $\mathbf{A}_n$ have only one "1" per row for each of its I_n rows, the column vector $\mathbf{a}_{i_n}^{(n)}$ must be one of the R *unit* column vectors $\boldsymbol{\varepsilon}_1, \boldsymbol{\varepsilon}_2, \ldots \boldsymbol{\varepsilon}_R$. Thus f_{i_n} need be evaluated only for those R permissible unit vectors—entailing exactly R evaluations. Thus $\mathbf{A}_n$ as a whole can, in view of the separability of the assumed loss function, be optimized via a total of $I_n R$ such evaluations—a quite manageable search process, since it is *linear* in the size $(I_n R)$ of the matrix (as opposed to being proportional to $2^{I_n R}$, as would be true in the case of a totally general binary matrix and a *non*-separable loss function).

The conditional OLS, LAD (or other) estimation of the remaining matrices can be implemented either rowwise or in a "one-dimension at a time" elementwise manner, as with unconstrained CANDCLUS/CANDECOMP, depending on the nature of the parameters, constraints, or other factors. For purposes of the approach described here, conditional estimates of each of the other $N - 1$ other matrices must be given, and treated as fixed for reestimation of $\mathbf{A}_n$. This enables using the elementwise separability property described earlier for CANDCLUS to estimate overlapping cluster structures, or other discrete structures, for some modes and also, if desired, to implement continuous optimization of parameter matrices for other modes with (say) nonnegativity constraints. In the case of a matrix $\mathbf{A}_k$ with continuous parameters, and in which OLS or WLS estimation is being done, $\mathbf{A}_k$ may be estimated *matrix*wise via OLS regression as in CANDECOMP, or via a straightforward generalization of this to WLS matrixwise estimation via WLS regression. If the LAD (or WLAD) criterion is to be used, the elementary discrete LAD (or WLAD) procedure must be used, on an iterated "one-dimension at a time" elementwise basis.

It should be noted that other forms of constrained overlapping cluster structures can also be estimated (conditionally) on a rowwise basis. For example, if one wanted to constrain a cluster structure to one in which each object or other entity corresponding to a level of a specified mode/way is contained in no more than C other clusters, the search could be restricted to those binary R-vectors having C or fewer 1's. The additive decomposability of the objective function leads to a very simple algorithm for this problem. First, go through all components of a particular row vector, allowing each component to be 0 or 1 on an unconstrained elementwise basis. Then, if the number of 1's is C or less, you're finished (for that row vector). If not, choose the C components associated with the C largest reductions (or "differentials") in the objective function being minimized. (This step requires storing these differentials for each such unit component, followed by a sorting algorithm aimed at choosing the C largest absolute differentials among them.) This algorithm would enable choosing the optimal vector for the $i_n th$ row of $\mathbf{A}_n$ in considerably fewer than $\binom{R}{C}$ operations—the specific number of operations being dependent on the data and other parameters' current estimates. Other constraints are possible (e.g., that each object be in *exactly* C clusters, or that the R-vectors be restricted to a predefined subset of all possible binary R-dimensional vectors). Any constraints that can be defined in terms of the set of permissible binary R-vectors for each row of a specific parameter matrix (even if a different set for each row is specified, so long as these constraints are independent of parameter values in the other rows) can be imposed in this manner.

While we have only roughly outlined the constrained CANDCLUS approach, at least one special case merits particular attention. This is the special case involving two-mode two-way data, in which one mode is modeled by partitions and the other by continuous parameters. The particular special case can easily be shown to be equivalent to the well-know clustering procedure know as K-means (where $K = R$, in this case) if an OLS, and K-medians if a LAD, criterion is used. We might mention, as a related model of interest, the case of unconstrained CANDCLUS fit to such two-mode data with one mode modeled by *overlapping* clusters and the other by continuous parameters. This model/method we have called "overlapping K-centroids" (Chaturvedi, Carroll, Green and Rotondo, 1994) elsewhere.

While we have dealt here only with the CANDCLUS model with constraints making the clustering (for one or more modes) a partition, this entire approach generalizes in a straightforward manner to a similarly constrained version of MUMCLUS, at least via OLS or WLS optimization. As discussed by Carroll and Chaturvedi (1995), estimation of the continuous parameters in the core array via LAD, WLAD or other additively decomposable fit criteria is not, in general, an easily implementable computational problem leading to a closed form solution. We shall not explore this last class of models and methods more fully, however, in the present paper.

3. CONCLUSION

The general CANDCLUS and MUMCLUS models have been defined and discussed, including references to previously published work on unconstrained estimation of each using various additively decomposable loss functions as fit criteria, based on a separability property originally observed by Chaturvedi. An extension of the separability property from individual elements of various parameter matrices to (row) vectors of those matrices leads to a straightforward approach for estimating these models with the clusters corresponding to certain ways or modes being constrained to satisfy partitioning constraints. Using this property to impose certain other kinds of constraints on the clustering is also possible. One that is discussed entails constraining the objects to be contained in no more than a fixed number (C) of clusters.

Some special cases of CANDCLUS are discussed, including OLS and LAD estimation of the ADCLUS/INDCLUS models and a procedure generalizing K-means and K-medians to the case of overlapping clusters, called K-overlapping centroids clustering (which includes methods called overlapping K-means and overlapping K-medians as special cases), as well as some other potential applications (e.g., to fitting "hybrid" models entailing combinations of continuous and discrete dimensions for the same set of entities corresponding to one or more modes of the multiway data array). We anticipate many future applications of these and other specific special cases of constrained and unconstrained CANDCLUS/MUMCLUS models, some not yet even contemplated, to a wide variety of data analytic situations.

New algorithmic developments may further improve fitting procedures for this class of models, using various loss functions as fitting criteria—potentially reducing problems of merely local optima and slow convergence, and generally increasing computational efficiency so as to make dealing with the increasingly large data arrays arising in many practical data analytic situations much more feasible.

4. References

Carroll, J. D. (1976): Spatial, non-spatial and hybrid models for scaling, *Psychometrika*, **41**, 439–463.

Carroll, J. D. and Arabie, P. (in press): Multidimensional scaling, In: *Handbook of Perception and Cognition. Volume 3: Measurement, Judgment and Decision Making*, Birnbaum, M. H. (ed.), San Diego, CA: Academic Press.

Carroll, J. D. and Chang, J. J. (1970): Analysis of individual differences in multidimensional scaling via an N-way generalization of "Eckert-Young" decomposition, *Psychometrika*, **35**, 283–319.

Carroll, J. D. and Chaturvedi, A. (1995): A general approach to clustering and multidimensional scaling of two-way, three-way, or higher way data, In: *Geometric Representations of Perceptual Phenomena*, Luce, R. D. et al. (eds.), 295–318, Mahwah, NJ: Erlbaum.

Carroll, J. D. et al. (1994): K-means,K-medians and K-modes: Special cases of partitioning multiway data. (Paper presented at meeting of the Classification Society of North America, Houston, TX.)

Carroll, J. D. and Pruzansky, S. (1980): Discrete and hybrid scaling models, In: *Similarity and Choice*, Lantermann et al., (eds.), 108–139, Bern: Hans Huber.

Carroll, J. D. and Pruzansky, S. (1984): The CANDECOMP-CANDELINC family of models and methods for multidimensional data analysis, In: *Research Methods for Multimode Data Analysis*, Law, H. G. et al. (eds.), 372–402, New York: Praeger.

Chaturvedi, A. and Carroll, J. D. (1994): An alternating combinatorial optimization approach to fitting the INDCLUS and generalized INDCLUS models, *Journal of Classification*, **11**, 155-170.

Chaturvedi, A. et al. (1994): A feature based approach to market segmentation via overlapping K-centroids clustering. Manuscript submitted for publication.

Chaturvedi, A. et al. (1995): Two L_1 norm procedures for fitting ADCLUS and INDCLUS. Manuscript submitted for publication.

Chaturvedi, A. et al. (1996): Market segmentation via K-modes clustering. (Paper presented at American Statistical Association Conference, Chicago, IL.

De Soete, G. and Carroll, J. D. (1996): Tree and other network models for representing proximity data, In: *Clustering and Classification*, Arabie, P. et al. (eds.), 157–197, River Edge, NJ· World Scientific.

DeSarbo, W. S (1982): GENNCLUS: New models for general nonhierarchical clustering analysis, *Psychometrika*, **47**, 449–475.

ten Berge, J. M. F and Kiers, H. A. L. (1996): Some uniqueness results for PARAFAC2, *Psychometrika*, **61**, 123–132.

A Distance-Based Biplot
for Multidimensional Scaling of Multivariate Data

Jacqueline J. Meulman

Department of Data Theory, University of Leiden
P.O. Box 9555, 2300 RB Leiden, The Netherlands

Summary: Least squares multidimensional scaling (MDS) methods are attractive candidates to approximate proximities between subjects in multivariate data (Meulman, 1992). Distances in the subject space will resemble the proximities as closely as possible, in contrast to traditional multivariate methods. When we wish to represent the variables in the same display - after using MDS to represent the subjects - various possibilities exist. A major distinction is between linear and nonlinear biplots. Both types will be discussed briefly, including their drawbacks. To circumvent these drawbacks, a third alternative will be proposed. By expanding the optimal p-space (where p denotes the dimensionality of the subject space) into an m-dimensional space of rank p (with $m > p$), we obtain a coordinate system that is appropriate for the evaluation of the MDS solution directly in terms of the m original variables. The latter are represented graphically as vectors in p-space, and their entries as markers that are located on these vectors. The overall approach, including the analysis of mixed sets of continuous and categorical variables, can be viewed as a distance-based alternative for the graphical display of multivariate data in Gifi (1990).

1. Introduction

In the approach to Multivariate Analysis (MVA) applied in this paper, the variables are used to define an observation or measurement space in which the units are located according to their scores. The distances in this observation space are regarded as proximities to be approximated by distances between subject points in a low-dimensional representation space. If the m-dimensional observation space is denoted by $\mathbf{Q}$, giving coordinates for n points in m dimensions, the proximities between all pairs of subjects are given in the proximity matrix $D(\mathbf{Q})$, where $D(\bullet)$ is the Euclidean (also called Pythagorean) distance function. So the $n \times n$ matrix $D(\mathbf{Q})$ contains proximities $d_{ik}(\mathbf{Q})$ between subject i and k. Squared distances are given by:

$$D^2(\mathbf{Q}) = \mathbf{v}\mathbf{1}' + \mathbf{1}\mathbf{v}' - 2\mathbf{Q}\mathbf{Q}', \qquad (1)$$

where $\mathbf{v} = \text{vecdiag}(\mathbf{Q}\mathbf{Q}')$ is the n-vector containing the diagonal elements of $\mathbf{Q}\mathbf{Q}'$, and $\mathbf{1}$ is the n-vector of all 1's. Analogously, squared distances in the p-dimensional representation space $\mathbf{X}$ are defined by $D^2(\mathbf{X}) = \mathbf{v}\mathbf{1}' + \mathbf{1}\mathbf{v}' - 2\mathbf{X}\mathbf{X}'$, now with $\mathbf{v} = \text{vecdiag}(\mathbf{X}\mathbf{X}')$.

Approximation of a set of proximities by a set of distances in some low-dimensional space is usually identified as a multidimensional scaling (MDS) task. In Meulman (1986), following Gower (1966), it is shown that techniques of multivariate analysis, like principal components, canonical correlation and homogeneity analysis, are equivalent to MDS tasks applied to particular derived proximities when the so-called classical Torgerson-Gower approach to MDS (Torgerson, 1958; Gower, 1966) is used. A basic ingredient is the original Young-Householder (1938) process that transforms a squared distance matrix $D^2(\mathbf{Q})$ into an $n \times n$ scalar product matrix $\mathbf{Q}\mathbf{Q}'$, modified by locating the origin in the centroid of points through the use of the $n \times n$ centering operator $\mathbf{J} = \mathbf{I} - (\mathbf{1}\mathbf{1}'/\mathbf{1}'\mathbf{1})$, which gives

$$-1/2\mathbf{J}(D^2(\mathbf{Q}))\mathbf{J} = -1/2\mathbf{J}(\mathbf{v}\mathbf{1}' + \mathbf{1}\mathbf{v}' - 2\mathbf{Q}\mathbf{Q}')\mathbf{J} = \mathbf{Q}\mathbf{Q}'. \qquad (2)$$

(**I** is the $n \times n$ identity matrix; we assume that the variables q_j have zero mean.) The scalar product matrix $\mathbf{QQ'}$ is then approximated by another scalar product matrix of lower rank, using an objective function that can be written in the form:

$$\text{STRAIN}(\mathbf{X}) = ||\mathbf{QQ'} - \mathbf{XX'}||^2 = \text{tr} \, (\mathbf{QQ'} - \mathbf{XX'})'(\mathbf{QQ'} - \mathbf{XX'}), \quad (3)$$

so $|| \bullet ||^2$ denotes a least squares discrepancy measure. (The term STRAIN is used after Carroll and Chang, 1972.) In our case. because $\mathbf{XX'} = -1/2\mathbf{J}(D^2(\mathbf{X}))\mathbf{J}$, (3) can also be written as

$$\text{STRAIN}(\mathbf{X}) = (1/4)||\mathbf{J}(D^2(\mathbf{Q}) - D^2(\mathbf{X}))\mathbf{J}||^2, \quad (4)$$

so Torgerson-Gower scaling approximates double-centered squared distances. To find the coordinates in $\mathbf{X}$, first an eigenanalysis of $\mathbf{QQ'}$ is performed: $\mathbf{QQ'} = \mathbf{K\Lambda K'}$, where $\mathbf{K}$ is an $N \times t$ matrix containing t eigenvectors, Λ is a $t \times t$ diagonal matrix containing the ordered positive eigenvalues, and t denotes the rank of $\mathbf{Q}$ (for $t \le m$). The optimal solution in the Torgerson-Gower procedure for obtaining a p-dimensional $\mathbf{X}$ (for $p \le t$), is given by $\mathbf{X} = \mathbf{K}_p\Lambda_p^{1/2}$; thus, the eigenvectors are rescaled using the eigenvalues to give coordinates $\mathbf{X}$ for the units in p-space with dimensions that reflect differential saliences.

The multivariate case described above, with proximities $D(\mathbf{Q})$, is usually called Principal Coordinates Analysis (Gower, 1966); using the singular value decomposition $\mathbf{Q} = \mathbf{K\Lambda L'}$, it can easily be shown that the solution for $\mathbf{X}$ that would be obtained in a principal components analysis is equivalent to the solution for $\mathbf{X}$ obtained by minimizing (3) or (4). Using the same strategy, it was shown in Meulman (1986, 1992) that Multiple Correspondence Analysis (MCA), also called homogeneity analysis, can be viewed as a classical scaling technique as well. In MCA, categorical variables are analyzed, and each categorical variable h_j defines a binary indicator matrix $\mathbf{G}_j$ with n rows and l_j columns, where l_j denotes the number of categories. Elements h_{ij} then define elements g_{ir}^j as follows:

$$h_{ij} = r \longrightarrow g_{ir}^j = 1; \quad (5)$$
$$h_{ij} \ne r \longrightarrow g_{ir}^j = 0,$$

where $r = 1, ..., l_j$ is the running index to indicate the category number of variable j. In analogy with (3) and (4). MCA can be written as a classical MDS problem since its optimal solution for the subject scores $\mathbf{X}$ minimizes

$$\text{STRAIN}(\mathbf{X}) = 1/m\sum_{j=1}^{m}||\mathbf{G}_j(\mathbf{G}_j'\mathbf{G}_j)^{-1}\mathbf{G}_j' - \mathbf{XX'}||^2 \quad (6)$$

$$= 1/m\sum_{j=1}^{m}||\mathbf{J}(D^2(\mathbf{G}_j(\mathbf{G}_j'\mathbf{G}_j)^{-1/2}) - D^2(\mathbf{X}))\mathbf{J}||^2,$$

where proximities are derived simultaneously from all $\mathbf{G}_j$ separately. The columns of the indicator matrix $\mathbf{G}_j$ are divided by the square root of the marginals $\mathbf{G}_j'\mathbf{G}_j$; the latter operation defines the chi-squared metric. Finally, the proximities are approximated by Euclidean distances in $\mathbf{X}$. As before, in the classical scaling approach, $\mathbf{X}$ would not be normalized to represent the subjects in an orthonormal cloud, but instead the eigenvalues are used to give the representation space a certain shape, displaying the differential saliences.

In Meulman (1986, 1992) an alternative is proposed, which is to analyse multivariate data by minimizing a loss function that is directly defined on the distances (so it does not approximate distances through inner products). The history of least squares MDS methods can be followed from Shepard (1962), Kruskal (1964), Guttman (1968), Takane, Young, and De Leeuw (1977), De Leeuw and Heiser (1980), Ramsay (1982), a.o. Least squares MDS methods are traditionally applied to a given proximity matrix, whose proximities are then approximated through minimization of some least squares loss function that is defined on (transformations of) proximities and distances in a representation space $\mathbf{X}$. In the multivariate cases described above, we derive the proximities from the multivarate data. Then, in the distance-based modification of principal components analysis (distance-based PCA, for short), we minimize

$$\text{STRESS}(\mathbf{X}) = ||D(\mathbf{Q}) - D(\mathbf{X})||^2 \tag{7}$$

over $\mathbf{X}$. As before, $D(\bullet)$ is an $n \times n$ matrix with Euclidean distances between subjects, and $\mathbf{X}$ is the low-dimensional space in which distances should match the proximities as closely as possible. To minimize loss function (7), we have to use an iterative procedure since there is no analytic solution. In the present paper, the majorization algorithm for MDS has been used (e.g., see De Leeuw and Heiser, 1980; Groenen and Heiser, 1996). The majorization approach amounts to computing an update for $\mathbf{X}$ that reduces the value of (7) from a starting point $\mathbf{X}^\circ$ by:

$$\mathbf{X} = 1/nB(\mathbf{X}^\circ)\mathbf{X}^\circ. \tag{8}$$

Here, the $n \times n$ matrix $B(\mathbf{X}^\circ)$ is defined as

$$B(\mathbf{X}^\circ) = B^+(\mathbf{X}^\circ) - B^*(\mathbf{X}^\circ), \tag{9}$$

where the elements of the matrix $B^*(\mathbf{X}^\circ)$ are given by $b_{ik}^*(\mathbf{X}^\circ) = d_{ik}(\mathbf{Q})/d_{ik}(\mathbf{X}^\circ)$ if $i \neq k$ and $d_{ik}(\mathbf{X}^\circ) \neq 0$; otherwise $b_{ik}^*(\mathbf{X}^\circ) = 0$. The elements of the diagonal matrix $B^+(\mathbf{X}^\circ)$ are given by $b_{ik}^+(\mathbf{X}^\circ) = \mathbf{1}'B^*(\mathbf{X}^\circ)\mathbf{e}_i$, where $\mathbf{e}_i$ is the ith column of the identity matrix $\mathbf{I}$. Repeatedly computing the update $\mathbf{X}$ gives a convergent series of configurations.

A special feature of the Gifi-system is the possibility of differential treatment of variables in the analysis. For example, some variables may be treated as containing numerical scores, while others may be treated as nominal variables. The latter treatment is appropriate when a variable partitions the subjects into unordered classes. In distance-based PCA nominal treatment of variables is carried out as follows. First, the nominal variable $\mathbf{h}_j$ is replaced by a binary indicator matrix $\mathbf{G}_j$ with n rows and l_j columns, as above. Then, proximities are derived simultaneously from $\mathbf{Q}$ and $\mathbf{G}_j$:

$$\Delta(\mathbf{Q}; \mathbf{G}) = D(\mathbf{Q}; \mathbf{G}_j(\mathbf{G}_j'\mathbf{G}_j)^{-1/2}) \tag{10}$$

where $j = 1, ..., m$ is the running index to indicate the variables in the analysis that classify the subjects into groups (there may be more than one classifying variable, and then multiple indicator matrices should be created). As in MCA and homogeneity analysis, the columns of the indicator matrix $\mathbf{G}_j$ are divided by the square root of the marginals $\mathbf{G}_j'\mathbf{G}_j$ to give distances in the chi-squared metric. Finally, the proximities $\Delta(\mathbf{Q}; \mathbf{G})$ are approximated by Euclidean distances $D(\mathbf{X})$.

Meulman (1992) has shown that if we apply the classical scaling approach (as in Torgerson, 1958; Gower, 1966) to approximate $\Delta(\mathbf{Q}; \mathbf{G})$ then this results in a solution for $\mathbf{X}$ that is equivalent to the subject scores in Gifi's PCA, with numerical and nominal variables. (Again, apart from a scaling factor per dimension, displaying the differential saliences; PCA usually displays the subject points as an orthonormal cloud, with equal dispersions.)

2. Supplementary representation of variables in a MDS solution

The idea of joint representation of subjects and variables, which originates with Tucker (1960), was subsequently very succesfully applied in the analysis of preference data (Carroll, 1972), and has become well-known as the biplot (Gabriel, 1971); the classical reference to the basic idea of lower-rank approximation is Eckart and Young (1936). A recent book on biplots is Gower and Hand (1996). From the biplot point of view, principal components analysis can be regarded as a bilinear model (Kruskal, 1978). (A variable-oriented approach would regard PCA as the analysis of a correlation matrix.) In the bilinear model, we minimize

$$\sigma(\mathbf{X}) = ||\mathbf{Q} - \mathbf{X}\mathbf{A}'||^2, \tag{11}$$

over $\mathbf{X}$ and $\mathbf{A}$: the observed scores in the m-dimensional space $\mathbf{Q}$ are approximated by the inner product of the p-dimensional component scores $\mathbf{X}$ and component loadings $\mathbf{A}$ (with p much smaller than m). Graphically, in the biplot representation, subjects are represented as points, and variables as vectors, and the orthogonal projection of the subject points onto the variable vectors gives an approximation of the observed scores.

In the framework of distance-based PCA, as in (7), variables can be considered from an *internal* and an *external* perspective. From the internal perspective, their role is to provide the proximities between the subjects. From the external perspective, the variables can be used afterwards to study whether we can account for the structure among the subjects. The latter can be done through linear and nonlinear *external* biplot methods. We call these biplots external, because the variables are fitted into the subject space in a second step, while the subject points $\mathbf{X}$ are kept fixed. By contrast, in internal biplots, the subject points and the vectors representing the variables are found simultaneously.

2.1 Linear external biplots through multiple regression

A straightforward way to fit a set of variables in a given configuration is through "property fitting" (e.g., Carroll, 1972; also, see Meulman, Heiser and Carroll, 1987). For distance-based PCA, this amounts to the projection of a variable $\mathbf{q}_j$ into the space of the subjects $\mathbf{X}$ by the use of multiple regression. In the regression, the columns in $\mathbf{X}$ are the independent variables, and the weights obtained from the regression determine the coordinates for the variable $\mathbf{q}_j$ in the space $\mathbf{X}$. The optimal direction $\mathbf{a}_j$ for the vector representing variable $\mathbf{q}_j$ in the p-space $\mathbf{X}$ is thus found as

$$\mathbf{a}_j = (\mathbf{X}'\mathbf{X})^{-1}\mathbf{X}'\mathbf{q}_j. \tag{12}$$

Using $\mathbf{a}_j$ to represent the endpoint of the vector gives a linear biplot representation. This biplot is obtained, however, through the use of different rationales for fitting the subjects on the one hand and the variables on the other: the subject points are fitted through the use of least squares distance fitting in (7), and the vectors representing the variables through ordinary multiple regression in (12).

2.2 Nonlinear external biplots through unfolding

As a possible alternative, coherent, method, Meulman and Heiser (1993) proposed a least squares generalization of the so-called nonlinear biplot in Gower and Harding

(1988). The latter nonlinear biplot was developed to obtain nonlinear representations of variables in a space that is generated through a principal coordinates analysis. The procedure discussed in Meulman and Heiser (1993) can be described as follows. First, regard each variable as a series of $s = 1, ..., S$ supplementary points (a trajectory) in the space $\mathbf{X}$. In terms of the data, a supplementary point for variable j has coordinates $\mathbf{e}_{r_s}$ in $\mathbf{X}$ that are all equal to zero, except for the jth variable; so when $\mathbf{e}_j$ is the jth column of the $m \times m$ identity matrix $\mathbf{I}$, $\mathbf{e}_{r_s} = r_s \mathbf{e}_j$, where $min(\mathbf{q}_j) \leq r_s \leq max(\mathbf{q}_j)$. Next, for each supplementary point, the distance is calculated to the n original points in observation space. The vector with squared distances between supplementary point $\mathbf{e}_{r_s}$ and the subjects in $\mathbf{Q}$ is given by

$$d^2(\mathbf{e}_{r_s}; \mathbf{Q}) = \mathbf{v} + \mathbf{1}\mathbf{e}'_{r_s}\mathbf{e}_{r_s} - 2\mathbf{Q}\mathbf{e}_{r_s} = \mathbf{v} + \mathbf{1}r_s^2 - 2\mathbf{Q}\mathbf{e}_{r_s}, \tag{13}$$

with $\mathbf{v}$ the n-vector containing the diagonal elements of $\mathbf{QQ}'$. The kth element of $d^2(\mathbf{e}_{r_s}; \mathbf{Q})$ gives the squared distance between the sth supplementary point and the kth subject point in observation space, and will be written as $d^2(\mathbf{e}_{r_s}; \mathbf{q}_k)$, where $\mathbf{q}_k$ denotes the kth row of $\mathbf{Q}$. Mapping the trajectory for variable $\mathbf{q}_j$ involves the approximation of $d(\mathbf{e}_{r_s}; \mathbf{Q})$ by $d(\mathbf{y}_s; \mathbf{X})$, where $\mathbf{y}_s$ gives p-dimensional coordinates in the space of $\mathbf{X}$, for different values of $r_s, s = 1, ..., S$. Here S denotes a prechosen number, appropriate to cover the range of $min(\mathbf{q}_j)$ to $max(\mathbf{q}_j)$. Each supplementary point has to be mapped separately, and the coherent method with respect to least squares multidimensional scaling as in (7) is the use of

$$\text{STRESS}(\mathbf{y}_s) = ||d(\mathbf{e}_{r_s}; \mathbf{Q}) - d(\mathbf{y}_s; \mathbf{X})||^2, \tag{14}$$

to be minimized over $\mathbf{y}_s$ for given $\mathbf{Q}$ and $\mathbf{X}$. The loss function (14) represents the least squares external unfolding problem; it is called unfolding because it fits distances between two sets of points, $\mathbf{X}$ and $\mathbf{y}$, and it is called external because $\mathbf{X}$ is known and fixed. There is no closed-form solution for STRESS-based external unfolding as in (14), so the loss function has to be minimized iteratively. The latter can be done using the SMACOF framework for unfolding (Heiser, 1981; 1987). The points $y_1, ...y_s, ..., y_S$, mapped in $\mathbf{X}$ by minimizing (14), will in general not be on a straight line, because (14) represents a nonlinear mapping. Nonlinear biplot representations have interesting properties, but some of them are not yet fully understood and are currently under study (Groenen and Meulman, 1995). In most cases, nonlinear biplots are harder to interpret than linear ones, and therefore here a biplot is presented that is linear and, unlike regression, at the same time consistent with the criterion minimized in distance-based PCA.

3. The Distance-Based Biplot

The basic notion to arrive at a simple, linear, and coherent biplot is that distances $D(\bullet)$ are invariant under rotation, i.e., $D(\mathbf{X}) = D(\mathbf{XA}')$ if $\mathbf{A}'\mathbf{A} = \mathbf{I}$. Usually, a rotation matrix is of the order $p \times p$; here, however, we will consider a matrix $\mathbf{A}$ of order $m \times p$. This matrix will be labeled a *rotation-expansion* matrix, because the transformation preserves the distances in $\mathbf{X}$, and at the same time expands the representation space. Thus, the coordinates $\mathbf{X}$ in p-space are replaced by m-dimensional coordinates in the space $\mathbf{XA}'$ of rank p, and since $\mathbf{A}'\mathbf{A} = \mathbf{I}$

$$\text{STRESS}(\mathbf{X}) = ||D(\mathbf{Q}) - D(\mathbf{X})||^2 = ||D(\mathbf{Q}) - D(\mathbf{XA}')||^2. \tag{15}$$

To obtain the biplot coordinates, we have to minimize $||\mathbf{Q} - \mathbf{XA}'||^2$ over all $\mathbf{A}$ satisfying $\mathbf{A}'\mathbf{A} = \mathbf{I}$. This amounts to an orthogonal Procrustes problem of the order $p \times p$ that is solved as follows.

Define the singular value decomposition of the $m \times p$ matrix $\mathbf{Q'X}$ as $\mathbf{Q'X = K\Lambda L'}$, and the eigenvalue decomposition of the $p \times p$ matrix $\mathbf{X'QQ'X}$ as $\mathbf{X'QQ'X = L\Lambda^2 L'}$. Then the rotation-expansion matrix is found by

$$\mathbf{A = KL' = Q'XL\Lambda^{-1}L'}. \tag{16}$$

Now the m-dimensional coordinate system $\mathbf{XA'}$ can be used to evaluate the MDS solution directly in terms of the original variables, with the Pearson correlation coefficient as a natural measure of association. At the same time, the jth row in $\mathbf{A}$ (denoted by $\mathbf{a}_j'$) gives the coordinates to display the variable $\mathbf{q}_j$ in the space $\mathbf{X}$. The scores $\{q_{ij}\}$ themselves can be represented as well; the projected coordinates in the space $\mathbf{X}$ are given by $\mathbf{q}_j \mathbf{a}_j'/\mathbf{a}_j'\mathbf{a}_j$. The latter set of quantities are called single category coordinates in Gifi's (1990) approach to PCA. Therefore, the approach proposed here can be regarded as a distance-based alternative for Gifi's biplot display of multivariate data. The series of points $\mathbf{q}_j\mathbf{a}_j'/\mathbf{a}_j'\mathbf{a}_j$ are located on the vector that represents the variable $\mathbf{q}_j$, and these are usually called markers (Gabriel, 1971; Gower and Hand, 1996).

4. Material and Methods

The data that are used in the application of the distance-based biplot were collected by Van Strien and Van der Ham, at the Department of Psychiatry at Utrecht University Hospital (see Van der Ham, Meulman, Van Strien and Van Engeland, 1997). The data concern 16 variables that measure well-being at four points in time for 55 patients with eating disorders. The patients were diagnosed independently into four categories (using the DSM-III-R): 1. Anorexia Nervosa ($N_1 = 25$), 2. Anorexia with Boulimia Nervosa ($N_2 = 9$), 3. Boulimia Nervosa after Anorexia ($N_3 = 14$), and 4. Atypical Eating Disorder ($N_4 = 7$). The total number of patients $N = 55$, and data are available at four different time points, and the total number of observational units would be $4 \times 55 = 220$. However, there are a few subjects with missing observations at one of the four time points, so the actual number of observational units is $55 + 53 + 54 + 55 = 217$.

In addition to the 16 variables that measure well-being, two additional nominal variables are used, the first a variable (with four categories) that associates each observational unit with a point in time, and the second a variable that links each patient with a diagnosis (in one of the four eating disorder categories; the diagnosis was established before time point one, and does not change over time). In summary, the multidimensional scaling analysis is applied to a 217×217 matrix with proximities between the observational units, with the proximities derived from $16 + 4 + 4 = 20$ variables (time and diagnosis each have four categories). The analysis is intended to result in a subgrouping of the eating disorders in a longitudinal perspective.

The multidimensional scaling task was performed in two dimensions; the solution as a whole is judged on two different criteria. First, the correlation between the original variable $\mathbf{q}_j$ and the fitted $\mathbf{Xa}_j'$ is considered; variables were fitted into the subject space by the biplot method discussed in Section 3. The Pearson correlation coefficient is a natural goodness-of-fit measure for the biplot representation, since standard PCA maximizes the average squared correlation. So for the distance-based biplot, we can compare this particular goodness-of-fit measure with the optimum provided by standard PCA. Second, we consider classification of subjects on the basis of the Euclidean distances in two-dimensional space. Each subject is allocated to one of the four time category points and to one of the four diagnosis points. Specifically, the centroids of the subject points in a particular class define the associated category points. In addition to Euclidean distance, the allocation used the posterior probabilities, by employing Bayes rule and the a priori distribution over the categories. The resulting assignment is compared to the original time points and diagnostic group classification.

5. Results

The primary result of the analysis consists of the coordinates for the observational units in two-dimensional space; the graph displaying the cloud of points is given in Figure 1, top panel. The subjects have been labeled with their diagnosis. Visual inspection immediately suggests that the second dimension is related to the diagnostic categories of eating disorder. We see that the anorexia subjects (label 1) form a group, but patients with atypical eating disorder (label 4) form a subgroup. The latter patients can be considered as anorectic patients for whom the loss of weight is unknown or less than 15%. The second dimension separates the anorectic patients (classes 1 and 4) from the boulimic patients (classes 2: anorexia nervosa with boulimia nervosa, and 3: boulimia nervosa after anorexia). Having connected the diagnosis category points over the different points in time (by computing the appropriate centroids of subject points), it is clear that the first dimension displays the development in time. The variables are displayed in the bottom panel of Figure 1. Instead of displaying the variables and subjects together, it was chosen to display the variables with the group points, since this gives a more comprehensive biplot; groups of subjects are represented by their centroid, which is associated with a particular point in time and diagnosis.

The first thing to note is that all variables have a positive correlation with the first dimension; this means there is a general factor that correlates positively with all the variables. The second dimension separates the variables. We find three bundles of variables: Bingeing (4), Vomiting (5) and Purging (6) are clearly distinguished in the vertical direction of the graph; Preoccupation (15), Body Perception (16), Hyperactivity (7), Sexual Behavior (13), and Fasting (3) correlate most with the general factor, and Weight (1), Menstruation (2), Family Relations (8), Emancipation (9), Work/School record (11) and Sexual Attitude (12) form a third bundle of variables. Friends (10) and Mood (14) do not fit very well in the overall representation. The fit of the variables, as measured by the Pearson correlation between observed scores and fitted scores, is given in Table 1. We notice that distance-based PCA gives a very decent fit for the variables, compared to the maximum that could be obtained by applying standard PCA.

As described above, we compare the classification of the subjects on the basis of the results from both analyses with the original ones. With respect to time, the grouping is given in Table 2. Here, rows indicate the original time points, and columns the fitted time points. From the percentage correctly assigned subjects in the bottom row, we conclude that distance-based PCA performs better than standard PCA, although it is obviously hard to distinguish between consecutive time points, especially between 3 and 4. Next, we inspect assignment to the eating disorders categories. Results are given in Table 3, with the rows and columns ordered according to the subgroupings. It is clear that neither approach to PCA can distinguish between groups 1 and 4 on the one hand, and groups 2 and 3 on the other hand. On the whole, distance-based PCA performs better. This becomes more clear when we combine results from Table 3 in its bottom row: now standard PCA finds 83% and 91% correctly, while distance-based PCA obtains 92% and 97%.

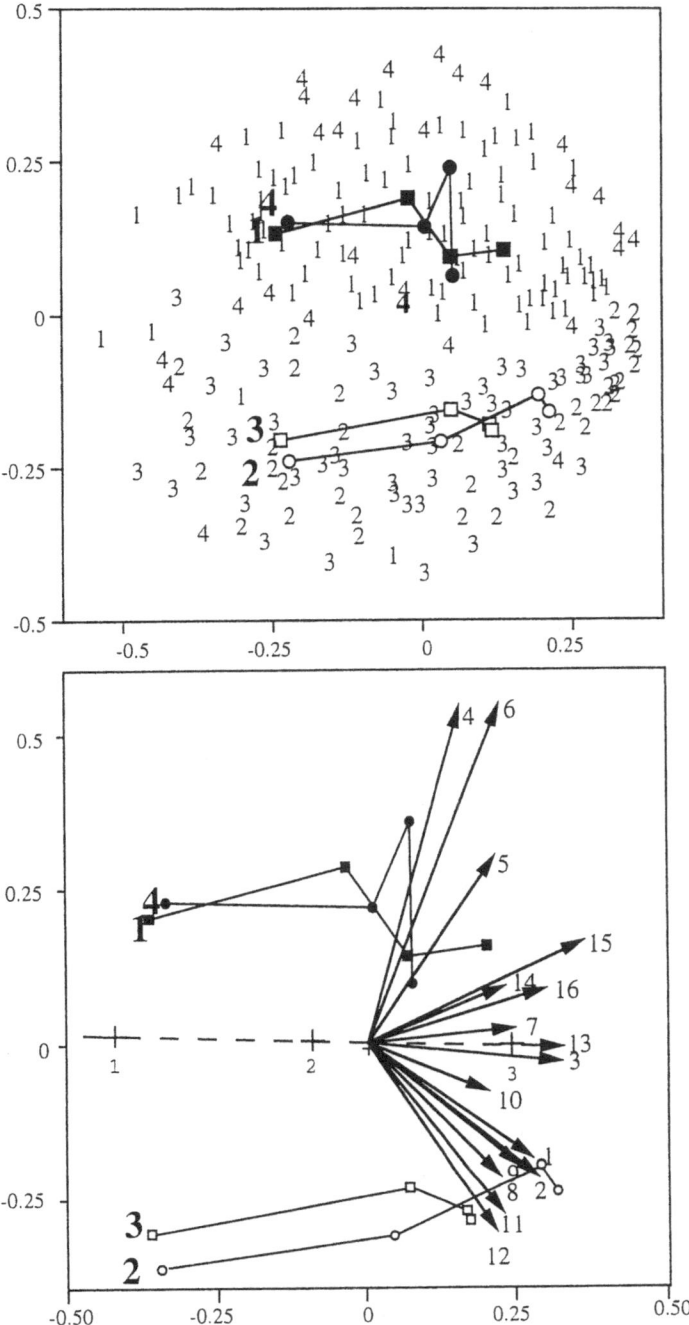

Figure 1. Top panel: Subjects represented in two-dimensional space. Labels 1: anorexia, 2: anorexia with boulimia, 3: boulimia after anorexia, 4: atypical eating disorder. Trajectories represent the diagnostic categories in time (from left to right). Lower panel: Trajectories, now displayed with the variables (described in Table 1). Markers represent the three categories for variable 13.

Table 1: Correlations between Observed Variables and Fitted
Variables in Standard and Distance-Based PCA

Variables	Standard PCA	D-Based PCA
Weight	0.69	0.68
Menstruation	0.72	0.72
Fasting	0.70	0.72
Bingeing	0.79	0.71
Vomitting	0.62	0.55
Purging	0.88	0.79
Hyperactivity	0.53	0.53
Family Relations	0.58	0.58
Emancipation	0.64	0.64
Friends	0.46	0.44
Work/School	0.60	0.63
Sexual Attitude	0.62	0.62
Sexual Behavior	0.70	0.71
Mood	0.47	0.50
Preoccupation	0.76	0.74
Body Perception	0.63	0.63
Mean	0.66	0.63

Tabel 2: Classification of Subjects
in Time Categories.

	Standard PCA				D-Based PCA			
	1	2	3	4	1	2	3	4
1	47	6	2	0	48	5	2	0
2	13	15	10	15	13	16	5	19
3	9	15	5	25	6	16	8	24
4	6	10	4	35	5	12	2	36
% Correct	.85	.28	.09	.64	.87	.30	.15	.65

Table 3: Classification of Subjects
in Diagnosis Categories

	Standard PCA				D-Based PCA			
	1	4	2	3	1	4	2	3
1	82	2	11	2	92	0	3	2
4	20	0	5	3	22	1	2	3
2	3	0	15	18	1	0	19	16
3	2	3	20	31	2	0	25	29
% Correct	.85	.00	.42	.55	.95	.04	.53	.52
Combined	0.83		0.91		0.92		0.97	

6. Monte Carlo Study

By definition, the distance-based biplot is outperformed by ordinary PCA with respect to Pearson correlations between observed and fitted variables. The differences
in our empirical example are small, however, and distanced-based PCA performs better with respect to the recovery of the original classification of subjects into groups.

Table 4: Correlations between True Scores and Fitted
Scores in Distance-based PCA compared to Standard PCA

Variables	D-Based PCA	Standard PCA
1	.80	.79
2	.81	.79
3	.79	.78
4	.78	.75
5	.77	.73
6	.73	.69
7	.68	.64
8	.65	.60
9	.66	.59
10	.57	.54
11	.51	.45
dim 1	.93	.92
dim 2	.66	.56
dim 3	.27	.20

To inspect these properties in a more general context, (replicated) artificial data were generated, with 75 subjects and 13 variables with a perfect representation in three dimensions. From this set, two variables were selected as partitioning variables, and five categories were created using an optimal discretization strategy. The remaining 11 variables were subjected to a fair amount of random error (with an average of 53%). The resulting set of variables was analyzed by distance-based PCA, and compared to standard PCA. In both cases, analyses were done in two dimensions. The number of replications was set to 100. Four criteria were inspected:

• 1. The fit per variable, as measured by the Pearson correlation between the true scores (without error) and the bilinear approximation.

• 2. The fit per dimension, as measured by the Pearson correlation between the true dimensions and the fitted dimensions.

• 3. The correct classification of subjects in two-space as compared to the original classes. The a priori distribution in the population was taken into account to compute the posterior probabilities.

• 4. The distances between the subjects in two-space as compared to the distances in true-space.

The results reported below were obtained by averaging the results after applying distance-based and standard PCA to the 100 samples of the artificially created structure described above. The first part of Table 4 gives the correlations between true scores and fitted scores; the second part reports on the results with respect to the original three dimensions (that are approximated in two-space; the correlations here are again obtained by applying the rotation-expansion strategy from Section 3, but now to the original dimensions; so if the original dimensions are denoted by $\mathbf{Z}$, we minimize $||\mathbf{Z} - \mathbf{XB'}||^2$ over all $\mathbf{B}$ satisfying $\mathbf{B'B} = \mathbf{I}$, and we compute the correlations between $\mathbf{Z}$ and $\mathbf{XB'}$). We notice that distance-based PCA performs better for each variable and each dimension separately. Results for the two classification variables were combined; the aggregated results are given in Table 5, where again rows indicate the original categories and columns the fitted categories. Except for the third category (75% versus 76% correct), distance-based PCA performs better; this effect is strongest for the extreme categories 1 and 5 (88% versus 78% and 87% versus 76% correct). Finally, the overall statistics are given in Table 6, confirming the superior performance of distance-based PCA over standard PCA.

Table 5: Classification of Subjects
in the Monte Carlo Study

	D-Based PCA					Standard PCA				
	1	2	3	4	5	1	2	3	4	5
1	.88	.11	.01	.00	.00	.78	.20	.02	.00	.00
2	.08	.81	.11	.01	.00	.05	.74	.19	.02	.00
3	.01	.12	.76	.10	.01	.00	.13	.75	.12	.00
4	.00	.01	.10	.80	.09	.00	.02	.18	.76	.05
5	.00	.00	.01	.13	.87	.00	.00	.02	.22	.76

Table 6: Summary of Results
of the Monte Carlo Study:
Overall Statistics

	D-Based PCA	Standard PCA
Distances	0.91	0.88
Variables	0.74	0.70
Dimensions	0.62	0.56
Classification	0.81	0.75

7. Discussion

A distance-based biplot was developed for a least squares MDS analysis of multivariate data. The fitted p-dimensional space is expanded into an m-dimensional space of rank p that directly represents the fitted variables, while distances between subjects are preserved. The biplot method was applied to a data concerning patients with various types of eating disorders. The variables obtained a decent fit, also when compared to standard PCA. The method was further studied in a Monte Carlo study. Here results were compared with respect to the true structure, and the method proposed performed better than standard PCA with respect to the criteria considered. The analysis allows nominal variables to be included; their categories are represented by centroids of subjects. The method could include optimal scoring of ordinal variables too. By representing nominal variables as points in the subject space, and the other variables as vectors in the same space, we have actually developed a triplot, with subjects, variables, and classes as its constituents.

References:

Carroll, J.D., and Chang, J.J. (1972): *IDIOSCAL (Individual differences in orientation scaling): A generalization of INDSCAL allowing IDIOsyncratic reference systems as well as an analytic approximation to INDSCAL*, Paper presented at the Psychometric Society Meeting, Princeton, NJ.

Carroll, J. D. (1972): Individual differences and multidimensional scaling, In: *Multidimensional scaling: Theory and applications in the behavioral sciences*, R. N. Shepard, A. K. Romney, and S. B. Nerlove (eds.), Vol. 1, 105-155, Seminar Press, New York and London.

De Leeuw, J., and Heiser, W.J. (1980): Multidimensional scaling with restrictions on the configuration, In: *Multivariate analysis*, P.R. Krishnaiah (ed.), , Vol. V, 501-522, North-Holland, Amsterdam.

Eckart, C. and Young, G. (1936): The approximation of one matrix by another of lower rank, *Psychometrika*, 1, 211-218.

Gabriel, K.R. (1971): The biplot graphic display of matrices with application to principal components analysis, *Biometrika*, **58**, 453-467.

Gifi, A. (1990): *Nonlinear multivariate analysis*, John Wiley and Sons, Chichester.

Gower, J.C. (1966): Some distance properties of latent roots and vector methods used in multivariate analysis, *Biometrika*, **53**, 325-338.

Gower, J.C. and Hand, D.J. (1996): *Biplots*, Chapman & Hall, London.

Gower , J.C., and Harding, S.A. (1988): Nonlinear biplots, *Biometrika*, **75**, 445-455.

Groenen, P.J.F. and Heiser, W.J. (1996): The tunneling method for global optimization, *Psychometrika*, **61**, (in press).

Groenen, P.J.F. and Meulman, J.J. (1995): *Joint nonlinear biplots through least squares MDS*, Paper presented at the 9th European Meeting of the Psychometric Society, Leiden.

Guttman, L. (1968): A general nonmetric technique for finding the smallest coordinate space for a configuration of points, *Psychometrika*, **33**, 469- 506.

Heiser, W.J. (1981): *Unfolding analysis of proximity data*, Dept. of Data Theory, Leiden.

Heiser, W.J. (1987): Joint ordination of species and sites: the unfolding technique, In: *Developments in numerical ecology*, P. Legendre and L. Legendre (eds.), 189-221, Springer, New York.

Kruskal, J.B. (1964): Multidimensional scaling by optimizing goodness of fit to a nonmetric hypothesis, *Psychometrika*, **29**, 1-28.

Kruskal, J.B. (1978): Factor analysis and principal components analysis: bilinear methods. In: *International encyclopedia of statistics*, W.H. Kruskal and J.M. Tanur (eds.), 307-330, The Free Press, New York.

Meulman, J.J. (1986): *A distance approach to nonlinear multivariate analysis*, DSWO Press, Leiden.

Meulman, J.J. (1992): The integration of multidimensional scaling and multivariate analysis with optimal transformations of the variables, *Psychometrika*, **57**, 539-565.

Meulman, J.J., and Heiser, W.J. (1993): Nonlinear biplots for nonlinear mappings. In: *Information and Classification: Concepts, Methods and Applications*, O. Opitz, B. Lausen, and R. Klar (Eds), 201-213, Springer Verlag, Berlin.

Meulman, J.J., Heiser, W.J., and Carroll, J.D. (1987): *PREPMAP-3 User's Guide*, Bell Telephone Laboratories, Murray Hill, NJ.

Ramsay, J.O. (1982): Some statistical approaches to multidimensional scaling data, *Journal of the Royal Statistical Society, Series A*, **145**, 285-312.

Shepard, R.N. (1962): The analysis of proximities: Multidimensional scaling with an unknown distance function I and II, *Psychometrika*, **27**, 125-140, 219-246.

Takane, Y., Young, F.W., and De Leeuw, J. (1977): Nonmetric individual differences multidimensional scaling: An alternating least squares method with optimal scaling features, *Psychometrika*, **42**, 7-67.

Torgerson, W.S. (1958): *Theory and methods of scaling*, Wiley, New York.

Tucker, L. R (1960): Intra-individual and inter-individual multidimensionality. In: *Psychological Scaling: Theory and Applications*, H. Gulliksen and S. Messick (eds.), Wiley, New York.

Van der Ham, Th., Meulman, J.J., Van Strien, D.C, and Van Engeland, H. (1997): *Empirically based subgrouping of eating disorders in adolescents: A longitudinal perspective, Britisch Journal of Psychiatry*, in press.

Young, G. and Householder, A.S. (1938): Discussion of a set of points in terms of their mutual distances, *Psychometrika*, **3**, 19-22.

Latent-class scaling models for the analysis of longitudinal choice data

Ulf Böckenholt[1]

[1] University of Illinois at Urbana-Champaign
Department of Psychology
603 E. Daniel Street
Champaign, Il 61820

Summary: A new class of multidimensional scaling models for the analysis of longitudinal choice data is introduced. This class of models extends the work by Böckenholt and Böckenholt (1991) who proposed a synthesis of latent-class and multidimensional scaling (mds) models to take advantage of the attractive features of both classes of models. The mds part provides graphical representations of the choice data while the latent-class part yields a parsimonious but still general representation of individual differences. The extensions discussed in this paper involve simultaneously fitting the mds latent-class model to the data obtained at each time point and modeling stability and change in preferences by autocorrelations and shifts in latent-class membership over time.

1. Introduction

Over the years, latent class models have proven to be a versatile tool for the analysis of panel, survey, or experimentally derived choice data. For example, numerous marketing applications demonstrate that these models can provide useful insights into market structure by simultaneously segmenting and structuring a market at a particular point in time or over time with the use of panel data. However, to some extent the interpretation of classification results obtained by these models is complicated by the fact that no information is provided about the perceptual space of the choice options or the decision process of a respondent. To overcome this limitation, Böckenholt and Böckenholt (1991) presented a synthesis of latent-class and multidimensional scaling (mds) models that takes advantage of the attractive features of both classes of models. The mds part determines the perceptual space of the choice options while the latent class part yields a parsimonious but still general representation of individual differences.

This paper introduces an extension of this approach for the analysis of stability and change in choice data collected at two time points. The new class of models represents persons and items in a joint space. Individual differences are captured by different vector termini or ideal point positions in a multidimensional space. Person-specific changes are modelled by allowing for switching among the latent classes over time. Changes in the perception of items are modelled by drifts in the item parameters. As a result, the proposed approach provides a much stronger test of the stability and validity of a multidimensional representation than possible on the basis of a data set collected at a single point in time. Moreover, a rich set of hypotheses can be tested to determine the locus of change in the time-dependent scaling results.

The remainder of this paper is structured as follows. First, parametric representations of ideal and vector models are reviewed. Next, a general modeling framework is presented for the analysis of choice data collected at two time points. Various special cases of the approach are derived. The paper concludes with an analysis of a sociometric choice data set.

2. Latent class scaling models

A natural and economic way of investigating choice behavior is to ask respondents to select the preferred items from a set of items. Coombs (1964) termed these data pick any when the set of items is unconstrained, and pick any/n when the set of items is fixed. In the latter case, a decision outcome may be written as $i, j, \bar{k}, \ldots, q$ indicating that options i, j, and q are chosen and option k is not chosen. Clearly, persons may differ strongly in their preferences for the various items (Takane, 1983). Latent-class models are well-suited to account for these individual preference effects by grouping respondents with high intragroup similarity and large intergroup differences (Lazarsfeld and Henry, 1968). Thus, preference variability is described by assigning judges to latent classes such that members of a latent class share the same choice probabilities. The unobserved classes are determined by invoking the principle of local independence which states that the latent-class membership variable accounts completely for any dependencies among the observed choices.

Let $\pi_{i|a}$ denote the conditional probability that item i is selected by members of class a. Under the local independence representation of the latent-class model the joint probability of observing the choice outcome $(i, j, \bar{k}, \ldots, q)$ is

$$\Pr(i, j, \bar{k}, \ldots, q) = \sum_a \pi_a \pi_{i|a} \, \pi_{j|a} \, (1 - \pi_{k|a}) \, \ldots \, \pi_{q|a} \tag{1}$$

where π_a represents the relative size or proportion of class a and $\sum_{a=1}^{S} \pi_a = 1$. Although the unconstrained latent-class model is well-suited to describe individual preference differences, it does not furnish information about the perception of the items and the response process of the respondents. However, this information can be extracted from the data by constraining the class-specific probabilities, $\pi_{i|a}$, to be a function of a scaling model. In particular, the ideal point and vector models may prove useful in supplying succinct graphical representations of the choice behavior of the respondents.

Böckenholt and Böckenholt (1991) showed that latent-class models and scaling models can be combined by setting $\pi_{i|a} = \Phi(z_{i|a})$ and Φ is the normal cumulative distribution function. The deviate $z_{i|a}$ is expressed as a function of an unfolding model or a vector model. For example, the ideal point model specifies that an item is selected when it is close to a person's ideal point position. As a result, for members of class a we may write

$$z_{i|a} = -\tau_i - \sum_{h=1}^{r} (\iota_{ah} - \beta_{ih})^2, \tag{2}$$

where τ_i is an item-specific threshold parameter, and ι_{ah} is the ideal point position of class a on dimension h. The location of item i on dimension h is denoted by β_{ih}. Thus, according to (2) the items' locations are perceived homogeneously in an r-dimensional space, however, the positions of the ideal points differ among the classes. The smaller the distance is between an item and a class' ideal point position, the more members of this class prefer the item.

Occasionally, a special case of the ideal point model, the vector model, may prove sufficient for describing individual differences in choice data. According to this model each class is characterized by a preference vector $\boldsymbol{\nu}_a = (\nu_{a1}, \nu_{a2}, \ldots, \nu_{ar})$. An item's projection onto the preference vector of a class determines its probability of being chosen,

$$z_{i|a} = \tau_i + \sum_{h=1}^{r} \nu_{ah} \beta_{ih} \tag{3}$$

where ν_{ah} is the h-th element of the preference vector ν_a.

By combining the vector or ideal point model with the latent-class representation we get

$$\Pr(i, j, \bar{k}, \ldots, q) = \sum_a \pi_a \Phi(z_{i|a})\, \Phi(z_{j|a})\, \Phi(-z_{k|a}) \ldots \Phi(z_{q|a}) \tag{4}$$

Clearly, when $z_{i|a}$ is unconstrained the latent class representation in (1) is obtained.

3. Modeling stability and change

For the investigation of stability and change in choice data consider the situation of N individuals measured at two time points. If preferences or attitudes are stable, both item and person parameters are time-homogeneous. In contrast, systematic response differences at both time points may indicate changes in the perception of (some of) the items and/or person-specific variability.

Within the framework of the latent-class scaling models, hypotheses about person-specific change can be tested by allowing for time-dependent switches among latent classes. Shifts among classes can be interpreted as shifts among preference states because each latent-class position corresponds to a particular preference state. As a result, the approach can account for both individual preference differences and preference changes over time.

Hypotheses about item-specific changes can be tested by letting the item location or threshold parameters vary over time. When the perceptual space of the items is not time-homogeneous, it is difficult to assess any changes in the person-specific parameters. Thus, although both hypotheses about the locus of change in choice data are of interest, the interpretation of the data is greatly simplified when item-specific changes are small and can be ignored.

In addition, to the analysis of preference-related effects, it is desirable to take into account "test-retest" effects (Hagenaars, 1990). In particular, when the time interval separating the two choice occasions is small, it is likely that choices among identical options by a person are not independent. Instead, outcomes of previous choices may strongly influence current choices. In this case the local independence representation is not appropriate and a more general approach that allows for correlated choices is needed.

Based on these considerations the latent-class mds model for two time points is written as

$$\Pr(i^{(1)}, j^{(1)}, \bar{k}^{(1)}, \ldots, q^{(1)}; i^{(2)}, j^{(2)}, \bar{k}^{(2)}, \ldots, q^{(2)}) = \sum_a \sum_b \pi_{ab}$$

$$\Phi_2(z_{i|a}^{(1)}, z_{i|b}^{(2)}; \rho_i)\ \Phi_2(z_{j|a}^{(1)}, z_{j|b}^{(2)}; \rho_j)\ \Phi_2(-z_{k|a}^{(1)}, -z_{k|b}^{(2)}; \rho_k)\ \ldots\ \Phi_2(z_{q|a}^{(1)}, z_{q|b}^{(2)}; \rho_q) \tag{5}$$

where $\Phi_2(z_{i|a}^{(1)}, z_{i|b}^{(2)}; \rho_i)$ is the bivariate normal distribution function with correlation coefficient ρ_i and upper integration limits $z_{i|a}^{(1)}$ and $z_{i|b}^{(2)}$. These limits are either unconstrained, or a function of the ideal point or vector models. For example, in the case of the unidimensional ideal point model

$$z_{i|a}^{(t)} = -\tau_i^{(t)} - (\iota_a - \beta_i^{(t)})^2. \tag{6}$$

The class size parameter π_{ab} refers to the probability of belonging to classes a and b at time points t and $t + 1$, respectively. Switching among class locations occurs whenever $\pi_{ab} > 0$ provided $a \neq b$.

Autodependencies of item i between two time points t and $t+1$ are represented by ρ_i. Note that this "test-retest" effect is not class-dependent. In contrast, dependencies between different items are accounted for by the latent class representation. Thus, by separately modeling between- and within-item dependencies we can distinguish individual difference effects and local associations in the choices.

Model (5) includes a variety of published models as special cases. When the correlation between choices among identical options is zero, and $z_{i|a}^{(t)}$ is unconstrained we obtain a class of latent change models originally proposed by Wiggins (1973). Special cases of this class include the latent Markov model (Langeheine and van der Pol, 1990; Poulsen, 1990) and the latent symmetry model (Böckenholt and Langeheine, 1996; Böckenholt, 1987). However, for testing the former model at least three time points are required.

Hypotheses of interest may be roughly classified as follows: (a) no change, (b) person-specific change, (c) item-specific change, and (d) test-retest effects. To test the no-change hypothesis, we set $z_{i|a}^{(1)} = z_{i|a}^{(2)}$ (for all items) and $\pi_{ab} = 0$ when $a \neq b$. Hypotheses regarding person-specific change can be tested by specifying a latent change mechanism. For example, under the symmetric change model we expect $\pi_{ab} = \pi_{ba}$. According to this constraint, an equal number of individuals switches from class a to class b as from class b to a. Changes in the perception of item i can be tested by constraining $\tau_i^{(1)} = \tau_i^{(2)}$ and $\beta_i^{(1)} = \beta_i^{(2)}$ for the ideal point and vector models. Similarly, test-retest effects for item i are investigated by setting $\rho_i = 0$.

4. Estimation and Model Testing

Parameter estimates of the time-dependent latent-class mds model can be obtained by using maximum likelihood methods. Under the assumption of random sampling of N subjects for the pick any/n task the log-likelihood function is written as

$$\ln L = c + \sum_u f_u \ln \{\Pr(\mathbf{y}_u)\} \tag{7}$$

where c is a constant, and f_u denotes the observed number of persons with selection vector $\mathbf{y}_u = (i^{(1)}, \ldots, q^{(1)}; i^{(2)}, \ldots, q^{(2)})$. The Expectation-Maximization (EM) algorithm is used for parameter estimation (Dempster, et al., 1977). The implementation of this algorithm is straightforward and not further discussed here because it is well documented in the literature. For example, Hathaway (1985) and Lwin and Martin (1989) provide a detailed discussion of the EM algorithm for estimating normal mixture models, and Böckenholt (1992) reviews methods for the computation of normal probabilities.

Large sample tests of fit are available based on the likelihood-ratio (LR) χ^2 statistic (G^2) and/or Pearson's goodness-of-fit test statistic (P^2). Asymptotically, if a latent-class mds model provides an adequate description of the data, then both statistics follow a χ^2-distribution with $(4^n - m - 1)$ degrees of freedom, where m refers to the number of parameters to be estimated. The LR test is most useful when the number of items is small. Otherwise, only a small subset of the possible choice patterns may be observed. In this case, it is doubtful that the test statistics will follow approximately a χ^2-distribution. However, useful information about a model fit may be obtained by inspecting standardized differences between the observed and expected model probabilities for certain partitions of the data (e.g., subsets of items). Although these residuals are not independent, a careful inspection of their direction and size is useful in identifying sources of systematic model misfits.

Nested latent-class mds models can be compared by computing the difference between their LR-statistics. This difference is asymptotically distributed as a χ^2-statistic with the degrees of freedom equal to the difference between the number of parameters in the unrestricted and restricted models. For example, the hypothesis of test-retest effects can be tested with n degrees of freedom by comparing the fits of Model (5) with and without correlation coefficients.

5. Sociometric Choices

This section presents the analysis of a small study with two items observed at two points in time. Although the number of items is too small for testing the mds part of the latent-class models, the data set is useful for demonstrating the importance of taking into account latent change and re-test effects in an analysis of longitudinal data.

In a sociometric choice investigation reported by Langeheine (1994) students were asked with respect to every classmate whether they would choose (C) or reject (R) this person on the basis of two criteria measuring interpersonal attractiveness. Criterion i is "share a table in case of a new seating arrangement" and criterion j is "share a tent if a class would go for a camping trip". Table 1 contains the results of this study for two measurement occasions separated by a one week interval. For example, the majority of responses (673) rejects class mates on the basis of the two criteria at both time points. For the most part the following results agree with the ones obtained by Langeheine (1994). The main difference between his and the analyses reported here is the application of Model (5).

Because the LR-test of the two-class no-change model,

$$\Pr(i^{(1)}, j^{(1)}; i^{(2)}, j^{(2)}) = \sum_{a=1}^{2} \pi_a \, \Phi(z_{i|a}^{(1)}) \, \Phi(z_{j|a}^{(1)}) \, \Phi(z_{i|a}^{(2)}) \, \Phi(z_{j|a}^{(2)}) \tag{8}$$

yields a $G^2 = 239.1$ with $df = 10$, it is justified to test whether the poor fit is a result of variability in the evaluation of the items over time, person-specific changes, or both. When allowing for switching among classes (i.e., $\pi_{ab} > 0$ when $a \neq b$) G^2 drops to 49.9 ($df = 8$) indicating a substantial latent change effect. In contrast, when, additionally, relaxing the assumption of time-homogenous item probabilities (i.e., $\Phi(z_{i|a}^{(1)}) \neq \Phi(z_{i|a}^{(2)})$), G^2 reduces by little to 49.7 ($df=4$). Clearly, item-specific effects can be ignored after allowing for person-specific change. Because of the short interval between the time points, it seems likely that the poor fit of the model is caused by strong autodependencies of the choices. Support for this hypothesis is obtained by inspecting the fitted frequencies of the latent change model in (the $\rho = 0$ column of) Table 1. In particular, the two choice patterns with identical choices are underpredicted. This result suggests that the fit of the model can be improved considerably by Model (5) because it allows for test-retest effects. The LR-test for this model with equal correlations for both items is $G^2 = 5.99$ ($df = 7$), and the estimated correlation coefficient is $\hat{\rho} = \hat{\rho}_i = \hat{\rho}_j = .49$. The predicted frequencies are given in (the $\hat{\rho} = .49$ column of) Table 1 and the estimated choice probabilities and class size parameters are listed in Table 2. We note that the class consisting of reject responses is twice as large as the class consisting of pick responses. However, the transition probabilities of both class are symmetric with $\hat{\pi}_{1|2} \simeq \hat{\pi}_{2|1}$.

Table 1: Observed and expected frequencies of sociometric choices

Choices				Obs. freq.	Exp. freq. $(\rho = 0)$	Exp. freq. $(\hat{\rho} = .49)$
$i^{(1)}$	$j^{(1)}$	$i^{(2)}$	$j^{(2)}$			
R	R	R	R	673	668.7	672.4
R	R	R	C	32	35.5	31.8
R	R	C	R	27	29.8	26.5
R	R	C	C	62	61.5	62.0
R	C	R	R	25	27.9	27.2
R	C	R	C	25	8.9	23.4
R	C	C	R	6	4.3	2.3
R	C	C	C	29	44.0	32.0
C	R	R	R	24	27.9	24.6
C	R	R	C	3	4.4	2.3
C	R	C	R	10	2.5	11.0
C	R	C	C	17	19.3	16.4
C	C	R	R	37	35.9	36.0
C	C	R	C	29	43.2	31.0
C	C	C	R	12	18.3	15.2
C	C	C	C	249	230.9	245.8

Table 2: Parameter estimates of Model (5)

Class	Size	Items		Transition Prob.	
	$\hat{\pi}_a$	i	j	$\hat{\pi}_{a|1}$	$\hat{\pi}_{a|2}$
1	.67	.04	.04	.90	.10
2	.33	.85	.93	.11	.89

6. Discussion

This paper presented latent-class scaling models for graphical representations of preference stability and change in choice data. The models provide a parsimonious framework for the analysis of individual taste differences. A rich set of hypotheses can be tested that examine the stability of the perceptual space of the items and shifts among different preference states for two time points. In addition, as illustrated in the application section test-retest effects can also be taken into account. Because the approach can be applied with minor modifications to different data types (i. e., rankings, first choices, preference ratings) it can be viewed as a general framework for obtaining graphical representations of preference changes.

Acknowledgements

This work was partially supported by the National Science Foundation (Grant No. SBR 94-09531).

References:

Böckenholt, U. (1992). Thurstonian models for partial ranking data. *British Journal of Mathematical and Statistical Psychology*, **45**, 31-49.

Böckenholt, U. (1997). Modeling time-dependent preferences: Drifts in ideal points. In: *Visualization of Categorical Data*, Greenacre, M., and Blasius, J. (eds.). Lawrence Erlbaum Press.

Böckenholt, U., and Böckenholt, I. (1991). Constrained latent class analysis: Simultaneous classification and scaling of discrete choice data. *Psychometrika*, **56**, 699-716.

Böckenholt, U., and Langeheine, R. (1996). Latent change in recurrent choice data. *Psychometrika*, **61**, 285-302.

Carroll, J. D., and Pruzansky, S. (1980). Discrete and hybrid scaling models. In E. D. Lantermann and H. Feger (eds.), *Similarity and Choice*. Vienna: Huber Verlag.

Coombs, C. H. (1964). *A theory of data*. New York: Wiley.

Dempster, A. P., Laird, N. H., and Rubin, D. B. (1977). Maximum likelihood from incomplete data via the EM algorithm. *Journal of the Royal Statistical Society*, **B39**, 1 - 38.

Hagenaars, J. (1990). *Categorical longitudinal data*. Newbury Park: Sage.

Hathaway, R. J. A. (1985). A constrained formulation of maximum-likelihood estimation for normal mixture distributions. *Annals of Statistics*, **13**, 795-800.

Langeheine, R. (1994). Latent variables markov models. In: A. von Eye, and C. C. Clogg (eds.) *Latent variables analysis*. Thousand Oaks: Sage.

Langeheine, R., and van de Pol, F. (1990). A unifying framework for Markov modeling in discrete space and discrete time. *Sociological Methods and Research*, **18**, 416-441.

Lazarsfeld, P. F., and Henry, N. W. (1968). *Latent structure analysis*. New York: Houghton-Mifflin.

Loewenstein, G. F., and Elster, J. (1992). *Choice over Time*. New York: Russell Sage Foundation.

Lwin, T. and Martin, P. J. (1989). Probits of mixtures. *Biometrics*, **45**, 721-732.

Poulsen, C. S. (1990). Mixed Markov and latent Markov modeling applied to brand choice data. *International Journal of Marketing*, **7**, 5-19.

Takane, Y. (1983). *Choice model analysis of the "pick any/n" type of binary data*. Handout at the European Psychometric and Classification Meetings, Jouy-en-Josas, France.

Wiggins, L. M. (1973). *Panel analysis: Latent probability models for attitude and behaviour processes*. Amsterdam: Elsevier.

Part VI

Multivariate and Multidimensional Data Analysis

- Multidimensional Data Analysis

- Multiway Data Analysis

- Non-Linear Modeling and Visual Treatment

Nonlinear Multivariate Analysis by Neural Network Models

Yoshio Takane

Department of Psychology, McGill University
1205 Dr. Penfield Ave., Quebec, H3A 1B1 CANADA

Summary: Feedforward neural network (NN) models approximate nonlinear functions that connect inputs to outputs by repeated applications of simple nonlinear transformations. By combining this feature of NN models with traditional multivariate analysis (MVA) techniques, nonlinear versions of the latter can readily be constructed. In this paper, we examine various properties of nonlinear MVA by NN models in two specific contexts: Cascade Correlation (CC) networks for nonlinear discriminant analysis simulating the learning of personal pronouns, and a five-layer auto-associative network for nonlinear principal component analysis (PCA) finding two defining features of cylinders. We analyze the mechanism of function approximations, focussing, in particular, on how interaction effects among input variables are captured by superpositions of sigmoidal transformations.

1. Introduction

Feed-forward neural network (NN) models and statistical models have much in common. The former can be viewed as approximating nonlinear functions that connect inputs to outputs. Many statistical techniques can be viewed as approximating functions (often linear) that connect predictor variables to criterion variables. It is thus beneficial to exploit various developments in NN models in nonlinear extensions of linear statistical techniques. There is one aspect of nonlinear transformations by NN models that is particularly attractive in developing nonlinear multivariate analysis (MVA). It allows joint multivariate transformations of input variables, so that interactions among them can be captured automatically in as much as they are needed for prediction. In this paper we examine various properties of nonlinear MVA by NN models in two specific contexts: Cascade Correlation (CC) networks for nonlinear discriminant analysis simulating the learning of personal pronouns, and a five-layer auto-associative network for nonlinear principal component analysis (PCA) recovering two defining attributes of cylinders. In particular, we analyze the mechanism of function approximations in these networks.

2. Cascade Correlation (CC) Network

NN models consists of a set of units, each performing a simple operation. Units receive contributions from other units, computes activations by summing the incoming contributions and applying prescribed (nonlinear) transformations (called transfer functions) to the summed contributions, and send out their contributions according to the activations and strengths of connections to other units. A network of such units can produce rather complicated and interesting effects. It can produce almost any kind of nonlinear effects and interactions among input variables by looking at examples that show specific input–output relationships.

The CC learning network is capable of dynamically growing nets (Fahlman & Lebiere, 1990). It starts as a net without hidden units, and it adds hidden units to improve its performance until a satisfactory degree of performance is reached. Thus, no *a priori* net topology has to be specified. Hidden units are recruited one at a time in

527

such a way that all pre-existing units are connectd to the new one. Input units are directly connected to output units (cross conneions) as well as to all hidden units. The cross connections capture linear effects of input variables. Hidden units, on the other hand, produce nonlinear and interaction effects among the input variables that are necessary to connect inputs to outputs in some tasks. When a new hidden unit is recruited, the connection weights associated with input connections are determined so as to maximize the correlation between residuals from network predictions at the particular stage and projected outputs from the recruited hidden unit, and are fixed throughout the rest of the learning process. This avoids the necessity of back-propagating error across different levels of the network, and leads to faster and more stable convergence. The weights associated with output connections are, however, re-estimated after each new hidden unit is recruited.

The CC algorithm constructs a net and estimates connection weights based on a sample of training patterns. For each input pattern, a unit in a trained net sends contributions to units it is connected to. A contribution is defined as the product of the activation for the pattern at the sending unit and the weight associated with the connection between the sending unit and the receiving unit. The receiving unit forms its activation by summing up the contributions from other units and applying the sigmoid transformation to the summed contribution. An activation is computed at each unit and for each input pattern in the training sample. Let a_1 denote an input pattern (a vector of activations at input units and bias, which acts like a constant term in regression analysis), and let w_1 represent the vector of weights associated with the connections from the input and bias units to hidden unit 1 (h_1). Then, the activation for the input pattern at h_1 is obtained by $b_1 = f(a_1' w_1) - .5$, where f is a sigmoid function, i.e., $f(t) = 1/\{1 + \exp(-t)\}$. Now h_1 as well as the input and bias units send contributions to h_2. The activation at h_2 is then obtained by $b_2 = f(a_2' w_2) - .5$. A similar process is repeated until an activation at the output unit is obtained, which is the network prediction for the output. In the training phase, connection weights are determined so that the network prediction closely approximates the output corresponding to the input pattern.

3. Two-Number Identification

The CC network algorithm was first applied to the two-number identification problem, in which there are two input variables, x_1 and x_2 (excluding the bias). Pairs of x_1 and x_2 are classified into group 1 (indicated by output variable y equal to .5) when the two numbers are identical, and are otherwise classified into group 2 (indicated by $y = -.5$). This is a simple two-group discrimination problem, but the function to be approximated is highly nonlinear, as can be seen in Figure 1(a). The problem is interesting because of its implication to real psychological problems; identifying two objects underlies many psychological phenomena, as exemplified by an example given in the next section.

One hundred training patterns, generated by facorially combining x_1 and x_2 varied systematically from 1 to 10 in the step of 1, were used in the training. The CC network algorithm constructed a network depicted in Figure 1(b). This net has three input units (including the bias), one output unit, and two recruited hidden units. Network predictions are computed in a manner described above. Figure 1(c) displays the function approximated by the CC net (the set of network predictions as a function of x_1 and x_2). The approximation looks quite good, although the ridge at $x_1 = x_2$ in the approximated function is not as "sharp" as in the original target function. This is due to the "crudeness" of the training sample. The minimum difference between two distinct numbers in the training sample is 1, so that the net was not required

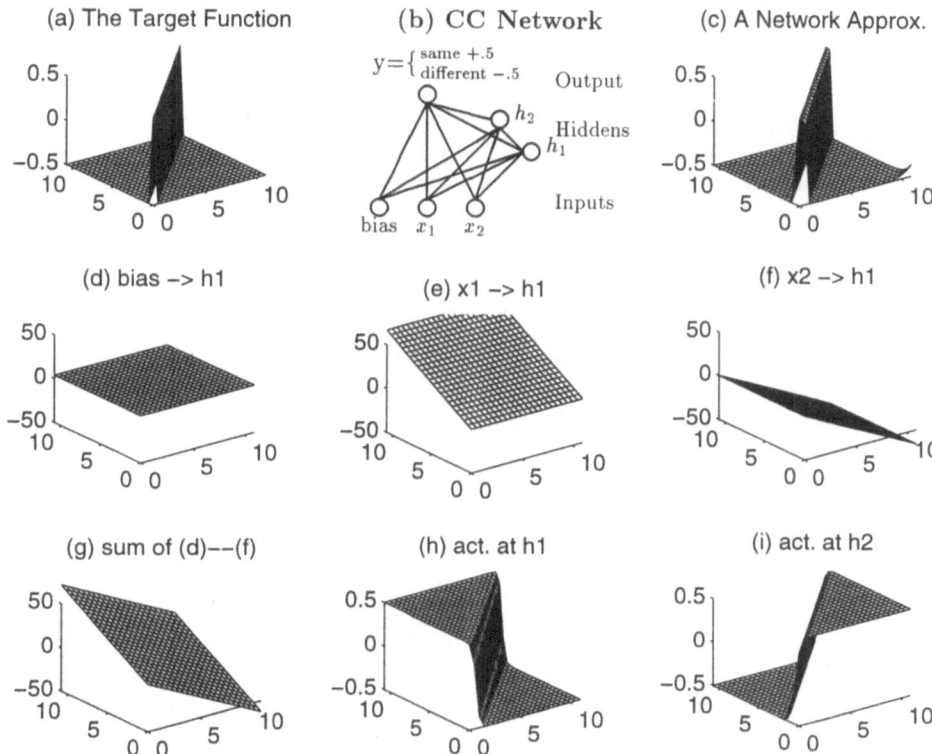

Figure 1: The mechanism of a function approximation for the two-number identification problem by CC network. (a) depicts the target function approximated by the CC network, (b), with the approximated function displayed in (c). (d) through (f) are contributions from three input units to h_1, which are summed to obtain (g). which is sigmoid-transformed to obtain the activation function (h) at h_1. The activation function at h_2 (i) is similarly derived. These activation functions are used to define contributions of the units to other units.

to discriminate beween two numbers whose differences are less than 1. The ridge in the approximated function can be made sharper if pairs of numbers with smaller differences are included in the training sample. Note that interpolations are done quite nicely. That is, although numbers like 5.5 were not included in the training sample, the identification involving such numbers are handled as expected. Extrapolation, on the other hand, seems a bit difficult, as indicated by a slight increase in function values toward the righthand side corner. Note that the target function involves a form of interaction between x_1 and x_2, where the word " interaction" is construed broadly; the meaning of a specific value on one variable, say x_1, depends on the value on the other variable, x_2.

It is interesting to see how the approximated function is bulit up and what roles the two hidden units play. Figure 1(d) through (f) present contributions of the three input units to h_1. As described above, contributions are defined as products of activations at the input units and the weights associated with the connections leading to h_1. The contributions are summed up (Figure 1(g)), and further sigmoid-transformed to obtain the activation function at h_1 (Figure 1(h)). It seems that h_1 is identifying if $x_1 \geq x_2$. The activation function at h_2 (Figure 1(i)) is similarly derived. Contributions now come from four units (three input units plus h_1). h_2 seems to be identifying if $x_2 \geq x_1$. The output unit (y) receives contributions from all other units. However, h_1 and h_2 seem to play particularly important roles. y stands out to take .5, when and only when input patterns satisfy both $x_1 \geq x_2$ and $x_2 \geq x_1$, but otherwise -.5. Interestingly, this is essentially how we prove $x_1 = x_2$ in mathematics.

4. Pronoun Learning

We were interested in the two-number identification problem because of its implication to a real psychological problem, that is, the learning of first and second person pronouns. When the mother talks to her child, me refers to the mother and you to the child. However, when the child talks to the mother, me refers to the child, and you to the mother. The child has to learn the shifting reference of these pronouns. There are three relevant input variables in this problem (excluding bias) and one output variable indicating me ($y = .5$) or you ($y = -.5$). The three input variables are speaker (sp), addressee (ad), and referent (rf). The rule (or the function) to be learned is: "Use me when the speaker and the referent agree (i.e., $y = .5$, when sp = rf)", and "use you when the addressee and the referent agree (i.e., $y = -.5$, when ad = rf)." The network should be able to judge which two of the three input variables agree in their values. The two-number identification problem is thus a prerequisite to the pronoun learning problem.

How children learn the correct use of these pronouns has been studied by Oshima-Takane (1988, 1992) and her collaborators (Oshima-Takane, et al., 1996). Simulation studies by CC networks have also been reported in Oshima-Takane, et al. (1995), and in Takane, et al. (1995). All previous simulation studies, however, presupposed the existence of only two pronouns, me and you. This severely limits the scope of these studies. In particular, the operating rule may not coincide with the one assumed above. That is, seemingly correct behavior can follow from rules other than the one described above. For example, a rule such as me if sp = rf and you otherwise, or you if ad = rf and me otherwise, works equally well so far as only me and you are considered. That is, ad = rf is equivalent to sp $\neq$ rf, and sp = rf is equivalent to ad $\neq$ rf when only me and you are to be distinguished.

We, therefore, first investigate what rule is in fact learned under the me–you–only condition. Forty training patterns were created by systematically varying the three input variables from -2 to 2 in the step of 1, and by discarding all but *me* and *you* patterns. Forty patterns were retained. (Remember that sp and ad cannot agree, and such patterns were also discarded.) The CC network algorithm recruited two hidden units to perform the task. The approximated function is depicted in Figure 2 in terms of ad on the y-axis and rf on the x-axis for nine different values of sp (-2, -1.5, -1. -.5, 0. .5, 1, 1.5, and 2). It looks like the output variable, y, takes the value of -.5, as it is supposed to (see the diagonal "ditch" observed in each graph), but it also takes the value of .5 in all other cases, including sp = rf and sp $\neq$ rf $\neq$ ad. This is correct for sp = rf, but not for sp $\neq$ rf $\neq$ ad. Remember that no training patterns were given for the latter and so it is quite natural that the net responded rather arbitrarily to the latter patterns. This implies, however, that pronouns other than *me* and *you* are necessary to learn the correct use of these two pronouns. That is, to learn to discriminate between sp = rf and sp $\neq$ rf $\neq$ ad, patterns involving other pronouns such as *he* and *she* have to be included in the training sample.

To verify the above assertion, another simulation study was conducted, this time, with pronouns other than *me* and *you* also included in the training sample. This condition, called the me–you–others condition, had 100 training patterns with 40 *me-you* patterns plus 60 *others* patterns. The net was trained to take the value of 0 ($y = 0$) when sp $\neq$ rf $\neq$ ad in addition to $y = .5$ when sp = rf and $y = -.5$ when ad = rf. Figure 3 shows the approximated function under this condition, which looks as it is supposed to. The task is appreciably more complicated than before, and the CC network algorithm recruited five hidden units to perform the task.

5. Five-Layer Auto-Associative Network

The next example pertains to a five-layer auto-associative network. A simplified version of this network is depicted in Figure 4(a). There are five layers of units including the input layer at the bottom and the output layer at the top. Units are interconnected between adjacent layers, but not within same layers or between nonadjacent layers. It is well known (e.g., Baldi & Hornik, 1989) that a three-layer neural network with linear transfer functions at both middle (hidden) and output layers has a rank reducing capability when the number of units in the hidden layer is smaller than both the number of input units and that of output units. This is a network version of (linear) reduced-rank regression (Anderson, 1951), also known as PCA with instrumental variables (Rao, 1964) and redundancy analysis (Van den Wollenberg, 1977). The usual (linear) PCA results when inputs and outputs coincide, as in Figure 4(a). The name "auto-associative" derives from the fact that this net attempts to reproduce X from input X with a reduced number of components (the number of units in the middle layer). The network version of PCA is not interesting in itself since there are more efficient and accurate algorithms to do linear PCA. It becomes interesting when the model is extended to nonlinear PCA by including two additional hidden layers with nonlinear transfer functions (most often, with sigmoidal transformations), one between the input layer and the middle layer, and the other between the middle layer and the output layer, resulting in a five-layer network. Layer 2 (hidden layer 1) and layer 4 (hidden layer 3) perform nonlinear input encoding and nonlinear output decoding, respectively. Unlike the CC network, the network topology (the number of layers, the number of units in each layer and how the units in different layers are connected) is *a priori* specified and fixed throughout the learnig process, in which only connection weights are adjusted using the backpropagation (BP) algorithm.

The five-layer auto-associative network was proposed (apparently independently) by several authors at about the same time (e.g., Irie & Kawato, 1989; Oja, 1991), and

532

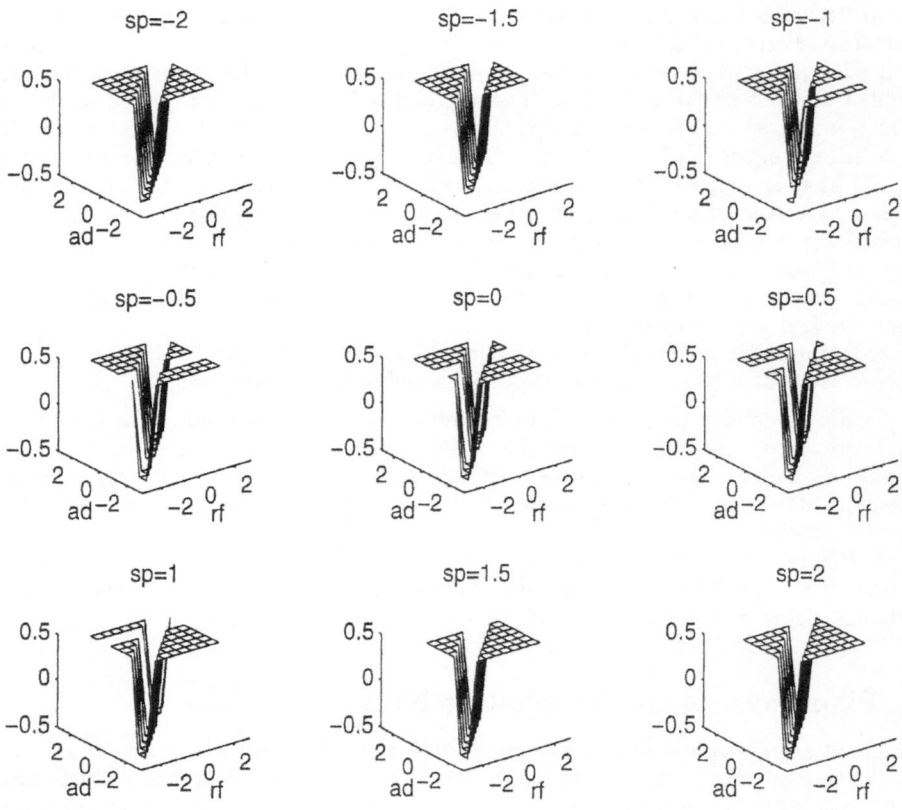

Figure 2: The approximation function for the pronoun learning problem obtained under the me-you-only condition. The function is depicted as functions of 'addressee' (y-axis) and 'referent' (x-axis) at several values of 'speaker'. Function values (z-axis) at ad=rf should indicate *you* ($y = +.5$), and those at sp=rf *me* ($y = -.5$), if the pronouns are correctly learned. The problem is that the function takes the assumed value for *me* even if sp≠rf≠ad for which no examples were given in the training. Discontinuities in the function correspond to points where sp=ad which never occurs.

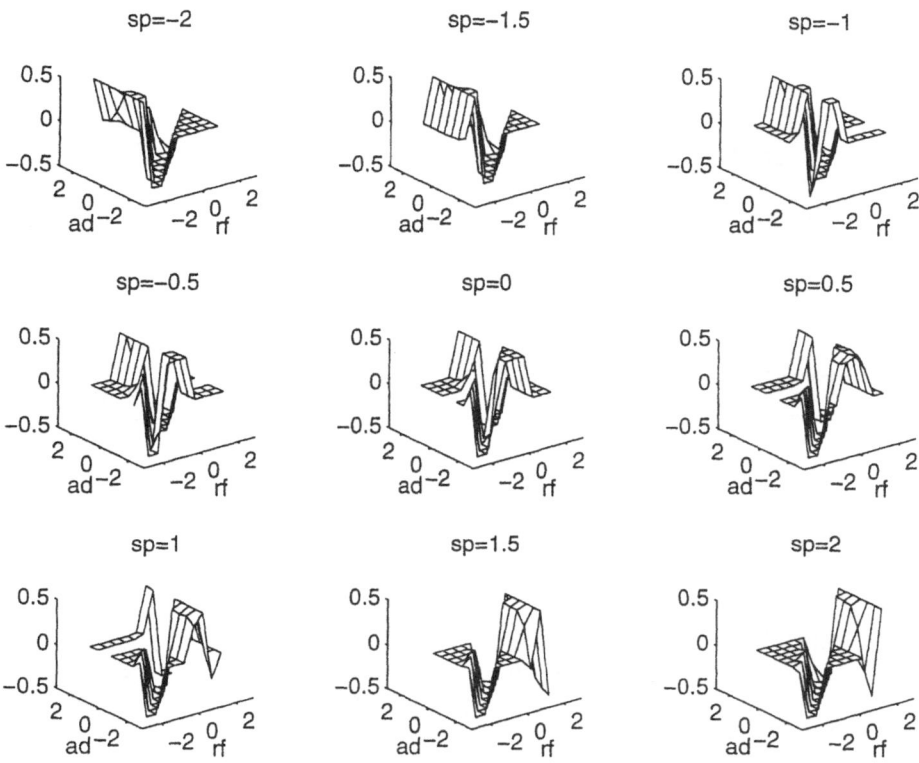

Figure 3: The same as Figure 2, but under the me-you-others condition. When the correct learning occurs, the function takes the value of *you* ($y = -.5$) if and only if ad=rf, the value of *me* ($y = +.5$) if and only if sp=rf, and $y = 0$ if and only if sp≠rf≠ad.

534

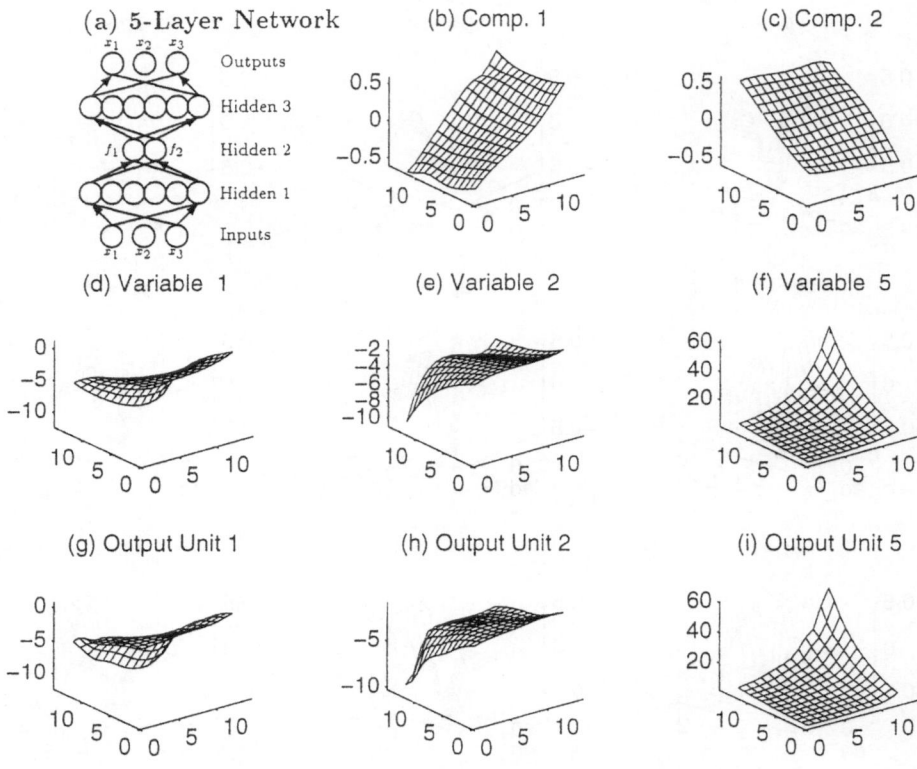

Figure 4: The mechanism of a function approximation in the five-layer auto-associative network. (a) depicts the basic construction of the network. (b) and (c) represent recovered components as functions of the original components ($ln\ a$ on the y-axis, $ln\ b$ on the x-axis). (d) through (f) display a sample of input functions (out of 12 altogether), and (g) through (i) recovered functions at the output units corresponding to (d)—(f).

has been applied to extracting components that determine facial expressions of emotion (DeMers & Cottrell, 1993) and internal color representation (Usui, et al., 1991). Takane (1995) examined recovery properties of nonlinear PCA by the NN models using several artificial data sets.

6. The Cylinder Problem

The example used to demonstrate nonlinear PCA by NN models was adapted from Kruskal & Shepard (1974), who generated a set of cylinders by systematically varying log altitude ($\ln a$) and log base area ($\ln b$) of the cylinders. These two variables serve as two assumed components to be recovered by nonlinear PCA. Kruskal & Shepard then measured the cylinders with respect to twelve variables which are all monotonic functions of $\ln a$ and $\ln b$: 1. altitude, 2. base area, 3. circumference, 4. side area, 5. volume, 6. moment of inertia, 7. slenderness ratio, 8. diagonal–base angle, 9. diagonal–side angle, 10. electrical resistance, 11. conductance, and 12. torsional deformability.

The training patterns used in the present study were generated in a similar way, except that 1) $\ln a$ and $\ln b$ were systematically varied from -.6 to .6 in the step of .1 to obtain 13 equally spaced levels, which were factorially combined to obtain 169 cylinders (as opposed to 30 prescribed cylinders in Kruskal & Shepard), and 2) after the same twelve variables were used to measure the cylinders, they were further transformed by an arbitrary linear transformation to define a completely new set of twelve variables, which may no longer be monotonic with either $\ln a$ or $\ln b$. Three examples of these variables are shown in Figure 4(d)—(f), as functions of $\ln a$ and $\ln b$. These variables are joint multivariate nonlinear transfomations of $\ln a$ and $\ln b$. Nonmetric PCA allowing only variablewise monotonic transformations is expected to have great difficuties in recovering the original components from such data. However, nonlinear PCA by means of a five-layer auto-associative network with 12 units in each of the the first and third hidden layers (this number is the same as the number of input units and that of output units) could almost perfectly recover the input data. The recovered data are shown in Figure 4(g)—(i) for the variables corresponding to those in Figure 4(d)—(f). The recovered variables at the output units look remarkably similar to the corresponding input variables, except that small wiggles are observed in the former. Figure 4-(b) and (c) give two recovered components plotted against $\ln a$ and $\ln b$. In both cases, recovered components are fairly linear with the original components.

It is interesting to see how input variables are approximated (recovered) at the ouput units with a reduced number of components in the middle layer (Hidden layer 2). Figures 5 and 6 display the activation functions created at hidden layers 1 and 3 (H_1 and H_3), respectively. The activation functions at H_1 were obtained by sigmoid transformations of linear combinations of input unit activations. They are in turn linearly combined to obtain the two recovered components at H_2. The recovered components are then linearly combined and sigmoid–transformed to obtain the activation functions at H_3. They were then linearly combined to obtain the approximated input functions at the output units.

7. Discussion

NN models present interesting perspectives to nonlinear multivariate analysis by allowing joint multivariate nonlinear transformations of input variables. In this paper, we highlighted the mechanisms of these transformations in two specific context: CC

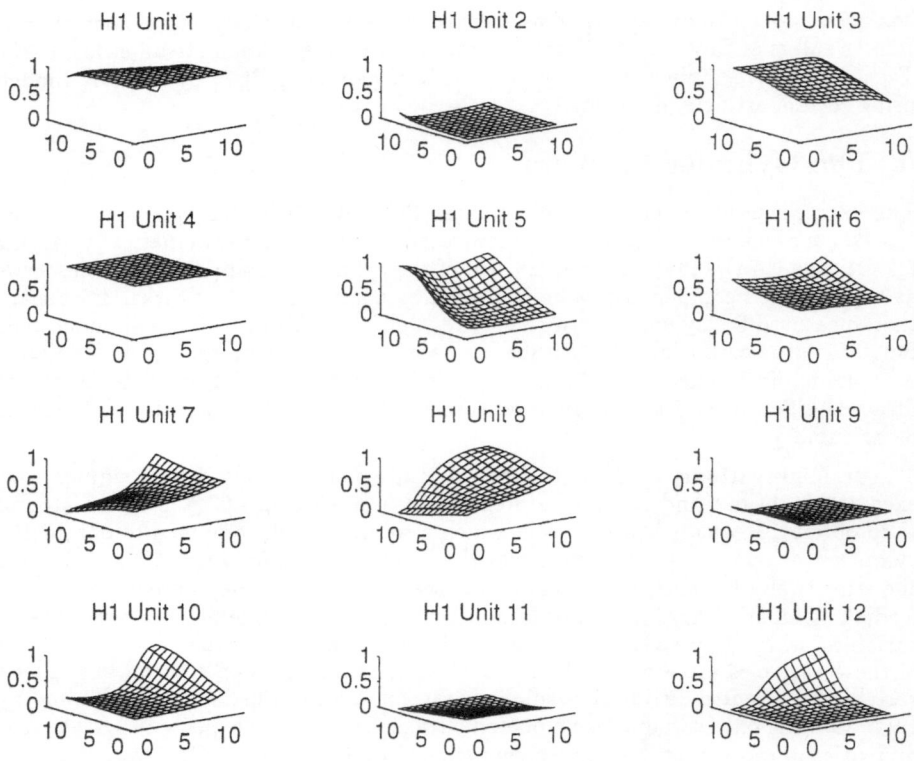

Figure 5: The activation functions created at units in hidden layer 1. These activation functions are linearly combined to obtain the activation functions ((b) and (c) of Figure 4), which are recovered component scores.

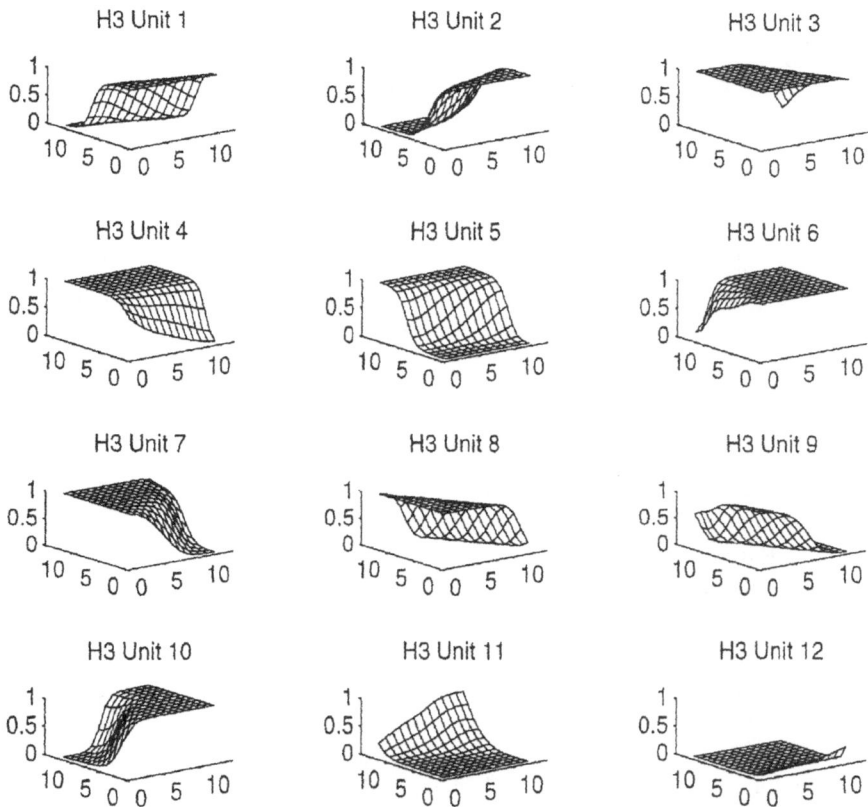

Figure 6: The same as Figure 5, but for hidden layer 3. These activation functions are linearly combined to obtain the output functions (activation functions at output units), some of which are given in (g)—(i) of Figure 4.

networks for nonlinear discriminant analysis and a five-layer auto-associative network for nonlinear PCA. In the present studies, no random errors were added in the data generation process. Investigating how the networks cope with random errors in the data is an important next step to evaluate the viability of the approach as a general method for developing nonlinear multivariate analysis techniques.

References

Anderson, T. W. (1951). Estimating linear restrictions on regression coefficients for multivariate normal distributions. *Annals of Mathematical Statistics*, **22**, 327–351.

Baldi, P. & Hornik, K. (1989). Neural network and principal component analysis: Learning from examples without local minima. *Neural Network*, **2**, 53–58.

DeMers, D., & Cottrell, G. (1993). Non-linear dimension reduction. In: *Neural Information Processing Systems 5*, Hanson, S. J. et al. (eds.), 580–587, Morgan Kaufmann, San Mateo, CA.

Fahlman, S. E., & Lebiere, C. (1990). The cascade correlation learning architecture. In: *Neural Information Processing Systems 2*, Touretzky, D. S. (eds.), 524–532, Morgan Kaufmann, San Mateo, CA.

Irie, B. & Kawato, M. (1989). Tasō pāseputoron ni yoru naibu hyōgen no kakutoku. [Acquisition of internal representation by multi-layered perceptron.] *Shingakugihō*, **NC89-15**.

Kruskal, J. B., & Shepard, R. N. (1974). A nonmetric variety of liner factor analysis. *Psychometrika*, **39**, 123–157.

Oja, E. (1991). Data compression, feature extraction, and autoassociation in feedforward neural networks. In: *Artificial Neural Networks*, Kohonen, T. et al. (eds.), 737–745.

Oshima-Takane, Y. (1988). Children learn from speech not addressed to them: The case of personal pronouns. *Journal of Child Language*, **15**, 94–108.

Oshima-Takane, Y. (1992). Analysis of pronomial errors: A case study. *Journal of Child Language*, **19**, 111-131.

Oshima-Takane, Y. et al, (1995). The learning of personal pronouns: Network models and analysis. *McGill University Cognitive Science Center Technical Report*, **2095**. McGill University.

Oshima-Takane, Y. et al. (1996). Birth order effects on early language development: Do second born children learn from overheard speech? *Child Development*, **67**, 621–634.

Rao, C. R. (1964). The use and interpretation of principal component analysis in applied research. *Sankhyā A*, **26**, 329–358.

Takane, Y. (1995). Nonlinear multivariate analysis by neural network models. *Proceedings of the 63rd Annual Meeting of the Japan Statistical Society*, 258–260.

Takane, Y. et al. (1995). Network analyses: The case of first and second person pronouns. *Proceedings of the 1995 IEEE International Conference on Systems, Man and Cybernetics*, 3594–3599.

Usui, S. et al. (1991). Internal color representation acquired by a five-layer neural network. In: *Artificial Neural Network*, Kohonen, T. et al. (eds.), 867–872.

Van den Wollenberg, A. L. (1977). Redundancy analysis: An alternative for canonical analysis. *Psychometrika*, **42**, 207–219.

Generalized Canonical Correlation Analysis with Linear Constraints

Haruo Yanai[1]

[1]Research Division
National Center for University Entrance Examination
2-19-23 Komaba, Meguro-ku, Tokyo 153, Japan

Summary: A method is introduced by imposing linear constraints upon parameters corresponding to more than two sets of variables. We call the method introduced here 'generalized canonical correlation analysis with linear constraint'. It covers canonical correlation analysis with linear constraints proposed by Yanai & Takane (1992) as its special case. Further, by employing dummy variables, our method turns out to be identical to the multiple correspondence analysis with linear constraints.

1. Introduction

Yanai & Takane (1992) proposed canonical correlation analysis with linear constraints by imposing linear constraints upon parameters corresponding to two sets of variables. In this paper, we extend the earlier results to the case where more than two sets of variables are available, and thereby derive general explicit solutions for canonical correlation analysis with more than two sets of variables by imposing linear constraints upon the parameters corresponding to the sets of variables. We call this method the generalized canonical correlation analysis with linear constraints (GCCAC). Furthermore, we show that for categorical data our method yields the multiple correspondence analysis with linear constraints (MCC),or the Quantification method of the third type (Hayashi,1952) with linear constraints which covers the canonical analysis of contingency tables with linear constraints presented by Bockenholt & Bockenholt (1990) as well.

2. Generalized Canonical Correlation Analysis (GCCA)

2.1 Formulation of GCCA

Suppose that for n subjects data on m sets of variables $x_1, x_2, \cdots, x_m$ were recorded (where x_i comprises p_i variables) and are represented in terms of the matrix $X = (X_1, X_2, \cdots, X_m)$. The sizes of the matrices $X_1, X_2, \cdots, X_m$ are $n \times p_1, n \times p_2, \cdots, n \times p_m$, respectively. Let $a' = ((a_1)', \cdots, (a_m)')$ be a vector of weight vectors for $X_1, X_2, \cdots, X_m$ with $a_i \in R^{p_i}$. We look for a weight vector a that maximizes the following criterion

$$\sum_{j=1}^{m}\sum_{i=1}^{m}(X_i a_i, X_j a_j)/\sum_{i=1}^{m}\|X_i a_i\|^2 = a'(X'X)a/a'D_{XX}a \to \max_{a} \qquad (1)$$

where $(\ ,\ )$ denotes the scalar product in R^n and

$$X'X = \begin{pmatrix} X_1'X_1 & X_1'X_2 & \cdots & X_1'X_m \\ X_2'X_1 & X_2'X_2 & \cdots & X_2'X_m \\ \cdots & \cdots & \ddots & \vdots \\ X_m'X_1 & X_m'X_2 & \cdots & X_m'X_m \end{pmatrix} \quad \text{and} \quad \begin{pmatrix} X_1'X_1 & O & \cdots & O \\ O & X_2'X_2 & \cdots & O \\ \cdots & \cdots & \ddots & \vdots \\ O & O & \cdots & X_m'X_m \end{pmatrix}$$

are block matrices of order $p \times p$ where $p = \sum_{i=1}^{m} p_i$. Further, the $p_i \times p_j$ matrices $X_j'X_j (j = 1, \cdots, m)$ are assumed to be nonsingular throughout the paper. Any solution of (1) is necessary a solution of the following eigenvalue problem:

$$(X'X)a = \lambda D_{XX} a. \tag{2}$$

Thus the maximum value of (1) is given by λ_1, the maximum eigenvalue of (2). If $X_1, X_2, \cdots, X_m$ are m sets of dummy variables, then (2) implies nothing but the eigen equation of multiple correspondence analysis (MCA) (see, Lebart et al.(1984)), or of the Quantification method of the third type of by Hayashi (1952).

We give an alternative equation to solve a in the following Theorem.

Theorem 1: The weight vectors $a_j \in R^{p_j}$ maximizing (1) can be obtained by solving:

$$_gPf = \lambda f \quad \text{with} \quad _gP = P_{X_1} + P_{X_2} + \cdots + P_{X_m} \quad \text{for} \quad \lambda \quad \text{and} \quad f \in R^n \tag{3}$$

where $P_{X_j} = X_j(X_j'X_j)^{-1}X_j'$ is the orthogonal projection matrix onto $S(X)$, the column subspace of X_j. From the eigenvector f of $_gP$, the weight vectors

$$a_j = (1/\lambda)(X_j'X_j)^{-1}X_j'f \quad \text{for} \quad j = 1, \cdots, m, \tag{4}$$

provide a solution of (2).

Proof: Putting $f = Xa$ and multiplying (2) from the left by XD_{XX}^{-1}, we get

$$(XD_{XX}^{-1}X')f = \lambda f \tag{5}$$

which is easily transformed into (3). Furthermore, by substituting $f = Xa$ on the left hand side of (2), we obtain $X'f = \lambda D_{XX} a$. This yields $a = (1/\lambda)D_{XX}^{-1}X'f$, which implies (4). (Q.E.D.)

An alternative method for deriving the eigenequations (3) and (4) is given by minimizing

$$Q_2(a) = \sum_{j=1}^{m} \|f - X_j a_j\|^2 \to \min_a \tag{6}$$

with respect to $a_1, \cdots, a_m$ for a given f. This shows that the minimum value of the $Q_2(a)$ defined above is attained as

$$Q_2(a^*) = mf'f - f'(P_{X1} + P_{X2} + \cdots + P_{Xm})f \tag{7}$$

when $a_j^* = (X_j'X_j)^{-1}X_j'f$. Now, minimizing (7) with respect to f leads to (3). We show an important property of $_gP$ in the following corollary.

Corollary 1: $0 \leq \lambda_j(_gP) \leq m(j = 1, \cdots, m)$

Proof: Observe that $I_n \geq P_{X_j}(j = 1, \cdots, m)$. which implies that $I_n - P_{X_j}$ is nonnegative definite. Then we have

$$mI_n \geq \sum_{i=1}^{m} P_{X_i} =_g P,$$

thus establishing the desired result due to Poincare separation theorem (PST). (Q.E.D.)

3. Generalized Canonical Correlation Coefficient

Corollary 2 : Let $\lambda_j = \lambda_j(_gP)(j = 1, \cdots, k)$ be j-th eigenvalue of (3) and f_j be the corresponding normalized eigenvector. Let us denote by

$$R_k(X_1, \cdots, X_m) = \sum_{i \neq j}^{m} (P_{X_i}f_k, P_{X_j}f_k)/((m-1)\sum_{i=1}^{m}\|P_{X_j}f\|^2) = \frac{\lambda_k - 1}{m - 1} \tag{8}$$

Then, we have

$$\frac{-1}{m-1} \leq R_k(X_1, \cdots, X_m) \leq 1 \quad \text{for} \quad k = 1, \cdots, \text{rank}(_gP). \tag{9}$$

Proof: First, multiply (3) by $f'P_{X_j}$ from the left and sum over j. We get

$$\sum_{i \neq j}^{m} (P_{X_i}f_k, P_{X_j}f_k)/(\sum_{i=1}^{m}\|P_{X_i}\|^2 f_k) = \lambda_k - 1.$$

By Corollary 1, we get (9). (Q.E.D.)

Observe that if the intersection of all m subspaces $S(X_j)(j = 1, \cdots, m)$ is not empty, then $R(X_1, X_2, \cdots, X_m)$ attains 1. Further, if $X_i'X_j = O$ for $i \neq j$ then $R(X_1, X_2, \cdots, X_m) = 0$. Thus, if $m = 2$, then $R(X_1, X_2)$ is the ordinary canonical correlation coefficient between X_1 and X_2. From the above, it can be justified that $R_j(X_1, \cdots, X_m)$ is the j-th generalized canonical correlation coefficient among $X_1, X_2, \cdots, X_m$.

4. Generalized Canonical Correlation Analysis with linear Constraints

Next, we consider maximizing (1) under some additional linear constraints on $a_j(j = 1, \cdots, m)$.

Theorem 2: Let $C_j(j = 1, \cdots, m)$ be matrices of orders $p_j \times s_j$(where $s_j > 0$). Maximizing (1) subject to the $s = \sum_{j=1}^{m} s_{x_j}$ linear constraints

$$C_j'A_j = 0 \quad (j = 1, \cdots, m) \tag{10}$$

yields eigen equations of the form

$$\sum_{j=1}^{m} (P_{X_j} - P_{X_j(X_j'X_j)^{-1}C_j})g = \lambda_c g \tag{11}$$

and

$$(\sum_{j=1}^{m} P_{X_j Q_{C_j}})f = \lambda_c f \quad \text{and} \quad Q_{C_j} = I - P_{C_j}. \tag{12}$$

Further, weight vectors a_j are given by

$$a_j = (X_j'X_j)^{-1}X_j'f - (X_j'X_j)^{-1}C_j(C_j'(X_j'X_j)^{-1}C_j)^{-1}C_j'(X_j'X_j)^{-1}X_j'f \quad \text{for} \quad j = 1, \cdots, m \tag{13}$$

Proof: The proof uses the following auxiliary lemma.

Lemma (Rao(1973)). Suppose that A and B are symmetric matrices of the same orders, and let B be nonsingular. Then, maximization of $a'Aa/a'Ba$ subject to $C'a = 0$ leads to the necessary condition for a:

$$(I - P)Aa = \lambda Ba \quad \text{where} \quad P = C(C'B^{-1}C)^{-1}C'B^{-1}. \tag{14}$$

Proof of Theorem 2: First, consider the $p \times p$ matrices

$$A = X'X = \begin{pmatrix} X_1'X_1 & X_1'X_2 & \cdots & X_1'X_m \\ X_2'X_1 & X_2'X_2 & \cdots & X_2'X_m \\ \cdots & \cdots & \ddots & \vdots \\ X_m'X_1 & X_m'X_2 & \cdots & X_m'X_m \end{pmatrix}, \quad B = \begin{pmatrix} X_1'X_1 & O & \cdots & O \\ O & X_2'X_2 & \cdots & O \\ \cdots & \cdots & \ddots & \vdots \\ O & O & \cdots & X_m'X_m \end{pmatrix}$$

and the $p \times s$ matrix

$$C = \begin{pmatrix} C_1 & O & \cdots & O \\ O & C_2 & \cdots & O \\ \cdots & \cdots & \ddots & \vdots \\ O & O & \cdots & C_m \end{pmatrix}.$$

Multiplying (14) from the left by XB^{-1} and expanding, we get the equation for λ and a:

$$(XB^{-1}X' - XB^{-1}C(C'B^{-1}C)^{-1}C'B^{-1}X')Xa = \lambda Xa,$$

thus leading, by $g = Xa$, to

$$\sum_{j=1}^{m}(P_{X_j} - P_{X_j(X_j'X_j)^{-1}C_j})g = \lambda_c g, \tag{15}$$

observing that $(XB^{-1}C)'(XB^{-1}C) = C'B^{-1}C$.

From the above, and using the Theorem 2.1 of Yanai & Takane(1992), we immediately have

$$(\sum_{j=1}^{m} P_{X_j Q_{C_j}})f = \lambda_c f \quad \text{where} \quad Q_{C_j} = I - P_{C_j}. \tag{16}$$

If $\text{rank}(X_j', C_j) = \text{rank}(X_j) + \text{rank}(C_j)$, then we can use

$$(P_{X_j} - P_{X_j(X_j'X_j)^{-1}C_j}) = P_{X_j Q_{C_j}} = P_{X_j}$$

which implies that adding to the constraint $C_j a_j = 0$ is meaningless under the condition stated above.

We term the method explained in this theorem 'Generalized Canonical Correlation Analysis with linear constraints (GCCAC)'

Corollary 3: Let $\lambda_j(X, C)$ be the j-th eigenvalue of (15 and 16). Then $\lambda_j(X, C) \leq \lambda_j(X)$ for all j.

Theorem 3 includes the result of the canonical correlation analysis with linear constraints by Yanai & Takane (1992) as its special case.

Suppose that the data matrices X_1 and X_2 contain only dummy variables. Further, let $N_{12} = X_1'X_2$, $D_1 = X_1'X_1$ and $D_2 = X_2'X_2$. Then maximizing

$$a'N_{12}b/(\sqrt{a'D_1a}\sqrt{b'D_2b}) \to \max_{a,b}$$

subject to $E'a = 0$ and $F'b = 0$ yields the singular value decomposition of

$$(I - E(E'D_1^{-1/2}E)^{-1}E'D_1^{-1/2})Y(I - F(F'D_2^{-1/2}F)^{-1}F'D_2^{-1/2})$$

where $Y = D_1^{-1/2}N_{12}D_2^{-1/2}$.

This is the canonical analysis of a contingency table with linear constraints by Bockenholt & Bockenholt (1990), which is here subsumed as a special case of GCCAC with $m = 2$ and X_1 and X_2 two matrices of dummy variables.

5. Numerical Examples

We give three numerical examples demonstrating the usefulness of our method.

Example 1: Given $m = 3$ dummy matrices A, B and C of orders 8×2, we compute the eigenvalues of $_gP$ where $_gP = P_A + P_B + P_C$. Then we have $\lambda_1(_gP) = 3$, $\lambda_2(_gP) = \lambda_3(_gP) = \lambda_4(_gP) = 1$. and obtain generalized canonical correlation coefficients by the equation (8): $R_1(A, B, C) = 1$, $R_2(A, B, C) = R_3(A, B, C) = R_4(A, B, C) = 0$.

$$A = \begin{pmatrix} 1 & 0 \\ 1 & 0 \\ 1 & 0 \\ 1 & 0 \\ 0 & 1 \\ 0 & 1 \\ 0 & 1 \\ 0 & 1 \end{pmatrix}, \quad B = \begin{pmatrix} 1 & 0 \\ 1 & 0 \\ 0 & 1 \\ 0 & 1 \\ 0 & 1 \\ 0 & 1 \\ 1 & 0 \\ 1 & 0 \end{pmatrix} \quad \text{and} \quad C = \begin{pmatrix} 1 & 0 \\ 1 & 0 \\ 0 & 1 \\ 0 & 1 \\ 1 & 0 \\ 1 & 0 \\ 0 & 1 \\ 0 & 1 \end{pmatrix}$$

Example 2: Let $c_1 = (1, 2)$, $c_2 = (1, -2)$ and $c_3 = (2, 1)$ be constraint vectors. Then by means of the formula (15), we have $\lambda_{c1}(_gP) = 1.477$, $\lambda_{c2}(_gP) = 0.900$ and $\lambda_{c3}(_gP) = 0.623$ and the remaining eigenvalues are 0. Thus, from Examples 1 and 2, we have demonstrated Corollary 3 of Theorem 2, numerically. Further, the resulting weights are as follows.

weight	a_{11}	a_{12}	a_{21}	a_{22}	a_{31}	a_{32}
dimension 1	0.287	−0.144	0.361	0.181	−0.144	0.287
dimension 2	−0.300	0.150	0.000	0.000	−0.150	0.300
dimension 3	0.166	−0.083	−0.264	−0.132	−0.083	0.166
constraints	1	2	1	−2	2	1

Observe that $0.287 \times 1 + (-0.144) \times 2 = 0$, $0.361 \times 1 + 0.181 \times (-2) = 0$ and $-0.144 \times 2 + 0.287 \times 1 = 0$, thus establishing that the obtained weights satisfy the given constraints.

Example 3: As data X, we consider answers of 25 subjects to 11 categories of four items (see Table 1). Using these data we computed the eigenvalues and the corresponding weights for the categories by changing the constraints as in the following four ways:

case 1: without any constraint.

case 2: $c_1=(1,1,1)$, $c_2=(1,1)$, $c_3=(1,1,1)$, $c_4=(1,1,1)$

case 3: $c_1=(1,1,-1)$, $c_2=(1,-1)$, $c_3=(1,1,-1)$, $c_4=(1,1,-1)$

case 4: $c_1=(1,2,3)$, $c_2=(1,2)$, $c_3=(1,2,3)$, $c_4=(1,2,3)$

It follows that the maximum eigenvalue attains 4 in case 1. Thus, we showed the the weights of 11 categories corresponding to the second largest eigenvalue in case 1, and we showed the weights corresponding to the largest eigenvalues in each of the other three cases. As can be seen from Table 3, weights given to categories in any of the four items satisfy the constraints in the cases 2,3 and 4.

Table 1: Labels of 11 categories of four items

item	1) age group	2) sex	3) married status	4) numbers of children
category 1	20-39	male	single	none
category 2	40-59	female	married	one and two
category 3	over 60	———	alternatives	more than three

Further, we showed the eigenvalues of the four cases in Table 4. It is shown that the eigenvalue of the case 1 is larger than any of the other three eigenvalues obtained from (3). The result also demonstrates the validity of Corollary 3.

6. Discussions

Our method proposed in this paper is said to be the general method of multivariate analysis for finding appropriate linear combinations by imposing linear constraints upon parameters corresponding to many sets of variables. It is to be noted here that our method covers canonical correlation analysis, canonical discriminant analysis, principal component analysis, multiple correspondence analysis and also Quantification method of the third type, taking linear constraints into consideration. In

Table 2: Data of 25 subjects of the four items

item	1 2 3 4	item	1 2 3 4	item	1 2 3 4
subject 1	3 1 2 3	subject 10	1 2 2 3	subject 19	2 2 2 2
subject 2	1 2 3 3	subject 11	3 2 1 1	subject 20	1 2 2 2
subject 3	2 1 2 2	subject 12	3 2 3 2	subject 21	1 1 2 1
subject 4	2 1 2 3	subject 13	2 2 3 3	subject 22	2 2 2 3
subject 5	1 1 2 1	subject 14	3 1 2 3	subject 23	3 1 1 1
subject 6	3 2 3 3	subject 15	2 2 2 2	subject 24	1 2 2 3
subject 7	1 2 2 2	subject 16	1 2 2 3	subject 25	1 2 3 3
subject 8	3 2 1 1	subject 17	2 2 2 3		
subject 9	3 2 3 1	subject 18	2 2 3 3		

Table 3: Weights of 11 categories corresponding to the second maximum eigenvalue of the four cases

	case 1	case 2	case 3	case 4
1-1	0.070	-0.073	0.139	-0.099
1-2	0.145	-0.122	0.134	-0.191
1-3	-0.223	0.195	0.273	0.160
2-1	-0.095	0.104	0.198	0.080
2-2	0.037	-0.104	0.198	-0.040
3-1	-0.459	0.295	0.097	0.412
3-2	0.077	-0.141	0.169	-0.117
3-3	0.031	-0.153	0.266	-0.059
4-1	-0.304	0.257	0.127	0.311
4-2	0.123	-0.113	0.115	-0.079
4-3	0.084	-0.143	0.242	-0.051

Table 4: eigenvalues of the four cases

component	1	2	3	4	5	6	7	8
case 1	4.000	2.162	1.142	1.093	0.997	0.632	0.456	0.197
case 2	2.203	2.203	1.350	0.900	0.605	0.533	0.295	0.000
case 3	3.785	1.501	1.040	0.505	0.108	0.043	0.018	0.000
case 4	2.042	1.477	1.200	0.982	0.686	0.394	0.220	0.000

our paper, we have not given any comment how to find the appropriate constraints for each of the methods. More often, natural forms of constraints may appear from specific empirical questions posed by the investigators concerned in the problem of their fields. With such formulations, one specifies the space in which the original parameter should lie, and then proceeds to test the hypothesis by means of an appropriate statistics. It should be noted, however, that little work has been done on various kinds of multivariate analysis. In this viewpoint, how to test the hypothesis stated in (10) provides an interesting problem to be tackled in the future.

References:

Yanai,H. & Takane,Y.(1992): Canonical correlation analysis with linear constraints, *Linear Algebra & its Applications*, **176**, 75–89

Bokenholt, U. & Bockenholt,I.(1990): Canonical analysis of contingency tables with linear constraints, *Psychometrika*, **55**, 633–639

Lebart, L. Morineau, A.K. & Warwick, M.(1984): Multivariate *Descriptive Statistical Analysis*, John Wiley & Sons, New York

Rao,C.R. & Yanai,H.(1979): General definition of a projector, its decomposition and application to statistical problems, *J. of Statistical Planning and Inference*, **3**, 1–17

Rao, C.R.(1973): *Linear Statistical Inferences and its Applications (second Edition)*, John Wiley & Sons, New York

Hayashi,C. (1952): On the prediction of phenomena from qualitative data from the mathematico-statistical point of view, *Annals of the Institute of Statistical Mathematics*, **3**, 69–96

Principal Component Analysis Based on a Subset of Variables for Qualitative Data

Yuichi Mori[1], Yutaka Tanaka[2] and Tomoyuki Tarumi[2]

[1] Kurashiki City College
160 Kojima Hieda-cho
Kurashiki 711, Japan

[2] Department of Environmental and Mathematical Sciences
Okayama University
2-1-1, Tsushima-naka, Okayama 700, Japan

Summary: A principal component analysis based on a subset of variables is proposed by Tanaka and Mori (1996) to derive principal components which are computed as linear combinations of a subset of quantitative variables but which can reproduce all the variables very well. The present paper discusses an extension of their modified principal component analysis so that it can deal with qualitative variables by using the idea of the alternating least squares method by Young et al.(1978). A backward elimination procedure is applied to find a suitable sequence of subsets of variables. A numerical example is shown to illustrate the performance of the proposed procedure.

1. Introduction

Consider a situation where we wish to make a small dimensional rating scale to measure latent traits. On the one hand, from the validity aspect all the variables should be included in the stage of constructing the rating scale. On the other hand, the number of variables should be as small as possible in the stage of application of the rating scale from the practical aspect . Thus we meet the problem of variable selection in principal component analysis (PCA).

Suppose we observe quantitative variables and wish to make a small dimensional rating scale by applying a PCA–like method. Furthermore, we wish to obtain the rating scale or a set of principal components (PCs) in such a way that it is based on only a subset of variables but it represents all the variables very well. If we can find such PCs, we may say that those PCs provide a multidimensional rating scale which has high validity and is easy to be applied practically. To do this Tanaka and Mori (1996) proposed a method of PCA based on a subset of variables using the idea of Rao(1964)'s PCA of instrumental variables and called it a "modified PCA" (M.PCA), while the problems of variable selection in the ordinary PCA have been studied by Jolliffe (1972, 1973, 1986), Robert and Escoufier (1976), McCabe (1984) and Krzanowski (1987a, 1987b) among others.

In this paper we extend M.PCA so that it can deal with qualitative data with unordered or ordered categories. Namely we perform both quantification of qualitative data and M.PCA at the same time, and study the performance of our method numerically by applying it to a real data set. For analyzing qualitative data with a PCA approach, a lot of techniques have been proposed, e.g., De Leeuw (1982, 1984), De Leeuw and Van Rijckevorsel (1980), Israels (1984), and Young et al. (1978). We use Young et al.(1978)'s PCA in which the alternating least squares (ALS) method is utilized. We call this qualitative PCA based on a subset of variables a "qualitative modified PCA" (QM.PCA) to distinguish it from M.PCA and others.

2. PCA based on a subset of qualitative variables

2.1 Quantification of qualitative data

Let X be a data matrix with n observations and p variables (items), where the j-th ($j = 1, \ldots, p$) variable x_j has c_j possible values (c_j categories) labeled 1, ..., c_j. To quantify X we use an indicator matrix G_j ($n \times c_j$) for each x_j. The columns of G_j are dummy variables which have scores 0 and 1, score 1 indicating the category that the observation belongs to. Then, as an optimally scaled matrix of X we make an $n \times p$ quantified matrix Y whose j-th column is given by $y_j = G_j a_j$, where a_j is a score vector for categories of x_j. The scores a_j are assigned so that the criterion of PCA is maximized under the constraints that all y_j's have zero means and unit variances (which will be described in 2.3).

2.2 Modified PCA for quantitative data matrix Y

Suppose we have obtained a quantitative data matrix Y by quantifying the qualitative variables. We wish to represent this Y by r PCs as well as possible, where r is preassigned and the PCs are linear combinations of a subset of variables of Y. As discussed by Tanaka and Mori (1996), to derive such PCs we can use PCA of instrumental variables proposed by Rao (1964) by assigning the subset of variables as instrumental variables, when variables are quantitative.

We wish to make r linear combinations $Z = Y_1 W$, which reproduce the original p variables as well as possible, where Y_1 is a subset of Y with q ($1 \le q \le p$) variables and W is a $q \times r$ ($1 \le r \le q$) weight matrix for the q variables. Here a_j ($j = 1, \ldots, q$) and W are determined in such a way that y can be predicted as well as possible by means of linear functions of z. Thus the predictive efficiency is maximized for y by using a linear predictor in terms of z.

Let the covariance matrix of $Y = (Y_1, Y_2)$ be $S = \begin{pmatrix} S_{11} & S_{12} \\ S_{21} & S_{22} \end{pmatrix}$. The residual covariance matrix of y after subtracting the best linear predictor is expressed as

$$S_{res} = S - S_1' W (W' S_{11} W)^{-1} W' S_1 = S - S_{Reg}, \tag{1}$$

where $S_1 = (S_{11}, S_{12})$. Then the problem becomes to maximize S_{Reg}. If it is formulated as the maximization problem of $tr(S_{Reg})$ among other possibilities, a generalized eigenvalue problem

$$[(S_{11}^2 + S_{12} S_{21}) - \lambda S_{11}] w = 0 \tag{2}$$

is obtained. Let the q eigenvalues of (2) be ordered from the largest to the smallest as $\lambda_1, \lambda_2, \ldots, \lambda_q$ and the associated eigenvectors be denoted by $w_1, w_2, \ldots, w_q$. Then, the solution is expressed as $W = (w_1, \ldots, w_r)$, and the maximized value of the criterion $tr(S_{Reg})$ is given by $\sum_{i=1}^{r} \lambda_i$, or the proportion of the original variations explained by the r PCs is given by

$$P = \sum_{i=1}^{r} \lambda_i / tr(S). \tag{3}$$

In QM.PCA this proportion P indicates the average squared multiple correlation between each of the original variables and the r PCs, since all y_j's are standardized in the quantification process.

Thus we apply the following two-stage procedure of variable selection as a practical strategy to find PCs which are based on a small number of variables but represent all the variables very well.

A. Initial fixed-variable stage

Assign q variables to subset Y_1, usually $q := p$, and solve the eigenvalue problem (EVP) (2). Looking carefully at the eigenvalues and the proportions P in (3), determine the number r of PCs to be used.

B. Variable selection stage (backward elimination)

B-1: Based on the results of stage A, start with q variables – r PCs model.

B-2: Remove each one of the q variables in turn, solve the q EVP (2) with $q-1$ dimensional matrices, and find the best subset of size $q-1$, that is, the one for which the proportion P in (3) is the largest. Remove the corresponding variable from the present subset of q variables and put $q := q - 1$.

B-3: If both P and the number of variables in Y_1 are larger than preassigned values, go back to *B-2*. Otherwise stop.

2.3 Estimation of category scores and variable weights

Before performing M.PCA we have to estimate the unknown parameters a_j and then make optimally scaled variables. To do this we use the idea of PRINCIPALS (Young et al., 1978) which is an extension of ordinary PCA to the situation where the variables may be measured at a variety of scale levels. It is used to estimate both a_j and W simultaneously. Here, we use all variables in this estimation phase, i.e., $q := p$.

Let

$$\hat{Y} = ZB, \tag{4}$$

where Z $(= YW$ in this phase) is an $n \times r$ matrix of n component scores and B is a $r \times p$ coefficient matrix. We wish to obtain B so that $\hat{Y}$ predicts Y as well as possible, where Y contains unknown a_j's and Z contains unknown W. The optimization criterion is expressed as

$$\theta = tr(Y - \hat{Y})'(Y - \hat{Y}). \tag{5}$$

Based on the ALS principle, unknown parameters are estimated as follows.

(Step 0) Determine initial values Y, which is assigned as category scores provided externally or assigned as random numbers. Standardize Y columnwise.

(Step 1) Apply M.PCA to the data matrix Y, that is, solve EVP (2) and obtain W. Successively obtain Z and $B = (Z'Z)^{-1}Z'Y$.

(Step 2) Evaluate θ. If the improvement in fit from the previous iteration to the present iteration is negligible, stop.

(Step 3) From Z and B compute $\hat{Y}$ by eq.(4). Obtain the matrix of optimally scaled data Y which gives the minimum θ in (5) for the fixed $\hat{Y}$. The optimal scaling of data is performed for each variable separately and independently. Standardize the optimally scaled data and go back to *Step 1*.

Steps 1 through *3* are iterated until convergence is obtained.

3. A numerical example

We illustrate the proposed method by applying it to the data gathered for the purpose of making a rating scale to measure the seriousness of mild disturbance of consciousness (MDOC) due to head injury and other lesions (Sano et al., 1977, 1983). The data set consists of 87 individuals and 25 variables (test items, four points scale).

Tab. 1 Process of removing variables.

Step	q	P	P_q	Removed Variable	P_o
0	23	0.69389	1.00000	–	0.69389
1	22	0.69355	0.99304	V9	0.69314
2	21	0.69323	0.98847	V17	0.69276
3	20	0.69285	0.97941	V10	0.69200
4	19	0.69238	0.97238	V3	0.69070
5	18	0.69183	0.96316	V16	0.68965
6	17	0.69135	0.95196	V12	0.68853
7	16	0.69036	0.93944	V11	0.68733
8	15	0.68933	0.91520	V21	0.68454
9	14	0.68837	0.89372	V7	0.68212
10	13	0.68734	0.88519	V19	0.67778
11	12	0.68602	0.87106	V4	0.67560
12	11	0.68448	0.83876	V23	0.67551
13	10	0.68244	0.81823	V2	0.67449
14	9	0.68019	0.80063	V5	0.67133
15	8	0.67743	0.78617	V13	0.66767
16	7	0.67412	0.77004	V20	0.66034
17	6	0.67035	0.74620	V22	0.66114
18	5	0.66264	0.71730	V1	0.65993
19	4	0.65145	0.68579	V6	0.64770
20	3	0.63371	0.64599	V8	0.63340
21	2	0.60734	0.60734	V15	0.60734

P_q is P value using all q eigenvalues obtained by EVP (2).
P_o is P value computed by the ordinary PCA of the selected variables.

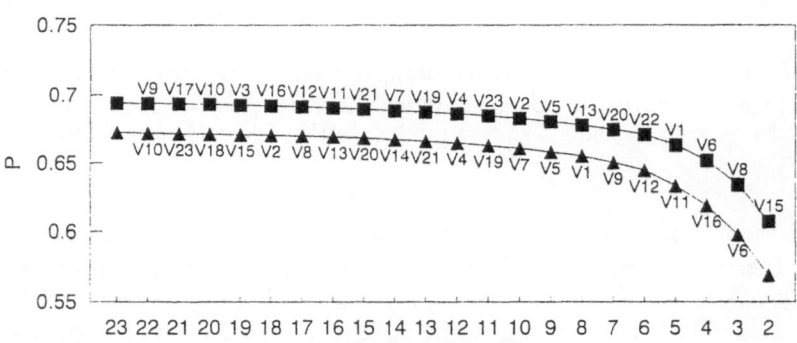

Fig. 1: Process of removing variables (QM.PCA and M.PCA).
(■ indicates the result of QM.PCA which means M.PCA of Y and ▲ indicates that of M.PCA which is applied to X as quantitative variables.)

These data were originally analyzed by Sano, et al. (1977, 1983) to make a rating scale for MDOC, and later by Tanaka and Kodake (1981) and Tanaka (1983) using principal factor analysis with variable selection functions. All of these studies adopted a 23 variables – 2 factors model, since two of 25 variables were thought not to be important, and analyzed the 23 variables as quantitative ones using scores 0, 1, 2 and 3 or 0 and 3 according to its grade of disturbance.

Now let us apply QM.PCA to this data set. We denote this data set by X.

At first indicator matrices G_j are generated from X and are transformed into initial quantified vectors y_j by assigning original values of x_j to score vectors a_j ($j = 1$, ..., 23). After normalizing y_j's so that they have zero means and unit variances, we apply the ordinary PCA to the initial Y. The eigenvalues and the cumulative proportions are $\lambda_1 = 13.878$ (60.34%) $> \lambda_2 = 1.579$ (67.20%) $> \lambda_3 = 0.9580$ (71.37%) $> \lambda_4 = 0.8456$ (75.05%) $> \lambda_5 = 0.7432$ (78.28%) $> \cdots$. Looking at these values and referring the previous studies we decided to extract two PCs. Starting with this initial Y and $r = 2$, scores a_j's and weights W are estimated iteratively by means of ALS method. Y obtained at the final iteration is an optimally quantified matrix of qualitative X. The eigenvalues and the cumulative proportions of the quantitative variables Y are $\lambda_1 = 14.458$ (62.86%) $> \lambda_2 = 1.502$ (69.39%) $> \lambda_3 = 0.9401$ (73.48%) $> \lambda_4 = 0.7788$ (76.86%) $> \lambda_5 = 0.6948$ (79.88%) $> \cdots$. These proportions are larger than those obtained by the ordinary PCA applied to the original X.

Next, we apply M.PCA to this optimally scaled variables Y. The process of removing variables is shown in Tab.1 and the change of P values are visualized in Fig.1 (■). They illustrate that the change of P value is very small until 11 or 12 variables are removed. Even when the number of variables is reduced drastically to 6, the reduction of P is less than 2.5%.

Fig.2 shows scatter plots of the correlation loadings which are defined as the correlations between the original variables and the (varimax-rotated) derived PCs and which play important roles for the interpretation of PCs. Three scatter plots correspond to the cases of (a) $q = 23$, (b) $q = 11$ and (c) $q = 6$. Black circles in (b) and (c) indicate the selected variables. Comparing these scatter plots, it is observed that the configurations are almost the same between (a) and (b), but they are slightly different between (a) and (c). This fact indicates that the meanings of the PCs do not change when 12 variables are removed and they change only slightly when 17 variables are removed. We can say that PCs based on the selected variables have almost the same information as PCs based on the whole ones.

Comparing with the result of M.PCA applied to the original X, Fig.1 shows that QM.PCA provides higher values of P at every step. This difference is due to the effect of optimal scaling by ALS method.

To illustrate how our method selects variables from clusters of variables, the selected variables are marked on the dendrogram in Fig.3 obtained by cluster analysis of Y using standardized squared Euclidean distances and the furthest neighbor method. As illustrations 11 variables and 6 variables selected by QM.PCA are marked below the variable numbers in Fig.3. The dendrogram suggests that there exist three clusters which are composed of 11, 2 and 10 variables, respectively. It is noted that the selected variables are distributed in two major clusters in a well-balanced manner. The reason why no variables are selected from the smallest cluster may be explained in such a way that variables No.21 and No.23 in this cluster are located near the origin in the scatter plots of the correlation loadings (Fig.2) and therefore they do not play important roles for composing these PCs.

Finally let us compare our method with other variable selection methods. On the basis of the previous studies we consider the following methods.

(1) *Method of regression analysis*: Remove one variable which has the largest multiple correlation with the remaining variables successively.

(2) *Jolliffe(1972. 1973. 1986)'s method based on PCA*: Apply PCA to all p variables. Looking at loadings associated with the p-th largest eigenvalue, remove one variable which has the largest loading. Next, looking at loadings associated with the $(p-1)$-th largest eigenvalue, remove one variable among the remaining ones in the same way, and so on. Iterate the process until the number of removed variables is $p - q$.

(3) *Method based on cluster analysis*: Form q clusters of variables using a method of

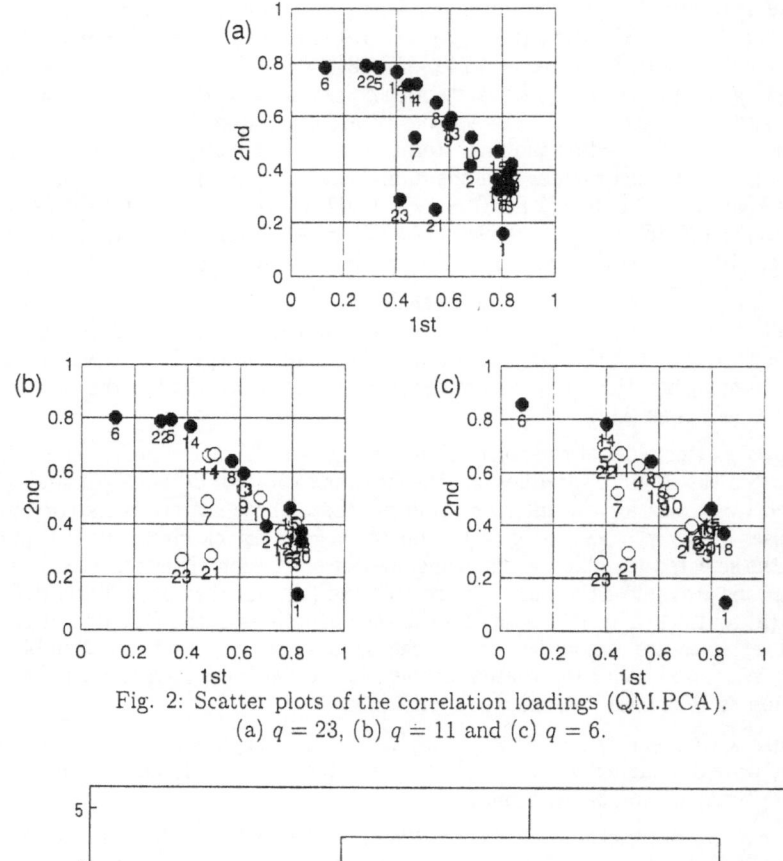

Fig. 2: Scatter plots of the correlation loadings (QM.PCA).
(a) $q = 23$, (b) $q \doteq 11$ and (c) $q = 6$.

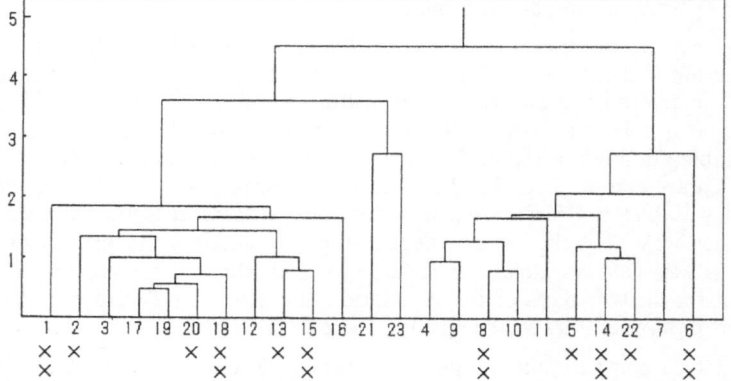

Fig. 3: Dendrogram of 23 variables (Furthest neighbor method with standardized squared Euclidean distances). Eleven and six variables selected by QM.PCA are marked by $\times$ below the variable number.

cluster analysis and select one variable from each cluster.

(4) *Method using Robert and Escoufier(1976)'s RV-coefficient (1)*: Find $q-1$ variables among q variables, which have the closest configuration of the PC scores to that based on q variables, and remove the corresponding variable. The *RV*-coefficient is used as the criterion to evaluate the closeness between the two configurations.

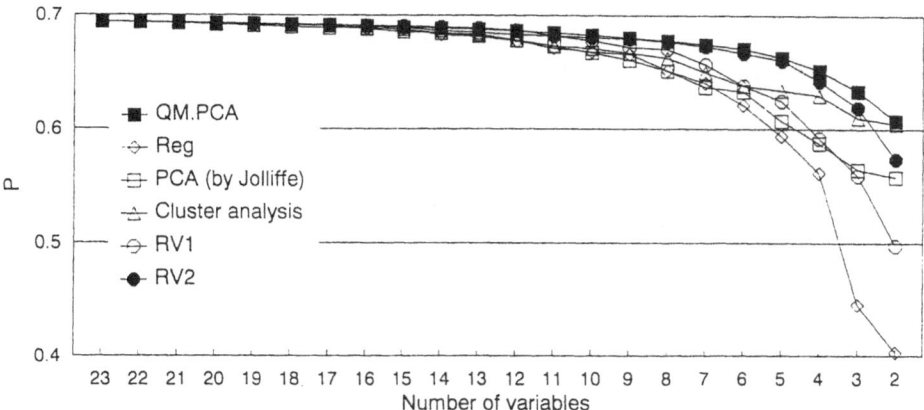

Fig. 4: Comparison of P obtained by applying various selection methods to Y.

(5) *Method using Robert and Escoufier(1976)'s RV-coefficient (2)*: Find $q-1$ variables among q variables, which have the closest configuration of the modified PC scores to that of the original variables Y in terms of the RV-coefficient. In this case the RV-coefficient is computed as $RV = \left\{ \sum_{j=1}^{r} \lambda_j^2 / tr(S^2) \right\}^{1/2}$ where λ_j's are obtained by solving EVP (2) (see, Tanaka and Mori, 1996).

Fig.4 indicates the changes of P for each selection method. P values at each step are computed using Y_1 obtained by each method as the instrumental variables and Y as the main variables. The reason why we use the formulation based on instrumental variables to compute P is to make fair comparisons, because P values obviously become smaller if the ordinary PCA is applied to the selected variables (this fact is observed in Tab.1, that is, P values in the last column, which are computed by the ordinary PCA of the selected variables, are always smaller than those in the third column). Fig.4 shows that the proposed method selects a subset of variables which explains the original variables better than other methods.

4. Concluding remarks

Modified PCA (M.PCA), which was proposed by Tanaka and Mori (1996) to derive PCs which are formulated as linear combinations of a subset of variables but which represent all the variables very well, is extended so as to deal with qualitative data by combining a method of optimal scaling and M.PCA. Performance of the proposed method (QM.PCA) is studied numerically by applying it to a real data set. From the results of the numerical study we can say the followings:

1) The proportion of variance explained by a specified number of PCs in QM.PCA is larger than in M.PCA without optimal scaling in every step of variable selection. It suggests the usefulness of optimal scaling in analyzing qualitative data.

2) In our numerical example the average proportion P of the variance of all the variables explained by PCs does not change much by omitting at most 17 among 23 variables. Also the loadings of PCs based on all the variables ($q = 23$) and on two subsets of variables ($q = 11$ and $q = 6$) are almost the same. These facts suggest that we can construct a multidimensional rating scale which is based on a small number of variables but which has almost the same information as the case using all the variables.

554

3) Comparison is made among the performances of our method and other variable selection methods in PCA. Our method is superior to the other methods in the sense that the proportion P of our method is larger than those of the other methods.

4) To study how our method select variables the selected variables are marked in the scatter plots and the dendrogram for cluster analysis of the variables. It seems that the variables are selected from the major clusters in a well-balanced manner.

References:

De Leeuw, J. (1982): Nonlinear principal component analysis. *COMPSTAT 1982*, 77-86, Physica-Verlag, Vienna.

De Leeuw, J. (1984): The Gifi system of nonlinear multivariate analysis. *Data Analysis and Informatics III*, Diday, E. et al. (eds.), 415-424, Elsevier Science Publishers, North-Holland.

De Leeuw, J. and Van Rijckevorsel, J. (1980): HOMALS & PRINCALS: Some generalization of principal components analysis. *Data Analysis and Informatics*, Diday, E. et al. (eds.), 415-424, North-Holland Publishing Company.

Israels, A. Z. (1984): Redundancy analysis for qualitative variables. *Psychometrika*, **49**, 331-346.

Jolliffe, I. T. (1972): Discarding variables in a principal component analysis. I. Artificial data. *Applied Statistics*, **21**, 160-173.

Jolliffe, I. T. (1973): Discarding variables in a principal component analysis. II. Real data. *Applied Statistics*, **22**, 21-31.

Jolliffe, I. T. (1986): *Principal component analysis*. Springer-Verlag, New York.

Krzanowski, W. J. (1987a): Selection of variables to preserve multivariate data structure, using principal components. *Applied Statistics*, **36**, 22-33.

Krzanowski, W. J. (1987b): Cross–validation in principal component analysis. *Biometrics.*, **43**, 575-584.

McCabe, G. P. (1984): Principal Variables. *Technometrics*, **26**, 137-144.

Rao, C. R. (1964): The use and interpretation of principal component analysis in applied research. *Sankhya*, A, **26**, 329–58.

Robert, P. and Escoufier, Y. (1976): A unifying tool for linear multivariate statistical methods: the RV-coefficient. *Applied Statistics*, **25**, 257-265.

Sano, K. et al. (1977): Statistical studies on evaluation of mind disturbance of consciousness: Abstraction of characteristic clinical pictures by cross-sectional investigation. *Sinkei Kenkyu no Shinpo*, **21**, 1052-1065 (in Japanese).

Sano, K. et al. (1983): Statistical studies on evaluation of mind disturbance of consciousness. *Journal of Neurosurg*, **58**, 223-230.

Tanaka, Y. and Kodake, K. (1981): A method of variable selection in factor analysis and its numerical investigation. *Behaviormetrika*, **10**, 49-61.

Tanaka, Y. (1983): Some criteria for variable selection in factor analysis. *Behaviormetrika*, **13**, 31-45.

Tanaka, Y and Mori, Y. (1996): Principal component analysis based on a subset of variables: Variable selection and sensitivity analysis. *American Journal of Mathematics and Management Sciences*, Special Volume (to appear).

Young, F. et al. (1978): The principal components of mixed measurement level multivariate data: An alternating least squares method with optimal scaling features. *Psychometrika*, **43**, 279-281.

Missing Data Imputation in Multivariate Analysis

Mario Romanazzi

Department of Statistics, "Ca' Foscari" University of Venice
2347 S. Polo, 30125 Venice, Italy

Summary: A new imputation method for incomplete data is suggested, to be used with non-random multivariate observations. The principle is to fill the gaps in a data set so that the partial complete-data configuration and the total filled-in configuration are similar, according to a matrix correlation coefficient. The optimality criteria are Escoufier's RV and Procrustes normalized statistic. Three examples are illustrated.

1. The method

Suppose p numerical variables have been observed on n units, but only n_1 units have complete data, whereas the other $n_2 = n - n_1$ have incomplete data. Let X be the $n \times p$ data matrix, with some empty cells, and let X_1 be the $n \times p_1$ submatrix including the $p_1 \geq 1$ variables with complete data. It is assumed that the variables (columns of X_1 and X) are centered with respect to the means. We suggest estimating the missing values in such a way that X_1 and X are as similar as possible, according to a matrix correlation coefficient. Crettaz De Roten and Helbling (1991) also use a matrix correlation criterion, but they consider an *a priori* partition of the variables in two groups, or try to maximize the match between X_1 and the submatrix of the incomplete-data variables.

A geometrical interpretation can be given. X_1 and X correspond to two constellations representing the same n points in p_1- and p-dimensional Euclidean space, respectively. The constellation associated with X_1 is fixed, whereas in the constellation corresponding to X, relative positions of points and overall shape change for each selection of values to substitute for missing values in the empty cells of X. Our method amounts to fixing X at the position of maximum similarity with X_1.

Obviously, a definition of "similar configurations" must be given. In multivariate analysis there are two standard equivalence criteria: congruence up to linear transformations, typical of canonical correlation and multivariate regression, and congruence up to orthogonal transformations, as in principal component and factor analysis. In this work the match of X_1 and X is evaluated under orthogonal congruence. It is well-known that a maximal invariant is the pair of inner-product matrices $X_1 X_1^T$, $X X^T$ that are comparable by Escoufier's RV

$$RV(X_1, X) = \frac{\operatorname{tr} X_1 X_1^T X X^T}{\sqrt{\operatorname{tr}(X_1 X_1^T)^2 \operatorname{tr}(X X^T)^2}} = \frac{\operatorname{tr} X_1^T X X^T X_1}{\sqrt{\operatorname{tr}(X_1^T X_1)^2 \operatorname{tr}(X^T X)^2}},$$

or Lingoes-Schönemann normalized Procrustes statistics (Seber, 1984)

$$LS(X_1, X) = \frac{\{\operatorname{tr}(X_1^T X X^T X_1)^{1/2}\}^2}{\operatorname{tr}(X_1 X_1^T) \operatorname{tr}(X X^T)} = \frac{\{\operatorname{tr}(X_1^T X X^T X_1)^{1/2}\}^2}{\operatorname{tr}(X_1^T X_1) \operatorname{tr}(X^T X)}.$$

Up to the normalization factor appearing in the denominator, RV and LS are different functions of the singular values of the matrix $X_1^T X$, whose elements are the covariances of the observed variables.

Suppose there are $q \geq 1$ missing values $\theta_1, \ldots, \theta_q$, forming the vector $\theta = (\theta_1, \ldots, \theta_q)^T$. With no loss of generality, they can be thought to be pertinent to the last n_2 units and last p_2 variables. Thus X can be partitioned as

$$X = (\, X_1 \quad X_2 \,) = \begin{pmatrix} X_{11} & X_{12} \\ X_{21} & X_{22} \end{pmatrix},$$

where X_j is an $n \times p_j$ matrix and X_{ij} is an $n_i \times p_j$ matrix, $i, j \in \{1, 2\}$. The $p \times 1$ vector of the means is

$$\bar{x}_n = \begin{pmatrix} \bar{x}_n^{(1)} \\ \bar{x}_n^{(2)} \end{pmatrix},$$

with $\bar{x}_n^{(1)}$ ($\bar{x}_n^{(2)}$) denoting the vector of the means of the first p_1(last p_2) variables. For $j \in \{1, 2\}$, it is useful to interpret $\bar{x}_n^{(j)}$ as the weighted mean of $\bar{x}_1^{(j)}$ and $\bar{x}_2^{(j)}$, the group averages computed from first n_1 and last n_2 units. The relevant expression is

$$\bar{x}_n^{(j)} = \frac{n_1}{n}\bar{x}_1^{(j)} + \frac{n_2}{n}\bar{x}_2^{(j)} = \bar{x}_1^{(j)} + \frac{n_2}{n}(\bar{x}_2^{(j)} - \bar{x}_1^{(j)}) = \bar{x}_2^{(j)} - \frac{n_1}{n}(\bar{x}_2^{(j)} - \bar{x}_1^{(j)}).$$

Hence the submatrices in X can be written as

$$\begin{aligned} X_{11} &= A - \frac{n_2}{n}1_{n_1}a^T, \quad X_{12} = C - \frac{n_2}{n}1_{n_1}\beta^T, \\ X_{21} &= B + \frac{n_1}{n}1_{n_2}a^T, \quad X_{22} = \Delta + \frac{n_1}{n}1_{n_2}\beta^T \end{aligned}$$

where A, B, C and Δ are column-centered matrices, $a = \bar{x}_2^{(1)} - \bar{x}_1^{(1)}$ and $\beta = \bar{x}_2^{(2)} - \bar{x}_1^{(2)}$. Since X_1 is the submatrix of X corresponding to the p_1 complete-data variables, A, B, C and a are free of θ, whereas $X_2 \equiv X_2(\theta)$, $\Delta \equiv \Delta(\theta)$ and $\beta \equiv \beta(\theta)$ depend on the missing values to be estimated. Thus the matrix $X \equiv X(\theta)$ is a function of θ, and it makes sense to look for the vector θ^* which gives RV (LS) its maximum value. This amounts to choosing, among all possible matrices $X(\theta)$, the most similar one to X_1, according to RV (LS). Putting $\Gamma = (X_1^T X_1)^2$, $\gamma = \operatorname{tr}\Gamma$, $\Phi(\theta) = X_1^T X_2(\theta)X_2(\theta)^T X_1$, $\phi(\theta) = \operatorname{tr}\Phi(\theta)$, $\psi(\theta) = \operatorname{tr}\{X_2(\theta)^T X_2(\theta)\}^2$, since $X_1^T X(\theta)X(\theta)^T X_1 = \Gamma + \Phi(\theta)$ and $\operatorname{tr}(X^T X)^2 = \gamma + 2\phi(\theta) + \psi(\theta)$, leads to

$$RV(X_1, X) \equiv RV(\theta) = \frac{\gamma + \phi(\theta)}{\sqrt{\gamma\{\gamma + 2\phi(\theta) + \psi(\theta)\}}},$$

$$LS(X_1, X) \equiv LS(\theta) = \frac{\{\operatorname{tr}(\Gamma + \Phi(\theta))^{1/2}\}^2}{\operatorname{tr} X_1^T X_1 \{\operatorname{tr} X_1^T X_1 + \operatorname{tr} X_2(\theta)^T X_2(\theta)\}}.$$

Several properties of $RV(\theta)$ and $LS(\theta)$ can be noted.

Remark 1. Since $\operatorname{tr}(X^T X)^2 \geq \operatorname{tr}(X_1^T X_1)^2 > 0$, $RV(\theta)$ is an ∞ times differentiable function of $\theta \in \mathbb{R}^q$.

Remark 2. A sufficient condition for $X_1^T X(\theta)X(\theta)^T X_1$ to be positive definite is that $X_1^T X_1$ is positive definite. In this case $\{X_1^T X(\theta)X(\theta)^T X_1\}^{1/2}$ is an analytic function of θ, which implies that $LS(\theta)$ is too. The behaviour of $LS(\theta)$ in the singular case is beeing investigated.

Remark 3. $RV(\theta) = LS(\theta) = 0 \Leftrightarrow X_1^T X(\theta) = (\, X_1^T X_1 \quad X_1^T X_2(\theta) \,) = 0_{p_1, p}$. But

$X_1^T X_1 \neq 0_{p_1}$, thus $RV(\theta)$ and $LS(\theta)$ are positive for all $\theta \in \mathbb{R}^q$.

Remark 4. $RV(\theta) = LS(\theta) = 1$ iff there exists a column-orthogonal matrix $Q(\theta)$ $p_1 \times p_2$ and a scalar δ such that $X_2(\theta) = \delta X_1 Q(\theta)$.

Remark 5. If $\Phi(\theta)$ does not depend on θ, $LS(\theta)$ attains its maximum value at the θ-vector where $\operatorname{tr} X_2(\theta)^T X_2(\theta)$ is minimized. Then the optimal value for all missing elements in the j-th incomplete-data variable is the mean of observed data for that variable. If $\Phi(\theta)$ does not depend on θ and all the data corresponding to last n_2 units and last p_2 variables are missing, $RV(\theta)$ satisfies the same property.

Remark 6. If $p_1 = p_2 = q = 1$, then

$$RV(\theta) = \frac{s_{11} + (s_{12}(\theta))^2/s_{11}}{\sqrt{s_{11}^2 + 2(s_{12}(\theta))^2 + (s_{22}(\theta))^2}}, \qquad LS(\theta) = \frac{s_{11} + (s_{12}(\theta))^2/s_{11}}{s_{11} + s_{22}(\theta)},$$

where $s_{11} = \operatorname{var} X_1$, $s_{12}(\theta) = \operatorname{cov}(X_1, X_2)$ and $s_{22}(\theta) = \operatorname{var} X_2$. It can be shown that the value θ_{LS}^* which maximizes $LS(\theta)$ satisfies

$$\theta_{LS}^* - \bar{x}_2(\theta_{LS}^*) = \frac{s_{12}(\theta_{LS}^*)}{s_{11}} \cdot (x_{n,1} - \bar{x}_1) \cdot [LS(\theta_{LS}^*)]^{-1},$$

whereas the value θ_{RV}^* which maximizes $RV(\theta)$ satisfies

$$\theta_{RV}^* - \bar{x}_2(\theta_{RV}^*) = \frac{s_{12}(\theta_{RV}^*)}{s_{11}} \cdot (x_{n,1} - \bar{x}_1) \cdot \frac{s_{22}(\theta_{RV}^*) + (s_{12}(\theta_{RV}^*))^2/s_{22}(\theta_{RV}^*)}{s_{11} + (s_{12}(\theta_{RV}^*))^2/s_{11}}.$$

Therefore, in the optimal two-dimensional configuration including the estimated point $(x_{n,1}, \theta^*)$, θ^* is the ordinary linear regression prediction, with the slope adjusted by a positive factor that depends on the strength of the linear relation between X_1 and X_2 and, for $RV(\theta)$, the relative magnitudes of s_{11} and $s_{22}(\theta^*)$.

2. Optimization of RV and LS

To obtain maximum similarity between X_1 and $X(\theta)$, the optimization problems $\max_{\theta \in R^q} RV(\theta)$, or $\max_{\theta \in R^q} LS(\theta)$, must be solved. The two functions vary in the $(0, 1]$ interval, if $X_1^T X_1$ is positive definite they are continuous and differentiable for all $\theta \in R^q$, but the search for the extrema might be complicated by the existence of local extrema. A detailed discussion of the behaviour of $RV(\theta)$ in the particular case $q = 1$ is given in Romanazzi (1995). When p and q are not very high (tentatively, $p \leq 10$, $q \leq 4$, in the case of $RV(\theta)$), the optimizations are conveniently performed by symbolic computation, using for instance Mathematica or MathLab. Otherwise, *ad hoc* algorithms based on steepest descent methods must be designed. The gradient vector of $RV(\theta)$ is (Romanazzi, 1995):

$$RV^{(1)}(\theta) = \frac{RV(\theta)}{2\{\gamma + \phi(\theta)\}\{\gamma + 2\phi(\theta) + \psi(\theta)\}}\{2[\phi(\theta) + \psi(\theta)]\phi^{(1)}(\theta) - [\gamma + \phi(\theta)]\psi^{(1)}(\theta)\}.$$

Here $\partial \operatorname{vec} \Delta(\theta)/\partial \theta \equiv \Delta^{(1)}$ is the $q \times n_2 p_2$ matrix of the partial derivatives of the elements of $\operatorname{vec} \Delta(\theta)$ with respect to $\theta_1, \ldots, \theta_q$ and $\partial \beta(\theta)/\partial \theta \equiv \beta^{(1)}$ is the $q \times p_2$ matrix of the partial derivatives of the components of $\beta(\theta)$ with respect to $\theta_1, \ldots, \theta_q$. $\Delta^{(1)}$ and $\beta^{(1)}$ are constant matrices no longer depending on $\theta_1, \ldots, \theta_q$. The expressions of $\phi^{(1)}(\theta)$ and $\psi^{(1)}(\theta)$ turn out to be

$$\phi^{(1)}(\theta) = 2\{\Delta^{(1)}[(I_{p_2} \otimes BB^T)\,\mathrm{vec}\,\Delta(\theta) + \frac{n_1 n_2}{n}(I_{p_2} \otimes B)\,\mathrm{vec}\,a\beta(\theta)^T + \mathrm{vec}\,BA^TC]$$
$$+\frac{n_1 n_2}{n}\beta^{(1)}[\frac{n_1 n_2}{n}(a^Ta)\beta(\theta) + \Delta(\theta)^TBa + C^TAa]\},$$

$$\psi^{(1)}(\theta) = 4\{\Delta^{(1)}[(\Delta(\theta)^T\Delta(\theta) + C^TC + \frac{n_1 n_2}{n}\beta(\theta)\beta(\theta)^T) \otimes I_{n_2}]\,\mathrm{vec}\,\Delta(\theta)$$
$$+\frac{n_1 n_2}{n}\beta^{(1)}[\Delta(\theta)^T\Delta(\theta) + C^TC + \frac{n_1 n_2}{n}(\beta(\theta)^T\beta(\theta))\,I_{p_2}]\beta(\theta)\}.$$

If $X_1^TX_1$ is positive definite, the gradient vector of $LS(\theta)$ is (Romanazzi, 1995):

$$LS^{(1)}(\theta) = \frac{LS(\theta)}{\mathrm{tr}[\Gamma + \Phi(\theta)]^{1/2}\,\mathrm{tr}[X(\theta)^TX(\theta)]} \times$$
$$\{\mathrm{tr}[X(\theta)^TX(\theta)][\Delta^{(1)}(\Phi_\Delta^{(1)}(\theta))^T + \frac{n_1 n_2}{n}\beta^{(1)}(\Phi_\beta^{(1)}(\theta))^T]\,\mathrm{vec}[\Gamma + \Phi(\theta)]^{-\frac{1}{2}}$$
$$-2\,\mathrm{tr}[\Gamma + \Phi(\theta)]^{1/2}[\Delta^{(1)}\,\mathrm{vec}\,\Delta(0) + \frac{n_1 n_2}{n}\beta^{(1)}\beta(0)]\},$$

where

$$\Phi_\Delta^{(1)}(\theta) = \frac{\partial\,\mathrm{vec}\,\Phi(\theta)}{\partial\,\mathrm{vec}\,\Delta(0)}$$
$$= B^T\Delta(0) \otimes B^T + (B^T \otimes B^T\Delta(0))K_{n_2,p_2} + (B^T \otimes A^TC)K_{n_2,p_2}$$
$$+\frac{n_1 n_2}{n}(a\beta(\theta)^T \otimes B^T + B^T \otimes a\beta(\theta)^T),$$

$$\Phi_\beta^{(1)}(\theta) = \frac{\partial\,\mathrm{vec}\,\Phi(\theta)}{\partial\beta(0)}$$
$$= \frac{n_1 n_2}{n}(a\beta(0)^T \otimes a + a \otimes a\beta(0)^T) + a \otimes A^TC + A^TC \otimes a$$
$$+a \otimes B^T\Delta(0) + B^T\Delta(0) \otimes a.$$

Here K_{n_2,p_2} denotes the commutation matrix of the order $n_2 p_2$, transforming $\mathrm{vec}\,\Delta(\theta)$ into $\mathrm{vec}(\Delta(0)^T)$.

Unfortunately, explicit expressions of θ_{RV}^* and θ_{LS}^* are not available: even in the simplest case — RV coefficient, $q = 1$ — the optimal value of θ is a root of a fourth-degree equation. This is an important drawback in comparison with imputation based on linear regression.

The optimization of $RV(\theta)$ and $LS(\theta)$ is computationally feasible and the results are sensible if the number of missing values is small. In particular, at least one variable must have complete data. With large matrices scattered with missing values we suggest the following iterative procedure.

Step 0. Substitute a naive initial estimate (*e.g.*, the mean) for each missing value.

Step 1. For $k = 1, \ldots, q$, determine the optimal value of θ_k, according to RV or LS, keeping the other θ_j's, $j \neq k$, fixed at their previous values.

Step 2. Iterate step 1 until the configuration of points no longer changes, *i.e*, $|\theta_k^{(i)} - \theta_k^{(i-1)}| < \varepsilon_k$, $k = 1, \ldots, q$.

While the general method determines the optimal values of $\theta_1, \ldots, \theta_q$ simultaneously, the iterative procedure tries to locate the optimum θ-vector optimizing the

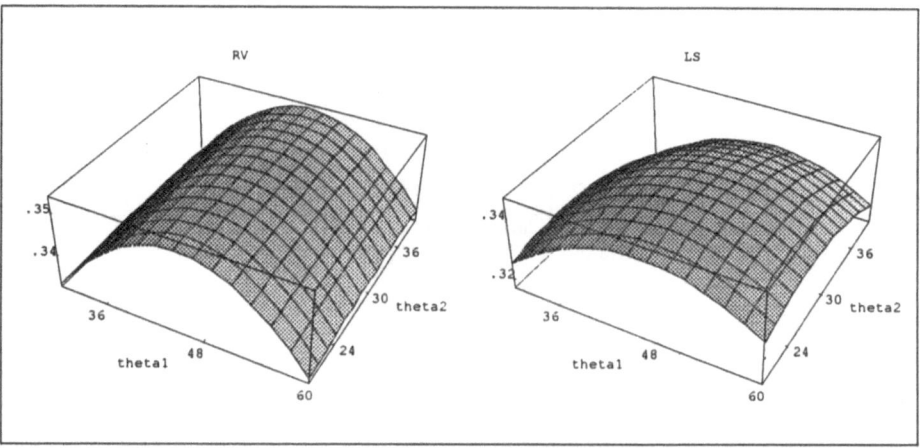

Figure 1: Graphs of $RV(\theta)$ and $LS(\theta)$.

θ_k's one at a time. This implies that it will only find a sub-optimal configuration $X(\overline{\theta}_{RV})(X(\overline{\theta}_{LS}))$ such that $RV(\overline{\theta}_{RV})(LS(\overline{\theta}_{LS})) \leq RV(\theta^*_{RV})(LS(\theta^*_{LS}))$.

3. Numerical illustrations

The following examples illustrate the behaviour of matrix correlation imputation in three situations which are fairly different both in the characteristics of the data and the number of missing values. A comparison with alternative procedures — imputation of means, multiple regression — is also given.

Example 1. The data set includes $p = 4$ measures of school performance of a sample of $n = 85$ students of the faculty of sociology at the university of Trento. The measures are (1) final secondary-school mark, (2) average score in the university examinations, (3) and (4) mathematics and sociology examination scores, respectively. Variable 1 is measured on a sixty-points scale, with 36 and 60 corresponding to pass-mark and maximum mark, respectively. The other variables are measured on a thirty-points scale, with 18 and 30 corresponding to pass-mark and maximum mark. In the data set there are $q = 2$ missing values, since one student did not record the final secondary-school mark (the other data are $x_2 = 28$, $x_3 = 21$, $x_4 = 28$) and another one did not record the average examination score (the other data are $x_1 = 42$, $x_3 = 25$, $x_4 = 28$).

The means of the variables (standard deviations in parentheses) for the $n_1 = 83$ complete-data students are $\overline{x}_1 \doteq 49.6$ (6.2), $\overline{x}_2 \doteq 26.6$ (1.49), $\overline{x}_3 \doteq 25.1$ (3.19), $\overline{x}_4 \doteq 27.7$ (2.06). The pair-wise linear correlations between the variables are positive and low; the minimum correlation is $r_{1,3} \doteq .203$, the maximum correlation is $r_{2,4} \doteq .558$ and the average correlation is .363.

Imputation of the means over complete cases gives $\theta_{AV,1} \doteq 49.6$, $\theta_{AV,2} \doteq 26.6$. The predictions derived from multiple regression are $\theta_{MR,1} \doteq 51.2$ (average examination

560

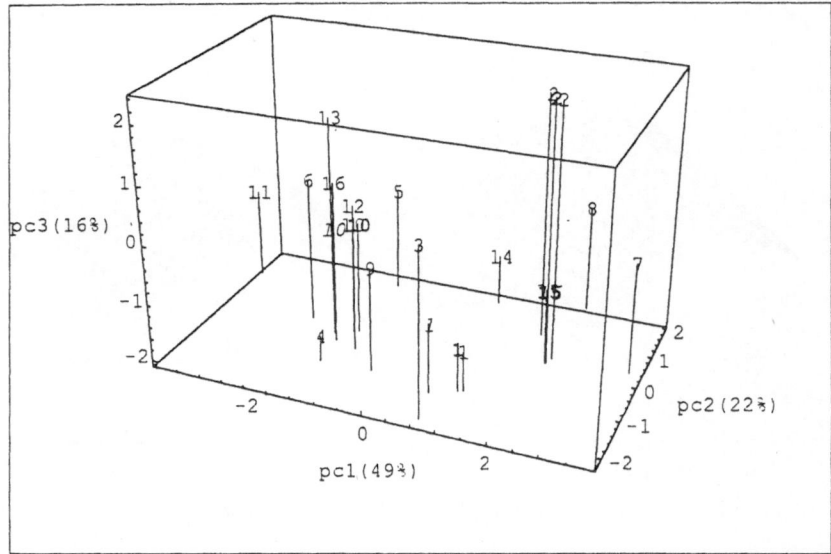

Figure 2: Principal components of crimes data for 16 American cities (numbers correspond to cities); bold (italic) numbers are cities with discarded data estimated by optimization of LS (multiple regression).

score, mathematics and sociology scores as predictors), $\theta_{MR,2} \doteq 25.7$ (final secondary-school mark, mathematics and sociology scores as predictors). The optimal θ-vectors according to RV and LS are $\theta_{RV}^* \doteq (45.5, 34.5)^T$, $\theta_{LS}^* \doteq (46.8, 26.0)^T$. The optimal values of RV and LS are rather low ($RV_{max} \doteq .353$, $LS_{max} \doteq .346$), which suggests that the optimal four-dimensional configuration $X(\theta_{RV}^*)$ $(X(\theta_{LS}^*))$ is not very similar to the two-dimensional configuration of the complete-data variables X_1. Moreover, Figure 1 reveals that the surface described by $RV(\theta)$ in a neighbourhood of θ_{RV}^* is almost flat in the direction of θ_2: this means that the value of $\theta_{RV,2}^*$ can be altered without a substantial reduction in the value of $RV(\theta)$. Predictions derived from multiple regression should also be interpreted with caution, since the squared multiple correlations are low (.156 in the estimation of θ_1, .480 in the estimation of θ_2). It is clear that the estimates of θ_1 vary according to the imputation method, whereas the estimates of θ_2 are similar for all of the methods except optimization of RV. (However, $\theta_{RV,2}^* \doteq 34.5$ is *not* valid, since the maximum score is 30; this surprising result can be due to the fact that $RV(\theta)$ is almost constant in the direction of θ_2.)

The discrepancy between multiple regression and matrix correlation estimates of θ_1 follows from the "global" character of the first method and the "local" character of the second one. Multiple regression uses mainly the positive correlation between dependent and explanatory variables (in particular, average examination score) in the complete data set, thus producing a value somewhat greater than the mean. On the contrary, optimization of RV (LS) looks for the value of θ_1 for which the position of the missing-data student is as near as possible to the positions of the complete-data students with similar values of the average examination score and mathematics and sociology scores. Inspection of the data shows that this value is about 46.

Example 2. To explore the validity of iterative optimization of $RV(\theta)$ and $LS(\theta)$, we discarded, at random, $q = 4$ values from the city-crimes data set described by

Discarded	Imputation Methods			
Values	AV	MR^*	RV^{**}	LS^{**}
$x_{1,3} = 106$	277	293 (.719)	142 (237)	137 (139)
$x_{2,6} = 669$	1064	952 (.888)	858 (853)	810 (1004)
$x_{10,4} = 226$	211	275 (.745)	351 (199)	212 (232)
$x_{15,4} = 148$	211	145 (.745)	256 (188)	147 (143)

Table 1: Estimates of discarded values from city-crimes data (AV: average; MR: multiple regression; RV, LS: optimization of RV or LS; $*$: squared multiple correlation in brackets; $**$: results from iterative optimization in brackets).

Everitt (1984). The variables are crime rates for $p = 7$ different crimes in $n = 16$ American cities. The minimum, maximum and average (absolute) correlations are $r_{1,6} \doteq .050$, $r_{2,4} \doteq .772$ and $.403$, respectively. The scatter plot of the first three principal components of the standardized data is shown in Figure 2. The first component is interpreted as a size factor, with high scores associated with low levels of crime. The second is a shape factor, contrasting the first four variables ("violent" crimes) to the last three ("non-violent" crimes); high values correspond to a prevalence of "non-violent" crimes. The third component is easily interpreted, being highly correlated with variable no. 7 (auto-theft).

In Table 1 we record the discarded values and their matrix correlation imputations, obtained by direct and iterative optimization. Results from imputation of means and multiple regression are also given. Iterative estimates derived from RV are better than global estimates in three cases out of four. Iterative estimates derived from LS are very near to the global ones, except $x_{2,6}$. Global and local optima are almost equal ($RV(\theta^*_{RV}) \doteq .921$, $RV(\bar{\theta}_{RV}) \doteq .920$, $LS(\theta^*_{LS}) \doteq .774$, $LS(\bar{\theta}_{LS}) \doteq .771$), this maybe indicating, as in Example 1, that the surfaces $RV(\theta)$ and $LS(\theta)$ are not very steep in a neighbourhood of θ^*. Finally, the results in Table 1 suggest that, in this data set, matrix correlation imputation (in particular, optimization of LS) might be superior to multiple regression imputation. This is confirmed by the scatter plot in Figure 2 where, together with the complete-data units no. 1, 2, 10 and 15, we represent as supplementary points the same units with discarded values replaced by imputations derived from optimization of LS and multiple regression. It is clear that the supplementary points corresponding to LS are closer to the true ones than those corresponding to multiple regression.

Example 3. We consider the sons data discussed by Seber (1985): the $n = 25$ units form a compact cluster of points and the 4 variables are positively correlated. The minimum, maximum and average correlations are $r_{2,3} \doteq .693$, $r_{3,4} \doteq .839$ and $.732$, respectively. We also consider an artificial data set with $n = 15$, $p = 4$, whose characteristics mimic the city-crimes data . The minimum, maximum and average (absolute) correlations are $\mid r_{1,2} \mid \doteq \mid -.126 \mid$, $\mid r_{2,3} \mid \doteq \mid -.889 \mid$, and $.387$, respectively.

To assess the behaviour of imputation methods, all possible pairs of points are discarded in turn from the configuration of $\mathbf{X_3}$ and $\mathbf{X_4}$, and the omitted values are estimated by averages over complete data, multiple regressions using $\mathbf{X_1}$ and $\mathbf{X_2}$ as predictors, optimizations of RV and LS. As a summary of the results, the "sampling" distribution of the sum of the squared deviations between true values and estimates and the percentage of best results, $i.e.$, minimum squared errors, are computed. According to the frequency distributions of squared errors in Table 2, the best results

(a) Squared Error	Frequency AV	MR	RV	LS	(b) Squared Error	Frequency AV	MR	RV	LS
0 ⊢ 50	.12	.16	.06	.10	0 ⊢ 5	.01	.20	.16	.24
50 ⊢ 100	.14	.20	.15	.24	5 ⊢ 10	.03	.16	.19	.14
100 ⊢ 150	.17	.15	.12	.16	10 ⊢ 15	.04	.22	.25	.31
150 ⊢ 200	.06	.20	.13	.13	15 ⊢ 20	.08	.18	.13	.11
200 ⊢ 300	.12	.17	.28	.21	20 ⊢ 30	.24	.14	.10	.05
300 ⊢ 500	.21	.12	.21	.15	30 ⊢ 50	.42	.09	.16	.14
≥ 500	.18	≈ 0	.05	.01	≥ 50	.18	.01	.01	.01
% Best Results	30	27	10	33	% Best Results	10	19	26	45

Table 2: Distributions of squared errors and percentages of best results (symbols as in Table 1). (a) Sons data, 300 trials. (b) Artificial data, 105 trials.

are given by multiple regression for the sons data, by the optimization of LS for the artificial data. In both cases, optimization of LS produces the highest percentage of minimum squared errors.

4. Concluding remarks

Imputation of values to non-observed data requires a criterion that is subjectively, if not arbitrarily, chosen. Our criterion is to fill the gaps in such a way that the partial complete-data configuration and the total "filled-in" configuration are as similar as possible according to a measure of matrix correlation. We confined the attention to coefficients invariant under separate orthogonal transformations of the matching configurations, but other choices are possible, such as coefficients that are invariant under linear transformations. The computational effort is not negligible - even in the simplest cases optimization routines are necessary - but this drawback is bearable if the aim is a truly multivariate solution, using all the available information.

Preliminary results suggest that when data is homogeneous with strong linear relations between complete-data variables and those with missing values, then imputation based on linear regression is superior. Conversely, when the units belong to different groups and there are no important linear relations, the matrix correlation method is more reliable. In these situations, LS is often better than RV.

5. References

Crettaz De Roten F. and Helbling, J.-M. (1991): Une estimation de données manquantes basée sur le coefficient RV. *Revue de Statistique Appliquée*, **39**, 47-57.

Everitt, B. S. (1984): *An introduction to latent variable models*. Chapmann and Hall.

Romanazzi, M. (1995): Missing values imputation and matrix correlation. *Quaderni di Statistica e Matematica applicata alle Scienze Economico-Sociali*, **15**, 41-59.

Seber, G. A. F. (1984): *Multivariate Observations*. Wiley.

Recent Developments in Three-Mode Factor Analysis: Constrained Three-Mode Factor Analysis and Core Rotations

Henk A.L. Kiers

Department of Psychology
University of Groningen
Grote Kruisstraat 2/1
9712 TS Groningen
The Netherlands

Summary: A review is presented of some recent developments in three-mode factor analysis, that are all aimed at reducing the difficulties in interpreting three-mode factor analysis solutions. First, variants of three-mode factor analysis with zero constraints on the core are described, and attention is paid to algorithms for fitting these models, as well as to uniqueness of the representations. Next, various methods for rotation of the core to simple structure are discussed and related to two-way simple structure rotation techniques. In the concluding section, new perspectives for simplification of the interpretation of three-mode factor analysis solutions are discussed.

1. Introduction

1.1 Three-way data and three-way methods

Three-way data are data associated with three entries. Three-way data are collected in various disciplines. For instance, in the behavioral sciences, three-way data may consist of scores of a set of individuals on a set of variables at different occasions. As another example, in spectroscopy, three-way data are obtained when measuring absorbed energy at various absorption levels on various mixtures of substances that have been exposed to various sorts of light emission.

Several methods have been proposed for the analysis of three-way data (for overviews see Law, Snyder, Hattie & McDonald, 1984; Coppi & Bolasco, 1989; Carlier et al., 1989; Kiers, 1991). The most popular methods are probably Three-Mode Factor Analysis (3MFA; Tucker, 1966; Kroonenberg & De Leeuw, 1980) and CANDE-COMP/PARAFAC (Carroll & Chang, 1970; Harshman, 1970; Harshman & Lundy, 1984). In 3MFA the data x_{ijk} are modelled by

$$x_{ijk} = \sum_{p=1}^{P} \sum_{q=1}^{Q} \sum_{r=1}^{R} a_{ip} b_{jq} c_{kr} g_{pqr} + e_{ijk}, \tag{1}$$

$i=1,...,I, \ j=1,...,J, \ k=1,...,K$, where a_{ip}, b_{jq}, c_{kr} are elements of the matrices **A**, **B**, and **C** of orders I by P, J by Q, and K by R, and the additional parameters g_{pqr} denote the elements of the P by Q by R so called "core array". The matrices **A**, **B**, and **C** can be considered component matrices for "idealized subjects" (in A), "idealized variables" (in B), and "idealized occasions" (in C), respectively. The elements of the core indicate how the components from the different modes interact. The model is fitted to the data by minimizing the sum of squared error terms. In this minimization, the component matrices can be taken columnwise orthonormal without loss of fit, since orthonormalizations of **A**, **B**, and **C** can be compensated for by transformations of the core. In fact, any nonsingular transformation of the component matrices can be compensated by applying the inverse transformation to the core. As a consequence, the model parameters have a great deal of transformational freedom, comparable to the rotational indeterminacy in factor analysis.

The CANDECOMP/PARAFAC model is given by

$$x_{ijk} = \sum_{r=1}^{R} a_{ir} b_{jr} c_{kr} + e_{ijk}, \tag{2}$$

$i=1,\dots,I,\ j=1,\dots,J,\ k=1,\dots,K$, where $a_{ir},\ b_{jr},\ c_{kr}$ again denote elements of matrices **A**, **B**, and **C**. As has been noted by Carroll and Chang (170, p. 312), the PARAFAC model can be considered as a version of the 3MFA model where the core is constrained to be "superdiagonal" (which implies that g_{pqr} is unconstrained if $p=q=r$ and g_{pqr} is constrained to 0 otherwise). It follows that, if $P=Q=R$, the 3MFA fit is always at least as good as the PARAFAC fit, because the 3MFA model uses not only the superdiagonal elements of the core, but also the off-superdiagonal elements, which may considerably enhance the fit. In contrast to the 3MFA model, the parameters in the CANDECOMP/PARAFAC model are (usually) unique, up to scalar multiplications and permutations.

To increase insight in the difference between the two models and in the role of the core in the 3MFA mode, we use a (simplified) tensorial description of the two models. Considering **x** as a vectorized version of the modelled three-way array, and **e** as a vector with error terms, the 3MFA model can be written as

$$\mathbf{x} = g_{111}(\mathbf{a}_1 \phi \mathbf{b}_1 \phi \mathbf{c}_1) + g_{112}(\mathbf{a}_1 \phi \mathbf{b}_1 \phi \mathbf{c}_2) + \dots + g_{211}(\mathbf{a}_2 \phi \mathbf{b}_1 \phi \mathbf{c}_1)$$

$$+ \dots + g_{PQR}(\mathbf{a}_P \phi \mathbf{b}_Q \phi \mathbf{c}_R) + \mathbf{e}, \tag{3}$$

where $\mathbf{a}_p \phi \mathbf{b}_q \phi \mathbf{c}_r$ denotes the triple tensor product of column p of **A**, column q of **B**, and column r of **C**. This tensor product is a vectorized version of the three-way array computed from all possible products of elements from the three different vectors. Specifically, it contains the products $a_{ip} b_{jq} c_{kr}$ for $i=1,\dots,I,\ j=1,\dots,J,\ k=1,\dots,K$, ordered such that index i runs slowest and index k runs fastest; the vector **x** is composed analogously from the elements x_{ijk}. In the same notation, the CANDECOMP/PARAFAC model can be written as

$$\mathbf{x} = \mathbf{a}_1 \phi \mathbf{b}_1 \phi \mathbf{c}_1 + \dots + \mathbf{a}_R \phi \mathbf{b}_R \phi \mathbf{c}_R + \mathbf{e}. \tag{4}$$

It can now be seen easily that the essential difference between the 3MFA model and the CANDECOMP/PARAFAC model consists of the fact that the latter contains only a subset of the triple tensor products of the former. Specifically, the PARAFAC model contains only the triple products for which $p=q=r$. Another difference between the descriptions in (3) and (4) is that in (3) all triple tensor products are weighted by an element of the core matrix, whereas such weights in (4) are missing. This difference, however, is nonessential, because in (4) these weights can be understood to be subsumed in the columns of **A**, **B** or **C**.

1.2 Problems in interpreting three-mode factor analysis solutions

Above, we have described the CANDECOMP/PARAFAC model and the 3MFA model. The latter method, 3MFA, is one of the earliest generalizations of ordinary (two-mode) principal components analysis (PCA), but, as we see from (1) and (3), the generalization is considerably more complicated than PCA itself. This is first of all because it employs components for three rather than two modes. In addition, it involves a three-way core array that describes the relations between all factors of all three modes. The interpretation of the results is a somewhat burdensome enterprise, especially because of the need to take into account all core elements. It would therefore be attractive if many core elements could be ignored. Another complicating aspect is that the solution is rotationally undetermined. This implies that a solution is equivalent to infinitely many rotated versions of this solution, and part of the process of interpreting one's results is to choose *which* rotated solution must be interpreted. These complications of the 3MFA model have already been recognized by Tucker (1966), and have since evoked several proposals for handling these problems. In these proposals, the CANDECOMP/PARAFAC method played an important role. This is because the CANDECOMP/PARAFAC model can be seen as the extremely simplified version of the three-mode factor analysis model where the core is superdiagonal. Because of this simplification, CANDECOMP/PARAFAC solutions are much easier to interpret. An added benefit is that CANDECOMP/PARAFAC solutions are unique, hence no choice has to be made between various rotated versions of the solution. The main disadvantage of CANDECOMP/PARAFAC is that it is much more restrictive than the 3MFA model, and may therefore more often fail to fit the data well. For this reason, researchers have not abandoned the 3MFA model, and from the seventies on, procedures have been proposed for simplifying the core of a 3MFA solution.

1.3 Early proposals for simplification of the core

Kroonenberg (1983, p.58) has observed that if the 3MFA core can be simplified into a core with diagonal core planes, then the 3MFA model has the same form as the CANDECOMP/PARAFAC model (which, as we saw above, is considerably less complicated than the 3MFA model); it should be noted that the CANDECOMP/PA-RAFAC model thus turns out to be equivalent to two different forms of the 3MFA model, that is, not only the 3MFA model with a superdiagonal core, but also a different 3MFA model employing a core with diagonal frontal planes. Kroonenberg (1983, Chapter 5) has mentioned and demonstrated that procedures for diagonalizing the frontal planes of a core that have been investigated in the related context of the IDIOSCAL model (Cohen, 1974, 1975; MacCallum, 1976; De Leeuw & Pruzansky, 1978) apply equally to the 3MFA model. Thus, the first attempts at simplifying the 3MFA core consist of diagonalizing the frontal planes of the core by means of nonsingular transformations. By applying nonsingular transformations to the core, and the inverse transformations to **A**, **B**, and **C**, the fit of the model is unaffected. In this way, the simplicity of the CANDECOMP/PARAFAC model is approximated, without losing the better fitting properties of 3MFA.

In practice, diagonalization of the core planes cannot be expected to work perfectly. Therefore, the obtained cores have only approximately diagonal core planes, and cannot be reduced to CANDECOMP/PARAFAC representations. This diagonalization loses part of its motivation. What one can expect to find is a core with relatively high elements on the diagonals of the frontal planes, and relatively small elements elsewhere. In this way possibilities for further simplification may well be overlooked. Some of the more recent developments have specifically aimed at simplifications of the core that go beyond simplification to plane diagonal form. On the one hand, approaches have been proposed that impose a (usually) very simple structure on the core, and search among the thus constrained models for models that fit almost as well as the unconstrained 3MFA model, thus attaining great simplicity at small cost. On the other hand, procedures have been developed for rotations of the core that lead to simpler forms than those consisting of approximately diagonal planes, thus enhancing simplicity at no cost whatsoever. In Section 2, the constrained 3MFA approaches will be discussed; the rotations to simpler cores are the subject of Section 3.

2. Constrained Three-mode Factor Analysis

As has been mentioned in the introduction, the CANDECOMP/PARAFAC model can be seen as a variant of the 3MFA model in which the core is constrained to be superdiagonal. It has also been mentioned that the CANDECOMP/PARAFAC model often fits considerably more poorly than the 3MFA model. This observation has motivated the study of methods in between the two models: In "Constrained 3MFA" (Kiers, 1992; Rocci, 1992) the core is constrained to have certain elements equal to zero, but usually not so many that it yields the CANDECOMP/PARAFAC model. In this way, a class of new models is generated, that has a greater fitting potential than CANDECOMP/PARAFAC has, and yet is not as intricate as the full three-mode factor model.

Obviously, the more elements of the core are constrained to zero, the poorer the fit of the model, and hence Constrained 3MFA (C3MFA) seems merely to offer a set of compromise methods, in which the benefit of a simpler solution can only be attained at the cost of a poorer fit. Fortunately, in practice, the costs of using a considerably simpler solution are usually low. Experience with C3MFA indicates that a fit almost as well as that of C3MFA can often be attained by models with only a few more triple product terms than the CANDECOMP/PARAFAC model. In fact, it can be proven that the core of a 3MFA model can always be constrained to have a considerable number of zero elements *without loss of fit*. For instance, for P by Q by R cores with $P=QR-1$, Murakami, Ten Berge and Kiers (1996) have shown that at least $QR(QR-2)-(R-2)$ elements can be constrained to zero without loss of fit. For example, in a 5x3x2 core array as many as 24 of the 30 elements can be constrained to zero. The possibility of costless constraining core elements in other situations is still under study, but experience suggests that usually (far) more than half of the elements can be constrained to zero without affecting the fit.

As has been seen above, an important feature of C3MFA is that it offers models that are considerably more simple than the full 3MFA model and still fit nearly as well. An added property of *some* C3MFA models is that they give unique solutions just as CANDECOMP/PARAFAC. In particular, Kiers, Ten Berge and Rocci (in press) have shown that C3MFA employing 3x3x3 cores yield unique solutions if the cores are constrained to have zeros in the positions indicated by 0 in the core, the frontal planes of which are given below:

$$
\begin{matrix}
Y & 0 & 0 \\
0 & 0 & X \\
0 & X & 0
\end{matrix}
\qquad
\begin{matrix}
0 & 0 & X \\
0 & Y & 0 \\
X & 0 & 0
\end{matrix}
\qquad
\begin{matrix}
0 & X & 0 \\
X & 0 & 0 \\
0 & 0 & Y
\end{matrix}.
\qquad (5)
$$

The elements indicated by X may or may not be constrained to 0; the elements indicated by Y must be unconstrained and nonzero. This is only one class of models for which uniqueness has been proven. Practical experience has suggested that models employing other sets of constraints on 3x3x3 cores or that employ smaller core sizes are nonunique (except, of course, the models corresponding to CANDECOMP/PA-RAFAC). For models employing larger cores, to the author's knowledge no uniqueness results are available as yet.

3. Rotation of the Core to Simple Structure

An alternative to constraining a three-mode factor model to have many zeros in the core, is to rotate the core (in all three directions) such that it has many (near) zero elements, and hence is easy to interpret as well. An advantage of rotating rather than constraining the core is that rotation *never* affects the fit of the solution. A disadvantage is that it need not yield a solution that is as simple as desired. Two types of approaches have been proposed recently. On the one hand, approaches for rotation to a *fixed* simple form have been considered, thereby varying on the earlier approaches by Kroonenberg (1983). On the other hand, attempts to rotation to *arbitrary* simple structures have been made, following the approach to simple structure rotations developed for PCA.

3.1 Rotation of the core to a fixed simple structure

Just as Kroonenberg (1983), Kiers (1992) searched for methods for rotation of the core such that the core approximates a core associated with the CANDECOMP/PA-RAFAC model. However, contrary to Kroonenberg's approach, Kiers suggested rotating the core to approximate superdiagonality (rather than diagonality of the core planes). The rationale behind this approach is twofold. On the one hand, the approximation is expected to yield more small sized elements, simply because it explicitly aims at finding more elements close to zero. On the other hand, a 3MFA model employing a superdiagonal core is more directly related to the CANDECOMP/PARA-FAC model than one employing a core with diagonal planes. When using the former, the component matrices are directly related to those of CANDECOMP/PARAFAC; when using the latter, one of the matrices can only be obtained after a transformation. Especially when only orthonormal rotations of the core are considered, the latter (usually nonorthonormal) transformation makes the relation with the CANDECOMP/ PARAFAC model more indirect and complicated.

Kiers (1992) studied procedures for both oblique and orthonormal rotation to superdiagonality. Two of the procedures for oblique rotation turned out to behave poorly in that they frequently led to degenerate solutions, in which the transformation matrices tended to singular matrices, and hence inverse transformations could not be computed sensibly anymore. A third procedure for oblique rotation, which behaved better, was deemed uninteresting because it imposed unwanted restrictions on the transformation matrices. A procedure for oblique rotation that does not share these problems has been devised only recently as a variant of a procedure for rotation to arbitrary simple structure (Kiers, 1995, see below).

The procedure for orthonormal rotation proposed by Kiers (1992) behaved well computationally. In a simulation study with randomly rotated superdiagonal cores of sizes 2x2x2, 3x3x3 and 4x4x4 (ten of each), the method recovered all 30 superdiagonal cores. Further properties of this rotation procedure, like the derivation of theoretical bounds to the "superdiagonalizability" of core arrays, have been studied by Henrion (1993). Practical experience with the method, however, often indicated that superdiagonality was not well approximated, and that the resulting rotated core was sometimes far from simple. Further research, therefore, mainly aimed at rotation of cores to an *arbitrary* simple structure.

3.2 Rotation of the core to an arbitrary simple structure

Rather than rotating the core to an a priori specified form, one may choose to rotate the core to an unspecified simple structure. One such procedure was employed by Murakami (1983) who rotated the frontal core planes to simple structure by means of varimax rotation applied to the transpose of the supermatrix containing all core planes next to each other. In Murakami's case, the core could be rotated in only one direction. The general situation where the core can be rotated in all three directions such that some kind of overall simple structure is attained was first dealt with by Kruskal (1988). He proposed a method called "tri-quartimax" rotation, which is a procedure for maximizing a combination of normalized quartimax functions applied to the supermatrices consisting of the frontal, lateral and horizontal planes, respectively, of the core. Kruskal proposed to maximize this function over oblique rotations, thus ignoring the fact that quartimax was originally proposed as a criterion for orthonormal rotation. Therefore, there is little reason to expect that the properties of "ordinary" quartimax carry over to tri-quartimax. Unfortunately, no published information is available on the performance of the method, nor on its implementation.

Kruskal's (1988) idea has been an important point of departure for Kiers' (in press) "three-mode orthomax" rotation procedure. "Orthomax" (Jennrich, 1970) is a general family of simple structure criteria containing the well-known varimax (Kaiser, 1958) and quartimax (Carroll, 1953; Ferguson, 1954; Neuhaus & Wrigley, 1954; Saunders, 1953) criteria as special cases. Kiers' proposal consists of the application of the orthomax criterion to the supermatrices consisting of the frontal, lateral and horizontal planes of the core, thereby using a generalization of Kruskal's (1988) criterion. In contrast to Kruskal's method, three-mode orthomax is restricted to orthonormal rotations, thus respecting the orthonormality of the 3MFA component matrices. Because of the latter fact, the method should be used with care: If the rotated core is not as simple as one would like, one should take into account that further simplicity might be attainable upon relaxing the orthonormality restriction. It seems, however, that in practice the method often gives quite simple solutions for the core, despite the restriction of orthonormality on the rotations.

Three-mode orthomax is not just a single method, but a class of methods. First, different methods arise from different choices of the simplicity criterion (e.g., quartimax, varimax, or other orthomax criteria). Second, three-mode orthomax allows for rotation of the core in all three directions simultaneously, but also, if desired, in two or only one direction. Similarly, the criterion can measure simplicity of the core in all three directions or a subset of those. The quartimax criterion always measures simplicity in all three directions, because it operationalizes amount of simplicity by the sum of fourth powers of all elements. Clearly, the order in which these fourth powers are considered is irrelevant. For other criteria, the situation is quite different. For instance, varimax measures simplicity as a sum of variances of columns of squared loadings. In three-mode varimax these columns can be constituted in three different ways, depending on the mode of interest. For instance, varimax in direction A takes the variances of the columns that each contain the elements related to one of the entries of mode A (e.g. "idealized individual"). Hence, varimax in direction A aims at large variation between the elements corresponding to the same idealized individual. Varimax simplicity in direction A does not imply varimax simplicity in direction B, because large variances of squared loadings could be found per entry of mode A, even when elements related to one entry in mode B are all equal. Finally, it is possible to attach different weights to the criteria used for simplicity in direction A, B, and C. Thus three-mode orthomax is a very flexible approach. Despite this flexibility, the procedure for optimizing the three-mode orthomax criterion is relatively simple: It consists of iteratively updating estimates for the three rotation matrices by applying an ordinary orthomax procedure to a supermatrix computed from the original core and the current values for the rotations. The method is somewhat sensitive to local optima, but because the algorithm converges quickly, using several restarts is an adequate and feasible way of dealing with this problem.

For practical applications of three-mode orthomax it is useful to have some guidelines for which method to choose from the class of methods. First one has to decide in which directions the core will be rotated. Of course, the more rotational freedom is employed, the simpler the core can get. However, sometimes rotational freedom in one or two directions may be exploited for rotating the component matrix rather than the core to a desirable form. Next, one has to choose among the simplicity criteria. Kiers (in press) reports a simulation study from which one might conclude that effects of using different simplicity criteria tend to be small. Although in some situations the use of quartimax should be disadvised for its tendency to yield a general factor (as is well-known for the two-way case), in general quartimax and varimax behave similarly, and it seems safest to start with varimax, but to keep in mind that, if this does not yield desirable results, other options should be considered.

As an example of application of three-mode varimax, consider the cores reported in Table 1. The first core is an unrotated 3MFA solution (details of which are to be found in Kiers, in press). The second core results from three-mode varimax applied to the unrotated core. It can be seen that the varimax rotated core is considerably more simple than the unrotated core: Most of its elements are extreme; it has far fewer medium sized elements (in the intervals [-.5,-.2] or [.2,.5] say) than the unrotated core.

Table 1: Unrotated exemplary core and three-mode varimax rotated core

Frontal planes of unrotated core								
2.07	0.06	0.21	0.07	0.14	-0.26	0.04	0.34	-0.25
-0.19	-0.60	-0.39	0.99	0.08	-0.60	-0.38	0.62	-0.36
-0.12	0.59	0.29	0.13	-0.05	-0.13	-0.14	0.22	-0.16

Frontal planes of three-mode varimax rotated core								
2.11	0.01	-0.03	0.00	0.33	-0.06	0.01	0.01	0.00
0.03	-0.02	0.38	0.14	0.97	-0.06	0.97	0.13	0.59
0.00	-0.14	0.84	0.12	0.06	-0.02	0.17	0.11	0.09

Three-mode orthomax can indirectly also be used for oblique core rotations, for instance, by combining them with normalization operations, in analogy to Harris and Kaiser's (1964) orthoblique approach. Recently, Kiers (1995) proposed a different procedure for oblique simple structure rotation of the core. He developed a straight-forward generalization of the SIMPLIMAX procedure for oblique rotation of a loading matrix to an optimal simple target (Kiers, 1994). Specifically, in three-mode SIMPLIMAX, oblique rotation matrices for all three modes are found in such a way that the m (a number to be specified in advance) smallest elements of the rotated core have a minimal sum of squares (σ). The technique thus aims at simple structure in a very explicit way. For example, when a 3x3x3 core is analyzed by SIMPLIMAX with $m=20$, the method finds a core in which 20 elements are optimally close to zero (as expressed by the fact that the method minimizes its sum of squares), whereas the other seven elements will be relatively large. *How* close to zero the smallest elements are depends on the choice for m. If m is chosen very small, then the method will often succeed in setting all m elements to exactly 0, but as m increases, the small values (or at least their sum of squares) will increase as well. Because of this trade-off relationship one should apply SIMPLIMAX with different values for m, and search the solution which has sufficiently many small elements that are sufficiently close to zero. It should be noted that in three-mode SIMPLIMAX it is, just as in three-mode orthomax, possible to rotate over only one or two modes, rather than all three.

The method is computationally much less efficient than three-mode orthomax. The algorithm for three-mode SIMPLIMAX consists of iterative application of the two-mode SIMPLIMAX procedure, applied to supermatrices of frontal, lateral or horizontal planes of the current rotated core matrix. This procedure turns out to be very sensitive to local optima, and hence requires many starts from different starting positions (say 200), which makes it relatively inefficient (e.g., using about one hour for one full analysis of 200 runs on an 486 66MHz pc). Hence, better starting procedures, or approaches for avoiding local optima are called for. On the other hand, in practice the cores are usually rather small, and not many different values of m need to be tested, as in the empirical, and, as far as size is concerned, rather typical example reported below.

Table 2: Three-way SIMPLIMAX applied to a core reported by Kroonenberg (1994)

Original Core (Frontal planes next two each other)					
24	-26	-5	-2	1	1
18	11	9	-3	-3	0
-2	-12	15	-1	-6	7

Core rotated to 13 zeros		$\sigma = 0.000$			
35.4	0	0	**3.1**	0	0
0	**24.0**	0	0	0	0
0	0	**18.9**	0	0	**12.4**

Core rotated to 14 zeros		$\sigma = 2.958$			
35.5	0.0	0.0	0.9	0.3	0.0
0.0	**23.5**	0.0	0.2	-1.4	0.0
0.0	0.0	**18.8**	0.0	0.0	**11.3**

Core rotated to 15 zeros		$\sigma = 10.967$			
33.9	0.0	-0.0	1.9	-0.6	0.0
0.0	**25.7**	-0.1	-0.5	-2.5	0.0
0.0	0.0	0.4	-0.0	0.0	**21.6**

Core rotated to 16 zeros		$\sigma = 276.355$			
36.6	0.0	-0.7	-0.2	0.7	-1.4
0.0	**25.8**	1.4	0.7	2.7	-0.4
0.4	0.8	-9.3	2.2	-2.2	12.9

The core array used in the present example was reported by Kroonenberg (1994), and is based on a three-mode factor analysis on scores of 82 subjects measured on five variables (pertaining to performance and drunkenness) at eight occasions (at which different doses of alcohol had been administered to them). The present 3x3x2 core array has been rotated here by means of three-way SIMPLIMAX. We used the values $m=12,...,16$. For $m=12$ and $m=13$ three-way SIMPLIMAX found a solution in which the smallest m elements were zero, up to the accuracy implied by the convergence criterion used. In fact, it can be proven that any 3x3x2 cores can be transformed into a core with as many as thirteen exactly zero elements (Ten Berge, 1995), which explains what we found here. Hence, the only nontrivial applications of SIMPLIMAX are those with $m=14$, $m=15$ and $m=16$. In each complete SIMPLIMAX analysis we used 200 random starts. The rotated cores, as well as the values of the function σ (the sums of smallest squared core elements) are given in Table 2, with the $(18-m)$ highest elements in bold face. It can be seen that the core rotated towards 14 zeros indeed gives only four important core elements, the others being about ten times as small or smaller. Even in the core rotated to 15 zeros the high values are at least eight times as large as the small values. However, in the core rotated to 16 zeros the smallest elements were no longer negligible compared to the high values. It can hence be concluded that this 3x3x2 core can be simplified tremendously, and, in fact, the main relations between components can be described in three or four terms. Therefore, even though this procedure may destroy some of the simplicity of the component matrices (Kroonenberg's matrix was obtained after 'simplicity' rotations of two of the component matrices), the gain in simplicity of the core is worth considering.

The algorithms used in SIMPLIMAX can also be used for rotations to a target in which the positions of the small elements are fixed. In this way, the procedure can also be used for superdiagonalization, and hence a new technique has come available for superdiagonalization by means of oblique rotation.

4. Discussion

In the present paper, an overview has been given of recent developments concerning simplification of the core. Simplification of the core has been proposed as an aid for interpreting a 3MFA solution. The interpretation of a 3MFA solution usually starts by giving interpretations to the components for the three modes, and next, the relations between these modes, as reflected in the core, are considered. The developments discussed here all focused on simplifying the core, none aimed at simplifying the interpretation of the component matrices. In fact, methods yielding simplified cores may yield component matrices that are hard to interpret. In other words, from the former situation where component matrices could be interpreted easily, but the core made the results rather complex, we are now at the other extreme where the relations are made simple, but the component matrices may be rather complex. One way to deal with this problem, as suggested by Kiers (1995, in press) is to consider rotation of the core in only one or two directions, and not affect those component matrices that are very important in interpretation, and for which simple interpretations are available. Of course, such approaches do not always work. On the one hand, it is possible that for all three component matrices simple interpretations are available, which one does not wish to disturb; on the other hand, it is possible that using only one or two rotation directions no longer leads to a simple core. However, even in those situations it is conceivable that a solution exists in which the core as well as the component matrices are reasonably simple. A way to find such solutions would be to define and optimize criteria that combine simplicity of the component matrices and simplicity of the core. Because of the interdependence of the rotation of the core and of the component matrices, optimization of such combined simplicity criteria seems far from trivial.

An alternative position can be taken as well: Methods that simplify the core can be seen as methods that simplify the *structure* of the model, just as CANDECOMP/PA-RAFAC is a simpler model than 3MFA. The fact that the components themselves are not related in a simple way to the original individuals, variables, occasions, or whatever, can be deemed less disturbing. For instance, it can be deemed acceptable that, when a 3x3x3 core is reduced to only 4 nonnegligible elements, the 4 ensuing tensor product terms are somewhat complicated to interpret. The alternative of trying to grasp up to 27 interactions between (more simple) components does not seem more attractive. Using a few tensor product terms based on more complex components is a way of moving most of the interactions into the components, and once these are conceptualized, the model becomes easy to grasp.

One new development discussed here concerned constrained 3MFA methods. Probably the most surprising finding was that the core can be constrained to have very many zero elements at no or very little cost. This has led to the study of procedures for reducing a core to a maximally parsimonious core without loss of fit. This study has been and will be aided by the availability of three-way SIMPLIMAX. By means of three-way SIMPLIMAX one can search how many elements of a particular core can be set to zero by means of oblique rotations (and hence without loss of fit). In this way, three-way simplicity rotation can be used to obtain useful C3MFA models.

References:

Carlier, A., Lavit, Ch., Pagès, M., Pernin, M.O. and Turlot, J.C. (1989): Analysis of data tables indexed by time: a comparative review. In: *Multiway data analysis*, Coppi, R. and Bolasco, S. (Eds.), 85-101, Amsterdam, Elsevier Science Publishers.

Carroll, J.B. (1953): An analytic solution for approximating simple structure in factor analysis, *Psychometrika*, **18**, 23-38.

Carroll, J.D. and Chang, J.-J. (1970): Analysis of individual differences in multidimensional scaling via an n-way generalization of "Eckart-Young" decomposition, *Psychometrika*, **35**, 283-319.

Cohen, H.S. (1974): *Three-mode rotation to approximate INDSCAL structure (TRIAS)*, Paper presented at the Psychometric Society Meeting, Palo Alto.

Cohen, H.S. (1975): *Further thoughts on three-mode rotation to INDSCAL structure, with jackknifed confidence regions for points*, Paper presented at U.S.-Japan seminar on Theory, Methods and Applications of Multidimensional Scaling and Related Techniques. La Jolla.

Coppi, R. and Bolasco, S. (Eds.) (1989): *Multiway data analysis*, Amsterdam, Elsevier Science Publishers.

De Leeuw, J. and Pruzansky, S. (1978): A new computational method to fit the weighted Euclidean distance model, *Psychometrika*, **43**, 479-490.

Ferguson, G.A. (1954): The concept of parsimony in factor analysis, *Psychometrika*, **19**, 281-290.

Harris, C.W. and Kaiser, H.F. (1964): Oblique factor analytic solutions by orthogonal transformations, *Psychometrika*, **29**, 347-362.

Harshman, R.A. (1970): Foundations of the PARAFAC procedure: models and conditions for an "explanatory" multi-mode factor analysis, *UCLA Working Papers in Phonetics*, **16**, 1-84.

Harshman, R.A. and Lundy, M.E. (1984): The PARAFAC model for three-Way factor analysis and multidimensional scaling, In: *Research methods for multimode data analysis*, Law, H.G., Snyder, C.W., Hattie, J.A. and McDonald, R.P. (Eds.), 122-215, New York, Praeger.

Henrion, R. (1993): Body diagonalization of core matrices in three-way principal components analysis: Theoretical bounds and simulation, *Journal of Chemometrics*, **7**, 477-494.

Jennrich, R.I. (1970): Orthogonal rotation algorithms, *Psychometrika*, **35**, 229-235.

Kaiser, H.F. (1958): The varimax criterion for analytic rotation in factor analysis, *Psychometrika*, **23**, 187-200.

Kiers, H.A.L. (1991): Hierarchical relations among three-way methods, *Psychometrika*, **56**, 449-470.

Kiers, H.A.L. (1992): TUCKALS core rotations and constrained TUCKALS modelling, *Statistica Applicata*, **4**, 659-667.

Kiers, H.A.L. (1994): SIMPLIMAX: Oblique rotation to an optimal target with simple structure, *Psychometrika*, **59**, 567-579.

Kiers, H.A.L. (1995): *Three-way SIMPLIMAX for oblique rotation of the three-mode factor analysis core to simple structure*, Manuscript submitted for publication.

Kiers, H.A.L. (in press): Three-mode Orthomax rotation, *Psychometrika*.

Kiers, H.A.L., ten Berge, J.M.F. and Rocci, R. (in press): Uniqueness of three-mode factor models with sparse cores: The 3x3x3 case, *Psychometrika*.

Kroonenberg, P.M. (1983): *Three-mode principal component analysis: Theory and applications*, Leiden, DSWO press.

Kroonenberg, P.M. (1994): The TUCKALS line: A suite of programs for three-way data analysis, *Computational Statistics and Data Analysis*, **18**, 73-96.

Kroonenberg, P.M. and De Leeuw, J. (1980): Principal component analysis of three-mode data by means of alternating least squares algorithms, *Psychometrika*, **45**, 69-97.

Kruskal, J.B. (1988): *Simple structure for three-way data: A new method intermediate between 3-mode factor analysis and PARAFAC-CANDECOMP*, Paper presented at the 53rd Annual Meeting of the Psychometric Society, Los Angeles, June 27-29.

Law, H.G., Snyder, C.W., Hattie, J.A. and McDonald, R.P. (Eds.)(1984): *Research methods for multimode data analysis*, New York, Praeger.

MacCallum, R.C. (1976): Transformations of a three-mode multidimensional scaling solution to INDSCAL form, *Psychometrika*, **41**, 385-400.

Murakami, T. (1983): Quasi three-mode principal component analysis - A method for assessing factor change, *Behaviormetrika*, **14**, 27-48.

Murakami, T., Ten Berge, J.M.F. and Kiers, H.A.L. (1996): *A class of core matrices in three-mode principal components analysis which can be transformed to have a majority of vanishing elements*, Manuscript submitted for publication.

Neuhaus, J.O. and Wrigley, C. (1954): The quartimax method: An analytic approach to orthogonal simple structure, *British Journal of Mathematical and Statistical Psychology*, **7**, 81-91.

Rocci, R. (1992): Three-mode factor analysis with binary core and orthonormality constraints, *Journal of the Italian Statistical Society*, **3**, 413-422.

Saunders, D.R. (1953): *An analytic method for rotation to orthogonal simple structure*, Research Bulletin, RB 53-10, Princeton, New Jersey, Educational Testing Service.

Ten Berge, J.M.F. (1995): *How sparse can core arrays get: The 3x3x2 case*, Unpublished note.

Tucker, L.R. (1966): Some mathematical notes on three-mode factor analysis, *Psychometrika*, **31**, 279-311.

Tucker2 as a Second-order
Principal Component Analysis

Takashi Murakami

School of Education, Nagoya University
Furo-cho, Chikusa-ku, Nagoya
464-01, Japan

Summary: Statistical properties of the Tucker2 (T2) model, a simplified version of three-mode principal component analysis (PCA), are investigated aiming at applications to the study of factor invariance. The T2 model is derived as a restricted form of second-order PCA in the situation comparing component loadings and component scores across occasions. Several statistical interpretations of coefficients obtained from the least squares algorithm of T2 are proposed, and several aspects of T2 are shown to be natural extensions of characteristics of classical PCA. A scale free formulation of T2 and a new derivation of the algorithm for the large sample case are also shown. The relationship with a generalized canonical correlation model is suggested.

1. Introduction

Consider a set of data collected by administering an inventory consisting of p items to n subjects on m occasions with or without changing the conditions. In this situation, we may be interested in the comparisons between factor loadings of the same items (variables) and between factor scores of the same subjects on different occasions. This is the factor invariance problem.

In the present article, we will investigate how to use the *Tucker2* (T2) model, a simplified version of three-mode principal component analysis (three-mode PCA; Tucker, 1966; Kroonenberg & De Leeuw, 1980), as a tool for the study of factor invariance. While T2 is a very general model of multidimensional analysis of three-way data, we will specify the manner of preprocessing of input data and of transformation of output coefficients which are appropriate to the restricted purpose. In Section 2, we will recapitulate several formulations of classical PCA, and distinguish two solutions; PCA-1 and PCA-2. We will prefer PCA-1 because of several favorable aspects; rotational freedom, statistical convenience in interpretations of coefficients, and the scale free derivation. In Section 3, we will formulate T2 as a restricted second-order PCA, and will show that the least squares solution obtained by the TUCKALS2 algorithm (Kroonenberg & De Leeuw, 1980) has almost all properties of classical PCA, which will be listed in Section 2. We will find two classes of solutions; T2-1 and T2-2, neither are the same as the standard formulation. We will conclude the T2-1 is more favorable due to the similar convenient properties to PCA-1 while T2-2 is also useful for straightforward derivation of the algorithm for large sample data.

2. Classical Principal Component Analysis Revisited

2.1 Two solutions of classical PCA

There are many ways to formulate classical PCA (e.g. Ten Berge & Kiers, 1996). We will start from the approximation of the matrix of input data by the factor analytical model, namely, the minimization of the following function,

$$f(F, A^*) = \|Z - FA^{*\prime}\|^2, \tag{1}$$

575

where Z is an $n \times p$ matrix of input data, F, an $n \times q$ matrix of component scores, and A^*, a $p \times q$ matrix of component loadings. For the simplicity, we will assume that $n > p$, and the rank of Z is equal to p. In addition, we assume that Z is centered and standardized; $Z'\mathbf{1} = \mathbf{0}$, and diag $Z'Z = I$ (rather than diag $Z'Z = nI$). Hence $R = Z'Z$ is the correlation matrix between input variables.

The constraint is imposed on F such that the problem is to minimize (1) subject to $F'F = I$. (We will use a superscript asterisk to denote that no constraints will be imposed on the matrix.) The minimum will be attained by the use of the singular value decomposition (SVD), $Z = PDQ'$, where P is the $n \times p$ column-wise orthonormal matrix, D, the $p \times p$ diagonal matrix of singular values arranged in descending order, and Q, the $p \times p$ orthonormal matrix. Then, the solution is given by

$$F = P_q T, \quad \text{and} \quad A^* = Q_q D_q T, \tag{2}$$

where P_q contains the first q columns of P, Q_q contains the first q columns of Q, D_q is the upper left $q \times q$ submatrix of D, T is an $q \times q$ arbitrary orthonormal matrix, and q is the number of components satisfying $q < p$ (cf. Ten Berge, 1993, pp. 35-36). The formulation can be seen to be a kind of regression problem where columns of F are predictor variables and elements of A^* are regression coefficients to predict columns of Z. The matrix A^* in this sense is sometimes called the *pattern* matrix.

Because constraints are essentially inactive in this problem (Ten Berge, 1993, p. 45 etc.), we can change the constraints without loss of optimality; if we minimize

$$f(F^*, A) = \|Z - F^* A'\|^2, \tag{3}$$

subject to $A'A = I$. The solution will be given by

$$F^* = P_q D_q U, \quad \text{and} \quad A = Q_q U, \tag{4}$$

where U is an arbitrary square orthonormal matrix. In this formulation, it is usual to set $U = I$ otherwise U destroys the orthogonality of columns of F^*.

We will call the class of formulations PCA leading to (2) *PCA*-1, and call one producing (4) *PCA*-2. Solutions of PCA with various criteria and constraints result in either of them as long as the analysis is based on centered and standardized data matrix Z. Of course, we can transform a solution to the other through $F^* = F D_q$, and $A = A^* D_q^{-1}$ if T=U=I, and $F^* A' = F A^{*'}$ irrespective of T and U.

2.2 Statistical meanings of PCA

We will examine the statistical meanings of PCA in some detail. First, we will return to (1) and (3). By the use of regression theory, we obtain

$$A^* = Z'F. \tag{5}$$

and

$$F^* = ZA, \tag{6}$$

irrespective of F and A as long as they satisfy the orthonormal constraints. Formally, these equations represent a kind of *dual relations* (Nishisato, 1994, p.102). In addition, they have some interesting interpretations; (5) defines the *structure* matrix, namely, the correlation matrix between input variables and (standardized) component scores while (6) defines the component scores as linear composites of input variables by the use of A as the matrix of weights. These interpretations are not only useful in applications but also give us some important insights on PCA.

For example. we can see that both F^* and F are matrices of linear composites of input variables, namely, columns of F^* and F exist in the column space of Z although the fact is not explicit in (1) and (3). Hence, we know that the minimum of the function $f(F, A^*)$ in (1) is equal to the minimum of

$$f(V, A^*) = \|Z - ZVA^{*\prime}\|^2, \tag{7}$$

where V is the $p \times q$ matrix of weights constrained as $V'RV = I$, and given by $V = Q_r D_r^{-1} T$ (Ten Berge & Kiers, 1996). This can be much simplified in PCA-2; substituting (6) into (3), we have

$$f(A) = \|Z - ZAA'\|^2. \tag{8}$$

in which the matrix A plays the two roles; the weight matrix and the pattern matrix.

By the use of (8). we can rewrite the minimization problems into the maximization problems. After some operations using $A'A = I$, we have

$$f(A) = \mathrm{tr}\,R - \mathrm{tr}\,A'Z'ZA = p - \mathrm{tr}\,A'RA, \tag{9}$$

which means that the PCA-2 solution is also obtained by the maximization of $\mathrm{tr}\,A'RA$ under the constraint of $A'A = I$. This is the most popular formulation of classical PCA in the literature because it directly leads to the well-known solution of $A = K_q$, where K_q is the matrix of eigenvectors associated with q largest eigenvalues of R.

Next, substituting (5) into (1), we obtain

$$f(F, A^*) = \mathrm{tr}\,R - \mathrm{tr}\,F'ZZ'F = p - \mathrm{tr}\,A^{*\prime}A^*. \tag{10}$$

This implies that PCA-1 can be interpreted as the maximization of sums of squares of elements of structure matrix, $\mathrm{tr}\,A^{*\prime}A^*$.

Ten Berge and Kiers (1996) called the formulation based on the maximization of $\mathrm{tr}\,A'RA$ *Hotelling* PCA, and the approximation of a data matrix by the factor model like (1) *Pearson* PCA. Their classification does not perfectly overlap with our PCA-1 and -2. For example, they classify (3) as Pearson PCA. They prefer Pearson PCA to Hotelling PCA due to its full rotation freedom including oblique one, elegance of the idea of least squares predictive efficiency, and convenience of orthonormal component scores (only if one uses the constraint of PCA-1, and does not perform the oblique rotation.) We can expect that the loading matrix A^* is the pattern matrix and the structure matrix simultaneously, and the property does not change after orthogonal rotation because F is orthonormal. A squared element of A^* is the size of explained variance by the component in the variable, and the row sum of them is the squared multiple correlation for each variable, and so forth. However, the matrix of weights, V in (7) has slightly complicated relationship with A^*; $V = A^*(A^{*\prime}A^*)^{-1}$ (cf. Ten Berge, 1986, p. 36).

PCA-2 has other kinds of good properties; A plays roles both of the weight and pattern matrix shown in (8), and the straightforward derivation of algorithm as the eigenproblem of R. However, resistance for rotation seems to be a shortcoming of this solution. Moreover, we will add that the scale-free formulation of PCA will be formulated on the basis of PCA-1 below.

2.3 Scale free formulation of classical PCA

We have used the centered and standardized data array Z without any excuse so far. The choice of scale transformations is a crucial problem because there is no simple formula, for example, to transform the result of SVD of raw score matrix X into that of Z. Because variables do not share the same measurement unit and origin in many applications, centering and standardizing are practically useful. We want a scale free formulation of PCA to justify it.

Meredith and Millsap (1985) proposed the scale free formulation justifying the standardization based on the maximization of sum of squared multiple correlations. We will extend it to justification of the centering.

Let us define the matrix $A^\dagger$ as

$$A^\dagger = D_S^{-1/2} X'F, \tag{11}$$

where X is the matrix of raw input data. and D_S is the diagonal matrix of variances of columns of X. We will introduce one more constraint, $F'1 = 0$ as well as $F'F = I$. If we define the $n \times n$ matrix, $J = I - 11'$, then $F'1 = 0$ means $F = JF$ (Ten Berge, 1993, p.66), hence $X'F = X'JF$. Because $X'J = (X - 1\bar{x}')'$, we obtain that $A^\dagger = Z'F$, where $Z' = D_S^{-1/2}(X - 1\bar{x}')'$. As a result, the elements of $A^\dagger$ are correlation coefficients between variables and components although variables are not centered in (11), and we know that $A^\dagger$ is equal to A^* in (5).

Now, we have justified not only the preprocessing of centering and standardization but also constraints of PCA-1. In other words, we can use classical PCA of standardized variables with orthogonal rotation as the maximization of $\mathrm{tr}A^{\dagger\prime}A^\dagger$ under the constraints of $F'1 = 0$ and $F'F = I$. Our result may not be so impressive because it looks to depend solely on the *word* correlation. However, the same rationale will produce the result which is not necessarily trivial on T2 in the next section.

Here, we will list the formulae to obtain final output of PCA-1 to compare them with those for T2: The loading matrix is given by

$$A^* = K_q \Lambda^{1/2} T, \tag{12}$$

where Λ is the diagonal matrix of q largest eigenvalues of R, K_q. the matrix of corresponding eigenvectors as before, and T, an arbitrary orthogonal matrix determined, say, Varimax method. Then, the matrix of component scores are obtained by

$$F = ZA^*(A^{*\prime}A^*)^{-1}, \tag{13}$$

which is equivalent to ZV in (7). These equations are the same as those resulting from the algorithm of PCA of "factor analysis with unit diagonals" in many standard statistical packages.

2.4 Asymmetric role of scores and coefficients

Gifi (1990, pp. 49-50) distinguished *multivariate analysis* (MVA) from *multidimensional analysis* (MDA) based on the asymmetric role of rows (subjects) and columns (variables) of the $n \times p$ matrix of input data. While the latter treats the matrix as an arbitrary rectangular matrix, the former uses it as p random variables and regards rows of the matrix as a random sample of size n, even when no probabilistic model is assumed explicitly. Their definition of MVA is the study on "systems of random variables or random samples from such systems." We accept the standpoint of Gifi, and treat PCA as MVA in his sense.

The asymmetry of rows and columns of a matrix of data is extended to the asymmetry of two matrices obtained by PCA, for example, F and A^* in (1). While F is the matrix of q derived random variables, A^* is the matrix of some coefficients with the definite statistical meaning rather than values of random variables. As was mentioned before, elements of A^* have two implications; coefficients of correlation and coefficients of regression. While coefficients themselves will be objective of interpretations, individual values on variables per se are not usually interpreted, especially in large sample cases. Properties of components are interpreted through the coefficient matrix and the relationships with external information.

Our treatment of PCA reflects these asymmetries. For example, we are willing to rotate the result of PCA-1 because the orthogonal rotation keeps the orthonormality of derived variables.

The reason why we have emphasized the asymmetric treatment is that T2 had been formulated as a method of MDA rather than MVA in Gifi's sense. In the sequel, we will not necessarily follow the standard symmetric notations and treatments of three-mode PCA as in, for example, Kroonenberg (1983) because we consider that the factor invariance problem commonly involves the asymmetry.

3. T2 as a second-order PCA

3.1 Derivation of the T2 model as a second-order PCA

Let Z_k $(k = 1, \ldots, m)$ be the $n \times p$ matrix of data on k-th occasion, namely, the k-th frontal plane of a three-mode data array in the terminology of three-mode analysis. From the standpoint of asymmetric role of rows and columns of the matrix of data mentioned above, we consider that the three-mode array consists of mp random variables. We will postpone to specify the manner of centering and standardization of Z_k, but we assume that $\sum_{k=1}^{m} Z_k' Z_k = mI$. We also assume that $n > mp$ and that all mp columns of data are linearly independent.

The simplest way of applying classical PCA to these matrices may be the separate analysis of p variables on each occasion such as

$$Z_k \simeq F_k A_k^{*\prime} \qquad k = 1, \ldots, m, \tag{14}$$

where F_k is an $n \times q$ $(q < p)$ matrix of component scores, and A_k^* is an $p \times q$ matrix of component loadings. (We will adhere to PCA-1.) It looks easy to attain the two aims mentioned in the introduction, namely, comparisons between factor loadings and factor scores obtained on different occasions. To do so, one can compare matrices of loadings on different occasions directly, or compute any of indices of coefficients of congruence between them (e.g. Ten Berge, 1986), and compute correlation coefficients between component scores. However, there are several problems in these methods. First, the comparisons are not so easy when m and q are large. Second, rotation which can be performed in each condition separately may bring the indeterminacy to any indices for comparisons, for example, correlation coefficients between component scores. Third, there may be much redundancy in mq columns of loadings and scores which makes estimates of coefficients unstable and interpretations of results confusing.

A simple way to partially avoid these difficulties is applying PCA-1 to an $mn \times p$ matrix obtained by juxtaposing all frontal planes vertically:

$$Z_k \simeq F_k A^{*\prime} \qquad k = 1, \ldots, m, \tag{15}$$

where A^* is a $p \times q$ common loading matrix, and we assume that $p > q$. (Here, we regard the data array as a sample of size nm on p variables temporarily.) Although (15) is more parsimonious than (14), some of mq columns of F_k's can remain redundant. Hence, we will apply PCA-1 again to the $n \times mq$ matrix obtained by juxtaposing F_k's horizontally. (We return to a sample of size n);

$$F_k \simeq GC_k^{*'} \qquad k = 1, \ldots, m, \tag{16}$$

where C_k^* is a $q \times r$ matrix of *second-order loadings* and G is an $n \times r$ matrix of second-order component scores. From the relationship of ranks of matrices, it follows that $r \leq mq$, and $q \leq mr$. By substituting (16) into (15), we can get the equation having the same form as T2; $Z_k \approx GC_k^{*'}A^{*'}$. This is a kind of second-order PCA with the equality restriction on first-order loadings (Bloxom, 1984), but $\approx$ does not mean the least squares approximation of the model. We will obtain the three matrices simultaneously in the least squares sense by minimizing

$$f(G, C_k, A^*) = \sum_{k=1}^{m} \|Z_k - GC_k'A^{*'}\|^2, \tag{17}$$

subject to $G'G = I$, and $\sum_{k=1}^{m} C_k C_k' = mI$. We can impose constraints on G and C_k because the model of T2 is the product of three matrices. We will call the minimization of (17) T2-1 problem. The original problem for T2 by Kroonenberg and De Leeuw (1980) with different constraints from T2-1 will be introduced later.

One can distinguish the change of loadings from the change of scores by checking the pattern appeared in C_k unless the changes are not so drastic ones. Hence T2-1 can be used a tool for the study of factor changes on the descriptive level. That is illustrated in Murakami (1983; pp.31-34), and we will not repeat it here. We will only point out that the simple structure attained by orthogonal rotation in (43) is crucial for such interpretations.

Next, we redefine the *first-order components* in (15) as

$$F_k = GC_k' \qquad k = 1, \ldots, m, \tag{18}$$

where G and C_k are given by the minimization of (17) rather than the heuristic solutions in (16). Correspondingly, we will call A^* the matrix of the *first-order loadings*. F_k defined in (18) has some convenient aspects: First, it is orthonormal in the sense of $m^{-1}\sum_{k=1}^{m} F_k'F_k = I$. Second, as will been shown, the formula similar to (5),

$$A^* = m^{-1}\sum_{k=1}^{m} Z_k'F_k, \tag{19}$$

will give the basis for a scale free formulation of T2. Third, another analogous equation is derived immediately from (18) by the use of $G'G = I$;

$$C_k = F_k'G \qquad k = 1, \ldots, m, \tag{20}$$

which means that C_k is a kind of structure matrix, whose elements are covariances between the first-order components and the second-order components.

There may be another approach to the second-order PCA of three-mode data; we can derive it through the definition of the linear composites. First, we define the matrix of the first-order composites for each occasion such as

$$\ddot{F}_k^* = Z_k A \qquad k = 1, \ldots, m. \tag{21}$$

The weights held constant across occasions such as used in (21) are sometimes called the *stationary weights* (cf. Meredith & Tisak, 1982). Next, let us define the matrix of the second-order composites as

$$G^* = \sum_{k=1}^{m} \ddot{F}_k^* \tilde{C}_k. \tag{22}$$

Then, we can formulate the second-order PCA problem in the similar way to Hotelling PCA as the maximization of the following function under the constraints of $A'A = I$, and $\sum_{k=1}^{m} \tilde{C}_k' \tilde{C}_k = I$;

$$g(A, \tilde{C}_1, \tilde{C}_2, \ldots, \tilde{C}_m) = \text{tr} \sum_{k=1}^{m} \sum_{l=1}^{m} \tilde{C}_k' A' R_{kl} A \tilde{C}_l, \tag{23}$$

where $R_{kl} = Z_k' Z_l$. $\tilde{C}_k$ should be distinguished from C_k because they differ in the direction of constraints. We will call the maximization of (23) T2-2 problem.

As will be shown later, the relationship between G in (17) and G^* in (22) is simple. However, the relationship between the first-order composites $\ddot{F}^*$ and the first-order component F_k defined in (18) is somewhat complicated and must be defined separately because the former spans an mq dimensional subspace in mp columns of Z_k's, but the latter exists only in an r dimensional column space of G. (Note that $r \leq mq$.)

3.2 Equivalence of two formulations of second-order PCA to T2

Kroonenberg and De Leeuw (1980) defined TUCKALS2 as the algorithm minimizing the following function;

$$f(G, C_1^*, \ldots, C_m^*, A) = \sum_{k=1}^{m} \| Z_k - G C_k^{*'} A' \|^2 \tag{24}$$

where G is the $n \times r$ orthonormal matrix, C_k^*, the $q \times r$ frontal plane of a three-mode core matrix, and A, the $p \times q$ orthonormal matrix. This is the original T2 formulation.

Analogous to the case of (1) and (3), we can change the constraints without loss of optimality, hence we can also consider the minimization problem of (17) and of

$$f(G^*, \tilde{C}_1, \ldots, \tilde{C}_m, A) = \sum_{k=1}^{m} \| Z_k - G^* \tilde{C}_k' A' \|^2 \tag{25}$$

subject to $\sum_{k=1}^{m} \tilde{C}_k' \tilde{C}_k = I$, and $A'A = I$.

Assuming that we have the solution of (24), define

$$\Lambda = \sum_{k=1}^{m} C_k^* C_k^{*'}, \quad \text{and} \quad \Delta = \sum_{k=1}^{m} C_k^{*'} C_k^*. \tag{26}$$

Kroonenberg and De Leeuw (1980) showed that both Λ and Δ are the diagonal matrices of eigenvalues of positive definite matrices. Then,

$$A^* = A \Lambda^{1/2} T, \quad \text{and} \quad C_k = T' \Lambda^{-1/2} C_k^* U, \tag{27}$$

and

$$G^* = G\Delta^{1/2}, \quad \text{and} \quad \tilde{C}_k = C_k^* \Delta^{-1/2}, \tag{28}$$

where T and U are arbitrary orthonormal square matrices. We did not introduce rotational freedom to (28) for the same reason as in the case of PCA-2. It is easy to verify that matrices defined above produce the same optimum in (17) and (25) as that of (24), and C_k and $\tilde{C}_k$ satisfy their corresponding constraints.

Similar to the case of classical PCA, we can also convert the minimization problems to the maximization ones. On the one hand, by applying regression theory to (17), and noting that $\sum_{k=1}^{m} C_k G' G C_k' = mI$, we obtain

$$A^* = m^{-1} \sum_{k=1}^{m} Z_k' G C_k', \tag{29}$$

which is equivalent to (19), and can be regarded as an extension of the counterpart in classical PCA, (5). Substituting (29) into (17), we have the formula of the same form as (10);

$$f(G, C_1, C_2, \ldots, C_m, A^*) = \text{tr} \sum_{k=1}^{m} R_{kk} - m\text{tr}A^{*\prime}A^* = mp - m\text{tr}A^{*\prime}A^* \tag{30}$$

On the other hand, using (25), and $\sum_{k=1}^{m} A'\tilde{C}_k'\tilde{C}_k A = I$, we obtain

$$G^* = \sum_{k=1}^{m} Z_k A \tilde{C}_k. \tag{31}$$

which is equal to (22), and can be regarded as an extension of (6). Substituting this into (25), we also obtain the formula of a natural extension of (8);

$$f(A, \tilde{C}_1, \tilde{C}_2, \ldots, \tilde{C}_p) = \| Z_k - (\sum_{l=1}^{m} Z_l A \tilde{C}_l) \tilde{C}_k' A' \|^2, \tag{32}$$

where the matrix $A\tilde{C}_k$ has two roles; as a weight matrix and a pattern matrix. It is easy to confirm that the minimization of (32) is equivalent to the T2 problem, the maximization of (23), because (32) can be written as $mp - \text{tr} \sum_{k=1}^{m} \sum_{l=1}^{m} \tilde{C}_k' A' R_{kl} A \tilde{C}_l$. We also point out that the relationship between (29) and (31) is analogous to the dual relationship between (5) and (6).

As was in the case of classical PCA, implications of the two solutions are remarkably different. We prefer T2-1 to T2-2 for the same reason why we prefer PCA-1 to PCA-2; rotational freedom, convenient properties of A^*, and the (origin- and) scale free formulation derived below notwithstanding the several attractive properties of T2 such as (32).

3.3 Scale free formulation of T2-1

We will reformulate an origin- and scale free version of the second-order PCA on the basis of T2-1 in the same way in classical PCA. We will start from the redefined structure matrix of the first-order composites;

$$A^\dagger = m^{-1}D_S^{-1/2}\sum_{k=1}^{m} X_k'GC_k', \tag{33}$$

where X_k is the matrix of raw input data on k-th occasion, and D_S is the diagonal matrix of variances of variables which are centered in such a way as to transform $A^\dagger$ into the correlation matrix.

The process is almost the same as in the case of classical PCA in Section 2.3. First, we will add one constraint that $G'1 = 0$, which means $G = JG$. Therefore, we have $X_k'G = (X_k - 1\bar{x}_k')' G$ where $\bar{x}_k = n^{-1}X_k'1$. Hence, we know that D_S must be defined as $D_S = m^{-1}\text{diag}\sum_{k=1}^{m}(X_k - 1\bar{x}_k')'(X_k - 1\bar{x}_k')$,

This suggests that a sufficient condition for obtaining G such that $A^\dagger$ defined in (33) has the interpretation of the structure matrix is the transformation

$$Z_k' = D_S^{-1/2}(X_k - 1\bar{x}_k')', \tag{34}$$

which satisfies that $Z_k'1 = 0$, and $\sum_{k=1}^{m} Z_k'Z_k = mI$, and we have $A^\dagger = A^*$, where A^* is given in (29).

Although the above discussion may look almost trivial, we should consider that there are some other possible methods of preprocessing. First, one seemingly plausible transformation $Z_k' = D_S^{-1/2}(X_k - 1\bar{x}')'$, where $\bar{x} = (mn)^{-1}\sum_{k=1}^{m} X_k'1$, and $D_S = \text{diag}\sum_{k=1}^{m}(X_k - 1\bar{x}')'(X_k - 1\bar{x}')$, are not a sufficient condition to make $A^\dagger$ be a structure matrix. Second, another transformation, $Z_k' = D_{S_k}^{-1/2}(X_k - 1\bar{x}_k')'$, where $D_{S_k} = \text{diag}(X_k - 1\bar{x}_k')'(X_k - 1\bar{x}_k')$, the standardization for each k, is also plausible. This possibility is not precluded but somewhat spurious notwithstanding Murakami(1983)'s early recommendation.

We will not assert that the preprocessing (34) is universally valid. Theoretical and empirical studies to find better methods of preprocessing (e.g. Kroonenberg, 1983) are meaningful for the vast class of applications. Our conclusion is limited to the study of factor comparisons which we define in 3.1.

3.4 The algorithm

As the sample size n is usually very large comparing to mp in the study of factor invariance, the iterative algorithm based on R_{kl} is more convenient than that on Z_k. Murakami(1983) derived such an algorithm from TUCKALS2 through the algebraic manipulations. A very straightforward derivation of an improved version is possible on the basis of T2-2 criterion in (23).

First, we will assume that A is given. We define the $n \times mp$ data matrix Z by arranging frontal planes next to each other as $Z = [\, Z_1\ Z_2\ \ldots\ Z_m]$, and compute the $mp \times mp$ covariance matrix as $R = Z'Z$. Next, we define the $mq \times mq$ matrix,

$$H = (I \otimes A)'R(I \otimes A), \tag{35}$$

where $\otimes$ denotes the Kronecker product, and also define the mp by q matrix, $\tilde{C} = [\ \tilde{C_1}'\ \tilde{C_2}'\ \dots\ \tilde{C_m}'\ ]'$. Then we can rewrite (23) as

$$g(\tilde{C}) = \tilde{C}'H\tilde{C}, \tag{36}$$

where the constraint is $\tilde{C}'\tilde{C} = I$. Columns of $\tilde{C}$ will be given as the eigenvectors associated with the r largest eigenvalues of H. (For the simplicity, we assumed that all the eigenvalues are distinct.)

Next, we assume that $\tilde{C}$ is given. In addition, we also assume that we have a set initial values of elements of A. (It should be one given in the previous iteration in an ALS process.) We will rewrite (23) into

$$g(A) = \mathrm{tr}A' \sum_{k=1}^{m} \sum_{l=1}^{m} R_{kl} A \tilde{C}_l \tilde{C}_k', \tag{37}$$

and consider the singular value decomposition

$$\sum_{k=1}^{m} \sum_{l=1}^{m} R_{kl} A \tilde{C}_l \tilde{C}_k' = P\Lambda Q', \tag{38}$$

and let

$$B = PQ'. \tag{39}$$

Using Schwartz inequality, Ten Berge (1988) proved that

$$\mathrm{tr}\tilde{C}'(I \otimes B)'R(I \otimes B)\tilde{C} \geq \mathrm{tr}\tilde{C}'(I \otimes A)'R(I \otimes A)\tilde{C}, \tag{40}$$

or $g(B) \geq g(A)$. This means that the process replacing A by B increases the criterion monotonically. Therefore, we can alternate the eigendecomposition of H and the SVD in (38) until convergence is attained.

Clearly, when the algorithm converges,

$$g(A, \tilde{C}) = \mathrm{tr}\Delta = \mathrm{tr}\Lambda \tag{41}$$

holds, and we can use this for a criterion of convergence.

As the step for A in the algorithm described here performs the singular value decomposition of the matrix of p by q, it is much better than that in Murakami(1983) which needs the eigendecomposition of the p by p matrix. However, more careful studies will be necessary to compare the efficiency with that of the new algorithm using Gram-Schmidt orthogonalization (Kiers, et al., 1992).

Finally, we will list formulae to complete the analysis of T2-1. First, we will obtain the first-order loading matrix;

$$A^* = m^{-1/2}A\Lambda^{1/2}T, \tag{42}$$

where T is an orthogonal matrix which should be determined to attain the simple structure of A^*. Next, we will have the second-order loading matrix by

$$C_k = m^{1/2}T'\Lambda^{-1/2}\tilde{C}_k\Delta^{1/2}U, \tag{43}$$

where U is also an orthogonal matrix determined to reach the simple structure of C. If necessary, the second-order component scores are obtained by

$$G = \sum_{k=1}^{m} Z_k A^* C_k (\sum_{l=1}^{m} C_l' A^{*'} A^* C_l)^{-1}. \tag{44}$$

Eq. (42) and (44) show the apparent similarities to their counterparts for PCA-1, (12) and (13)

3.5 Use of the first-order composites

In analyzing the three-mode data, one may want to evaluate the correlations of components between occasions. For the purpose, the covariances between the first-order components defined in (18), $F'_k F_l = C_k C'_l$, may be useful, and it is easily computed on only the second-order loading matrix. However, elements in the matrix often overestimate the correlations because columns of the first-order components exist in the r dimensional subspace as is mentioned in 3.1. Hence, the first-order composites defined in (21) may be appropriate for because they are the linear combinations of input variables, and columns of them spans an mq dimensional space.

Although one can compute the correlation coefficients between the first-order composites directly, we can transform them into variables with unit variances which are mutually orthogonal in advance. In addition, we can maximize the congruence of the transformed first-order composites with the first-order components, which is conceptually in the same level as the first-order composites. That is, we will define the composites,

$$\ddot{F}_k = Z_k V \qquad k = 1, \ldots, m, \tag{45}$$

which satisfy $\sum_{k=1}^{m} \ddot{F}'_k \ddot{F}_k = mI$ and minimize $\sum_{k=1}^{m} \|F_k - \ddot{F}_k\|^2$. Through somewhat complicated operations using eigenequation of H and the rationale for the orthogonal Procrustes method (Cliff, 1966), we obtain

$$V = \ddot{A}(\ddot{A}'\bar{R}\ddot{A})^{-1/2} T, \tag{46}$$

where $\ddot{A} = A\Lambda^{-1/2}$, $\bar{R} = m^{-1} \sum_{k=1}^{m} R_{kk}$ and T is the orthogonal matrix in (42) attaining the simple structure of A^*. We can obtain the covariances of first-order composites between occasions by $V'R_{kl}V$.

One may wonder what criteria V optimizes because it does not produce the composites with maximum variances while A in (15) does it. Eq. (36) suggests that A in (17), on the basis of which V is defined, maximizes the sum of the r largest eigenvalues of H. As H is the covariance matrix rather than the correlation matrix, the criterion consists of the maximization of variances of composites and correlations between composites. The meaning is somewhat ambiguous. However, it is interesting to point out that the criterion is almost equal to SUMCOV, which is a generalized canonical analysis of more than three sets of variables with the stationary weights (Meredith & Tisak, 1982) if one uses the constraints $A'\bar{R}A = I$ instead of $A'A = I$. (A slight difference is that Meredith and Tisak employ the successive formulation instead of our simultaneous one.)

Because of the simple relationship between G and G^* in(28), we can write

$$G = \sum_{k=1}^{m} Z_k V W_k, \tag{47}$$

where $W_k = T'(\ddot{A}'\bar{R}\ddot{A})^{1/2}\Lambda^{1/2}\tilde{C}_k\Delta^{-1/2}U$. Substituting (47) into (17), we obtain

$$f(V, W_1, W_2, \ldots, W_p, A^*, C_1, C_2, \ldots, C_p) = \sum_{k=1}^{m} \|Z_k - (\sum_{l=1}^{m} Z_l V W_l)C'_k A^{*'}\|^2, \tag{48}$$

which is an extension of (7), and completes our list of the parallel relationships between classical PCA and T2.

3.6 Concluding remarks

As long as one considers classical PCA to be a rank q approximation of data matrix via truncated SVD, it is simple enough. However, if one utilizes the indeterminacy of the model to transform the output matrix of coefficients into the weight, pattern, and structure matrix with orthogonal or oblique rotations, it is far from simple. Similarly, Tucker2 is also simple as a symmetric decomposition of a three-way array. We have reformulated Tucker2 in multiple ways such that it can be seen as a natural extension of classical PCA which is surprisingly full of statistical implications. The multiplicity of transformations of output coefficients and scores from T2 with orthogonal rotation as well as possibilities of several formulations with different criteria is expected to facilitate interpretations of complex associations between variables in three-mode array.

Acknowledgment

The author is obliged to anonymous reviewers for their many helpful comments.

References:

Bloxom. B. (1984): Tucker's three-mode factor analysis model. In: *Research Methods for Multimode Data Analysis*, Law, H.G. et al (eds.), 104-120, Praeger Publishers, New York.

Cliff, N. (1966): Orthogonal rotation to congruence, *Psychometrika*, **31**, 33-42.

Gifi, A. (1990): *Nonlinear Multivariate Analysis*. Chichester: Wiley.

Kiers, H.A.L. et al. (1992): An efficient algorithm for TUCKALS3 on data with large number of observation unit. *Psychometrika*, **57**, 415-422.

Kroonenberg, P.M. (1983): *Three-mode Principal Component Analysis*, DSWO Press. Leiden.

Kroonenberg, P.M. and De Leeuw. J. (1980): Principal component analysis of three-mode data by means of alternating least squares algorithm, *Psychometrika*, **45**, 69-97.

Meredith, W. & Millsap, R.E. (1985): On component analysis. *Psychometrika*, **50**, 495-507.

Meredith, W. and Tisak, J. (1982): Canonical analysis of longitudinal and repeated measures data with stationary weights, *Psychometrika*, **47**, 47-67.

Murakami, T. (1983): Quasi three-mode principal component analysis: A method for assessing the factor change, *Behaviormetrika*, **14**, 27-48.

Nishisato, S. (1994): *Elements of Dual Scaling: An Introduction to Practical Data Analysis*, Lawrence Erlbaum, Hilsdale.

Ten Berge, J.M.F. (1986): Some relationship between descriptive comparisons of components from different studies, *Multivariate Behavioral Research*, **21**, 29-40.

Ten Berge. J.M.F. (1988): Generalized approach to the MAXBET problem and the MAXDIFF problem, with applications to canonical correlations. *Psychometrika*, **53**, 487-494.

Ten Berge, J.M.F. (1993): *Least Squares Optimization in Multivariate Analysis*. DSWO Press. Leiden.

Ten Berge. J.M.F. & Kiers, H.A.L. (1996): Optimality criteria for principal component analysis and generalizations, *British Journal of Mathematical and Statistical Psychology*, **49**, 335-345.

Tucker, L.R. (1966): Some mathematical notes on three-mode factor analysis. *Psychometrika*, **31**, 279-311.

Parallel Factor Analysis with Constraints on the Configurations: An overview

Pieter M. Kroonenberg[1] and Willem J. Heiser[2]

[1] Department of Education, Leiden University
[2] Department of Data Theory, Leiden University
Wassenaarseweg 52, 2333 AK Leiden, The Netherlands

Summary: The purpose of the paper is to present an overview of recent developments with respect to the use of constraints in conjunction with the Parallel Factor Analysis PARAFAC model (Harshman, 1970). Constraints and the way they can be incorporated in the estimation process of the model are reviewed. Emphasis is placed on the relatively new triadic algorithm which provides a large number of new ways to use the PARAFAC model.

1. Introduction

The PARAFAC model is a data-analytic model for three-way data, in which each way represents a different mode (three-mode data), for example, subjects (mode 1) have scores on semantic differential scales (mode 2) under several conditions (mode 3), or in case of a three-way analysis of variance design the mean yield of several varieties of maize (mode 1) planted in several locations (mode 2) during several years (mode 3).

In most applications to date the full model has been used, sometimes with provisions for missing data. However, it is possible to include constraints on the configurations of the components in the model. In this paper an overview is given of constraints that have been proposed in this context and the practical relevance of such constraints is illustrated. Moreover, attention will be paid to ways of fitting the model to include constraints. To this end the recently developed alternatives to the basic algorithm will be reviewed, in particular, the triadic or component-wise estimation.

2. The PARAFAC model

The Parallel factor analysis model (PARAFAC) was formulated by Harshman (1970; Harshman and Lundy, 1984a,b) and in a different context by Carroll and Chang (1970). It can be considered as the three-way generalization of both component analysis and the singular value decomposition.

The standard formulation of the model is (seen as a generalization of component analysis)

$$x_{ijk} = \sum_{s=1}^{S} a_{is} b_{js} c_{ks} + e_{ijk} \tag{1}$$

with $i = 1, .., I$, $j = 1, .., J$, and $k = 1, .., K$. The a_{is}, b_{js}, and c_{ks} are the elements of the components $\mathbf{a}_s$, $\mathbf{b}_s$, and $\mathbf{c}_s$, respectively, and the e_{ijk} are the errors of approximation. Note that each component depends only on one the indices i, j, k and that each component s is present in all three ways. An alternative formulation of the model (seen as a generalization of the singular value decomposition)

$$x_{ijk} = \sum_{s=1}^{S} \lambda_s a_{is} b_{js} c_{ks} + e_{ijk} \tag{2}$$

where the vectors (components) $\mathbf{a}_s$, $\mathbf{b}_s$, and $\mathbf{c}_s$, have lengths equal to 1 (or mean squares equal to 1), and the scale factors λ_s can be considered the three-way analogues of the singular values.

In many applications, especially in the social and behavioral sciences, no *a priori* models exist for three-way data, so that three-way models such as PARAFAC have to be used in an exploratory fashion. The only exception known to us can be found in the field of event-related potentials and has been formulated by Möcks (1988). In particular, Möcks derives what he calls the *topographic component model* based on biophysical considerations and continues to show that his model is identical to the PARAFAC model.

In the physical sciences, explicit models for physical processes occur frequently, and with respect to three-way and higher data several examples can be found in chemistry. Smilde, Van der Graaf, and Doornbos (1990) discuss a model for the multivariate calibration of reversed phase chromatographic systems which is identical to the PARAFAC model, and Leurgans and Ross (1992) discuss several three- and multimode models in spectroscopy. For instance that discuss a model in which the light emission measured is separately linear (1) in the number of photons absorbed, (2) in the fraction of photons absorbed that lead to emission at a particular wavelength (3) for several concentrations of light emitting entities.

In contrast with several other three-way models, the PARAFAC model is an identified model, so that after estimation the parameters of the model cannot be changed without affecting the fit of the model to the data. In particular no transformations of the components are possible without loss of fit. The identifiability is a great help in evaluating and interpreting solutions, and it is this feature which makes the model extremely relevant in those cases where an a priori model for the data is available. A PARAFAC analysis is in that case not a search for a good fitting model, but a method to obtain identified estimates for the parameters.

3. Constraints

Substantive reasons, parsimony, or modelling considerations may require constraints on the parameters. In particular, the following situations may be considered, which we will discuss in turn. (1) Orthogonality of components, (2) non-negativity of components, (3) linear constraints including design variables for the components, (4) fixed components, (5) order constraints on the components, (6) mixed measurement levels, (7) missing data. It should be noted that all constraints lead to a loss of fit, but the constrained solution may be compared with an unconstrained one to assess the importance of the constraints.

Constraints can enter the problem of finding a solution for the parameter estimates in essentially two ways, i.e. as constraints on the parameters, cases (1)-(5), or as constraints on possible transformations of the data (6) and (7). An example of the latter is optimal scaling in which case optimal transformations for the data are sought given the measurement level of the variables, simultaneously with optimal parameter estimates for the model. Another example is the estimation of the model in the presence of missing data, in which case either the missing data are 'ignored' via a 0-1 weight matrix for the data (the model is fitted around the missing data), or the missing data are estimated simultaneously with the model in a form of expectation-maximization algorithm.

3.1 Orthogonality constraints

This type of constraints requires that the components within a particular mode are pairwise orthogonal, i.e. they form an orthogonal base in the space of the components. In principle, restricting one of the modes to orthogonality is sufficient to obtain a partitioning of the total variability by components (see e.g. Kettenring, 1983). The orthogonality restriction is not necessarily a logical constraint for the PARAFAC model. After all, the parallel proportional profile principle which lies at the heart of the model, refers to a property which is specified for a triplet (a_s, b_s, c_s), and does not refer to a projection of, say the rows, into a lower-dimensional space. (Similarly, confirmatory factor analysis is not concerned with a projection into a lower-dimensional space either.) A more extensive discussion of this issue can be found in Franc (1992, pp. 151-155), who discusses the differences between a model which consists of the sum of rank-one tensors (PARAFAC), and a model which projects the data matrix from the Cartesian product space $\mathcal{R}^{I \times J \times K}$ into a lower dimensional product space $\mathcal{R}^{P \times Q \times R}$, where P, Q, and R are the number of components of the three ways in the three-mode principal component model defined by Tucker (1966).

Harshman and Lundy (1984a) suggested to use orthogonality constraints to avoid so-called degenerate (non-converging) solutions. A detailed treatment of degeneracy is beyond the scope of this review paper, and the reader is referred to Harshman and Lundy (1984b, pp. 271-281), Kruskal, Harshman, and Lundy (1989), and for tests to detect degeneracy to Krijnen and Kroonenberg (submitted).

Harshman and Lundy (1984a) also discuss fitting covariance and similarity data by PARAFAC, and show that this indirect fitting (indirect with respect to fitting the PARAFAC model to the raw data) implies that one mode (usually that of the subjects or generally the data generators) is orthogonal.

3.2 Non-negativity constraints

Non-negativity is very natural constraint to impose on components in several cases, especially when quantities expressed in the components cannot be negative. In spectroscopy components cannot have negative elements because negative values for absorption, emission or concentration are nonsensical. Similarly, in the latent class model, tackled with three-way methods by Carroll, De Soete, and Kamensky (1992), the estimated probabilities should be nonnegative. As a final example, it is known that subtests of intelligence tests generally correlate positively which each other, because all measure intelligence in some way. Krijnen and Ten Berge (1992; Krijnen, 1993, chap. 4) make a case for imposing nonnegativity constraints on the components of PARAFAC when analyzing intelligence tests. This is particularly useful to avoid so-called contrast components on which, for instance, the numerical tests have negative coefficients and the verbal tests positive ones.

3.3 Linear constraints including design variables

Design present for the subjects. Suppose that subjects come into specific groups, and that not their individual values but their group means are interesting. Carroll, Pruzansky, and Kruskal (1980) showed that such constraints can be handled by first averaging the original data according to the design, followed by a three-way analysis on the condensed data. They also warned that on interpretational grounds this does in general not seem a good procedure for the subject mode in the INDSCAL model. DeSarbo, Carroll, Lehmann, and O'Shaughnessy (1982) employ such linear constraints in their paper on three-way multivariate conjoint analysis, and also take up the issue of the appropriateness of restrictions on the subject mode.

Facet or factorial designs for variables. Tests and questionnaires are sometimes constructed according to a facet or factorial design and as above these variables may be combined and then subjected to a standard PARAFAC analysis.

A priori clusters on the variables. Another type of constraint occurs when variables belong to certain explicitly defined a priori clusters and this is to be evident via a simple structure on the components. In the three-way case one might like to fit such a constraint via the application of the PARAFAC model. The details of such approach has been worked out by Krijnen (1993, chap. 5).

A posteriori clusters on variables. Rather than knowing the clusters beforehand, a PARAFAC solution might be desired in which optimal non-overlapping clusters are searched at a prespecified number of clusters. In practice, several numbers of clusters may be tried to find a solution which is optimal in some sense (Krijnen, 1993, chap. 6).

3.4 Fixed components

Constant component(s). Suppose one is uncertain whether the data at hand really exhibit individual differences in all components. One notices differences in the first component of the subjects, but the higher ones might not be very large. One could consider a three-way analysis with the last component fixed at a constant value and compare the fit of the solutions. In case of a negligible difference one can conclude that the differences were not worth bothering about. In the extreme case of no individual difference whatsoever, one will most likely end up with averaging over subjects. One example of this might be the study of the question whether semantic differentials are subject to individual differences.

Inclusion of external information. Most studies are part of a research tradition, so that almost always earlier results exist using the same type of data and/or the same variables. Thus one might have a configuration available for one of the modes, and one might be interested in the question whether the older information is comparable to the information contained in the new three-way data set. By fixing the configuration of the mode one can estimate the other parameters in the model within the context of the already available external information. By comparing the restricted with the unrestricted model one may assess the differences between the present and the previous study. For an example using the Tucker3 model (Tucker, 1966; Kroonenberg and De Leeuw, 1980) see the study by Van der Kloot and Kroonenberg (1985).

Another view of the same situation might be that external (two-way) information on continuous variables is available. For instance, personality information is available for the subjects, and it is desired to explain the structure of the analysis in terms of these external variables. Within the analysis-of-variance context such procedures have sometimes been called factorial regression. A discussion of this approach for two-way case as well as references to examples can be found in Van Eeuwijk, Denis, and Kant (1995).

Fixing components for modelling purposes. It is possible to decide for each component in a PARAFAC model whether one, two of three ways should have constant values for this component. By doing this, one can, for instance, perform a three-way analysis of variance with the PARAFAC model. At present, we are working on a viable way of carrying out three-way analysis of variance with multiplicative terms for the interactions via a PARAFAC model as suggested by De Leeuw in Kroonenberg (1983, p. 141).

3.5 Order constraints

Order constraints for autocorrelated data. When one of the modes in the three-way array is a time node or two of the modes contain spatial coordinates, it may be desirable to use this design information during the analysis. One way to do this would be by requiring that the levels of a component obey certain order restrictions. One might also try to take this further and require equal intervals, but then one more or less returns to the previous section.

3.6 Constraints to create discrete models

Discrete and hybrid clustering. Carroll and co-workers have developed procedures to perform clustering on two-mode three-way similarity data by using constraints of the components. This approach demands that certain components can only have a limited number of integer values. Like in the previous section this is an example of using constraints to fit specific models. Details can be found in Carroll and Chaturvedi (1995).

3.7 Optimal scaling

As mentioned above, optimal scaling concerns constraints on the data rather than on the model. Detailed discussions of optimal scaling can be found in Gifi (1990), and the earlier references mentioned there in. The fundamental idea is that the measurement level of a variable is specified by the transformations which may be applied to the values of that variable without changing the meaning of the variable and its interpretation. For example, any monotone transformation of the values of an ordinal variable will maintain the ordering of the values and thus their basic meaning. Sands and Young (1980) presented an optimal scaling version of the PARAFAC model as well as the algorithm to compute its solution, but no further applications of this approach are known (see, however, some comments by Harshman and Lundy, 1984, pp. 188-191).

3.8 Constraints for single elements

Missing data. Missing data may lead to constrained estimation problems. They can be handled within the PARAFAC model either by a kind of EM algorithm estimating the model and the missing data in turn, or by differentially weighting elements of the three-way data array, for instance, by weighting every valid data point with 1 and each missing data point by 0.

Equality constraints. In certain applications it might be interesting to demand or test whether certain elements in a component are equal to one another, where it is not known beforehand what the size of these equal elements should be. Such constraints clearly operate at the level of the individual parameters in the model, and are thus model constraints.

4. Algorithms

In this section we will give an overview of several algorithms proposed for the PARAFAC model and comment on the way constraints can be handled. In particular, we will concentrate on the standard or so-called CP algorithm and the triadic algorithm.

4.1 Standard (unconstrained) algorithm

The by-now standard algorithm was independently worked out by Harshman (1970) who called it the PARAFAC algorithm and by Carroll and Chang (1970) who called it the CANDECOMP algorithm. Following Kiers and Krijnen (1991), we will refer to it as the *CP algorithm*. As shown in Ten Berge, Kiers and Krijnen (1993) in the standard CP algorithm the estimates for the parameter matrices are found row by row (even though this is not always evident from its formulation), and thus the algorithm may be called a *row-wise algorithm* for the PARAFAC model. The discrepancy function (objective function, loss function) which lies at the basis of most algorithms is

$$\Phi(\mathbf{A}, \mathbf{B}, \mathbf{C}) = \sum_{i=1}^{I} \sum_{j=1}^{J} \sum_{k=1}^{K} \left(z_{ijk} - \sum_{s=1}^{S} a_{is} b_{js} c_{ks} \right)^2 \qquad (3)$$

The basic solution proposed by Carroll and Chang (1970) and Harshman and Lundy (1970) which at a later date has also been independently worked out by several other authors (e.g. Möcks, 1988), consists of fixing two of the parameters matrices, solving for the third one via (multivariate) regression, performing this procedure for each permutation of the component matrices in turn, and repeating the whole procedure until convergence. Also Hayashi and Hayashi (1982) presented an algorithm which show similarities to the standard one. Technically such the CP algorithm is straightforward to implement, but it turns out that the algorithm does not always converge, due to the mismatch of data and model. This situation is called degeneracy and details can be found in Harshman and Lundy (1984b), Kruskal, Harshman, and Lundy (1989), and Krijnen and Kroonenberg (submitted). As an aside the work of Pham and Möcks (1992) should be mentioned, as they prove the consistency and asymptotic normality of the least squares estimators.

Variants. A variant on the basic algorithm was proposed by Kiers and Krijnen (1991). They showed by rearranging the calculations and by operating on the (multivariable-multioccasion) covariance matrix rather than the raw data, that the computation time becomes independent of the number of observations, but that at each iteration step the results are the same as the standard algorithm. Moreover, by operating on the covariance matrix, the storage space does not increase with the number of observations, and multivariable-multioccasion covariance matrices from the literature can be directly used. A deficiency of the algorithm is that it cannot handle missing data other than during the building of the covariance matrix.

Using the assumption of normality for the errors, Mayekawa (1987) developed a maximum likelihood procedure for indirect fitting (see also section 3.2), i.e. first the raw data are converted covariance matrices per occasion (with possibly different samples per occasion). Note that the covariances are different from the Kiers and Krijnen (1991) proposal, because they use only a single (multivariable-multioccasion) covariance matrix, and thus necessarily assume repeated measurements.

Weights for error terms. Several authors (Harshman and Lundy, 1984b, pp. 242ff.; Carroll, De Soete and Kamensky, 1992; Heiser and Kroonenberg, 1994) considered a generalization of the discrepancy function by including weights for the error terms e_{ijk}

$$\Psi(\mathbf{A}, \mathbf{B}, \mathbf{C}) = \sum_{i=1}^{I} \sum_{j=1}^{J} \sum_{k=1}^{K} w_{ijk} \left(z_{ijk} - \sum_{s=1}^{S} a_{is} b_{js} c_{ks} \right)^2 \qquad (4)$$

The addition of weights changes the details of the algorithm in several places but not its basic character. The seemingly small change has considerable consequences in the ability of the algorithms to handle certain kinds of constraints. Harshman and Lundy (1984b) presented the weighted version of the standard CP algorithm in the context of reweighting or scaling of error terms to achieve equal error variances but seemed to look only at diagonal weight matrices (i.e. $w_{ijk} = w_i$).

Orthogonality. Because the basic algorithm operates on rows or whole matrices at a time, in principle only constraints can be handled which operate on rows or entire component matrices. As the least squares steps for each of the three parameter matrices are independent, one may use different constraints on different component matrices as long as at each step the discrepancy function is decreased. Orthogonality constraints are an example of this but they are generally handled by including the constraints in the discrepancy function via Lagrange multipliers, and thus changing fundamentally the characteristics of the discrepancy function.

Weights on components. Several constraints can be incorporated via weight matrices for the component matrices (rather than the errors) and the estimation can then proceed via the standard algorithm or with specialized algorithms. This leads to a very general discrepancy function

$$\Psi(\mathbf{A}, \mathbf{B}, \mathbf{C}) = \sum_{i=1}^{I} \sum_{j=1}^{J} \sum_{k=1}^{K} w_{ijk} \left(z_{ijk} - \sum_{s=1}^{S} (p_{is} a_{is})(q_{js} b_{js})(r_{ks} c_{ks}) \right)^2 , \qquad (5)$$

where $\mathbf{P} = (p_{is})$, $\mathbf{Q} = (q_{js})$, and $\mathbf{R} = (r_{ks})$, are, generally known, weight matrices for the components. Krijnen (1993, chap. 5) uses such, binary, weights as a priori cluster constraints. Moreover, Krijnen (1993, chap. 6) also proposed an algorithm for optimally clustering the coordinates of the variables using explicitly the row-wise character of the standard algorithm, in which the binary weights had to be determined. Carroll, Pruzansky, and Kruskal (1980) showed that if a factorial design or linear constraints are specified on the components there is no need for a special algorithm because one may first reduce the data matrix according to the design and then use the basic algorithm to solve for the parameter estimates (see also Franc (1992, pp.188ff.). The latter author also discusses the inclusion of different metrics for the component matrices and shows that this situation can be handled by rewriting the basic equations and then using the standard algorithm. Such a procedure is analogous to the use of the ordinary singular value decomposition to solve the singular value decomposition with weighted metrics in correspondence analysis (see e.g. Greenacre, 1983, p. 40).

Nonnegativity. Nonnegativity constraints require an adaptation of the basic algorithm to solve the regressions at each step. One may use a non-negative least squares (NNLS) procedure such as described in Lawson and Hanson (1974, pp. 158–161) and discussed or referred to in Krijnen and Ten Berge (1992), Durrell, Lee, Ross, and Gross (1990), Carroll, De Soete, and Kamensky (1992). The basic principle of the NNLS procedure is to set negative coordinate values to zero during iterations. Paatero and Tapper (1994) take a different approach and use a penalty function whose value is dependent on the size of the negative element. Paatero's procedure – Positive Matrix Factorization – is as of yet only fully described for the two-way case, but a program with documentation (Paatero, 1996) exists which includes the estimation of the parameters of the nonnegative PARAFAC model as well.

4.2 Triadic algorithm or component-wise algorithm

Recently, *column-wise algorithms* have been put forward which can handle a larger number of constraints. A column-wise algorithm for symmetric matrices was first worked out in unpublished notes by Ten Berge (1986) for the specific purpose to fit the INDSCAL model, and it was published in Ten Berge, Kiers, and Krijnen (1993). In another unpublished manuscript Carroll (1987) produced the same or similar algorithm for the same purpose and a full column-wise algorithm alternative for the CP algorithm was published in Carroll, De Soete and Kamensky (1992). Again independently in yet another unpublished manuscript Heiser and Kroonenberg (1994) also proposed the column-wise algorithm or triadic algorithm, as they called it.

In the standard algorithm the discrepancy function is solved by optimizing over a complete parameter matrix of a mode holding the other two fixed. However, it is possible to rearrange the computations yet again by making the number of components the outer loop and computing for each component s first the component $\mathbf{a}_s$, then $\mathbf{b}_s$, followed by $\mathbf{c}_s$, while holding the other two components fixed. This may be accomplished by using the following development of the discrepancy function

$$
\begin{aligned}
\Phi(\mathbf{a}_s, \mathbf{b}_s, \mathbf{c}_s) &= \left(\sum_{i=1}^{I} \sum_{j=1}^{J} \sum_{k=1}^{K} \left\{ \left[z_{ijk} - \sum_{s' \neq s}^{S} a_{is'} b_{js'} c_{ks'} \right] - a_{is} b_{js} c_{ks} \right\} \right)^2 \\
&= \left(\sum_{i=1}^{I} \sum_{j=1}^{J} \sum_{k=1}^{K} \left\{ \bar{z}_{ijk} - a_{is} b_{js} c_{ks} \right\} \right)^2
\end{aligned}
\tag{6}
$$

Note that after the s-th components have been estimated the components $s+1$ (or 1) will be next after the new component s have been incorporated in $\bar{z}_{ijk}$.

Variants. The triadic algorithm opens the way to including many more types of constraints and it also allows the development of special variants of the PARAFAC model using these constraints. In particular, such features as structured components as in time series, design matrices for the modes, nonnegativity and equality constraints, etc. can be easily handled in this context, and different constraints may be imposed on different components. However, for including orthogonality constraints again a change of the discrepancy function is necessary. In terms of special models using constraints, Carroll, De Soete, and Kamensky (1992) used a constrained version of the columnwise algorithm to estimate Lazarsfeld latent class model, and Carroll and Chaturvedi (1995) developed both a clustering procedure and a hybrid clustering procedure for two-mode three-way similarity data. Finally, the estimation of analysis of variance with multiplicative components for all interactions seems to come within reach (Heiser and Kroonenberg, in preparation).

Other triadic algorithms. Yoshizawa (1988) discussed at a theoretical level an algorithm to solve the PARAFAC model which produces orthogonal triads, and which shows considerable similarity with an algorithm proposed by Denis and Dhorne (1989) for the same purpose, the latter, however, are able to completely decompose a three-way array with their method. No specific details about feasibility or practical relevance of these algorithms seems to be known, nor have they been used to incorporate constraints.

5. Programs

The standard algorithm has been included in (FORTRAN) programs such as CANDECOMP (Carroll and Chang, 1970) and PARAFAC (Harshman and Lundy, 1994), of which the latter program is probably the most extensive one as it includes many special features relevant for the analysis of three-mode data.

Most of the authors who have contributed to the further development of PARAFAC have written their own programs, primarily in Matlab, Splus, or other matrix-based languages. The standard algorithm for the PARAFAC model, as well as the nonnegativity and orthogonal variants, have also been included in the analysis package for three-way data 3WAYPACK (Kroonenberg, 1994, 1996), and the triadic or componentwise version will be in the next version of 3WAYPACK. This package also includes other three-way models, such as those proposed by Tucker (1966, 1972).

6. References

Carroll, J.D. (1987): New algorithm for symmetric CANDECOMP. Unpublished manuscript. AT&T Bell Laboratories, Murray Hill, NJ.

Carroll, J.D. and Chaturvedi, A. (1995): A general approach to clustering and multidimensional scaling of two-way, three-way, and higher way data. *Geometric representations of perceptual phenomena: Papers in honor of Tarow Indow on his 70th birthday*, Luce, R.D. et al. (eds.), Erlbaum. Mahwah, NJ.

Carroll, J.D. and Chang, J.-J. (1970): Analysis of individual differences in multidimensional scaling via an N-way generalization of "Eckart-Young" decomposition. *Psychometrika*, **35**. 283–319.

Carroll, J.D., De Soete, G., and Kamenski, A.D. (1992): A modified CANDECOMP algorithm for fitting the latent class model: Implementation and evaluation. *Applied Stochastic Models and Data Analysis*, **8**, 303–309.

Carroll, J.D., Pruzansky, S., and Kruskal, J.B. (1980): CANDELINC: A general approach to multidimensional analysis of many-way arrays with linear constraints on parameters. *Psychometrika*, **45**, 3–24.

Denis, J.B. and Dhorne, T. (1989): Orthogonal tensor decomposition of 3-way tables. In: *Multiway data analysis*, Coppi, R. and Bolasco, S. (eds.), 31–38, Elsevier, Amsterdam.

DeSarbo, W.S., Carroll, J.D., Lehmann, D.R., and O'Shaughnessy, J. (1982). Three-way multivariate conjoint analysis. *Marketing Science*, **1**, 323–350.

Durrell, S.R., Lee, C.-H., Ross, R.T., and Gross, E.L. (1990): Factor analysis of the near-ultraviolet absorption spectrum of plastocyanin using bilinear, trilinear, and quadrilinear models. *Archives of Biochemistry and Biophysics*, **278**, 148–160.

Franc, A. (1992): Étude algébrique des multitableaux: Apports de l'algèbre tensorielle. Unpublished PhD thesis, Université de Montpellier II, France.

Gifi, A. (1990): *Nonlinear multivariate analysis*, Wiley, Chicester. UK.

Greenacre, M.J. *Theory and applications of correspondence analysis*, Academic Press, London.

Harshman, R.A. (1970): Foundations of the PARAFAC procedure: Models and contributions for an "explanatory" multi-modal factor analysis. *UCLA Working Papers in Phonetics*,16, 1–84. [Also available as University Microfilms, No. 10,0085].

Harshman, R.A. and Lundy, M.E. (1984a): The PARAFAC model for three-way factor analysis and multidimensional scaling. In: *Research methods in multimode data analysis*, Law, H.G. et al. (eds.). 122–214, Praeger, New York.

Harshman, R.A. and Lundy, M.E. (1984b): Data preprocessing an the extended PARAFAC model. In: *Research methods in multimode data analysis*, Law, H.G. et al. (eds.), 216–284, Praeger, New York.

Harshman, R.A. and Lundy, M.E. (1994): PARAFAC: Parallel factor analysis. *Computational Statistics and Data Analysis*, **18**, 39–72.

Hayashi, C. and Hayashi, F. (1982): A new algorithm to solve PARAFAC-model. *Behaviormetrika*. **11**, 49–60.

Heiser, W.J. and Kroonenberg, P.M. (1994): Dimensionwise fitting in Parafac-Candecomp with missing data and constrained parameters. Unpublished manuscript, Department of Data Theory, Leiden University, Leiden.

Kettenring, J.R. (1983): Components of interaction in analysis of variance models with no replications. In: *Contributions to statistics: Essays in honor of Norman L. Johnson*, Sen, P.K. (ed.), North-Holland, Amsterdam. Kiers, H.A.L. and Krijnen, W.P. (1991): An efficient algorithm for PARAFAC of three-way data with large numbers of observation units. *Psychometrika*, **56**, 147–152.

Krijnen, W.P. (1993): *The analysis of three-way arrays by constrained PARAFAC methods*, DSWO Press, Leiden.

Krijnen, W.P. and Kroonenberg, P.M. (submitted): Detecting degeneracy when fitting the PARAFAC model.

Krijnen, W.P. and Ten Berge, J.M.F. (1992): A constrained PARAFAC method for positive manifold data. *Applied Psychological Measurement*, **16**, 295–305.

Kroonenberg, P.M. (1983): *Three-mode principal component analysis: Theory and applications*, DSWO Press, Leiden.

Kroonenberg, P.M. (1994): The TUCKALS line: A suite of program for three-way data analysis. *Computational Statistics and Data Analysis*, **18**, 73–96.

Kroonenberg, P.M. (1996): 3WAYPACK *user's manual*, Leiden University, Leiden.

Kroonenberg, P.M. and De Leeuw, J. (1980): Principal component analysis of three-mode data by means of alternating least squares algorithms. *Psychometrika*, **45**, 69–97.

Kruskal, J.B., Harshman, R.A., and Lundy, M.E. (1989): How 3-MFA can cause degenerate PARAFAC solutions, among other relationships. In: *Multiway data analysis*, Coppi, R. and Bolasco, S. (eds.), 115–122, Elsevier, Amsterdam.

Lawson, C.L. and Hanson, R.J. (1974): *Solving least squares problems*, Prentice Hall, Englewood Cliffs, NJ. Leurgans, S.E. and Ross, R.T. (1992): Multilinear models: Application in spectroscopy (with discussion). *Statistical Science*, **7**, 289–319.

Mayekawa, S.-I. (1987). Maximum likelihood solution to the PARAFAC model. *Behaviormetrika*, **21**, 45–63.

Möcks, J. (1988): Decomposing event-related potentials: A new topographic components model. *Biological Psychology*, **26**, 129–215.

Paatero, P. and Tapper, U. (1994): Positive matrix factorization: A non-negative factor model with optimal utilization of error estimates of data values. *Environmetrics*, **5**, 111–126.

Paatero, P. (1995): User's guide for positive matrix factorization programs PMF2.EXE and PMF3.EXE, Department of Physics, University of Helsinki.

Pham, T.D. and Möcks, J. (1992). Beyond prinicpal component analysis: A trilinear decomposition model and least squares estimation. *Psychometrika*, **57**, 203-215.

Sands, R. and Young, F.W. (1980): Component models for three-way data: ALSCOMP3, an alternative least squares algorithm with optimal scaling features. *Psychometrika*, **45**, 39–67.

Smilde, A.K., Van der Graaf, P.H., and Doornbos, D.A. (1990): Multivariate calibration of reversed-phase chromatographic systems. Some designs based on three-way data analysis. *Analytica Chimica Acta*, **235**, 41-51.

Ten Berge, J.M.F. (1986): Three notes on three-way analysis. Paper presented at the Workshop on TUCKALS and PARAFAC, Leiden University, July 2.

Ten Berge, J.M.F., Kiers, H.A.L., and Krijnen, W.P. (1993): Computational solutions for the problem of negative saliences and nonsymmetry in INDSCAL. *Journal of Classification*, **10**, 115–124.

Tucker, L.R. (1966): Some mathematical notes on three-mode factor analysis. *Psychometrika*, **37**, 279–311.

Tucker, L.R. (1972): Relations between multidimensional scaling and three-mode factor analysis. *Psychometrika*, **37**, 3–27.

Van der Kloot, W.A. and Kroonenberg, P.M. (1985): External analysis with three-mode principal component analysis. *Psychometrika*, .

Van Eeuwijk, F.A., Denis, J.-B., Kang, M.S. (1995): Incorporating additional information on genotypes and environments in models for two-way genotype by environment tables. In: *Genotype by environment interaction: New perspectives.* M.S. Kang and H.G. Gaugh Jr. (eds.), CRC-Press, Boca Raton, USA.

Yoshizawa, T. (1988): Singular value decomposition of multiarray data and its applications. In *Recent developments in clustering and data analysis*, C. Hayashi et al. (eds.), Academic Press, New York.

Acknowledgement

The research of the first author was financially supported by a grant from the Nissan Fellowship Programme of the Netherlands Organization for Scientific Research (NWO).

Regression Splines for Multivariate Additive Modeling

Jean-François Durand

Probabilités et Statistique
Université Montpellier II
Place Eugène Bataillon
34095 Montpellier, France

Unité de Biométrie
ENSAM-INRA-UM II
9 Place Pierre Viala
34060 Montpellier, France

Summary: Four additive spline extensions of some linear multiresponse regression methods are presented. Two of them are defined in this paper and their properties are compared with those of two other recently devised methods. Dimension reduction aspects and quality of the regression are discussed and illustrated on examples.

1. Introduction

Let $(x_1, \ldots, x_p)$ be a set of predictors related to a set of responses $(y_1, \ldots, y_q)$, all measured on the same n individuals with sample data matrices X $(n \times p)$ and Y $(n \times q)$. The goal of this paper is to present methods for multivariate additive modeling and data reduction which integrate regression splines in their settings. Piecewise polynomials or splines are extensively used in statistics and data analysis, see for example (Ramsay 1988), (Gifi 1990), (Hastie and Tibshirani 1990) and (Durand 1993). Spline transformations of the explanatory variables presented in Section 2, non necessarily monotonic in contrast to (Ramsay 1988), are conjointly used with orthogonal projections on either such smoothed predictors or "synthetic" explanatory variables called additive components. Spline functions are attractive due to their appealing local sensitivity to data and orthogonal projectors preserve some linear properties in the considered nonlinear methods. Section 3 outlines that problems arise with least-squares splines (Eubank 1988) when the number of predictors is large. Scarcity of data in multivariate setting may cause harm in additive modeling by least-squares splines, thus providing a point in favour of dimension reduction.

Four multiresponse additive regression models are considered, all presenting dimension reducing aspects. Because of their similar scope of applicability, the three first methods are compared on a "running example". In Section 4, the regression on Additive Spline Principal Components (ASPCs henceforth) is based on a new definition of ASPCs. Additive principal components have been recently defined by Donnell et al. (1994) who explore the low end of the component spectrum for detecting concurvities in additive models. Here, large uncorrelated ASPCs are used not for condensing in an additive fashion the X sample matrix since the predictors are transformed, but rather for predicting linearly the response data set. The second method, Partial Least Squares regression via additive splines referred to as ASPLS (Durand and Sabatier 1994), is summarized in Section 5. This method differs from the preceding in that dimension reduction is processed at the same time as the regression is computed. When predictor and response data sets are identical, ASPLS components are called self-ASPLS components. In Section 6, such self-ASPLS components are interpreted as an additive summary of the original predictor matrix and used for regression purpose. Finally, Principal Component Analysis with Instrumental Variables (Durand 1993) whose scope of applicability differs from that of the preceding methods, is presented in Section 7 with applications to simple regression and to nonlinear Discriminant Analysis.

2. Spline transformations of the predictors

For simplicity, let us choose the same kind of spline functions for transforming the predictors: we take K interior knots in which piecewise polynomials of order m are required to join end to end so that $r = m + K$ is the dimension of the spline space. A spline function $s^i(\bullet)$ used for transforming the predictor x_i is a linear combination of normalized B-spline basis functions $\{B_l^i(\bullet)\}_{l=1,\ldots,r}$

$$s^i(x_i) = \sum_{l=1}^{r} a_l^i B_l^i(x_i) , \tag{1}$$

see (De Boor 1978) for computational and mathematical properties of B-splines. The ith column of $\mathbf{X}$, denoted $\mathbf{X}^i$, is replaced by $\mathbf{X}^i(\mathbf{a}^i)$ linearly depending on $\mathbf{a}^i$ through $\mathbf{X}^i(\mathbf{a}^i) = \mathbf{B}^i \mathbf{a}^i$, where $\mathbf{B}^i$ is the $n \times r$ coding matrix of $\mathbf{X}^i$, and $\mathbf{a}^i$ is the vector of the r spline coefficients. The matrix $\mathbf{X}$ is now being transformed in a $n \times p$ matrix $\mathbf{X}(\mathbf{a})$ which is a function of the spline vectors. This matrix is columnwise denoted

$$\mathbf{X}(\mathbf{a}) = [\mathbf{X}^1(\mathbf{a}^1)| \ldots |\mathbf{X}^p(\mathbf{a}^p)] . \tag{2}$$

In order to make $\mathbf{X}^i(\mathbf{a}^i)$ centered independently of the values of the spline coefficients, B^i is column centered with respect to D, a $n \times n$ diagonal matrix of weights for the observations. The knot sequence $\{\xi_1^i, \ldots, \xi_{2m+K}^i\}$ used for transforming the ith predictor is written as

$$\min_{k=1,\ldots,n} X_k^i = \xi_1^i = \ldots = \xi_m^i < \xi_{m+1}^i < \ldots < \xi_{m+K}^i < \xi_{m+1+K}^i = \ldots = \xi_{2m+K}^i = \max_{k=1,\ldots,n} X_k^i .$$

When $m \geq 2$, particular spline coefficients called *nodal coefficients* (Durand 1993) and given by $a_l^i = mean(\xi_{l+1}^i, \ldots, \xi_{l+m-1}^i)$, keep the ith predictor invariant which gives $\mathbf{X}^i(\mathbf{a}^i) = \mathbf{X}^i$. The existence of such spline coefficients implies firstly that the additive spline model as defined in section 3 can effectively take account of possibly linear relationships and secondly, that the different iterative algorithms can reasonably be initialized.

3. Additive models and least-squares splines

In multiresponse regression, an additive spline model is a fit of the form

$$\hat{y}_j = \sum_{i=1}^{p} f_i^j(x_i), \qquad j = 1, \ldots, q , \tag{3}$$

where the coordinate function $f_i^j(x_i)$ is a spline function as defined in (1) with spline coefficients stored in $\hat{a}_i^j$. In matrix notation, the jth column of the $n \times q$ model matrix $\widehat{Y}$ is the sum of the smoothed sample predictors

$$\widehat{Y}^j = \sum_{i=1}^{p} B^i \hat{a}_i^j = \sum_{i=1}^{p} X^i(\hat{a}_i^j) . \tag{4}$$

We will note whether or not an additive method is associated with a multivariate linear smoother (Hastie and Tibshirani 1990) defined by $\widehat{Y} = SY$, where the smoother matrix S does not depend on Y.

Spline coefficients can be chosen to minimize the mean squared error $\|Y - BA\|_D^2$, where $\|X\|_D^2 = trace(X'DX)$, $B = [B^1| \ldots |B^p]$ and A is the $pr \times q$ matrix of spline coefficients. Linear multiple regression on the $n \times pr$ design matrix B provides the so called *least-squares spline model* (Eubank 1988) associated with the smoother matrix $S = P_B$, where P_B is the D-orthogonal projector on $span(B)$. However, using least squares splines may cause problem when no sufficient points are available for fitting surfaces in high dimensional spaces. Our O.E.C.D. "running" data set is typical of such unlucky context since only eighteen points are needed for fitting an additive surface in a space of fourteen dimensions ($n = 18$, $p = 13$). The scope of applicability of this method concerns data sets with large samples of few predictors since scarcity of data is increased because the number of independent variables (the columns of the design matrix B) is getting larger.

The aim of the paper is to present four methods for additive modeling that include dimension reducing aspects in their settings. All are based on the column centered design matrix $X(a)$ defined by (2). In the different models, $\hat{a}_i^j$ is expressed as a linear combination of M optimal spline vectors, all associated with the same ith predictor transformed at the different stages of the methods (M is the *dimension* of the model, that is, the number of additive components or latent variables). Model dimension M may be considered a *tuning parameter* and can be checked out in the same way as in linear methods by using cross-validation. When it is computationally feasible, cross-validation is more important in nonlinear modeling because the risk of overfitting is increased due to the greater flexibility of splines. Other tuning parameters allow to choose what type of splines is to be used: the order of the polynomials, the number and position of the knots. Choosing few well located knots generally suffices in multiresponse regression but finding their optimal number and location is a difficult problem so that giving the optimal answer is beyond the scope of this paper. To summarize, the model defined by (3) and (4), belongs to the family of nonparametric additive models depending on the aforesaid tuning parameters.

4. Regression on additive spline principal components

Additive principal components have been recently defined by Donnell et al. (1994) to detect instability in additive regression models due to *concurvity* between the predictors. They explore the low end of the principal components spectrum, neglecting the use of the largest components for dimension reduction because plots of such components cannot be interpreted as a projection of the "same" data set. This insightful remark sets us the problem of what is really going on when using the largest components in dimension reduction not for purpose of exploratory data analysis but rather for multivariate additive modeling.

The definition of additive spline principal components (ASPCs) presented in the next section differs from that of Donnell et al. in these points: first, only finite-dimensional function spaces are used here, namely, piecewise polynomials of space dimension $r = m + K$. Second, uncorrelation of ASPCs is not obtained through orthogonal sets of functions but constructed step by step through an orthogonalizing procedure. Finally, ASPCs are not deduced from the eigenanalysis of some variance matrix, but instead, are considered as mutually uncorrelated solutions of a maximum variance optimization problem.

4.1 Additive spline principal components

Some finite-dimensional nonlinear principal component extensions (Ramsay 1988, Gifi 1990), are based on the Eckart-Young theorem that constructs low rank approximations of a matrix. Others are directly deduced from the eigenanalysis of some covariance matrix (Donnell et al. 1994). Here, we follow a nested approach motivated by the fact that taking spline coefficients equal to nodal coefficients leads to *common principal components*.

DEFINITION. The kth ASPC is the linear composite of $X(a)$, uncorrelated with the previous components, which has the largest variance. Denoting $c = X(a)u$, the objective function $f(a^1, \ldots, a^p, u) = var(c)$ is maximized subject to

$$
\begin{aligned}
\|u\|^2 &= 1, \\
var(X^i(a^i)) &= var(X^i), \quad i = 1, \ldots, p, \\
cov(c, c^j) &= 0, \quad j = 1, \ldots, k-1.
\end{aligned}
$$

The last constraint is omitted when $k = 1$. Writing $(a^{1,k}, \ldots, a^{p,k}, u^k)$ as an optimal argument of f and $X(a^{(k)}) = [B^1 a^{1,k} | \ldots | B^p a^{p,k}]$ as an optimal matrix, the kth ASPC becomes $c^k = X(a^{(k)})u^k$. Note that X as well as all coding matrices B^i are column centered with respect to D in order to make c^k centered for arbitrary spline coefficients.

The kth additive principal function may be defined as $C^k(x_1, \ldots, x_p) = \sum_{i=1}^p \phi^{i,k}(x_i)$ with $\phi^{i,k}(x_i) = u_i^k s^{i,k}(x_i)$, where $s^{i,k}$ is the optimal spline function used for transforming the ith predictor. A crucial problem is the choice of M, the number of components. Here we have no total variance decomposition theorem because data sets are changing when successive ASPCs are computed. However the sequence $var(c^1), \ldots, var(c^M)$, cannot increase because nested optimizations are considered (an exception to this rule may nevertheless occur when local optima are reached by the algorithm). The only question is then of setting a stopping rule, that is, on estimating whether $var(c^k)$ is "small". Since uncorrelated ASPCs are constructed for regression purpose only, one can pragmatically check out the goodness-of-fit for different model dimensions, see Section 4.3.

4.2 ASPCs from the O.E.C.D. predictors

Our "running example" is a data set from econometrics, that consists of eighteen observations (countries) and eighteen variables analyzed by Bertier and Bouroche (1975). More precisely, thirteen predictors characterizing the eighteen countries have been measured to explain five consumer responses, see Table 1.

Table 1. O.E.C.D. variables

Predictors		Responses	
POP	population	CAL	calories per capita and per day
DENS	density per km^2	LODG	number of lodgings per 1000 cap.
POPG	population growth	ELEC	electricity consumption
AGRF	% of farming & fishing population	EDUC	public expenditure for education
INDU	% of industrial population	TV	number of TV sets per 1000 cap.
GNP	gross national product per capita		
GDPA	% of GNP for agriculture		
FCF	fix capital formation		
RR	running receipts		
OFR	official reserves (million $)		
DR	discount rate		
IMP	importations (million $)		
EXP	exportations (million $)		

Variables are centered and standardized with equally distributed weights $(D = 18^{-1}I_{18})$ and for all the competing regression methods we will compare, the 13 predictors are transformed by B-spline functions of degree 1 (order 2) with 2 equally spaced interior knots. In this section, twelve ASPCs have been computed whose variances are in Table 2.

Table 2. ASPC's variance for the O.E.C.D. data set

ASPC	1	2	3	4	5	6	7	8	9	10	11	12
var	7.15	6.62	4.76	3.52	3.02	2.48	1.99	1.82	1.73	1.59	1.09	0.98

In contrast to common principal components, we observe the fact, shared by all the examples we have studied, that the sequence of variances is mildly decreasing. We are now ready to address the problem of what to do with the ASPC's D-orthogonal basis: the next section presents a simple way of selecting the components that best explain the responses.

4.3 Regression on additive spline principal components

Denoting by C the $n \times M$ matrix of M selected ASPCs, the regression on additive spline principal components, in short ASPCR, is geometrically defined as the orthogonal projection of Y on $span(C)$, the linear space spanned by the columns of C. The multivariate additive model is given by $\widehat{Y} = P_C Y$ where P_C is the projection matrix on $span(C)$. Obviously ASPCR is a multivariate linear smoothing procedure that enters the additive framework of Section 3 and any modeled response may be expressed as the sum of p coordinate functions of the predictors separately, each function being a particular spline function.

Choosing M ASPCs can be checked out by examining the fit of nested models. Table 3 shows the R^2 values of the O.E.C.D. responses according to different model dimensions. It is clear that some ASPCs may be deleted, for instance. the second and the three last.

Table 3. R^2 of the responses for nested ASPCR models

Mod. dim. ASPC	CAL	LODG	ELEC	EDUC	TV	% of total Y-variance
1	0.056	0.005	0.084	0.034	0.608	15.74
2	0.105	0.062	0.091	0.080	0.608	18.92
3	0.118	0.101	0.188	0.482	0.700	31.78
4	0.378	0.103	0.358	0.727	0.821	47.74
5	0.378	0.373	0.359	0.744	0.823	53.54
6	0.443	0.463	0.360	0.746	0.828	56.80
7	0.615	0.718	0.671	0.781	0.904	73.78
8	0.629	0.732	0.693	0.814	0.905	75.46
9	0.769	0.838	0.714	0.832	0.905	81.16
10	0.865	0.838	0.737	0.837	0.929	84.12
11	0.880	0.844	0.785	0.853	0.952	86.28
12	0.923	0.871	0.827	0.878	0.966	89.30

Finally one can reduce the model dimension M by choosing ASPCs that best explain the responses. Here ASPCs 1, 3, 4, 5, 7 and 9 seem to provide a good trade-off between goodness of fit and dimension reduction. Table 4 presents the results for the nested corresponding models and the goodness-of-fit may be compared with that of the model dimension 6 in Table 3 (73.04% against 56.80%).

Table 4. R^2 of the responses for nested selected ASPCR models

Mod. dim.	ASPC	CAL	LODG	ELEC	EDUC	TV	% of total Y-variance
1	1	0.056	0.005	0.084	0.034	0.608	15.74
2	3	0.068	0.044	0.181	0.436	0.700	28.58
3	4	0.329	0.047	0.352	0.681	0.820	44.58
4	5	0.329	0.317	0.352	0.698	0.822	50.36
5	7	0.501	0.571	0.664	0.732	0.898	67.32
6	9	0.641	0.678	0.685	0.750	0.898	73.04

Responses TV and EDUC are well reconstituted by the six dimensions model whereas CAL, LODG and ELEC are rather badly approximated. Let us now pay more attention on selecting the predictors of main influence for the variable EDUC which will be our "running response" for comparing different models. Figure 1 shows coordinate function plots for the six main variables of the ASPCR model. The influence of a predictor on a response is here measured by the range of the transformed data marked by their corresponding numbers: Germany (1 D), Austria (2 A), Belgium (3 B), Canada (4 CND), Denmark (5 DK), Spain (6 E), USA (7 USA), Finland (8 FI), France (9 F), Greece (10 G), Ireland (11 IRL), Italia (12 I), Japan (13 J), Norway (14 N), the Netherlands (15 NL), Portugal (16 P), England (17 UK), Sweden (18 S). As a confirmatory point, note that EDUC is modeled in the same fashion when all ASPCs are used (same influential predictors with similar function shapes).

5. Additive splines for Partial Least Squares regression

This method, whose multivariate linear version is due to Wold et al. (1983), has been extended to additive spline multiresponse regression by Durand and Sabatier (1994). It mainly differs from ASPCR in that components or latent variables are constructed *at the same time* as the regression is computed, and not before, thus hopefully providing components with better explanatory potential. The example of O.E.C.D. data will actually

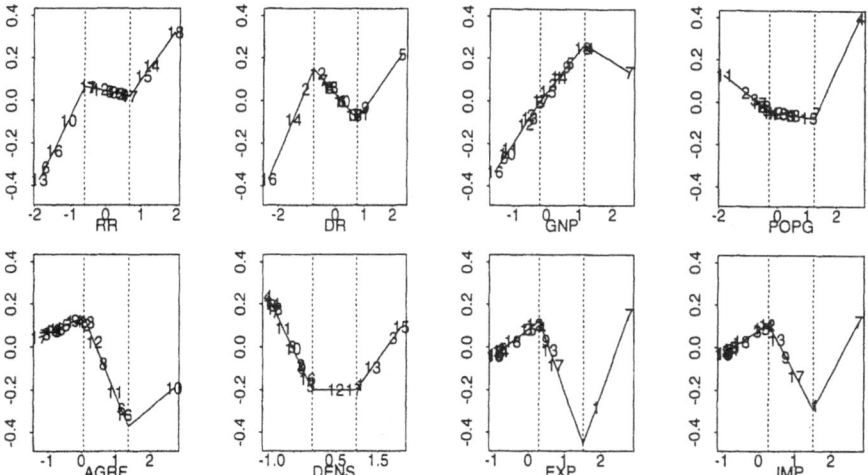

Figure 1: *Coordinate function plots of the main predictors (in decreasing order from left to right) for EDUC modeled by ASPCR. The dotted vertical lines indicate the position of the knots.*

illustrate that a smaller number of components are needed in ASPLS (Additive Spline Partial Least Squares) than in ASPCR for explaining the same amount of Y-variance.

5.1 Mathematical background

DEFINITION. Using the standard notation of linear PLS, see for example (Frank and Friedman 1993), the kth ASPLS components are the linear composites $t = X(a)w$ and $u = F_{k-1}c$, with $F_0 = Y$, which maximize the objective function $f(a^1, \ldots, a^p, w, c) = w'X(a)'DF_{k-1}c = cov(t, u)$ subject to

$$
\begin{aligned}
\|w\|^2 = \|c\|^2 &= 1, \\
var(X^i(a^i)) &= var(X^i), \quad i = 1, \ldots, p, \\
cov(t, t^j) &= 0, \quad j = 1, \ldots, k-1.
\end{aligned}
$$

In the same way as ASPCs are defined, the last constraint is omitted when $k = 1$. Writing $(a^{1,k}, \ldots, a^{p,k}, w^k, c^k)$ as an optimal argument of f and $X(a^{(k)}) = [B^1 a^{1,k} | \ldots | B^p a^{p,k}]$ as an optimal matrix, the kth ASPLS components become $t^k = X(a^{(k)})w^k$ and $u^k = F_{k-1}c^k$.

The final part of step k of ASPLS consists of updating F_k as the residual regression of F_{k-1} onto t^k,

$$F_k = F_{k-1} - P_{t^k}F_{k-1} . \tag{5}$$

It must be noted that fixing spline coefficients equal to nodal coefficients in the optimization problem above, does not lead to linear PLS components except for t^1 and u^1. Moreover, the linear PLS property of reconstructing the predictor matrix is obviously not preserved in ASPLS whose aim is only to provide an additive approximation to the response variables. As a consequence of (5), the ASPLS model is given by

$$Y = F_0 = \sum_{k=1}^{M} \widehat{Y}_k + F_A = \widehat{Y} + F_A , \tag{6}$$

where $\widehat{Y}_k = P_{t^k}F_{k-1}$ is the kth partial model matrix of rank 1. Model dimension M can be determined by cross-validation. More pragmatically, the fact that the t^ks are mutually uncorrelated implies the additive decomposition of the total Y-variance

$$trace(Y'DY) = \|Y\|_D^2 = \sum_{k=1}^{M} \|\widehat{Y}_k\|_D^2 + \|F_A\|_D^2 . \tag{7}$$

Table 5. R^2 of the responses for nested ASPLS models

Mod. dim.	CAL	LODG	ELEC	EDUC	TV	% of total Y-variance
1	0.054	0.101	0.370	0.463	0.874	37.26
2	0.534	0.547	0.398	0.579	0.881	58.82
3	0.722	0.576	0.423	0.820	0.889	68.64 `
4	0.814	0.648	0.692	0.820	0.902	77.57

To determine the ASPLS model dimension, one can easily measure the part of each component in the reconstruction of the total response variance. It can be shown (Durand and Sabatier 1994) that (6) enters the additive framework (4). No linear smoother matrix can

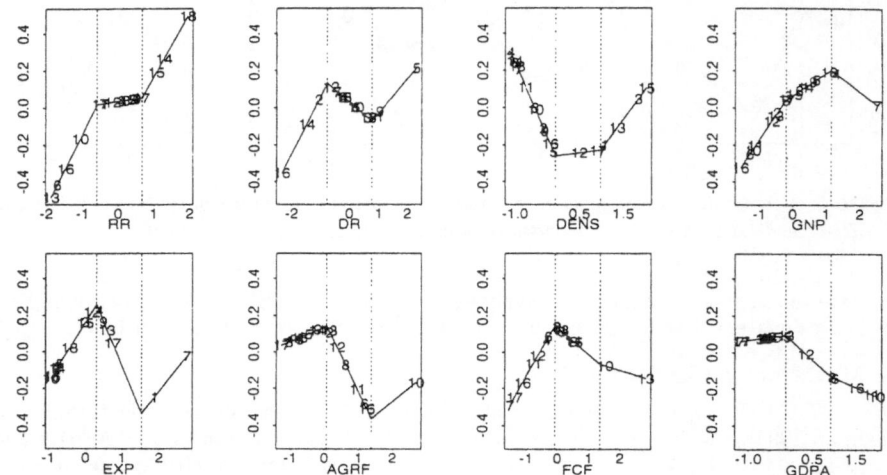

Figure 2: *Influential coordinate function plots (in decreasing order from left to right) for EDUC modeled by ASPLS. The dotted vertical lines indicate the position of the knots.*

be associated with model (6) since Y does not enter linearly in the expression of $\widehat{Y}$. One can also consult the latter paper for computational aspects of ASPLS which are based on normal equations just like the ASPC's algorithm of Section 4.2.

5.2 ASPLS applied on O.E.C.D. data

Going back to the example, Table 5 shows R^2 response values for up to 4 model dimensions. Four components provide a better fit than the six selected ASPCR components. Moreover,

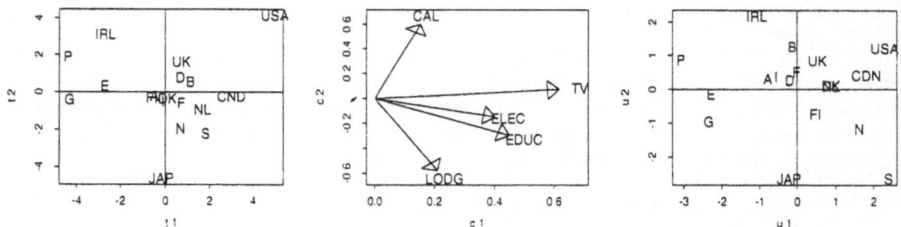

Figure 3: *The (t_1, t_2) display is similar to a first principal component scatterplot used to explain the responses summarized by (c_1, c_2) and (u_1, u_2) plots.*

response EDUC, whose eight coordinate functions of main influence are presented in Figure 2, is similarly modeled by ASPLS than by ASPCR: predictors RR, DR, DENS, GNP, EXP and AGRF mainly participate in the additive fit. Because responses are projected on components, a particular aspect of ASPLS reducing properties is the possibility of "naming" predictor components by their capability in explaining the responses. In Figure 3, Axis 1 summarizes the opposition between strong and weak consumer countries, while axis 2 contrasts countries according to responses LODG and CAL.

6. Self-ASPLS components for exploratory analysis and additive modeling

An appealing idea is that of using ASPLS components to approximate the X data matrix itself. When predictors are taken for responses, $\widehat{X}$ computed by (6) can be seen as an additive spline summary of rank M for X. This approximation cannot be better than that provided by the optimal Eckart-Young theorem, here the objective function to be maximized is $cov(t, u)$. Components $t_1, \ldots, t_A$, referred to as the self-ASPLS components, may be used as an exploratory tool. In fact, self-ASPLS components are close to linear principal components and Figure 4 illustrates the comparison between self-ASPLS and common principal components both computed from the predictors of the O.E.C.D. data.

We observe great similarities in component plots. Note however that six self-ASPLS com-

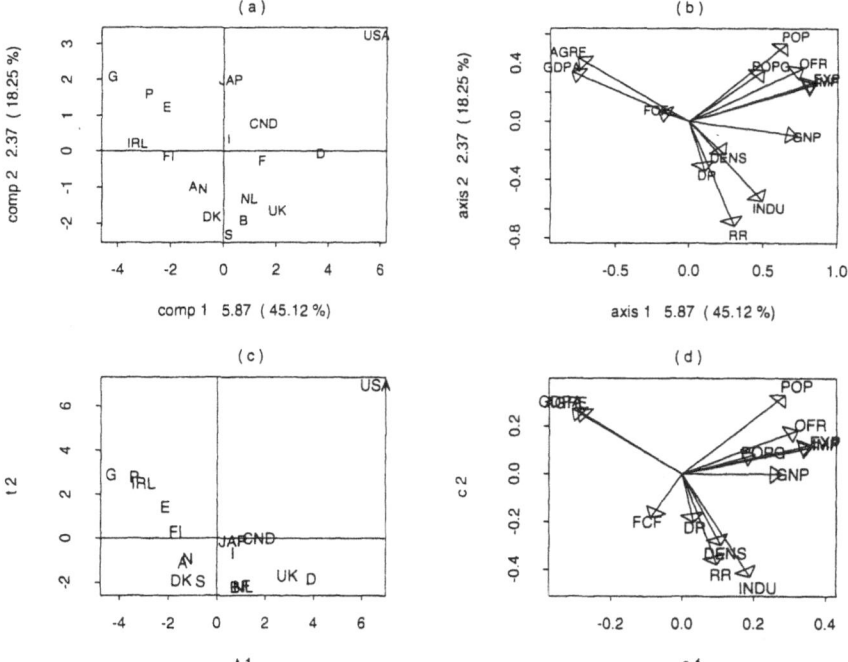

Figure 4: *Comparison between common (a),(b) and self-ASPLS (c),(d) principal component plots.*

ponents explain respectively 43.98, 14.94, 12.08, 7, 5.26 and 4.11 percents (total 87.37%) of the total variance against 45.12, 18.25, 11.91, 8.49, 7.52 and 3.80 percents (total 95.09%) for common principal components.

Linear regression on self-ASPLS components provides an additive model in the same way ASPCR does in Section 4.3. Table 6 shows R^2 values for the O.E.C.D. responses regressed on six explanatory self-ASPLS components. Figure 5 displays coordinate function plots of the influential predictors on the "current" response EDUC. For all the models and methods studied, we observe a great stability in the prediction since variables DR, RR, GNP, DENS and AGRF all chiefly intervene with similar transformations in the model.

Table 6. R^2 of the responses for six nested self-ASPLS component regression models.

Mod. dim.	CAL	LODG	ELEC	EDUC	TV	% of total Y-variance
1	0.047	0.018	0.106	0.072	0.671	18.28
2	0.052	0.084	0.107	0.245	0.676	23.28
3	0.338	0.091	0.327	0.465	0.821	40.84
4	0.427	0.376	0.466	0.700	0.861	56.60
5	0.466	0.545	0.475	0.701	0.864	61.02
6	0.570	0.606	0.605	0.757	0.875	68.26

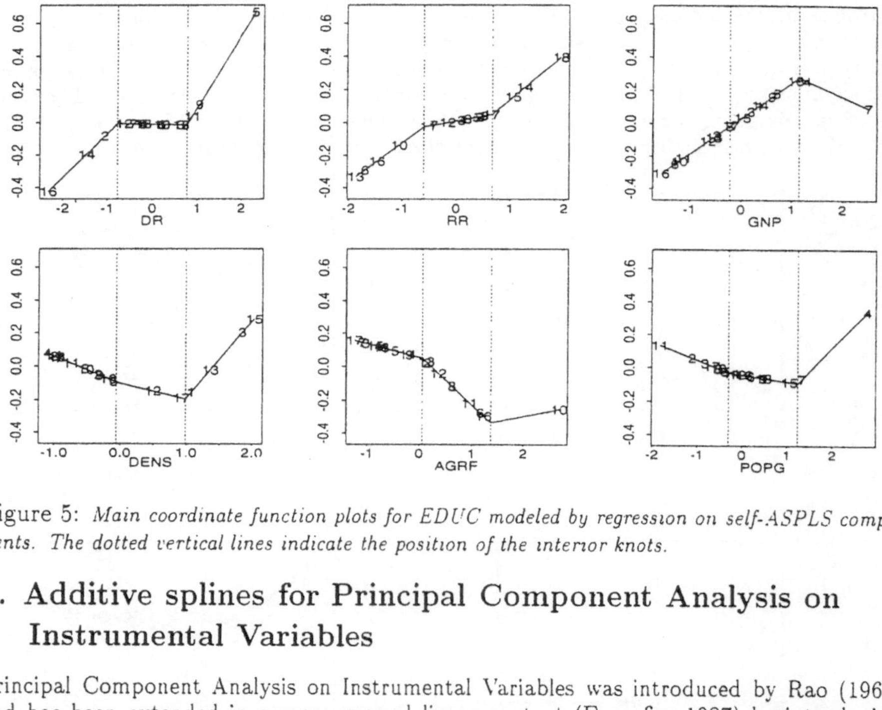

Figure 5: *Main coordinate function plots for EDUC modeled by regression on self-ASPLS components. The dotted vertical lines indicate the position of the interior knots.*

7. Additive splines for Principal Component Analysis on Instrumental Variables

Principal Component Analysis on Instrumental Variables was introduced by Rao (1964) and has been extended in a more general linear context (Escoufier 1987) by introducing adapted metrics in principal component analysis. A nonlinear additive spline extension to this approach, referred to as ASPCAIV, can be found in (Durand 1993). This method differs from those of Sections 4, 5 and 6 in that dimension reduction (here, linear PCA) occurs *after* a multivariate linear smoothing procedure has been applied. The associated smoother matrix is here $P_{X(a)}$ which is to be compared with that of the least-squares regression spline model of Section 3. Although projections are processed on spaces of smaller dimensions, ASPCAIV is sensitive to scarcity of data and the O.E.C.D. example does not provide sufficient number of samples to interpret the model. To illustrate the performance of the method, two examples are shown in Section 6.2. The first one presents a simple regression based on 200 observations. In the second, two predictors are related to three responses, all measured on 250 items.

7.1 Mathematical background

DEFINITION. Let Q be a $q \times q$, positive definite matrix that defines a metric for computing Euclidian distances between objects in the response space $\mathbb{R}^q$. The aim of the

ASPCAIV method is first to find a $p \times p$ metric $\overline{R}$ and a vector of spline coefficients $\overline{a}$ that minimize the objective function $f(R, a) = trace\,[YQY'D - X(a)RX(a)'D]^2$. Then, dimension reduction is done (if needed), by solving the eigenanalysis of $X(\overline{a})\overline{R}X(\overline{a})'D$.

The objective function can be interpreted as a measure of discrepancy between eigenmatrices for components associated with the predictor and response sets of variables. For fixed a, an optimal metric is explicitly given by

$$R = [X(a)'DX(a)]^+ [X(a)'DY] Q [Y'DX(a)] [X(a)'DX(a)]^+ , \tag{8}$$

but the reverse is false: for fixed R, no explicit optimal a is available. An iterative algorithm is needed which alternates a step of computing the best R for fixed a, with a step of gradient descent with respect to a. Due to (8), when a local optimum is obtained, solving the eigenproblem for $X(\overline{a})\overline{R}X(\overline{a})'D$ leads to the same solutions as for $\widehat{Y}Q\widehat{Y}'D$, where

$$\widehat{Y} = P_{X(\overline{a})}Y. \tag{9}$$

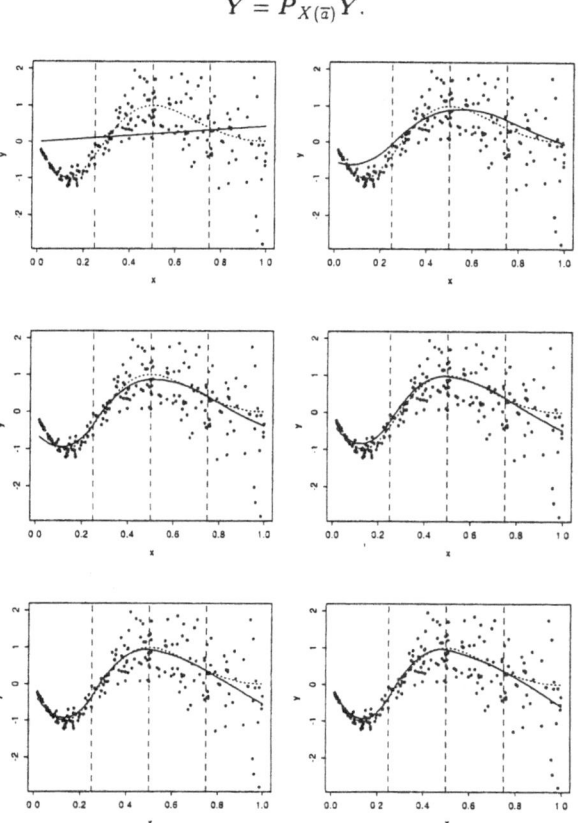

Figure 6: *Evolution of the smoother along with the 6 first steps of the method applied on the example based on 200 (x_i, y_i) observations (dots), $y_i = sin(2\pi(1 - x_i)^2) + x_i\epsilon_i$, with x_i uniform on $[0, 1]$ and ϵ_i standard normal. The signal (dashed), the spline smooth (solid), degree 2 with 3 knots.*

As a consequence, ASPCAIV is a multivariate linear smoothing procedure associated with the smoother projection matrix $P_{X(a)}$. Figure 6 presents the evolution of the scatterplot smoother along with the first steps of the method applied to the example from the S-PLUS (1991) user's manual in chapter 18 (here $Q = [1]$, and splines are of degree 2 with 3 equally spaced knots). Note that the initializing choice of nodal coefficients consists of computing the common linear regression at the first step of the method.

7.2 Application to additive spline discriminant analysis

By (9), ASPCAIV provides a competing model for classical additive models (Hastie and Tibshirani 1990). However, to make use of ASPCAIV's abilities, the metric Q can be different from the identity and specifically chosen according to specific multivariate regressions. For example, it can be shown that taking $Q = (Y'DY)^{-1}$ leads to considering the method as the best canonical correlation analysis between Y and any smoothed predictor matrix defined by (2). More precisely, the optimization problem becomes

$$trace \left(P_Y P_{X(\overline{a})} \right)^2 \geq trace \left(P_Y P_{X(a)} \right)^2, \quad \text{for arbitrary } a.$$

By taking the indicator matrix of classes for the Y sample matrix, a direct application of this result concerns nonlinear Discriminant Analysis (Durand 1992, 1993, Hastie et al 1994). Note that, here, linear Discriminant Analysis is obtained at the first step of the algorithm (because nodal spline coefficients are used) and that discriminating by additive spline variables can only improve linear results. Discriminant variables are deduced from principal components by using a scaling factor (Escoufier 1987). Denoting $G = (Y'DY)^{-1}Y'DX(\overline{a})$, the matrix whose rows G_i are the centroids of the classes defined by $X(\overline{a})$ and Y, the classification rule is: The object x to be classified, is transformed in t (considered as a row vector) by using optimal B-spline functions, and then affected to class j if

$$(t - G_j) \left[X(\overline{a})'DX(\overline{a}) \right]^+ (t - G_j)' = \min_{i=1,...,q} (t - G_i) \left[X(\overline{a})'DX(\overline{a}) \right]^+ (t - G_i)'. \quad (10)$$

As in linear Discriminant Analysis, other metrics can be chosen for the geometrical affectation rule (10). However, in order that spline transformations could make sense, the user has to verify that each observation lies within the range of the corresponding variable in the training sample.

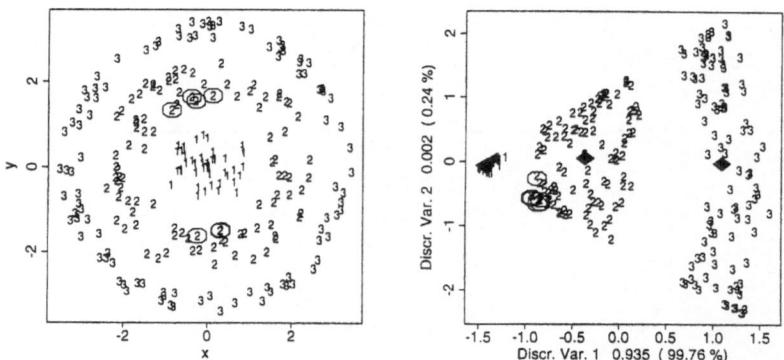

Figure 7: *Misclassified items are marked with circles and centroids with black squares.*

The application illustrating the performance of the method is based on three class, two dimensional data (x, y). Simulated data are generated as follows: the three class distributions are uniform on the annuls centered at the origin with respective extreme radii $(0, 1)$, $(1.5, 2.5)$ and $(3, 3.5)$. Number of items are 50 for the first group and 100 for the others so that the training sample matrix X is 250×2 and Y, the indicator matrix of classes, 250×3. Usual linear and quadratic discriminant methods perform poorly on this data set while discrimination by additive spline discriminant variables provides good results: Figure 7 presents seven misclassified items, marked with circles, in both (x, y) and two first discriminant variables. Only one discriminant variable is needed for separating the classes and the close to one first eigenvalue yield well separated groups.

8. Conclusion

In this paper we have compared and illustrated various additive multiresponse regression methods from the point of view of prediction as well as interpretation. It is well known that in case of extreme collinearity in predictors, interpreting linear regression coefficients is dangerous. ASPCAIV is an additive modeling method in which dimension reducing aspects occur *after* the regression is computed. Therefore, its scope of application is that of predictor data sets with much more observations than variables. The possibility of choosing different metrics leads us to view ASPCAIV as a unifying framework for additive extensions to some two data blocks linear methods.

In some chemometrics, econometrics or sociometrics applications, the number of predictors exceeds or equals approximately the number of observations and nonlinear methods presenting dimension reduction stages *before* or *at the same time* prediction is processed, are to be preferred. The ASPLS method, the regression on ASPCs as well as on self-ASPLS components, all construct a set of uncorrelated components which are additive functions of the predictors. In ASPLS, linear regression on such components occurs as soon as they are constructed, thus generally providing a more parsimonious additive model in the sense that less additive components are needed for explaining the same amount of the response variance. A great stability has been found in interpreting the different additive models on the O.E.C.D. data: the choice of a set of influential predictors is in a large way independent of the method and coordinate spline function shapes are similar.

References

Bertier, P. and Bouroche, J.-M. (1975), *Analyse des données multidimensionnelles*, Paris: PUF.

De Boor, C. (1978), *A practical guide to splines*, New York: Springer.

Donnell, D. J. et al. (1994), Analysis of additive dependencies and concurvities using smallest additive principal components (with discussion), *The Annals of Statistics*, 4, 1635-1673.

Durand, J. F. (1992), Additive spline discriminant analysis, in *Computationnal Statistics, Vol.1*, (Y. Dodge and J. Whittaker, eds.), Physica-Verlag, 144-149.

Durand, J. F. (1993), Generalized principal component analysis with respect to instrumental variables via univariate spline transformations, *Computational Statistics & Data Analysis*, 16, 423-440.

Durand, J. F. and Sabatier, R. (1994), Additive splines for PLS regression, *Tech. Rept. 94-05*, Unité de Biométrie, ENSAM-INRA-UM II, Montpellier, France. In press in *Journal of the American Statistical Association*.

Escoufier, Y. (1987), Principal components analysis with respect to instrumental variables, *European Courses in Advanced Statistics*, University of Napoli, 285-299.

Eubank, R. L. (1988), *Spline smoothing and nonparametric regression*, New York and Basel: Dekker.

Frank, I. E., and Friedman, J. H. (1993), A statistical view of some Chemometrics regression tools (with discussion), *Technometrics*, 35, 109-148.

Gifi, A. (1990), *Nonlinear multivariate analysis*, Chichester: Wiley.

Hastie, T. and Tibshirani, R. (1990), *Generalized additive models*. London: Chapman and Hall.

Hastie, T. et al. (1994), Flexible discriminant analysis by optimal scoring, *Journal of American Statistical Association*, 89, 1255-1270.

Ramsay, J. O. (1988), Monotone regression splines in action (with discussion), *Statistical Science*, 3, 425-461.

Rao, C. R. (1964), The use and the interpretation of principal component analysis in applied research, *Sankhya A*, 26, 329-356.

Wold, S. et al. (1983), The multivariate calibration problem in chemistry solved by the PLS method, *Proc. Conf. Matrix Pencils*. Ruhe, A. and Kagstrom, B. (Eds), Lecture notes in mathematics, Heidelberg: Springer Verlag, 286-293.

Bounded Algebraic Curve Fitting for Multidimensional Data Using the Least-Squares Distance

Masahiro Mizuta[1]

[1] Division of Systems and Information Engineering, Hokkaido University
N.13, W.8, Kita-ku, Sapporo-shi
Hokkaido 060, Japan

Summary: Linear regression or smoothing techniques are not adequate for curve fitting, in cases in which neither variable can be designated as the response. We present a new method for fitting bounded algebraic curve to multidimensional data using the least-squares distance between data points and the curve. Numerical examples of the proposed method are also shown.

1. Introduction

In data analysis process, we can sometimes investigate data structures with methods of curve fitting. Curves can be represented by explicit functions, parametric functions, or implicit functions. Curves represented by explicit functions reveal influences of other variables on one variable, like regression analysis. Curves by parametric functions give latent order of data and are depicted easily with computer graphics. Curves by implicit functions, i.e., *Algebraic curves* show relations with variables.

Many researchers proposed curve fitting methods with algebraic curves. Particularly, it is worthwhile to notice that Keren et al.(1994), and Taubin et al.(1994) independently developed the algorithms for *bounded* algebraic curve. However, most studies are based on *approximate distances* between data points and the algebraic curve. We have developed a method to find algebraic curve that minimizes the sum of squares *exact distances*. In this article, we propose a method to find *bounded* algebraic curves based on exact distances.

2. Algebraic curve fitting and bounded algebraic curve

A p-dimensional algebraic curve is the set of zeros of k-polynomials $\boldsymbol{f}(\boldsymbol{x}) = (f_1(\boldsymbol{x}), \cdots, f_k(\boldsymbol{x}))$ on $\boldsymbol{R}^p$,

$$Z(\boldsymbol{f}) \;=\; \{\boldsymbol{x} : \boldsymbol{f}(\boldsymbol{x}) = \boldsymbol{0}\}.$$

We restrict ourselves $p = 2$ and $k = 1$ without a loss of generality hereafter:

$$Z(f) \;=\; \{(x,y) : f(x,y) = 0\}. \tag{1}$$

2.1 Approximate distance and exact distance

The distance from a point $\boldsymbol{a} = (\alpha, \beta)$ to the curve $Z(f)$ is usually defined by

$$
\begin{aligned}
\mathrm{dist}(\boldsymbol{a}, Z(f)) \;&=\; \inf(\| \, \boldsymbol{a} - \boldsymbol{y} \, \|: \boldsymbol{y} \in Z(f)) \\
&=\; \inf_{x,y} \{ \sqrt{(x-\alpha)^2 + (y-\beta)^2} \; ; f(x,y) = 0 \}.
\end{aligned}
\tag{2}
$$

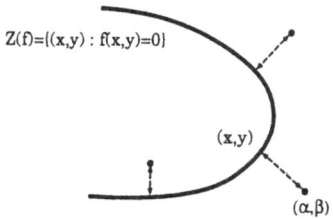

Fig. 1: Distance between curve $Z(f)$ and point (α, β).

It was said that the distance between a point and the algebraic curve cannot be computed by direct methods. So, Taubin proposed *an approximate distance* from $\boldsymbol{a}$ to $Z(f)$ (Taubin (1991)). The point $\hat{\boldsymbol{y}}$ that approximately minimizes the distance $\|\boldsymbol{y} - \boldsymbol{a}\|$, is given by

$$
\begin{aligned}
\hat{\boldsymbol{y}} &= \boldsymbol{a} - (\nabla f(\boldsymbol{a})^T)^+ f(\boldsymbol{a}) \\
&= (\alpha, \beta) - \frac{1}{\|\nabla f(\boldsymbol{a})\|^2} \left(\frac{\partial f}{\partial x}, \frac{\partial f}{\partial y}\right)|_{(\alpha,\beta)} f(\boldsymbol{a}),
\end{aligned} \tag{3}
$$

where $(\nabla f(\boldsymbol{a})^T)^+$ is the pseudoinverse of $\nabla f(\boldsymbol{a})^T$. The distance from $\boldsymbol{a}$ to $Z(f)$ is approximated by

$$
\text{dist}(\boldsymbol{a}, Z(f))^2 \approx \frac{f(\boldsymbol{a})^2}{\|\nabla f(\boldsymbol{a})\|^2}. \tag{4}
$$

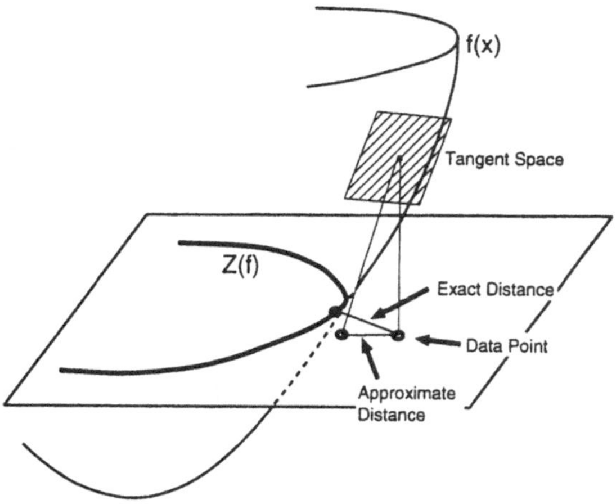

Fig. 2: Approximate distance and distance.

Kriegman and Ponce(1990) have shown that elimination theory can be used to construct a closed-form expression for the exact distance, but the amount of computation is impractical.

We have developed a method to calculate the exact distance with a technique of constrained optimization (Mizuta (1995)). Let $\boldsymbol{x} = (x, y)$ be the nearest point on $Z(f)$ to a point $\boldsymbol{a} = (\alpha, \beta)$. The point $\boldsymbol{x} = (x, y)$ can be calculated with an augmented Lagrangian function (Hestenes (1969)):

$$
Q(\boldsymbol{x}, \mu, r) = (x - \alpha)^2 + (y - \beta)^2 + \mu f(x, y) + \frac{1}{2} r f(x, y)^2
$$

where μ is a Lagrangian parameter and r is a penalty parameter. The algorithm proceeds with the following steps:

Initialization: Set $\mu_0 = 1, r_0 = 1, k = 0$.
 Set x_0 the initial points.
Step 1 Minimize $Q(x, \mu_k, r_k)$ with respect to x.
 Put them x_k.
Step 2 Stop if the value of $f(x, y)^2$ is below some threshold.
Step 3 $r_{k+1} = cr_k$, where c is a constant, greater than 1.
Step 4 $\mu_{k+1} = \mu_k + r_k f(x, y)$
Step 5 $k = k + 1$, and return to **Step 1**.

2.2 Methods for curve fitting

Taubin (1991) presented the algorithm to find the algebraic curve such that the sum of *approximate* squares distances between data points and the curve is minimum.

We have already developed the algorithm to find the algebraic curve with exact squares distance (Mizuta (1995)). Although algebraic curves can fit the data very well, they usually contain points far remote from the given data set. In 1994, Keren et al.(1994) and Taubin at al.(1994) independently developed the algorithms for *bounded* (closed) algebraic curve with approximate squares distance. We will introduce the definition and properties of bounded algebraic curve in the next subsection.

2.3 Bounded algebraic curve

We call $Z(f)$ is *bounded* iff there exists a constant r such that $Z(f) \subset \{x : \| x \| < r\}$. For example, it is clear that $Z(x^2 + y^2 - 1)$ is bounded, but $Z(x^2 - y^2)$ is not bounded.

Keren et al.(1994) defined $Z(f)$ to be *stably bounded* if a small perturbation of the coefficients of the polynomial leaves its zero set bounded. An algebraic curve $Z((x-y)^4 + x^2 + y^2 - 1)$ is bounded, but not stably bounded because $Z((x-y)^4 + x^2 + y^2 - 1 + \varepsilon x^3)$ is not bounded for any $\varepsilon \neq 0$.

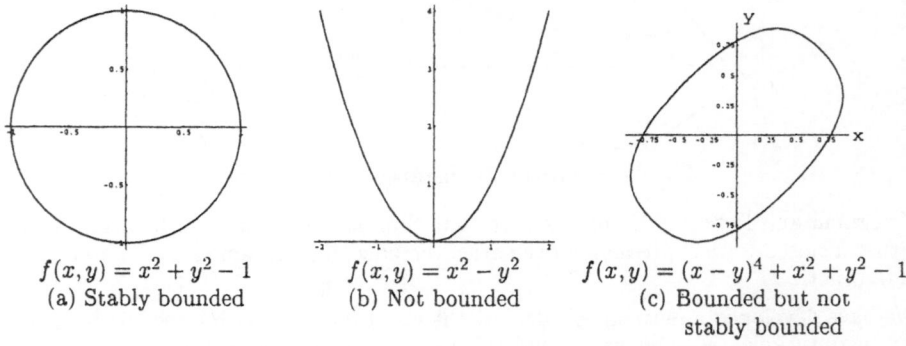

$f(x, y) = x^2 + y^2 - 1$
(a) Stably bounded

$f(x, y) = x^2 - y^2$
(b) Not bounded

$f(x, y) = (x - y)^4 + x^2 + y^2 - 1$
(c) Bounded but not stably bounded

Fig. 3: Examples on bounded and stably bounded

Let $f_k(x, y)$ be the form of degree k of a polynomial $f(x, y)$: $f(x, y) = \sum_{k=0}^{d} f_k(x, y)$.

The leading form of a polynomial $f(x, y)$ of degree d is defined by $f_d(x, y)$. For example, the leading form of $f(x, y) = x^2 + 2xy - y^2 + 5x - y + 3$ is $f_2(x, y) =$

$x^2 + 2xy - y^2$.

Lemma: For an even positive integer d, any leading form $f_d(x,y)$ can be represented by $\boldsymbol{X}A\boldsymbol{X}^T$. Where A is a symmetric matrix and $\boldsymbol{X} = (x^{\frac{d}{2}}, x^{\frac{d}{2}-1}y, \cdots, xy^{\frac{d}{2}-1}, y^{\frac{d}{2}})$.

Remark: Symmetric matrix A is not unique. For example,

$$x^4 - x^2y^2 + y^4 = (x^2, xy, y^2)\begin{pmatrix} 1 & 0 & -\frac{1}{3} \\ 0 & -\frac{1}{3} & 0 \\ -\frac{1}{3} & 0 & 1 \end{pmatrix}(x^2, xy, y^2)^T$$

$$= (x^2, xy, y^2)\begin{pmatrix} 1 & 0 & -\frac{3}{4} \\ 0 & \frac{1}{2} & 0 \\ -\frac{3}{4} & 0 & 1 \end{pmatrix}(x^2, xy, y^2)^T$$

Theorem (Keren et al. (1994)): The $Z(f)$ is stably bounded iff d is even and there exists a symmetric positive definite matrix A such that

$$f_d(x,y) = (x^{\frac{d}{2}}, x^{\frac{d}{2}-1}y, \cdots, xy^{\frac{d}{2}-1}, y^{\frac{d}{2}})A(x^{\frac{d}{2}}, x^{\frac{d}{2}-1}y, \cdots, xy^{\frac{d}{2}-1}, y^{\frac{d}{2}})^T.$$

3. Bounded algebraic curve fitting based on exact distance

We must show two steps to find the bounded algebraic curve fitted to a given data. In the first step, we use the method to calculate the distance from a given point to a given algebraic curve already described in the section 2. The next step is to find the bounded curve that minimizes the sum of the squares distances from the data.

We parametrize a set of polynomials that induce (stably) bounded algebraic curve. In general, a polynomial f of degree p with q parameters can be denoted by $f(a_1, \cdots, a_q, x, y)$, where $a_1, \cdots, a_q$ are parameters of the polynomial.

For example, *all of the polynomials* of degree 2 can be represented by

$$f(a_1, a_2, \cdots, a_6, x, y) = A^T X,$$

where $X = (1, x, y, x^2, xy, y^2)^T, A = (a_1, a_2, ..., a_6)^T$.

For *stably bounded algebraic curves* of degree 4,

$$f(a_1, \cdots, a_{16}, x, y) = (x^2, xy, y^2)B^2(x^2, xy, y^2)^T + (a_7, \cdots, a_{16})(1, x, y, \cdots, y^3)^T$$

where,

$$B = \begin{pmatrix} a_1 & a_2 & a_3 \\ a_2 & a_4 & a_5 \\ a_3 & a_5 & a_6 \end{pmatrix}.$$

Let $(\alpha, \beta)(i = 1, 2, \cdots, n)$ be n data points in the plane. The point in $Z(f)$ that minimizes the distance from (α_i, β_i) is denoted by $(x_i, y_i)(i = 1, 2, \cdots, n)$. The sum of squares of distances is

$$R = \sum_{i=1}^{n} R_i,$$

where $R_i = (x_i - \alpha_i)^2 + (y_i - \beta_i)^2$.

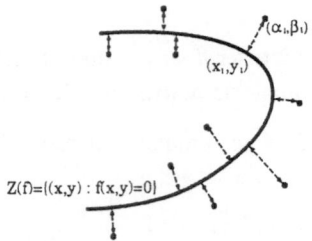

Fig. 4: The sum of squares of distances.

We can minimize R with respect to the parameters of polynomial f with *Levenberg-Marquardt Method*. The method requires the partial derivatives of R with respect to a_j:

$$\frac{\partial R_i}{\partial a_j} = 2\left((x_i - \alpha_i)\frac{\partial x_i}{\partial a_j} + (y_i - \beta_i)\frac{\partial y_i}{\partial a_j}\right). \tag{5}$$

The only thing left to discuss is a solution for $\frac{\partial x_i}{\partial a_j}$ and $\frac{\partial y_i}{\partial a_j}$. Hereinafter, the subscript i is omitted.

By the derivative of the both sides of
$f(a_1, \cdots, a_q, x, y) = 0$ with respect to a_j $(j = 1, \cdots, q)$, we obtain

$$\frac{\partial f}{\partial x}\frac{\partial x}{\partial a_j} + \frac{\partial f}{\partial y}\frac{\partial y}{\partial a_j} + \frac{df}{da_j} = 0, \tag{6}$$

where $\frac{df}{da_j}$ is the differential of f with a_j when x and y are fixed.

Because (α_i, β_i) is on the normal line from (x_i, y_i), then $(y - \beta)\frac{\partial f}{\partial x} - (x - \alpha)\frac{\partial f}{\partial y} = 0$. By the derivative of the both sides with respect to a_j, we obtain

$$\frac{\partial f}{\partial x}\frac{\partial y}{\partial a_j} + (y - \beta)\left(\frac{\partial^2 f}{\partial x^2}\frac{\partial x}{\partial a_j} + \frac{\partial^2 f}{\partial x \partial y}\frac{\partial y}{\partial a_j} + \frac{\partial^2 x}{\partial a_j \partial x}\right)$$
$$-\frac{\partial f}{\partial y}\frac{\partial x}{\partial a_j} - (x - \alpha)\left(\left(\frac{\partial^2 f}{\partial x \partial y}\frac{\partial x}{\partial a_j} + \frac{\partial^2 f}{\partial y^2}\frac{\partial y}{\partial a_j} + \frac{\partial^2 f}{\partial a_j \partial y}\right)\right) = 0. \tag{7}$$

The equations (6) and (7) are simultaneous linear equations in two variables $\frac{\partial x}{\partial a_j}$ and $\frac{\partial y}{\partial a_j}$. So, we can get $\frac{\partial x_i}{\partial a_j}$ and $\frac{\partial y_i}{\partial a_j}$.

4. Numerical examples

Two examples are provided of bounded versus unbounded curve fitting.

The first example data is two dimensional artificial data of size 40 that lies in the neighborhood of an asteroid. We set the result of GPCA (*Generalized Principal Components Analysis*: Gnanadesikan (1977), Mizuta (1983)) for an initial curve, and search for fitting an algebraic curve and a bounded algebraic curve of degree 4 with the proposed method (Fig. 5). The sum of squares of distances R is 0.088 in the case of bounded fitting. This value is greater than the value in the case of bounded fitting ($R = 0.026$). But the figure 5 shows that the bounded algebraic curve reveals a suitable outline of the data points.

The second is three-dimensional data of size 210. The 210 points almost lie on a closed cylinder (Fig. 6 (a)). We also apply the method to the data with an algebraic curve and a bounded algebraic curve of degree 4 (Fig. 6 (b), (c)). The value of R is 1.239 in the case of bounded fitting and the value of R is 0.924 in the case of unbounded fitting. The unbounded algebraic surface reproduces the structure of the closed cylinder and the bounded surface shows a global shape of the data points.

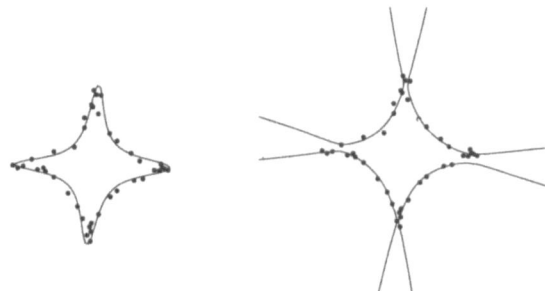

Fig. 5: Asteroid data: bounded (left) and unbounded (right)

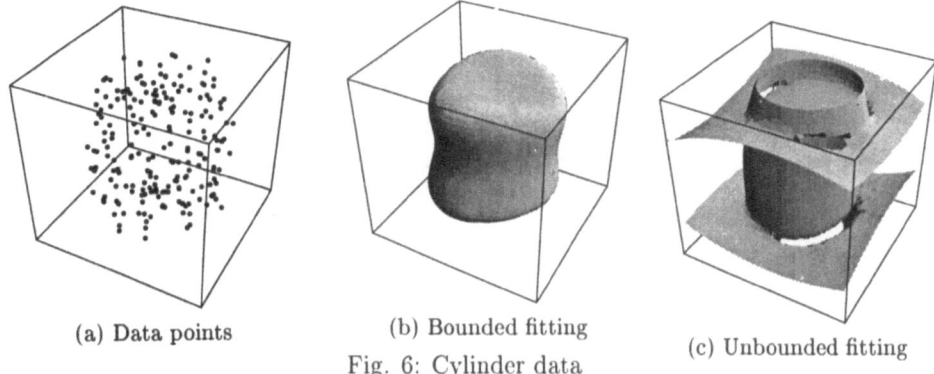

(a) Data points   (b) Bounded fitting   (c) Unbounded fitting

Fig. 6: Cylinder data

5. Concluding remarks

In the article, we do not mention the curve fitting in the 3-dimensional space:

$$Z(\boldsymbol{f}) = \{(x, y, z) : f_1(x, y, z) = 0 \text{ and } f_2(x, y, z) = 0\}.$$

However, extending the proposed method is not difficult.

Taubin (1994) proposed the approximate distance of order k and presented algorithms for rastering algebraic curves. The proposed algorithm for exact distance can also be used for rastering algebraic curves.

References:

Gnanadesikan, R. (1977). *Methods for Statistical Data Analysis of Multivariate Observations*. John Wiley & Sons.

Hestenes, M.R. (1969). Survey paper: Multiplier and gradient methods, *J. of Optimization Theory and Applications*, **4**, 303-320.

Keren, D., Cooper, D. and Subrahmonia, J. (1994). Describing complicated objects by implicit polynomials, *IEEE trans. Patt. Anal. Machine Intell.*, **16**, 1, 38–53.

Kriegman, D. J. and Ponce, J. (1990). On recognizing and positioning curved 3-D objects from image contours, *IEEE Trans. Patt. Anal. Machine Intell.*, **12**, 12, 1127–1137.

Mizuta, M. (1983). Generalized principal components analysis invariant under rotations of a coordinate system, *J.Japan Statist. Soc.*, **14**, 1–9.

Mizuta, M. (1995). A derivation of the algebraic curve for two-dimensional data using the least-squares distance, *In Data Science and Its Application*, Hayashi, C. et al. (eds.), 167–176, Academic Press, Tokyo.

Taubin, G.(1991). Estimation of planar curves, surfaces, and nonplanar space curves defined by implicit equations with applications to edge and range image segmentation, *IEEE Trans. Patt. Anal. Machine Intell.*, **13**, 11, 1115–1138.

Taubin, G.(1994). Distance Approximations for Rastering Implicit Curves, *ACM Trans. on Graphics*, **13**, 1, 3–42.

Taubin, G., Cukierman, F., Sullivan, S., Ponce, J. and Kriegman, D. J. (1994). Parameterized families of polynomials for bounded algebraic curve and surface fitting. *IEEE trans. Patt. Anal. Machine Intell.*, **16**, 3, 287–303.

Using the Wavelet Transform for Multivariate Data Analysis and Time Series Analysis

Fionn Murtagh[1], Alexandre Aussem[2]

[1] Faculty of Informatics, University of Ulster
Magee College, Londonderry BT48 7JL, Nth. Ireland
Email fd.murtagh@ulst.ac.uk Web www.infm.ulst.ac.uk/~fionn

[2] Université Blaise Pascal, Clermont-Ferrand II
ISIMA, Campus des Cezeaux, BP 125 63173 Aubière Cedex, France
Email: alex@sp.isima.fr

Summary: We discuss the use of orthogonal wavelet transforms in multivariate data analysis methods such as clustering and dimensionality reduction. Wavelet transforms allow us to introduce multiresolution approximation, and multiscale nonparametric regression or smoothing, in a natural and integrated way into the data analysis. Applications illustrate the powerfulness of this new perspective on data analysis.

1. Introduction

Data analysis, for exploratory purposes, or prediction, is usually preceded by various data transformations and recoding. In fact, we would hazard a guess that 90% of the work involved in analyzing data lies in this initial stage of data preprocessing. This includes: problem demarcation and data capture; selecting non-missing data of fairly homogeneous quality; data coding; and a range of preliminary data transformations.

The wavelet transform offers a particularly appealing data transformation, as a preliminary to data analysis. It offers additionally the possibility of close integration into the analysis procedure as will be seen in this article. The wavelet transform may be used to "open up" the data to de-noising, smoothing, etc., in a natural and integrated way.

2. Some Perspectives on the Wavelet Transform

We can think of our input data as a time-varying signal, e.g. a time series. If discretely sampled (as will almost always be the case in practice), this amounts to considering an input vector of values. The input data may be sampled at discrete wavelength values, yielding a spectrum, or one-dimensional image. A two-dimensional, or more complicated input image, can be fed to the analysis engine as a rasterized data stream. Analysis of such a two-dimensional image may be carried out independently on each dimension, but such an implementation issue will not be of further concern to us here. Even though our motivation arises from the analysis of ordered input data vectors, we will see below that we have no difficulty in using exactly the same approach with (more common) unordered input data vectors.

Wavelets can be introduced in different ways. One point of view on the wavelet transform is by means of filter banks. The filtering of the input signal is some transformation of it, e.g. a low-pass filter, or convolution with a smoothing function. Low-pass and high-pass filters are both considered in the wavelet transform, and their complementary use provides signal analysis and synthesis.

3. The Wavelet Transform Used

Manuscripts must not exceed **12 pages** for invited lectures, a prizewinner's speech, and invited sessions; and **8 pages** for contributed sessions. The following discussion is based on Strang (1989), Bhatia et al. (1996) and Strang and Nguyen (1996). Our task is to consider the approximation of a vector x at finer and finer scales. The finest scale provides the original data, $x_N = x$, and the approximation at scale m is x_m where usually $m = 2^0, 2^1, \ldots 2^N$. The incremental detail added in going from x_m to x_{m+1}, the detail signal, is yielded by the wavelet transform. If ξ_m is this detail signal, then the following holds:

$$x_{m+1} = H^T(m)x_m + G^T(m)\xi_m \qquad (1)$$

where $G(m)$ and $H(m)$ are matrices (linear transformations) depending on the wavelet chosen, and T denotes transpose (adjoint). An intermediate approximation of the original signal is immediately possible by setting detail components $\xi_{m'}$ to zero for $m' \geq m$ (thus, for example, to obtain x_2, we use only x_0, ξ_0 and ξ_1). Alternatively we can de-noise the detail signals before reconstituting x and this has been termed wavelet regression (Bruce and Gao, 1994).

Define ξ as the row-wise juxtaposition of all detail components, $\{\xi_m\}$, and the final smoothed signal, x_0, and consider the wavelet transform W given by

$$Wx = \xi = [\xi_{N-1} \ldots \xi_0 x_0]^T \qquad (2)$$

The right-hand side is a concatenation of vectors. Taking $W^T W = I$ (the identity matrix) is a strong condition for exact reconstruction of the input data, and is satisfied by an orthogonal wavelet transform. The important fact that $W^T W = I$ will be used below in our enhancement of multivariate data analysis methods. This permits use of the "prism" (or decomposition in terms of scale and location) of the wavelet transform.

Examples of these orthogonal wavelets, i.e. the operators G and H, are the Daubechies family, and the Haar wavelet transform (Press et al., 1992; Daubechies, 1992). For the Daubechies D_4 wavelet transform, H is given by

$$(0.4829629131, 0.8365163037, 0.2241438680, -0.1294095226)$$

and G is given by

$$(-0.1294095226, -0.2241438680, 0.8365163037, -0.4829629131).$$

Implementation is by decimating the signal by two at each level and convolving with G and H: therefore the number of operations is proportional to $n + n/2 + n/4 + \ldots = O(n)$. Wrap-around (or "mirroring") is used by the convolution at the extremities of the signal.

4 .Wavelet-Based Multivariate Data Analysis: Basis

We consider the wavelet transform of x, Wx. Consider two vectors, x and y. The squared Euclidean distance between these is $\|x - y\|^2 = (x - y)^T(x - y)$. The squared Euclidean distance between the wavelet transformed vectors is $\|Wx - Wy\|^2 = W^T W(x-y)^T(x-y)$, and hence identical to the distance squared between the original vectors. For use of the Euclidean distance, the wavelet transform can replace the original data in the data analysis. The analysis can be carried out in wavelet space rather than direct space. This in turn allows us to directly manipulate the wavelet transform values, using any of the approaches found useful in other areas. The results based on the orthogonal wavelet transform exclusively imply use of the

Euclidean metric, which nonetheless covers a considerable area of current data analysis practice.

Note that the wavelet basis is an orthogonal one, but is not a principal axis one (which is orthogonal, but also optimal in terms of least squares projections). Wickerhauser (1994) proposed a method to find an approximate principal component basis by determining a large number of (efficiently-calculated) wavelet bases, and keeping the one closest to the desired Karhunen-Loève basis. If we keep, say, an approximate representation allowing reconstitution of the original n components by n' components (due to the dyadic analysis, $n' \in \{n/2, n/4, \ldots\}$), then we see that the space spanned by these n' components will not be the same as that spanned by the n' first principal components.

5. Wavelet Filtering or Wavelet Regression

Foremost among modifications of the wavelet transform coefficients is to approximate the data, progressing from coarse representation to fine representation, but stopping at some resolution level m. As noted above, this implies setting wavelet coefficients $\xi_{m'}$ to zero when $m' \geq m$.

Filtering or non-linear regression of the data can be carried out by deleting insignificant wavelet coefficients at each resolution level (noise filtering), or by "shrinking" them (data smoothing). Reconstitution of the data then provides a cleaned data set. A practical overview of such approaches to data filtering (arising from work by Donoho and Johnstone at Stanford University) can be found in Bruce and Gao (1994, chapter 7). For other model-based work see Starck et al. (1995).

6. Examples of Multivariate Data Analysis in Wavelet Space

We used a set of 45 astronomical spectra. These were of the complex AGN (active galactic nucleus) object, NGC 4151, and were taken with the small but very successful IUE (International Ultraviolet Explorer) satellite which was still active in 1996 after nearly two decades of operation. We chose a set of 45 spectra observed with the SWP spectral camera, with wavelengths from 1191.2 Å to approximately 1794.4 Å, with values at 512 interval steps. There were some minor discrepancies in the wavelength values, which we discounted: an alternative would have been to interpolate flux values (vertical axis, y) in order to have values at identical wavelength values (horizontal axis, x), but we did not do this since the infrequent discrepancies were fractional parts of the most common regular interval widths. Fig. 1 shows a sample of 20 of these spectra. A wavelet transform (Daubechies 4 wavelet used) version of these spectra was generated, with a number of scales generated which was allowed by dyadic decomposition. An overall $0.1\,\sigma$ (standard deviation, calculated on all wavelet coefficients) was used as a threshold, and coefficient values below this were set to zero. Spectra which were apparently more noisy had relatively few coefficient values set to zero, e.g. 31%. More smooth spectra had up to over 91% of their coefficients set to zero. On average, 76% of the wavelet coefficients were zeroed in this way. Fig. 2 shows the relatively high quality spectra re-formed, following zeroing of wavelet coefficient values.

620

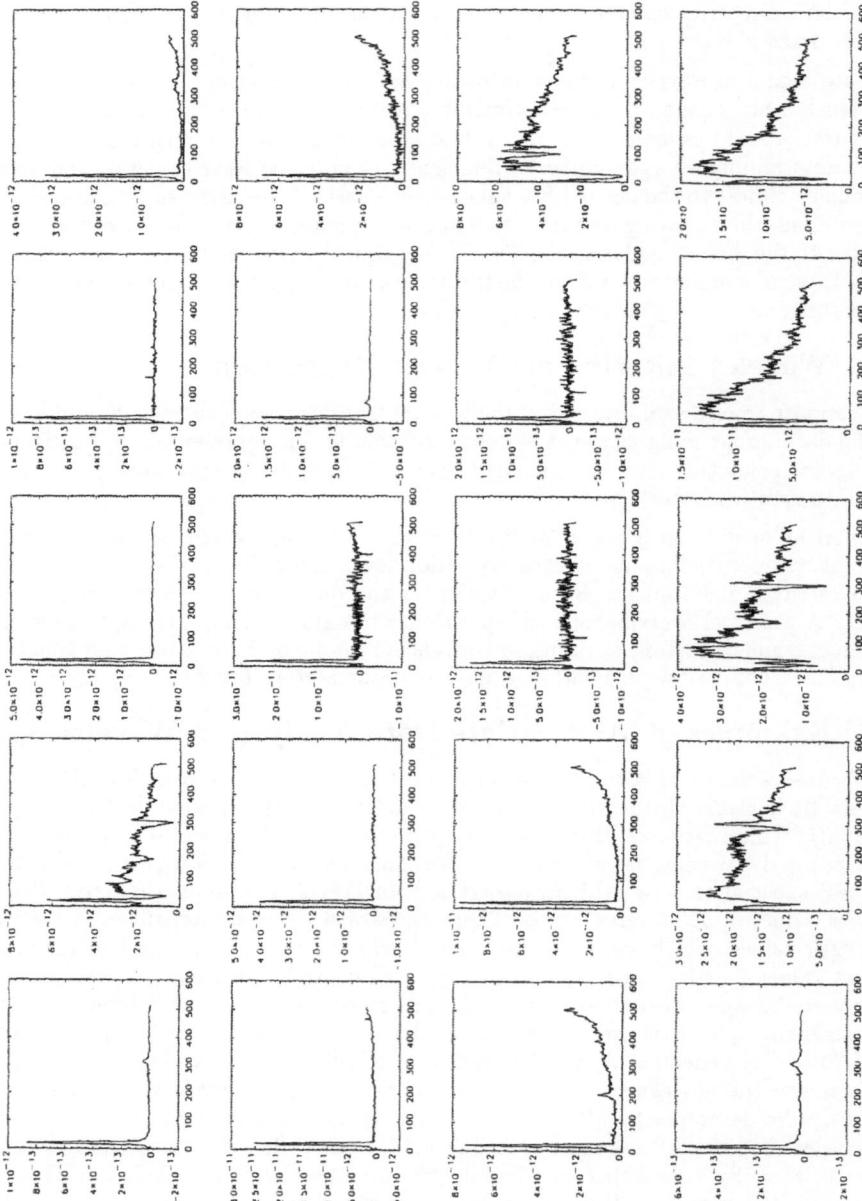

Figure 1: Sample of 20 spectra (from 45 used) with original flux measurements plotted on the y-axis.

The Kohonen "self-organizing feature map" (SOFM; Murtagh and Hernández-Pajares, 1995) was applied to this data. A 5 × 6 output representationalgrid was used.

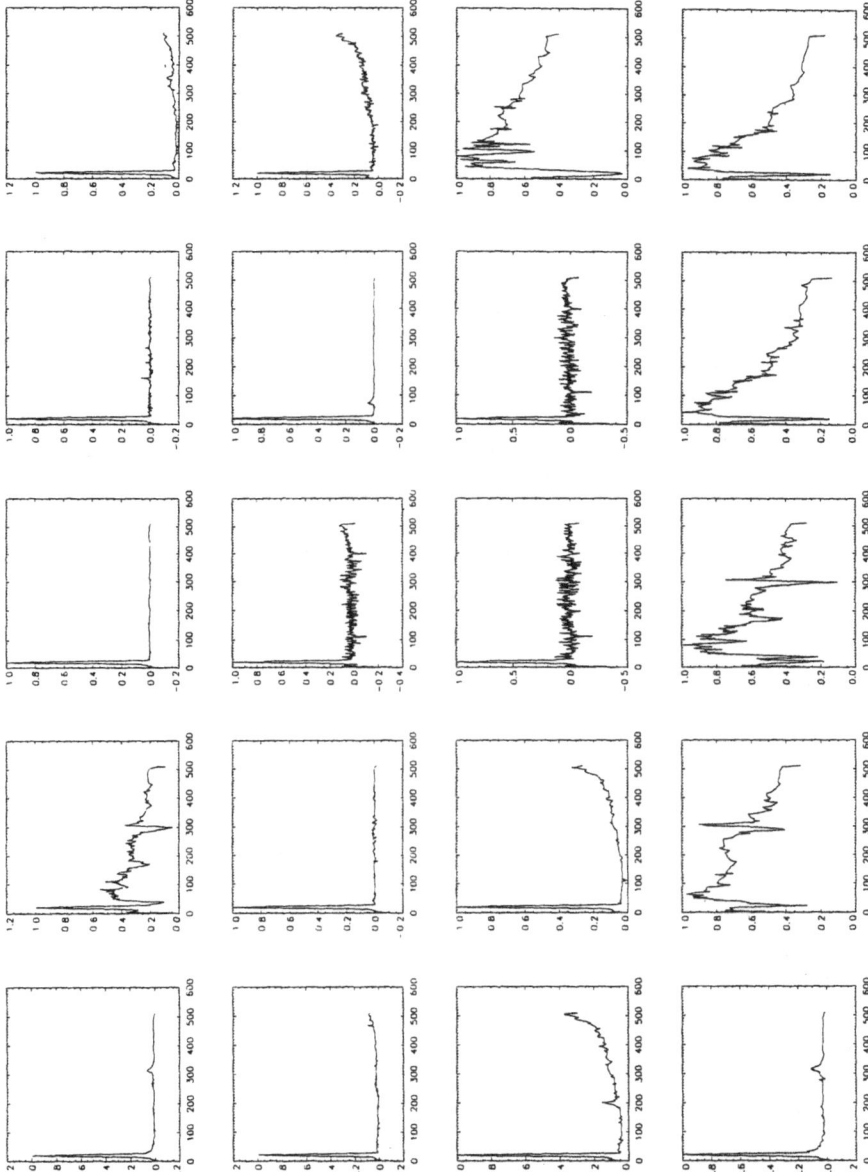

Figure 2: Sample of 20 spectra (as in previous Fig.), each normalized to unit maximum value, then wavelet transformed, approximately 75% of wavelet coefficients set to zero, and reconstituted.

In wavelet space or in direct space, the assignment results obtained were identical. With 76% of the wavelet coefficients zeroed, the result was very similar, indicating that redundant information had been successfully removed. This approach to SOFM construction leads to the following possibilities:

1. Efficient implementation: a good approximation can be obtained by zeroing most wavelet coefficients, which opens the way to more appropriate storage (e.g. offsets of non-zero values) and distance calculations (e.g. implementation loops driven by the stored non-zero values). Similarly, compression of large datasets can be carried out. Finally, calculations in a high-dimensional space, R^m, can be carried out more efficiently since, as seen above, the number of non-zero coefficients may well be $m'' \ll m$ with very little loss of useful information.

2. Data "cleaning" or filtering is a much more integral part of the data analysis processing. If a noise model is available for the input data, then the data can be de-noised at multiple scales. By suppressing wavelet coefficients at certain scales, high-frequency (perhaps stochastic or instrumental noise) or low-frequency (perhaps "background") information can be removed. Part of the data coding phase, prior to the analysis phase, can be dealt with more naturally in this new integrated approach.

A number of runs of the k-means partitioning algorithm were made. The exchange method, described in Späth (1985) was used. Four, or two, clusters were requested. Identical results were obtained for both data sets, which is not surprising given that this partitioning method is based on the Euclidean distance. For the 4-cluster, and 2-cluster, solutions we obtained respectively these assignments:

1232131144411143113431331411214122222221121114

1222111111111111111111111111112111122222221121111

The case of principal components analysis was very interesting. We know that the basic PCA method uses Euclidean scalar products to define the new set of axes. Often PCA is used on a variance-covariance input matrix (i.e. the input vectors are centered); or on a correlation input matrix (i.e. the input vectors are rescaled to zero mean and unit variance). These two transformations destroy the Euclidean metric properties vis-à-vis the raw data. Therefore we used PCA on the unprocessed input data. We obtained identical eigenvalues and eigenvectors for the two input data sets.

The eigenvalues are similar up to numerical precision:

| 1911.217163 | 210.355377 | 92.042099 | 13.908587 | 7.481989 |
| 2.722113 | 2.304520 | | | |

| 1911.220703 | 210.355392 | 92.042336 | 13.908703 | 7.481917 |
| 2.722145 | 2.304524 | | | |

The eigenvectors are similarly identical. The actual projection values are entirely different. This is simply due to the fact that the principal components in wavelet space are themselves inverse-transformable to provide principal components of the initial data.

Various aspects of this relationship between original and wavelet space remain to be investigated. We have argued for the importance of this, in the framework of data coding and preliminary processing. We have also noted that if most values can be set to zero with limited (and maybe beneficial) effect, then there is considerable scope for computational gain also. The processing of sparse data can be based on an "inverted file" data-structure which maps non-zero data entries to their values. The inverted file data-structure is then used to drive the distance and other calculations. Murtagh (1985, pp. 51–54 in particular) discusses various algorithms of this sort.

7. An Isotropic Redundant Wavelet Transform

It is common in pattern recognition to speak of "features" when what is intended are small density perturbations in feature space, small glitches in time series, etc. Such "features" may include sharp (edge-like) phenomena which can be demarcated using wavelet transforms like the orthogonal ones described above. Sometimes the glitches which are of interest are symmetric or isotropic. If so, a symmetric wavelet may be more useful. The danger with an asymmetric wavelet is that the wavelet itself may impose artifacts.

The "à trous" (with holes) algorithm is such an isotropic wavelet transform. It does not have the orthogonality property of the transform described earlier. The French term is commonly used, and arises from an interlaced convolution which is used instead of the usual convolution (see Shensa, 1992; Holschneider et al., 1989; see also Starck and Bijaoui, 1994; and Bijaoui et al., 1994). The algorithm can be described as follows: (i) smoothing p times with a B_3 spline – hence Gaussian-like, but of compact support; (ii) the wavelet coefficients are given by the differences between successive smoothed versions of the signal. The latter provide the detail signal, which (we hope) in practice will capture small "features" of interpretational value in the data. The following attractive additive decomposition of the data follows immediately from the design of the above scheme:

$$c_0(k) = c_p(k) + \sum_{i=1}^{p} w_i(k) \qquad (3)$$

The set of values provided by c_p provide a "residual" or "continuum" or "background" Adding w_i values to this, for $i = p, p - 1, \ldots$ gives increasingly more accurate approximations of the original signal. Note that no decimation is carried out here, which implies that the size or dimension of w_i is the same as that of c_0. This may be convenient in practice: cf. next section. It is readily seen that the computational complexity of the above algorithm is $O(n)$ for an n-valued input, and the storage complexity is $O(n^2)$.

8. Wavelet-Based Forecasting

In experiments carried out on the sunspots benchmark dataset (yearly averages from 1720 to 1979, with forecasts carried out on the period 1921 to 1979: see, e.g., Tong, 1990), a wavelet transform was used for values k up to a time-point k_0. One-step-ahead forecasts were carried out independently at each w_i. These were summed to produce the overall forecast (cf. the additive decomposition of the original data, provided by the wavelet transform). An interesting variant on this was also investigated: this variant was that there was no need to use the same forecasting method at each level, i. We ran autoregressive, multilayer perceptron and recurrent connectionist networks in parallel, and kept the best results indicated by a cross-validation on withheld data at that level. We found the overall result to be superior to working with the original data alone, or with one forecasting engine alone. Details of this work can be found in Aussem and Murtagh (1996).

9. Conclusion

The results described here, from the multivariate data analysis perspective, are very exciting. They not only open up the possibility of computational advances but also provide a new approach in the area of data coding and preliminary processing.

The chief advantage of these wavelet methods is that they provide a multiscale decomposition of the data, which can be directly used by multivariate data analysis methods, or which can be complementary to them.

A major element of this work is to show the practical relevance of doing this. It has been the aim of this paper to do precisely this in a few cases. Finding a symbiosis between what are, at first sight, methods with quite different bases and quite different objectives, requires new insights. Wedding the wavelet transform to multivariate data analysis no doubt leaves many further avenues to be explored.

Further details of the experimentation described in this paper, details of code used, and further information, can be found in Murtagh (1996).

References:

Aussem, A. and Murtagh, F. (1996): Combining neural network forecasts on wavelet-transformed time series, *Connection Science*, in press.

Bhatia, M., Karl, W.C. and Willsky, A.S. (1996): A wavelet-based method for multiscale tomographic reconstruction, *IEEE Transactions on Medical Imaging*, 15, 92–101.

Bijaoui, A., Starck, J.-L. and Murtagh, F. (1994): Restauration des images multi-échelles par l'algorithme à trous, *Traitement du Signal*, 11, 229–243.

Bruce, A. and Gao, H.-Y. (1994): *S+Wavelets User's Manual*, Version 1.0, Seattle, WA: StatSci Division, MathSoft Inc.

Daubechies, I. (1992): *Ten Lectures on Wavelets*, Philadelphia: SIAM.

Holschneider, M., Kronland-Martinet, R., Morlet, J. and Tchamitchian, Ph. (1989): A real-time algorithm for signal analysis with the help of the wavelet transform, in J.M. Combes, A. Grossmann and Ph. Tchamitchian (eds.), *Wavelets: Time-Frequency Methods and Phase Space*, Berlin: Springer-Verlag, 286–297.

Murtagh, F. (1985): *Clustering Algorithms*, Würzburg: Physica-Verlag.

Murtagh, F. and Hernández-Pajares, M. (1995): The Kohonen self-organizing feature map method: an assessment, *Journal of Classification*, 12, 165–190.

Murtagh, F. (1996): Wedding the wavelet transform and multivariate data analysis, *Journal of Classification*, submitted.

Press, W.H., Teukolsky, S.A., Vetterling, W.T. and Flannery, B.P. (1992): *Numerical Recipes*, 2nd ed., Chapter 13, New York: Cambridge University Press.

Shensa, M.J. (1992): The discrete wavelet transform: wedding the à trous and Mallat algorithms, *IEEE Transactions on Signal Processing*, 40, 2464–2482.

Späth, H. (1985): *Cluster Dissection and Analysis*, Chichester: Ellis Horwood.

Starck, J.-L. and Bijaoui, A. (1994): Filtering and deconvolution by the wavelet transform, *Signal Processing*, 35, 195–211.

Starck, J.-L., Bijaoui, A. and Murtagh, F. (1995): Multiresolution support applied to image filtering and deconvolution, *Graphical Models and Image Processing*, 57, 420–431.

Strang, G. (1989): Wavelets and dilation equations: a brief introduction, *SIAM Review*, 31, 614–627.

Strang, G. and Nguyen, T. (1996): *Wavelets and Filter Banks*, Wellesley, MA: Wellesley-Cambridge Press.

Tong, H. (1990): *Non Linear Time Series*, Oxford: Clarendon Press.

Wickerhauser, M.V. (1994): *Adapted Wavelet Analysis from Theory to Practice*, Wellesley, MA: A.K. Peters.

Visual Manipulation Environment for Data Analysis System

Masahiro Mizuta[1], Hiroyuki Minami[2]

[1] Division of Systems and Information Engineering, Hokkaido University
N.13, W.8, Kita-ku, Sapporo-shi
Hokkaido 060, Japan

[2] Department of Information and Management Science, Otaru University of Commerce
3-5-21, Midori, Otaru-shi
Hokkaido 047, Japan

Summary: Most statistical softwares utilize graphical facility to display data mainly, not to construct and execute analysis.
We had developed the data analysis system with visual manipulation and improve it on the UNIX platform with Tcl/Tk, the interface builder available on many architectures and operation systems. We offer its features and introduce some examples on our environment.

1. Visual Environment for Statisticians

We statisticians are familiar with graphical environments. We pick up, detect many features within observations through many analysis methods and plots. Most statistical softwares have also much graphical facility. For example, S-Language(Becker *et al.*(1988)), SAS/GRAPH have functions not only to display but to rotate, print data with various ways. Graphical environments provide us more potential for EDA(Tukey(1977)). We can see data from various viewpoints with computer softwares. For example, xgobi may tell us over what we expected with the function "GrandTour."

In computer science, "Visualization" is one of the most attractive keywords. What we introduced above is the effect by "Visualization of Data." "Visualization" is useful not only to reveal data but to deal with analysis processes. A statistician is often forced to apply many methods to the same data in his/her analysis since he/she wants to find the most reasonable result. A lot of results he/she made, however, may perplex him/her. If he/she can view an analysis flow (made of some statistical methods) on the graphical display, he/she can easily find the relations between data and applied methods. It is called "Visualization of Execution."

As we described, "Visualization" has much power then we try to build "Visual Environment," which synthesizes both "Visualization" merits. In short, our target is to build the system which can view, display data and construct analysis flows and execute them in the same environment. In the system, if we can add and execute some statistical methods in an analysis flow, we can easily change some methods or apply another procedure to the original data or intermediate ones.

We had already developed the prototypes and now improve it on the workstation with Tcl/Tk environment for its interoperability. In addition, this idea is also effective for novices in statistics but they have less knowledge on statistics then may make a meaningless statistical flow. We have already recognized this problem and tried to add a supporting module which makes use of technique on knowledge engineering.

In this paper, we introduce the system and some additional modules with examples.

2. Concept and Overview of Our System

Shu(1988) proposes a classification of visualizations according to their usage.

1. Visualization of Data or Information about Data

2. Visualization of a Program and/or Execution

3. Visualization of Software Design

What we use on analyses are classified into 1 and 2. Then, we have tried to apply the concept of 3 to a data analysis system. The process of making softwares is similar to that of making data analysis flows. We regard data analysis procedure as a flow of data. Our concept is shown in Figure 1 and flows of whole analysis processes are shown in Figure 2.

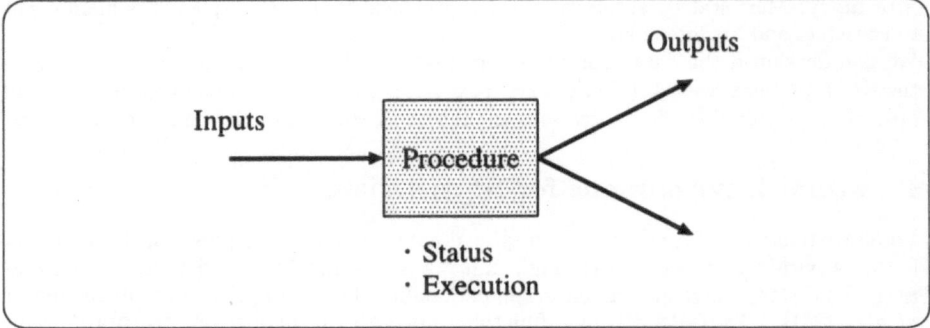

Fig. 1: Basic concept

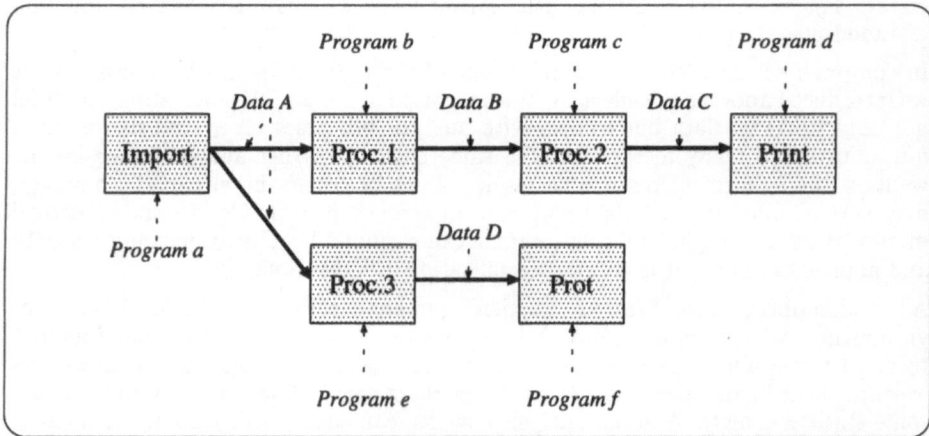

Fig. 2: Analysis flows

An arrow stands for a datum and a box is a (statistical) method. The leftmost box stands for the import procedure for an observation and the others are methods which are applied to the input. The data flow through many arrows and boxes.

"Flow Chart" which shows a control of a program is used in software development and similar to our concept. It is really useful when we check the control in a program but cannot map it into a real program since we cannot find the input and the output from this chart.

The flows we construct can execute as procedures directly since they show flows of data.

2.1 Overview of our system

We made a prototype of a data analysis system based on data flow diagram with PC(MS-DOS)(Mizuta (1989)). The prototype was ported into a workstation(X Window system on UNIX) in 1991.

Recently, the progress of computer environments has been accelerated. Then, we decide to rebuild our system, based on the environment available in many computer platforms.

Tcl/Tk is a kind of interface builder in computer graphical environment. It is developed on X Window originally and ported many platforms. Now it is available on MS-Windows, Windows95, WindowsNT and so on. It can handle interface parts (button, menu, bar, etc.) and assign its function easily. The characteristics are suitable for our system then we make a new prototype with Tcl/Tk.

3. How to build analysis flow and its examples

Figure 3 is a snapshot of the system at the first phase. The user first clicks "Put procedure" button then the menu of procedures appears. The items in this menu are sorted by procedure's purposes or type of data.

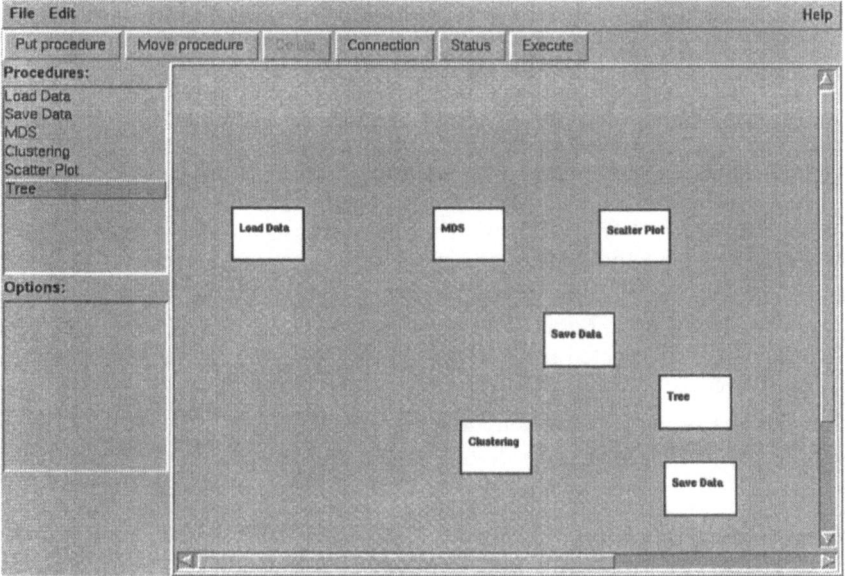

Fig. 3: First phase

The selected procedure box appears in the view follows the movement of pointing device(mouse). He/She can put it any place as you like. If he/she decides its position, click the left button of the mouse then the box is fixed. Figure 4 shows the situation some boxes are put.

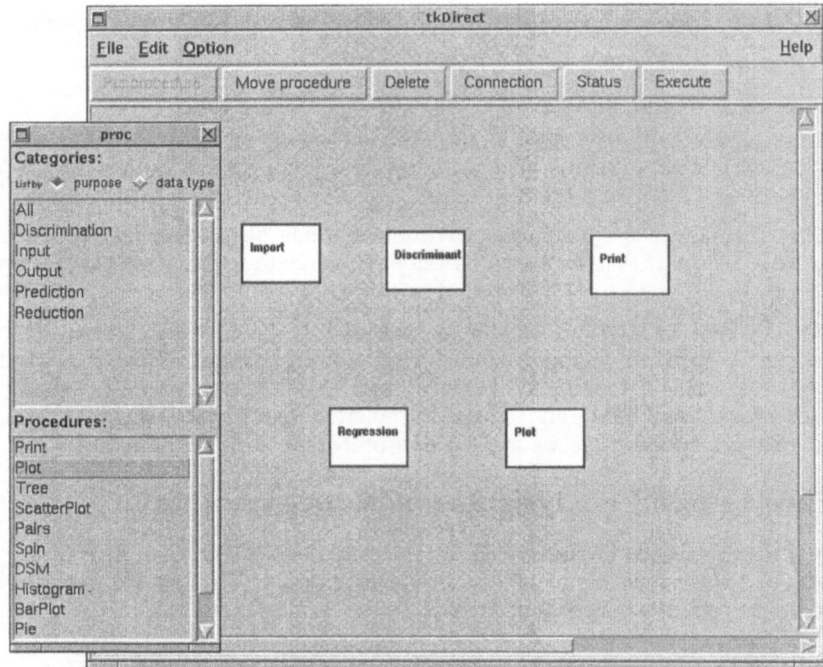

Fig. 4: Placed methods

Next, the user makes connections between boxes to construct a flow. He/She clicks a source box then an arrow appears. The start point is fixed to the source box and the other follows the movement of its pointing device. He/She clicks on the other box then a source box and the other are connected. The view in Figure 5 shows two streams. The upper flow is to do discriminant analysis and print results. The lower is to do regression analysis and plot results.

Now the user can really do analysis. He/She can click the rightmost box of a flow. A procedure box is executed if the input is calculated or ready to import. If not, the procedure which is expected to make the input is triggered. In short, if the rightmost box is activated, all boxes on the flow are executed recursively.

Figure 6 is the result of regression analysis. The user can get the result through the plot with the points and the regression line.

3.1 Parameter handling

Some methods may have parameters: for example, a parameter in specifying the dimension for execution MDS procedure. The system has a facility to set them on the same screen. Figure 7 shows that the dimension of MDS is 2, do output the eigen value in addition to the result and use an additive constant in the analysis.

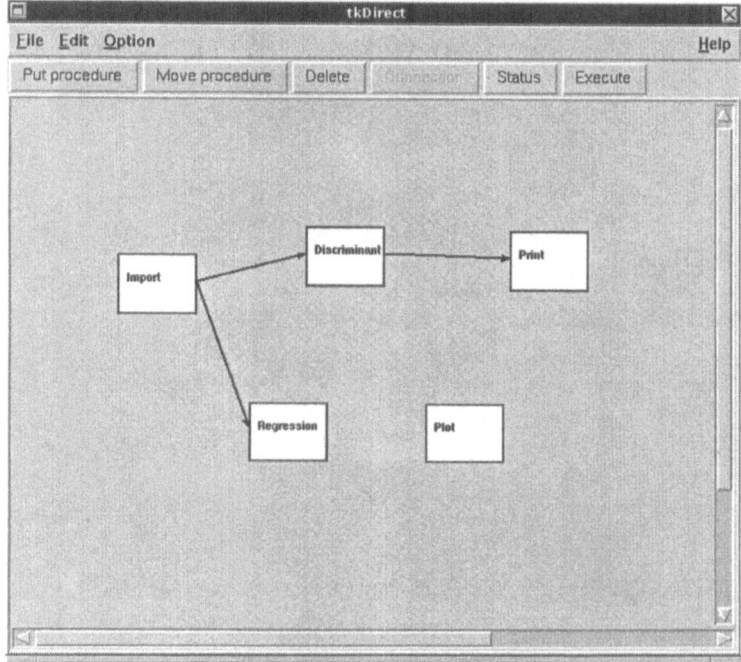

Fig. 5. Two flows

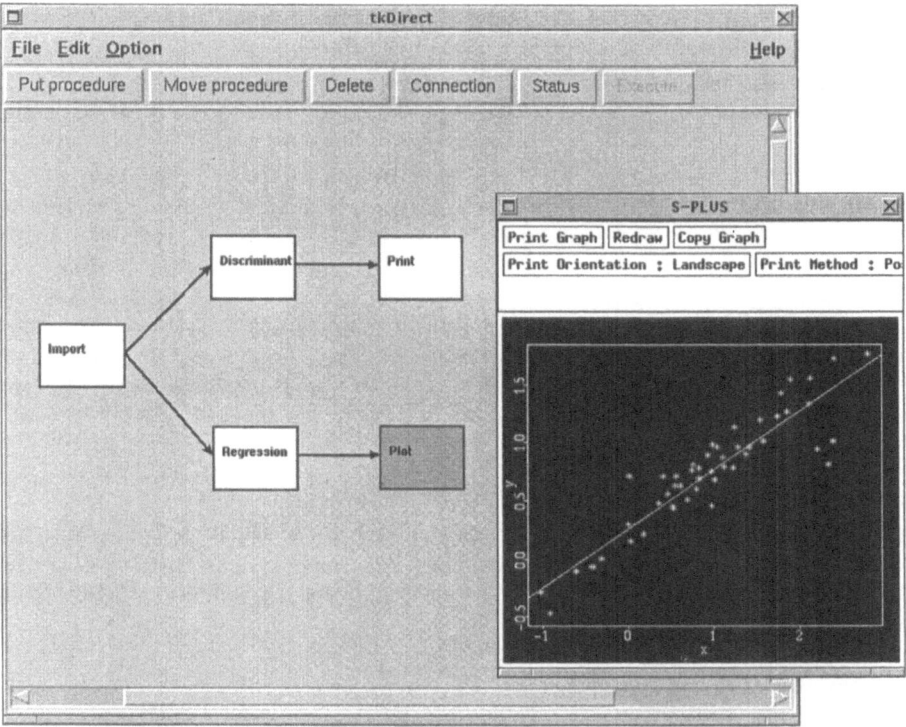

Fig. 6: Results(with plot)

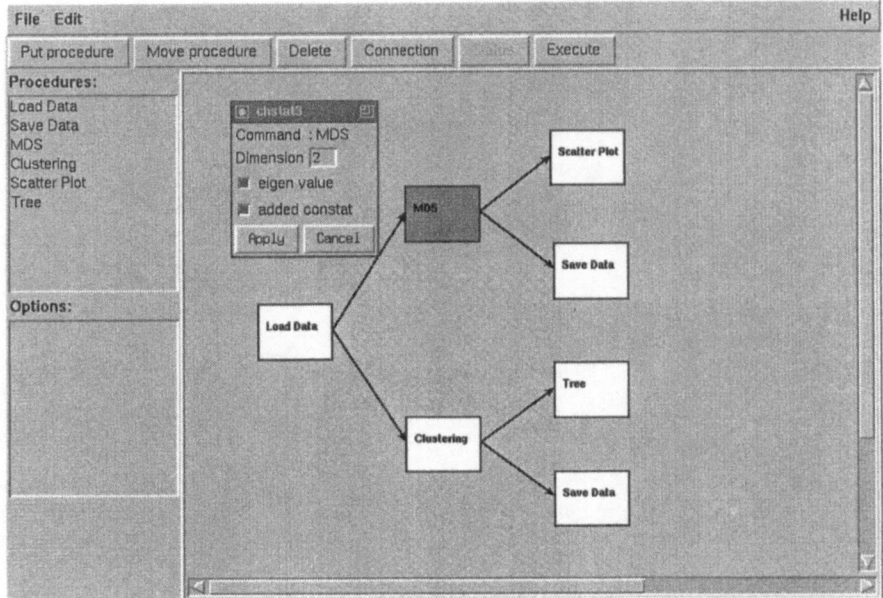

Fig. 7: Parameter handling(MDS)

4. Details and Applications

This system is developed on some independent routines.

As introduced already, the interface parts are written with Tcl/Tk (Ousterhout(1994)). As a statistical engine, we adopt S Language. These are familiar under UNIX workstations.

Each engine, however, is developed independently thus we can add another statistical engine (*i.e.* your original statistical library) as a method under the restriction that its input device is "stdin(standard input)" and the output is "stdout(standard output)."

A new function can also attach to this system easily. For example, we are about to add a new function "Validity Check," which verifies if the connection between statistical procedures is reasonable or not. If inappropriate, the routine suggests us the suitable method. How to realize this module is to check the input attribute of a procedure and the attribute of data as the input. If both are consistent, the connection must be reasonable. If not, the system tries to find another procedure which satisfies its condition. If a suitable procedure is not found, the module searches another procedure which can convert the attribute of data into the input one.

This module (written with Prolog) is implemented on MS-DOS version and works well. In addition, we have already developed a knowledge supporting system of data analysis (Minami *et al.* (1993)) as the extension of "Validity Check." We are going to synthesize the system and this module.

5. Concluding Remarks

The "Macro" feature may be effective for the system. The flow except an import box can be regarded as one (synthesized) procedure then we can consider it as another one method. To realize this feature is not so hard and we will offer it soon.

Our concept is not enough for all statistical methods since it forces that the result of procedures provides by value mainly. As you know, some methods output their result as a function or formula. The current version of our system cannot handle these datatypes directly. The concept of "Object-Oriented" is important to extend kinds of datatypes. Tcl/Tk is based on its concept. If we make all parts of the system object-oriented, we may handle formulas and functions same as values.

This idea has an influence on the module "Validity Check" and supporting modules. The essence of them is to check the types of data, input and output attributes on procedures. If procedures are managed with object-oriented style, it is easy to make some checker routine since the routine has only to check their classes are valid.

Acknowledgment

We would like to thank Mr. Kikuchi (Graduate School of Information Engineering, Hokkaido University) for his programming support.

A part of this work was supported by a Grant-in-Aid for Scientific Research from the Ministry of Education, Science, Sports, and Culture of Japanese Governments.

References:

Becker, R.A., Chambers, J.M. and Wilks, A.R. (1988). *The New S Language*, Wadsworth & Brook/Cole Advanced Books & Software, Pacific Grove.

Minami, M., Mizuta, M. and Sato, Y.(1993). A Knowledge Supporting System for Data Analysis. *Journal of the Japanese Society of Computational Statistics*, **6**, 1, 85–97.

Mizuta, M. (1990). Data Analysis System with Visual Manipulations, *Bulletin of the Computational Statistics of Japan*, **3**, 1, 23–29 (in Japanese).

Ousterhout, J. K. (1994). *Tcl and the Tk Toolkit*, Addison Wesley.

Shu, N. C. (1988). Visual Programming. Van Nostrand Reinhold Company.

Tukey, J.W.(1977). *Exploratory Data Analysis*, Addison-Wesley.

Human Interface for Multimedia Database with Visual Interaction Facilities

Toshikazu Kato

Electrotechnical Laboratory (ETL), AIST, MITI
1-1-4, Umezono, Tsukuba Science City, 305 Japan

Summary: This paper describes visual interaction mechanisms for image database systems. The typical mechanisms for visual interactions are query by visual example (QVE) and query by subjective descriptions (QBD). The former includes a similarity retrieval function by showing a sketch, and the latter includes a sense retrieval function by learning user's personal taste. We modeled user's visual perception process by four levels; a physical level, a physiological level, a visual psychological level, and visual cognition level. These models are automatically created by image analysis and statistical learning, are referred as multimedia indexes to database systems.

1. Introduction

"A picture is worth a thousand words." A human interface plays an important role in a multimedia information system. For instance, we request a content based visual interface in order to communicate visual information itself to and from a multimedia database system, Iyenger and Kashyap (1988), Grosky and Mehrotra (1989). The algorithms of multimedia operations have to suit user's subjective viewpoint, such as a similarity measure, a sense of taste, etc. Thus, we have to provide flexible interaction mechanism to design a multimedia human interface.

We expect a multimedia database system to manage multimedia data themselves, such as image data, as well as alphanumeric data. We also expect it to provide a human interface to accomplish flexible man-machine communication in a user-friendly manner. Then, what are needed in multimedia interaction? We can summarize the essential needs in multimedia interaction as follows, Kato, et al. (1991).

(a) Visual query on pictorial domain: We need to communicate multimedia data to and from the database in a user-friendly manner. For instance, we would like to show image data itself as a pictorial key to retrieve some visual information from database systems.

(b) Subjectivity of judging criteria: We want to adjust database operations as well as database schemata to each of our subjective views. In case of similarity retrieval, we would like to get some suitable candidates according to our subjective measures where the measures may differ with each individuals.

(c) Interpretation between multimedia domains: Some of the multimedia queries should evaluate multimedia data on the different domains. In case of content based retrieval, we expect the system to retrieve some image data by describing their contents as text data.

We can answer these needs by a multimedia human interface with our visual interaction facilities. Let us show the general framework of visual interaction by typical user's query requests in our applications. Our basic ideas are QVE (query by visual example) and QBD (query by subjective description). Multimedia interaction requires interpreting the contents of multimedia information in order to operate from user's subjective viewpoint. Thus, interpretation algorithms have to suit the perception processes of each user. Such processes belong to a subjective human factor. On our multimedia human interface, the system refers to the object model on multimedia information and the user model on the perception process to operate from user's subjective viewpoint.

2. Intelligent Visual Interaction

2.1 Conventional Approach to Visual Interface

Several experimental image database systems have been proposed to provide visual interfaces. The QPE system provides a schema of pictorial data in a graphic form as well as in a tabular form, Chang and Fu (1980). While this system shows the data in a graphic form, its query style is only a substitute for the query languages on alphanumeric data. In the icon-based system, icons and their two dimensional strings are referred to as pictorial keys to image data, Chang, et al. (1988). A user can specify the target images by placing icons on the graphic display as a visual query. The system evaluates only two dimensional strings of icons in its retrieval process. Therefore, it is difficult to perform similarity retrieval according to the subjective measure of the user. The hypermedia system provides an indexing mechanism for multimedia data in a uniformed style, Yankelovich, et al. (1988). Although this system enables a kind of subjective indexing, its process fully owes much to the user's effort on defining many links.

We originally want to process visual information and its contents according to our subjective views. On this point, although the above systems use graphic devices to show schema, icons and guidelines, their facilities are not enough to perform such interaction in a user-friendly manner.

2.2 Computational Models of Visual Perception Process

In human-to-human communication, we can exchange not only pictorial data themselves as objective information but also emotion, personal taste, and so on, as subjective information. The latter type communication, we call "kansei-oriented communication," is rather important to express and to understand personal opinions well.

As a working hypothesis, we have developed an artificial sense model on visual perception process.

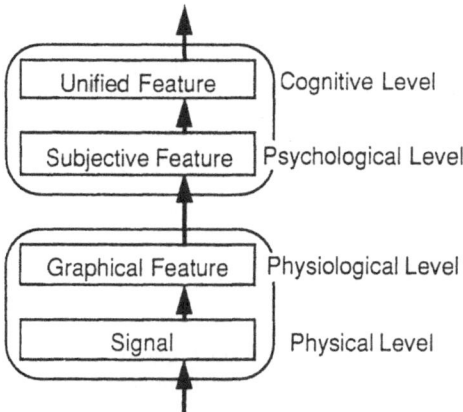

Fig. 1: Conceptual schema of visual perception process

(1) Physical level interaction: A picture may often remind us similar images or related pictures. This process is a kind of similarity and associative retrieval of pictorial data by physical level interaction with pictorial database.
(2) Physiological level interaction: As we can recognize hand-written characters as well as printed block characters, some graphical features have almost the same values to distinguish the category from the others. Early stage of mammal vision mechanism extracts this sort of graphical features such as intensity levels, edge, contrast, correlation, spatial frequency, and so on. Visual perception may depend on such

graphical features.

(3) Psychological level interaction: A user may wish to see some graphic symbols which give him the similar impression from his view. We have to notice that the criteria for similarity belongs to a subjective human factor. Although human beings have anatomically the common organs, each person may show different interpretation in classification and similarity measure. It means each person has his own weighting factors on graphical features. The system should evaluate similarity according to his subjective criterion. Therefore, the system should analyze and learn the subjective similarity measure on the images with each user. Graphical features are mapped into subjective features by the weighting factors.

(4) Cognitive level interaction: We often have difference impressions, even when viewing the same painting. Each person may also differently give a unique interpretation even viewing the same picture. It seems each person has his own correlation between concepts and graphical features and/or subjective features. The system should evaluate a subjective description according to his criterion. Therefore, the system should analyze and learn the correlation between the subjective descriptions and the images with each user.

A user model is needed to operate visual information based on the subjective viewpoint of each user. We have to develop a simple learning algorithm to adjust the criteria for each user.

3. Query by Visual Example at Physiological Level

This chapter describes the visual perception models and algorithms for query by visual example (QVE), i.e. similarity retrieval on objective criteria.

3.1   Visual Perception Model for Graphic Symbols

We have been developing an image database system called TRADEMARK. The TRADEMARK database is a collection of graphic symbols, Kato, et al. (1991). These figures are protected as intellectual property. At a Patent Office, an examiner compares each figure with tens of thousands of existing registered graphic symbols. It is a burdensome task that can be avoided if the image database system accepts query by visual example (QVE) using sketch retrieval facility. This is the essential problem of an image database system.

Let us discuss the technical problems associated with QVE. We can point out these problems:

(i) The pixelwise pattern matching is quite a time consuming task.

(ii) We can not give a pictorial key which is exactly the same with the original design of the symbol.

(iii) Our visual impressions of graphic symbols are psychologically ambiguous. The similarity measure may differ with each of us.

We have assumed an image model using several kinds of graphic features (GF) from our psychological experiments and from the recent knowledge on visual physiology. We can simulate the process of visual perception of graphic symbols with a GF space. These features are primitive ones in the early stage of our human vision system. The TRADEMARK system refers to these GF vectors as the pictorial index of graphic symbols.

(1) Spatial distribution of the gray level *Gray8, Edge8:* The distribution of black pixels represents the outline of the graphic symbol. For this purpose, the graphic symbol is divided into 8×8 square meshes. *Gray8* denotes the number of black pixels m_{ij} in each mesh.

$$Gray8 = \{m_{ij}\} \quad (0 \le i \le 7, 0 \le j \le 7)$$

Similarly, we defined *Edge8* with the contour of the graphic symbol.

(2) Spatial frequency *RunB/W, RunW'*: The spatial frequency measures the complexity of graphic symbols. *RunB/W* approximates the frequency by the run-length distribution of each rectangle mesh. Here, the figure is divided into four horizontal meshes as well as four vertical meshes. Similarly, we defined *RunW'* without distinguishing the black and white runs.

(3) Local correlation measure and local contrast measure *Corr4, Cont4*: The local correlation and the local contrast show the spatial structure, such as the regularity of arrangement of partial figures.

$$
\left.
\begin{aligned}
Corr4 &= \{m_{ij} \times m_{i'j'}\} \\
Cont4 &= \frac{m_{ij} - m_{i'j'}}{m_{ij} + m_{i'j'}}
\end{aligned}
\right\} (0 \le i \le 4, 0 \le j \le 4)
$$

Where, m_{ij}, $m_{i'j'}$ are adjacent meshes. These parameters are defined on 4×4 square meshes.

[Alg. 1] GF space for graphic symbols
(1) Analyze the layout of a document image to extract graphic symbols. Normalize the image size of the graphic symbols.
(2) Calculate the GF vector p_i with each graphic symbol.

3.2   Sketch Retrieval on GF space

We may expect that the neighboring graphic symbols in a GF space will have a similar shape. For example, a fine copy and its rough sketch are neighbors in GF space. Therefore, the system can retrieve similar graphic symbols by comparing their GF vectors.

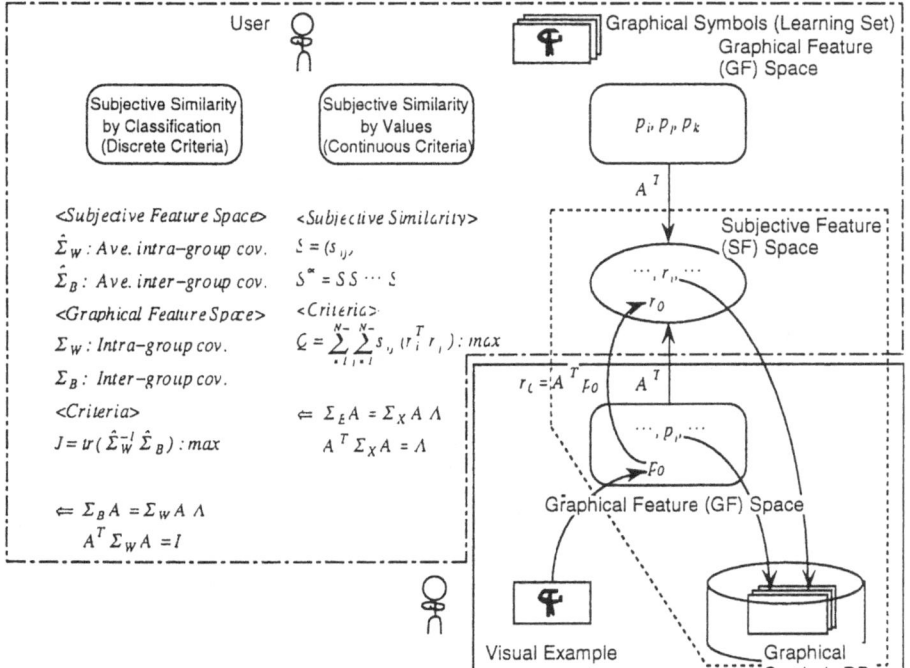

Fig. 2:   Overview of Query by Visual Example mechanism

Let us show the sketch retrieval algorithm for graphic symbols. Fig. 2 shows the outline of the whole QVE mechanisms. In Fig. 2, the sketch retrieval process is enclosed by solid lines.

[Alg. 2]   Sketch retrieval on GF space
(1)   Normalize the image size of the sketch, i.e. the visual example.
(2)   Calculate the GF vector p_0 of the sketch.
(3)   Calculate the distance d_i between the sketch p_0 and the graphic symbols in the database p_i

$$d_i = \sum_{k=1}^{K} (w_k \| p_i - p_j \|)$$

Where, k and w_k mean the GF vector and its weight factor.
(4)   Choose the graphic symbols in the ascending order of d_i.

3.3   Experimental Results of Sketch Retrieval

Let us show our experimental results of sketch retrieval algorithms for graphic symbols and full color paintings. Fig. 3 shows an example of sketch retrieval. A user has written down a sketch shown as "your visual example" in the QVE window in Fig. 3. The system searches for the most similar graphic symbols on the pictorial index comparing their GF vectors. The QVE window also shows the candidates for similar graphic symbols in descending order of priority. The first candidate is the original design of the rough sketch. We can also find similar graphic symbols in Fig. 3.

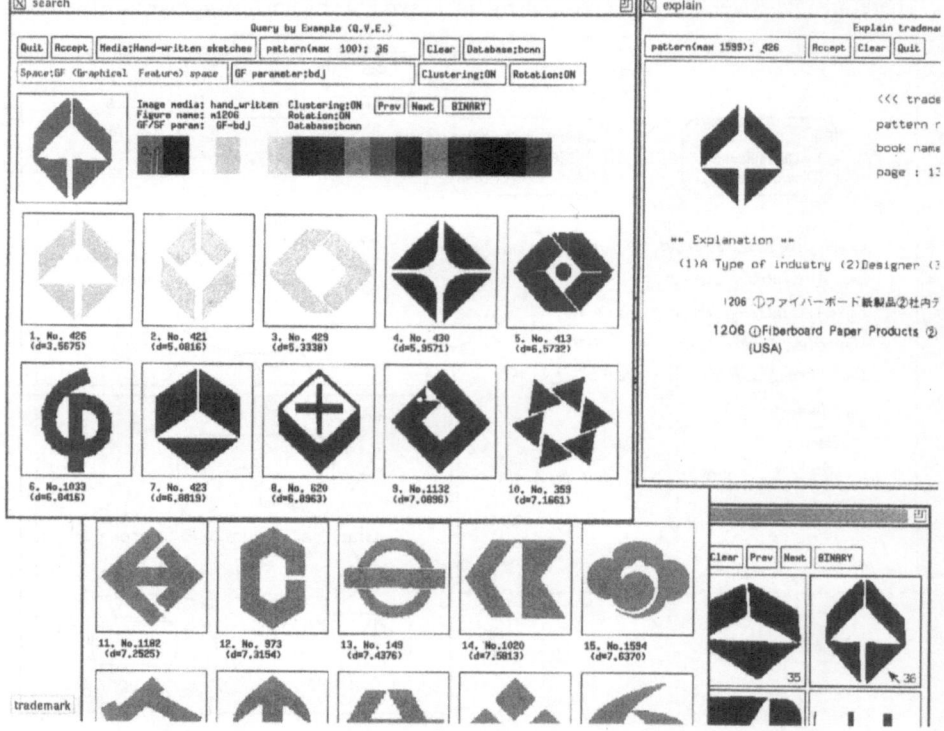

Fig. 3: Sketch retrieval of graphic symbols by showing a rough sketch

We have evaluated this algorithm in an experiment in which we showed fair copies, hand-written sketches and rough sketches with every 100 visual example. (Currently, the TRADEMARK database manages about 2,000 graphic symbols.) We have tested the recall ratio. Here, the recall ratio shows the rate of retrieval of the original graphic symbol among the best ten candidates For a fair copy, the system had an almost 100% recall ratio among the first ten candidates, using the GF features. Even for the rough sketches, it had about 95% recall. We may conclude that our GF features satisfy the requirements for a robust image model for sketch retrieval.

4. Query by Visual Example at Psychological Level

This chapter describes another aspect to query by visual example (QVE), i.e. similarity retrieval on subjective criteria.

4.1 Personal View Model for Graphic Symbols

Remember that the human being does not rely on only geometric process for determining shape similarity. Psychological processes also relate to it. Therefore, we have to consider a psychological aspect of similarity retrieval. Such similarity differs with each user even for the same figures. Therefore, the system should learn the subjective similarity measure as a personal view model of each user.

A user classifies the test samples from the database into several clusters judging similarity. The system extracts the GF vector of each graphic symbol. We need a subjective feature (SF) space which reflects the subjective similarity measure. We can construct such an SF space by the discriminant analysis. The discriminant analysis is one of the multivariate analysis methods to evaluate the classification. The algorithm to construct the SF space and the personal index is as follows. The learning process is shown enclosed by dotted dashed lines in Fig. 2

[Alg. 3]   Learning subjective similarity measure (SF space)
(1)   Choose appropriate graphic symbols from the database to make the learning set P. The user classifies the graphic symbols into several clusters without overlapping.
(2)   Normalize the image size. Calculate the GF vector p_k of each graphic symbol $k \in P$. (The GF vector is the same one in the sketch retrieval algorithm.)
(3)   Apply the discriminant analysis to the clustering result by the user. The linear mapping A is given by solving the following eigenvalue problem.

$$\left. \begin{array}{l} \Sigma_B A = \Sigma_W A \Lambda \\ A' \Sigma_W A = I \end{array} \right\}$$

Where, Σ_B and Σ_W denote inter-group and intra-group covariance matrixes of GF vectors, respectively. A' means the transposed matrix of A, and Λ is an eigenvalue vector. Thus, we can define the SF space with the user.

$$r_k = A' p_k .$$

Where, r_k is the SF vector of the graphic symbol k.
(4)   Calculate the SF vectors with every graphic symbol in the database.

$$r_k = A' p_k .$$

We will refer to the SF space of r_k as the personal index. Note that we do not have to examine the similarity of all the graphic symbols in the database. Once the system has learned the linear mapping A, it can automatically construct the personal index only from the GF vectors. This algorithm reduces the personnel expenses for indexing.

4.2 Similarity Retrieval on SF space

We may expect that the neighboring graphic symbols in an SF space will give a similar impression from the user's view. Just the same with sketch retrieval, the user shows a sketch with which he wants to see the similar figures. The system can retrieve similar graphic symbols by comparing their SF vectors on the personal index. Then, the system shows suitable candidates. The algorithm for similarity retrieval is as follows, which is also shown enclosed by dotted lines in Fig. 2.

[Alg. 4]   Similarity retrieval on SF space

(1)   Apply the linear mapping A to the GF vector p_0 of the sketch.

$$r_0 = A' p_0.$$

(2)   Choose the neighboring graphic symbols p_i on the personal index as the candidates for similarity retrieval.

(3)   Calculate the distance d_i between the sketch r_0 and the graphic symbols r_i on the personal index.

$$d_i = \| r_i - r_0 \|.$$

(4)   Choose the graphic symbols in the ascending order of d_i.

4.3   Experimental Results of Similarity Retrieval

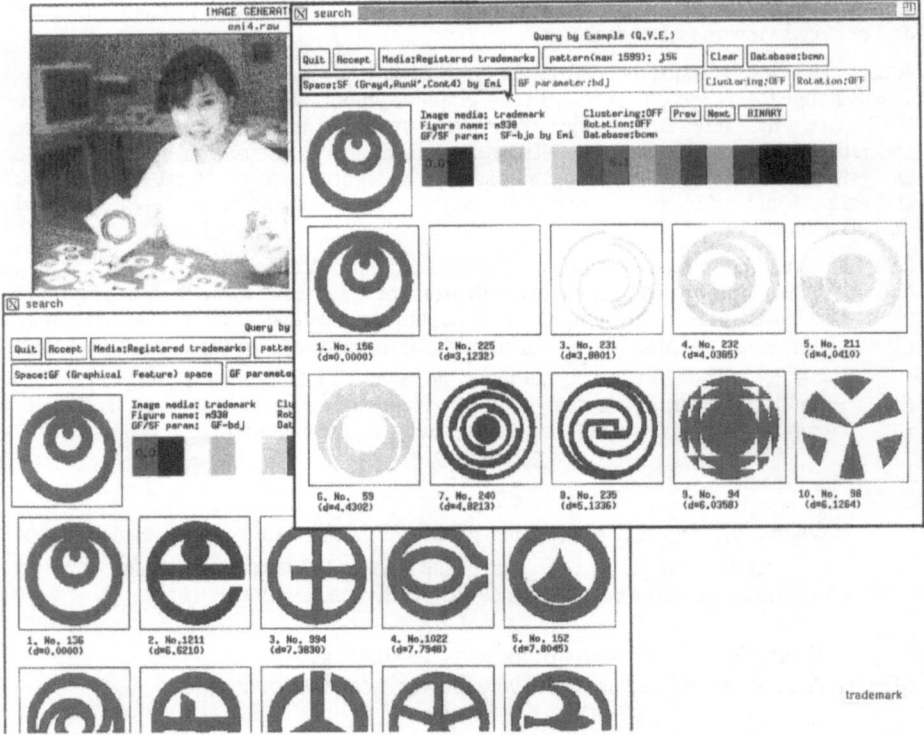

Fig. 4: Example of similarity retrieval
The upper QVE window and the lower one show the result of similarity retrieval on the SF space of the user and that of sketch retrieval on the GF space.

Fig. 4 shows an example of similarity retrieval. The upper QVE window in this figure shows the ten candidates for similarity retrieval on the SF space, while the lower one shows the ten candidates on the GF space. The second to the eighth candidates on the SF space have matched with the classification by this user. The system could not retrieve these candidates in the sketch retrieval on the GF space, since their graphic features differ from those of the visual example.

We have evaluated the learning algorithm and the similarity retrieval algorithm in an experiment with eleven users. In this experiment, we used 230 graphic symbols for the samples out of 2000. The system had at least one similar graphic symbol more than 98% recall ratio among the first ten candidates. We may conclude that our SF spaces satisfy the subjective similarity measure of each user.

5. Query by Subjective Description at Cognitive Level

This chapter gives a more complex visual interaction algorithm for full color paintings. A user has only to show several words in a query by subjective description (QBD) after constructing his visual perception model as the user model.

5.1 Full Color Paintings and Artistic Impressions

In conventional database systems, keyword indexes have been used to retrieve some paintings which give us certain impressions. A user describes his request with a combination of such keywords, which are assigned by the indexer to each painting. Even though a keyword thesaurus is available, this approach has the following problems.
(a) The indexer has to assign several keywords to each painting in the database, which is a laborious work.
(b) The indexer assigns such keywords according to his personal view, which affects the user's query. Even when viewing the same painting, such descriptions may differ with persons, nations and cultural backgrounds.
(c) While a keyword thesaurus is useful for enlarging the vocabulary of the user's query, the operations can be defined only on the text data domain.

Our approach aims to unify text data and image data to describe the contents of an image. Thus, we have to model how a person feels certain impressions when viewing a painting. Art critics view paintings from several aspects, such as motif, general composition and coloring. Chijiiwa (1983) reported that the dominant impression generated by paintings is coloring. This report suggests that there is a reasonable correlation between the coloring and the words in the reviews.

We have been developing an electronic art gallery called ART MUSEUM, Kato, et al. (1991). The ART MUSEUM is a collection of full color paintings. The ART MUSEUM system provides QBD facility for sense retrieval. A user can retrieve full color paintings by presenting some words on his personal taste.

The ART MUSEUM system has the personal index on unified features (UF), which are derived from graphic features (GF) on color and subjective features (SF) on impression words.

We can parameterize the coloring of a painting by the distribution of the RGB intensity value in the subpictures.

[Alg. 5] Pictorial index (GF space) for coloring features
(1)   Divide a painting into 4×4 subpictures to approximate the combination and the arrangement of colors.
(2)   Calculate the distribution of the RGB intensity value in the subpictures as the GF vector p_i with each graphic symbol.

We also need a subjective criterion on artistic impression. A user answers his artistic impressions on sample paintings as the weight vector of adjective words. (Currently, the adjectives are restricted up to about 30.)

[Alg. 6] Inquiry for artistic impression (SF space)
(1)   Choose appropriate paintings from the database to make the learning set P.

(2)   The user describes his impressions as the weight of the adjectives a_k to each painting $k \in P$.

5.2   Personal View Model for Artistic Impressions

Let us show the algorithm for learning a personal view model for artistic impressions. We, of course, can not directly compare the subjective words in a query and the coloring features of a painting, since they are on the different domains.

Remember that the subjective descriptions of each user are related with the coloring of paintings. We may expect that the set of words and the parameterized coloring feature correlates with each other. The ART MUSEUM system analyzes such correlation between the different domains. We will regard the correlation as the personal view model for the user. We can construct a unified feature (UF) space on this model to compare the subjective words and coloring features.

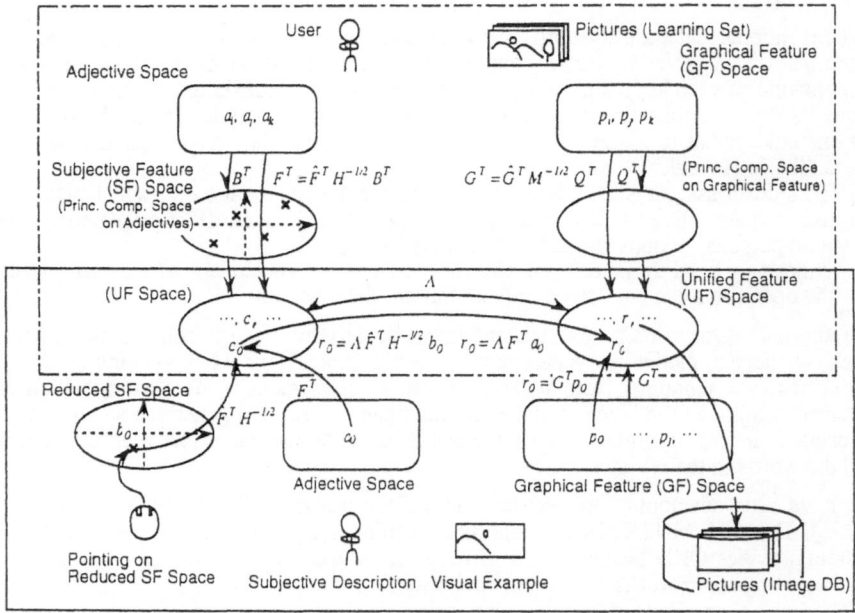

Fig. 5: Overview of QBD mechanism

We can construct such a UF space by the canonical correlation analysis. The algorithm to construct the UF space and the personal index is as follows. (It is also shown enclosed by dotted dashed lines in Fig. 5).

[Alg. 7]   Learning artistic impression measure
(1)   Apply the canonical correlation analysis to the result of the inquiry by the user. The linear mappings F and G make their correlation maximum:

$$\left.\begin{array}{l} c_k = F\, a_k \\ r_k = G\, p_k \end{array}\right\}$$

Where, F', G' mean the transposed matrix of F, G, respectively.
(2)   Calculate the UF vectors of paintings in the database from the following formula.

$$r_i = G\, p_i \,.$$

We will refer to the UF space of r_i as the personal index of the user model. Note that we do not have to assign the adjectives a_i to every painting in the database. Once the system has learned the linear mappings F and G, it can automatically construct the personal index only from the GF vectors. This is a labor-saving algorithm for indexing.

5.3 Sense Retrieval on UF space

We may expect that the neighboring paintings in UF space will give similar impressions of coloring to the user.

The user has only to show several words in QBD. The system evaluates the most suitable coloring for the words according to the personal view model. Then, the system can provide paintings of suitable coloring. The algorithm for sense retrieval on coloring is as follows. (It is shown enclosed in solid lines in Fig. 5.)

[Alg. 8]   Sense retrieval on UF space
(1)   Apply the linear mappings F and Λ to the adjective vector a_0 of a subjective description in the user's query.
$$r_0 = \Lambda F \, a_0 \, .$$
Where Λ is the regression by the diagonal matrix of canonical correlation coefficients.
(2)   Choose neighboring paintings of the image r_i on the UF space as candidates for sense retrieval.

The UF space gives a criterion to evaluate the text data and the image data by their contents. The UF space enables us to operate a multimedia query which has multimedia data of different domains in its parts. Therefore, this algorithm corresponds to a multimedia join on text data and image data.

Note that the sense retrieval algorithm evaluates the visual impression in UF space. Therefore, we can retrieve paintings without assigning keywords to every painting; this reduces the labor cost of indexing.

For other applications of the UF space, we can retrieve paintings that give us a similar impression by showing a painting as a visual example. We can also infer the suitable keywords for simulating the user's personal view, using the inverse mappings as follows.
$$a_0 = F^{-1} \Lambda^{-1} G \, p_0 \, .$$

5.4 Experiments on Sense Retrieval

We have experimented with the learning algorithm and the sense retrieval algorithm on our ART MUSEUM system. In this experiment, we adjusted the UF space according to the personal view of female students. The learning algorithm is applied to the average answers of female students. This is a user-group model on artistic impressions.

Fig. 6 shows an example of sense retrieval. This figure shows the best eight candidates for the adjectives; "romantic, soft and warm". These paintings roughly satisfied the personal view of the subjects. We may conclude that the UF personal index on UF space reflects a personal sense of coloring.

642

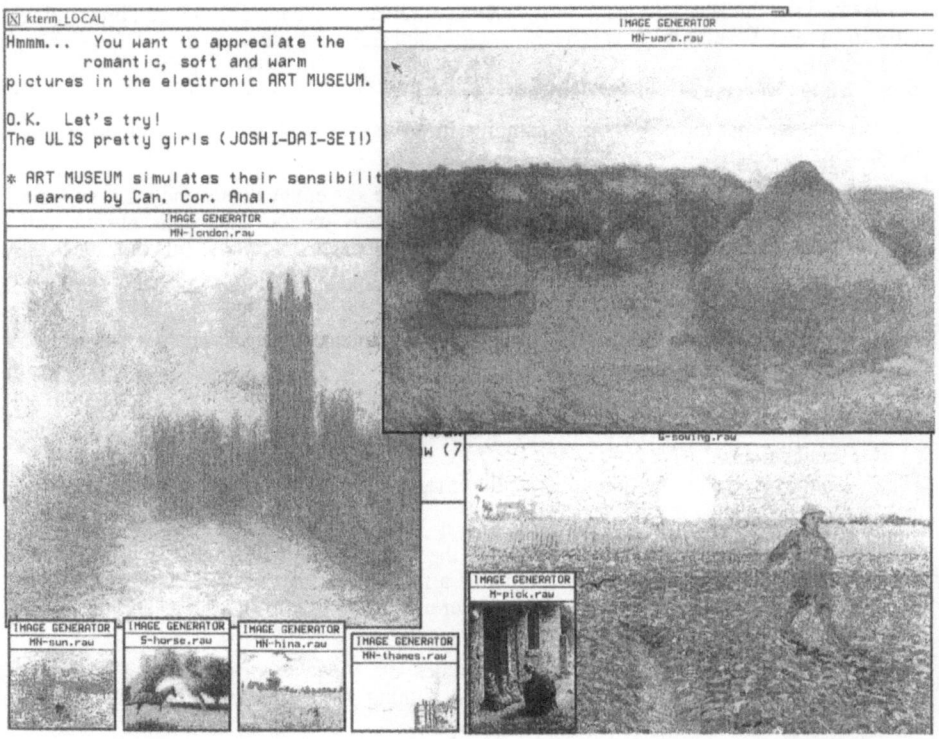

Fig. 6: Example of sense retrieval in QBD
The best eight paintings in the database appear for the words "romantic",
"soft" and "warm". The words in the query and colorings of the paintings
are evaluated on a UF space of female students.

6. Summary

We proposed the concept of cognitive view mechanism which interprets the multimedia
data. We have developed the algorithms for sketch retrieval, similarity retrieval and sense
retrieval to support visual interaction. The sketch retrieval algorithm accepts visual data as
an example. The similarity retrieval algorithm and the sense retrieval algorithm adjust their
interpretation criteria to the personal view of each user. We have shown the fundamental
method for evaluating a personal view in multimedia information systems. These
algorithms are implemented and tested in our experimental multimedia database systems,
TRADEMARK and ART MUSEUM. These functions formed visual interaction in a user-
friendly manner.

Our research gives the guiding principle to the cognitive view mechanism for multimedia
information systems. The methods in this paper are basis for user-centered human
interface designing and multimedia information understanding.

Acknowledgments:

The author would like to thank the colleagues in Electrotechnical Laboratory, especially
Dr. Akio Tojo, Dr. Toshitsugu Yuba, Dr. Kunikatsu Takase, Dr. Hideo Tsukune, Mr.
Koreaki Fujimura and Mr. Takio Kurita for their support in this research.

The author would also thank to the students from University for Library and Information Science (ULIS) and Tsukuba University and visitors from private companies.

References:

Chang, N. S. and Fu, K. S. (1980): Query-by-Pictorial Example, *IEEE Trans. on Software Engineering,* **SE-6**, 6, 519-524.

Chang, S. K., et al. (1988): An Intelligent Image Database System", *IEEE Trans. on Software Engineering,* **SE-14**, 5, 681-688.

Chijiiwa, H. (1983): Chromatics, Fukumura Printing Co., 128-163.

Grosky, W. I. and Mehrotra, R. (eds.) (1989): Image Database Management, *COMPUTER (special issue),* **2 2**, 12, 7-71.

Iyenger, S. S. and Kashyap, R. L. (1988): Image Databases, *IEEE Trans on Software Engineering (special selection),* **SE-14**, 5, 608-688.

Kato, T., et al. (1991): A Cognitive Approach to Visual Interaction, *Proc. of Multimedia Information Systems MIS'91,* 271-278.

Yankelovich, N., et al. (1988): Intermedia: The Concept and the Construction of a Seamless Information Environment, COMPUTER, **2 1**, 1, 81-96.

Part VII

Case Studies of Data Science

- Social Science and Behavioral Science

- Management Science and Marketing Science

- Environmental, Ecological, Biological, and Medical Sciences

Proposition of a new Paradigm for a scientific Classification for Leadership Behavior Research Perspective

Jyuji Misumi

Institute of Social Research,

Institute of Nuclear Safety System, Incorporated

Keihanna Plaza, 1-7

Hikaridai, Seika-cho,

Soraku-gun, Kyoto 619-02,

Japan

Summary: This is our approach to the behavioral science of leadership, comprising a behavioral morphology and a behavioral dynamics, each with its specifically and generality. In this sense, it is different from current social science research. The later lacks the category of general behavioral morphology and cross-disciplinary perspective. A balance must be struck among the four areas of the above-mentioned paradigm, if there is to be productive interdisciplinary research.

1.The Limitations of Traditional Leadership Types

The behavioral classification concerning leadership types employed from common usage has the following imitations.

First, it is unidimensional-democratic/ autocratic, conservative/progressive, liberal /authoritarian, hawk/dove, employee-centered/ production-centered-rather than multidimensional.

Second, the terms used in this classification have multiple meanings in common usage, which makes them difficult to operationalize.

Third, these terms are heavily value-laden.

Fourth, the categories are used on historical concepts and functional concepts.

2.PM Leadership Concept

As remedy, we developed the leadership PM concept which 1)allows multidimensional analysis, 2) can be operationally defined, 3)is itself value-neutral, 4)makes possible experimental research and statistical studies.

Measuring leadership which is very much a group phenomenon, requires a group functional concept like PM concept(Misumi, 1985).

In the concept of PM,P stands for performance and represents the kind of leadership that is oriented towards achievement of the group's goal and problem solving. Being an abbreviation of maintenance,M stands for the kind of leadership that is oriented   towards the group's self-preservation or maintenance and strengthening of the group process itself. These two conceptual elements(P and M)are similar to Bale's(1953) "task leader" and "emotional leader".

The concept of PM is a constructive concept to classify and organize the factors obtained from leadership at different levels. It is not merely a descriptive concept for the factors obtained from factor analysis, but is at a higher level of abstraction. Because of being abstract, PM concept applies not only to industrial organizations, but also to many other social groups, P does not concern production only but also more general group goals or problem solving tasks. This is what principally distinguishes it from Blake

Mouton's(1964) model.
In the case of PM concept, we   consider P and M to be two axes on which the level of each type can be measured(high or Low),thus obtaining four distinct types of leadership (see figure 1).The validity of these four PM types was proved using correspondence analysis which was first developed by Guttman(1950) and later by Hayashi(1956).

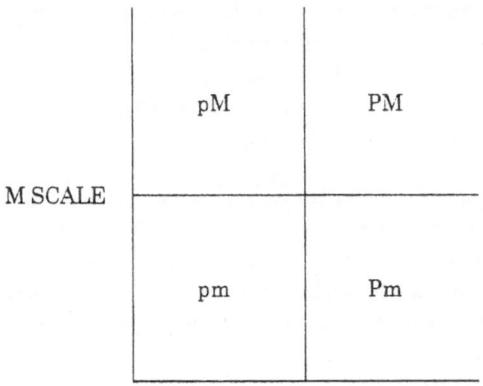

Fig.1.Conceptual representation of 4 patterns of PM leadership behavior
(Misumi,J.1984)

3.A New Research Paradigm:
Searching for Differences and Similarities

As a group phenomenon, leadership pertains to all the fields of social science. Consequently, for a better comprehension of leadership, an interdisciplinary perspective is indispernsable. PM theory, more than any other leadership theory, is the result of abroad interdisciplinaly research program(Misumi & Hafsi, 1989; Misumi & Perterson, 1987).
This interdisciplinary orientation would have been difficult, if not impossible, without the existence of an adequate research paradigm.
One of the principal characteristics of PM Leadership Theory is, as indicated by the paradigm if Fig.2, to apprehend the study of leadership in terms of two principal perspectives:(a)behavioral morphology and (b)behavioral dynamics.

	situation	
	General	Specific
Behavioral-morphology dimension	General behavioral morphology	Specific behavioral morphology
Behavioral-dynamics dimension	General behavioral dynamics	Specific behavioral dynamics

Fig.2.paradigm of science of leadership behavior.(Misumi,J.1984)

4.Behavioral Morphology of Leadership

The morphological approach of behavioral morphology consists in indentifying, describing, naming, and categorizing the forms of leadership in both general and specific settings. This distinction is based on the idea that there are both universal and particularistic leadership situations.

5.Behavioral Dynamics of Leadership

The behavioral morphology approach alone is, however, not sufficient to apprehend the leadership phenomenon. We need also a complementary approach that helps ascertain the causal laws that govern or determine the effectiveness of leadership.

Like behavioral morphology, behavioral dynamics can also be further subdivided into general behavioral dynamics and specific behavioral dynamics, depending on the degree of abstraction being considered(specific or general).

In the behavioral dynamics area, field research using quantitative behavioral methods has been complemented by laboratory research.

This is our approach to the behavioral science of leadership,comprising a behavioral morphology and a behavioral dynamics, each with its specificity and generality. In this sense, it is different from current social science research. The latter lacks the category of general behavioral morphology and cross-disciplinary perspective. A balance must be struck among the four areas of the above mentioned paradigm,if there is to be productive interdisciplinary research.

6.Some Empirical Results

Our research on the PM model consisted of both field surveys in different kinds of organizations and laboratory studies. Regarding measurement in the field, we found that evaluation by subordinates of their superiors was more valid than evaluation by superiors, peers or self .We,therefore, had subordinates evaluate the leadership of their superiors on the P and M dimensions.

To determine the level of P and M leadership for each subject, we first calculated the mean score of al subjects on each item of the two dimensions (P and M). As discussed by Misumi(1985), these P and M items, represented in Table 1, are the results of factor analysis. A leader whose score in P and M, is, for example, higher than the mean, is thought to provide a leadership of PM-type. A leader whose score is higher than the mean only in P dimension, is classified as providing a P-type(or Pm-type) leadership. When a leader's score is higher than the mean only in M dimension, he is referred to as a M-type(pM-type).When a leaders obtains a score lower than the mean in both dimensions, he is thought to provide a leadership of pm-type. This results in our final four-type classification:PM,P,M and pm.

To test the validity and reliability of these leadership categories in industrial organizations, we examined their relationship with some objective and cognitive variables such as productivity, accident rate, rate of turnover, job satisfaction, satisfaction with compensation, sense of belongingness to company and labor union, team work meetings quality, mental hygiene, and performance norms. More than 300,000 subjects were surveyed. As indicated in Table 2, of the four types PM-type was found to provide the best results, and pm-type the worst. In the long run, M-type ranks second, and in the short run, P-type ranks second.

Table 1

Factor Loading's of Main Items on Leadership(Misumi, 1984)

Items	Factor loadings		
	I	II	III
59.Make subordinates work to maximum capacity	.687	-.017	-.203
57.Fussy about the amount of work	.670	-.172	.029
50.Fussy about regulations	.664	-.072	.001
58.Demand finishing a job within time limit	.639	.070	.065
51.Give orders and instructions	.546	.207	.198
60.Blame the poor job on the employee	.528	.113	-.121
74.Demand reporting on the progress of work	.466	.303	.175
86.Support subordinates	.071	.780	.085
96.Understand subordinates' viewpoint	.079	.775	.229
92.Trust subordinates	.024	.753	-.003
109.Favor subordinates	.067	.742	-.050
82.Subordinates talk to their superior without any hesitation	-.026	.722	.059
101.Concerned about subordinates' promotion, pay-raise, and so forth	.147	.713	.134
88.Show consideration for subordinates' personal problems	.132	.705	.150
94.Express appreciation for job well done	.058	.651	.129
104.Impartial to everyone in work group	-.143	.644	.164
95.Ask subordinates' opinion of how on-the-job problems should be solved	.049	.643	.121
85.Make efforts to fill subordinates' request when they request improvement of facilities	.110	.606	.333
81.Try to resolve unpleasant atmosphere	.233	.538	.338
87.Give subordinates jobs after considering their feelings	-.276	.478	.457
76.Work out detailed plans for accomplishment of goals	.229	.212	.635
75.No time is wasted because of inadequate planning and processing	.038	.333	.614
70.Inform of plans and contents of the work for the day	.254	.278	.607
52.Set time-limit for the completion of the work	.319	.299	.554
53.Indicate new method of solving the problem	.251	.489	.479
56.Show how to obtain knowledge necessary for the work	.295	.492	.472
61.Take proper steps for an emergency	.360	.451	.305
69.Know anything about the machinery and equipment subordinates are in charge of	.255	.304	.458

Table 2

The Summary of Comparison of the Effectiveness of 4 Patterns of P-M Leadership Behavior on Various Kinds of Factors of Work Group (the figures of this table show the ranking of effectiveness in each factor) (Misumi,J., 1984)

		\multicolumn{4}{c}{Pattern of leadership behavior}
		PM	M	P	pm
Productivity	Long term	1	2	3	4
	Short term[a]	1	3	2	4
Accidents[b]	Long term	1	2	3	4
	Short term[a]	1	3	2	4
Turn over		1	2	3	4
Group norm for high performance	Long term	1	2	3	4
	Short term[a]	1	3	2	4
Job satisfaction		1	2	3	4
Satisfaction with salaries		1	2	3	4
Team work		1	2	3	4
Evaluation of work group meeting		1	2	3	4
Loyalty (belongingness) to	Company	1	2	3	4
	Labor union	1	2	3	4
Communication		1	2	3	4
Mental hygiene (excessive tension and anxiety)[c]		1	2	3	4
Hostility to supervisor[d]		1	2	3	4

a Including the data obtained by laboratory studies.
b Smaller figures indicate lower rate of accidents or turn over.
c Smaller figures indicate less tension and anxiety
d Smaller figures indicate less hostility to supervisor.

It is noteworthy that this order of effectiveness is not limited to businesses only, but is the same for teachers(Misumi, Yoshizaki & Shinohara, 1977), government offices (Misumi, Shinohara & Sugiman, 1977), sports coaches(Misumi, 1985) and religious groups(Kaneko,1986).

References:

Bales,R.F.(1953): The equilibrium problem in small groups. In *working papers in the theory of action*, ed. Parsons,T.,Bales,R.F.& Shils,E.A. Glencoe,III: Free Press.

Blake,R.R., & Mouton, J.S.(1964): *The managerial grid.* Houston: Gulf.

Guttman,L.(1950): Chaps. 2,3,6, and 8. In Stuffer.S.(Ed.), *Measurement and prediction.* Princeton:Princeton University Press.

Hayashi,C.(1956): Theory and examples of quantification(II). *Proceedings of the Institute of Statistics and Mathematics* 4,19-30.

Kaneko,S.(1986): Religious consciousness and behavior of religious a adherents' representatives. *Kyoto Survey Research center of Jodo-shin-syu,* 65-86.

Misumi,J.,(1984): *The behavioral science of leadership*(Second Edition). Yuhikaku

Misumi,J.,& Hafsi,M.(1989): La theorie de leadership de PM(Performance-Maintenance):une approach Japonaise de l'etude scintifique de leadership. *Bulletin de Psychologie,* Tome XLII,392,727-736.

Misumi,J.,& Peterson,M.F.(1987): Developing a Performance-Maintenance (PM) Theory of leadership. *Bulletin of the Faculty of Human Sciences,* Osaka University,13,135-170.

Misumi,J., Shinohara,H., & Sugiman,T.(1977): Measurement of leadership behavior of administrative managers and supervisors in local government offices. *The Japanese Journal of Educational and Social Psychology,*16,2,77-98.

Misumi,J.,Yoshizaki,S.,& Shinohara,S.(1977): The study on teacher's leadership: its measurement and validity. *The Japanese Journal of Educational and Social Psychology,*25,2,157-166.

Structure of Attitude toward Nuclear Power Generation

Kiyoshi Karube[1]   Chikio Hayashi[2]   Shinichi Morikawa[3]

[1]Teikyo University
359 Ohtsuka, Hachiohji
Tokyo 192-03, Japan

[2]Institute of Statitical Mathematics
4-6-7 Minami-Azabu, Minato-ku
Tokyo 106, Japan

[3]Institute of Nuclear Safety System, Incorporated
Keihana Plaza, 1-7 Hikaridai, Senka-cho, Soraku-gun
Kyoto 619-02, Japan

Summary: How does the Japanese general public respond to the planned future progress of nuclear power generation? Research and surveys on the Japanese national character, which have been carried out in Japan for over 40 years, confirm that despite considerable change, the "core" of the Japanese national character continues to be firmly preserved. What consideration, therefore, should be given to such national character in connection with nuclear power generation, when nuclear power generation is expected to continue in future, and even higher-level developments in nuclear power technology are likely to be pursued, both quantitatively and qualitatively?
In the present research project, data gathered via opinion survey are analyzed to elucidate how Japanese general public attitudes toward nuclear power generation, whether favorable or unfavorable, and the Japanese national character, which governs various aspects of societal life, are interrelated within an attitude space, and how subjects can be classified and divided therein.

1. Data collection and analysis

Data were collected in an opinion survey addressed to 1,138 subjects, 18 to 79 years of age, selected by probability sampling in the area of Kansai. The questionnaire included questions concerning ideas and feelings about various matters that constitute attitude toward nuclear power generation (e.g. images of nuclear power, understanding of nuclear power, attitude toward energy and environmental issues, sense of anxiety, sensitivity to risk, views of science and civilization that can influence the above, etc); social and political attitude closely related to the above; and archetypal Japanese characteristics (e.g. tendency toward moderate opinions, sense of trust, concept of typical Japanese leadership, attitude toward supernatural beings such as ghosts, psychological relationship with superstition, etc.), as well as questions concerning how electric power companies should deal with the general public.

Hayashi's Quantification Method III (pattern classification) is used in the analysis. When the maximum correlation coefficient is obtained in calculation, using numeric assigned to respondents and respective reply categories, and taking into account all reply categories, those categories generally present multi-dimensional category scores, the corresponding respondent scores also being multi-dimensional. Categories of proximate category scores are placed proximately in the attitude space, indicating their close correlation. Since respective dimensions of respondent scores serve as attitude scales that are independent from each other, respondents concentrated on the same scale can be classified into one group.

2. Findings

Replies to a group of closely correlated questions, such as questions concerning attitudes toward nuclear power generation, were analyzed first. When a group of questions was found to form a clear-cut scale, a category defined by the value range of the scale was regarded as representative of replies to the group of questions, in the same manner as in other question categories. Replies to almost all questions presented in the survey were analyzed, yielding the following findings.

The first axis in the attitude space divides the subjects (respondents) into those who are completely indifferent and those who are not. Subjects with high respondent scores on this axis (inflection point on or above the distribution curve) can be distinguished as the indifferent group. The number of subjects in this group accounts for 13% of all subjects.

The second axis divides the subjects into those who are strongly positive toward nuclear power generation and those who are not, while the third axis divides the subjects into those who are strongly negative toward nuclear power generation and those who are not. Respondent scores show that strongly positive accounts for 11%, and strongly negative for 9%.

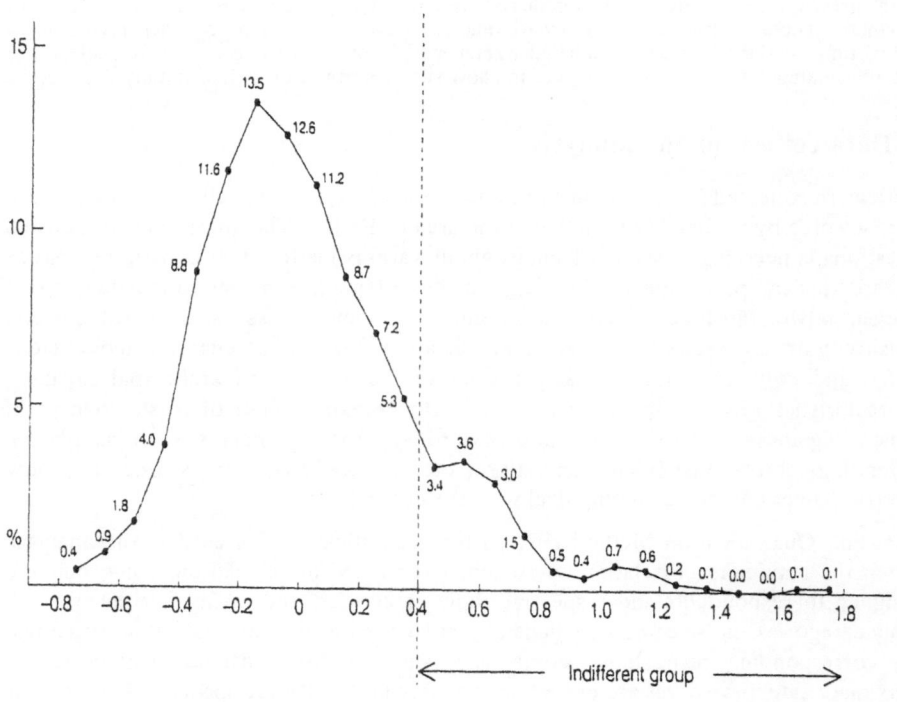

Fig. 1 I-axis Sample Score Distribution

The straight line formed by the "positive", "moderate" and "negative" categories which runs diagonally on the plane demarcated by the second and third axis, becomes, when projected onto the parallel line passing through the original point, a scale that divides the three categories. The subjects are classified on this scale into groups that account for 12%, 50%, and 5%, respectively, of the total.

The positions of item categories that correspond to these groups, the item categories of the Japanese national character, and other item categories indicate that respondents in the indifferent, strongly positive, and strongly negative groups have nothing to do with typical Japanese sentiments. (Tab.1, Tab.2)

It is believed that for the majority of respondents, which excludes people in these groups, somewhat Japanese style communication can be effective in promoting nuclear power generation.

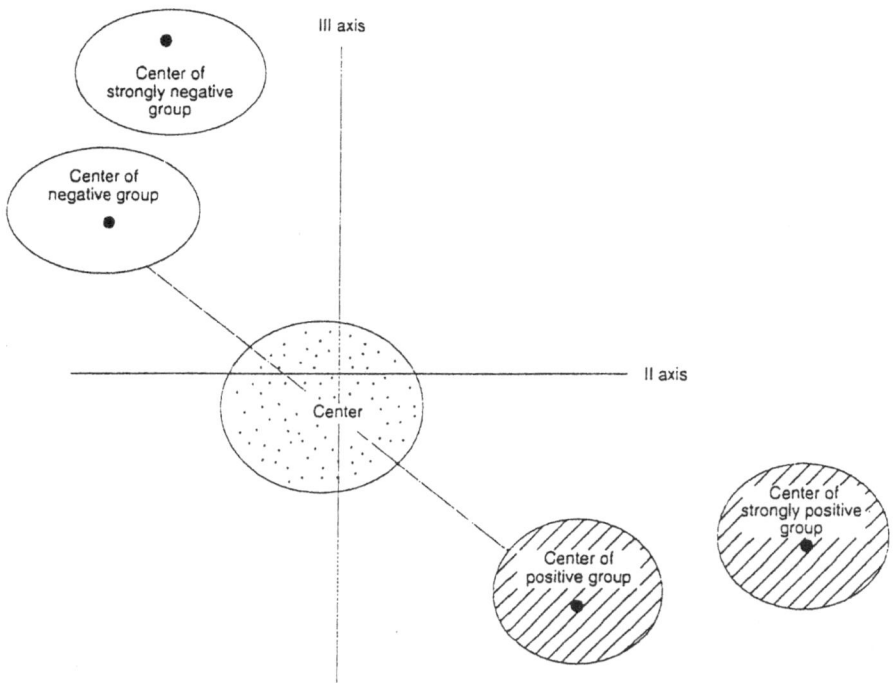

Fig.2 II-and III-axis Relationship (Model presentation)

Tab.1 Characteristics of Subjects with Positive and Negative Attitude toward Nuclear Power Generation

Characteristics	Common Characteristics
Strongly positive Positive views of science and civilization Few centrist replies (tendency to express opinions clearly) Not very interested in accidents **Strongly negative** Development of aircraft, etc./:not useful Do not like typical Japanese leadership Consider most dangerous/:diseases, drug hazards, malpractice, nuclear power, nuclear power generation(plant), radioactive contamination, war, environmental contamination and natural environmental destruction Negative views of science and civilization Consider Japan's power generation capacity sufficient Strong interest in environmental issues	Sufficient knowledge about positive and negative effects of nuclear power generation Little interest in supernatural beings Little influence of superstition Consider crime, bullying and inter-personal trouble most dangerous
Negative Usefulness of aircraft, etc./: moderate Importance of aircraft, etc./: moderate Images of nuclear power/:radioactivity, breakdown, environmental contamination, accident, explosion Widely used power generation method in Japan/: hydro Regarding nuclear power generation, wish to know more about past incidents and accidents, treatment of waste materials, regional development, impact of radioactivity, necessity of nuclear power generation, difference from atomic bomb Consider socialism good Fearful of impact of nuclear power plant accidents that can last for generations Moderate replies to many questions Main cause of nuclear power accident/: inadequate operation , manual, equipment system failure... Sense of anxiety/: moderate and strong Interest in accident/: moderate and strong Somewhat negative views of science and civilization Moderate view of typical Japanese leadership Evaluation of ideologies/: depend on times and situation Images of nuclear power/: danger	Strong distrust of others Cause of nuclear power plant accidents: operation error Remember Chernobyl accident very well Insufficient knowledge of positive and negative effects of nuclear power generation
Positive Consider democracy and capitalism good Little anxiety Trust in others very much	Strong interest in politics Male Japan's power generation capacity/: insufficient Consider energy issues important University graduate 60 years of age or older

(Continue)

(Continued)

Japan's power generation capacity/: appropriate
Somewhat positive and moderate view of science and
  civilization
Strong influence of superstition
Strong interest in supernatural beings
Chernobyl accident/:somewhat remember, do not remember
Widely used power generation method/: nuclear
Consider natural disaster most dangerous
Consider typical Japanese leadership favorable
Images of nuclear power/: electricity, power, power generation,
  energy, fuel, resources
Somewhat weak interest in environmental issues
Usefulness of aircraft, etc./: very useful
Importance of aircraft, etc./: very important
Fearful of nuclear power plant accident/: exposure to
  radioactivity above permissible levels

Tab.2 Category Score and Weighted Sum of Respondents
(Sample weight =pop/respondents in each of urban and provincial strata)

	N(4676)	Axis-1	Axis-2	Axis-3
Attitude toward nuclear power generation				
1 Strongly negative	608	-0.245	-0.537	4.430
2 somewhat negative	1192	0.922	-1.570	0.8φ6
3 Moderate	1141	0.713	-0.405	-0.621
4 Somewhat positive	1122	0.235	0.550	-1.853
5 Strongly positive	613	-2.226	4.350	-1.433
Image of nuclear power				
1 Energy, fuel, resource	331	-0.443	1.397	-0.760
2 Electricity, power, power generation	1801	-0.310	0.613	-0.504
3 War, atomic bomb, nuclear weapon	1339	-0.242	-0.209	0.148
4 Accident, explosion	805	-0.849	-0.747	0.844
5 Radioactivity, breakdown, environmental contamination	858	-0.874	-0.635	1.460
6 Danger, fear	267	0.566	-1.049	-0.159
7 Other	648	0.287	0.148	-0.140
Understanding of nuclear power				
(1) Power generation method in Japan				
1 Thermal	1653	-0.875	0.720	0.395
2 Hydro	1106	0.447	-0.564	1.056
3 Nuclear	1546	0.492	-0.011	-1.323
4 Other	371	2.233	0.197	0.575

(Continue)

658

(Continued)

(2) Know about positive/negative effects of nuclear power generation?				
1 Yes	700	-2.537	3.061	1.591
2 No	3052	0.734	-0.700	-0.580
3 Neither	924	0.189	0.669	0.699
(3) What you wish to know about nuclear power generation				
1 Mechanism	1416	-0.870	-0.580	-0.412
2 Necessity	1351	-0.162	-1.718	-0.260
3 Economic efficiency	692	-0.461	-1.317	-0.635
4 Safety	3467	-0.144	-0.526	-0.273
5 Past incidents and accidents	1153	-1.228	-0.867	0.509
6 Disaster prevention system	2919	-0.480	-0.564	-0.220
7 Impact of radioactivity	2818	-0.197	-1.025	-0.152
8 Treatment and disposal of spent fuel and waste materials	2678	-0.899	-0.496	0.331
9 Difference from atomic bomb	974	-0.011	-2.268	-0.545
10 Regional development in localities around nuclear power plants	570	-0.986	-1.136	-0.916
11 Nothing in particular	305	5.194	5.684	1.321
(4) Remember Chernobyl accident?				
1 Very well	2883	-1.022	0.340	0.837
2 A little	1222	1.042	-0.440	-1.485
3 No	571	4.045	0.316	-1.067
(5) Main cause of accident				
1 Equipment system failure	2692	-0.398	-0.286	0.270
2 Operator error(human error)	1896	-0.618	0.106	1.009
3 Inadequate management system	3308	-0.179	-0.030	-0.156
4 Inadequate operation manual	1191	-0.753	-0.396	0.524
(6) Impact of accident				
1 Death or injury in power plant or local community	341	1.964	2.318	-1.153
2 Radioactivity exposure above permissible levels	575	0.809	1.822	-1.853
3 Radioactive contamination spreading to other parts of the world	1608	-0.392	-0.087	-0.100
4 Impact on future generations	2047	-0.134	-0.565	0.691
Energy, environmental issues, sense of anxiety, sensitivity to risk				
(1) Energy issues				
1 Very important	2036	-1.493	0.852	0.455
2 Important	2318	0.893	-0.849	-0 791
3 Not important	303	4.932	2.549	2.588
(2) Japan's power generation capacity				
1 Sufficient	619	-0.268	-0.118	2.267
2 Almost sufficient	670	0.042	-0.690	-0.746
3 Exactly appropriate	1520	1.147	0.064	-0.561
4 Almost insufficient	1449	-0.284	0.273	-0.203
5 Insufficient	386	-1.844	1.567	0.196

(Continue)

(Continued)

(3) Environmental issues			
1 Very interested	1001	-1.581	-1.679	2.181
2 Interested	1580	-0.881	-0.305	-0.389
3 Slightly interested	1592	0.969	0.540	-1.418
4 Not very interested	503	4.117	3.830	1.345
(4) Fear (Sense of anxiety)			
1 Little	2052	0.508	0.802	-0.389
2 Moderate	1849	-0.145	-0.104	0.607
3 Much	775	-0.176	-1.070	-0.433
(5) Interest in accident (Sensitivity to risk)			
1 Little	1027	2.438	2.954	-0.636
2 Moderate	2174	-0.537	-0.216	0.500
3 Much	1475	-0.473	-1.316	-0.301
(6) Most dangerous thing in social life			
1 Traffic accident	1976	-0.382	0.599	-0.195
2 Natural disaster	831	-0.011	0.256	-1.167
3 Environmental pollution, destruction, abnormal weather	247	-1.766	-1.337	2.400
4 Fire	244	0.174	-1.588	-1.717
5 Crime, bullying, interpersonal trouble	46	-0.676	2.340	4.727
6 War	77	0.444	0.444	2.926
7 Nuclear power(generation), radioactive contamination	54	-0.436	-1.981	3.282
8 Disease, drug hazards, malpractice	101	-0.423	-1.631	3.744
Views of science and civilization			
1 Very negative	563	-0.409	-1.675	2.769
2 Somewhat negative	862	-0.479	-0.639	0.187
3 Slightly negative	1288	1.180	0.239	0.166
4 Moderate	738	0.859	-0.489	-0.777
5 Somewhat positive	957	-0.454	1.287	-0.901
6 Very positive	268	-1.635	3.504	-1.902
Social and political attitude			
(1) No. of items considered important (importance of aircraft etc.)			
1 0 item	604	3.055	1.459	1.862
2 1 item	1118	0.530	-0.809	1.949
3 2 items	1703	-0.410	-0.685	0.078
4 3 items	1251	-0.880	1.448	-2.755
(2) No. of items considered useful (usefulness of aircraft etc.)			
1 0-1 item	623	2.740	0.822	3.941
2 2 items	1415	-0.019	-1.114	1.656
3 3 items	2638	-0.395	0.639	-1.823
(3) Interest in political affairs			
1 Very interested	877	-1.984	1.707	1.321
2 Somewhat interested	2009	-0.410	-0.060	-0.245
3 Not interested	1749	1.811	-0.405	-0.412

(Continue)

(Continued)

(4) Ideology				
1 Democracy and capitalism considered good	1884	-1.256	1.408	-0.349
2 Depend on time and situation	1843	1.146	-0.715	0.035
3 Socialism considered good	949	0.940	-0.750	0.613
Japanese national characteristics				
(1) Tendency toward moderate opinions				
1 0-4 items (Tend for express opinions clearly)	462	-2.689	3.702	-0.818
2 5-12 items	2792	-0.566	-0.132	-0.189
3 13 or more items	1422	2.435	-0.505	0.328
(2) Scale of sense of trust				
1 0 item	1316	0.449	0.398	1.099
2 1 item	1544	0.808	-0.577	-0.385
3 2-3 item (Strong distrust of others)	1816	-0.661	0.546	-0.475
(3) Typical Japanese leadership				
1 0-2 items (In favorable to Japanese style of leadership)	412	1.268	0.884	3.837
2 3-6 items	2480	0.617	-0.483	0.018
3 7 or more items (Favorable)	1784	-0.793	0.816	-0.917
(4) Scale of interest in supernatural beings				
1 0-3 items (Little interest)	627	1.566	2.861	2.705
2 4-9 items	2300	0.099	-0.065	0.124
3 10 or more items (Much interest)	1749	-0.326	-0.584	-1.139
(5) Superstition believable?				
1 0-2 items (Little influence)	653	-0.288	2.324	3.615
2 3-6 items	1797	-0.231	-0.184	-0.025
3 7-8 items (Much influence)	2226	0.558	-0.253	-1.045
Demographics				
(1) Sex				
1 Male	2214	-0.784	1.486	0.653
2 Female	2462	0.964	-1.083	-0.592
(2) Age				
1 18-29 years old	1121	0.972	-0.092	-0.533
2 30-39 years old	967	-0.044	-0.170	-0.389
3 40-59 years old	1937	-0.323	0.078	0.455
4 60 years old or above	651	0.334	1.138	0.126
(3) Education				
1 Elementary/secondary school graduate	664	1.410	0.436	0.490
2 High school	2467	0.467	-0.441	-0.415
3 University	1505	-1.023	0.918	0.410
(4) Residence				
1 Urban	4000	0.030	0.046	0.017
2 Provincial	676	0.768	0.649	-0.115
R(correlation)		0.335	0.282	0.265

References:

Hayashi, C. and Morikawa, S. (1995)   National Character and Communication.   INSS Report May, 1995

Research Concerning the Consciousness of Women's Attitude toward Independence

Setsuko Takakura

Tokyo International University
2509 Matoba, Kawagoe-shi
Saitama-ken 350-11, Japan

Summary: Applying Quantification Method III(Correspondence analysis) to the results of our survey we found a pattern of thinking as regards the consciousness of women's independence. The majority of the samples (female university graduates) consider "women's independence" from three points of view: financial independence, psychological strength and family role. It is said that financial independence is a very important aspect for women's independence, but more than half of the samples value psychological independence rather than financial independence, even if they consider that, concerning men's independence, both financial power and psychological strength are necessary conditions for it.

1. Outline of Survey

Recently what do Japanese women understand by the expression "women's independence", and what is the relation with the other items: consciousness of liberty. equality between the sexes, or happiness, autonomy, identity, etc.? What are the obstructions to their independence? The purpose of this research is to clarify these subjects, (we conducted a mail survey*).

We have chosen 7 universities and 5 junior colleges situated within Tokyo and surrounding areas. As the population we have determined female graduates of these universities and junior colleges in the years: 1958, '67, '75, '81, '86, '91. The number of samples and the number of effective responses are as table-1.

Table-1

Number of samples and number of effective responses

graduation year	'58	'67	'75	'81	'86	'91	TOTAL
7 universities	188 132 (74%)	188 98 (58%)	258 130 (58%)	258 109 (53%)	316 137 (52%)	315 132 (37%)	1523 738 (56%)
5 colleges	160 98 (64%)	160 89 (59%)	220 106 (53%)	221 91 (49%)	372 153 (45%)	367 132 (37%)	1500 669 (48%)
TOTAL	348 230 (69%)	348 187 (58%)	478 236 (49%)	479 200 (51%)	688 290 (48%)	682 264 (41%)	3023 1407 (52%)

number on the left : number of samples
number on the right : number of effective responses   ( ) : ratio of effective responses

We obtained 1,407 effective responses in total.

2. Results

2.1 outline

We will show the results of the responses to the two main questions. (The numbers are the percentage of the responses.)

Regarding the consciousness of independence:

Q11. Do you think that you are independent now? (no dependency upon anyone)

1.sufficiently independent   (12%)   2.just independent   (46%)
3.not sufficiently independent (24%)   4.hardly independent   (13%)
5.D.K.   (5%)

Regarding the degree of satisfaction with the level of one's own independence:

Q13. What do you think of the level of your independence?

1.satisfied   (27%)   2.unsatisfied, but have no choice   (33%)
3.unsatisfied, then seeking some other way   (27%)   4.nothing, D.K.   (13%)

We show in Fig.-1 some aspects of the responses 1 and 2 in Q.11, according to several categories: university or junior college, year of graduation, occupation, marital status; and we show in Fig.-2 the response 1 in Q.13, according to the same categories.

Q.11 responses:1,2 (%)

Q.13 response:1(satisfied) (%)

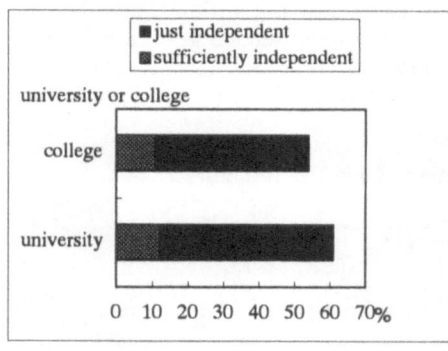

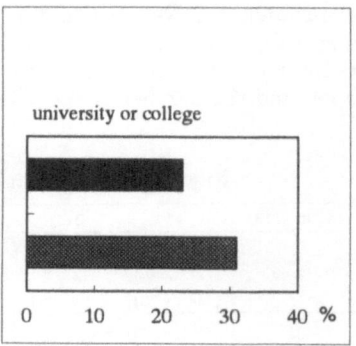

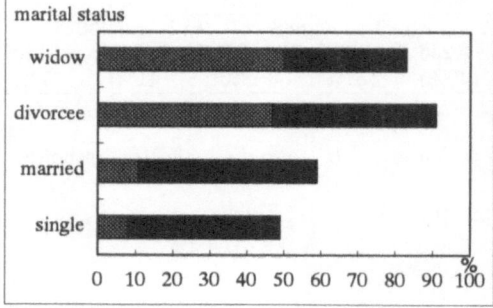

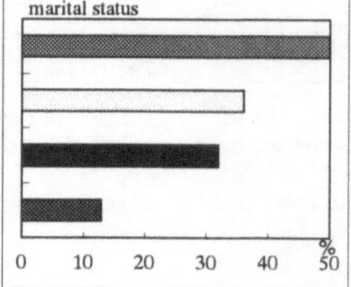

Fig.-1-1   Fig.-2-1

Q.11 responses:1,2 (%)   Q.13 response:1(satisfied) (%)

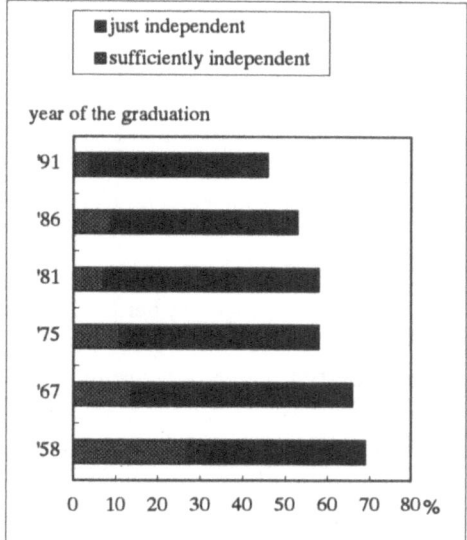

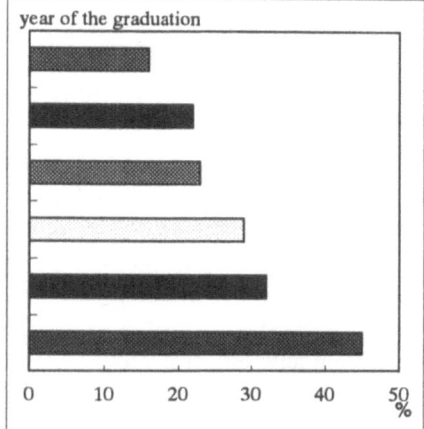

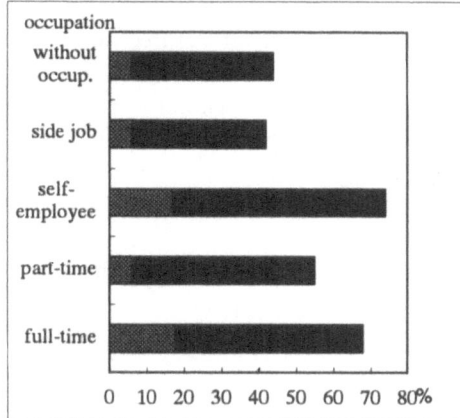

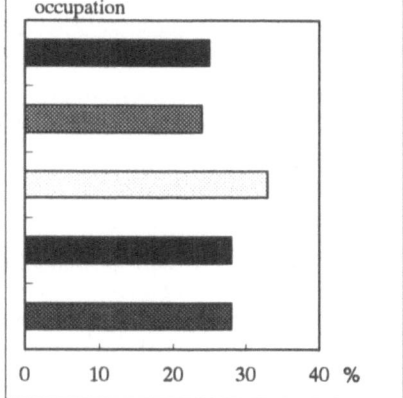

Fig.-1-2   Fig.-2-2

From these figures we understand that the highest degree of consciousness of independence was as follows: by age, among the oldest samples; by profession, among full-time workers and the self-employed; by marital status, among divorcees and widows.

In order to ascertain the fundamental structure of women's consciousness regarding independence we asked 3 questions: Q.2, Q15, Q.29. These questions are as follows. (The numbers are the percentage of the responses.)

Q.2 Some examples of women's life-styles are listed below. Please mark the types that attract you. (no more than 2)

1. fashionable single career woman able to make full use of her abilities in her profession (6%)

2. woman with a full-time profession living with husband (without children) (9%)

3. married woman with a child/children who devotes her energies to her profession, entrusting housework and child care to somebody else as necessary (25%)

4. woman with a full-time profession who shares partial charge of housework with husband and child/children (37%)

5. woman with a full-time profession who does housework and child care without help from anybody else (5%)

6. woman whose priority is taking care of husband and child/children and who works part-time locally during the daytime as long as she has time to spare (11%)

7. woman who takes care of husband and child/children and participates actively in some activity which contributes to society (30%)

8. woman who depends on the income of husband, and who efficiently deals with housework and child care and then participates actively in various free-time activities of her own choice (21%)

9. woman who while preserving good judgement and making efforts to extend her knowledge, and although depending on the income of her husband, successfully takes care of the family and domestic management and acts as the mainstay of the family (25%)

Q.29 When you hear the expression "women's independence" how do you feel about it? Please mark any of the following categories which correspond to your feelings. (as many as you like)

#. unattractive (4%)	R. impressive (27%)
S. self-confident (54%)	T. financially independent (56%)
U. vibrant (44%)	V. active (54%)
#. selfish (2%)	W. trendy (18%)
X. self-reliant (39%)	Y. self-assertive (24%)
#. too busy (3%)	Z. equality between the sexes (32%)

Q.15 How important do you think each of the following items is, from the point of view of women's independence and men's independence ?

F: women's independence				Item	M: men's independence			
very import-ant	a lit-tle import-ant	not very import-ant	not at all import-ant		very import-ant	a lit-tle import-ant	not very import-ant	not at all import-ant
45	46	8	0	A. being able to support oneself financially	78	18	1	0
46	47	5	0	B. having one's own con-victions	62	32	4	0
56	41	1	0	C. acting according to one's own principles	70	26	1	0
55	42	2	0	D. being able to select one's own way of life	68	28	2	0
34	54	10	0	E. being able to manage a family budget	22	51	23	0
36	52	10	0	F. being able to produce a harmonious family atmosphere	28	57	12	0
31	51	15	1	G. protecting and suppor-ting the family	40	45	11	1
21	51	26	1	H. having a profession	69	25	3	0
4	44	47	4	I . participating actively in local social activities	6	41	46	4
14	62	20	1	J . taking an interest in political and economic affairs	37	50	9	1
34	56	8	0	K. being able to coope-rate with others	36	54	7	0
39	55	5	0	L. respecting the view-point of others	42	52	4	0
24	57	16	1	M. being able to accomp lish one's aims	36	49	11	1
56	39	3	0	N. being able to look after oneself	42	48	7	0
31	49	18	1	O. being able to do house-work	12	49	35	2
33	46	17	2	P. being able to take care of children	10	46	38	3
10	43	42	2	Q. supporting the family financially	66	26	4	1

(The numbers are the percentage of the responses of each category.)

2.2. The application of Quantification Method III(Correspondence analysis)

We applied Quantification Method III (Correspondence analysis) to the response of Q15-F (except item-I) regarding women's independence (we have taken notice of the response 1(very important) (Fig.-3); and in the same way, to the response of Q15-M regarding men's independence (Fig.-4).

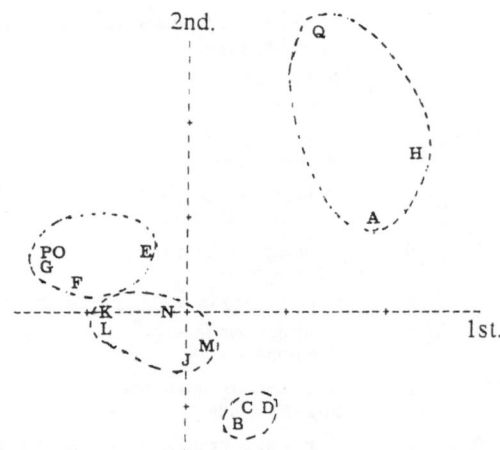

Fig.-3 Q.15-F: concerning women's independence

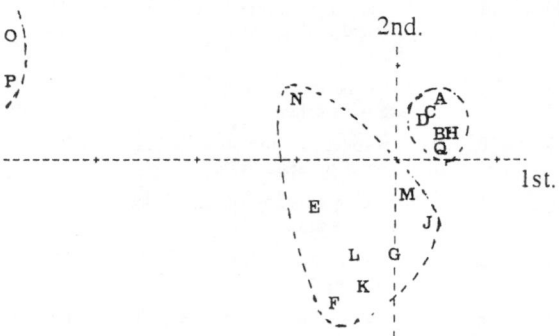

Fig.-4 Q.15-M: concerning men's independence

As a result of these figures we were able to classify the items as follows:

Q15F (Fig.3)		Q15M (Fig.4)	
concerning women's independence		concerning men's independence	
group of items	meaning	group of items	meaning
A,H,Q	financial power	A,H,Q,B,C,D	financial power and
B,C,D	psychological strength		psychological strength
J,K,L,M,N	human relations etc.	E,F,G,J,K,L,M,N	some role in the family and human relations etc.
P,O,E,F.G	role in the family	O,P	ability of doing housework

This classification reveals that the samples (female graduates) consider women's independence from four different aspects: financial power, psychological strength, human relations and role in the family; while, concerning men's independence they consider almost only two aspects: the aspect of financial and psychological strength and the aspect of human relations etc.. (The item-O and the item-P are situated very far from the other items; it means that the samples who marked "very important" for items O and P are few and heterogeneous.)

We apply also Quantification Method III to all the responses of Q.2, Q.29, and Q.15-F. (Fig.-5). We understand that the samples who regard financial power as very important items for women's independence chose the categories 1 or 2 in Q.2, which is the opposite of the samples who chose the categories 6,7,8,9 in Q.2; the latter choosing also "W" in Q.29 seem conservative; it seems that they do not consider "independence" positively. The samples who chose O,P,G,K,L,F,E,N, in the Q.15-F respect the family role and they chose type-5 in Q.2, it seems that they consider this kind of life-style as an ideal.

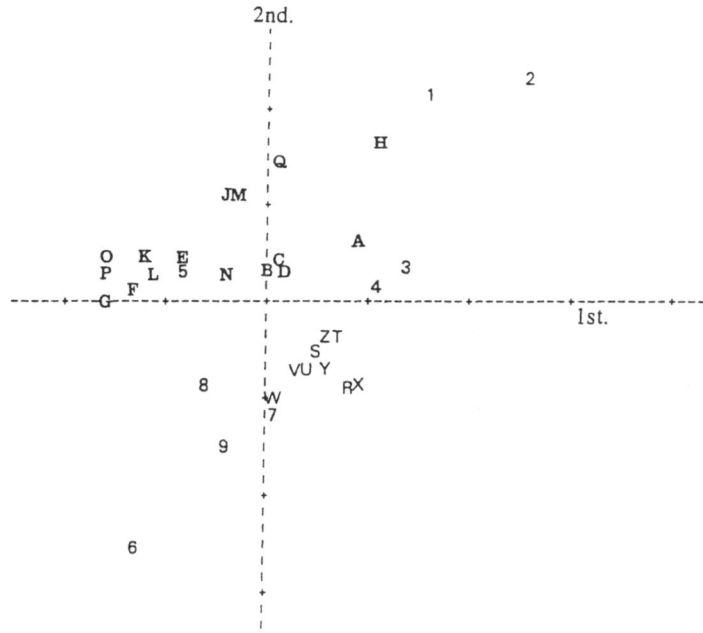

Fig.-5   Q.2: (1 -- 9);   Q.29: (R -- Z);   Q.15-F: (A -- Q)

People say that financial power and psychological strength are very important for independence. We found here that women's (female graduates) opinions of independence are different according to whether they are considering women's independence or men's independence. That is, concerning men's independence they consider that the two strengths, financial and psychological, are important, while concerning women's independence some samples (female graduates) attach importance to financial power and other some samples to psychological strength.

Concerning the women's independence we have taken the items A,H,Q (which mean

668

financial power); dividing two categories: "very important" and the others, we applied again Quantification Method III. Then, taking the first 'eigen-vector' as the out-side criterion we applied the multiple regression analysis for the purpose of finding the effective factors; we have taken as predictor factors: kind of school, year of the graduation, marital status, occupation, response to Q.11 and response to Q.13. We found from this result that the occupation (having one or not) is the remarkable effective factor, and the kind of schools (university or junior college) and the response to Q.11 are quite effective factors. We have tried this same method concerning the items B,C,D (which mean the psychological strength). We found as effective factors (to discriminate the samples who attach importance to these items for women's independence and the other samples who do not) the kind of schools (university or junior college), the response to Q.11 and the response to Q.13.

3. Conclusion

After these analyses we can say that the consciousness of female graduates toward independence is not so high; especially the consciousness of young graduates is lower than that of older graduates. The consciousness of the samples who are widows or divorced is strong; most of them having an occupation, retain financial independence. Most of the samples who have husbands do not consider financial power as a very important factor for women's independence, they attach importance to the psychological aspect rather than the financial aspect. The great part of young women graduates who are not yet married have not enough consciousness of independence and are not satisfied with the level of their independence. It is particularly interesting that we found by the method of the Quant.-3, the difference in women's (graduates) way of thinking between women's independence and men's independence. There were many discussions on the important factors: financial power and psychological strength regarding independence. We found that the samples who support the former aspect and the samples who support the latter aspect are not necessarily the same. The former are many among the samples who have an occupation, the latter are many among the samples who graduated from junior college are confident of the situation of their own independence. We obtained in this way some new views concerning the consciousness of women's attitude toward independence.

References

Garaudy, R. (1981): Pour l'avèment de la femme. Albin Michel

Moen, Y. (1992): Women's Two Roles. Auburn House

Sodei, T. et al. (1992): Research Concerning the Women's Attitude toward Independence, Bureau of Citizens and Cultural Affaires, Tokyo metropolitan Government (in Japanese)

Watanabe, K. (1990): The Concept of "Jiritsu" in Psychology, The Journal of the Faculty of Integrated Arts and Social Sciences, Japan Women's University, 1, 189-206 (in Japanese)

Acknowledgment:

We conducted this survey with the help of a grant in aid of The Tokyo Women's Foundation, and this study was carried out under the ISM Cooperative Research Program (95-ISM-CRP A-34).

A Cross-National Analysis of the Relationship between Genderedness in the Legal Naming of Same-Sex Sexual/Intimate Relationships and the Gender System

Saori Kamano

Institute of Statistical Mathematics
4-6-7 Minami-Azabu, Minato-ku
Tokyo 106 Japan

Summary: In this paper, I approach the naming of same-sex sexual/intimate relationships from a social constructionist perspective and undertake a theoretically informed cross-sectional cross-national analysis. Classifying the countries in terms of the genderedness in the legal naming of same-sex sexual/intimate relationships, I explore first, how various socio-politico-economic factors are related to the differentiations of the countries in the naming of same-sex sexual/intimate relationships; and, second, how such differentiations in naming are linked to the gender system. I undertake correlation analyses and logistic regression analyses to examine the these linkages.

1. Introduction

Feminist scholarship has noted that it is ultimately a gender issue as to whether and how same-sex sexual/intimate relationships are named or seen as a category apart from other types of relationships (see, e.g., Connell 1987). The very conceptual possibility of "same-sex sexual/intimate relationships" (abbreviated as SSSIR) as an identifiable type of relationships hinges on the viability of gender as a social category. In other words, it is only when "gender" or the social categories of "men" and "women" operate as a social divider and when heterosexuality is assumed can sexual/intimate relationships involving people of the same-sex be constructed as an identifiable and distinguishable category of relationships. Extant historical and anthropological studies have documented that such "naming" of same-sex sexual/intimate relationships varies overtime and across societies (see, e.g., Durberman, et al. 1980; D'Emilio and Freedman 1988). However, such documentation of the variation awaits more rigorous theorizing as well as systematic analysis, which I will attempt in this study by focusing on one aspect of naming--the "genderedness" of the naming of SSSIR--and take a first step toward a more theoretical and rigorous exploration of its cross-national variation.

I will first lay out my argument regarding the genderedness of the naming of SSSIR and the gender system before discussing how genderedness is operationalized and how countries are differentiated in this respect by socio-politico-economic factors. I will next consider whether and how genderedness in naming is related to the rigidity of gender categories and the level of gender inequality.

2. Theoretical linkage between the genderedness in the naming of SSSIR and the gender system

Since the naming of SSSIR is basically a gender issue, one can expect it to be related to the gender system of a society. Specifically, I argue that the rigidity of gender categories--which partially constitute the gender system--produces differences in how men and women are generally treated in a society, leading to gender differences in the naming of SSSIR. Highly rigid gender categories mean that men and women are treated and conceptualized separately and differently in a society. Similarly, the level of gender inequality is expected to contribute to whether or not men's and women's SSSIR are named in the same way. Gender inequality means that men and women are evaluated differently and unequally for who they are and what they do. In virtually all modern societies, men are granted more importance in the system; whatever men do receives more attention whereas what women do tends to be disregarded. I argue that such a tendency is expected to be more prominent in societies with a higher level of gender inequality. Rigidity and inequality in gender system are analytically distinct but inseparable empirically. I therefore expect the cross-national pattern of the naming of SSSIR to be as follows: the more rigid and/or unequal the gender system, the more likely the naming of SSSIR is gendered--i.e. only men's SSSIR (and not those of women's) are named.

3. Coding of Legal Naming of SSSIR

I operationalize legal naming as the codification of SSSIR in the law of each country. I use as data the description of laws regarding SSSIR for 117 countries compiled in "Country-by-Country Survey" in *ILGA Pinkbook* (Tielman and de Jonge 1988). I first classify legal statements of the 88 countries where SSSIR are named to identify the gender of the subject referenced: men only; women only; men and women; and gender unspecified person. I then construct a dummy variable (MENREF) differentiating countries that reference men exclusively in naming SSSIR from those referencing gender-unspecified persons and/or both men and women. Such coding identifies countries in which "men" are singled out in the naming of SSSIR and contrast them with all other cases. Table 1 shows the classification of countries by the genderedness in the naming of SSSIR, by geographic region.

Table 1: Genderedness in the Naming of SSSIR, by Geographic Region

MENREF Code: Description	Africa	N. America	C.& S. America	Asia	Europe	Oceania	All
0: countries with at least one statement referencing SSSIR of both men and women or gender-unspecified persons	14 (73.7)	2 (100)	11 (91.7)	12 (63.2)	23 (82.1)	5 (62.5)	67 (76)
1: countries in where statements reference men's SSSIR exclusively	5 (17.9)	0 (0)	1 (4.2)	7 (36.8)	5 (17.9)	3 (37.5)	21 (24)
N %	19 (100)	2 (100)	12 (100)	19 (100)	28 (100)	8 (100)	88 (100)

The table indicates that 21 out of 88 countries, or 24% of the countries considered here, name only men's SSSIR while 67 countries, or 76%, do so for SSSIR of both genders and/or persons without gender specification. Examination by each geographic region indicates that the tendency to name only men's relationships is stronger in Asian countries and Oceanic countries (about 40%). The proportion of countries naming only men's SSSIR is particularly low in Central and South American countries--only 1 out of 12 countries references only men's SSSIR.

4. Correlation analyses of the relationships between socio-politico-economic factors and the genderedness in the legal naming of SSSIR

Having demonstrated cross-national variation of genderedness in naming, I now explore, through correlation analyses, how the cross-national patterns of the genderedness in naming is related to various socio-politico-economic factors which affect a range of social phenomena. The factors considered include: (a) political systems, as indicated by democratic system vs. non-democratic system (DEMOC) and communist system vs. non-communist system (COMM); (b) the level of legalization in a society, as indicated by the number of institutional areas under jurisdiction (LEGAL), (c) the level of economic development, as indicated by GNP per capita (LGNPCP); (d) history of colonization, as indicated by whether or not a country has been colonized by Britain (COLBRI) and whether or not a country has been colonized by France (COLFRN); religious system, as indicated by the percentages of population affiliated with Protestantism (PCTPROT), Catholicism (PCTCATH) and Islam (PCTISLM) respectively; and, the strength of the linkage between sex and procreation as indicated by the percentage of married women using contraceptives (CONTRA). All the indicators are measured in the 1980's, except for the level of legalization of a society which is measured in 1978. (See Kamano (1995) for details of the data sources.)

The correlation coefficients between the genderedness in the naming of SSSIR and each of the socio-politico-economic factors are presented in Table 2 below.

Table 2: Correlation Analyses of Socio-Politico-Economic Factors and Genderedness in the Naming of SSSIR

	Correlation Coefficient (Pearson's R)	Meaning of the Findings:
DEMOC: democracy/non-democracy	.053	not significant
COMM: communist/non-communist	-.009	not significant
LEGAL: # of institutional areas under jurisdiction	.082	not significant
LGNPCP: natural log of GNP per capita	-.322***	More economically developed, Less likely to exclusively reference men
COLBRI: former British colony	.401***	Former British colony, More likely to exclusively reference men
COLFRN: former French colony	-.177*	Former French colony, Less likely to exclusively reference men
PCTCATH: % Catholic population	-.096	not significant
PCTPROT: % Protestant population	-.134#	(Higher % of Protestant people, Less likely to exclusively reference men)
PCTISLM: % Muslim population	-.128	not significant
CONTRA: % of married women using contraceptives	-.143#	(Higher % of Contraceptives use, Less likely to exclusively reference men)

#: p<.10; *: p < .05; **: p < .01; ***: p < .005

The correlation analyses show that economically more developed countries and former French colonies tend to belong to the group of countries referencing both men and women or "persons" in the naming of SSSIR, while former British colonies are more likely grouped among countries exclusively referencing men in naming such relationships, as indicated by the statistically significant coefficient (at .05 level) of -.322, -.177, and .401 respectively. Using a more generous criterion of .10 level, one can also conclude that countries with a higher percentage of Protestant population and countries with a higher percentage of married women using contraceptives are countries which name SSSIR of both men and women or without gender specification.

Embedded in these political, social and economic structures might also be differences in the gender system, which might underlie the observed differences in the particular gender group(s) referenced. For example, it might be the level of rigidity and inequality in gender categories divide countries varying in economic development and which differ in the naming of SSSIR. In the next section, I directly examine the how the pattern of the naming of SSSIR is related to the gender system.

5. Logistic regression analysis of the effect of gender system on the genderedness in the legal naming of SSSIR

To explore how the cross-national pattern of the genderedness in the naming of SSSIR is linked to the gender system, I undertake a logistic regression analysis with genderedness in the legal naming of same-sex sexual/intimate relationships as a dependent variable and three measures of the level of gender inequality and the rigidity of gender categories as explanatory variables, controlling for the socio-politico-economic variables I discussed above. The rigidity of gender categories and the level of gender inequality are indicated by the gender segregation in occupation and education, the ratio of women-to-men in labor force participation rate, and the legal rights of women. I expect the following cross-national pattern: countries with a more rigid and more unequal gender system will tend to name only men's SSSIR, rather than those of both men and women or gender-unspecified persons. Table 3 shows the results of logistic regression analysis.

Among the three indicators of the rigidity of gender categories and gender inequality, the statistically significant logit coefficient of one measure of gender segregation is consistent with the expected pattern. The observed effect of this variable indicates that rigidity and inequality in the gender system are related to the differentiation among countries in terms of the genderedness in the naming. More specifically, societies where men and women are more segregated (unequally evaluated and treated) in occupation and education tend to reference men exclusively in the naming of SSSIR, while societies where men and women are less segregated tend to either explicitly reference both men and women or do not reference either gender. The level of gender inequality and the rigidity of gender categories as measured by women's participation in the labor force (relative to men's) and women's legal rights do not seem to differentiate the countries in whether they reference men exclusively or reference both men and women or gender-unspecified persons in the naming of SSSIR.

Table 3: Logistic Regression Analysis of the Effect of Gender System on the Genderedness in the Naming of SSSIR (Dependent Variable: **MENREF**: 1 =Men only; 0=Men and Women and/or "Persons")

	Logit Coefficient[i] (std. error)	exp(b) [ii]	ΔP [iii] (P=.5)
EXPLANATORY VARIABLES			
SEGREG: gender segregation in occupation and education	1.98 (1.07)*	7.27	.495
FMLFP: ratio of women-to-men in labor force participation rate	3.08 (4.72)	21.7	.770
EQLAW legal equality of women	-1.43 (1.24)	.238	.358
CONTROL VARIABLES [iv]			
DEMOC: democracy/non-democracy	-2.30 (.928)**	.100	.575
COMM: communist/ non-communist	-1.50 (1.35)	.224	.375
LEGAL: # of institutional areas under jurisdiction	.021 (.024)	1.02	.005
LGNPCP: natural log of GNP per capita	-.555 (.754)	.574	.139
COLBRI: former British colony	-.166 (.820)*	.190	.041
COLFRN: former French colony	4.95 (24.2)	140.8	1.23
PCTCATH: % Catholic population	-.032 (.019)#	.969	.008
PCTPROT: % Protestant population	-.084 (.047)#	.920	.021
PCTISLM: % Muslim population	-.030 (.027)	.970	.008
CONTRA: % of married women using contraceptives	.085 (.076)	1.09	.021
FERTILE: total fertility rate [v]	.154 (.079)*	1.17	.038

-2 x Log Likelihood = 33.6 (p=.994)
Model χ^2=42.7 (p<.001)
Goodness of Fit=95.0 (p<.001)   N=72

#: p<.10;   *:p < .05;   **: p < .01;   ***: p < .005;
(i): log [P/(1-P)];   (ii): P/(1-P);   (iii): The change in probability of referencing men only (ΔP) at the level where the effects of X_b are maximized (where P=.5);   (iv): Correlation analyses (results not shown here) among the explanatory variables and control variables have shown that the variables are not highly correlated with one another (<.400), with expections of those among CONTRA, LGNPCP and FERTILE (>.700). It is possible that such high correlation among these three variables to have affected the overall results of the logistic regression; (v) used as a "control variable" for CONTRA which intended to indirectly indicate the extent to which sex and procreation is linked.

Apart from the link of the gender system to the cross-national patterns of the genderedness of the naming of SSSIR, statistically significant coefficients of some socio-politico-economic factors show the following patterns: (a) democracy increases the likelihood of naming SSSIR for both or unspecified genders; (b) countries with a history of British colonization tend to name both men's and women's SSSIR ,while countries not colonized by Britain tend to name only men's relationships; (c) the higher the total fertility rate, the less likely SSSIR are named exclusively for men; and, (d) the higher the proportion of Catholic and Protestant population, the more likely are men exclusively referenced.

The observed effect of democracy on gender reference in the naming of same-sex sexual/intimate relationships is consistent with conventional wisdom. If one grants that democratic countries are more "inclusive," then it follows that democratic countries either reference both men and women or do not make any distinctions by gender in the naming of SSSIR. If Britain has imposed its law which exclusively references men onto its colonies, and that these colonies have kept the law in its original form in the 1980's, then one would expect a history of British colonization (COLBRI) to have a positive effect on the genderedness in naming (Dynes 1990). However, the effect observed here is negative. The key to understanding the unexpected effect might lie in understanding the post-colonial developments in former British colonies. One can conjecture that Britain might have imposed its legal and cultural paradigm on its colonies which, upon independence, also show a greater tendency to reverse these imposed legal and cultural norms. Testing these ideas requires a separate analysis. For the present purpose, it is sufficient to note that presence/absence of British colonization divides countries in how they name SSSIR.

The positive effect of total fertility rate means that countries with a higher fertility rate name only men's SSSIR, while the countries with a lower fertility rate name SSSIR of both men and women or gender-unspecified persons. The observed effect here might support my argument linking a higher level of gender inequality and a rigidity of gender categories to a stronger tendency to name only men's SSSIR. It is generally the case that the total fertility rate is higher in societies with more rigid and unequal gender categories, which in turn increases the likelihood of the exclusive reference to men in the naming of SSSIR.

The negative effect of the percentage of population affiliated with Catholicism and Protestantism indicates that the higher the proportion of Catholics and Protestants in the population, the more likely that same-sex sexual/intimate relationships are named exclusively for men. Given that the formal teachings of Judeo-Christianity tend to address men exclusively, this finding is perhaps readily comprehensible.

6. Conclusion

In this paper, I focused on the genderedness in the naming of same-sex sexual/intimate relationships, differentiating between countries that exclusively reference men in naming and those which reference both genders or gender-unspecified "persons". I argued that the gender system is one of the important factors differentiating the two types of the naming of SSSIR: men are referenced exclusively in the naming of same-sex sexual/intimate relationships in societies with rigid and unequal gender categories. I reasoned that societies with rigid gender categories conceptualize and treat men and women as two distinctive groups, resulting in differences in the naming of men's and women's same-sex sexual/intimate relationships. Similarly, societies with a higher level of gender inequality by definition privilege men over women and grant the former more visibility and prominence. It is therefore expected that men tend to be referenced exclusively in the naming of same-sex sexual/intimate relationships in these type of societies.

After presenting a coding scheme of the legal naming of same-sex sexual/intimate relationships, I discussed the results of correlation analysis of the genderedness of naming and various socio-politico-economic factors, which showed that countries which are former French colonies, Protestant, and economically more developed belong mostly to the group that references both men and women or persons in the naming of SSSIR. In contrast, countries which are former British colonies belong largely to the group referencing only men in the naming of SSSIR. Furthermore, I undertook a logistic regression analysis to examine the linkage between gender references in naming and the rigidity of gender categories and the level of gender inequality, controlling for the effects of socio-politico-economic factors. The anticipated pattern of the genderedness in the naming of SSSIR in relation to the rigidity and inequality of gender categories was borne out by the finding that the more rigid the gender categories are, the stronger the tendency to exclusively reference men, rather than both genders or gender-unspecified "persons," in naming SSSIR. More importantly, the analyses presented here affirm that aspects of the gender system are important in differentiating among societies which name SSSIR differently.

References:

Connell, R. W. (1987): *Gender and Power*. Stanford University Press, Stanford.

D'Emilio, J. and Freedman, E. B. (1988): *Intimate Matters: A History of Sexuality in America*. Harper and Row, New York.

Durberman, M. B. et al. (eds.) (1980): *Hidden from History: Reclaiming the Gay and Lesbian Past*. New American Library, New York.

Dynes, W. R. (ed.). (1990): *Encyclopedia of Homosexuality*. Garland, New York and London.

Kamano, S. (1995): *Same-Sex Sexual/Intimate Relationships: A Cross-National Analysis of the Interlinkages among Naming, the Gender System, and Gay and Lesbian Resistance Activities*. Ph.D. Dissertation, Stanford University.

Tielman, R. and de Jonge, T. (1988): A worldwide inventory of the legal and social situation of lesbians and gay men. In *ILGA Pink Book: A Global View of Lesbian and Gay Liberation and Oppression*, Tielman, R. and van der Veen, E. (eds.), 183-242, Utrecht University, Utrecht.

A Constrained Clusterwise Regression Procedure for Benefit Segmentation

Daniel Baier

Institute of Decision Theory and Operations Research
University of Karlsruhe; Post Box 69 80;
76128 Karlsruhe; Germany

Summary: A new procedure for benefit segmentation using clusterwise regression is presented. Constraints on the model parameters ensure that the derived benefit segments can be easily attached to single competing products under consideration. The new procedure is compared to other one-stage and two-stage procedures for benefit segmentation using data from the European air freight market.

1. Introduction

Conjoint analysis is the label attached to a popular research tool for measuring buyers' tradeoffs among competing multiattributed products (see, e.g., Green, Srinivasan (1990) for a review): First, respondents are asked for preferential judgments w.r.t. a set of attribute-level-combinations (stimuli) which serve as (hypothetical) product descriptions. Then, the observed response data are analyzed using regressionlike estimation procedures at the individual level. The resulting so-called part-worths (estimated preferences for attribute-levels) are later used to predict responses w.r.t. a set of competing products and—assuming, e.g., that each individual chooses the product with the highest predicted preference—shares of choices or market shares.

A research purpose served by many commercial applications of conjoint analysis is the identification and understanding of so-called benefit segments (see, e.g., Green, Krieger (1991), Wittink, Vriens, Burhenne (1994)): The observed response data are used in order to identify groups of buyers having similar preferences. For this purpose, various procedures have been proposed and successfully applied during the last years. Some of these procedures are so-called two-stage procedures where the individual part-worth estimates are used in an unrelated secondary stage as an input for clustering techniques. Others are so-called one-stage procedures (see, e.g., Kamakura (1988), DeSarbo, Oliver, Rangaswamy (1989), Wedel, Kistemaker (1989), Wedel, Steenkamp (1989), (1991), DeSarbo, Wedel, Vriens, Ramaswamy (1992), and Baier, Gaul (1995) for sample procedures or Wedel, DeSarbo (1994) for a review) where segment-specific part-worth functions and segment-membership indicators are simultaneously estimated using generalizations of well-known clusterwise regression procedures (see, e.g., Bock (1969), Späth (1983), DeSarbo, Cron (1988)). In both cases, the resulting parameter estimates are then used to predict preferences and choices at the segment level w.r.t. a set of competing products.

However, a major shortcoming of these procedures consists in the (missing) link between the computational derivation of segment-specific model parameters and the prediction of choices for the competing products: No guarantee is provided that each competing product is chosen by at least one benefit segment, a fact that—when analyzing real markets—could lead to the (surprising) situation that established products in these markets are predicted to have no buyers. For this reason, a new one-stage procedure for benefit segmentation is proposed where constraints are implemented which ensure that each competing product is selected by at least one segment.

2. A constrained clusterwise regression procedure

2.1 Model formulation

Let be i an index for N respondents, t an index for T segments or homogeneous groups of respondents, j an index for n stimuli, $\tilde{j}$ an index for $\tilde{n}$ ($\tilde{n} \leq T$) competing products, v an index for V attributes, and w an index for W_v levels of attribute v.

The data are (binary) profile data $B_{111}, \ldots, B_{nVW_V}$ for the stimuli, $\tilde{B}_{111}, \ldots, \tilde{B}_{\tilde{n}VW_V}$ for the competing products (where B_{jvw} and $\tilde{B}_{\tilde{j}vw}$ indicate whether stimulus j and product $\tilde{j}$ has level w for attribute v ($=1$) or not ($=0$)), and response data $y_{11}, \ldots, y_{nN}$ (where y_{ji} describes the observed preference value for stimulus j obtained from respondent i).

The model parameters are the segment-membership indicators $h_{11}, \ldots, h_{TN}$ (where h_{ti} denotes whether respondent i belongs to segment t ($=1$) or not ($=0$)) and segment-specific part-worths $u_{111}, \ldots, u_{TVW_V}$. With the help of the segment-specific response estimates for the stimuli and the products

$$u_{jt} = \sum_{v=1}^{V} \sum_{w=1}^{W_v} B_{jvw} u_{tvw} \quad \forall j, t \quad \text{and} \quad \tilde{u}_{\tilde{j}t} = \sum_{v=1}^{V} \sum_{w=1}^{W_v} \tilde{B}_{\tilde{j}vw} u_{tvw} \quad \forall \tilde{j}, t, \tag{1}$$

the least squares loss function

$$Z = \sum_{i=1}^{N} \sum_{j=1}^{n} (y_{ji} - \sum_{t=1}^{T} h_{ti} u_{jt})^2 \tag{2}$$

with

$$\tilde{u}_{\tilde{j}\tilde{j}} \geq \tilde{u}_{\tilde{j}'\tilde{j}} \quad \forall \tilde{j}, \tilde{j}', \qquad \sum_{t=1}^{T} h_{ti} = 1 \quad \forall i, \qquad h_{ti} \in \{0,1\} \quad \forall t, i \tag{3}$$

is minimized. The constraints in formula (3) ensure that segment $\tilde{j}$ chooses competing product $\tilde{j}$ resp. that each competing product is selected by at least one segment (Note that for obvious reasons index $\tilde{j}$ is also used for segments.), and that the segmentation scheme is non-overlapping.

It should be mentioned that with $\tilde{n}=0$ the standard versions of Wedel, Kistemaker's (1989) as well as Baier, Gaul's (1995) (non-overlapping) clusterwise regression procedures are contained in the model formulation. (In this case the inequality constraints in formula (3) are omitted.)

2.2 Algorithm

For estimation of the model parameters, an exchange algorithm is proposed as given in Tab. 1: In the initialization phase we start with a segmentation matrix $\mathbf{H}$, segment-specific response estimates for the stimuli $\mathbf{U}$, and segment-specific response estimates for the competing products $\tilde{\mathbf{U}}$ so that the constraints in formula (3) are fulfilled. Additionally, the initial loss function value is computed. In the iteration phase we repeatedly test, whether an exchange of a respondent from one segment to another improves the loss function without violating the constraints in formula (3). If so, a new loss function value is calculated. If not, the exchange is cancelled (using the variables $\tilde{h}_1, \ldots, \tilde{h}_T$ for restoration). In the final phase segment-specific part-worth estimates are computed.

{Initialization phase:}

Set $\dot{\mathbf{B}} := \begin{pmatrix} 1 & B_{111} & \cdots & B_{11(W_1-1)} & \cdots & B_{1V1} & \cdots & B_{1V(W_V-1)} \\ \vdots & \vdots & & \vdots & & \vdots & & \vdots \\ 1 & B_{n11} & \cdots & B_{n1(W_1-1)} & \cdots & B_{nV1} & \cdots & B_{nV(W_V-1)} \end{pmatrix}$ and

$\ddot{\mathbf{B}} := \begin{pmatrix} 1 & \tilde{B}_{111} & \cdots & \tilde{B}_{11(W_1-1)} & \cdots & \tilde{B}_{1V1} & \cdots & \tilde{B}_{1V(W_V-1)} \\ \vdots & \vdots & & \vdots & & \vdots & & \vdots \\ 1 & \tilde{B}_{\tilde{n}11} & \cdots & \tilde{B}_{\tilde{n}1(W_1-1)} & \cdots & \tilde{B}_{\tilde{n}V1} & \cdots & \tilde{B}_{\tilde{n}V(W_V-1)} \end{pmatrix}$

Choose $\mathbf{H}^{(0)}$ and compute $\mathbf{C} := (\dot{\mathbf{B}}'\dot{\mathbf{B}})^{-1}\dot{\mathbf{B}}'\mathbf{Y}\mathbf{H}^{(0)'}(\mathbf{H}^{(0)}\mathbf{H}^{(0)'})^{-1}$, $\tilde{\mathbf{U}}^{(0)} := \ddot{\mathbf{B}}\mathbf{C}$
s.t. $\tilde{u}_{\tilde{j}\tilde{j}}^{(0)} \geq \tilde{u}_{\tilde{j}'\tilde{j}}^{(0)} \, \forall \tilde{j}, \tilde{j}'$.

Set $\mathbf{U}^{(0)} := \dot{\mathbf{B}}\mathbf{C}$, $Z^{(0)} := \sum_{i=1}^{N}\sum_{j=1}^{n}(y_{ji} - \sum_{t=1}^{T}h_{ti}^{(0)}u_{jt}^{(0)})^2$, $s := 0$ and choose $\epsilon > 0$.

{Iteration phase:}

Repeat   Set $s := s+1$, $\mathbf{H}^{(s)} := \mathbf{H}^{(s-1)}$, $Z^{(s)} := Z^{(s-1)}$.
       For $i := 1$ to N do
         For $t := 1$ to T do
           Begin Set $\tilde{h}_{t'} := h_{t'i}^{(s)} \, \forall t'$, $h_{t'i}^{(s)} := 0 \, \forall t' \neq t$, $h_{ti}^{(s)} := 1$.
          Set $\mathbf{C} := (\dot{\mathbf{B}}'\dot{\mathbf{B}})^{-1}\dot{\mathbf{B}}'\mathbf{Y}\mathbf{H}^{(s)'}(\mathbf{H}^{(s)}\mathbf{H}^{(s)'})^{-1}$, $\mathbf{U}^{(s)} := \dot{\mathbf{B}}\mathbf{C}$, $\tilde{\mathbf{U}}^{(s)} := \ddot{\mathbf{B}}\mathbf{C}$.
          If $((\tilde{u}_{\tilde{j}\tilde{j}}^{(s)} \geq \tilde{u}_{\tilde{j}'\tilde{j}}^{(s)} \, \forall \tilde{j}, \tilde{j}')$ and $(\sum_{i=1}^{N}\sum_{j=1}^{n}(y_{ji} - \sum_{t=1}^{T}h_{ti}^{(s)}u_{jt}^{(s)})^2 < Z^{(s)}))$
            Then.   Set $Z^{(s)} := \sum_{i=1}^{N}\sum_{j=1}^{n}(y_{ji} - \sum_{t=1}^{T}h_{ti}^{(s)}u_{jt}^{(s)})^2$
            Else   Set $h_{t'i}^{(s)} := \tilde{h}_{t'} \, \forall t'$.
         End.
Until $Z^{(s-1)} - Z^{(s)} < \epsilon$.

{Final phase:}

Set $\mathbf{C} := (\dot{\mathbf{B}}'\dot{\mathbf{B}})^{-1}\dot{\mathbf{B}}'\mathbf{Y}\mathbf{H}^{(s)'}(\mathbf{H}^{(s)}\mathbf{H}^{(s)'})^{-1}$

Set $u_{tvw} := \begin{cases} \frac{c_{1t}}{V} + c_{(1+(W_1-1)+\cdots+(W_{v'-1}-1)+w)t} & \text{if } w \neq W_v \\ \frac{c_{1t}}{V} & \text{else} \end{cases} \quad \forall t, v, w.$

Tab. 1: An algorithm for constrained clusterwise regression

Note that in the algorithm for constrained clusterwise regression $\dot{\mathbf{B}}$ is the dummy-coded design matrix for the stimuli with elements

$$\dot{B}_{jl} = \begin{cases} 1 & \text{if } l = 1 \\ B_{jv_0w_0} & \text{else,} \quad \begin{aligned} v_0 &= \max\{v|l>1 + (W_1\text{-}1) + \cdots + (W_{v\text{-}1}\text{-}1)\}, \\ w_0 &= l - (1 + (W_1\text{-}1) + \cdots + (W_{v_0\text{-}1}\text{-}1)), \end{aligned} \end{cases} \tag{4}$$

and that $\ddot{\mathbf{B}}$ is defined in the same way as the dummy-coded design matrix for the competing products. The algorithm uses some computational simplifications concerning parameter estimation when the segmentation matrix $\mathbf{H}$ with elements h_{ti} is additionally known: In this case, we get segment-specific response estimates for the stimuli $\mathbf{U} = \dot{\mathbf{B}}\mathbf{C}$ with elements u_{jt} and for the competing products $\tilde{\mathbf{U}} = \ddot{\mathbf{B}}\mathbf{C}$ with elements $\tilde{u}_{\tilde{j}t}$ by using

$$\mathbf{C} = (\dot{\mathbf{B}}'\dot{\mathbf{B}})^{-1}\dot{\mathbf{B}}'\mathbf{Y}\underbrace{\mathbf{H}'(\mathbf{H}\mathbf{H}')^{-1}}_{=: \mathbf{G}} \tag{5}$$

where the elements of matrix $\mathbf{G}$ can be easily computed via

$$g_{it} = \begin{cases} \frac{1}{N_t}, & \text{if } h_{ti} = 1 \\ 0, & \text{else}, \end{cases} \quad \forall t, i \quad \text{with} \quad N_t = \sum_{i=1}^{N} h_{ti} \quad \forall t. \tag{6}$$

3. Comparisons

The new constrained clusterwise regression procedure (in the following referred to as CCR) was empirically compared to other one-stage and two-stage procedures for benefit segmentation. For comparisons, data from the European air freight market were used (see Baier, Gaul (1995) for a more detailed discussion of the data) which describe preferences of 150 respondents w.r.t. 18 hypothetical product descriptions for an over-night parcel service with house-to-airport delivery and European destination collected from responsibles for parcel delivery of German companies with more than 25 air freight parcels per month within Europe. The reduced design of the 18 stimuli w.r.t. the attributes 'collection time', 'delivery time', 'transport control', 'agency type', and 'price' (for a 10kg parcel) and descriptions of six competing products selected according to Baier, Gaul (1995) are given in Tab. 2 and 3.

	collection time	delivery time	transport control	agency type	price
stimulus 1	16:30	10:30	active	airline company	160DM
stimulus 2	16:30	10:30	passive	airline company	200DM
stimulus 3	16:30	13:30	active	integrator	200DM
stimulus 4	16:30	13:30	passive	integrator	240DM
stimulus 5	16:30	12:00	active	forwarding agency	160DM
stimulus 6	16:30	12:00	active	forwarding agency	240DM
stimulus 7	17:30	13:30	active	airline company	160DM
stimulus 8	17:30	13:30	active	airline company	240DM
stimulus 9	17:30	12:00	passive	integrator	160DM
stimulus 10	17:30	12:00	active	integrator	200DM
stimulus 11	17:30	10:30	active	forwarding agency	200DM
stimulus 12	17:30	10:30	passive	forwarding agency	240DM
stimulus 13	18:30	12:00	passive	airline company	200DM
stimulus 14	18:30	12:00	active	airline company	240DM
stimulus 15	18:30	10:30	active	integrator	160DM
stimulus 16	18:30	10:30	active	integrator	240DM
stimulus 17	18:30	13:30	passive	forwarding agency	160DM
stimulus 18	18:30	13:30	active	forwarding agency	200DM

Tab. 2: 18 stimuli for data collection in the European air freight market

	collection time	delivery time	transport control	agency type	price
product A	17:30	13:30	passive	integrator	160DM
product B	17:30	13:30	active	integrator	240DM
product C	16:30	10:30	passive	integrator	200DM
product D	17:30	10:30	passive	forwarding agency	200DM
product E	16:30	12:00	passive	integrator	160DM
product F	18:30	13:30	active	integrator	200DM

Tab. 3: Six products for simulations in the European air freight market

Besides the new procedure, two-stage procedures using Ward- or k-Means-clustering (referred to as Ward or k-Means) as well as one-stage procedures using Kamakura's (1988) hierarchical clusterwise regression (HCR), Wedel, Kistemaker's (1989) clusterwise regression (CR) with Baier, Gaul's (1995) iterative minimum-distance algorithm, Wedel, Steenkamp's (1989) fuzzy clusterwise regression (FCR), or DeSarbo, Wedel, Vriens, Ramaswamy's (1992) latent class metric conjoint analysis (LCA) were applied to the conjoint data in order to derive 1- to 8-segment solutions. For CCR, competing products were selected in the fixed order from Tab. 3.

Afterwards, the observed preference values for the stimuli at the individual level were compared to the respective segment-specific response estimates using different fit measures like, e.g., DeSarbo et al.'s (1992) variance accounted for (VAF)

$$\text{VAF} = 1 - \frac{\sum_{i=1}^{N}\sum_{j=1}^{n}(y_{ji} - \sum_{t=1}^{T} h_{ti}u_{jt})^2}{\sum_{i=1}^{N}\sum_{j=1}^{n}(y_{ji} - \bar{y})^2} \quad \text{with} \quad \bar{y} = \sum_{i=1}^{N}\sum_{j=1}^{n} y_{ji}/(Nn) \tag{7}$$

(also applicable to fuzzy clustering), Kamakura's (1988) predictive accuracy index (PAI), Green, Helson's (1989) 1st-choice hits, average product moment correlations (Prod.mom.corr.) and average mean sums of errors (MSE) as well as average values for Kendall's τ. Tab. 4 and 5 show the results based on 50 random starting solutions. (Underline denotes best performance.) $T=6$ was selected for further comparisons since there are six competing products for simulations.

T	Ward	k-Means	HCR	CR	FCR	LCA	CCR
1	0.2413	0.2413	0.2413	0.2413	0.2413	0.2412	0.2413
2	0.4641	0.4641	0.4616	0.4645	0.4752	0.4294	0.4577
3	0.5311	0.5299	0.5257	0.5329	0.5329	0.4856	0.5320
4	0.5800	0.5787	0.5831	0.5867	0.5755	0.5590	0.5867
5	0.6140	0.6154	0.6273	0.6323	0.6164	0.5607	0.6322
6	0.6541	0.6538	0.6585	0.6645	0.6337	0.6377	0.6353
7	0.6813	0.6821	0.6886	0.6923	0.6442	0.6528	0.6639
8	0.7095	0.7119	0.7106	0.7142	0.6539	0.6479	0.6964

Tab. 4: Summary of VAF-values for different numbers of segments T

	Ward	k-Means	HCR	CR	FCR	LCA	CCR
Prod.mom.corr.	0.7992	0.7992	0.8054	0.8078	0.7865	0.7818	0.7883
MSE	6.2258	6.2321	6.1479	6.0388	6.5934	6.5213	6.5650
1st-choice hits	0.8333	0.8533	0.8333	0.7933	0.8400	0.8133	0.7867
Kendall's τ	0.6489	0.6437	0.6544	0.6565	0.6494	0.6315	0.6366
VAF	0.6541	0.6538	0.6585	0.6645	0.6337	0.6377	0.6353
PAI	0.3579	0.3582	0.3534	0.3472	0.4833	0.3856	0.3771

Tab. 5: Summary of values according to various fit measures ($T=6$)

Even a first glance at Tab. 4 and 5 reveals that the best one-stage procedure (CR) performs better (with the exception of $T=1$ or $T=2$) than the other one-stage and two-stage procedures. CCR, the new procedure, competes well in this context with only minor deteriorations w.r.t. various fit measures. However – to be honest – one should mention that this behavior heavily depends on the selected competing products in the market. (If, e.g., some benefit segments are not satisfied by the available products in the market, an application of CCR should lead to inferior classification results compared to an application of CR.)

attribute	level	segm. 1 (14.7%)	segm. 2 (37.3%)	segm. 3 (8.7%)	segm. 4 (14.7%)	segm. 5 (16.7%)	segm. 6 (8.0%)
collection time	16:30	0.123	0.768	0.000	0.273	0.000	0.022
	17:30	0.050	0.359	0.529	0.424	0.018	0.054
	18:30	0.000	0.000	0.617	0.000	0.014	0.000
delivery time	10:30	0.043	0.068	0.172	0.129	0.061	0.608
	12:00	0.025	0.036	0.108	0.056	0.004	0.427
	13:30	0.000	0.000	0.000	0.000	0.000	0.000
transport control	active	0.048	0.043	0.067	0.106	0.491	0.030
	passive	0.000	0.000	0.000	0.000	0.000	0.000
agency type	airline company	0.123	0.002	0.021	0.000	0.080	0.095
	integrator	0.049	0.000	0.059	0.082	0.250	0.000
	forwarding agency	0.000	0.005	0.000	0.200	0.000	0.078
price	160DM	0.663	0.117	0.084	0.141	0.182	0.217
	200DM	0.313	0.059	0.021	0.047	0.085	0.115
	240DM	0.000	0.000	0.000	0.000	0.000	0.000
		prod. E	prod. E	prod. F	prod. D	prod. F	prod. D

Tab. 6: Standardized segment-specific part-worth functions and most preferred products of the six-segment solution from CR (VAF=0.6645)

attribute	level	segm. 1 (15.3%)	segm. 2 (12.7%)	segm. 3 (29.3%)	segm. 4 (17.3%)	segm. 5 (14.7%)	segm. 6 (10.7%)
collection time	16:30	0.093	0.050	0.776	0.089	0.611	0.000
	17:30	0.126	0.061	0.367	0.196	0.357	0.385
	18:30	0.000	0.000	0.000	0.000	0.000	0.494
delivery time	10:30	0.025	0.085	0.091	0.350	0.044	0.139
	12:00	0.034	0.007	0.056	0.228	0.000	0.082
	13:30	0.000	0.000	0.000	0.000	0.015	0.000
transport control	active	0.053	0.497	0.052	0.126	0.042	0.112
	passive	0.000	0.000	0.000	0.000	0.000	0.000
agency type	airline company	0.119	0.014	0.000	0.000	0.009	0.092
	integrator	0.052	0.203	0.011	0.008	0.000	0.154
	forwarding agency	0.000	0.000	0.008	0.147	0.019	0.000
price	160DM	0.667	0.153	0.069	0.181	0.283	0.100
	200DM	0.298	0.060	0.037	0.091	0.129	0.047
	240DM	0.000	0.000	0.000	0.000	0.000	0.000
		prod. A	prod. B	prod. C	prod. D	prod. E	prod. F

Tab. 7: Standardized segment-specific part-worth functions and most preferred products of the six-segment solution from CCR (VAF=0.6353)

In order to demonstrate advantages of the new procedure, Tab. 6 and 7 show the six-segment solutions with part-worth functions and most preferred competing products from CR and CCR. (The attribute-levels of the most preferred products are underlined.) The part-worth functions are standardized in the usual way that the segment-specific part-worth estimates for the least preferred attribute-levels are fixed to 0 and that the response estimates for the combinations of the most preferred attribute-levels (not necessarily the most preferred product) sum up to 1.

Using this standardization, the (relative) importances of the single attributes are given by the part-worth estimates for the most preferred attribute-levels: So, e.g., from Tab. 6 we can see that segment 1 is highly price-sensitive (attribute 'price' accounts for 66.3% of the overall preference), that ('16:30', '10:30', 'active', 'airline company', '160DM') is its most preferred attribute-level-combination (with a segment-specific response estimate of 1), that 'product E' with attribute-level-combination ('16:30', '12:00', 'passive', 'integrator', '160DM') is its most preferred competing product (with a response estimate of 0.860), and that 'product E' is also the most preferred product of segment 2 where early collection times are most important. However, only the competing products 'product D' to 'product F' can be attached to benefit segments. The utilities for 'product A' to 'product C' are for all segments inferior to the utilities for 'product D', 'product E', or 'product F', s.t. under the usual assumptions these products are predicted to have no buyers.

This shortcoming is overcome in the six-segment solution from CCR as given in Tab. 7: Here, segments with similar interpretations as in the CR solution can be found: e.g., a highly price-sensitive cluster (segment 1) and two clusters focusing on an early collection time (segment 3 and 5). Nevertheless, each product can be attached to a segment. So, e.g., segment 1 now prefers 'product A' whereas 'product E' is preferred by segment 5. (Note that segment 1 in the CR solution also had a fairly high response estimate of 0.769 for 'product A'.)

CCR: CR:	segm. 1 (15.3%)	segm. 2 (12.7%)	segm. 3 (29.3%)	segm. 4 (17.3%)	segm. 5 (14.7%)	segm. 6 (10.7%)	
segm. 1 (14.7%)	19	0	0	0	3	0	prod. E
segm. 2 (37.3%)	0	0	42	0	14	0	prod. E
segm. 3 (8.7%)	1	0	0	0	0	12	prod. F
segm. 4 (14.7%)	2	1	2	12	5	0	prod. D
segm. 5 (16.7%)	1	18	0	2	0	4	prod. F
segm. 6 (8.0%)	0	0	0	12	0	0	prod. D
	prod. A	prod. B	prod. C	prod. D	prod. E	prod. F	

Tab. 8: Cross-tabulation of segment-membership

The cross-tabulation in Tab. 8 shows that only few reallocations were necessary for achieving this easier interpretation: Segment 1 of both solutions consists mainly of the same respondents whereas segment 2 of the CR solution was divided up into the two CCR-segments 3 and 5.

4. Conclusions and Outlook

A new procedure for benefit segmentation has been presented, where segment-specific constraints ensure, that competing products can be easily attached to specific segments, a feature that simplifies interpretation. Comparisons so far show that the new procedure also competes well with other one-stage or two-stage procedures w.r.t various fit measures. The proposed procedure can be easily modified in order to support other research purposes. So, e.g., if the most preferred attribute-level-combinations of additional ($j > n$) benefit segments were treated as candidates for new promising products, the procedure could be used for simultaneous product design and benefit segmentation. Another possible extension could be the integration of two-mode clustering in this context (see, e.g., the model formulations in Wedel, Steenkamp (1991) or Baier, Gaul, Schader (1996)). In this case, the procedure could be used for simultaneous market structuring and benefit segmentation.

References:

Baier, D., Gaul, W. (1995): Classification and Representation Using Conjoint Data, In: *From Data to Knowledge:*, Gaul, W., D. Pfeifer (eds.), Berlin, Springer, 298-307.

Baier, D., Gaul, W., Schader, M. (1996): Two-Mode Overlapping Clustering With Applications to Simultaneous Benefit Segmentation and Market Structuring, To appear in: *Classification, Data Analysis and Knowledge Organization*, Klar, R., Opitz, O. (eds.), Berlin, Springer.

Bock, H. H. (1969): The Equivalence of Two Extremal Problems and its Application to the Iterative Classification of Multivariate Data, In: *Report on the Conference Medizinische Statistik*, Forschungsinstitut Oberwolfach.

DeSarbo, W. S., Cron, W. L. (1988): A Maximum Likelihood Methodology for Clusterwise Regression, *Journal of Classification*, 5, 249-282.

DeSarbo, W. S., Oliver, R., Rangaswamy, A. (1989): A Simulated Annealing Methodology for Clusterwise Linear Regression, *Psychometrika*, 54, 707-736.

DeSarbo, W. S., Wedel, M., Vriens, M., Ramaswamy, V. (1992): Latent Class Metric Conjoint Analysis, *Marketing Letters*, 3, 273-288.

Green, P. E., Helson, K. (1989): Cross-Validation Assessment of Alternatives to Individual-Level Conjoint Analysis: A Case Study, *Journal of Marketing Research*, 26, 346-350.

Green, P. E., Krieger, A. M. (1991): Segmenting Markets with Conjoint Analysis, *Journal of Marketing*, 55, 20-31.

Green, P. E., Srinivasan V. (1990): Conjoint Analysis in Marketing: New Developments with Implications for Research and Practice, *Journal of Marketing*, 54, 3-15.

Kamakura, W. A. (1988): A Least Squares Procedure for Benefit Segmentation with Conjoint Experiments, *Journal of Marketing Research*, 25, May, 157-167.

Späth, H. (1983): Cluster-Formation und -Analyse, Oldenbourg, München.

Wedel, M., DeSarbo, W. S. (1994): A Review of Recent Developments in Latent Class Regression Models, In: *Advanced Methods of Marketing Research*, Bagozzi, R. P. (ed.): Basil Blackwell, Cambridge, MA, 352-388.

Wedel, M., Kistemaker, C. (1989): Consumer Benefit Segmentation Using Clusterwise Linear Regression, *International Journal of Research in Marketing*, 6, 45-49.

Wedel, M., Steenkamp, J.-B. E. M. (1989): A Fuzzy Clusterwise Regression Approach to Benefit Segmentation, *International Journal of Research in Marketing*, 6, 241-258.

Wedel, M., Steenkamp, J.-B. E. M. (1991): A Clusterwise Regression Method for Simultaneous Fuzzy Market Structuring and Benefit Segmentation, *Journal of Marketing Research*, 28, November, 385-396.

Wittink, D. R., Vriens, M., Burhenne, W. (1994): Commercial Use of Conjoint Analysis in Europe: Results and Critical Reflections, *International Journal of Research in Marketing*, 11, 41-52.

Application of Classification and Related Methods to SQC Renaissance in Toyota Motor

Kakuro Amasaka

TQM Promotion div.Toyota Motor Corporation
1, Toyota-cho, Toyota, Aichi 471, Japan

Summary: To capture the true nature of making products, and in the belief that the best personnel developing is practical research to raise the technological level, we have been engaged in SQC promotion activities under the banner "SQC Renaissance". The aim of SQC promoted by Toyota is to take up the challenge of solving vital technological assignments, and to conduct superior QCDS research by employing SQC in a scientific, recursive manner. To this end, we must build Toyota's technical methods for conducting scientific SQC as a key technology in all stages of the management process, from product planning and development through to manufacturing and sales. Especially, multivariate analysis resolves complex entanglements of cause and effect relationships for both quantitative and qualitative data. A part of application examples are reported below, focusing on the cluster analysis as a representative example.

1. Toyota's SQC Renaissance for Making the Most of Staff

These days, a review of the changes in the environment surrounding the manufacturing industry indicated an ever greater necessity for corporate efforts to amplify and capitalize upon the technical prowess of young engineering staff who bear the main burden of these times. To capture the true nature of products, and in the belief that the best personnel developing is practical research to raise the technological level, we have gained a new awareness of Statistical Quality Control (SQC) as a behavioral science. In recent years, the entire company has been engaged in SQC promotion activities as shown in Fig. 1, under the banner "SQC Renaissance" (Amasaka 1993). Specifically, the objectives are that all members including the engineering staff and the management should seek to reach excellent solutions for technical problems and that they should realize practical achievement by both improving the proprietary technologies and management technologies through SQC practices. Another aspect of the objective is to develop the SQC promotion cycle activities in which the SQC practices result in practical and full development of SQC education that facilitates effective development of human resources, which, in turn, will be reflected in the performance of operations (Amasaka and Azuma 1991).

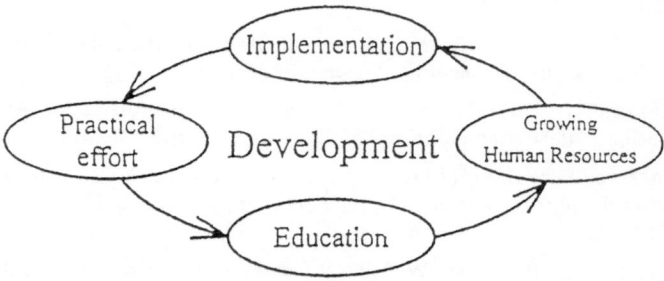

Fig.1   Schematic Drawing of Company-wide SQC Promotion Cycle Activities

If we are to make products that can satisfy our customers, we must work under the most appropriate conditions for raising work quality and minimizing problems. If staff keep a careful watch over their work and apply SQC properly, SQC can assist them in remedying work processes and effectively raise the quality of work (Amasaka and Yamada 1991).

2. SQC for improving Technology

The aim of SQC promoted by Toyota is to take up the challenge of solving vital technological assignments, and to conduct superior Quality, Cost and Delivery (QCD) research by employing SQC in a scientific with the exhibition of insight, inductive problem-solving methodology in addition to engineer's deductive work methods.

This SQC goes beyond reactive technological assignments, to solve proactive technological ones that must be anticipated. This is not a matter of merely performing analytically-oriented SQC in the form of statistical analysis, but is a scientific application of SQC at all stages from problem construction, assignment setting, through to the achievement of objectives, and entails the planning of surveys and experiments to ascertain the desirable scenarios, and devises approaches for tackling problems. When engineers and managers place value on logical thinking, they can resolve the cause and effect relationships of the gap between theory and practice, and can obtain new facts and knowledge for improving proprietary technology and managerial techniques. Rather than ending up with one-off solutions and partial solutions, they can create technology for general solutions, which in turn leads to improve product quality (Amasaka 1995).

By operating this way, a wide variety of SQC practical reports should be utilized as guidelines that can contribute to building up of wealth of engineering technologies as well as support for handing down and developing engineering technologies (Kamio and Amasaka 1992).

Based on this view point, we propose new Schematic Drawing of Scientific SQC that all departments make a Superior QCDS Research at each step of business process, as shown in Fig. 2.

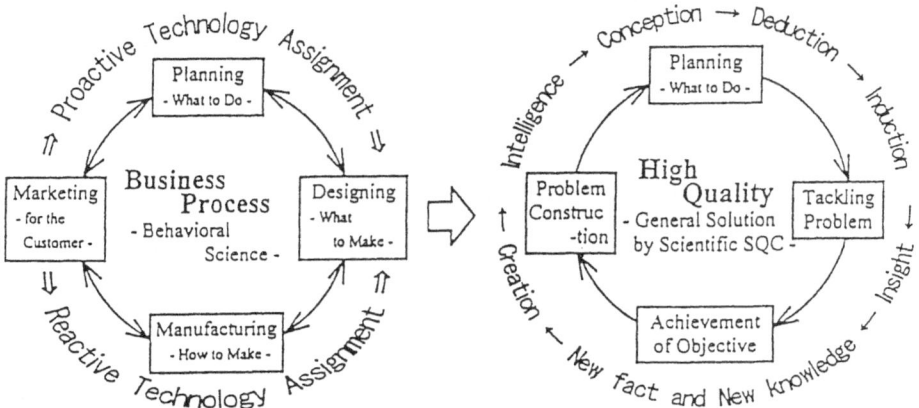

2-1 Business Process for Customer Science   2-2 Scientific SQC for Improving Technology

Fig.2   New Schematic Drawing of Scientific SQC to Conduct of Superior QCDS Research

686

3. SQC Established as a Toyota technical Methodology

In order to be able to provide customer-oriented, attractive products, it is important that we implement customer science that deftly reflects the feelings and voices of customers in the products we make. To this end, we must build Toyota's technical methods for conducting scientific SQC as a key technology in all stages of the management process, from planning and design through to manufacturing and marketing as shown in Fig. 2.

3.1 SQC as a core of the technical methods

Using proprietary technologies and acquired knowledge, SQC resolves complex entanglements of cause and effect relationships for both quantitative and qualitative data. Hence, it is a highly convenient method for technological analyses for improving proprietary technology. New seven tools for Total Quality Control (N7) and basic SQC methods enable full support for designing the experimental process and the analytical process, thus making it possible to analyze technology rapidly and with error-free thinking.

In addition by capitalizing upon proprietary technology, the use of the multivariate analysis method enables 70% to 80% of the mileage required to go before finding solution for a problem to be covered. Combined with SQC methods such as design of experiment, the remaining distance can be covered effectively.

The detailed practical reports shown in the references (Amasaka et al. 1992) (Amasaka et al. 1993) (Amasaka et al. 1994) (Amasaka et al. 1996a) (Amasaka and Sakai 1996) (Kusune and Amasaka 1992) (Takaoka and Amasaka 1991), unravel complex entanglements of cause and effect relationships, by capitalizing on SQC methods such as N7, multivariate analysis and design of experiment in effective combination with the physical and scientific methodology. The use of SQC methods brings about the outstanding achievement on the jobs, using analysis of sources of variation, modeling for prediction and control (Amasaka et al. 1993) (Amasaka et al. 1994) (Kusune and Amasaka 1992) (Takaoka and Amasaka 1991), and concurrent application with neural network as the new technical method (Amasaka et al. 1996a). Moreover, IT (Information Technology) needed for production control (Amasaka and Sakai 1996), the equipment diagnostic technology (Amasaka et al. 1992) are capitalized on as scientific support.

These methodologies for technological problem-solving have been established as new SQC methodology and Toyota's technical method for improving quality works done by engineering staff and managers. Fig. 3 shows a conceptual diagram of new SQC methodology.

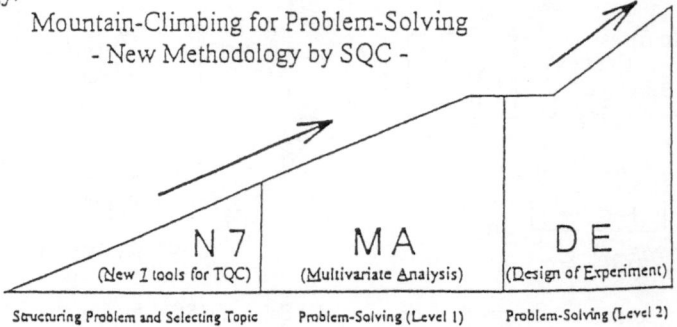

Fig.3   Schematic Drawing of Toyota's Technical Methods
for Conducting Scientific SQC in Toyota

3.2 Using multivariate analysis as the core of scientific SQC practices

Upstream product technology departments are required to quickly develop advanced design technology. Large amounts of data cannot be obtained from individual R&D projects, but lots of the collected trial and experimental data have excellent features in terms of status and condition control. For this reason, even time-series data collected from similar dead and buried in the R&D projects can be used for gaining scientific insights into the cause and effect relationships through multivariate analysis that makes use of proprietary technology. Thus we are able to build logical quantitative models for analysis of source of variation, prediction and control (Amasaka et al. 1996a) (Takaoka and Amasaka 1991).

Midstream manufacturing preparation departments are also required to rapidly develop new production technology and control systems. Comparatively large amounts of data can be obtained, although they stem from R&D, trials, and attempts using mass-production facilities, and rather than being planned, they tend to be of the trial and error variety, changing over time series.

The soundness of these data is not as clear as those of data from upstream departments, but they are data gathered on the spot by engineers, and are reliable in terms of the status and conditions pertaining at the time of collection. Hence, they lend themselves to the application of multivariate analysis using proprietary technology, so that technological knowledge and new facts can be probed for use in raising the technological level (Amasaka and Sakai 1996) (Kusune and Amasaka 1992).

In the downstream manufacturing departments, the important task is to improve manufacturing technology for making stable, high-quality products. Although the reliability of the collected data is not as clear as for the upstream and midstream departments, a great deal of raw data can be obtained if they are collected in a purposeful and planned fashion.

Then, by taking advantage of the manufacturing engineers' unique insights into the actual status of the site and the product, multivariate analysis can be performed to analysis of sources of variation, to extract factors that may aid in quality improvement, and to control processes at the optimum level, thus contributing to improvement of process capability (Amasaka et al. 1992) (Amasaka et al. 1993) (Amasaka et al. 1994).

In this way, multivariate analysis can be used for flexible analysis of technology even with varying quantities of diverse data collected in the past. It has taken root not only as an analytic method of SQC specialists, but also as Toyota's technical methods for skillfully raising the work quality of engineers (Amasaka et al. 1996b) (Amasaka et al. 1996c) (Amasaka and Kosugi 1991) (Amasaka and Maki 1991).

Recently, we have developed and provided our staff with friendly SQC analysis software to be used from personal computers (Amasaka et al. 1995a) (Amasaka and Maki 1992). In all types of technological fields, multivariate analysis has become the core of the SQC practices in applying scientific use of the technical field, and practical research is progressing.

A part of application examples are reported below, focusing on the cluster analysis as a representative example.

4. Examples of Applying Multivariate Analysis Method Focusing on Cluster Analysis

The research examples outlined are "Analysis of sources of variation in vehicle rusting", both belonging to the product and production technology arenas, and "Latent structure of engineers' attitudes to good inventions and patents", a theme impinging on the technological development and control areas. All these research examples started out with cluster analysis to unravel the complex technical assignments, and show that the application of a combination of different multivariate analysis methods has brought about the expected results.

4.1 "Latent structure of engineers' attitudes to goodness of inventions and patents"

Acquiring "good patents" that enable intelligent properties in possession to improve the corporate quality, has become more important measures for allowing permanent business activities. Hence, good patents recognized by managers and technical staff (here-in-after referred to as engineers), containing the contents of both invention and right, are clarified in terms of quality and required to conduct researches aimed of encouraging the acquisition for strong and wide patents. Therefore, this paper selects the representative examples that grasp the conceptual structure of the "good patents" admitted by the engineers (Amasaka et al. 1996d) (Amasaka and Ihara 1996) (Ihara and Amasaka 1996).

4.1.1 Characteristic Classification of Common Images of "Good Patents" and Grasping their Structural Concept

A good patent is said that the invention and right leads to the benefit of own company while allowing the company to maintain its main business and affecting others . The patent information available currently gives only simple statistical values, not studied about qualitative analysis by probing into the conscious region of engineers as inventors. In this connection, the survey aimed of objectifying the subjective structural concept of "good patent" is employed by grasping the engineers' latent status-quo recognition interpreted as the language information. Regarding to the survey's process (Amasaka et al. 1995b), free opinions are collected from engineers in advance about "what a good patent is, while considering the recent environmental situations." Those collected opinions are grouped and arranged into the key words by employing an affinity chart method and a relation chart method in the cooperation with the special department for the patent application. Then the key words are largely classified into "content of inventive technique" and "content of patent right." Respective contents are summarized into a format of 11 questions as outlined in Table 1 sheet. The survey by the questionnaire in marking the Table 1 sheet, one of given answers for selection, is conducted to a total of 97 persons selected from among those with experience of patent acquisition in seven departments from Product Technology: Research "b", Design "c", Development "d", Technical Administration "a", Production Engineering "e", "f", and "g", as shown in Fig. 4.

Fig. 4.1 shows the grouping by shared recognition of the "good patent" through the cluster analysis. The results are classified mainly into three major clusters ((1), (2), and (3)). The figure shows the percentage of the total number of persons who answer the questionnaire and those of each cluster of the seven departments. The figure indicates that the percentage is dispersed by the department. The questionnaire results are now analyzed with factor analysis by probing into the commonly shared structural concept of the "good patent." Fig. 4.2 shows the result of the analysis obtained from the scatter diagram of factor loading. Varimax method allows the reading of the strength of the right and the engineering development capability standing opposite to each other on axis 1 and the commercialization and profit-minded on axis 2. The result of this analysis confirms that there exists in the seven departments four types of engineers who attach importance to commercialization, profit-minded, the strength of the right, or the engineering development capability. Analysis of the similar scatter diagram for the factor scores reveals an interesting new knowledge that each engineer has his own structural concept of the "good patent."

And the summaries of these two analytical results allow us to interpret, as shown in the lower section of Fig. 4.1, that group (1) is A: the realist group (that places importance on practical invention based on Toyota's engineering capability and on the right of practical effect); group (2) is B: the advance group (that places importance on advanced invention superior to competitors or specific right competitors are eager to get); and group (3) is C: the futuristic group (that places importance on the prospective invention on the construction stage and the right to lead competitors with internationally). For example, research section "b" mostly consists of the futuristic group while Technical Administration "a" is largely composed of the realist group and so on for the rest of sections that reflect logical result of analysis. Regarding to this results, it is found that respective departments are conscious and in need of well-balanced activities for "good patent" as they understand what to expect from and what role to play in executing their job. We have thus obtained valuable results (of the analysis) which constitute the preparation for our future patent strategy.

Table 1   Example : " What is a good patent ? "

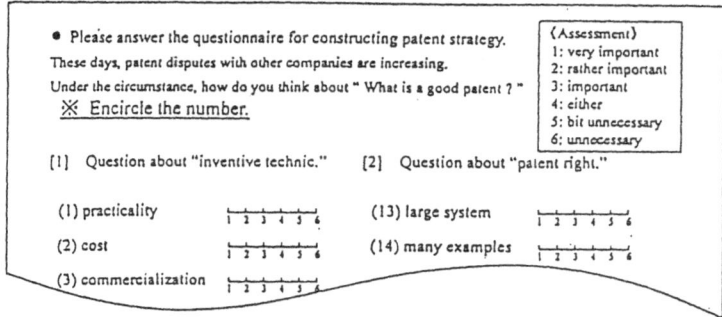

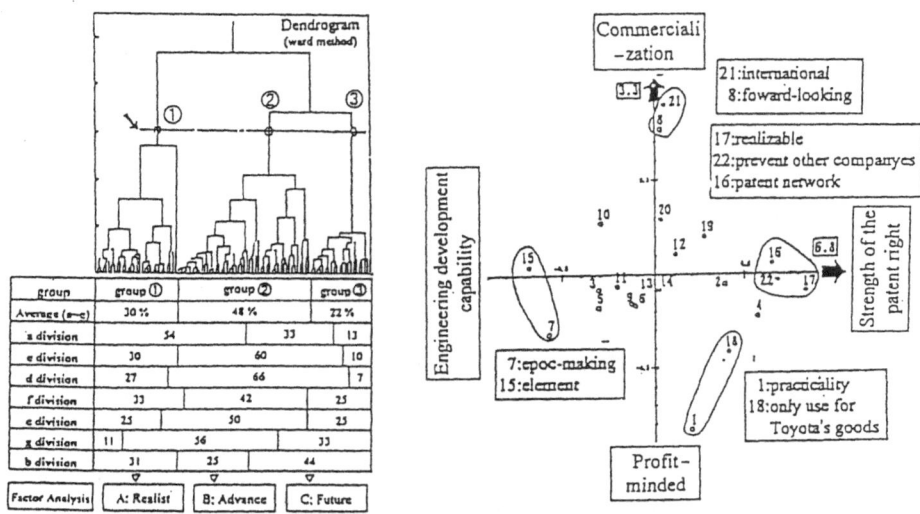

Fig. 4-1   Cluster Analysis   Fig. 4-2   Factor Analysis

Fig.4   Patterning and common understanding of good patents

4.1.2 Grasping Linkage of Recognition between "Good Invention" and "Good Right"

As mentioned in the proceeding section, a "good patent" has two sides of invented technique and subsequent right. To determine characteristic key words for respective sides as recognized by the realist, advance, and futuristic groups classified as the results of cluster and factor analyses, and obtain the degrees of mutual linkage among them, canonical correlation analysis is conducted. Fig. 5 shows the result of the analysis. It is found from the main component relation scatter diagram in the figure that most of the engineers belonging to A: the realist group can be related to the profit-minded in terms of the content of invented art and to the effective-minded in terms of the content of the right by the key words that each group is required to have. Likewise, the figure reads that the B: the advance group and C: the futuristic group are configured with the pioneer-minded and the innovative-minded, and the competitive-minded and the monopoly-minded respectively. This result of analysis can be judged logicalty from the viewpoint of empirical technique. We are able to obtain a verification to the effect that such cause and effect relationships can be a qualitative general evaluation index for a patent application by weighting respective key words as the multiple regression equation. At present, this process is in the middle of transfer to the implementation stage as a regular task.

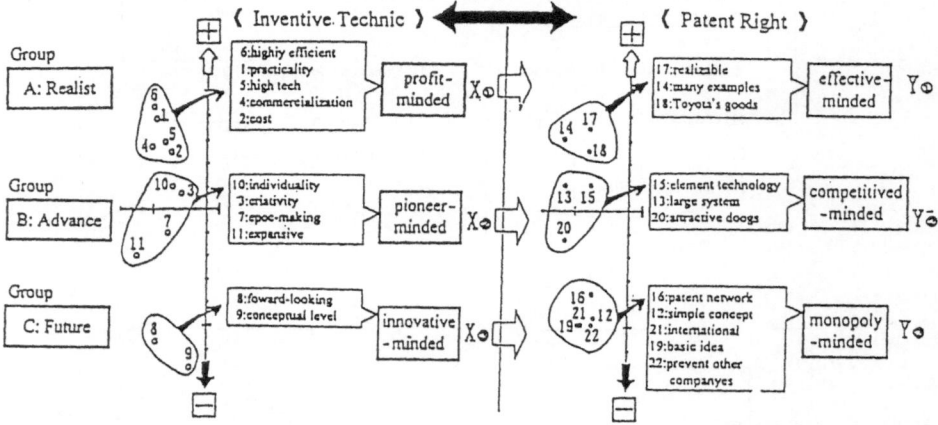

Fig. 5 The Relate of Inventive Technic and Patent Right   (canonical correlation analysis)

This case (1) shows a scientific application of SQC exactly as seen in Fig. 2 by upgrading the quality of job on the business process stage in a proactive engineering area. It is thus judged that the application effect of multivariate analysis as the core of Toyota Technical Method has been verified.

4.2 "Analysis of sources of variation for preventing vehicles' rusting"

One of the subjects for technological development of higher quality, longer life vehicles is the quality assurance of anti-rusting of the body. From the engineering viewpoint of anti-rusting of vehicle body, it involves consideration for structural design, adoption of anti-rusting steel, adoption of local anti-rusting processing (such as the application of wax, sealer, etc.), and paint design (conversion treatment and improvement of electrodeposition coating, etc.). On the implementation stage of the research, these anti-rusting measures are adopted singularly or in combination of multiple measures as may be required by the construction of subject section or corrosion factors. In order for us to proceed with advanced and timely QCDS study activities, it is necessary to conduct variable factor analysis of vehicle's anti-rusting and the optimization. In this sense, SQC centering around the multivariate analysis has much to contribute as a scientific approach.

In this connection, this paper describes a characteristic case for study (Amasaka et al. 1995c).

4.2.1 Anti-rusting Methods and How to Outline their Characteristics

To establish a superior quality assurance system, it is important to set up the network of quality assurance on the job processing stage so as to raise reliability technology of whole the sectors including product planning, design, review, production engineering, process design, administration, inspection and so on. Such a quality assurance activity under the cooperation of all the sectors has been established as Toyota's QA network, where SQC plays an important role as the behavioral science for enhancing quality performance.

Toyota has been incorporating various anti-rusting methods to various sections of the vehicle. To outline the deployment of quality performance, the application of matrix diagram method is effective. For example, Table 2 outlines complex correlationship, between corrosive environmental factors of 72 sections of a vehicle divided for the ease of arrangement of anti-rusting measures and the manufacturing processes factors, and the respective anti-rusting methods adopted in an anti-rusting QA network table. This table is a summary of objective facts inductively arrested by engineers from multiple engineering sectors, which are then arranged deductively in the subjective point of view.

To make this table more effective, it is necessary to provide it with the ease of visual recognition so that it shows at a glance where Toyota stands in its present activities for rust prevention quality assurance. Table 2 enables staff and engineers to make the engineering judgment more accurately, subsequently contributing much to the strategic decision making on the part of the management.

Table 2 Anti-corrosion Q.A. Matrix

Part		Category Rest Prevent				
Large Classificati	Small Classification	No	Electoro-deposited Coating	Corrosion Prevented Steel Sheet	Adhesive	W S
Heming of Shell Parts	Lower part of door	1		1	1	
	Lower part of laggage	2		1	1	
	Lower part of fuel filler lid	3		1		
	Door, The others	4		1	1	
Shell Parts * R/F	Hood * Lock R/F	5	1	1		
	Door * Lock R/F	6		1		
	Door * Side Protection Bar	7	1	1		
	Back Door * Lock R/F	8		1		
Front Floor	General Surface	9		1		
	Under R/F	10		1		
	Exhaust Pipe R/F					

72 parts — Every part of body shell

28 categories — Anti-corrosion factor / Process influence factor / Corrosion factor etc. } Yes: 1, No: blank

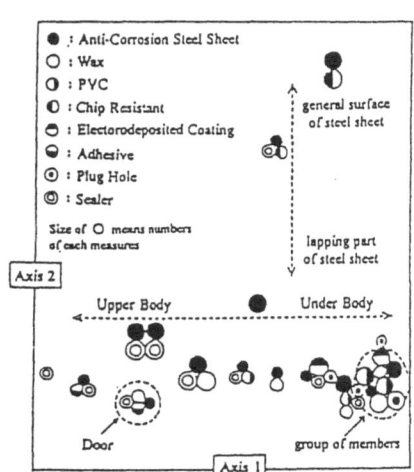

● : Anti-Corrosion Steel Sheet
○ : Wax
◑ : PVC
◐ : Chip Resistant
◓ : Electorodeposited Coating
◒ : Adhesive
◉ : Plug Hole
◎ : Sealer

Size of ○ means numbers of each measures

Axis 2

general surface of steel sheet

lapping part of steel sheet

Upper Body   Under Body

Door   group of members

Axis 1

(Scatter diagram of individual-score by quantification method type3)

Fig. 6   Characteristic of Toyota's Anti-Corrosion Measures

To proceed with such an aim and make it much easier to understand complex entanglements of cause and effect relationships, they will be summarized visually as shown in Fig. 6 by using quantification method type III. From a scatter diagram of vehicle section (axis I x axis II), it is apparent that Toyota's anti-rusting measures are taken mainly against the steel sheet joint portions and that the local anti-rusting processing and many other methods tend to be adopted as the measures reach the underbody sections. The diagram indicates that the combination of several types of anti-rusting methods is applied to the doors and underbody members positioned in the upperbody sections.

Adoption of such an analytical method makes it easy to evaluate the vehicles of competition, enabling the reactive bench marking for additional advantages. In addition, insight into proprietary technologies with the knowledge acquired from the result of this analysis enables us to grasp proactively our responses to the future quality assurance including the trends of anti-rusting measures and techniques of competition that reflect their thought on the target quality and the counter-marketability of anti-rusting materials.

4.2.2 Factorial Analysis Method for an Optimal Anti-Rusting of Vehicle Bodies

This subsection takes up the door hemming sections to which application of anti-rusting technique meets particular difficulties. Spraying of snow melting salts during winter to prevent the roads from freezing allows the filtration of salt water, a corrosive factor, to the hemmed joint portions of door outer-panels and inner-panels of rust prevention steel sheet. To prevent this, wax and/or sealer are adopted or the joint portions of the door outer-panels and inner-panels are sealed with adhesive agent for the dual purpose of adhesion and anti-rusting.

To evaluate the performance of various anti-rusting measures, we conduct a market monitor test using actual vehicles, in-house accelerated corrosion test, and the accelerated corrosion test on the bench using testpieces and/or parts. For any of these tests, it is imperative to conduct analysis of sources of variation for the optimization of anti-rusting of the body.

Example indicated in Table 3 outlines the test data showing the result of analysis on the relationship between the (combined) rust prevention specifications of a testpiece and the depth of corrosion. A dendrogram can be generated as shown in Fig. 7 by subjecting the above data to a cluster analysis. This dendrogram is used to hierarchically outline the degree of effect from the factors of anti-rusting measures by grouping the experiment numbers of a corrosion test. Moreover, analysis using quantification method type I enables us to verify quantitative degree of effect from the factors of anti-rusting measures by category-scores and the size of partial correlation coefficient. From this diagram, it is possible to grasp the effect of anti-rusting steel sheets used to the door hemming sections and the validity of quantitative effect of local anti-rusting processing A, B, and C from the engineering point of view, which in concurrent application with the analytical results of other testing methods using actual vehicles enables the optimization of the vehicle body rust prevention.

We have applied comprehensive and technical insight into these results of analyses. And by adopting controllable factors effective for anti-rusting measures in addition to the environmental factors in consideration of market environments, we have been able to verify deftly the factorial effects with the application of the design of experiment. We have thus succeeded in the embodiment of production with good QCDS performance through the optimization of the vehicle body rust prevention.

Table 3 Experiment Data of Test Piece

Anti-corrosion specification (combination)* and Result

Factor and Level					Corrosion depth	
X1	X2	X3	X4	X5	outer	inner
1	1	2	2	2	0.15	0
1	1	1	2	2	0.10	0
2	1	2	1			

X1: type of steel (outer)   1: steel sheet

2: anti-corrosion steel sheet level 1

X2: type of steel (inner)   1: anti-corrosion steel sheet level 1

2: anti-corrosion steel sheet level 2

X3: spot anti-corrosion steel process A (1:no 2:yes)

X4: spot anti-corrosion steel process B (1:no 2:yes)

X5: spot anti-corrosion steel process C (1:no 2:yes)

*: 21 combinations

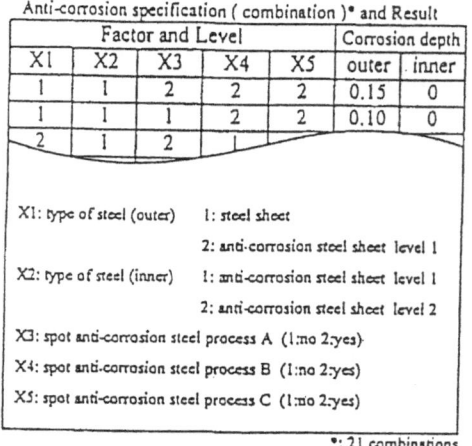

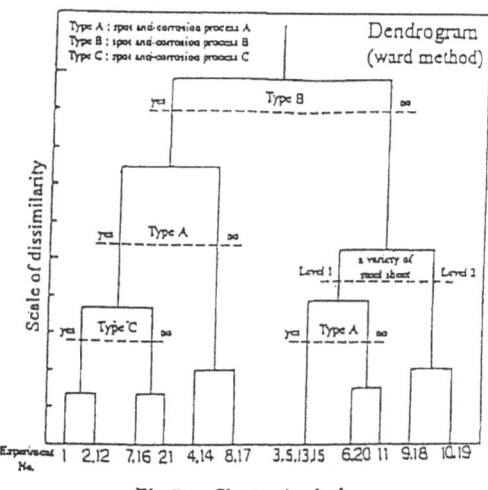

Fig.7   Cluster Analysis

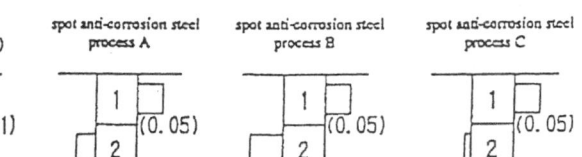

( ): category-scores , [ ]: partial correlation coefficient

Fig.8   Influence of Anti-corrosion Measues by Quantification Method (type 1)

It is judged that this case (2) too constitutes a new methodology of mountain-climbing for Problem-Solving effectively using SQC in concurrent application of multivariate analysis with N7 and design of experiment. We think that here too it has been verified that the multivariate analytical method forms the core of the Toyota Technical Method for scientific implementation of SQC as observed in Fig. 2 and 3.

5. Conclusion

We have been able to verify the following from the two demonstrative cases for studies as above mentioned: In combination with N7 and design of experiment, various multivariate analysis starting from the cluster analysis have been established as the core of technology-advancing SQC from a mere statistical analysis that tends to place unbalanced importance on analysis.

We think that we have verified that the multivariate analysis offers a great application effect as the core of the SQC methods in connection with the construction of the presently advocated Toyota Technical Method for the scientific implementation of SQC and the enhancement of the effectiveness of called SQC.

In unraveling confound situations and complex entanglements of cause and effect relationships, one of the multivariate analysis' methods cluster analysis enables them to clarify and organize visually and logically. On this view point, engineers' abilities with latent pursuit-minded and new ideas-minded are enhanced and improved. It is considered that this new technological approach has great insight to unravel and pursue complex problem-assignments appropriately. The author appreciates valuable teaching and comments from those people concerned.

References:

Amasaka, K. (1993): "SQC Development and Effects at TOYOTA," *(in Japanese) QUALITY, JSQC (Journal of the Japanese Society for Quality Control)*, 23, 4, 47–58.

Amasaka, K. (1995): "A Construction of SQC Intelligence System for Quick Registration and Retrieval Library, – A Visualized SQC Report for Technical Wealth –," *Springer Lecture Notes in Economics and Mathematical Systems*, 318–336.

Amasaka, K. et al.(1992): "A Method on Equipment Diagnosis of Grinder," *(in Japanese) QUALITY, JSQC (Journal of the Japanese Society for Quality Control)*, The 42th Technical Conference, 37–40.

Amasaka, K. et al.(1993): "A Study of Quality Assurance to Protect Plating Pants from Corrosion by SQC, – Improvement of Grenading Roughness for Rod Piston by Centerless Grinding –," *(in Japanese) QUALITY, JSQC (Journal of the Japanese Society for Quality Control)*, 23, 2, 90–98.

Amasaka, K. et al. (1994): "Consideration of effieientical counter measure method for Foundry, – Adaptability of defects control to Casting Iron Cylinder Block –," *(in Japanese) JSQC (Journal of the Japanese Society for Quality Control)*, The 47th Technical Conference, 60–65.

Amasaka, K. et al. (1995a): "Aiming at Statistical Package using in the Job Process," *(in Japanese) JSQC (Journal of the Japanese Society for Quality Control)*, The 25th Annual Technical Conference, 3–6.

Amasaka, K. et al. (1995b): "A study of Questionnaire Analysis of the Free Opinion, –The Analysis of Information Expressed in Words Using N7 and Multivariate Analysis Together–," *(in Japanese) JSQC (Journal of the Japanese Society for Quality Control)*, The 50th Technical Conference, 43-46.

Amasaka, K. et al. (1995c): "The Q.A. Network Activity for Prevent Rusting of Vehicle by Using SQC," *(in Japanese) JSQC (Journal of the Japanese Society for Quality Control)*, The 50th Technical Conference, 35–38.

Amasaka, K. et al. (1996a): "A Study on Estimating Vehicle Aerodynamics of Lift, – Combining the Usage of Neural Networks and Multivariate Analysis –," *(in Japanese) The Institute of Systems Control and Information Engineers*, 9, 5, 229–237.

Amasaka, K. et al. (1996b): "Influence of Multicollinearity and Proposal of New Method of Variable Selection, –A Study of Applied Multiple Regression Analysis for Analysis of Source of Valuation–," *(in Japanese) JIMA (Japan Industrial Management Association)*, 46, 6, 573–584.

Amasaka, K. et al.(1996c): "A Study on Validity of the BN method for Variable Selection, – A Study of Applied Multiple Regression Analysis for Analyzing Source of Variation Factors (Part II) –," *(in Japanese) JIMA (Japan Industrial Management Association)*, 47, 4, 248–256.

Amasaka, K. et al. (1996d): "An Investigation of engineers' Recognition and Feelings about Good Patens by New SQC Method," *(in Japanese) JSQC (Journal of the Japanese Society for Quality Control)*, The 52th Technical Conference, 17–24.

Amasaka, K. and Azuma, H. (1991): "The Practice SQC Education at TOYOTA, – For Growing Human Resource and Practical Effort –," *(in Japanese) QUALITY, JSQC (Journal of the Japanese Society for Quality Control)*, 21, 1, 18–25.

Amasaka, K. and Ihara, M. (1996): "Latent Structure of Goodness-of-invention," *Springer Lecture Notes in Economics and Mathematical Systems*, 348–353.

Amasaka, K. and Kosugi, T. (1991): "Application and Effects of Multivariate Analysis in TOYOTA," *(in Japanese) The Behavior Metric Society of Japan, The 19th Annual Conference*, 178–183.

Amasaka, K. and Maki, K. (1991): "Application of Multivariate Analysis for the Attraction of Manufacturing Vehicles," *(in Japanese) The Behavior Metric Society of Japan, The 19th Annual Conference*, 190–195.

Amasaka, K. and Maki, K. (1992): "Application of SQC Analysis Soft in Toyota," *(in Japanese) QUALITY, JSQC (Journal of the Japanese Society for Quality Control)*, 22, 2, 79–85.

Amasaka, K. and Sakai, H. (1996): "Improving the Reliability of Body Assembly Line Equipment," *The International Journal of Reliability and Safety Engineering*, 3, 1, 11–24.

Amasaka, K. and Yamada, K. (1991): "Re-evaluation of Present QC concept and Methodology in Autoindustry, – Deployment of SQC Renaissance in Toyota –," *(in Japanese) Total Quality Control*, 42, 4, 13–22.

Ihara, M. and Amasaka, K. (1996): "Factor Analysis for Selected Observations," *Springer Lecture Notes in Economics and Mathematical Systems*, 354–361.

Kamio, M. and Amasaka, K. (1992): "Collection of Activity Example Using SQC Method to Improve Engineering Technologies," *(in Japanese) JSA (Japanese Standards Association) NAGOYA QC Research Group*.

Kusune, K. and Amasaka, K. (1992): "The Statistical Analysis of the Springback for Stamping Parts with longitudinal Curvature," *(in Japanese) QUALITY, JSQC (Journal of the Japanese Society for Quality Control)*, 22, 4, 24–30.

Takaoka, T. and Amasaka, K. (1991): "Derivation of Statistical Equation for Fuel Consumption in S.I.Enginenes," *(in Japanese) QUALITY, JSQC (Journal of the Japanese Society for Quality Control)*, 21, 1, 64–69.

Application of Statistical Binary Tree Analysis to Quality Control

Atsushi Ootaki

Department of Precision Engineering
School of Science and Technology
Meiji University
Higashi-mita 1-chome, Tama-ku
Kawasaki 214-71, JAPAN
E-mail:ootaki@isc.meiji.ac.jp

Summary: The purpose of this paper is to show how statistical binary tree analysis would be applied in the field of QC. Analysis of discrimination of uncollectible issue of credit loan in sales management is shown, by using Classification And Regression Trees (CART).

1. Introduction

Quality of Japanese industrial products has been improved by modern quality control(QC) which is introduced into Japanese industries after the World War II. Especially, application of statistical techniques to QC (SQC) is one of the reasons why it helps us to improve quality of products. QC has spread not only to manufacturing department but also to almost all of departments in a company to improve quality of products, services and jobs as total quality control (TQC), which is known as a management tool of Japanese continuous quality improvement.

One of the basic concepts of SQC is how well we decompose many causes affecting variations of quality into each cause without confound. This means how many causes are well classified. To do this, engineers relating to quality planning, design, improvement of products and services apply many kinds of statistical techniques such as control chart, statistical test and estimation, design of experiments and multivariate statistical analysis. Especially, accompanying advance of computer during recent decade, they well apply multivariate statistical techniques such as multiple regression analysis, discriminant analysis and cluster analysis to specify and classify some causes affecting quality of products and services. They sometimes feel that the result of analysis by such techniques does not always bring them a useful information of problem solving since statistical linear model differs from the model that they estimate with their professional knowledge.

On the other hand, statistical binary tree analysis gives us a basic concept of stratification and/or classification. As shown in Fig. 1, the process is binary because parent nodes are always split into exactly two child nodes and is recursive because the process can be repeated by treating each child node as a parent. It gives us the same as a naturally simple thinking way of human being like an answer, "yes" or "no" according to successive questions. As anyone who would like to apply the methodology into real world can make clear that successive causes are followed until a terminal node of which the contents are sufficiently homogeneous, it is easier for him/her to understand the result.

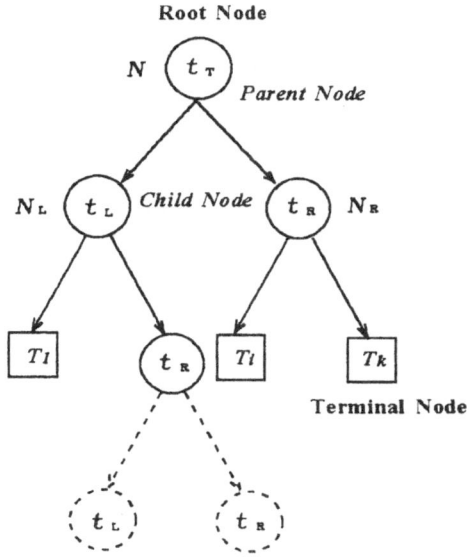

Fig. 1: Binary tree.

2. Data set used

This paper shows how to apply the statistical binary tree analysis in the field of QC. To do this, I will show an example: improvement of uncollectible issue of credit loan which is one of quality characteristics of sales management of construction machine. In the analysis, 47 variables including 31 categorical variables and 16 numerical variables are employed as shown in Fig. 2. The data set has 448 cases including 123 of uncollectible credit loan and 325 of collectible loan. The response variable is the name "(19) judge" which means the judgment; whether a case of credit loan is collectible or not.

3. Overview of binary tree by CART and analysis of the process

To analyze the loan problem, I employed Classification And Regression Trees (CART) developed by Breiman, Friedman, Olshen and Stone(1984). CART methodology is technically known as binary recursive partitioning and its computer program (1994) provides powerful means of carrying out the procedures. CART works in both classification and regression problem as shown in the name.

The earliest work of binary tree structure was proposed by Morgan and Sonquist(1963) as Automatic Interaction Detection (AID). The major difference between CART and AID lies in pruning and estimation process, that is, growing and honest estimation.

% DATA SPECIFICATIONS FOR CART.

Variables = 47   % must be specified first.

Names   % Characters on a line following % are ignored.

(1) juridic	(2) assess	(3) nominatn	(4) age	(5) dealings
(6) capital	(7) establish	(8) employee	(9) current	(10) anngross
(11) grading	(12) stabilit	(13) credit	(14) debt	(15) payrisk
(16) deposit	(17) undertak	(18) sufficnt	(19) judge	(20) unredeem
(21) interest	(22) install	(23) evenpay	(24) marks	(25) annincom
(26) selfcap	(27) district	(28) trade	(29) bkcredit	(30) stocks
(31) payable	(32) develomt	(33) agent	(34) fix	(35) brand
(36) competit	(37) heavymec	(38) machine	(39) complete	(40) rumor
(41) life	(42) mgntable	(43) mood	(44) native	(45) dishonor
(46) jump	(47) creditrn *			

Categories

```
( 1)  5 (.2)  2( 3)  5 ( 5)  2( 9)  2(12)  3 (19)  2
(23)  4 (25)  6(26)  6 (27)  6(28)  6(29)  6  (30)  7
(31)  7 (32)  9(33)  7(34)  7(35)  9  (36)  7(37)  6
(38)  6 (39)  4(40)  7 (41)  4  (42)  7(43)  7(44)  5
(45)  4 (46)  4(47)  5 *
```

Missing

```
( 3)  9 ( 7)  9( 9)  9 (19)  9(47)  9( 5)  9 (23)  9( 1)  9
( 2)  9 (12)  9 *
```

Fig. 2: Variables and data specifications.

The key elements of CART analysis are:

(1) rules for splitting each nodes into a tree
(2) deciding when a tree is complete, and
(3) assigning each terminal node to a class outcome
(or predicted value for regression.)

3.1 Splitting rules

Figure 3 is a part of the result of analysis by CART software.

The first topic addresses the method CART uses to select its questions for splitting nodes. The process is considerably simplified because CART always asks questions that have a "yes" or "no" answer.

For example, the questions might be:

Is ANNGROSS ≤ 950.0 millions yen ?
Is the rank of ANNINCOM ≤ 3.500?
(see the second line and the part of "Competitor" in Fig. 3.)

Where ANNGROSS denotes the variable name of annual gross sales amount and ANNINCOM the rank of annual sales income.

In this output CART answers:

"Node 1 was split on variable ANNGROSS. "
"A case goes left if variable ANNGROSS ≤ 9.500"

NODE INFORMATION

Node 1 was split on variable ANNGROSS
A case goes left if variable ANNGROSS <= 9.500
Improvement = 0.093   C. T. = 0.098

Node	Cases	Class	Cost
1	448	1	0.500
2	200	2	0.884
9	248	1	0.160

	Number Of Cases			Within Node Prob.		
Class	Top	Left	Right	Top	Left	Right
1	325	103	222	0.950	0.884	0.984
2	123	97	26	0.050	0.116	0.016

Surrogate	Split	Assoc.	Improve.
1 ANNINCOM s 2.500		0.927	0.089
2 BKCREDIT s 1,2,6		0.324	0.060
3 HEAVYMEC s 0,1,2		0.245	0.056
4 EMPLOYEE s 5.500		0.242	0.049
5 JURIDIC s 4,5		0.203	0.024

Competitor	Split	Improve.
1 ANNINCOM	2.500	0.089
2 DISTRICT	3.500	0.084
3 NOMINATN	1,2	0.072
4 ESTABLSH	44.500	0.071
5 CURRENT	1	0.067

Fig. 3: Node information for the first split.

In each case the question is used to split a node by sending the yes answers to the left child node and the no answers to the right child node. In the credit loan data, 200 cases go the left node; 248 cases go the right node. The cases in the left node contains 103 cases of Class 1 for collectible loan and 97 of Class 2 for uncollectible loan; the cases in the right node contains 222 of Class 1 for collectible loan and 26 of Class 2 for uncollectible loan.

CART's method is to look at all possible splits for all variables included in the analysis. For example, it would consider splits on ANNGROSS at one-hundred-billion, two-hundred-billions, etc. All the way through the highest annual gross sales amount observed in the data. It would then do the same for the rank of annual sales income which is the variable name ANNINCOM, and for all other variables as well. Since there are at most 448 different values for each variable in this data set. Any problem will have a finite number of candidate splits and CART will conduct a brute force search through them all by converting a continuous variable to an ordered categorical variable with quantile of data. If a categorical variable is nominal, CART selects the best of all combination of $_iC_2$, where i denotes number of categories.

3.2 Choosing a split: Measure of goodness-of-split criterion

CART's next activity is to rank order each splitting rule on the basis of a goodness-of-split criterion. One criterion commonly used is a measure of how well the splitting rule separates the classes contained in the parent node. The goodness-of-split criterion is defined as follows:

$$i(t) = \sum_{i \neq j} C(i \mid j)p(i \mid t)p(i \mid t)\pi(i) \qquad (1)$$

where, $C(i \mid j)$: misclassification cost which classify j as i
$\quad \pi\ (i)$: prior probability of class i
$\quad p(j \mid t)$: proportion of cases in node t which belong to class j.

This criterion is called the Gini index of diversity as a measure of node impurity. In practice, the following improvement of impurity by a split variable is employed as criterion for selection of split variable:

$$\Delta i(t) = i(t) - p_L i(t_L) - p_R i(t_R) \qquad (2)$$

where, $i(t)$: Gini index of parent node
$\quad i(t_L)$: Gini index of the left child node.
$\quad i(t_R)$: Gini index of the right child node.
$\quad p_L, p_R$: proportion of cases which go from parent node to the left node
$\qquad$ and/or the right node.

In the example, the improvement of the best split variable, ANNGROSS was 0.093 while the improvement of the best competitor variable, ANNINCOM which appears in the part of labelled "Competitor" was 0.089. Thus, the goodness-of- split criterion of the variable, ANNGROSS is higher than that of the variable, ANNINCOM which is the best competitor.

3.3 Prior probability

Generally, CART assumes that the relative population frequencies of the dependent variable classes are approximately equal, regardless of their distribution in the learning data set. If uncollectible loan occurs frequently as like as 27% of this data set, any company will go bankrupt. To change this operating assumption, the PRIOR command specifies that the sample proportion of the dependent variables do represent the population proportions. Any population distribution can be specified with the PRIOR command. In the analysis it is assumed that the prior probability of uncollectible credit loan occurs 0.05.

3.4 Misclassification cost

CART measures misclassification costs as the probability of misclassification. In doing so, CART has been implicitly treating all classification errors equally. Although the assumption of equal unit costs for all errors is often appropriate, there will be circumstances when non-unit costs are needed to describe a decision problem. For example, in some credit loan problem, the aim is to separate high risk of uncollectible loan in credit loans. In the example presented here, it is assumed that the expenses of collection to an uncollectible loan are paid about 10 times of that of a collectible loan.

3.5 Class assignment

Once a best split is found, CART repeats the search process for each child node, continuing recursively until further splitting is impossible or stopped for some other reason. Splitting will be impossible if only one case remains in a particular node or if all the cases in that node are exact copies of each other (on predictor variables). CART also allows splitting to be stopped for several other reasons, including that a node has too few cases. The default for this lower limit is 5 cases, but may be set higher or lower to suit a particular analysis.

Once a terminal node is found we must decide how to classify all cases falling within it. One simple criterion is plurality rule: the group with the greatest representation determines the class assignment. Thus, as shown in Fig. 3, the left-most node has 88.4% collectible cases, so the entire node is classified as collectible loan. However, if some cases in the left-most node was determined as uncollectible cases, the left node has 88.4% misclassification cost. Similarly, the right-most terminal node is 98.4% collectible cases, so all cases falling into that node are classified as collectible cases. CART goes a step further: Because each node has the potential for being a terminal node, a class assignment is made for every node whether it is terminal or not. The rule of class assignment can be modified from simple plurality to account for the costs of making a mistake in classification and to adjust for over-or under-sampling from certain classes.

3.6 Pruning trees

For example, if the splitting is carried out to the point where each terminal node contains only one data case, then each node is classified by the case it contains, and estimation error gives zero misclassification rate. However in virtually all applied statistics, less complex models are easier to understand and, for a given data set, can usually be estimated with greater precision.

Now we concern with two main issues: getting the right sized tree and getting more accurate estimate of the probability of misclassification or of the true expected misclassification cost.

Preference for simplicity can be seen in the model selection criteria proposed for conventional least squares regression such as adjusted R-squared, the Akeike information criteria(AIC), etc. All of them devise a trade-off between the error sum of squares and the number of parameters in the model. Analogous concept has been developed for CART tree. A natural measure of the complexity of a tree is the number of terminal node it contains, and misclassification rate estimated from a model is an accuracy measure that always improves as trees get larger. CART employes the following cost compexity measure for any tree:

$$Cost\ Complexity = Estimation\ of\ Misclassification\ Rate \\ + \alpha(Number\ of\ Terminal\ Nodes)$$

where, α is the penalty imposed per additional terminal node.

If the penalty parameter α, also known as the complexity parameter, is set to 0, the largest possible tree will always have the lowest cost complexity. This is because the cost-complexity criterion uses the estimates of misclassification rate to measure cost. On the other hand, if α is set sufficiently high, for example, it set to infinity, a tree with only a root node will be preferred because it is the smallest tree. Values of between infinity and 0 will pick out different trees, with selected trees becoming larger as a moves towards 0.

To see how the complexity parameter works, let's take another look at the tree summary information from CART run shown in Fig. 4. In this output the right-most column, labelled "Complexity Parameter," contains the value of α that would make that tree optimal. We can see that α starts out at a value of 0.099 for the tree with just one terminal node and declines to 0 by the time we reach the largest tree, with thirty-five terminal nodes.

To use the cost-complexity formula we first need to convert the relative costs into their absolute equivalents. We do this by multiplying the relative costs by the initial misclassification cost, which is 0.5 in this run. Thus, the tree with one terminal node has cost equal to $1.000 \times 0.500 = 0.500$ and the tree with two terminal nodes has cost $0.605 \times 0.500 = 0.3025$. Therefore the cost complexities for the two trees are:

(One terminal node): Cost Complexity = 0.500 + 1α

(Three terminal nodes): Cost Complexity = 0.3025 + 3α

The value of α that equalizes the two cost complexities is 0.099, and therefore an α of 0.099 is sufficient to make the smaller tree preferable. We can successively get the complexity parameters.

Instead of attempting to decide whether a given node is terminal or not, CART proceeds by growing trees until it is not possible to grow them any further. Once it has generated what we call a maximal tree, it examines smaller trees obtained by pruning away branches of maximal tree. The important point is that CART trees are always grown bigger than they need to be and are then selectively pruned back. Unlike other systems, CART does not stop in the middle of the tree-grown process, even if the tree appears to be optimal at that point, because important branches of the most accurate tree might be missed.

TREE SEQUENCE

Dependent variable: JUDGE

Terminal Tree Nodes		Cross-Validated Relative Cost	Resubstitution Relative Cost	Complexity Parameter
1	35	0.722 +/- 0.060	0.248	0.000
7	21	0.685 +/- 0.059	0.286	0.003
8	20	0.685 +/- 0.059	0.292	0.003
9	15	0.644 +/- 0.058	0.323	0.003
10	12	0.654 +/- 0.058	0.351	0.005
11*	11	0.612 +/- 0.057	0.363	0.006
12	9	0.643 +/- 0.058	0.392	0.007
13	7	0.650 +/- 0.058	0.447	0.014
14**	5	0.650 +/- 0.058	0.504	0.014
15	3	0.731 +/- 0.060	0.605	0.025
16	1	1.000 +/- 0.000	1.000	0.099

Initial misclassification cost = 0.500

Initial class assignment = 1

Fig. 4: Tree summary information.

3.7 Best pruned subtree: Test sample/Cross-validation

Once the maximal tree is grown and various sub-tree are derived from it, the best tree is determined by testing each for its error rate or cost. When there are sufficient data the simplest method is to divide the sample into learning and test sub-samples. However, many problems will not have sufficient data to allow a separate test sample. The tree-growing methodology is data intensive and requires many more cases than classical regression.

When data are in short supply, CART employs the computer-intensive technique of cross validation and the default is 10 cross validation. The results of the 10 mini-test samples are then combined to form error rates for trees of each possible size; these error rates are applied to the trees based on the entire learning sample. This complex process brings a set of reliable estimates of independent predictive accuracy of the tree. The middle column in Fig. 4, labelled "Cross-Validated Relative Cost," contains the relative misclassification rates. The first value is the average of cross-validated relative error and the second value is the standard error of cross-validated relative cost. We can see that the minimum cross-validated relative error is 0.612 of 11-terminal nodes. Let us notice that the maximal tree does not always have the minimal value of cross-validated relative error though the maximal tree has the minimal value of resubstitution relative cost, which means estimates of relative misclassification rate.

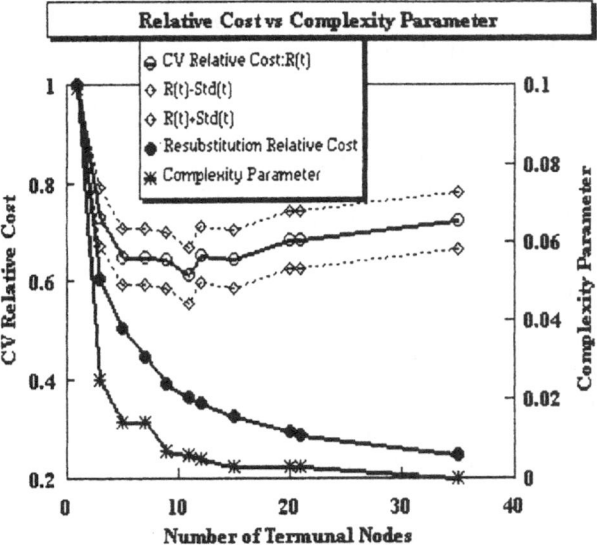

Fig. 5: Cross-validated relative cost.

The characteristics are fairly rapid initial decrease followed by a long, flat valley and then a gradual increase for larger trees shown in Fig. 5.

CART employs the following 1 SE rule for selecting the right sized tree to:

1. Reduce the instability
2. Choose the simplest tree whose accuracy is comparable to minimal misclassification rate.

In addition,

3. Less complex model is easier to understand and is prefered in all applied statistics.

$$\hat{R}(T_{k_1}) \leq \hat{R}(T_{k_0}) + SE(\hat{R}(T_{k_0})) \tag{3}$$

where T_{k_0} denotes the tree with the minimal misclassification cost and T_{k_1} denotes the tree having the minimal nodes within 1-SE rule.

In the analysis cross-validated relative cost of the tree with 5 terminal nodes is 0.650 and it is within 1-SE rule from theminimal cross-validated relative cost of the tree with 11 terminal nodes.

4. Application of CART to credit loan of construction machine

Finally, as shown in Fig. 6, the options for analysis are employed out of those for which CART software(1994) prepares.

It is assumed that prior probability of uncollectible credit loan occurs 0.05 and expenses of collection to uncollectible loan are paid about 10 times of collectable loan. As the result of the analysis, the tree having the minimal cross-validated relative error is selected by CART as shown in Fig. 7.

The following split variables are employed out of 46 candidate variables excluding one of 47 variables to construct the classification tree.

X1: Annual gross sales amount: (10)"angross"
X2: Undertaken cost: (17)"undertak"
X3: Deposit: (16)"deposit"
X4: Install: (22)"install"
X5: Nominate rank: (3)"nominatn"
X6: Rank in district: (27)"district"
X7: Number of installment plan: (22) install
X8: Number of own heavy machines: (37)"heavymec"
X9: Amount of debt: (14)"debt"

The cross-validated misclassification cost (probability) of the minimal tree which misclassifies a case of Class 2 for uncollectible loan into a case of Class 1 for collectible loan is 0.317 while the opposite cost is 0.175 shown as Fig. 8. The costs by cross-validation are the same as the optimal tree within the 1-SE rule having 5 terminal nodes unlike the costs by learning sample. Also the cross-validated costs of the trees having 9 and 7 terminal nodes within the 1-SE rule are the same as the above. The difference between the minimal cost tree and the optimal tree is shown in Fig. 9 and Fig. 10.

In conclusion the minimal cost tree having 11 terminal nodes are employed because the variable X5 to X9 are considerably important in business unlike variable importance in the analysis which is defined as a measure of a variable's ability to mimic the chosen tree and to play a role as a surrogate for the best splitting variable.

```
% OPTION SPECIFICATIONS FOR CART.
% Characters after % are ignored.
% Refer to the document USING THE CART PROGRAMS for syntax rules.
% The response variable must be identified first.
Response  =  19  Classes =   2  % Variable Name: JUDJE
Split      = GINI       % GINI possibly with altered priors
Priors     = thus       0.9500     0.5000E-01  *  % prior probabilities

Cost       = thus       % misclassification costs
   ( 1/ 1)     0.     10.00
   ( 1/ 2)  1.000        0.   *
Delete     - (11,  ) *          % delete the following variables
Secondary               % Secondary options
   atom         =   5  % minimum node size to split
   sample       =  1000  % larger nodes are subsampled
   nsurrogates  =  10,  5 % surrogates used, printed
   ncompetitors =   5    % competitors printed
   ntrees       =  100   % print errors for this many trees
   learnsize    =  20000  % maximum size of learning sample
   nodemax      =   750  % number of nodes in largest tree
   catmax       =  1000  % maximum categorical splits
   linmax       =    0  % maximum linear combination splits
   depthmax     =   750  % maximum depth  of tree *
Cross          10     % -fold cross validation *
```

Fig. 6: Option parameters for analysis.

```
                              Total Cases = 448
                                          | X1   960 billions yen
                                +---------1---------+
                    X2  25 thou. |                  | X9   949 thou.
                         +---------2------+     +-----9------+
               X3  40 thou. |                |   |         | X8 in (1,4,5)
                     +------3-------+                   +--10--+
           X4  24.5 |               |                   |      |
                 +--------4------+
           X5  3.5 |              |
            +---------5---------+
     X6     |                   |
    +-------6-------+
    |               |X7 in (1,2)
           +--------7--------+
           |X8 in (1,2,4,5) |
         +---8---+
         |       |

Terminal Regions
   1    2    3       4    5    6    7     8     9      10    11
   ⊜    ⊜    ×       ×    ×    ×    ⊜     ⊜     ⊜      ⊜     ×        Class
  19    6    2       0    8   16   15    39   203     16    21        1 (⊜)
   2    0    6      10   12   62    2     3    11      4   112        2 (×)
0.99 1.00 0.79    1.00 0.83 0.63 0.98  0.99  0.99   0.97  0.57
```

Fig. 7: Classification tree diagram of the minimal cost tree.

MISCLASSIFICATION BY CLASS

Minimal Cost Tree: Terminal Nodes = 11

Class	Prior Prob.	|-- CROSS VALIDATION --|			|-- LEARNING SAMPLE --|		
		N	N Mis-Classified	Cost	N	N Mis-Classified	Cost
1	0.950	325	57	0.175	325	37	0.114
2	0.050	123	39	0.317	123	18	0.146
Total	1.000	448	96		448	55	

Optimal Tree: Terminal Nodes = 5

Class	Prior Prob.	|-- CROSS VALIDATION --|			|-- LEARNING SAMPLE --|		
		N	N Mis-Classified	Cost	N	N Mis-Classified	Cost
1	0.950	325	57	0.175	325	39	0.120
2	0.050	123	39	0.317	123	34	0.276
Total	1.000	448	96		448	73	

Fig. 8: Comparison of cross-validated relative error between the minimal cost tree and the optimal tree.

CLASSIFICATION TREE DIAGRAM

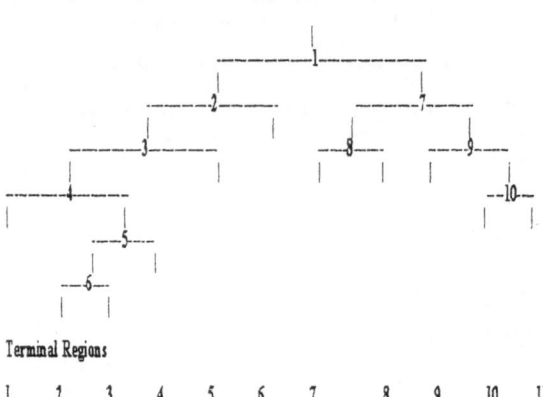

Terminal Regions

1   2   3   4   5   6   7   8   9   10   11

Fig. 9: The tree with 11-terminal nodes with minimal cross-validated cost.

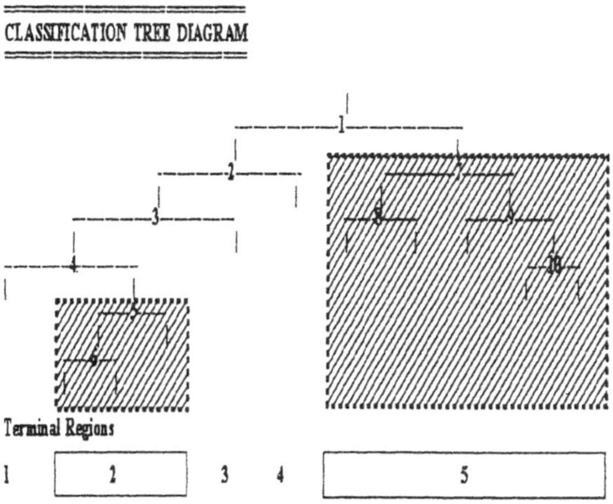

CLASSIFICATION TREE DIAGRAM

Fig. 10: The optimal tree with 5-terminal nodes.

References

Breiman, L., Friedman, J. H., Olshen, R. A., Stone, C. J. (1984): *Classification and Regression Trees*, Wadsworth.

California Statistical Software(1994): *CART Version 1.310*, CalStat, Inc.

Morgan, J. N., Sonquist, J. A. (1963): Problems in the analysis of survey data, and proposal, *JASA*, **58**, 415–434.

Steinberg, D., Colla, P.(1992): *CART*, SYSTAT, Inc.

Analysis of Preferences for Telecommunication Services in Each Area

Tohru UEDA[1] and Daisuke SATOH[2]

[1]Seikei University
3-3-1 Kichijoji-Kitamachi, Musashino-Shi
Tokyo 180, Japan

[2]NTT Telecommunication Networks Laboratories
3-9-11 Midori-Cho, Musashino-Shi
Tokyo 180, Japan

Summary: A method of combining conjoint analysis and regression analysis is proposed in order to analyze not only individual consumer preferences, but also preference tendencies on an aggregate basis, for example, in various geographic areas. We apply this method to new telecommunication services.

1. Introduction

Conjoint analysis [Luce and Tukey (1964)] has been used in marketing research to measure consumer preferences. It is a practical set of methods for predicting consumer preferences for multi-attribute options in a wide variety of product and service contexts. When developing new products and services, it is an effective method to determine service characteristics. As for the evolution of conjoint analysis in marketing research, see reviews by Green and Srinivasan (1978, 1990), Wittink and Cattin (1989), and Wittink et al (1994).

In this paper, we apply conjoint analysis to telecommunication services. We divide telecommunication services into two classes: services that are independent of subscriber networks and services that depend on subscriber networks. In the former case we can use conjoint analysis to determine service characteristics because we can regard typical consumer preference as reflecting general consumer preferences. Ueda (1994) has recently applied conjoint analysis to existing telecommunication services (voice mail). This kind of service offers a good opportunity to apply conjoint analysis.

In the latter case we must analyze overall consumer preferences in service areas because the services depend on subscriber networks. Moreover, when we have several service area candidates, we must forecast demands in order to build subscriber networks economically. It has been difficult to use conjoint analysis to identify preference tendencies on an aggregate basis and especially in specific geographic areas because it has mainly focused on individual consumers. Conjoint analysis alone is insufficient to measure consumer preferences for services such as cable television (CATV) in service areas.

In this paper we propose a method of identifying likely preferences in various areas. To do this we combine conjoint analysis with regression analysis. This combination enables us to analyze preference tendencies in specific geographic areas. We apply our method to new telecommunication services.

2. Application of conjoint analysis to telecommunication services

We assume that new telecommunication services (cable television services) have four attributes or factors that will influence consumer preference: video on demand (V.O.D) service, telephone service, registration fee, and monthly charge.

Here we consider the number of possible alternatives. There are two alternatives for each of the two attributes: the V.O.D service and the telephone service, because each is either present or not. Four registration fees are being considered: 20,000 yen, 40,000 yen, 60,000 yen, and 80,000 yen. Four monthly charges are being considered: 3,000 yen, 4,000 yen, 5,000 yen, and 6,000 yen. Consequently, a total of $2 \times 2 \times 4 \times 4 = 64$ alternatives would have to be tested if we were to array all possible combinations of the four attributes.

Table 1: Combinations for CATV service

	V.O.D	Telephone service	Registration fee (yen)	Monthly charge (yen)
A	Yes [2]	Yes [2]	80,000 [1]	4,000 [3]
B	No [1]	Yes [2]	80,000 [1]	3,000 [4]
C	No [1]	Yes [2]	60,000 [2]	4,000 [3]
D	No [1]	No [1]	60,000 [2]	3,000 [4]
E	Yes [2]	Yes [2]	40,000 [3]	6,000 [1]
F	Yes [2]	No [1]	40,000 [3]	5,000 [2]
G	Yes [2]	No [1]	20,000 [4]	6,000 [1]
H	No [1]	No [1]	20,000 [4]	5,000 [2]

Table 1 shows an orthogonal array that involves 8 of the 64 possible combinations that we wish to test in this case. As an example, we give tentative points, which are the numbers in square brackets in Table 1, to each alternative for each of the four attributes. Each combination has a total of 8 points. Therefore we cannot judge that one is clearly superior to another one.

We make inquiries about the preference-ranking of the choice-set of the 8 combinations in Table 1. We assume that every respondent would prefer a lower price and more services.

We consider the following utility function for combination i

$$U_i = a_1\delta_1(i) + a_2\delta_2(i) - a_3\log x_3(i) - a_4\log x_4(i),\tag{1}$$

where $\delta_j(i)$ and $x_h(i)$ are defined by

$$\delta_j(i) = \begin{cases} 1: & \text{if combination } i \text{ has attribute } j \quad (j = 1,2), \\ 0: & \text{otherwise}, \end{cases}\tag{2}$$

$$x_h(i): \text{the value of attribute } h \text{ for combination } i \quad (h = 3,4).\tag{3}$$

Respondents' preference for a lower price and more services is expressed by the following restrictions:

$$a_k \geq 0 \quad (k = 1,2,3,4).\tag{4}$$

3. Sampling

We conduct a survey in two areas of Japan, getting 389 respondents from one area and 200 from the other. In Table 2 the area denotes a major city in Japan. The respondents in each area are composed of random and purposive sampling. Although there are very few subscribers to CATV service in those areas, their opinions are very important, so we also intentionally select respondents who are CATV subscribers.

Table 2: Sampling

Area	Random sampling	Purposive sampling	Total
A	300	89	389
B	162	38	200

4. Estimation Method

Various algorithms for conjoint analysis have been proposed, such as MONANOVA [Kruskal (1965)], TRADE-OFF [Johnson (1973)], LINMAP [Srinivasan and Shocker (1973)], and RANKLOGIT [Ogawa (1987)]. LINMAP differs from the others in that it uses linear programming whereas the other approaches use nonlinear optimization. The use of linear programming enables LINMAP to obtain global optimum parameter estimates, while the other approaches cannot be guaranteed to achieve global optimums.

Satoh and Ueda have discovered two problems with LINMAP solutions.

1: Even if there is a set of solutions that expresses the preference data perfectly, LINMAP cannot always generate it. Instead it produces a set that expresses only the partial rankings.

2: LINMAP cannot necessarily produce a set of solutions that matches an analyst's inferences from observed data.

Satoh and Ueda have proposed an improvement of LINMAP[Satoh and Ueda]. Thus we applied it to the new telecommunication services in Table 1. The algorithm is composed of two steps as follows. STEP1 is LINMAP and STEP2 is a new additional part.

STEP 1 (LINMAP):

$$\min \sum_{i=1}^{n-1} \gamma_i, \tag{5}$$

subject to

$$U_{\hat{1}} - U_{\hat{2}} = (x_{\hat{1}} - x_{\hat{2}})a + \gamma_1 \geq 0,$$
$$U_{\hat{2}} - U_{\hat{3}} = (x_{\hat{2}} - x_{\hat{3}})a + \gamma_2 \geq 0,$$
$$\cdots$$
$$U_{n-1} - U_{\hat{n}} = (x_{n-1} - x_{\hat{n}})a + \gamma_{n-1} \geq 0,$$
$$U_{\hat{1}} - U_{\hat{n}} = (x_{\hat{1}} - x_{\hat{n}})a = 1,$$
$$a \geq 0 \quad , \quad \gamma_1, \cdots, \gamma_{n-1} \geq 0,$$

where n is the total number of combinations, a is the vector whose components are the utility parameters a_i for every attribute, and U_h and x_h are respectively the utility and the combination vector of the combination ranked hth by the respondent.

STEP 2:

$$\min(\varepsilon_l + \varepsilon_u), \tag{6}$$

subject to

$$(x_1 - x_2)a \quad + \quad \gamma_1 \leq \frac{1}{n-1} + \varepsilon_u,$$

$$(x_2 - x_3)a \quad + \quad \gamma_2 \leq \frac{1}{n-1} + \varepsilon_u,$$

$$\cdots$$

$$(x_{n-1} - x_n)a \quad + \quad \gamma_{n-1} \leq \frac{1}{n-1} + \varepsilon_u,$$

$$(x_1 - x_2)a \quad + \quad \gamma_1 \geq \frac{1}{n-1} - \varepsilon_l,$$

$$(x_2 - x_3)a \quad + \quad \gamma_2 \geq \frac{1}{n-1} - \varepsilon_l,$$

$$\cdots$$

$$(x_{n-1} - x_n)a \quad + \quad \gamma_{n-1} \geq \frac{1}{n-1} - \varepsilon_l,$$

$$(x_1 - x_n)a \quad = \quad 1,$$

$$\gamma_{min} \quad = \quad \sum_{i=1}^{n-1} \gamma_i,$$

$$a \geq 0, \quad \gamma_1, \quad \cdots \quad, \gamma_{n-1} \geq 0,$$

$$0 \leq \varepsilon_l \leq \frac{1}{n-1} \quad, \quad 0 \leq \varepsilon_u \leq \frac{n-2}{n-1},$$

where γ_{min} is the optimal solution of the objective function (5).

5. Analysis

Here is the sum of the range of every respondent's partworth for each attribute in Table 3 and Figs. 1 and 2.

Table 3: Sum of partworths

Area	V.O.D.	Phone	Registration fee	Monthly charge
A	100.11	57.51	69.66	62.99
B	50.00	30.58	25.53	22.20

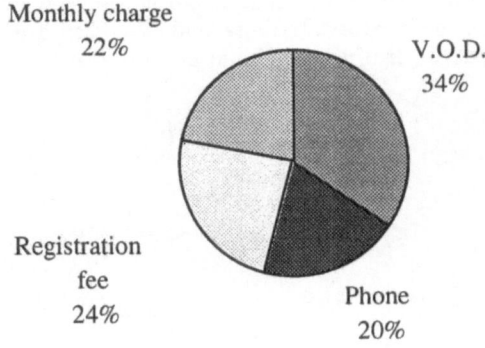

Figure 1: Sum of partworths in area A

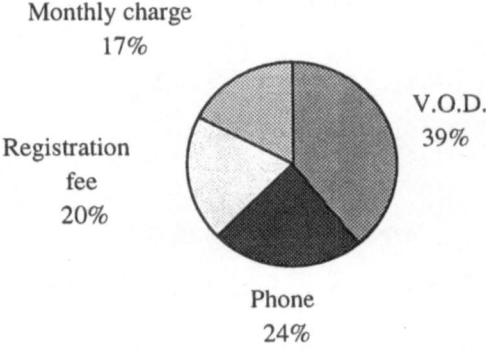

Figure 2: Sum of partworths in area B

6. Preference-ranking of choice-set combinations in each area

Let a_j^p denote the relative importance of attribute j to decision-maker p. Let $\tilde{U}_i^p$ denote the utility of combination i for decision-maker p through conjoint analysis. $\tilde{U}_i^p$ is expressed as

$$\tilde{U}_i^p = a_1^p \delta_1(i) + a_2^p \delta_2(i) - a_3^p \log x_3(i) - a_4^p \log x_4(i). \tag{7}$$

We assume that $\tilde{U}_i^p$ can be related to decision-maker p's characteristic θ_j^p as follows:

$$\tilde{U}_i^p = \sum_j b_j^i \theta_j^p + \varepsilon^p, \tag{8}$$

where ε^p is the disturbance term and θ_j^p is defined by

$$\theta_j^p = \begin{cases} 1 : & \text{if decision-maker } p \text{ has characteristic } j, \\ 0 : & \text{otherwise.} \end{cases} \qquad (9)$$

We obtain the value of b_j^i through multiple regression analysis.

To get a preference-ranking of the choice-set combinations in area A, we aggregate θ_j^p in area A as follows:

$$\overline{U_i(A)} = \sum_{p \in A} \sum_j b_j^i \theta_j^p = \sum_j b_j^i \sum_{p \in A} \theta_j^p. \qquad (10)$$

Thus, we can obtain $\overline{U_i(A)}$ for combination i in area A if we know the value of $\sum_{p \in A} \theta_j^p$. The factors we chose as explanatory variables θ_j^p are shown in Appendix I.

Moreover, we can obtain the relative importance of attribute $\overline{a_k(A)}$ in area A through multiple regression analysis by using conjoint analysis or the multiple regression equation

$$\overline{U_i(A)} = \overline{a_1(A)}\delta_1(i) + \overline{a_2(A)}\delta_2(i) - \overline{a_3(A)}\log x_3(i) - \overline{a_4(A)}\log x_4(i), \qquad (11)$$

where $\overline{U_i(A)}$ is the criterion variable and $\delta_j(i)$ and $x_h(i)$ are explanatory variables, which are defined by Eqs. (2) and (3), respectively.

Regression analysis gives coefficients b_j^i. Hence we can obtain a preference-ranking

Table 4: Coefficient of determination

	Coefficient of determination
A	0.3199
B	0.3526
C	0.3387
D	0.3671
E	0.3525
F	0.3702
G	0.3539
H	0.3111

of choice-set combinations in various areas if the data of explanatory variables in Appendix I are available in those areas. Moreover, we can estimate the relative importance of an attribute in various areas. The estimation, however, has low accuracy, as shown in Table 4.

7. Conclusion

We have applied conjoint analysis to new telecommunication services and investigated a method of obtaining a preference-ranking of choice-set combinations in various areas. We have chosen this elaborate method over other ones because it should enable us to determine the relative importance of an attribute in various areas and to choose a new service that most consumers in those area prefer to another service. Although the results had low accuracy due to a lack of effective explanatory variables, this method will be effective if appropriate ones are found. Further studies will be needed to transform our estimation into a forecast of the demand[Ueda et al (1995)] for new telecommunication services in various areas.

Appendix I: The factors used as explanatory variables

1: Are you male or female? (2)

2: How old are you? (9)

3: What is your relationship to the head of your household? (4)

4: What does your head of household do? (8)

5: What kind of company does your head of household work at? (9)

6: What is the annual income of your household? (11)

7: What kind of house do you live in? (4)

8: How many rooms does your house have? (8)

9: How many television sets are there in your house? (4)

10: How many hours per day do you watch television on average? (10)

11: How many videocassette recorders do you have? (5)

12: How many rental videocassettes do you rent per month? (7)

13: How many personal computers are there in your household? (4)

14: Do you have video game machine A? (2)

15: Do you have video game machine B? (2)

16: Do you have video game machine C? (2)

17: Do you have video game machine D? (2)

18: Do you have video game machine E? (2)

19: Do you have video game machine F? (2)

20: Do you have video game machine G? (2)

21: Do you have video game machine H? (2)

22: Do you have video game machine I? (2)

23: Do you have video game machine J? (2)

24: Do you get satellite broadcast TV in your house? (2)

25: Do you get CATV? (2)

26: Would you pay an additional charge for pay channels on CATV if they were interesting? (2)

Note that these questions were designed for Japanese respondents and might be inappropriate in other countries.

The number in parentheses is the number of categories.

References:

Green, P.E. and Srinivasan, V. (1978): Conjoint Analysis in Consumer Research: Issues and Outlook, *Journal of Consumer Research*, **5**. 103-123.

Green, P.E. and Srinivasan, V. (1990): Conjoint Analysis in Marketing: New Developments with Implications for Research and Practice, *Journal of Marketing*, **54**, 3-19.

Johnson, R.M. (1975): A Simple Method for Pairwise Monotone Regression, *Psychometrika*, **40**, 2, 163-168.

Kruskal, J.B. (1965): Analysis of Factorial Experiments by Estimating Monotone Transformations of the Data, *Journal of the Royal Statistical Society*, Series B, **27**, 251-263.

Luce, R.D. and Tukey, J.W.(1964): Simultaneous conjoint measurement: A new type of fundamental measurement, *Journal of Mathematical Psychology*, **1**, 1-27.

Ogawa, K. (1987): An Approach to Simultaneous Estimation and Segmentation in Conjoint Analysis, *Marketing Science*, **6**, 1, 66-81.

Satoh, D. and Ueda, T.: to be submitted.

Srinivasan, V. and Shocker, A.D. (1973): Estimating the Weights for Multiple Attributes in a Composite Criterion Using Pairwise Judgments, *Psychometrika*, **38**, 473-493.

Ueda, T. (1994): Analysis of Preferences for Services Based on Conjoint Analysis, *Singaku ron*, J77-B-I, **9**, 542-549, (in Japanese).

Ueda, T. et al. (1995): A method of forecasting demand for new telecommunication services, *9th European Meeting of the Psychometric Society*, 123.

Wittink, D. and Cattin, P. (1989): Commercial Use of Conjoint Analysis: An Update, *Journal of Marketing*, **53**, 91-96.

Wittink, D. et al. (1994): Commercial Use of Conjoint in Europe: Results and Critical Reflections, *International Journal of Research in Marketing*, **11**, 41-52.

Effects of End-Aisle Display and Flier
on the Brand-Switching of Instant Coffee

Akinori Okada

Department of Industrial Relations
School of Social Relations
Rikkyo (St. Paul's) University
3 Nishi Ikebukuro
Toshima-ku, Tokyo 171, Japan

Summary: Brand-switching data among instant coffee brands were analyzed by a nonmetric asymmetric multidimensional scaling (Okada and Imaizumi, 1987) to identify effects of the end-aisle display and of the flier. Two-dimensional solutions show that the end-aisle display of the brand is in general not effective to induce switching to the brand and is vulnerable against switching to other brands, and that for some brands the flier of the brand is effective to induce switching from similar brands to the brand and is defensive against switching to other brands, but that for some brands the flier is not effective to induce switching to the brand and is vulnerable against switching to similar brands.

1. Introduction

After several asymmetric multidimensional scaling (MDS) models and procedures have been introduced (Zielman and Heiser, 1996), asymmetric MDS has been utilized to analyze various sorts of data such as attraction relationships (Chino, 1978; Collins, 1987), journal citations (Chino, 1978, 1990; Weeks and Bentler, 1982), word associations (Chino, 1990; Harshman et al., 1982; Zielman and Heiser, 1993), telephone communication (Okada, 1989), intergenerational occupational mobility (Okada, 1988a), foreign trade (Chino, 1978), marriages among ethnic groups (Zielman, 1991), or data from various areas of psychology and of sociology.

One of the most important areas for applying asymmetric MDS seems to be marketing research. Brand switching data have been analyzed by asymmetric MDS, because asymmetries in brand switching might have a relationship with the differences in attractiveness among brands (DeSarbo and De Soete, 1984; Zielman and Heiser, 1996). Asymmetric MDS has been used to analyze brand switching data among car categories or among soft drink brands (DeSarbo and Manrai, 1992; DeSarbo, et al., 1992; Harshman, et al., 1982; Okada, 1988b; Zielman, 1991). In the present study, brand switching data among instant coffee brands are analyzed by a nonmetric asymmetric MDS (Okada and Imaizumi, 1987) to investigate the effects of the end-aisle display and of the flier (a pamphlet or circular for mass distribution issued by a supermarket informing of sales) on the brand switching.

2. Data

The brand switching data analyzed in the present study was derived from scanner data of about 5,000 instant coffee purchases made in 1993 by a panel which consists of 796 households who frequently came to a super market. Eleven instant coffee brands which were analyzed in the present study (10 brands and other instant coffee brands which were treated as the 11-*th* brand) are represented in Table 1. These brands include three types of instant coffee; freeze-dried instant coffee (type a in Table 1), regular instant coffee (type b in Table 1), and ones which are already mixed with sugar and cream or which are already packed in a plastic or paper cup (type c in Table 1). They also include brands of Nestle which dominates in the Japanese instant coffee market and brands of Ajinomoto General Foods which is a joint venture between Ajinomoto and General Foods.

Tab. 1: Eleven Instant Coffee Brands.

brand	type*	abbreviation
1 Nescafe Goldblend 150g	a	NGB150
2 Nescafe Goldblend 100g	a	NGB100
3 Nescafe Excella 250g	b	NEX250
4 Ajinomoto General Foods Maxim 100g	a	AMX100
5 Ajinomoto General Foods Maxim 2 cup	c	AMX2cup
6 Nescafe Cappuccino	c	NCP
7 Ajinomoto General Foods Blendy 250g	b	ABL250
8 UCC The Blend 144 100g	a	UCC100
9 Nescafe Excella 100	b	NEX100
10 Ajinomoto General Foods Maxim 30g	a	AMX30
11 others	**	others

*a: freeze-dried instant coffee
b: regular instant coffee
c: already mixed with sugar and cream or already packed in a plastic or paper cup
** Others are not a single brand, and consist of different types of instant coffee brands.

Seven brands were purchased both when there was and when there was not an end-aisle display of each of them at the supermarket. Others, the 11-*th* brand, were not purchased when there was an end-aisle display of any of them. If we distinguish a purchase of a brand when there was an end-aisle display of that brand (the brand with the end-aisle display) from a purchase when there was not an end-aisle display of that brand (the brand without the end-aisle display) as two different items, we have 18 items or brands. A switching matrix among 18 items or brands was calculated for each household. The sum of 796 matrices was derived to construct the switching matrix among 18 items or brands for the panel. Table 2 shows the 18 x 18 switching matrix, whose (j,k) element represents the frequency of switching from items or brands j to k, which is called the end-aisle display data.

Six brands were purchased both when a flier accompanied and when did not accompany each of the brands. If we distinguish a purchase of a brand when the flier issued by the supermarket accompanied that brand (the brand with the flier) from a purchase when the flier did not accompany the brand (the brand without the flier) as two different items, we have 17 items or brands. Others were not purchased when a flier accompanied any of them. Table 3 shows the switching matrix among 17 items or brands which is called the flier data.

3. Method and analysis

The nonmetric asymmetric MDS (Okada and Imaizumi, 1987) was applied to the 18 x 18 table of the end-aisle data and the 17 x 17 table of the flier data respectively. The asymmetric MDS represents each item or brand in a multidimensional

Tab 2: Switching Matrix among 18 Items or Brands (with/without the End).

from	\<to\> 1	2	3	4	5	6	7	8	9	10	11	12	13	14	15	16	17	18
1 NGB150 w/o end	70	76	26	34	8	5	7	12	1	3	1	4	5	3	1	5	2	47
2 NGB150 with end	65	92	39	71	8	21	15	23	1	3	5	6	10	9	1	7	5	52
3 NGB100 w/o end	26	24	33	24	4	4	10	12	0	2	2	0	2	1	0	2	3	23
4 NGB100 with end	27	37	39	25	4	6	11	51	0	9	4	0	2	6	0	3	2	47
5 NEX250 w/o end	9	13	7	5	84	59	2	1	0	0	1	7	12	0	0	29	1	22
6 NEX250 with end	10	13	8	4	60	88	0	10	2	4	2	5	8	3	0	18	1	26
7 AMX100 w/o end	11	22	8	14	3	1	22	18	1	5	2	2	2	15	0	3	6	45
8 AMX100 with end	11	31	47	12	5	4	38	65	0	1	3	0	4	7	2	6	2	51
9 AMX2 w/o end	1	1	1	0	0	0	0	1	39	0	1	0	1	3	0	0	2	19
10 NCP w/o end	4	6	5	6	3	1	4	7	0	187	21	0	2	1	0	3	6	73
11 NCP with end	3	1	4	4	4	1	1	5	0	15	18	0	0	0	0	0	1	31
12 ABL250 w/o end	1	5	0	3	8	8	0	4	0	1	0	6	18	0	0	1	0	21
13 ABL250 with end	11	13	5	6	19	19	3	9	0	2	4	20	41	3	0	2	1	43
14 UCC100 w/o end	6	14	9	15	0	1	12	31	1	1	0	1	5	36	0	3	3	49
15 UCC 100 with end	1	0	1	1	0	0	0	0	0	0	0	0	0	1	0	0	0	2
16 NEX150 w/o end	6	5	3	3	19	27	1	6	1	2	2	3	1	0	1	51	3	28
17 AMX30 w/o end	0	2	4	2	1	3	8	8	2	5	3	1	0	7	0	0	51	38
18 others	37	63	54	36	25	30	27	79	16	69	35	23	35	48	2	25	32	768

Tab. 3: Switching Matrix among 17 Items or Brands (with/without the Flier).

from	to																
	1	2	3	4	5	6	7	8	9	10	11	12	13	14	15	16	17
1 NGB150 w/o flier	279	22	139	11	30	6	43	11	2	9	2	23	2	14	11	7	97
2 NGB150 with flier	2	0	20	0	1	5	1	2	0	1	0	0	0	0	1	0	2
3 NGB100 w/o flier	99	2	116	1	12	3	55	27	0	14	1	4	0	6	3	4	66
4 NGB100 with flier	13	0	4	0	3	0	1	1	0	2	0	0	0	1	2	1	4
5 NEX250 w/o flier	28	4	17	1	160	52	3	2	1	1	1	25	1	1	37	1	35
6 NEX250 with flier	12	1	6	0	50	29	3	5	1	5	0	6	0	2	10	1	13
7 AMX100 w/o flier	56	2	60	3	10	1	86	13	1	9	0	7	0	22	5	8	80
8 AMX100 with flier	16	1	16	2	1	1	32	12	0	2	0	1	0	2	4	0	16
9 AMX2cup w/o flier	2	0	1	0	0	0	1	0	39	1	0	1	0	3	0	2	19
10 NCP w/o flier	12	1	17	1	7	1	17	0	0	227	8	1	1	1	3	7	97
11 NCP with flier	1	0	0	1	1	0	0	0	0	6	0	0	0	0	0	0	7
12 ABL250 w/o flier	25	2	12	2	34	13	11	3	0	6	0	78	5	2	3	1	60
13 ABL250 with flier	3	0	0	0	3	4	2	0	0	1	0	2	0	1	0	0	4
14 UCC100 w/o flier	19	2	24	2	1	0	28	15	1	1	0	6	0	37	3	3	51
15 NEX150 w/o flier	10	1	5	1	33	13	5	2	1	4	0	4	0	1	51	3	28
16 AMX30 w/o flier	2	0	6	0	2	2	15	1	2	8	0	1	0	7	0	51	38
17 others	96	4	88	2	38	17	80	26	16	100	4	52	6	50	25	32	768

Euclidean space as a point and a circle (in a two-dimensional space), a sphere (in a three-dimensional space) or a hypersphere (in a four- or higher-dimensional space) centered at that point. A configuration, which consists of points and circles (spheres, hyperspheres), represents both symmetric and asymmetric proximity relationships among items or brands. The Euclidean distance between two points imbedded in a configuration corresponds to the symmetric switching between two items or brands represented by those two points, and the difference of two radii of the circles centered at those two points corresponds to the asymmetric switching from one item or brand to the other.

We regard the number of purchases of item or brand k switched from item or brand j as the similarity from items or brands j to k, and the number of purchases of item or brand j switched from item or brand k as the similarity from items or brands k to j. Let s_{jk} be the similarity from items or brands j to k, and s_{kj} be the similarity from items or brands k to j. s_{jk} is not necessarily equal to s_{kj}. The similarity is assumed to be monotonically decreasingly related with m_{jk}. m_{jk} is defined by Equation (1)

$$m_{jk} = d_{jk} - r_j + r_k, \tag{1}$$

and m_{kj} is defined by Equation (2)

$$m_{kj} = d_{jk} - r_k + r_j. \tag{2}$$

Figure 1 represents items or brands j and k geometrically in a two-dimensional space. Item or brand j is represented by a point and a circle of radius r_j centered at that point, and item or brand k is represented by a point and a circle of radius r_k centered at that point. The length of the arrow directing from objects or brands j to k (from left to right) in Figure 1 shows m_{jk}, and the length of the arrow directing from objects or brands k to j (from right to left) in Figure 1 shows m_{kj}.

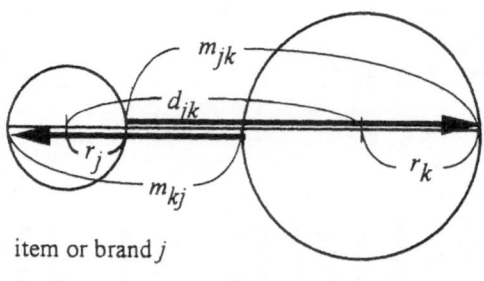

item or brand j

item or brand k

Fig. 1: Geometric Representation of Items or Brands j and k
in a Two-Dimensional Space.

When a point is in a central part of a configuration, that point is on average near to other points. The item or the brand represented by the point in a central part of a configuration is likely to be switched to/from other items or brands. When a point is in a periphery of a configuration, that point is on average near to some points representing similar items or brands but is rather distant to points representing dissimilar items or brands. The item or the brand represented by a point in a periphery of a configuration is likely to be switched to/from similar items or brands, but it is unlikely to be switched to/from dissimilar items or brands.

The radius of a circle (sphere or hyper sphere) centered at a point represents the vulnerability of the item or the brand represented by that point in switching. When the radius of a circle centered at a point is small, the item or the brand represented by that point is unlikely to be switched to others and is likely to be switched from others. When the radius of a circle centered at a point is large, the item or the brand represented by that point is likely to be switched to others and is unlikely to be switched from others. The larger the radius is, the more vulnerable the item or brand is.

A nonmetric algorithm based on Kruskal's nonmetric MDS (Kruskal, 1964) is used to derive a configuration (coorinates of points representing items or brands and radii) where m_{jk} is monotonically decreasingly related with the frequency of switching. By using the algorithm, the configuration which minimizes the badness-of fit measure S (Equation (3.1) of Okada and Imaizumi, 1987) for a given dimensionality is derived.

In the analysis the maximum dimensionality of the analysis is determined, and in each of the maximum dimensional through unidimensional spaces the configuration which minimizes S is derived. In the maximum dimensional space, an initial configuration of items or brands and their radii is derived by processing the similarities among items or brands using a factor analytic procedure. In a lower dimensional space, an initial configuration based on the principal components of the higher dimensional result is used. The initial configuration is improved by the steepest descent method iteratively to minimize S. The iteration is stopped, when (a) S is smaller than the stopping criterion (0.00001), (b) the improvement of S is smaller than the stopping criterion (0.00001), or (c) the number of iterations becomes larger than the maximum number (70).

Each data were analyzed by using maximum dimensionalities of seven through four. We had one S in seven-dimensional space, two S in six-dimensional space, three S in five-dimensional space, and four S in each of four- through unidimensional spaces. In six-dimensional space, the smaller S was chosen as the minimized S in six-dimensional space. In five-dimensional space, the smallest S of three S was chosen as the minimized S. In each of four- through unidimensional spaces, the smallest S was chosen as the minimized S in that dimensional space. Then we have the minimized S in each of seven- through unidimensional spaces.

4. Results

The minimized S in five- through unidimensional spaces for the end-aisle data were 0.232, 0.265, 0.309, 0.367 and 0.471, and those for the flier data were 0.204, 0.252, 0.303, 0.377 and 0.499. These figures and the interpretation of configurations suggest that we choose the two dimensional result as the solution for each of the two data sets.

Figure 2 shows the two-dimensional solution for the end-aisle display data. (The configuration was visually rotated so that the interpretation of the rotated configuration seems as clear as possible.) Each item or brand is represented as a point and a circle. The bold circle represents the brand with the end-aisle display, and the light circle represent the brand without the end-aisle display. For most of the brands, two

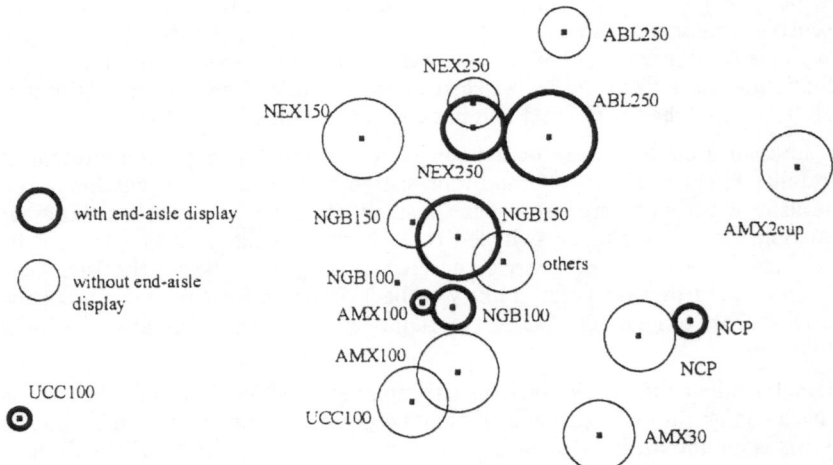

Fig. 2: Two-Dimensional Configuration of 11 Instant Coffee Brands
with/without the End-Aisle Display.

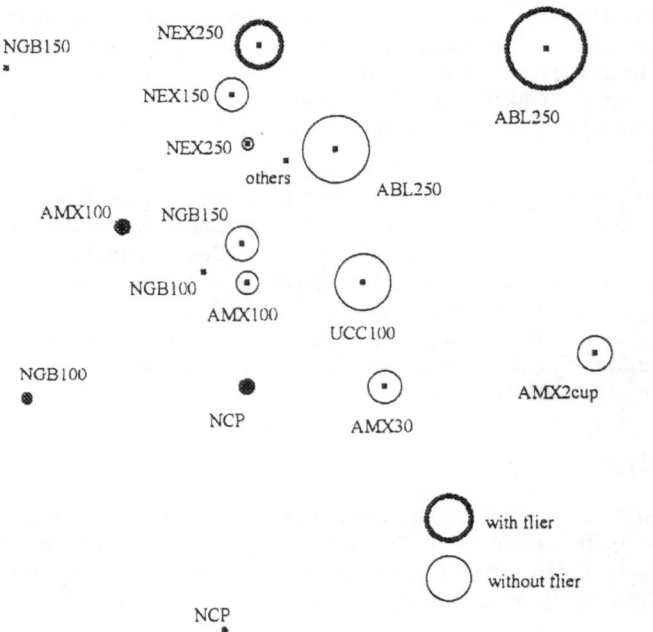

Fig. 3: Two-Dimensional Configuration of 11 Instant Coffee Brands
with/without the Flier.

points representing the same brand with and without the end-aisle display are closely located in the configuration. The vertical dimension seems to differentiate regular instant coffee brands from freeze-dried and ones already mixed with sugar and cream or already packed in a plastic or paper cup, and the horizontal dimension seems to represent the difference between brands of Nestle and those of Ajinomoto General Foods (Okada and Genji, 1995).

Figure 3 shows the two-dimensional solution for the flier data. (The configuration was also visually rotated.) The bold circle represents the brand with the flier, and the light circle represent the brand without the flier. Two points representing the same brand with and without the flier are also closely located in the configuration. The two dimensions look like to have the same meaning of those in Figure 2.

The obtained configuration derived from the analysis of the end-aisle data shows that most of the brands with the end-aisle display are in the central part of the configuration and that most of the brands without the end-aisle display are in the periphery of the configuration. Most of the brands with the end-aisle display located in the central part of the configuration have larger radii than those of the same brands without the end-aisle display located in the periphery of the configuration. The obtained configuration derived from the analysis of the flier data shows that most of the brands with the flier are in the periphery of the configuration and that most brands without the flier are in the central part of the configuration. Some of the brands with the flier located in the periphery of the configuration have larger radii and some of them have smaller radii than those of the same brands without the flier located in the central part of the configuration.

5. Discussion

We would like to focus our attention on those brands which are represented as two different items in a configuration; one with the end-aisle display or the flier and the other without the end-aisle display or the flier. Combining the location of a point (central part or periphery of a configuration) and the radius of a circle (smaller or larger), we can classify these brands into four categories (a) to (d) shown below (Okada and Genji, 1995). Characterization of a brand in each of four categories is accompanied.

(a) A brand with the end-aisle display or the flier has a smaller radius than the same brand without the end-aisle display or the flier and is located in rather a central part of a configuration, while that brand without the end-aisle display or the flier having a larger radius is located in rather a periphery of the configuration.
With the end-aisle display or the flier, a brand is likely to be switched from the same brand without the end-aisle display or the flier as well as from other brands, and is unlikely to be switched to the same brand without the end-aisle display or the flier as well as to other brands. Without the end-aisle display or the flier, a brand is unlikely to be switched from the same brand with the end-aisle display or the flier as well as from other brands, and is likely to be switched to the same brand with the end-aisle display or the flier as well as to other similar brands.

(b) A brand with the end-aisle display or the flier has a larger radius than the same brand without the end-aisle display or the flier and is located in rather a central part of a configuration, while that brand without the end-aisle display or the flier having a smaller radius is located in rather a periphery of the configuration.
With the end-aisle display or the flier, a brand is unlikely to be switched from the same brand without the end-aisle display or the flier as well as from other brands,

and is likely to be switched to the same brand without the end-aisle display or the flier as well as to other brands. Without the end-aisle display or the flier, a brand is likely to be switched from the same brand with the end-aisle display or the flier as well as from other similar brands, and is unlikely to be switched to the same brand with the end-aisle display or the flier as well as to other brands.

(c) A brand with the end-aisle display or the flier has a smaller radius than the same brand without the end-aisle display or the flier and is located in rather a periphery of a configuration, while that brand without the end-aisle display or the flier having a larger radius is located in rather a central part of the configuration.
With the end-aisle display or the flier, a brand is likely to be switched from the same brand without the end-aisle display or the flier as well as from other similar brands, and is unlikely to be switched to the same brand without the end-aisle display or the flier as well as to other brands. Without the end-aisle display or the flier, a brand is unlikely to be switched from the same brand with the end-aisle display or the flier as well as from other brands, and is likely to be switched to the same brand with the end-aisle display or the flier as well as to other brands.

(d) A brand with the end-aisle display or the flier has a larger radius than the same brand without the end-aisle display or the flier and is located in rather a periphery of a configuration, while that brand without the end-aisle display or the flier having a smaller radius is located in rather a central part of the configuration.
With the end-aisle display or the flier, a brand is unlikely to be switched from the same brand without the end-aisle display or the flier as well as from other brands, and is likely to be switched to the same brand without the end-aisle display or the flier as well as to other similar brands. Without the end-aisle display or the flier, a brand is likely to be switched from the same brand with the end-aisle display or the flier as well as from other brands, and is unlikely to be switched to the same brand with the end-aisle display or the flier as well as to other brands.

As mentioned earlier, seven brands were purchased both when there was and when there was not the the end-aisle display, and six brands were purchased both when there was and when there was not the flier. For the end-aisle display data, these seven brands are classified into categories (a) to (d) as shown in Table 4. For the flier data, these six brands are classified as shown in Table 5.

Tab. 4: Classification of the Seven Brands for the End-Aisle Display Data.

radius of a brand with the end-aisle display	location of a brand with the end-aisle display	
	central part	periphery
smaller radius	(a) AMX100	(c) NCP UCC100
larger radius	(b) NGB150 NGB100 NEX250 ABL250	(d) none

Five of the seven brands with the end-aisle display were in the central part of the configuration (categories (a) or (b) in Table 4), and four of the five had larger radii when they were with the end-aisle display than without the end-aisle display ((b) in Table 4), suggesting that the end-aisle display is in general not effective to induce switching from other brands and is vulnerable against switching to other brands. All six brands with the flier were in the periphery of the configuration ((c) or (d) in Table 5). Three of the six brands had smaller radii when they were with the flier than without the flier ((c) in Table 5), suggesting that the flier is effective to induce switching from similar brands as well as from the same brand without the flier and is defensive against switching to other brands. The other three had larger radii when they were with the flier than without the flier ((d) in Table 5), suggesting that the flier is not effective to induce switching from other brands and is vulnerable against switching to other similar brands.

Tab. 5: Classification of the Six Brands for the Flier Data.

	location of a brand with the flier	
radius of a brand with the flier	central part	periphery
smaller radius	(a) none	(c) NGB150 AMX100 NCP
larger radius	(b) none	(d) NGB100 NEX250 ABL250

Four of the six brands with the flier were always accompanied with the end-aisle display. For these four brands (NGB150, NGB100, AMX100, and ABL250), the effect of the flier actually means the effect of the end-aisle display and the flier. For the two brands (NEX250 and NCP), the effect of the flier means the mixture of the effect of the flier alone and the effect of the flier and the end-aisle display. To separate the effect of the end-aisle display and the effect of the flier, a switching matrix was constructed by treating each of the four brands above mentioned as three different items; (1) an item with the end-aisle display and the flier, (2) an item with the end-aisle display alone and (3) an item without the end-aisle display nor the flier, and by treating each of the two brands above mentioned as four different items; (1) an item with the end-aisle display and the flier, (2) an item with the end-aisle display alone, (3) an item with the flier alone and (4) an item without the end-aisle display nor the flier.

The resultant table was analyzed by the asymmetric MDS. Obtained results seem to suggest adopting the two-dimensional configuration as the solution. The two dimensions of the configuration have the same meaning of those in Figures 2 and 3. For the three brands (NGB150, NGB100, and ABL250) of the four mentioned above, the comparison between the brand with the end-aisle display alone and the same brand without the end-aisle display nor the flier tells that these three brands are classified in the same category (b) as shown in Table 4. For AMX100, the comparison shows that this brand is classified into (c) not (a) as shown in Table 4 (locations were reversed). Although locations of two items, one representing AMX100 with the end-aisle display alone and the other representing AMX100 without the end-aisle display nor the flier, were reversed, two items were closely located. It seems that the effect of the end-aisle display alone is almost the same as the mixture of the effect of the end-aisle display alone and the effect of the end-aisle display and the flier. For these four brands, the flier was always accompanied with the end-aisle display, and it seems impossible to separate the effect of the flier and that of the end-aisle display.

For NEX250 and NCP, the comparison between the brand with the end-aisle display alone and the same brand without the end-aisle display nor the flier shows that NEX250 is classified into (c) not (b) as shown in Table 4 (both radii and locations were reversed), and that NCP is classified in the same category (c) as shown in Table 4. Although locations and radii of two items; one representing NEX250 with the end-aisle display alone and the other representing NEX250 without the end-aisle display nor the flier, were reversed, two items were closely located and the difference of two radii was small. The comparison between the brand with the flier alone and the same brand without the end-aisle display nor the flier shows that NEX250 is classified in the same category (d) as shown in Table 5, and that NCP is classified into (d) not (c) as shown in Table 5 (radii were reversed). Although radii of two items; one representing the radius of NCP with the flier alone and the other representing NCP without the end-aisle display nor the flier, were reversed, the difference of two radii was small. It seems to suggest that the effect of the end-aisle display alone is almost the same as the mixture of the effect of the end-aisle display alone and the effect of the end-aisle display and the flier, and that the effect of the flier alone is almost the same as the mixture of the effect of the flier alone and the effect of the end-aisle display and the flier. For NEX250 and NCP, the item representing the brand with the end-aisle display and the flier was located between the item representing the brand with the end-aisle display alone and the item representing the brand with the flier alone, and had a radius whose length was between the two radii of these two items, suggesting that the interaction of the end-aisle display and of the flier are rather small.

Acknowledgment

The author would like to express his gratitude to Professor Dr. Wolfgang Gaul of the University of Karlsruhe to his helpful comments and suggestions given to him at his presentation on the IFCS-96 meeting. He also wishes to express his appreciation to Dr. Takeshi Moriguchi of The Distribution Economics Institute of Japan for providing him with the data. The author is indebted to H. A. Donovan for his helpful advice concerning English.

References

Chino. N. (1978): A Graphical Technique for Representing the Asymmetric Relationships between N Objects, *Behaviormetrika*, **No. 5**, 23-40.

Chino. N. (1990): A Generalized inner product model for the analysis of asymmetry, *Behaviormetrika*, **No. 27**, 25-46.

Collins. L.M. (1987): Deriving sociograms via asymmetric multidimensional scaling, In: *Multidimensional Scaling: History, Theory, and Applications*, Young, F.W. et al. (eds.), 179-196, Lawrence Erlbaum Associates, Hillsdale, NJ.

DeSarbo. W.S., and De Soete, G. (1984): On the use of hierarchical clustering for the analysis of nonsymmetric proximities, *Journal of Consumer Research*, **11**, 601-610.

DeSarbo, W.S. et al. (1992): TSCALE: A new multidimensional scaling procedure based on Tversky's contrast model. *Psychometrika*, **57**, 43-69.

DeSarbo, W.S., and Manrai, A.K. (1992): A new multidimensional scaling methodology for the analysis of asymmetric proximity data in marketing research, *Marketing Science*, **11**, 1-20.

Harshman, R.A. et al. (1982): A model for the analysis of asymmetric data in marketing research, *Marketing Science*, **1**, 205-242.

Kruskal, J.B. (1964): Nonmetric multidimensional scaling: A numerical method, *Psy-

chometrika, **29**, 115-129.

Okada, A. (1988a): An analysis of intergenerational occupational mobility by asymmetric multidimensional scaling, In: *The Many Faces of Multivariate Analysis: Proceedings of the SMABS-88 Conference Vol. 1*, Jansen, M.G.H. et al. (eds.), 1-15, RION, Institute for Educational Research, University of Groningen, Groningen.

Okada, A. (1988b): Asymmetric multidimensional scaling of car switching data. In: *Data, Expert Knowledge and Decisions*, Gaul, W. et al. (eds.), 279-290, Springer-Verlag, Berlin.

Okada, A. (1989). Asymmetric multidimensional scaling: Theory and application, *The Japanese Journal of the Acoustical Society of Japan*, **45**, 131-137. (in Japanese)

Okada, A., and Genji, K. (1995). Brand switching of instant coffee and the effect of end-aisle display, *Communications of the Operations Research Society of Japan*, **40**, 448-501. (in Japanese)

Okada, A., and Imaizumi, T. (1987): Nonmetric multidimensional scaling of asymmetric proximities, *Behaviormetrika*, **No. 21**, 81-96.

Weeks, D.G., and Bentler, P.M. (1982): Restricted multidimensional scaling models for asymmetric proximities, *Psychometrika*, **47**, 201-208.

Zielman, B. (1991): *Three-way scaling of Asymmetric Proximities* (RR-91-01), Department of Data Theory, University of Leiden.

Zielman, B., and Heiser, W.J. (1993): Analysis of asymmetry by a slide vector model, *Psychometrika*, **58**, 101-114.

Zielman, B., and Heiser, W.J. (1996): Models for asymmetric proximities, *British Journal of Mathematical and Statistical Psychology*, **49**, 127-146.

On the classification of environmental data in the Bavarian Environment Information System using an object-oriented approach

Erich Weihs

Bavarian State Ministry for State Development and Environmental Affairs
P.O. 810140 D 81901 Munich, Germany, Email Umwelt.Bayern@t-online.de

Summary: In hardly any other area is the availability of regional development data of such great importance as it is in the sector of environmental protection and conservation. It is therefore the goal of every environmental information system to provide relevant data collections for legislative bodies and for the daily execution of administrative tasks. In this context, environmental information systems are mainly represented by the organisational association of data collections by specialist information systems. Organisational association, because the required data should be accessible, but must remain with the authorities responsible for the specialist information. The current technology of data processing supports these requirements placed in divided processing.

The proof of where which data can be found and processed under which qualitative and quantitative conditions is the core of the system. The required 'common denominator' is the classification and the determination of the common vocabulary and its relations (=thesaurus) as a metalanguage, in order to be able to secure comparable and combinable research results regarding specialist information. This therefore means for information systems that the 'language' and the 'grammatics' of the data used in the said information system is clearly defined and known to the user. The dialogue with the user is organised and supported on the basis of these language patterns. For this reason, the importance of a thesaurus and of a data model as the basis of a common language is given particular weight.

Due to the fact that the references, as described above, are based on extremely varied contents and regulations of how the data is to be dealt with, it would appear justifiable to classify these points of view as being aspects of the object.

The content classification of data according to the topic, i.e. specialist origins and possible usability, is the aspect pertaining to the structuring of the data. The question of the technical procedural position of the data, within the data bank, points the way to an object-related classification and storage of data. Using a simple mathematic model, the following sections are intended to demonstrate,

- that the object-related classification of data as a content-independent aspect is an important supplement to the data itself and
- several possibilities of an object-orientated client server technology will be pointed out.

1. Introduction

One of the most important preconditions for thought is the comparative classification of terms to categories. It is only in this way that the remembrance of experiences can develop to become an aid in life. As means of categorisation, we have at our disposal, in addition to emotionally experienced images, language and script, also the ability to quantify. Finally however, even quantitative expressions of orders require their linguistic expression. The entirety of these experiences of orders and their cross-references to the model of reality is a part of what we mean when we claim to be informed (Wenzlaff 1991).

'Information' is therefore a subjective recognition - or imagination - of excerpts of reality. Subjects of information can include individuals or collective groups, i.e. organisations. 'Intersubjective' communication of data may, for example, take place by means of data transfer. Due to the fact that the recognition and formation of imagination is always bound to the existence of thought patterns, the registration of information (i.e. by acknowledging data) is the supplement of the subject's model of reality. Should the data not add to this model of reality in any way, then it will not be regarded by the recipient subject as being information (Whorf 1976, Herrschaft

1996). This statement is equally true for classification schemes and rules.

In this context, 'data' is to be understood as being definite values of statements, whose meaning (explicit or implicit - i.e. according to recognised agreement) has been determined. The meaning of a document can be made explicit through texts, tabular diagrams, legends or in files and data banks, thesaurii, etc. In this context, definite does not mean that it is a determined value in a mathematical sense. Even a parameter of a probability distribution has a definite value.

A classification system is therefore only relevant, if the data it contains make possible an extension of the user's subjective model of reality (Feyerabend 1976). Therefore, the order of the system to the research and the classification of the data according to the preconditions described above must be accessible.

In the sector of environmental conservation, linguistically-based order systems are of considerable importance: the development and use of existing data for environmental topics is receiving more and more recognition (KoEU, 1990). In this context, the establishment of verification and navigation systems - often described using the partially misleading title of metainformation systems - as the core of an environmental information system (compare LABO, 1994, p.15) is being accounted for increasingly.

Due to the fact that the majority of the available environmental data is regionally related, then, in addition to the subject-related search, regionally-related search into town names, certain regional units and co-ordinates are required and must be taken into consideration in classification systems (Weihs 1993, LABO 1994 p.21-22)

In this way, it is possible to carry out regional search asing both an order term, as well as the co-ordinates of a freely-defined area. The regionally-related access is necessary because the majority of the information is to be evaluated on the basis of area structures (i.e. natural regional units, such as river valleys, lakes, rivers and canals, aquatic and terrestrial terms or town names).

The basis for the indexing and search is the thesaurus, acting as a common 'vocabulary of language', for example for:

- the data model and the method catalogue of the system
- the data and method verification in the specialist information system
- the regulation of the access paths to the data stock
- a bibliographic, topical map with relevant reference (catalogue of maps)
- the subject and fact data
- the specialist integration of various and separate data stock
- filing of procedures, correspondence

Typically, the expression b_n

$$b_n \in B, n = 1, 2, \ldots j, \ldots k, \ldots, n^+ \quad \text{with} \tag{1}$$

$$b_j \neq b_k \in B, B = \text{final sum of the expression} \tag{2}$$

as a defined succession of symbols (data in the above mentioned sense) is not revocably and definitely based on only one object $o_{i,n}$. If b_n is homonymous (i.e. 'field' as a expression for agricultural land or a data section, see diagram 2) then

$$b_n \Leftarrow \{o_{g,n}\} , \; g=1,2, \; ..., \; g^+ \text{ with } g^+ = \text{ the number of all objects } o_{g,n} \text{ to the term } b_n \quad (3).$$

Correspondingly, we treat objects in the same way that, from the point of view of different users, are observed under different aspects: for example, 'pasture' has a very different meaning, depending on whether it is considered from the point of view of land use, or of natural preservation. Based on the same aspect (i.e. natural preservation versus land use), various objects can be observed in the same way. We use the factor g^* to denote the number of all possible aspects. We will continue from here without limitation from an aspect-related point of view, due to the fact that the homonyms, according to (3), are included. As we will see, *the introduction of the aspects mean a further classification, independent of the subject-logical classification.*

The relation expressed by (3) is not revocably definite (i.e. b_n = 'field', unlike in table 1 for agriculture and/or data-bank)

b_j means that there can be terms with equal rights g^+ (synonyms, i.e. 'field' and 'field') for one object:

$$b_j \rightarrow S_j = \{s_{i,j}\}, \text{ with } s_{i,j} \; i =1, 2, \; .., \; m, \; .. \; , \; i_j^+; \; s_{1,j} \neq s_{m,j}, \; j \neq m \text{ and } s_{i,j} \notin B \text{ and} \quad (4)$$

$$S_j \cap S_h = 0 \; \text{ with } S_j, \; S_h \subset S \text{ and } S \subset B \quad (5)$$

$$S = \{S_j\} \text{ and} \quad (6)$$

$$B^* = B \cup S \quad (7)$$

The relation stated by (4) is rarely transitive in a logical linguistic sense, i.e. for the term $n = b_1 \rightarrow$ 'field', according to table 1 under the aspect of 'land use' $s_{i,k} = b_3 \rightarrow$ 'pasture' it is defined in a synonymous manner, under the aspect 'data bank' b_2 'variable'. Thus, *in this case*, the chain of synonyms $s_{i,1} \equiv b_k \equiv s_{i,k}$ leads to the erroneous relation of synonyms 'pasture' $\equiv$ 'variable'.

It is a particularly frequent occurrence in the environmental sector to find a whole series of homonyms, i.e. one term denoting several different objects, which, amongst other things, result from the point of view of the use of various specialist terminologies, hereafter known as aspects (of a specialist language). Unlike in other areas of knowledge, no uniform use of language exists in this sector (Feyerabend, 1976), which would enable a generally valid expression and allocation of environmental influences. Quite the contrary, environmental questions are cross-section questions that affect many established areas of knowledge. Every knowledge or specialist sector will categorise according to its aspect (i.e. table 1, field to the land use versus the data processing) and according to its model of reality. The consideration of the various aspects in one thesaurus is an intention of the ideas presented here.

This then means for the process of indexing that data is not simply given a catchword or a definition because a finite number of preconditions (=allocation of catchwords, codes, etc.) have been fulfilled, rather only if the habits relating to the expression, which make information out of the data, are taken into consideration.

The homonymicity of the terms in a formulated system of order, which we denote as the thesaurus T, result in this generally not being of a transitive nature, i.e. these deficiencies are circumnavigated by redundant structures. Due to the fact that the transitive systems of order offer considerable advantages in the retrieval and in the object-orientated procedural technology, the model of a transitive thesaurus is the subject of discussion and will also be used in the environmental information system.

Initially, the preconditions for a consistent thesaurus T are stated for just one aspect. Subsequently, the required extensions for a consistent 'multi-lingual' thesaurus T are formulated according to several aspects: the classification and the research terms must correspond with the (specialist) linguistic understanding of the user according to various aspects $g = 1, 2, .., g^+$.

Term level				Aspect a_g			
n/j	b_n	$s_{i,n}$	b_j	$g = 1$	2	3	4 (g^*)
Type	Expression	Synonyms	Related terms	Land-Use	Agriculture	Environmental Protection	Data-Base
1/	Field			X	X	X	X
/3			Pasture	X			
		Variable					X
		Pasture		X			
2/	Variable						X
		Field					X
3/	Pasture			X	X	X	
/1			Field		X		
		Field		X			

Table 1: Relations of expressions

2 The Thesaurus

The ordering and evaluation of information is essentially a precondition for successful and, above all, for efficient work in all areas of knowledge and application. The drafting of classification systems and classifications is an essential method in creating order with regards to one certain criterion, i.e. regaining information, for objects which require classification. Therefore, in the process of regaining information, one must not always base all considerations on the terms stored in the data, as one is able to make use of the order created by the classification (i.e. research with the aid of generic terms, catchwords, etc.). Conversely, the classification can be made considerably easier by means of automation, provided that this contains the terms used by the thesaurus or has derived its terms from the thesaurus (compare diagram 2, Weihs 1992).

2.1 The Hierarchical Principle

The thesaurus T is an ordered sum of expressions B^*, which forms an open system for the specialist- and/or problem-orientated classification and ordering of terms; as a classification system, it is striving for the revocable, definite allocation of expressions b_n to the objects o_j This means that each terms is contained in the thesaurus T once and only once and only refers to one object o_j (excluding the demands of (3)):

$$o_n \Leftrightarrow b_n \text{ with } b_n \in B^* = \text{sum of all terms, } n = 1,2,3,j, ..., n^+ \text{ and} \tag{8}$$

$$b_n \neq b_j \text{ with } n \neq j. \tag{9}$$

Due to (1) through (7), the transitive, synonymous expressions are contained in B^*.

A definite allocation of the expressions b_n to the k^+ categories (Wersig, 1982) is defined as being the order system.

$$B^*_{2k} = R \, (B^*_k) \qquad \text{with } \{b_n, s_n\} \subset B^* \tag{10}$$

which satisfy the preconditions (11), (12).

$$B^*_1 = B^* \tag{11}$$

$$B^*_k \setminus B^*_{2k+1} \cap B^*_{2k+1} = \{0\}, \quad B^*_k \setminus B^*_{2k+1} \cup B^*_{2k} = B^*_k \text{ with } k = 1, 2, ...k^+ \text{ Categories} \tag{12}$$

Due to the fact that (1) and (2) ensure that a term will appear once and once only in the sum B^*, and therefore is included in only one subset. The allocations of the synonyms to the categories according to (4) $b_j \Rightarrow \{a_{j,i}\}$, is also definite according to the preconditions (5) and (6), because $A \cap B = 0$. (10) through to (12) define a strictly hierarchical, transitive system of order. The rule R, which, according to (10), must be defined for the ordering of the terms, is established on the basis of subject logical determined preconditions, determined between these as super, i.e. sub-orders of category forms. Nevertheless, the literature has examples of promising, mathematical-statistical principles for *the extraction* of thesaurii from texts or empirical material (see p.e. Bock 1993).

2.2   The Extended Principle

The equation (3) demands the consideration of the homonyms, because that is the point when the allocation between object and expression is no longer revocable. Conversely, (9) is the precondition for a logical subject set of rules according to (10) for the division of expressions in categories. We will therefore extend (8) to the effect that we can allocate in a definite sense each of the $h \geq 1$ varied objects with identical expression b_n (i.e. field, compare table 1) to a certain aspect $g = 1, ,..,u, .., g^*$:

$$o_{h,n} \Leftrightarrow a_g \text{ and } o_{h,n} \Leftrightarrow b_{n,g} \qquad \text{with } b_{n,g} \in B_g = \text{sum of all terms of the aspect } g \tag{13}$$

$$b_{n,r} \neq b_{m,u} \text{ and } r \neq u, \text{ and } S_{j,s} \cap S_{h,u} \supseteq 0 \tag{14}$$

The definition of the synonymous terms is extended in accordance with (4):

$$b_{j,g} \Rightarrow S_{j,g} = \{s_{j,g}\}, \text{ with } s_{i,j,g} \text{ } i = 1, 2, .., m, ..., i^+; \text{ } s_{i,j} \neq s_{m,j}, i \neq m \text{ and} \tag{15}$$

$$S_g \cap B_g = 0 \text{ but } S_g \cap B_t \supseteq 0 \tag{16}$$

The division rule is extended in an analogue manner according to (10):

$$B_{k,g} \setminus B_{2k+1,g} \cap B_{2k+1,g} = \{0\}; \, B_{k,g} \setminus B_{2k+1,g} \cup B_{2k,g} = B_{k,g} \text{ with } k = 1,2,.. \text{ } k^+ \text{ Categories} \tag{17}$$

with a division rule, based on the aspect, according to:

$$B_{2k,g} = \mathbf{R_g}\,(B_{k,g}) \text{ and} \tag{18}$$

$B_{k,g} \in C_g^+$: the $B_{k,g}$ of C_g^+ with a division rule, based on the aspect, according to: g. $\qquad$ (19)

In accordance with (19), it must be taken into consideration in the logical subject classification rules, that this is related to one and only the one related aspect a_g.

However, on the basis of (16), the following is also true:

$$B_{j,g} \cap B_{k,t} = H_{jk,gt} \supseteq 0, \tag{20}$$

$H_{jk,gt}$ =sum of the homonyms from the expressions of the categories j, k of the aspects g and t.

A pyramid structure, in accordance with Brito (1990), results from the inclusion of several languages r, u, v .... $\in \{g = 1, \dots, g^*\}$:

$$B_{k,g} \in C_g^+: \tag{21}$$

the $B_{k,g}$ of C_g^+ are the clusters of the pyramid structure (Brito, 1990) of the languages, i.e. aspects of $\{g^*\}$.

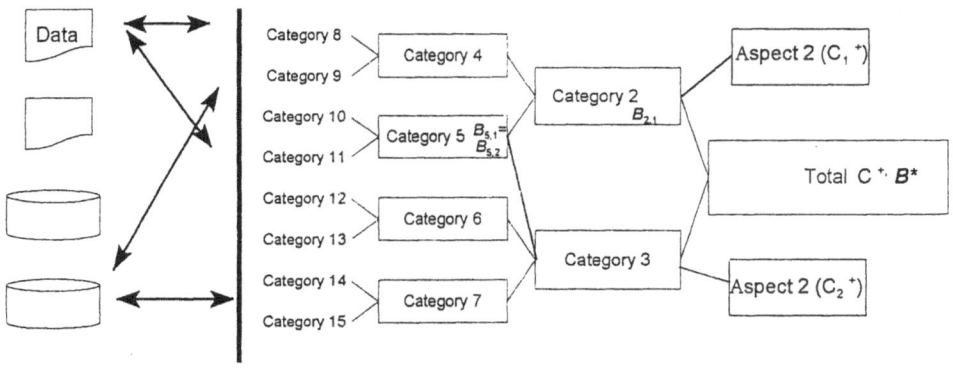

Reference by means of catchwording, indexing

Diagram 1: Indexing when dealing with several aspects

The sum total (universal sum) of the expressions of the thesaurus T of all expressions results from the terms and the synonyms allocated to the same.

$$B = B_t \cup B_u \cup B_v \dots = \bigcup Bg \text{ with (4) } b_{j,g} \Rightarrow S_{j,g} = \{s_{1j,g}\},\ 1 = 1,2,\dots,l^+ \tag{22}$$

The subject logical origins of a term is characterised by the affiliation to one (or several) aspects, (Schilling, 1991). Depending on which super, i.e. sub-orders have been permitted, the thesaurus will be, according to diagram 1, net-like, pyramidal or strictly hierarchical.

It is easily recognisable in diagram 2 that, through the introduction of the aspects, a $g^* + 1$ - dimensional variable area is stretched out. The aspects must not only be independent of each other and the categories in a Cartesian sense, they must also be independent of each other in a subject logical sense. The selection quantity, which is of interest for the retrieval of expressions, then results from the projection of the desired aspect area into the term level.

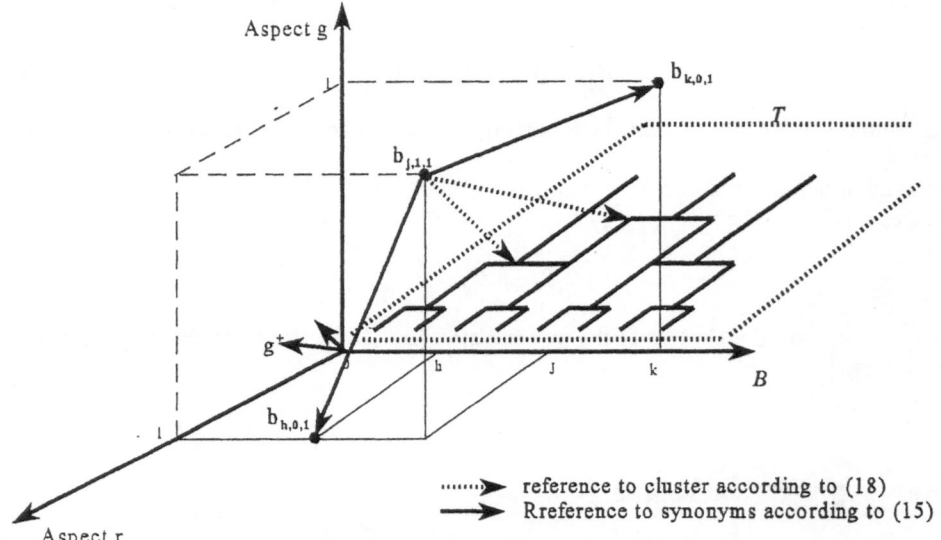

Diagram 2: Relations $g^+ + 1$ dimensionalen apect area

3. The Object-Orientated Principle

In an earlier part of this paper, we examined the preconditions which make possible the definition of a thesaurus T, which includes any number of aspects. *The introduction of the aspects makes possible the consideration of the homonymous expressions used in various specialist languages, i.e. specialist points of view.* In this context, the precondition was presupposed that each expression would appear once and only once in the thesaurus. The expressions are allocated to synonymous terms, depending on their aspect, and these are then equally components of the thesaurus. There is therefore an hierarchical clustering of terms within each aspect. The combination of various aspects then produces a pyramidal structure in the thesaurus. If one holds onto this precondition, it is possible to make a definite, non-redundant depiction of the expressions, simply by stretching out a $g^* + 1$ dimensional space. In accordance with the precondition, the aspects are now independent of each other. In as far as an aspect is then accurate for the expression b_n, the binary value 1 is commonly allocated, otherwise 0. The $g^* + 1$ th dimension makes reference to the term area. Due to the fact that, according to (2), each term appears only once, all changes related to the term (i.e. update of synonyms $\{g^+\}$, category affiliation $\{k\}$, references to aspects $\{s\}$) directly affect the relevant relations and objects. This is an essential precondition for the realisation of an object-orientated principle.

Our term object **b(n)** is defined by the vector of the variable area of the aspects, the categories of the thesuarii and the synonymous ring:

$$b(n) = (\{g^+\},\{s^+\},\{k^+\}) \tag{23}$$

The subject logical division of the point of view of the aspect from the classification determined by the contents makes possible an extended interpretation of the model: the aspect term can, without contradiction, be regarded as being a technical reference (address) to interfaces, such as HTML pages of the Internet, data-banks or methods. This therefore opens up a further area of use for a thesaurus of this kind.

References:

Bock, H.-H., Lenski W., Richter M.M. Eds. (1993): *Information Systems and Data Analysis*, Procedings of the 17th Anual Conference of the Gesellschaft für Klassifikation e.V., Berlin Springer

Brito, P., Diday, E. (1990): *Pyramidal representaton of symbolic objects*, Knowledge, Data Knowledge and Decision, Hamburg, Eds. Schader, Springer

Feyerabend, P. (1976): *Wider den Methodenzwang*, Suhrkamp, Frankft. 1976, 296f., p. 352ff..

Herrschaft, L.: *Zur Bestimmung eines medienspezifischen Informationsbegriffs in: Zeitschrift für Informationswissenschaft und -praxis*, Journal for Information Thgeory and Work, 47. Jahrg. Nr.3 Volume 47, No. 3 Nadoaw 47(3) S171 ff.

Kommission der Europäischen Gemeinschaft, KoEU (1990): *"Richtlinien über den freien Zugang zu Informationen über die Umwelt"*, Brüssel 1990

Bund Länderarbeitsgemeinschaft Bodenschutz LABO (1994): *Aufgaben und Funktionen von Kernsystemen des Bodeninformationssystems als Teil von Umweltinformationssystemen*, Umweltministerium Baden Württemberg, Stuttgart 1994

Nedobity, W (1989) *Ordnungsstrukturen für Begriffskategorien*, Studie zur Klassifikation, Bd 19(SK 19), Hrsg. Ges. f. Klassifikation e.V. Darmstadt 1989, p. 183 f. Opitz, O., Lausen, B., Klar (Eds.) (1992) Information and Classification, Springer Berlin New York

Schilling, P. (1991): *Variabler Thesaurus - eine Schlüsselfunktion für die zukünftige Informationsverarbeitung in einer Verwaltung* Konzeption und Einsatz von Umweltinformationssystemen, Brauer, W. im Auftrag der Gesellschaft für Informatik (GI)

Weihs, E. (1992): *On the Client-Server Conzept of Text Related Data*, Cognitive Paradigms in Knowledge Orga nisation, Sarada Ranganathan Endowment for Library Science, (Hrsg.), 452-459, Madras.

Weihs, E. (1993): *An approach to a Space Related Thesaurus*; Information and Classification, O.Opitz, B.Lausen, R.Klar (Hrsg), 469-476; Springer; Berlin, Heidelberg.

Weihs, E. (1993): *Datenbanken als Grundlage von Umweltinformationssystemen*, Tagungsunterlagen zur 17. Jahrestagung der Ges. für Klassifikation, Kaiserslautern.

Wenzlaff, B. (1991): *Vielfalt der Informationsbegriffe*, Nachrichten für Dokumentation 42, Heft 5/1991, 335-361, Weinheim.

Wersig, G. (1985): Thesaurus-Leitfaden: eine Einführung in das Thesaurus-Prinzip, *Theorie und Praxis*. - DGD-Schriftenreihe Bd.8; K.G. Saur; München.

Whorf, X. (1976): *Language, Thought and Reality*, MIT Press, 1956, p. 12

Cluster Analysis of Associated Words Obtained from a Free Response Test on Tokyo Bay

Shinsuke Suga[1], Ko Oi[1] and Sadaaki Miyamoto[2]

[1] National Institute for Environmental Studies
16-2, Onogawa, Tsukuba
Ibaraki 305, Japan

[2] Institute of Information Sciences and Electronics
University of Tsukuba, 1-1, Tennodai
Tsukuba, Ibaraki 305, Japan

Summary: This study is concerned with data analysis of associated words obtained from a free association test on Tokyo Bay. It is shown that cluster analysis of the words is an effective method to find respondents' concerns about the bay. Word clusters are considered to give structures of inseparable conceptions of the object of association. We analyze data from two survey areas near Tokyo Bay, i.e., one in a residential area and the other in a town where fisheries are primary industries. The data analysis shows that water pollution is a respondents' important concern in the two areas. Associated words showing various industries or development works are classified in some specific clusters. Further, we find a word cluster which indicates that Tokyo Bay is closely related to the lives of respondents engaged in fisheries.

1. Introduction

In the data analysis concerning the investigation in several environmental problems a variety of data are used. The authors have been analyzing the word data obtained from a questionnaire survey to find the environmental awareness of local residents. In the questionnaire survey respondents were asked to write down freely what they associated with a given stimulus word or a phrase. It is considered that people have a wide variety of conceptions about environmental problems. Thus, the questionnaire survey based on a free association is more useful to get satisfactory information about residents' concerns than the usual survey in which respondents find questions in the given list of individual items. We consider the classification of the words obtained from the free association test. In this aim, cluster analysis is applied to the associated words.

Applying the method of classification is useful to examine the awareness of local residents through the associated words in the following senses. First, discussing groups of classified words in the whole data is more practical than examining each word one by one. Second, a word cluster of associated words is considered to give the cognitive structures of the awareness, if an appropriate measure of the similarity is used. When people ponder on their living condition, for example, do they associate individual words, "convenience", "road", "quiet", etc., separately? On the contrary, as mentioned in Oi et al. (1986), they seem to rather recognize these items as a group of inseparable conceptions. A word cluster is considered to indicate some notion related to respondents' concerns.

The authors have analyzed word data obtained from various questionnaire surveys asking respondents to write down about living condition (Oi et al. (1988)), water side in general and Lake Kasumigaura (Suga et al. (1993)), and acoustic environment (Kondoh et al. (1993)). In the present paper, we examine local residents' awareness or images of Tokyo Bay. For data analysis the word data obtained from a free association test which was carried out in some regions near the inland sea are used. In the survey, respondents were asked to write down freely what they associated with a stimulus phrase "Tokyo Bay".

This paper is constructed as follows. In section 2, we summarize the questionnaire survey used in this study. In section 3, we describe the similarity measure for cluster analysis of word data. In section 4, the associated words used for cluster analysis are shown. Examining each of the words, especially written with a greater frequency, gives interesting information about Tokyo Bay that respondents conceive. Section 5 contains the results of cluster analysis and the interpretation of word clusters. We also describe characteristic residents' concerns in each survey area. In section 6, we discusses the efficiency of applying cluster analysis to the word data obtained from a free association. We conclude the paper with some remarks.

2. Questionnaire survey

The questionnaire survey based on a free association concerning this study was planned to find how people around Tokyo Bay evaluate the nearby sea area. In the survey, three stimuli, "sea", "Tokyo Bay", and "the new road across Tokyo Bay" were used for the free association. Respondents were asked to write down their association items in questionnaire sheets for each stimulus. In the present study, we analyze the words associated with "Tokyo Bay".

The survey was carried out in four areas. Two areas were in Kawasaki City in Kanagawa Prefecture, and other two areas in Kisarazu City in Chiba Prefecture. The two cities are on the opposite sides of Tokyo Bay each other. Data obtained from two areas nearer the bay in the cities are used in this study. One is a residential area, adjacent to a coastal industrial district in Kawasaki City, located about five kilometers from the edge of the bay. Another one is a rural area facing the bay on Kisarazu City side. We call the former one and the latter one Kawasaki and Kisarazu, respectively.

The questionnaires were mailed to 667 people in Kawasaki and 550 in Kisarazu selected from the residential map of each area by systematic sampling. The average ratio of recovery of questionnaires was 41%. About 45% of respondents in Kawasaki were office-workers. In Kisarazu about 60% of respondents were fisheries workers. In fact, fisheries are important industries in Kisarazu. The period of the survey was from February to March, 1993. The whole results concerning the survey are shown in Suga and Oi (1995).

3. Similarity measure of cluster analysis

To carry out cluster analysis of the words the following arrangement of answers is carried out. If phrases or sentences, written without any clear separation between words in Japanese, are given by a respondent, at first those are decomposed into words. Then nouns, adjectives and some verbs which are meaningful even after the decomposition in describing the environment are chosen as associated words. Those words are merged with words written down one by one from the beginning to make the set of associated words of a survey area. Among thus obtained associated words in the sets, those appearing in higher frequency are employed in the data analysis.

We use the similarity measure by Miyamoto and Nakayama (1980). The measure between two associated words is defined in the following way. Let $X = \{x_1, x_2, ..., x_n\}$ be a set of words each of which is found more than N times among the associated words from an area. Then let $Y = \{y_1, y_2, ..., y_m\}$ be a set of respondents each of which writes at least one word in X. Suppose a word x_i is written by a respondent y_k with the frequency p_{ik}, the similarity measure between the two words x_i and x_j is defined by

$$s(x_i, x_j) = \sum_{k=1}^{m} \min(p_{ik}, p_{jk}) / \sum_{k=1}^{m} \max(p_{ik}, p_{jk}) \qquad (1)$$

Clearly, $0 \leq s(x_i, x_j) \leq 1$. This measure shows that two words associated by more common respondents are more similar to each other. Thus

In this study, we set N=10 and N=7 for the data of Kawasaki and that of Kisarazu, respectively. Thus, 50 words and 53 words are analyzed for Kawasaki and Kisarazu, respectively. A computer package PAB developed by Miyamoto (1984) is employed. The method of average linkage between the merged groups is used.

Though other similarity measures may be used for classification of words, the measure defined by equation (1) is an effective one for considering respondents' concerns in the sense that the similarity between two words are measured based on common respondents' association. The efficiency of our measure will be shown in sections 5 and 6.

4. Words for data analysis

Table 1 shows the words used for cluster analysis in the order of the frequency of their appearance in each survey area. Those words are translated from associated words in the original Japanese language into English. Phrases in Table 1 are written in one word in the original language. Among whole respondents in the two survey areas, 221 respondents and 185 respondents answered the question about the stimulus "Tokyo Bay" in Kawasaki and in Kisarazu, respectively. In the table the words with an asterisk "Edo-mae", "sudate", and "Odaiba" are not translated because they cannot be found appropriate expressions in English. People often calls a dish made of marine products from Tokyo Bay or the marine products themselves "Edo-mae" in Japanese. The word "sudate" means a classical method of catching fish in a beach in the bay area. "Odaiba" is a tiny islet made at the end of the shogunate era in the 19th century to install a cannon.

Suga and Oi (1995) examined the words obtained from the survey and analyzed the frequency of each word. We summarize briefly the words in Table 1. We can find various images about Tokyo Bay which respondents conceive through the words. The words showing characteristics of Tokyo Bay are seen in each area. For example, the words "laver" and "short-necked clam" show marine products from the bay, while "reclamation" and "new road across Tokyo Bay" indicate industrial development in the bay area. Furthermore, "shell gathering" and "fishing" mean pastimes along a beach.

The words related to bad images are also seen in Table 1. The words "dirty" and "pollution" are written with a greater frequency, within the 20th place, in each area. Further, we can see "sludge" in each area. The words "death" and "red tide" are seen in Kisarazu, while "bad smell" in Kawasaki in Table 1. It indicates that the bad images related to pollution of Tokyo Bay are prominent in each survey area. We can see the words "fisheries", " fisherman", "place of fisheries", "fishing net", "marine products", "flat fish", "shellfish", and "trough shell" in Kisarazu. Most of these words are written with extremely small frequency in Kawasaki (see Suga and Oi (1995)). Greater frequency of the words showing marine products or fisheries is characteristic in Kisarazu.

5. Results of cluster analysis

5.1 Classification of words

Figures 1 and 2 show the results of cluster analysis of the associated words shown in Table 1 based on the similarity measure defined by equation (1). These figures show sketches of the actual dendrograms in which the similarity values are segmented into 25 classes in an interval [0, 1]. Figures 1 and 2 correspond to the results of Kawasaki and Kisarazu, respectively. In order to find respondents' concerns about Tokyo Bay we will discuss the interpretation of each word cluster by examining the words belonging to it. From the author's experiences, as for a word cluster containing many words it is often difficult to find its meaning. Thus, the whole word data are classified based on a dendrogram so that each cluster contains a suitable number of words for interpretation. On

Table 1 Associated words for data analysis

	Kawasaki		Kisarazu	
	word	frequency	word	frequency
1	sea	105	laver	75
	dirty	77	sea	74
	ship	73	short-necked clam	48
	reclamation	67	new road across Tokyo Bay	42
	laver	62	dirty	34
	clean	47	ship	32
	litter	43	old days	30
	water	41	reclamation	28
	fish	40	marine products	28
10	nature	34	litter	28
11	ferry	32	water	27
	fishing	30	fish	24
	old days	29	fisheries	23
	pollution	28	pollution	21
	sludge	28	life	21
	shell gathering	28	nature	20
	river	26	shell gathering	20
	Edo-mae*	25	Mt. Fuji	19
	industrial area	22	waste water	16
20	goby	21	tidal flat	16
21	human being	21	abundance	16
	Kawasaki	20	culture of marine products	15
	change	19	small	14
	tanker	18	shellfish	14
	industrial complex	18	trough shell	14
	seaside	17	change	14
	factory	17	clean	13
	life	16	death	13
	Haneda airport	16	factory	13
30	childhood	15	environment	13
31	Tokyo	15	ferry	12
	port	15	human being	12
	Japan	14	flat fish	11
	reclaimed land	14	fisherman	11
	landscape	14	development	11
	house boat	14	tasty	11
	new road across Tokyo Bay	13	goby	10
	short-necked clam	13	Tokyo	9
	Chiba	13	wind	9
40	fishing boat	12	Edo-mae*	9
41	bad smell	12	place of work	9
	culture of marine products	12	sudate*	8
	waterfront	11	fishing	8
	Kisarazu	11	oil	8
	color	11	industrial area	8
	bird	11	leisure	8
	overcrowding	10	work	8
	development	10	place of life	8
	Odaiba*	10	place of fisheries	7
50	Yokohama	10	construction	7
51			sludge	7
			fishing net	7
			reds tied	7

* The meaning is described in the text.

the other hand, some clusters containing a few words, say, less than three words, are also difficult to be considered.

We describe the classification procedure of the associated words by using the results in Kawasaki. In Figure 1, if the whole data are classified at level 0.054, then eight clusters A1 to A8 are obtained. Two clusters A2 and A3 are divided further because they contain too many words. Finally, 14 word clusters a1 to a14 are obtained. Clusters A4, A7, and A8 contain only one word. If such a cluster is formed at lower level, the word is considered to be associated independently of other words. In the same way, the whole data in Kisarazu are classified into 14 clusters b1 to b14 as shown in Figure 2.

5.2 Interpretation of word clusters

5.2.1 Clusters in Kawasaki

We consider the results in Kawasaki shown in Figure 1. As mentioned in section 1, the word clusters are considered to give the respondents' concerns or images about Tokyo Bay. In order to find the meaning of the clusters we examine the words belonging to each cluster. First, we consider cluster a1. The words "Tokyo", "Chiba", "Kawasaki", "Yokohama" are seen in a1. Each of these words shows the name of city located on the shore of Tokyo Bay. The formation of a1 indicates that the cognition concerning the cities shows an inseparable structure in the free association with Tokyo Bay.

Next, we consider clusters a2 and a4. It is clear that cluster a2 is characterized by the words showing the bad images of the sea area. Cluster a4 is considered to be an important one in the sense that all words in a4 are written with a greater frequency within the 10th place in Kawasaki (see Table 1). Such words are considered to show respondents' primal concerns, which depict an ordinary marine scenery related with pollution affairs dominant in this sea area.

Regarding those four clusters as one cluster A2 is more useful to find respondents' interests about Tokyo Bay. We should note that the words "litter", "bad smell", and "dirty" belong to A2. This indicates that an inseparable conception on the bay is composed of the bad images shown by those words. The dendrogram shows higher value of the similarity measure between "dirty" and the words with a greater frequency, "sea", "water", "ship", and "fish". Thus, we can find that the images of unclean water and dirty sea are the typical respondents' concerns about the bay.

Now, we consider A3. Although this cluster contains a rather large number of words, many of which are related to development works, various industries, and marine products. Further, we note that we can find few words concerning such things in other clusters. It shows that A3 gives a characteristic cognitive structure in the association. In order to find the meanings of association shown by those words more precisely, it is desirable to consider small clusters composing A3 separately.

Here, we consider three clusters a7 to a9 separately. Cluster a7 is characterized by the words concerning marine products. The cluster a7 indicates residents' recollection of the days before reclamation for heavy industries where the culture of laver, other marine products, and shell gathering were prevalent. Among the words in a8, we note "new road across Tokyo Bay", "ferry", and "Kisarazu". This cluster is considered to reveal respondents' interests about the new road being constructed to connect Kawasaki City and Kisarazu City across the bay. The word "ferry" shows the existing way of marine transport between the two cities which will not be used after the opening of the new road. The meaning of a9 is clear. We can find the image of the pollution caused by a coastal oil industry in the formation of the cluster. It is interesting that two words "house boat" and "fishing boat" belong to this cluster. Respondents may associate a scene of boating with a coastal industrial area as a background.

Finally, we consider a12. Three words "nature", "change", and "development" are characteristic. This cluster shows respondents' interest in the change of nature caused by various development works around the bay.

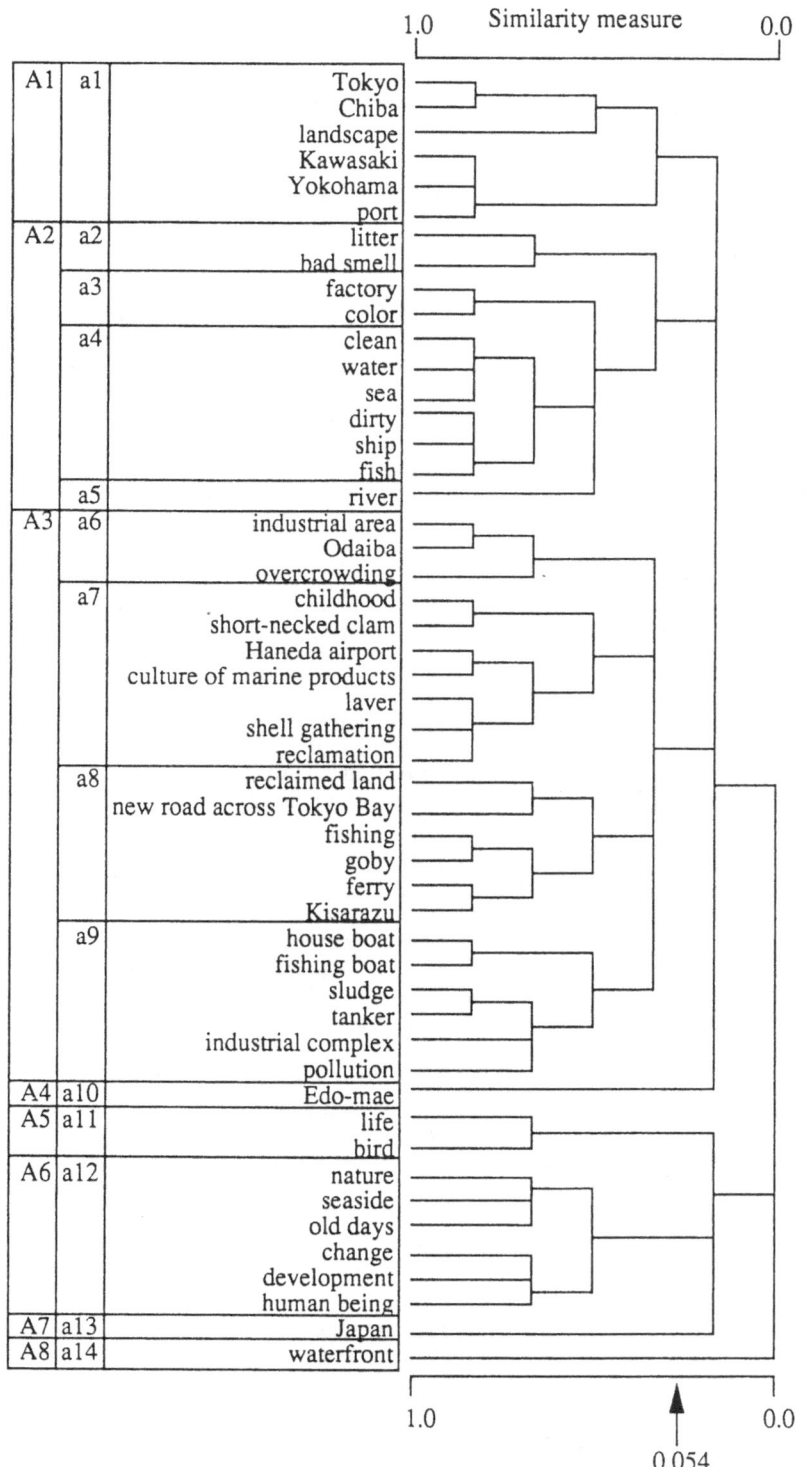

Figure 1 A sketch of the dendrogram showing the result of cluster analysis of word data in Kawasaki

742

5.2.2 Clusters in Kisarazu

Now, we consider the results in Kisarazu shown by Figure 2. The whole data are classified into seven clusters B1 to B7 at level 0.052. Two clusters B1 and B3 are further divided in the same way in the result of Kawasaki. Finally, we consider 14 word clusters b1 to b14.

First, we try to characterize B1 as one cluster. Many of the words showing marine products from Tokyo Bay and concerning fisheries belong to this cluster. It indicates that respondents' awareness related to such words shows an inseparable conception in the association in Kisarazu.

Next, we consider four clusters b1, b2, b5, and b6 separately. Cluster b1 is especially concerned with marine products from the bay. Though cluster b2 contains only three words, we should note that two words "tidal flat" and "sludge" belong to the same cluster. Actually, the conservation area of a natural tidal flat in the bay area exists near Kisarazu city. The formation of b2 reveals respondents' concerns about the conservation of the tidal flat area from water pollution. In cluster b5, we note three words "nature", "change", and "development" which are also found in a12 in Kawasaki. Thus, those two clusters have a common meaning each other. Examining all the words in b5 reveals that respondents' interest in the contrast between nature and development is relating to the environment and the abundance of nature in Kisarazu.

The words concerning fisheries are seen in b6. The result that the three words "sea", "life", and "fisheries" belong to one cluster shows that Tokyo Bay is closely related to the life of respondents in Kisarazu. This cluster is considered to be formed by the fact that about 60% of the respondents in this survey area engage themselves in fisheries.

Now, we consider two clusters b8 and b9. Although cluster b8 is composed of fewer words, it is clear that this cluster shows the bad images about water in Tokyo Bay. Various words related to pollution can be found in b9. The words "factory" and "oil" seem to show causes of pollution. The word "death" is used in phrases "death of sea" or "death of fish" etc., in the texts of answers. We can find that the pollution of the sea area is a serious problem for respondents in the fisheries community from this cluster.

Finally, we consider cluster b10. The dendrogram shows that the value of similarity measure between "Mt. Fuji" and "clean" is higher in b10. This indicates that respondents in Kisarazu make a positive evaluation about the landscape of Tokyo Bay coupled with Mt. Fuji. Actually, Mt. Fuji can be seen from the survey area Kisarazu. Further, we should note that the word "waste water" showing contrary images to "clean" belong to b10. Formation of this cluster seems to reveal that respondents in Kisarazu recognize the beautiful landscape of the bay and the bad images of waste water which show contrary images each other as an inseparable notion in the association of Tokyo Bay.

5.2.3 Common and different awareness between the two survey areas

We can find the clusters composed of the associated words relating to water pollution of Tokyo Bay in two survey areas. This indicates that water pollution is a respondents' important concern about the inland sea in the two areas. Two clusters a12 in Kawasaki and b5 in Kisarazu show respondents' concerns about the relation between nature and development. They also seem to show respondents' interest in the change of nature caused by various development works in Tokyo Bay. The associated words showing fisheries or individual names of marine products are classified into some specific clusters. Typical examples in each area are clusters a7 and b1.

In Kawasaki, the clusters containing words showing various industries or development works are found. Especially, cluster a9 which contains the words concerning a coastal oil industry is characteristic. As described in section 4, respondents in Kisarazu write various words related to fisheries and marine products from Tokyo Bay, and those words constitute some symbolic clusters. Among them, cluster b1 includes the words showing

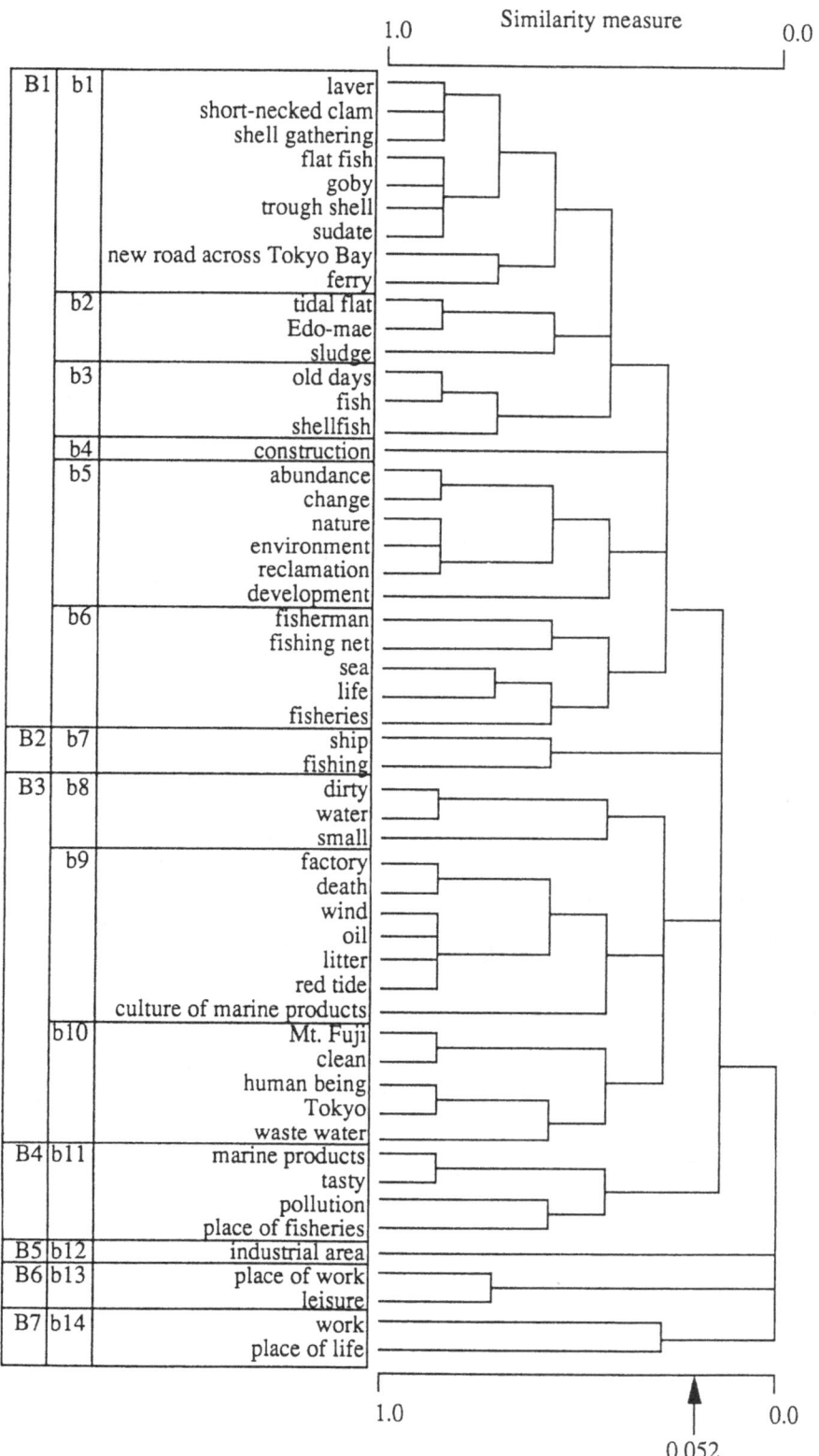

Figure 2 A sketch of the dendrogram showing the result of cluster analysis of word data in Kisarazu

several marine products. Formation of cluster b6 indicates that fisheries are not only parts of industries in Tokyo Bay but closely related to the lives of respondents themselves in Kisarazu.

6. Efficiency of cluster analysis

We may apply some other methods to the free response data except for the method based on cluster analysis. One is to examine each associated word in detail one by one. Indeed, the words especially written with a greater frequency in Table 1 give respondents' primal concerns about Tokyo Bay. We can also find characteristic respondents' concerns in each survey area by examining the difference of word frequencies in different survey areas. But that method dose not give the relationship between the associated words. Another way is to read each answer in detail. Knowing the contents of the actual answers is important for the data analysis in this study. This would be the best method for fewer number of answers. By using such method Oi et al (1994) analyzed the complaints caused by various pollution phenomena. However, it is not easy to integrate the contents of whole answers when we have to use more answers, say, more than 100 in each area as in this study.

As we have seen, classification of the associated words gives clear structures in a free association. Examination of words in each word cluster allows us to find residents' concerns without reading whole answers in detail. In order to find further interpretation of a cluster the authors often read the original answers and examine how the words belonging to the cluster are described in the actual answers. It is more useful to read original answers for a detailed interpretation of word clusters than to read them without the results of cluster analysis of words.

As mentioned in section 3, sentences or phrases appeared in the original answers are decomposed into words. Hence, those data are regarded as the sequence of words in which the order of the appearance of the words is closely related to the meaning of an answer. Miyamoto et al (1990) proposed a method, called a neighborhood method, of generating similarity measures between a pair of words obtained from long sentences in a free association. Suga et al (1994) applied the method to free response data about annoyance and trouble. The application of the method to the present data would reveal another aspect of the awareness of the respondents.

7. Concluding remarks

A free response test is useful for examining residents' concerns about several environmental problems directly. However, the analysis of such data is not as easy as that of the data obtained by a usual surveys in which respondents choose their answer from given items in a questionnaire. The method of classification we use in this study gives clear structures of the association with Tokyo Bay. Grasping such structures is important to discuss several issues about the bay in the future, for example development and conservation. It is not easy to reveal such structures through the data analysis based on a usual survey.

Acknowledgment: The authors would like to express our appreciation to the subjects for cooperating the survey.

References:

Kondoh, Y. et al. (1993): The acoustic environmental awareness of residents in high-rise apartment houses by the free response method, Proceedings of the Japan Society of Civil Engineers, **458**, 111-120 (in Japanese).

Miyamoto, S. and Nakayama, K. (1980): A hierarchical representation of citation relationship, IEEE Trans., Systems Man and Cybern., SMC-**10**, 899-903.

Miyamoto, S. (1984): Development of a Computer Program Package for Bibliometrics, Report of a research supported by the Grant in Aid for Fundamental Scientific Research of the Educational Ministry in fiscal 1983, (in Japanese).

Miyamoto, S. et al. (1990): Methods of digraph representation and cluster analysis for analyzing free association, IEEE Trans., Systems Man and Cybern., Vol. **20**, No. **3**, 695-701.

Oi, K. et al. (1986): Analysis of cognitive structures of environment of the local residents through word association method, Ecological Modelling, **32**, 29-41.

Oi, K. et al. (1988): The range and the structure of cognition of the living environment conceived by local residents, Proceedings of the Japan Society of Civil Engineers, **389**, 83-92 (in Japanese).

Oi, K. et al. (1994): Management of complaints caused by noise and other pollution phenomena filed by residents flowing into industrial areas, Proceedings of International Congress on Noise Control Engineering, Vol. **2**, 1141-1144.

Suga, S. et al. (1993): Study of the awareness of local residents on an expanse of water by a free association test and cluster analysis, Proceedings of the Japan Society of Civil Engineers, **458**, 91-100 (in Japanese).

Suga, S. et al. (1994): An application of a method of neighborhood to text of free response test on annoyance and trouble, Research Report from the National Institute for Environmental Studies, R-132-'94, 97-106 (in Japanese).

Suga, S. and Oi, K. (1995): A Survey of the Image of Sea through a Free Association Method, F-73-'95/NIES, National Institute for Environmental Studies (in Japanese).

Data Analysis of Deer-Train Collisions
in Eastern Hokkaido, Japan[1]

Keiichi Onoyama[1], Noboru Ohsumi[2], Naoko Mitsumochi[1], and Tsuyoshi Kishihara[1]

[1] Obihiro University of Agriculture and Veterinary Medicine
Inada-cho, Obihiro 080, Japan

[2] The Institute of Statistical Mathematics
4-6-7, Minami-Azabu, Minato-ku,
Tokyo 106, Japan

Summary: The data of 696 deer-train accidents which occurred on 330.95 km distance in eastern Hokkaido, Japan from April 1988 to March 1995 was statistically analyzed. Many of the accidents occurred at particular sites and night hours, which suggests the relation with the habitat and diel activity of deer. Relative densities of deer were estimated where the train runs were constant.

1. Background

One of the serious problems between human and animals is deer-train collisions. It includes the breakdown of or damage to trains, hence the disturbance to the train diagram, and the death or injury of deer. Although several studies were published on deer-car accidents (Allen and McCullough (1976), Schafer and Penland (1985), Waring et al. (1991), Reeve and Anderson (1993)), no actual data of deer-train accidents has been studied. Recently the number of accidents between the train and the Sika deer (*Cervus nippon yesoensis*) greatly increased from year to year in eastern Hokkaido, Japan.

We have produced a data set including a total of 696 cases of deer-train accidents from April 1987 to March 1995 (8 years) based on the driver reports of the Kushiro Branch of Hokkaido Railway Company. Determining the altitude and representative vegetation at 0.5 km distance along the Line on the basis of 1/50000 scale topographical maps by the National Geographical Survey Institute, Japan and 1/50000 scale actual vegetation maps (Environment Agency, 1988), we have created another data set consisting of the number of accidents per 0.5 km and environmental conditions. We present the results of statistical analysis on the data sets and an estimation of the relative densities of deer.

2. Statistics

Fig. 1 shows that the number of accidents increased year by year except for 1991. Since the number of train runs per year was almost the same, it is suggested the increase in the number of deer that passed across the railway track, hence the increase of the population of deer. Fig. 2 gives the hourly change in the number of accidents. Most of the accidents (79%) occurred between 16:00 to 23:00, when the deer activity is high. Fig. 3 presents the number of accidents per 10 km from Kamiochiai to Nemuro stations on the Nemuro

1: This study was in part carried out under the ISM Cooperative Research Program (95-ISM•CRP-A58).

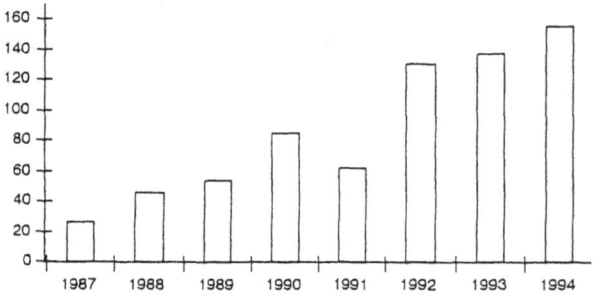

Fig.1: Yearly Change in the number of train-deer

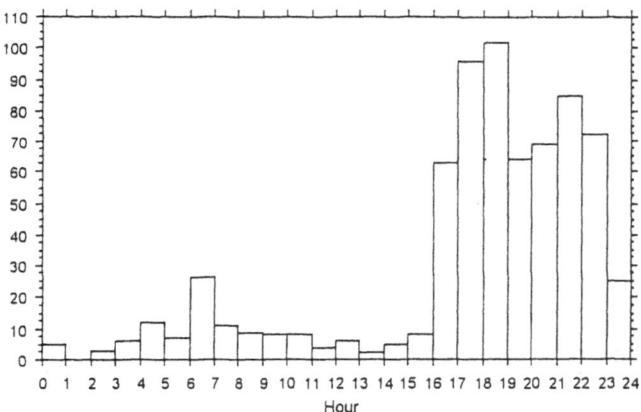

Fig. 2: Hourly change in the number of deer-train accidents

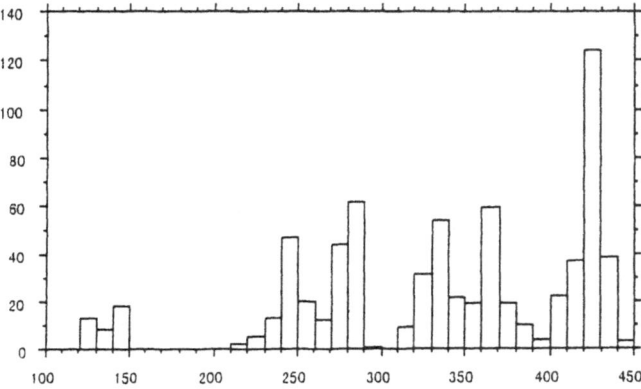

Fig. 3: Number of accidents per 10km. Numerals on abscissa are distances in km from the Takigawa station.

Line.   Many accidents occurred between Kushiro and Nemuro stations, where hunters reported that many deer lived.

Table 1 gives the number of cases in which numbers deer were found.   From January to April the number of deer was great and the mean ranged from 4.1 to 5.0.   However, the number of deer, which collided with trains ranged from 1 to 4 (0 means a nearmiss), and among these the cases of 1 occupied 94.1% (Table 2).   This situation is similar in every month.

Table 1: Number of cases in which numbers deer were found

Month										

	Number of deer found / case									
Month	1	2	3	4	5	6-9	0-19	20-	Total	Mean
Jan	16	12	9	10	10	6	6	1	70	4.1
Feb	17	15	14	1	7	9	6	3	72	4.7
Mar	12	18	17	10	6	11	8		82	4.2
Apr	6	1	5	3	5	9	4		33	5.0
May	12	8	4	2	3	4	0		33	2.8
Jun	4	6	1	1		1	1		14	3.0
Jul	17	13	9	2		0	0		41	1.9
Aug	14	13	4	3		2	0		36	2.1
Sep	19	17	3	5	1	4	1		50	2.6
Oct	33	25	7	5	5	1	0		76	2.0
Nov	47	23	12	5	6	14	0		107	2.6
Dec	24	15	11	10	5	7	3		75	3.1
Total	221	166	96	57	48	68	29	4	689	3.2

Tab. 2: Frequency distribution of the number of deer which collided with trains.

	Number of deer which collided with trains							Mean number of
Month	0	1	2	3	4	unknown	Total	deer when collided
Jan		63	6	1			70	1.11
Feb	7	58	6		1	1	73	1.14
Mar	5	68	5	1		4	83	1.09
Apr	4	25	2	2			33	1.21
May	2	31					33	1.00
Jun		12		1		1	14	1.23
Jul	2	34	3	1		1	41	1.13
Aug	1	35				1	37	1.00
Sep	4	45	2				51	1.04
Oct	5	68	1	1		2	77	1.04
Nov	2	100		1		5	108	1.02
Dec		71	4			1	76	1.05
Total	32	610	29	7	2	16	696	1.08

The distances at which drivers found deer ranged 0 to 300 m.   Fig. 4 is box plots for the distances in the daytime (8:00-16:00) and at night (20:00-24:00). Drivers found deer farther in front of them in the daytime than at night.   The mean distance at which drivers

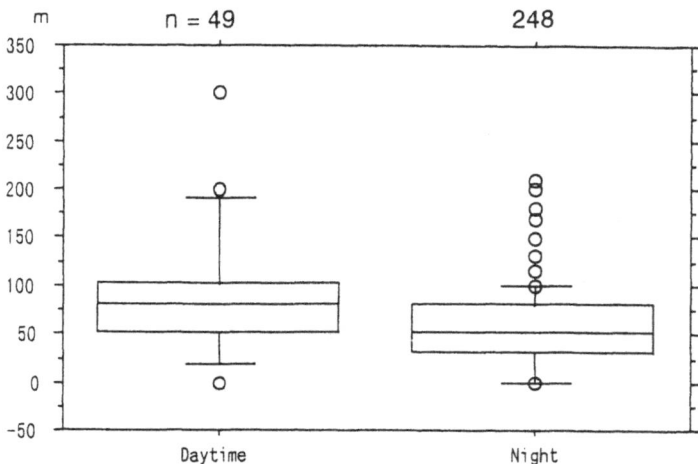

Fig. 4: Distances at which drivers found deer in the daytime and at night

found deer is in the daytime is 89 m with a standard deviation of 53 m in the daytime, but it is 56 m (*SD*=41 m) at night. Namely the distance in the daytime is on average 33 m longer than that at night.

Fig. 5 is box plots for the distances at which drivers found deer on 5 weathers. The distances were shorter in rainy or misty situations than in fine or cloudy ones. The difference in the means is about 18 m.

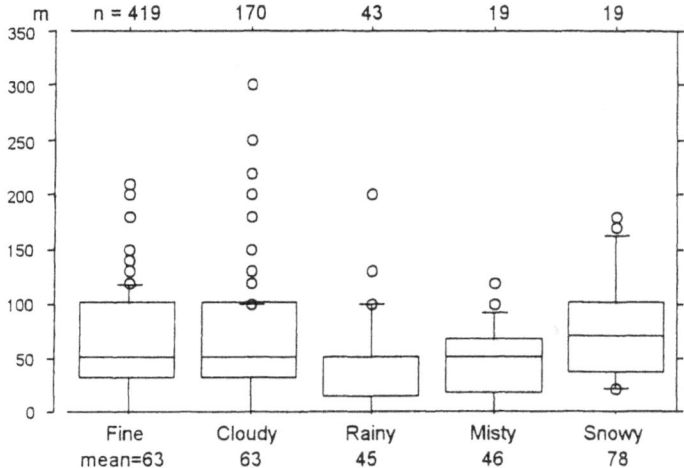

Fig. 5: Distance at which drivers found deer on each weather

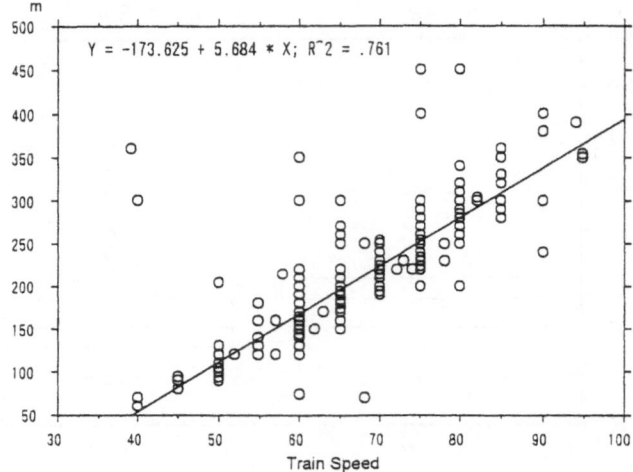

Fig. 6: Relation between train speed and distance to stop (ordinate)

Fig. 6 shows the relation between train speed and distance to stop.   As is expected, the greater the train speed is, the greater the distance becomes.

3. Relative densities of deer

Since the number of train runs was different between hours and stations, the number of accidents cannot be used as a direct indicator of the population density of deer. However, the 8 selected train runs (train numbers 5638, 5639, 5640, 5641, 5642, 5645, 5647, and 3644) between 16:00 and 22:00 together gave a total of 6 runs between Kushiro and Nemuro stations, also giving hourly runs nearly constant, so we may use the number of accidents on the 8 train runs as relative densities of deer along the railway.

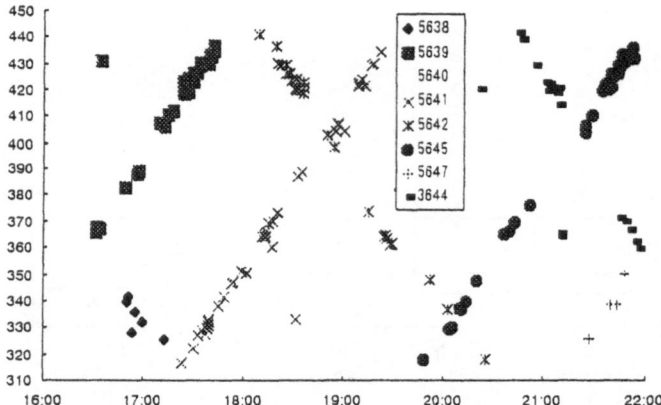

Fig. 7:   The accidents occurred on the 8 trains between Kushiro and
Nemuro stations. Ordinate: distance from the Takigawa Station in km.

Fig. 7 shows the accidents occurred on the 8 train runs, and Fig. 8 the relative densities per 2.5 km along the Nemuro Line between Kushiro and Nemuro stations. The relative density at the distances of 420 to 430 km was very high. It is suggested that there may be three or four large deer populations along the line between the two stations.

Hayashi's quantification method (type I) analysis performed on the data set of 0.5 km distance (270 cases) has revealed that the relative density of deer between Kushiro and Nemuro stations is related to the vegetation type and altitude category along the Nemuro Line. Since the multiple correlation coeficient is 0.468, other factors such as the deer's behavioral habit itself may be also related.

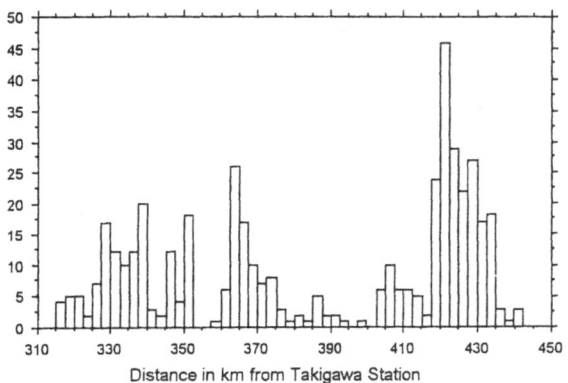

Fig. 8: Number of accidents per 2.5km on Nemuro Line

4. Acknowledgement

We would like to thank the Kushiro Branch of Hokkaido Railway Company for offering the material.

References:

Allen, R. and McCulough, D. (1976): Deer-car accidents in southern Michigan. *Journal of Wildlife Management,* **40,** 317-325.

Environment Agency. (1988): Actual vegetation map: the 3rd national survey on the natural environment (vegetation). Japan Wildlife Research Center, Tokyo.

Reeve, A. F. and Anderson, S. H. (1993): Ineffectiveness of Swareflex reflectors at reducing deer-vehicle collisions. *Wildlife Society Bulletin,* **21,** 127-132.

Schafer, J. A. and Penland, S. T. (1985): Effectiveness of Swareflex reflectors in reducing deer-vehicle accidents. *Journal of Wildlife Management,* **49,** 775-776.

Waring, G. H., Griffis, J. L. and Vaughn, M. E. (1991): White-tailed deer roadside behavior, wildlife warning reflectors, and highway mortality. *Applied Animal Behaviour Science,* **29,** 215-223.

Comparison of some numerical data between the *belisama* group of the genus *Delias* Hübner (Insecta: Lepidoptera) from Bali Island, Indonesia

Sadaharu Morinaka

The University of the Air
2-11 Wakaba , Mihama-ku, Chiba-ken, 261 Japan

Summary: In taxonomic study of closely related taxa, numerical data seem to be insufficiently used or discussed. In this study actual sizes of various parts in male genitalia and their ratios (to forewing length) are compared and discussed.   Generally ratios are considered to be useful for eliminating the environmental effects, as seen in the case that male genitalia of *Delias oraia* are proved clearly larger than those of *Delias belisama*. Meanwhile it is clarified through the discussion that there are such a character that is hardly effected by the environment, and that is therefore very important to discuss the speciation and biological evolution in related taxa.

1. Introduction

The description of biota is very important for various natural sciences because it is the basis of them, for instance taxonomy, phylogenetics, population genetics, ethology, evolutionary biology.   Since Linné, huge numbers of descriptions have been carried out but they are far from complete even now.   Many more biota await description.   As for descriptions of insects, we can see them in the journals of various learned societies, for instance, the Entomological Society of Japan, the Biogeographical Society of Japan, the Lepidopterological Society of Japan.   We can see non-numerical expressions for instance "antennae red, upper side of forewing blue and shining" in them.   We can see also numerical expressions such as "forewing 34 mm in average", but they are not well used in studies.

$$P = G + E \quad \text{(P: Phenotypic value, G: Genetic value, E: Environmental value)}$$

This formula (Kimura 1960) is well known in quantitative genetics. We know that insects, for instance, butterflies or moths when their larvae are not fed enough or live in severe conditions become small adults. Thus numerical data, for instance, the size of wings, is affected by various environmental factors, I consider.   Therefore I consider that numerical data is difficult to use for taxonomic studies.

Recently Lande (1976) or Lynch (1988) discussed the genetic model of evolution but hardly used actual data of organisms. Komatsu (1996) referred that variation in the size of genitalia was independent of that of general body but did not show actual data in his paper then.   I found that relative size of phallus (a part of male genitalia) had a distinct difference from those of other organs in the way of taxomic study and showed that it was peculiar and independent from the variety of sizes of other organs using actual data of two very closely related taxa (Morinaka 1996). And I referred that it had some important relation to the speciation and biological evolution (Morinaka 1996). In this paper I showed again that phallus had distinct difference from other organs using actual data and suggested it had very important meaning for the speciation and biological evolution between two very closely related taxa.

752

2. Materials

2.1 The genus *Delias* (Insecta: Lepidoptera, Pieridae)

The genus *Delias* Hübner, 1819 is a big genus which has more than two hundred species and belongs to Pieridae. The constituent species are distributed broadly from India to New Caledonia. Lots of them inhabit South East Asia and is also divergent on highlands of New Guinea Island. Their larvaes eat mistletoe plants and the adults have colorful wing markings on the undersides of the wings. These are known as remarkable characters of the genus *Delias*. Talbot divided this big genus to twenty groups (Talbot, 1928-1937). I have been studying Group 17 (*belisama* group). It has 15 species which are closely related to each other and distributed from Nepal to New Caledonia including Australia.

2.2 Materials

Delias belisama balina Fruhstorfer, 1910 (population from Bali Island) and *Delias oraia bratana* Kalis, 1941 (population from Bali Island) are used. Distributions of *Delias belisama* and its relatives are shown in Fig.1. *D. belisama* inhabits Sumatera, Jawa and Bali Island. On the other hand, *D. oraia* inhabits Lesser Sunda Islands for instance Lombok Island, Flores Island and also Bali Island beyond Wallace's line. They are closely related but clearly different species because they fly together in the mountains of Bali island, maintaining their separate identities. Males of both species are shown in Fig.2. Their wing markings are nearly identical and sometimes it is difficult to distinguish from each other.

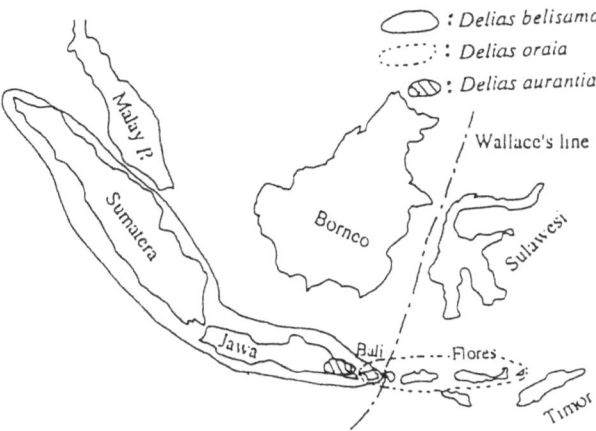

Fig. 1 Distribution of *Delias belisama* and its relatives

754

Fig. 2 Male adults of both species (A: *Delias belisama balina*, B: *D. oraia bratana*, upper: upperside, lower: underside)

3. Method and results

3.1 Method

As mentioned above comparison of wing markings between *D. belisama* and *D. oraia* is difficult. Therefore comparison of genitalia becomes to be rather important. In this study comparisons using various sizes of male genitalia are carried out. 35 males of *D. belisama balina* and 24 males of *D. oraia bratana* were measured for each sizes of male genitalia as Fig.3.

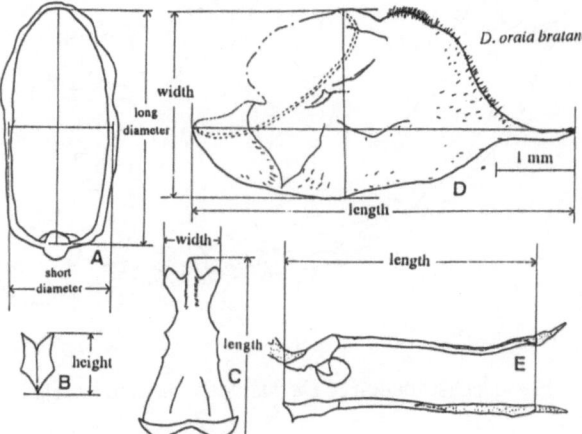

Fig. 3 Mesured portions of male genitalia. A: Ring (anterior), B: Juxta (posterior), C: Dorsum (dorsal), D: Valva (inner), E: Phallus (lateral and dorsal)

3.2 Results

The results are shown in Table 1. Generally sizes of *D. oraia* are larger than these of *D. belisama*. But the actual data is considered to include environmental effects. It is considered that originally one species existed in Bali Island and another species invaded there. Therefore it can be imagined that environments are not so suitable for the latter species. And also it is imagined that large individuals have large genitalia. Therefore correlation coefficients between sizes of genitalia and forewing length (=f.l.), and also regression equations are required. They are shown in Fig.4. and Table 2. And ratios of genital sizes/forewing length also shown in Table 2. It is clear that *D. oraia* is larger than *D. belisama* on some genital sizes of the same f.l. Therefore it is concluded that male genitalia of *Delias oraia* are larger than those of *Delias belisama*. But each value of the correlation coefficients has remarkable difference. Values of the dorsum and valva are large but the values of the phallus is markedly small.

	D.belisama balina (n=35) mean ± S.E.	D.oraia bratana (n=24) mean ± S.E.	Comparison
Forewing length	36.229 ± 0.328	37.500 ± 0.458	*belisama < oraia**
Abdominal length	16.057 ± 0.201	16.750 ± 0.235	*belisama < oraia**
Juxta length	0.662 ± 0.012	0.799 ± 0.016	*belisama < oraia***
Ring long diameter	2.901 ± 0.027	2.995 ± 0.048	*belisama < oraia*
Ring short diameter	1.465 ± 0.021	1.623 ± 0.032	*belisama < oraia***
Dorsum length	2.380 ± 0.027	2.593 ± 0.037	*belisama < oraia***
Uncus width	0.697 ± 0.008	0.757 ± 0.011	*belisama < oraia***
Valva length	4.440 ± 0.042	4.830 ± 0.061	*belisama < oraia***
Valva width	2.323 ± 0.021	2.530 ± 0.035	*belisama < oraia***
Phallus length	2.744 ± 0.021	3.082 ± 0.040	*belisama < oraia***

Table 1. Each sizes (mm) of male genitalia and comparison of both species.
(*:P<0.05, **: P<0.01 by t-test)

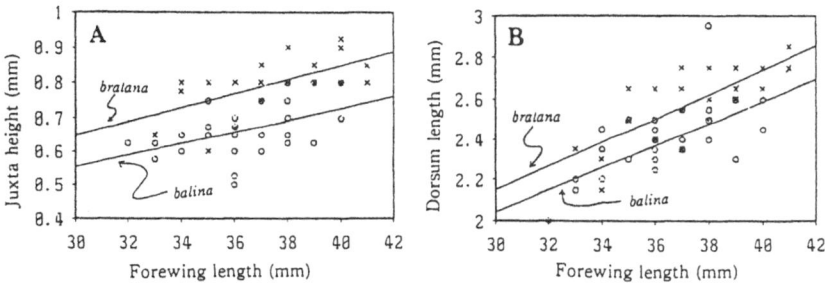

Fig. 4 Correlation of forewing length and juxta height (A) or dorsum length (B) in *Delias belisama balina* (○) and *D. oraia bratana* (×).

756

	Correlation coefficients and Regression equation				Genital sizes / forewing length (mean ± S E)	
	D.belisama balina (n=35)		D.oraia bratana (n=24)		D.belisama balina (n=35)	D.oraia bratana (n=24)
Juxta length	0.480*	Y=0.017**x + 0.033	0.597*	Y=0.020**x + 0.058	0.018 ± 0	0.022 ± 0
Ring long diameter	0.579*	Y=0.046**x + 1.250	0.344	Y=0.034 x + 1.716	0.080 ± 0.001	0.080 ± 0.001
Ring short diameter	0.190	Y=0.012 x + 1.045	0.190	Y=0.013 x + 1.154	0.041 ± 0.001	0.044 ± 0.001
Dorsum length	0.642*	Y=0.054**x + 0.441	0.762*	Y=0.059**x + 0.401	0.066 ± 0.001	0.069 ± 0.001
Uncus width	0.250	Y=0.006 x + 0.463	0.740*	Y=0.017**x + 0.124	0.019 ± 0	0.020 ± 0
Valva length	0.669*	Y=0.085 x + 1.351	0.704*	Y=0.090**x + 1.442	0.123 ± 0.001	0.129 ± 0.001
Valva width	0.745**	Y=0.048**x + 1.855	0.842*	Y=0.062**x + 0.194	0.064 ± 0	0.068 ± 0.001
Phallus length	0.386*	Y=0.025* x - 0.49	0.471	Y=0.035* x + 1.774	0.076 ± 0.001	0.083 ± 0.001

Table 2. Correlation coefficients between sizes of male genitalia and forewing length, and regression equations and each ratio to forewing length. (*:P<0.05, **: P<0.01 by t-test)

4. Discussion

Ratios are sometimes used for description because it is considered that they are comparatively constant and effects of environments are excluded. In this study it is concluded that male genitalia of *Delias oraia* are larger than those of *Delias belisama* using ratios. On the other hand I also found each value of the correlation coefficients has remarkable difference. Regression equations of forewing length and ratios (genital sizes/ forewing length) and correlation coefficients are shown in Fig. 5. In this correlation chart, the X axis is the forewing length and the Y axis is the ratio of two genital sizes to forewing length. Both lower regression equations show valva width and upper lines show phallus length. The ratio is constant in all forewing lengths in the lower graph, on the other hand, in the upper graph the ratio is not constant, reducing with the X axis. What does it mean? If environmental effects are constant to genitalia and forewing length, these regression lines have to be parallel to the X axis like the lower graph. In other words, this ratio has to be constant in all sizes of the forewing. Certainly the regression line of the valva is constant but that of the phallus is not, the ratio of phallus decreasing as forewing length increases. I consider that it means the phallus length is comparatively constant, independent of forewing length. In other words the environmental effects are not equal to phallus and forewing length. The sizes of these butterflies' phalli are rather constant and affected less than forewing length by the environment. I considered that such a character that is hardly affected by the environment as for phallus in this case, is most important to discuss the speciation and biological evolution.

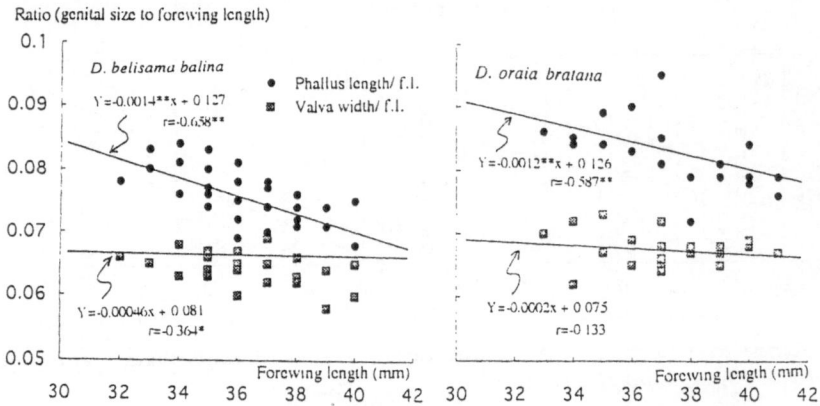

Fig. 5   Correlation of forewing length and ratios (Valva width and phallus length forewing length)

Acknowledgments

I greatly acknowledge Prof. Dr S. Sakai, Daito Bunka University, Saitama (The President of The Biogeographical Society of Japan), who recommended and encouraged to my talk in this IFCS-96 Conference. I wish to express his gratitude to Dr H. Mohri, Director General of National Institute for Basic Biology, Okazaki, and Prof. Dr T. Nakazawa, The University of The Air, Chiba for their kindly supporting and encouragement to my study. I also express my heartily thank to Dr N. Minaka, National Institute of Agro-Environmental Sciences, Tsukuba. He gave me much helpful advice, critically read and corrected the manuscript. I express my cordial thanks to Mr S. Sugi, Tokyo. He also critically read and corrected the manuscript.

References:

D'Abrera, B. (1986): Butterflies of the Oriental Region, Part 3. Lycaenidae and Riodinidae, Hill House, Victoria, Australia.

Kimura, M., 1960. An outline of Population Genetics, Baihûkan, Tokyo (In Japanese).

Komatsu, T. (1996): Morphometric study on stochastic processes and deterministic processes in evolution. Why do insect genitalia differentiate?, Fifth conference of International Federation of Classification Societies. Abstracts, 2, 212-215.

Lande, R. (1976): Natural selection and random genetic drift in phenotipic evolution, Evolution, 30, 314-334.

Lynch, M. (1988): The rate of polygenetic mutatio, Genetical research, 51, 137-148.

Morinaka, S. (1988): A Study on the *Belisama* Group of the Genus *Delias* from Bali, Indonesia (1), Tyo to Ga, 39, 2, 137-148 (In Japanese).

Morinaka, S. (1990): Ditto (2), - Comparison of Male Genitalia between *Delias belisama balina* and *D. oraia bratana* -, Tyo to Ga, 41, 3, 139-147 (In Japanese).

Morinaka, S. (1996): Comparison of some numerical data between the *belisama* group of the genus *Delias* Hübner (Insecta: Lepidoptera) from Bali Island, Indonesia. Fifth conference of International Federation of Classification Societies. Abstracts, 2, 302-305.

Morinaka, S. (1996): Comparison of some numerical data of genitalia relating to the speciation, 43rd Annual Meeting of the Lepidopterological Society of Japan, Abstracts, 50 (In Japanese).

Talbot, G. (1928-1937): A Monograph of the Pierine genus *Delias*, British Museum (Natural History), London.

Yagishita, A., S. Nakano and S. Morita. (1993): An illustrated list of the Genus *Delias* Hübner of the World, Nishiyama, Y. (ed), Khepera Publishers, Singapore (In Japanese).

Yata, O. and K. Morishita. (1981): Butterflies of the South East Asian Islands, vol. II. Pieridae, Danaidae, Tsukada, E. (ed), Plapac, Tokyo (In Japanese).

A method for classifying unaligned biological sequences

Tallur B.[1], Nicolas J.[1]

[1] IRISA, Campus Universitaire de Beaulieu,
Avenue de Gen. Leclerc, 35042 Rennes cedex, France

Summary : It is needless to emphasize the importance of classification of protein sequences in molecular biology. Various methods of classification are currently being used by biologists (Landès et al.1992) but most of them require the sequences to be prealigned – and thus to be of equal length – using one of the several multiple alignment algorithms available, so as to make the site-by-site comparison of sequences possible. Two LLA-based approaches for classifying prealigned sequences were already proposed (Lerman et al. (1994a)) whose results compared favourably with most currently used methods. The first approach made use of the "preordonnance" coding and the second one, the idea of "significant windows". The new directions of research leading to a clustering method free from this somewhat strong constraint were also suggested by the authors. The present paper gives an account of the recent developments of our research, consisting of a new method that gets round the sequence comparison problem faced with while dealing with unaligned sequences, thanks to the "significant windows" approach.

1. Introduction

The biological sequences are composed of strings of letters belonging to a finite size alphabet. In case of protein sequences, the alphabet $\mathcal{A}$ comprises 20 letters each one representing, respectively, one amino acid.

$$\mathcal{A} = \{ACDEFGHIKLMNPQRSTVWY\} \tag{1}$$

Similarity between amino acids is one of the important factors to be considered while computing the similarity between sequences. Several researchers have focussed their attention on this problem and put forward the similarity matrices (see, for example, Dayhoff et al. (1983), George et al. (1990), Risler et al. (1988), Lerman et al. (1994b)). The "profile matrix" considered by (Gribscov (1987)) is a particular case of stadardized similarity matrix between letters used in the second approach of classification proposed by Lerman et al. (1994a). The first approach of classification described by the latter is based upon the preordonnance coding and necessitates site-by-site comparison of sequences which is well adapted for aligned sequences. However, the high variation in sequence length renders the overall comparison of sequences very difficult and the site-by-site comparison altogether impossible in case of no prealignment. The "significant windows" approach for classifying unaligned sequences may be summarized as follows : A fixed-size window is made to slide along the sequences to be compared. Each window (i.e. subsequence delimited by each window position) of the shorter sequence is compared with the set of all windows of the longer one and, the "most significant window" – with respect to a proper null hypothesis of independence – is selected by means of beam search. Similarity index is then defined as a function of the similarities resulting from the site-wise comparison with the most significant window. This index depends on some parameters such as window size and window significance level whose values need to be chosen by the users, and also on some other parameters such as percentage of homologous sites that may be estimated from the data. Finally, the similarity index is standardized with respect to its observed distribution over the set of sequence pairs and hierarchical clustering using LLA program (Lerman et al. (1993)) is performed.

2. Similarity between sequences

The sequences being of unequal length, their site-by-site comparison is impossible and hence the pairwise comparison of sequences will be carried out by first selecting the "significantly comparable" windows and then using the site-by-site similarity between the latter for measuring the overall similarity between sequences. Let us consider two sequences of lengths L_1 and L_2 respectively with $L_1 < L_2$.

$$a_1, a_2, \ldots, a_{L_1}$$

and

$$b_1, b_2, \ldots, b_{L_1}$$

Let $D = ((D_{ij}))_{1 \leq i,j \leq 20}$ be a matrix of similarity scores associated with all possible letter pairs over a 20-letter alphabet $\mathcal{A}$ (such as for instance, Dayhoff's mutation data matrix (PAM 250)) where D_{ij} is the score of the letter pair (i, j), for $(i, j) \in \mathcal{A}$. Let $M = ((M_{ij}))_{1 \leq i,j \leq 20}$ be a "match matrix" where M_{ij} takes a value of 1 if the ith letter "matches" with the jth letter and 0 otherwise. We may build the match matrix in several ways and, in particular, by considering a classification of the set of letters. A window of fixed size l is made to slide along each of the sequences compared. Let us denote by w_{i1} the ith window of the first sequence (i.e. the subsequence $a_i, a_{i+1}, \ldots a_{i+l-1}$) and by w_{j2} the jth window of the second sequence (i.e. the subsequence $b_j, b_{j+1}, \ldots b_{j+l-1}$). The similarity score $S(w_{i1}, w_{j2})$ of a window pair (w_{i1}, w_{j2}) for $1 \leq i \leq (L_1 - l + 1)$ and $1 \leq j \leq (L_2 - l + 1)$ is defined as the sum of similarity scores of the corresponding letter pairs, as given by the matrix D. i.e.

$$S(w_{i1}, w_{j2}) = \sum_{k=0}^{l-1} D(a_{i+k}, b_{j+k}) \qquad (2)$$

But as a matter of fact, we will rather consider the standardized scoring matrix D^s instead of D (see Lerman et al. (1994a)).

2.1 Significant window pairs

Consider the window pair (w_{i1}, w_{j2}). The total number m of letter matches occurring in this pair is

$$m = \sum_{k=0}^{l-1} M(a_{i+k}, b_{j+k}) \qquad (3)$$

Under the null hypothesis that the letters in both windows are randomly and independently distributed, the random variable associated with the number m is distributed as a $Binomial(l, p)$ variable where p is the probability of one match occurring in the sequence pair. The parameter p may be estimated from the observed frequencies of letters in the sequences as follows :

$$\hat{p} = \sum_{(i,j)|M_{i,j}=1} freq(i, seq1).freq(j, seq2) \qquad (4)$$

where $(i, j) \in$ set of all letter pairs (a_i, b_j), $freq(i, seq1)$ is the relative frequency of the letter a_i in sequence1 and $freq(j, seq2)$ is the relative frequency of the letter b_j in sequence2. The window pair (w_{i1}, w_{j2}) is said to be *significant* or *significantly comparable* at level α if $m_0 > U$ where m_0 is the observed number of matches and U is an integer determined such that

$$Prob[m > U] \leq \alpha \qquad (5)$$

2.2 Determination of the significant region of comparison

Computation of the similarity measure between the given sequences is carried out according to an original idea that may be summarized as follows :

1. Determine the pertinent area (biologically speaking) in each of the sequences to which the search will be limited. As this information is generally not available, we have considered the two following solutions

   - The first one is based on two hypotheses : *(1)* the homology (or percentage of homologous sites among the sequences) h is either known or estimated from data, *(2)* the pertinent or biologically meaningful area is uniformly distributed over the sequence.

   - The second solution is obtained by using the *maximum predictable classification* (Lebbe and Vignes 1992, 1993) based upon the idea of local predictability of sequence areas from their contexts.

2. Compare – using the beam search technique – each window inside the pertinent area of the shorter sequence (i.e. sequence1) to all the windows inside the pertinent area of the longer one (i.e. sequence2).

3. Rule out the window pairs that are not *significantly comparable* by means of the binomial test described in section 2.1.

The significantly comparable sequence pairs alone will contribute to the overall similarity between sequences. For each window of sequence1, the best score associated with the window of sequence2 (among those selected as above) is considered and a "rough" similarity index between two given sequences is obtained by averaging it over all chosen window pairs. The standardized similarity index between sequences is then obtained by standardizing the "rough" index, first with respect to its empirical distribution over the set of window pairs and finally, over the set of all sequence pairs. The matrix of the probabilistic similarity index required by LLA is obtained by applying the standard normal cumulative distribution function as in case of prealigned sequences.

2.3 Aggregation criterion used in LLA method

The basic data required by the LLA method of hierarchical classification is the matrix of probabilistic similarity indices between the sequence pairs. To fix the ideas let us consider the set $\mathcal{O}$ of all sequences, and the probabilistic similarities between the sequences o_1 and o_2 given by the equation

$$P(o_1, o_2) = \Phi[Q_s(o_1, o_2)], \quad for \ \{o_1, o_2\} \in \mathcal{P}_2(\mathcal{O}) \qquad (6)$$

where Φ, Q_s and $\mathcal{P}_2(\mathcal{O})$ denote respectively the standard normal cumulative distribution function, standardized similarity index between sequences (defined in section 2.2) and the set of all possible sequence pairs. The algorithm builds a classification tree iteratively, by joining together at each step the two (or more in case of ties) most similar sequences or classes of sequences until all clusters are merged together. Thus the aggregation criterion that is maximized at each step or "level" of the algorithm

761

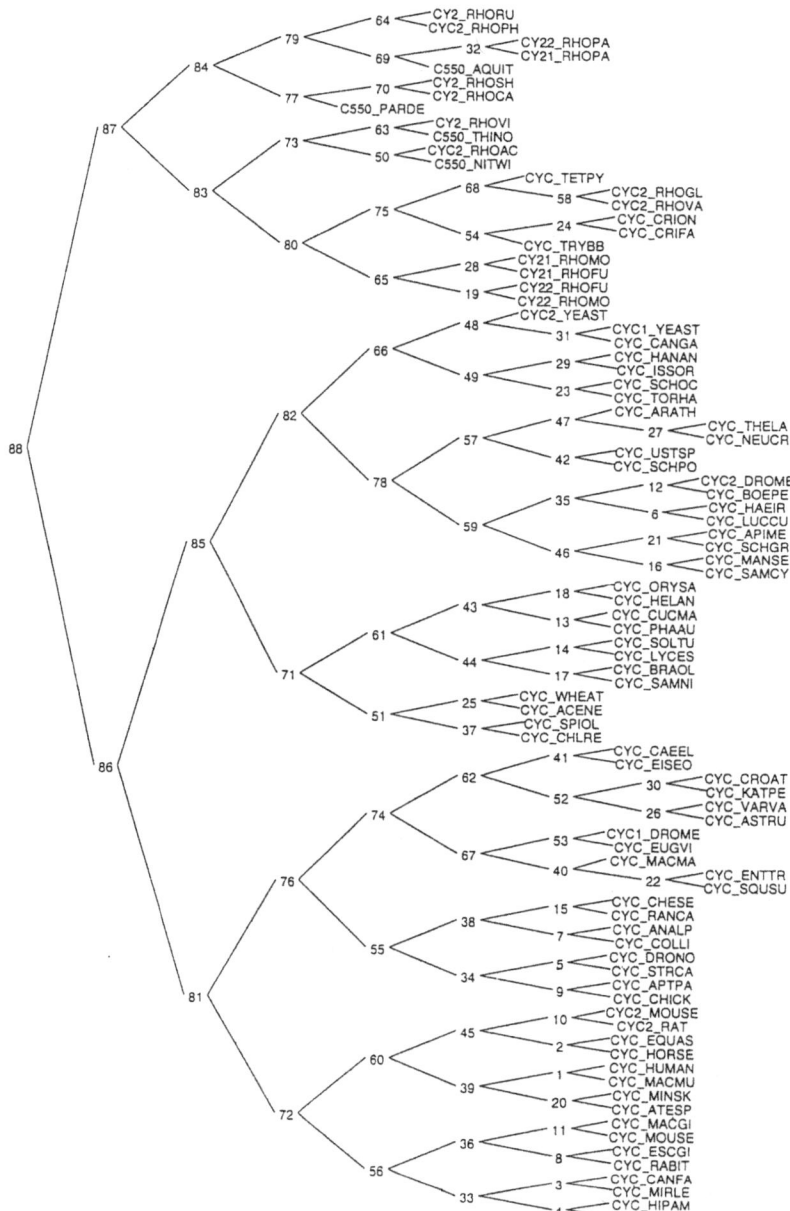

Figure 1: Classification of cytochrome sequences with the significant windows approach, Dayhoff matrix, window size = 7, $\alpha = 0.10$, percentage of homology = 70.

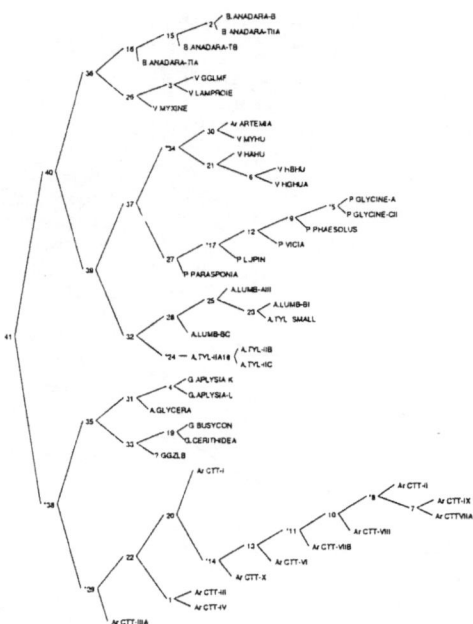

Figure 2: Classification of globin sequences with the significant windows approach, Dayhoff matrix, window size = 7, $\alpha = 0.10$, percentage of homology = 60.

is expressed as a similarity measure between two clusters. Suppose that C and D are any two arbitrary disjoint subsets (or clusters) of $\mathcal{O}$ comprising respectively r and s elements. Then a family of criteria of the "maximal link likelihood" is defined by the following measure of similarity between C and D

$$LL_\gamma(C, D) = [max\{P(c, d)/(c, d) \in C \times D\}]^{(r \times s)^\gamma} \quad 0 \leq \gamma \leq 1 \qquad (7)$$

In case of our data sets $\gamma = 0.5$ was found to yield the best results.

3. Applications

The experiments on the protein sequences belonging to *cytochrome* and *globin* families were carried out using different amino acid similarity matrices (e.g. Dayhoff et al. (1983), Risler et al. (1988), among others) and different values of the parameters such as window size, level of significance for window comparison and percentage of homology. The hierarchical classification method based on the LLA approach was used. Similar experiments on *aa-tRNA ligase* (also known as *aminoacyl-tRNA synthetase*) family of sequences were also conducted. The results were found to be rather unaffected by most of the above parameters whereas the choice of the relevent sequence areas retained for comparison was proved to be of utmost prevalence.

3.1 Sequences from cytochrome and globin families

A set of 89 sequences belonging to cytochrome family was classified using the similarity index described in section 2 and the LLA method of hierarchical classification. The most significant level was found to be the 86th where three main classes may be distinguished (see figure 1). Two of them group together the bacterial chromosomes and the third one is split into two subclasses corresponding respectively to plant and

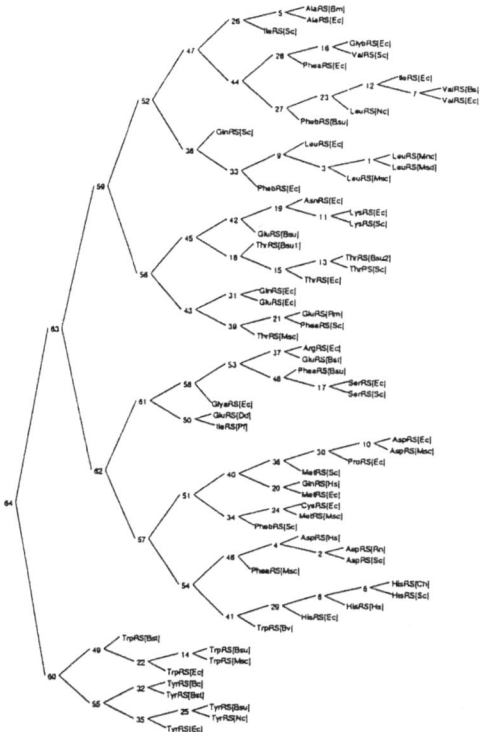

Figure 3: Classification of aatRNA ligases with the significant windows approach, Lerman's matrix, window size = 12, sequence area selection by maximam predictable classification.

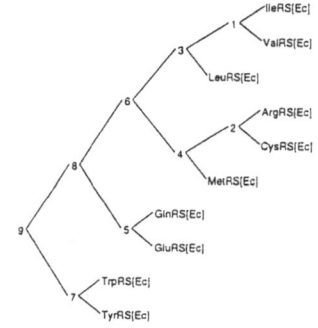

Figure 4: Classification of class I aatRNA from *E. coli* with Risler's matrix, window size = 9, sequence area selection by maximam predictable classification.

animal families. As for the globin family, a set of 42 sequences were classified and the figure 2 displays the corresponding classification tree. It may be observed that at 35th level of the tree – which is most significant according to Lerman's *global statistic* (Lerman et al. 1993) – 7 classes are clearly visible. Two of them are characterized by the vertebrates (V), the others being characterized by bivalves (B), plants (P), annelids (An), gastropods (G) and arthropods (Ar). All artropods but *artemia* are very clearly seperated from the rest of the species ; *artemia* is a miss which is classified among the vertebrates. Similarly, excepting *glycera* all the annelids are nicely put together and the bacterial hemoglobin *ggzlb*, which is very "neutral", is associated with the gastropods. It may also be noticed that the two classes of vertebrates are not joined quickly enough. Notwithstanding the above remarks, the results are globally quite satisfactory and are comparable to those produced by the best methods available.

3.2 Sequences from aa-tRNA ligase family

It is well known, from the biological standpoint, that the *ligases* may be considered as belonging to one of the two groups (see Eriani et al. (1990), Landès et al. (1995)). Class I comprising the sequences that recognize the amino acids *Met, Ile, Leu, Val, Cys, Arg, Gln, Glu, Tyr* and *Trp* seems to be most homogeneous wherein three subgroups may be distinguished : {*Met, Ile, Leu, Val, Cys, Arg*}, {*Gln, Glu*} and {*Tyr, Trp*}. The second group corresponding to the amino acids *Ser, Pro, Thr, Asp, Lys, His, Ala, Gly* and *Phe* is the least structured. The aim of our experiment was to validate our method by producing the results that are as close as possible to the biological knowledge todate by applying it to a test data set. The first test data set was made up of 65 sequences of *aa-tRNA ligases* belonging to various species and was particularly hard to classify due to the very high variation in the length of different amino acids for the same species on the one hand and in the length of the same amino acid for different species, on the other. For instance, for *E. Coli* the length of TrpRS was 334 and that of ValRS was 951, whereas the length of GlnRS was 554 for *E. Coli* and 809 for *Saccharomyces cerevisiae*. It was found that the selection of suitable sequence areas using the "maximum predictability classification" (Lebbe and Vignes (1993), Lebbe and Vignes (1992)) was particularly useful in this case. Figure 3 illustrates the hierarchical classification tree obtained by this method. The results are not fully satisfactory in that they do not agree perfectly with the biological classification given above. This is in fact explainable by a strong inter-species variation. The second test data set containing the sequences of class I *aa-tRNA ligases*, all belonging to a same species namely, *E.Coli* yielded much better results. Figure 4 shows clearly the three subgroups known to the biologists.

4. Conclusion and further research

A hierarchical classification method based on significant windows approach for classifying unaligned biological sequences has been presented and the results of the experiments with several not-easy-to-classify data sets have been described. On the whole, the results are close to the present knowledge of the phylogeny of corresponding species. In case of aa-tRNA ligases, it was shown that the quality of the results depends on the consideration of the biologically pertinent sequence areas as well as on the dgree of inter-species variation. Further direction of this research would be to improve the sensitivity of our method by a refinement of the clustering strategy on the one hand and to reduce the need for parameter tuning (window size, similarity matrix etc.), on the other.

References

Abe.K., Gita.N..(1982): Distances between strings of symbols: Review and remarks. *ICPR6, Munich.*

Barker W.C., Hunt L., George D.(1988): *Protein Seq. Data Anal., 1, 363.*

Dayhoff M.O., Barker W.C., Hunt L.T. (1983): *Mehods Enzymol., 91, 524–545.*

Dickerson R., Geis I. (1983): *Hemoglobin, Benjamin/Cummings, Menlo Park, CA.*

Eriani G., Delarue E., Poch O., Gangloff J., Moras D. (1990): Partitions of tRNA synthetases into two classes based on mutually exclusive sets of sequence motifs. *Nature, 347, 203–206.*

George D., Barker W., Hunt L. (1990): Mutation Data Matrix and Its Uses. *Mehods Enzymol., 183, 313–330.*

Gribscov M., Mclachlan A., Eisenberg D. (1987): Profile analysis: Detection of distantly related proteins. *Proc. Natl. Acad. Sci. 84, 4355–4358.*

Landès C., Hènault A., Risler J.L. (1992): A comparison of several similarity indices used in the classification of protein sequences: a multivariate analysis. *Nucleic Acids Research, 20, 3631–3637.*

Landès C., Perona J.J, Brunie S., Rould M.A., Zelwer C., Steitz T.A., Risler J-L. (1995): A structure-based multiple sequence alignment of all class I aminoacyl- tRNA synthetases. *Biochimie, 77, 194–203.*

Lebbe J., Vignes R. (1992): Sélection d'un sous-ensemble de descripteurs maximalement discriminant. *Troisième journées Symbolique-Numérique, université de Paris Dauphine.*

Lebbe J., Vignes R. (1993): Local predictability in biological sequences, algorithms and application. *Biochimie, 75, 371–378.*

Lerman I.C., Peter Ph., Leredde H. (1993): Principes et calculs de la méthode implantée dans le programme CHAVL (Classification Hiérarchique par Analyse de la Vraisemblance des Liens). *La revue de Modulad, Numéro 12, Dec 93.*

Lerman I.C., Nicolas J., Tallur B., Peter Ph. (1994a): Classification of aligned biological sequences. *New Approaches in Classification and Data Analysis, Springer Verlag, Berlin.*

Lerman I. C., Peter Ph., Risler J. L. (1994b): Matrices AVL pour la classification et alignement de séquences protéiques. *Publication IRISA No 866, IRISA, Rennes, France.*

Risler J.L., Delorme M.O., Delacroix H. and Hénault A. (1988): Amino acid substitutions in structurally related proteins: a pattern recognition approach. Determination of a new and efficient scoring matrix. *Journal of Molecular Biology, 204, 1019–1029.*

An Approach to Determine the Necessity of Orthognathic Osteotomy or Orthodontic Treatment in a Cleft Individual
-comparison of craniomaxillo-facial structures in borderline cases by roentogenocephalometrics-

Sumimasa Ohtsuka[1], Fumiye Ohmori[1], Kazunobu Imamura[1],
Yoshinobu Shibasaki[1] and Noboru Ohsumi[2]

[1]Department of Orthodontics, Showa University School of Dentistry
2-1-1 Kitasenzoku, Ohta-ku
Tokyo 145, Japan

[2]The Institute of Statistical Mathematics
4-6-7, Minami-Azabu, Minato-ku
Tokyo 106, Japan

Summary: At an early stage of growth, it seems nearly impossible to predict the treatment result which the malocclusion of a cleft patient could be corrected by surgical or orthodontic treatment. This study was done to get any clue to confirm the way which the treatment should be finished one or another of both plans for so called a "borderline case" by using a method of data analysis gained from roentgenocephalograms longitudinally. The subjects were unilateral cleft lip & palate patients, who were divided into two groups. One was the OPE group which are corrected by orthodontic treatment with orthognathic osteotomy, the other was the Non-OPE group which are corrected by orthodontics only. These cephalograms were used to evaluate some characteristics of maxillofacial structures. The results showed a possibility to identify the difference of the two groups by utilizing some parameters at an early stage of growth.

1. Introduction

Cleft patients have severe dental problems related to their abnormal facial structures, disturbed facial growth patterns and tooth anomalies, therefore their habilitation is needed from childhood to adulthood. Early orthodontic treatment is often indicated in order to change unfavorable growth pattern and to correct abnormal oral functions such as speech, mastication and swallowing. A majority of treatment objectives in some cases can be achieved through orthodontic treatment alone, while surgical treatment must be applied in the long run for others. At the adulthood, the combined approach between orthodontics and surgery such as the orthognathic osteotomy would be the best way for the one with severe maxillo-mandibular three dimensional disharmony which could not be treated by orthodontic treatment alone. However, the selection of orthodontic or surgical orthodontic treatment remains subjective in nature, which often result in forced long continuous orthodontic treatment on surgical case in the borderline cases. Besides, the decision for the surgery or not are multifactorial things which are related not only maxillo-mandibular relationship but also occlusion, soft tissue profile and the consent of the patient for the surgery. If we could judge the treatment plan for the surgical case earlier before the patient reaches maturity, we could avoid to force long term treatment of growth control which must be finally useless at the time of surgery. The earlier, the better.

This study was designed to investigate cephalometrically, on a longitudinal basis, the possibility of growth prediction for surgical-orthodontic treatment in the cleft patients as

766

early grwth stage as possible with the aid of the cephalometric analysis (Fig. 1).

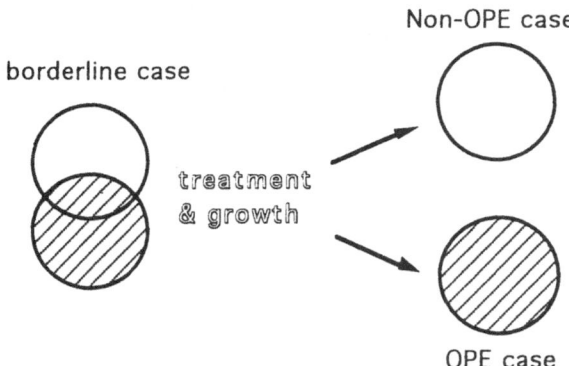

Fig. 1. Concept of the study
How to predict the orthognathic surgical case in its early growth stage?
What is some parameter to detect the surgical case?

2. Materials and methods

2.1 Subjects

The subjects were 67 Japanese unilateral cleft lip and palate (UCLP) patients (37 males and 30 females) at the Department of Orthodontics, Showa University Dental Hospital. Two orthodontic doctors who have the clinical experience in the treatment of the CLP patients over ten years have judged the 67 subjects to be **borderline cases** by the condition of malocclusion using dental cast model. Criteria of selection for the borderline cases were based on the severity of the malocclusion or treatment which all cases had light anterior crossbite before treatment at early mixed dentition. The cases which could be obviously made a diagnosis for surgical in the term of severe maxillo-mandibular relationship and would be favorable not to need the surgery in the future were omitted. Finally 33 of the patients had orthodontic treatment with osteotomy (**OPE group**) and 34 had only orthodontic treatment (**Non-OPE group**).

The lateral radiographic cephalograms, dental casts and hand-wrist radiograph were longitudinally assessed for both of the groups. The cephalograms were measured about cranio-facial structures at 3 stages which were the early mixed dentition, late mixed dentition and adult dentition according to the bone maturation by the hand-wrist radiograph. The maturity of bone as an index of growth was quantatively expressed by the percentages from 0 to 100 by the Ryokawa's method. The stage are as follows;
      stage A   early mixed dentition (Hellman dental stage IIIA), bone maturation=50~60% about 6 years old of age
      stage B   mixed dentition (dental stage IIIB), bone maturation=60~70% almost the time of adolescent growth initiation
      stage C   permanent dentition (dental stage IIIC), bone maturation=90~100% almost the time of bone growth completed, when it is clearly to make a treatment planning which should be corrected by orthodontics alone or with orthognathic sur-gery.

2.2 Cephalometric analysis

Cephalograms provide a quantitative medium for describing dynamic changes in the patient and for growth studies as for the dentofacial pattern in general.

Lateral cephalograms were taken with the same x-ray device and by a single technician. Focus median plane distance was 150 cm and film median plane distance was 10 cm with an enlargement of 10%. No correction was made for this radiographic enlargement, as it affected all the cephalograms of both groups in the same way.

In longitudinal cephalometric studies on growing subjects, reference line should be traced through craniofacial stable structures. Radiographs were traced and put the following landmarks which were identified or constructed:sella trucica (S), nasion (N), orbitale (Or), anterior nasal spine (ANS), point A (A), point B (B), pogonion (Pog), gnathion (Gn), menton (Me), gonion (Go), articulare (Ar), condylion (Cd), porion (Po), posterior nasal spine (PNS).

Reference planes were adopted as follows;
S-N plane (connects S to N), A-B line (connects A to B), Facial plane (N to Pog), FH plane (Po to Or), Ramus plane (Ar to the posterior border of the mandibular ramus), Y-axis (S to Gn), Mandibular plane (Me to the lower border of the mandible), Palatal plane (ANS to PNS). The definition of all these landmarks and planes were correspond to those given by Downs, Riedel and associates. The coordinates of each landmark for each cephalogram were recorded by means of a WACOM digitizer interfaced with a NEC PC-98Vm computer. The output values for each point were stored by coordinate representation on a disk for computer analysis. Nine linear and seventeen angular measurements were selected for quantitative cephalometric evaluation(Fig. 2,3).

Linear measurements for the assessment of
    cranial base dimensions: N-S.
    maxilarry dimensions: N-ANS, S'-Ptm', A'-Ptm'.
    mandibular dimensions: N-Me, Gn-Cd, Pog'-Go, Cd-Go.
Angular measurements for the assessment of
    maxillary dimensions: SNA (S-N-A).
    mandibular dimensions: SNB (S-N-B), SNP (S-N-Pog), mandibular plane angle, gonial angle, ramus inclination, Y-axis angle, facial plane angle.
    Intermaxillary relationship: ANB (A-N-B), A-B plane, convexity (N-A-Pog).

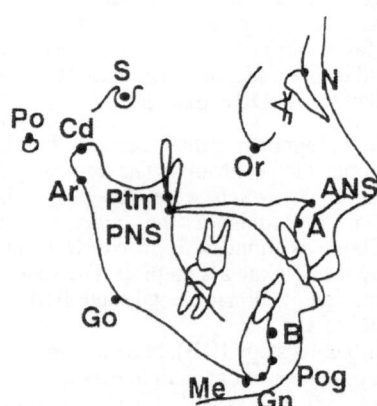

Fig. 2. Standard landmarks of roentogenocepharometrics

769

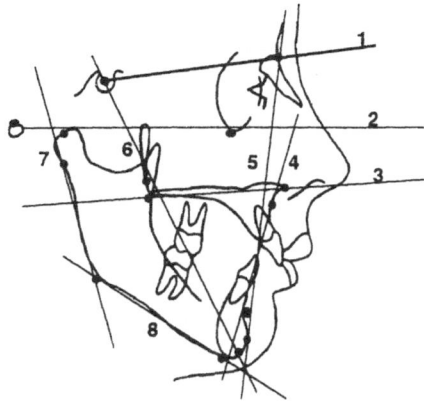

Fig. 3. Standard reference lines of roentogenocepharometrics
1, S-N plane (S to N). 2, FH plane (Po-Or). 3, Palatal plane (ANS to
PNS). 4, A-B plane (A to B). 5, Facial plane (N to Pog). 6, Y-axis (S to
Gn). 7, Ramus plane (Ar to construc-ted gonion). 8, Mandibular plane
(Me to constructed gonion).

2.3 Statistical evaluation

Cepalogram data were evaluated using StatView II and JUSE/MA1, the Statistical Package
for the Social Sciences designed for Apple and NEC compatible personal computers.

Initially, all the variable were tested for validity and robustness.Then, the morphological
difference was tested by means of a nonparametric test, Mann-Whitney Utest between
OPE and Non-OPE group at each stage. Moreover, the cephalometric data were analyzed
by a multivariate statistical approach, discriminant analysis. Fisher's type linear
discriminant analyses were carriied out, using treatment group (OPE or Non-OPE) as the
dependent variable at each growth stage. Then, some variables were selected as an
effective parameter to identify the Operation group.

3. Results

3.1 Descriptive statistics for the differences between the OPE and Non-OPE group

Table 1 summarizes the results of the nonparametric statistical comparison on the
differences between the OPE and Non-OPE groups. At the stage A in male, gonial angle
(Ar-Go-Me) exhibited significantly larger in OPE group ($p<0.001$); mandibular ramus
inclination angle (FH-Ramus plane) exhibited significantly smaller in OPE group
($p<0.05$). Whereas in female, SNB angle which represents the anterior limit of the
mandibular basal arch in relation to the anterior cranial base showed significantly larger in
OPE group ($p<0.01$); mandibular plane angle and Y-axis angle showed significantly
smaller in OPE group ($p<0.01$).

At the stage B in male, mandibular plane angle and gonial angle exhibited larger in OPE
group ($p<0.05$, $p<0.001$); ramus inclination angle, posterior position of maxilla to cranial
base (S'-Ptm') and mandibular ramus length (Cd-Go) appeared to be significantly smaller

in OPE group. In female, both SNB and Y-axis angle showed significantly difference as the same as stage A (p<0.01, p<0.05). The linear measurement for the assessment of mandibular ramus and total length of the mandible (Gn-Cd) showed significantly larger in OPE group (p<0.01, P<0.05).

At the stage C in male, Both SNB and gonial angle showed significantly larger in OPE group (p<0.05, p<0.001); ramus inclination and ramus length showed significantly smaller in OPE group (p<0.05). On the other hand, SNB and Y-axis angle showed significant difference between OPE and Non-OPE groups (p<0.05, p<0.01); mandibular total length and ramus length exhibited significantly larger in OPE group (p<0.05).

Table 1 Means values for variables which were indicated significantly morphological differences between OPE and Non-OPE group at each growth stage for male and female.

		male				female		
		OPE	Non	p		OPE	Non	p
stage A	Gonial	134.1	125.9	***	SNB	78.9	75.8	**
	Ramus	79.1	82.3	*	Mand P	30.7	35.2	**
					Y-axis	63.2	66.2	**
stage B	Mand P	33.9	30.5	*	SNB	78.1	75.0	**
	Gonial	132.2	125.3	***	Y-axis	64.1	67.3	*
	Ramus	81.7	85.2	*	Ramus	81.7	85.2	*
	S'-Ptm'	17.2	19.8	**	Gn-Cd	107.9	103.0	**
	Cd-Go	51.2	55.4	**	Cd-Go	51.0	48.2	*
stage C	SNB	76.0	72.8	*	SNB	78.7	74.9	*
	Gonial	130.8	124.6	***	Y-axis	63.8	67.6	**
	Ramus	81.8	85.8	*	Gn-Cd	116.0	110.1	*
	Cd-Go	57.9	61.8	*	Cd-Go	55.5	51.6	*

***p< 0.1% , **p< 1% and *p< 5% by Mann-Whitney test

3.2 Discriminant analysis of all cases

At each stage the discriminant analysis was done, however, there were some difference in the eligble variable and correct classified percentage. In male, variables which showed over 2.0 of F-ratio at stage A were Mandibular plane (MP), Gonial angle, ramus inclination and Cd-Gn. On the other hand, these were MP, Gonial angle, ramus inclination, N-ANS, S'-Ptm' at stage B, convexity, AB plane, SNP angle, SNB, ANB, gonial angle, ramus inclination, Pog'-Go and Cd-Gn at stage C. The same findings were observed in female samples. Therefore from the point of clinical view, elgible variables were selected commnly at each stage. For male, 3-factor model which is composed of gonial angle, ramus inclination and Cd-Gn was generated, giving 77.8, 83.3 and 83.8 percent correct classification of the OPE group at each growth stage. In the other hand, 4-factor model which is composed of Y-axis angle, SNB, ramus inclination and Cd-Gn was generated, giving 76.7, 74.1 and 78.6 percent correct classification of the OPE group in female (Table 2).

Table 2 Discriminant analysis generated by MAI
Percentage of case correctly classified OPE group for male and female.

	male	female
stage A	77.8	76.7
stage B	83.3	74.1
stage C	83.8	78.6
Predictive vaiable	Gonial angle Ramus inclination Cd-Go	Y-axis SNB Ramus inclination Cd-Gn

4. Discussion

Early treatment of the cleft patient's malocclusion has beeen generally recommended by many authors for the favorable results on growth and occlusal relationship. However, some disadvantages of the problems that might be encountered during the early dentition tretment are: (1)it may be not always that early treatment bring easier and better results, (2)patient's cooperation may deteriorate because of long periods of active treatment, and (3) family financing may also have an influence on the length and timing of treatment. If prediction of maxillo-facial relationship at the time of its growth completed could be done at early growth such as a childhood, it could be free from a wasted treatment and various sufferings with it. There are many reports about the criteria of treatment adaptation to skeletal Class III patients which should be selected orthodontic treatment or surgical orthodontic treatment. However, these subjects are almost for adults, there are few for growing young people, especially children. It may be the main reason that it is very hard to predict skeletal changes by growth and orthodontic treatment at the raky growth stage. On the other hand, variables to express the morphological differences in both groups got increased according to the raising the bone maturity. It suggested that growth prediction could be easier by aging. At the stage C which is almost completed growth, it seems eaier to make a diagnosis and treatment plannings what could be finished by orthodontic treatment alone or combined with orthognathic osteotomy on earth by the reason of stopped growth. In the present study a correct methodologic approach for evaluation of borderline case with malocclusions was then initiated.

The cephalometric analysis we applied was suitable for a geometric evaluation of maxillo-facial components. Several significant differences in craniofacial skeletal structures were found between the OPE group and Non-OPE group. The following results were obtained.

1. There were significant morphological differences of dentofacial complex, especially in the Mandible, between the OPE and Non-OPE group in both sexes. No significant difference in the maxillary components were noted in both groups.
2. The differences became more clearly with growth.
3. In the male OPE group samples were characterized at short and anterior position of the ramus with wide gonial angle, while dominant anterior growth direction of the mandible and its larger size in females (Fig. 4).
4. The morphological information from the mandible were available for determination of future surgical case.

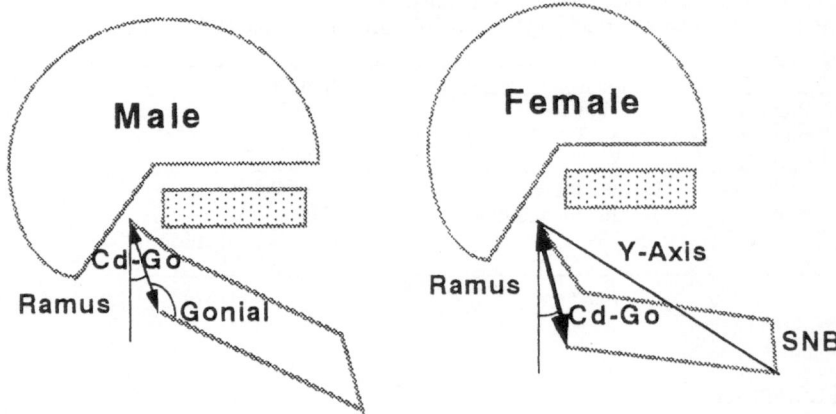

Fig.4. Schematic morphological characterisics in OPE case of male and female.
Male orthognathic surgical case has short ramus and downward rotated mandible with downward growth direction of its. On the other hand, larger mandible with its forward growth pattern is different from male one in female surgical case.

Comparing these data to the standard data which are gained from the subjects of normal occlusion by Iizuka and Ishikawa, the male gonial angle and ramus inclination showed larger than the standard. On the other hand, female Y-axis and ramus inclination showed almost same the standard except SNB which were larger by 2 degree. As they are substantively different from sample composed from size and criteria of growth stage, it coud not easily to compare the study and standard sample here.

In our opinion, a fundamental question arises from the obtained data. What were the reasons for morphological differences in the orthognathic surgical case between male and female? Moreover, why the differences were almostly found in the mandibular componets except by the maxillary components. There may be some important factors to be considered for the reason. One is the size of sampling in this study. The borderline case were selected from the poins of severity of maloocclusion with anterior and lateral crossbite. There is a data which indicated their similarity of malocclusion about inter-maxillary relationship in the term of ANB angle and SNA angle which shows no significant difference in both groups. In fact, there may be several morphological patterns for cleft patients. The male operation group shows the downward rotated of the mandible, the female cases show a typical skeletal class III which has overgrowth of the mandible relatively. The former is very difficult to correct anterior crossbite by the mandible backward rotation, since it makes the mandible more rotation result in a long face and shallow oberbite. On the other hand, the latter overgrowth of mandible is not easily controlled because of its size even if growth control would be begun from early growth stage. Treatment planning could be the other main cause to influence the results, since decision of treatment plannings may depend on some factors which are inter-maxillary relationship such as ANB angle, soft tissue profile from the point of the aesthetic sense, teeth movement in the orthodontic treatment and consent of the surgery by patient and parents. Therefore it could be said that bordeline case is multifactorial. That may be settled in the collecting more samples for borderline cases, which should be separated and examined by each factor.

The previous study suggested that there are some effective parameters between surgical and non-surgical orthodontic treatment for a cleft individual. The possibility of prediction for surgical case could be indicated in the earlier growth stage. After the patients had been treated for a period of time, from the first examination in mixed dentition, reexamination at about the initiation of puberal growth spurt on the hand-wrist radiographs may be of advantage in predicting future treatment procedures.

5. Concluding remarks

There were significant morphological differences in dentofacial complex, especially in the mandible, between OPE and Non-OPE group in both sexes. The differences became more and more clearly with growth. The male OPE group samples were characterized at short and anterior position of the ramus with wide gonial angle, while dominant anterior growth direction of the mandible and its large size are shown in females. The morphological information from the mandible were available for determination of future surgical case.

The previous study suggested that there are some effective parameters between surgical and non-surgical treatment for a cleft indvidual. The posssiblity of prediction for surgical case could be indicated in the early growth stage. We could not make our conclusion from the point of some problems by small sample, however, after this, we would like to collect more variable cases and examination them in detail. Although the growth prediction is really very hard things, we will do more in the future.

References:

Battagel J.M. (1994): The identification of Class III malocclusions by discriminant analysis, *European Journal of Orthodontics,* **16**, 71-80.

Cassidy D.W. et al. (1993): A comparison of surgery and orthodontics in "borderline" adults with Class II, Division1 malocclusions, *American Journal of Orthodontics and Dentofacial Orthopedics,* **104**, 455-470.

Downs, W.B. (1948) : Variation in facial relationships: their significance in treatment and prognosis, *American Journal of Orthodontics,* **34**, 812-840.

Iizuka T. and Ishikawa F. (1950): Points and landmark in head plates, *The Journal of Japan Orthodontic Society,* **16**, 66-75.

Sinclair P.M. et al. (1993): Combined surgical and orthodontic treatment, *In Contemporary orthodontics,* 2nd ed., Proffit W.R. (eds.), 607-631, Mosby Year Book, St. Louis.

Tollaro I., Baccetti T. and Franchi L. (1996): Craniofacial changes induced by early functional treatment of Class III malocclusion, *American Journal of Orthodontics and Dentofacial Orthopedics,* **109**, 310-318.

Valko R.M. (1968): Indications for selecting surgical or orthodontic correction of mandibular protrusion, *Journal of Oral Surgery,* **26**, 230-238.

(†)This study was in part carried out under the ISM cooperative Research Program (95-ISM•CRP-A54)

Data Analysis for Quality of Life and Personality

Kazue Yamaoka[1] and Mariko Watanabe[2]

[1]Department of Hygiene and Public Health, Teikyo University School of Medicine
2-11-1 Kaga, Itabashi-ku, Tokyo 173, JAPAN

[2]Department of Food Science, Showa Women's University Junior College
1-7 Taishido, Setagaya-ku, Tokyo 154, JAPAN

Summary: In order to examine the influence of Personality on Quality of Life (QOL) measurement, the questionnaire survey was conducted and the analysis was carried out. Using Hayashi's Quantification Method Type III, a structure of QOL questionnaire was confirmed and QOL score was calculated on the basis of the structure. The result of the analysis for the association between QOL and personality indicated that the QOL scores of the tolerable type subjects were greater than those of the intolerable subjects. Therefore, personality type was thought to be a possible confounding factor for a study related to QOL.

1. Introduction

In the last decade interest has increased in quality of life (QOL) measures in four broad health contexts: such as measuring the health of population, assessing the benefit of alternative uses of resources, and making a decision on treatment for an individual patient. Each context requires an assessment of the impact of ill health on aspects of everyday life of the individual. Increased attention is now being given to the importance of QOL, and methods of QOL measurement have accordingly been developed (Cox, et al., 1992). Yamaoka, et al. (1994) has developed the QOL20 Questionnaire for the measurement of non-disease specific QOL of a patient.

In the present study, we focused on the data analysis for the problems related to QOL measurement and it's association with personality type and special attention was given for the classification of subjective attitude measured by a questionnaire survey on the basis of structure analysis.

2. Study subjects and Method

2.1 Working Hypothesis

In general, there is a possibility that personality influences subjective responses to QOL questionnaire. Thus we focused on the influence of personality on QOL measurement. That is, the QOL20 (Yamaoka, et al., 1994) is a subjective measurement, it may be influenced by personality. In the present study, we hypothesized that the QOL20 scores may differ among personality types. Therefore, we classified personality type using

Eysenck Personality Questionnaire (EPQ, Shigehisa, 1989) described below.

2.2 Questionnaire
QOL20
In general, QOL measures were classified into three types: The first type is a performance score, e.g. an activity score. This type of measure is evaluated by a third party and Karnofsky Performance Status Scale (Karnofsky, 1949) and WHO Performance Status Scale (WHO, 1979) were established, for instance. The second type of evaluation employs objective data such as clinical findings. For this, nutritional parameters and duration of inpatient/outpatient stay, and so on were used. The third type is subjective evaluation by the patient. Although there is still no standard method for evaluating QOL, however, many trials have been carried out and FLIC (Schipper, 1985) and EORTC (Aaronson, et al, 1988), were proposed for QOL measurement in the West. Furthermore, a large number of studies have been performed on QOL questionnaires, little attention has been given developing a non-disease-specific QOL questionnaire for Japanese. Kobayahi, et al. (1994) summarized the problems in the investigation of QOL.

For the measurement of subjective QOL of patients, we developed the QOL20, which belongs to the third type and consists of 20 questions related to psychological, physiological, and environmental factors.

QOL20 was conducted under the working hypotheses that 1) The QOL of a patient is similar to that of a healthy person and that 2) QOL includes two main factors, i.e., "state of disease" (D) and "attitude toward disease" (F) and with the changes of these factors the QOL of a patient also changes. The scoring method has developed on the basis of the structure of the questionnaire. Because the structure of QOL recognized to be uni-dimensional, an additive scale, that is, the sum of number of responses, was used for the measurement of QOL. The symmetry of the positive score and negative score was not guarantied. This means that the distance in the configuration of the items were not equivalent among the items. In such a case, we thought that it is better not to use Likert Scale but to calculate the both positive score (QTP) and negative score (QTN).

Eysenck Personality Questionnaire ; EPQ
Personality type was classified using EPQ which consists of 25 questions with Likert scale with 4 categories (Shigehisa, 1989). Each question belongs to one of the dimensions such as extroversion (E), neuroticism (N), psychoticism (P), and dissimulation (L). Using the score for the target dimension (the score calculated as sum of Likert scales), we classified subjects into the personality type. That is, in case that the score is larger than the mean, it is classified into positive ("+") type and the other is classified into negative ("-") type. In this study, we concentrated on the tolerable type (E+,N-,P+), intolerable type (E-,N+,P-), and the other type. These types were classified using the above dimensions.

2.3 Subjects
Using the above two questionnaires, the survey was conducted on the parents of 145 students. A hundred and twenty-five males and 145 females responded to the questionnaires. We used the subjects who responded to both QOL20 and EPQ, and whose ages were between 40 to 65 years old. Thus the subjects used for the analysis

were 120 males and 128 females.

2.4 Statistical Method

In order to examine the validity of the construction of the QOL20, we analyzed the structure of the QOL20 using Hayashi's Quantification Method Type III (QIII) (Hayashi, 1952). The correlation between QOL20 scores and EPQ dimensions were examined using Spearman rank correlation coefficient. The difference of QOL20 scores among the personality types was examined by Kruscal Wallis test.

3. Results

3.1 Structure of QOL20

The structure of QOL20 was examined using QIII method and a scattergram of the category values corresponding to the maximum latent root and the category values corresponding to the second maximum latent root was figured. Although items were somewhat varied, however, it could be thought that a cup type curve like a uni-dimensional Guttmann's scale was reconfirmed (see Yamaoka, et al., 1994). Therefore, the scoring method, described in the above, was recognized to be valid.

3.2 QOL20 scores by personality types

The minimum value and the maximum value of the QOL20 positive scores for males were 1 and 20 and negative scores were -8 and 0, respectively. In case of female, those for positive scores were 0 and 20 and for negative scores were -8 and 0, respectively.

Spearman rank correlation coefficients between the positive and the negative QOL scores and EPQ dimensions are shown in Tab. 1. It was revealed that although the correlation coefficient between the positive score and the negative score of QOL20 were significant but they did not strongly correlate. Furthermore, Spearman rank correlation coefficients between QOL scores and EPQ dimensions were not so strong, too. However, the tendency for females were similar to that for males.

QOL20 scores by personality types are shown in *Tab. 2*. Results for males indicated that, in terms of a positive response tendency to the QOL20, the QOL20 scores of the tolerable type subjects were significantly greater than those of the intolerable type (p<0.01). The tolerable type, the other (intermediate type), the intolerable type made an order.

In terms of a negative response tendency to the QOL20, similar tendency was recognized (p<0.001). In case for females, although similar tendency was recognized, it was not significant.

Table 1. Spearman rank correlation coefficient between QOL scores and EPQ dimensions.

males: n=120

| | QOL20 | | EPQ dimensions | | |
	QTP	E	N	P	L
QTP	1.00	0.32**	-0.09	0.17	0.15
QTN	0.30***	0.10	-0.31***	0.14	0.13

females; n=128

| | QOL20 | | EPQ dimensions | | |
	QTP	E	N	P	L
QTP	1.00	0.14	-0.10	0.10	0.21*
QTN	0.31***	0.05	-0.27**	0.13	0.23**

E: extraversion、 N: neuroticism、 P: psychoticism、 L: dissimulation
*** $p<0.001$, ** $p<0.01$, * $p<0.05$

Table 2: Personality Types (Tolerable, Intolerable, Other) by EPQ and QOL20 scores.

Males (n=120)

QOL	Tolerable (17)	Intolerable (16)	Other (78)	χ^2 value
Positive	11.5 (6, 13)	5 (4, 7)	9 (7, 11)	9.42**
Negative	0 (-1, 0)	-2 (-6, -1)	-0.5 (-2, 0)	14.03***

Females (n=128)

QOL	Tolerable (18)	Intolerable (23)	Other (87)	χ^2 value
Positive	9 (7, 12)	7 (5.5, 9.5)	9 (7, 12)	5.81
Negative	0 (-2, 0)	-1 (-4.5, -0.5)	-0.5 (-2, 0)	4.16

Tolerable type: E+N-P+, Intolerable: E-N+P-,
The numbers of the table were Median(25%tile, 75%tile),
Kruscal Wallis test : ** $p<0.01$, * $p<0.05$

4. Conclusion

It is concluded that there is a possibility that the QOL20 scores of the tolerable type subjects were greater than those of the intolerable subjects. Therefore, personality type was thought to be a possible confounding factor for a study related to QOL. It is suggested that one must interpret carefully, as diagnostic measures, the QOL scores of the persons with "tolerable type" personality.

5. References

Aaronson NK, et al. (1988): A modular approach to quality of life assessment in cancer clinical trials. *Resent Results Cancer Res*, 111, 231-249.

Cox DR, et al (1992): Quality-of-life assessment: Can we keep it simple? *J R Statist Soc* A, 155, part 3: 353-393.

Kobayashi K, et al. (1994): Quality of life (QOL) studies in Japan-Problems in the investigation of QOL-. *Ann Cancer Res Ther*, 3(2), 83-89.

Karnofsky DA and Burchenal JH (1949): The clinical evaluation of chemotherapeutic agents in cancer. In: *Evaluation of Chemotherapeutic Agents* (MacLeod CM, ed.), Columbia Univ. Press, New York, 191-205.

Shigehisa T (1989): Behavioral regulation of dietary risk factor associated with stress-induced disease, in relation to personality and interpersonal behavior, in a sociocultural perspective. *Tokyo Kasei Gakuin University Journal*, 29, 25-45.

Schipper H, et al (1985): Measuring quality of life: Risks and benefits. *Cancer Treat Rep*, 69, 1115-1125.

World Health Organization (1979): WHO handbook for reporting results of cancer treatment. *Offset Publ.* No 48, WHO, Genove.

Yamaoka K, et al. (1994): A Japanese version of the questionnaire for quality of life measurement. *Ann Cancer Res Ther* 3, 45-53.